ISBN 978-1-331-91387-0
PIBN 10253261

This book is a reproduction of an important historical work. Forgotten Books uses
state-of-the-art technology to digitally reconstruct the work, preserving the original format
whilst repairing imperfections present in the aged copy. In rare cases, an imperfection in
the original, such as a blemish or missing page, may be replicated in our edition. We do,
however, repair the vast majority of imperfections successfully; any imperfections that
remain are intentionally left to preserve the state of such historical works.

# Similar Books Are Available from
# www.forgottenbooks.com

PHYSIOLOGICAL PLANT ANATOMY

MACMILLAN AND CO., Limited
LONDON · BOMBAY · CALCUTTA
MELBOURNE

THE MACMILLAN COMPANY
NEW YORK · BOSTON · CHICAGO
DALLAS · SAN FRANCISCO

THE MACMILLAN CO. OF CANADA, Ltd.
TORONTO

# PHYSIOLOGICAL
# PLANT ANATOMY

BY

## DR. G. HABERLANDT

PROFESSOR IN THE UNIVERSITY OF BERLIN
(FORMERLY IN THE UNIVERSITY OF GRAZ)

TRANSLATED FROM THE FOURTH GERMAN EDITION

BY

## MONTAGU DRUMMOND, B.A., F.L.S.

LECTURER IN PLANT PHYSIOLOGY, UNIVERSITY OF GLASGOW

WITH 291 FIGURES IN THE TEXT

MACMILLAN AND CO., LIMITED
ST. MARTIN'S STREET, LONDON
1914

573

*COPYRIGHT*

# TRANSLATOR'S PREFACE

SINCE its first appearance in 1884, Professor Haberlandt's *Physiologische Pflanzenanatomie* has undergone revision and amplification on three successive occasions. In its present form, therefore, this work may be assumed to embody the mature and considered views of its author, with regard to that section of botanical science which he has made peculiarly his own.

In order to retain, as far as possible, the fluency which is a prominent and agreeable feature of the original text, I have not scrupled to translate with some freedom, where necessary; at the same time, I have taken every care to ensure that the author's meaning should be faithfully reproduced throughout. The limited number of words and phrases for the insertion of which I am personally responsible,—for example the term **photosynthesis** on p. 262—are distinguished by being enclosed within square brackets.

I desire to record my indebtedness to my wife for assistance in proof-revision and in the compilation of the indices, and to Professor F. O. Bower, Sc.D., F.R.S., for much helpful advice and encouragement.

MONTAGU DRUMMOND.

GLASGOW, *1st January*, 1914.

# CONTENTS

## CHAPTER III.

## THE DERMAL SYSTEM.

## CHAPTER IV.

## THE MECHANICAL SYSTEM.

## CHAPTER V.

## THE ABSORBING SYSTEM.

# CONTENTS

# CONTENTS

## CHAPTER VIII.

## THE STORAGE SYSTEM.

# CHAPTER IX.

## THE AËRATING OR VENTILATING SYSTEM.

# CHAPTER X.

## THE SECRETORY AND EXCRETORY SYSTEMS.

## CHAPTER XI.

### THE MOTOR SYSTEM.

## CHAPTER XII.

### THE SENSORY SYSTEM.

## CHAPTER XIII.

### THE STIMULUS-TRANSMITTING SYSTEM.

## CHAPTER XIV

### SECONDARY GROWTH IN THICKNESS OF STEMS AND ROOTS.

# INTRODUCTION.

THE object of Physiological Plant Anatomy is twofold. It consists first, in the recognition of the physiological functions pertaining to the tissues of the plant and to the structural units, or cells, of which these tissues are composed ; and, secondly, in the discovery of the connection that exists between the several functions and the anatomical arrangements required for their proper performance. The study of the various cells and tissues of the plant body clearly demonstrates the fact that physiological activity depends upon the general structure of the organism and upon individual anatomical features, just as the special manner of action of every machine is the result of its particular mode of construction.

Physiological anatomy is an explanatory science, because it sets out to discover the relation between structure and function ; for the discovery of relations between distinct phenomena constitutes " explanation " in the scientific sense of the term. Since, however, any particular physiological function appears to the observer to be the aim and object of the correlated structural features, every demonstration of a connection between structure and function naturally assumes a teleological aspect.

The value of a teleological explanation depends entirely upon the philosophical attitude of its author. In some cases such an " explanation " merely serves as a pregnant mode of expressing the relation between structure and function, whereas an objective significance may be ascribed to it by those who recognise the existence of purpose as a final cause, not only in human actions, but also in all the workings of Nature.

The preceding definition of the scope of physiological anatomy at once raises the following question : Is there no exception to the rule —which almost amounts to a postulate from the anatomico-physiological standpoint—that each individual cell is endowed with some special function, and that a definite physiological value must be assigned to every morphological feature ?

There is no doubt that the rule in question holds good in the vast majority of cases. At the same time the adult organs of many plants certainly contain isolated cells, or even groups of cells, which must be regarded as functionless in the sense that they play no useful part in the general economy of the plant. Again, it may happen in the case of a particular cell-group or tissue that an individual morphological character cannot be "explained" with reference to a definite physiological function, or at any rate that the two cannot be brought into direct relation with one another. Such functionless units and unexplained features of the internal structure of plants are of various kinds ; their discrimination is a matter of considerable importance to the physiological anatomist, mainly because he is unable otherwise to determine the natural limits of his proper field of work.

The functionless condition may be due to **loss of function**. It is further necessary to distinguish between **ontogenetic loss of function,** which takes place during the development of the individual, and **phylogenetic loss of function**, which arises in the course of evolution of a species, a genus, or a family. A few examples will render the distinction more evident. When cork is formed, the primary dermal tissue or epidermis is ruptured and degenerates ; again, in many plants the stomata of the older leaves lose their power of adjustment and hence cease to regulate gaseous interchange ; lastly, in the older portions of secondary wood the water-conducting vessels become functionless by occlusion. All these instances illustrate loss of function of the ontogenetic type.

Less numerous are the cases in which a loss of function, initiated at some previous stage in the evolution of the species under consideration, has resulted in the phylogenetic reduction of a given organ or tissue. On becoming functionless, namely, an organ or a tissue inevitably undergoes—from causes which cannot conveniently be discussed here— a gradual reduction, which may end in complete atrophy. An illustration is provided by the more or less pronounced reduction of the parenchymatous bundle-sheaths which prevails in the leaves of certain species of *Euphorbia*. When typically developed, such bundle-sheaths are responsible for the removal of synthetic products from the leaf. In the *Euphorbias* referred to, this work is in part performed by the laticiferous system ; as a result, the bundle-sheaths are to a large extent relieved of their duties and accordingly undergo very obvious and extensive reduction.

Another instructive illustration of phylogenetic reduction is furnished by the functionless stomata which occur on the sporogonia of *Sphagnum* ; here guard-cells are still initiated, but the pore and the internal air-chamber both remain undeveloped. In this case the reduction of the stomata and the loss of function on the part of their guard-cells are secondary consequences of the phylogenetic reduction—or rather of the complete atrophy—which the photosynthetic system has undergone in the sporogonium ; owing to the disappearance of green tissue, the development of special ventilating organs in the shape of stomata has become superfluous.

Ontogenetic loss of function naturally cannot be regarded as a phenomenon of prime importance from the standpoint of physiological anatomy. When, namely, a particular organ or tissue ceases its physiological activity at a relatively early stage, as compared with other tissue-systems, it must do so for reasons which are in some way or other connected with its proper functions. The fact that loss of function may arise gradually in the course of evolution also in no wise affects the fundamental principles of the anatomico-physiological method. An organ or tissue which has undergone phylogenetic reduction must *ex hypothesi* have performed a definite function at some previous evolutionary stage, and must at that time have possessed an anatomical structure correlated with that function. Where reduction has not progressed too far, this former connection between structure and function may still be demonstrable. In cases of extreme reduction however, the existing morphological features may no longer be capable of interpretation in terms of their obsolete physiological value. In such instances it must always remain uncertain whether the reduced structure ever did perform any special function. Thus the gritty particles, composed of thick-walled stereides, which occur in the flesh of the Common Pear (and in other POMACEAE), have been regarded as relics of an ancestral protective shell. This interpretation may be quite correct ; but the thick-walled character of these stone-cells is not in itself a sufficient proof of their former mechanical function. For there is just as much to be said in favour of the view according to which these thick-walled stone-cells are devoid of any functional significance, and merely serve as repositories for superfluous cellulose material. Such ambiguous cases are probably rare.

Functionless characters sometimes arise in consequence of **change of function**, in which event they owe their existence partly to the influence of heredity. Numerous cases will be met with later on, in which the principal function of a given organ or tissue becomes altered in the course of phylogenetic—or it may be of ontogenetic—development, the change being accompanied by an appropriate modification of structure. It does not follow, however, that all the morphological features correlated with the original function necessarily undergo alteration in such cases. On the contrary, characters which are not incompatible with the newly acquired function are often inherited unchanged; owing to the change of function which has taken place such surviving features are physiologically valueless. An example will serve to render the argument more convincing.

In the genus *Aloe*, the leaf-margin is strengthened by thick-walled hypodermal mechanical cells of palisade-like form; transitional stages show that these palisade-stereides are phylogenetically derived from thin-walled chlorophyll-containing palisade-cells, that is to say, from typical photosynthetic elements. Now the palisade-like form of cells specialised for photosynthesis is closely related to their function; next to the presence of numerous chloroplasts, in fact, it is their most important anatomico-physiological characteristic. This feature is inherited unchanged when the photosynthetic cells of the leaf-margin of *Aloe* undergo the above-mentioned change of function, because the retention of this characteristic does not interfere with the newly assumed mechanical function. In respect of this new function, however, the palisade-like form of the cell is a useless feature, since it is by no means necessary that the mechanical cells employed for the purpose of strengthening the leaf-margin should be elongated at right angles to the surface of the leaf. As a matter of fact the protection of leaf-margins against the danger of laceration is far more frequently effected with the aid of fibrous strands, composed of cells that are elongated in a direction parallel to the leaf-surface.

A third category of functionless characters includes features which occur in association with some adaptive structure, their existence being determined by the particular mode of development of the latter. Very frequently, namely, the development of an adaptive arrangement involves the appearance of features which have no intrinsic physiological value, but which arise incidentally in connection with certain stages of a particular sequence of developmental processes. When a board is planed in a joiner's workshop spirally rolled shavings are produced; yet only a child would imagine these to be objects prepared for a definite purpose.

These useless **correlative characters** are of course not easy of recognition, since the discovery of such features generally implies a complete insight into the developmental mechanism of the adaptive structure with which they are correlated. An illustration is furnished by the stratification of thickened cell walls, which is so clearly displayed by many mechanical cells. Here the thickness of the membrane is the factor which is of physiological importance; the stratified condition, on the other hand, is valueless from the mechanical point of view, and merely represents an inevitable consequence of the particular mode of thickening adopted by the developing cell wall.

While the various kinds of functionless members and anatomical features that have been discussed so far complicate the task of the physiological anatomist, they do not fall altogether outside the scope of his subject, nor do they involve conceptions which conflict with its fundamental principles. It is otherwise, however, with such members and features as appear to have been devoid of physiological value *ab initio*. If purely individual variations be ignored for the present, and if attention be further restricted to undoubtedly hereditary characters, the question which first demands consideration relates to the occurrence or rather to the possibility of the existence of arrangements in the structure of an organism that are devoid of utility from their very first appearance. The answer to this question will depend largely upon the point of view from which one approaches the subject of evolution. It must be in the negative, if Charles Darwin's theory of natural selection be accepted; for if, as Darwin assumed, individual variations may take place *in any direction whatsoever*, then it is only owing to the natural selection which occurs in the struggle for existence that living beings are able to acquire a more complex organisation and thus gradually to rise to a higher plane of evolution. On this assumption a newly acquired morphological peculiarity can never be perpetuated by inheritance, unless it performs a definite service in the interests of the organism and hence proves useful in the struggle for existence. Utterly useless features, on the other hand, cannot possibly survive in these circumstances. From the point of view of natural selection all morphological characters are in the nature of useful adaptations, or at any rate must have fulfilled this condition at the time when they became "fixed."

If on the contrary one assumes with Nägeli that the progressive evolution of organisms is controlled by inherent tendencies—that is to say, by conditions located in the living substance itself—or, in other words, that the struggle for existence, instead of effecting the selection of useful characters, merely ensures the elimination of harmful or useless peculiarities, then the question propounded above must be answered in the affirmative, with certain reservations. For in the absence of the struggle for existence and of the unrestricted competition thereby implied, not only useful features—produced according to Nägeli by direct adaptation—but also all useless morphological peculiarities would be inherited, since both types of character develop under the compulsion of *internal* forces, as it were automatically. If we concede this point, we cannot deny the possibility of the inheritance of such structural features as are useless—without being actually harmful—and hence negligible in the struggle for existence, particularly where competition takes a relatively mild form owing to the specially favoured character of the environment, as may happen, for example, in an equably moist and warm tropical climate.

Even in its mildest form, however, the struggle for existence is active enough to prevent the perpetuation of a large proportion of useless morphological peculiarities in addition to the stock of useful adaptations with which every organism is provided. In fact it would not be going too far, altogether to deny the existence, in the internal organisation of plants, of structural units—whether single cells or entire tissues—which have always been functionless. On the other hand, the diversity which evidently characterises the detailed construction of organs of every type, is doubtless often nothing but the expression of an inherent " creative tendency." We are thus brought face to face with the problem of **alternative designs** (*Konstruktionsvariationen*), a phenomenon which is of widespread occurrence in the vegetable kingdom. An engineer may construct several bridges of the same span, of equal strength and at a similar cost, and follow a distinct plan in each case. Similarly, in the organisation of a plant, one and the same tissue-system may be capable of construction according to several distinct plans, none of which offers an appreciable advantage in comparison with the rest. Nature, as it were, takes a pleasure in ringing the changes indefinitely upon the possible variations of structural detail ; in this way arises the inconceivable diversity which prevails with regard to the details of anatomical structure. Not infrequently, indeed, a definite advantage does result from the choice of one among several possible methods of construction. Thus the isolation of the subepidermal fibrous strands which is characteristic of the haulms of the genus *Juncus* (and also of many CYPERACEAE) provides ample room for the intercalation of strips of photosynthetic tissue, and renders it easy for the haulm to assume the rôle of a photosynthetic organ. In many cases, however, it is a matter of indifference as far as the welfare of the plant is concerned, whether one plan of construction is selected rather than another ; it would, therefore, be futile to attempt an explanation of every case of alternative design on strictly utilitarian lines. The palisade-structure characteristic of specialised photosynthetic tissue, for instance, may be produced in two quite distinct ways, namely, either by the development of inwardly directed folds of the cell membrane, or else by the formation of complete partitions, each consisting of an entire cell wall. The two types of structure may be mutually exclusive as regards their occurrence within a group of closely related plants (*e.g.* in the family RANUNCULACEAE), or both may be found in the same individual ; there is, therefore, not the slightest reason to suppose that the two varieties of structure represent separate special adaptations. What the plant really requires in either case is the differentiation of a palisade-shaped photosynthetic element, and the special modes of construction employed for this purpose in different plants are adaptive only in so far as they tend to this common end. Similarly, most of the various (sectional) plans of arrangement embodied in the mechanical systems of inflexible organs are merely expressions of this tendency towards variation of design, and must, as far as their details are concerned, be treated as characters of purely morphological value.

All the phenomena with which the physiological anatomist is concerned partake of the nature of useful adaptations. They may be subdivided into two groups, which are, however, by no means sharply delimited from one another. In cases of **physiological adaptation** the morphological feature under consideration—whether it be a localised structure or an entire tissue-system—is intended to perform a physiological function that is to play a definite part in the internal economy of the plant. But there also exist numerous **ecological** (or **biological**) **adaptations**, which are developed in connection with the manifold requirements arising out of relations between the plant on the one hand, and environmental factors, such as climate, habitat, and the animal world, on the other.

Usually it is the physiological adaptations that mainly determine the structure of the several organs and tissue-systems, ecological adaptations being of minor importance in this respect. The photosynthetic system, for example, displays specialisation principally in connection with the physiological process of photosynthesis, as indeed its name implies. Its structure is such as to provide for the accommodation of the largest possible number of chloroplasts, and for the rapid removal of synthetic products. At the same time a certain amount of adaptation to external conditions, such as the intensity of illumination and the humidity of the atmosphere, is also discernible in the construction of the photosynthetic system. Another illustration is afforded by certain characteristics of the vessels that serve for the conduction of water. Here the presence of bordered pits, and, in fact, the peculiar mode of thickening of the cell-wall in general, is closely connected with a particular physiological process, namely, the ascent of sap. The number and the diameter of the vessels, on the other hand, are dependent upon ecological requirements, which in their turn may be determined by the nature of the habitat—if, for example, one compares a land plant with an aquatic species—or by some peculiarity of habit, as in the case of climbers. More rarely the appearance of a morphological feature is correlated mainly, or even entirely, with some ecological requirement; this condition is exemplified by the dermal system, the principal function of which is protection.

Very often structures originally devoted to purely physiological purposes become in particular species adapted to ecological ends, and accordingly undergo an appropriate change of form. This statement applies more especially to those ecological adaptations which control the relations between plants and animals. It would be a serious misconception to regard such adaptations as ecological in essence and origin. Take the case of the latex-tubes in the involucral bracts of *Lactuca*; these structures stand in open communication with short hairs, which exude an abundance of protective latex if they are injured (*e.g.* by insects). Here the laticiferous system has undoubtedly undergone a secondary specialisation which has nothing to do with its primary physiological function of conducting plastic materials. The same point is illustrated by the case of *oxalic acid*. This substance is a very common bye-product of vegetable metabolism in general and of protein-synthesis in particular. The free acid has a very poisonous action upon protoplasm, and is therefore rendered harmless by conversion into the almost insoluble crystalline *calcium oxalate*. Now, the crystals in question are often shaped like minute needles or darts, and are thus well adapted to provide " mechanical " protection against the attacks of insects and snails. Many other metabolic bye-products may thus secondarily assume an ecological importance. Physiological anatomy takes every adaptation into account, whether the latter be physiological or ecological in nature. But whenever the function of a particular structure has to be determined, full consideration should first be given to all the possibilities of physiological adaptation. Ecological adaptation should not be thought of unless it is found impracticable to bring the structure into relation with any genuinely physiological function, that is to say, with a process which constitutes an integral part of the internal vital activity of the plant. It is perhaps advisable at this stage once more to insist upon the fact that the two forms of adaptation cannot always be sharply separated from one another.[1]

The recognition of the fact that plants exhibit adaptation as regards their internal structure, is quite independent of the various suggestions and hypotheses which have been put forward in order to explain the origin of this adaptation. With respect to the latter point, one may follow Darwin in attributing the " purposeful " character of adaptation to the action of natural selection ; or one may share the belief of Lamarck and Nägeli in the existence of a direct accommodation which takes place without the intervention of natural selection ; or, finally, one may hold that the biologist is altogether unable to frame a satisfactory explanation of the purposeful aspect of nature, and thus relegate the question to the field of metaphysics.

In any case, however, it is impossible to deny the actual existence of purposeful adaptations. Physiological anatomy describes the adaptive features of the internal structure of plants, but is incapable of explaining their origin, and indeed does not profess to do so.

A few words must next be devoted to the general methods of investigation employed by the physiological anatomist. In many cases, an application of the methods of **comparative anatomy** suffices to disclose the physiological significance of a given tissue or arrangement, and to indicate the connection between its structure and its function. In the case of cork, for example, mere observation of the histological features of the cells and of their characteristic disposition at once leads to the conclusion that one is dealing with a protective dermal tissue. Similarly, the manner in which the photosynthetic cells of a leaf are linked up with the parenchymatous bundle-sheaths immediately suggests that the latter serve as paths for the translocation of synthetic products. Finally, the fact that the ventilating spaces of a leaf communicate with the outer atmosphere by means of the stomata can be determined by mere inspection.

Frequently, it is impossible to determine the principal function of a tissue with certainty from anatomical evidence alone. In these circumstances resort must be had to **experiment**. The thick-walled character of fibrous cells, for example, is not in itself sufficient proof that the elements in question are specialised for mechanical purposes ; but it is permissible to ascribe such a function to fibrous tissue when its great tensile strength has been demonstrated by experiment. Similarly, the existence of a photosynthetic tissue-system was not recognised until long after the chloroplasts had been shown to be organs of photosynthesis by means of physiological experiments. Again, the view held by the author, according to which the aleurone-layer of grasses is a glandular tissue engaged in the secretion of diastase, is largely founded upon experimental evidence.

The experimental method is also invaluable to the physiological anatomist as a means of confirming, of correcting, or of supplementing the results obtained by the methods of comparative anatomy. One or two examples will add force to this argument. The demonstration, by means of chemical tests and physiological experiments, of the fact that synthetic products travel in the parenchymatous bundle-sheaths, provides an experimental confirmation of conclusions previously arrived at on anatomical grounds. Similarly, microscopic investigation of one of the sensitive bristles of *Dionaea muscipula* (Ch. XII.) suggests that the structure in question is likely to suffer the greatest amount of deformation, when it is bent, in the constricted " hinge-area "; an actual trial at once demonstrates the correctness of this inference.

Again, quantitative experiments carried out in order to test the relative impermeability of cork towards water and gases throw additional light upon the protective function which has been attributed to corky tissues on the strength of their histological peculiarities. In this way, the results of anatomical investigation are both amplified and rendered more exact by the proper application of the experimental method. In fact, the utility of experiment from the point of view of the physiological anatomist largely depends upon the extent to which this method of research enables him to replace more or less speculative generalisations by well established "laws." No one, accordingly, who works in the field of physiological anatomy, can afford to dispense with experimental methods; at the same time, it will not do to overestimate the capabilities of this instrument of research. As a matter of fact, the experimental method is, in this branch of biology, subjected to two important restrictions.

In the first place, there are many questions concerning the relation between structure and function which do not admit of experimental treatment at all. How, for example, could the experimental method be applied with a view to determining the significance of such minute structures as the flanges of arm-palisade cells and the cuticular ridges of guard-cells, or of features like the parenchymatous shape of fibrous elements, the thickening of the edges of collenchymatous cells, the thin-walled character of tactile papillae, and many similar arrangements? The microscopic size of an object does not of course in itself place an insuperable obstacle in the way of experimental investigation. In the case of the stinging hairs of the Nettle, for example, it can readily be shown by experiment that the thin strip of the wall situated immediately below the head of the hair actually repre-sents the line of rupture. Again, the protoplast of an Algal cell of microscopic dimensions can be divided by plasmolysis into two portions, of which only one contains a nucleus; in this way, the difference of behaviour between nucleated and non-nucleated cyto-plasm can be observed. Chloroplasts even may be isolated and subjected to various tests in order to show that they do not lose their photosynthetic capacity in such circumstances. There are, however, just as many instances in which the physiological importance of microscopic features cannot be investigated by experimental means; where this is the case, it is permissible to base definite conclusions upon the data of comparative anatomy alone.

A further limitation of the experimental method—and one which is often not sufficiently appreciated—arises out of certain fundamental properties of the living organism. It is characteristic of physiological experiments that they expose an organism, or a portion thereof, to conditions which are always artificial and often unnatural. This statement applies more especially to the so-called **method of extirpation** which so frequently represents the only experimental resource at the disposal of the physiological anatomist. This method consists in the removal of a given organ or structure by means of an operation, the subsequent continuance of a particular function being regarded as proof of its connection with the excised structure, while cessation of the same function is interpreted in the opposite sense. This procedure may, however, easily lead to false conclusions for two distinct reasons. In the first place, a serious operation often gives rise to "shock," that is to say, to a condition of temporary or permanent general depression or paralysis; this effect may in reality be responsible for the lapse of a function, with which the extirpated organ is erroneously credited. It would clearly be dangerous to conclude, on the basis of an extirpation experiment, that a particular physiological process had ceased because the active organ had been removed, when it might be shock that was actually responsible for the observed cessation of activity. It is especially in connection with work upon irritability that errors of this kind are constantly liable to vitiate the results of experiment. Only careful study of the shock-phenomena themselves can enable the physiological anatomist to cope with this difficulty to some extent.

An equally serious cause of error in connection with extirpation experiments is the **faculty of self-regulation,** of which every organism is in some degree possessed. A mutilated organism generally "makes shift" to the best of its ability and is often able to transfer the duties of an excised organ to a different structure, and in this way to carry on the threatened function, not indeed with unimpaired vigour, but still to such an extent that the organism as a whole manages to survive. Neglect of this factor is responsible for the erroneous assumption that a structure necessarily *cannot* be the instrument of a particular function, if the latter persists in some degree after the structure in question has been excised. To take a concrete instance: Both anatomical and experimental data led the author to regard the so-called aleurone-layer of grasses as a glandular tissue which supplies the bulk of the diastase that is required for hydrolysis of the starch contained in the mealy portion of the endosperm. Other investigators have subsequently shown that an isolated grass-endosperm can, under certain conditions, hydrolyse its store of starch even after the aleurone-layer has been removed by an operation. But this result, obtained by the method of extirpation, by no means proves that the aleurone-layer plays no part in connection with the hydrolysis of starch under normal conditions; all it does is to demonstrate that the

starch-containing endosperm cells can, if necessary, themselves undertake the task of producing diastase, by an act of self-regulation.
Similarly no one would venture to conclude that the animal stomach
takes no share whatever in digestion, merely on the ground that the
small intestine can assume the digestive functions of the stomach when
the latter is removed by operation.

What has been stated concerning the method of extirpation, applies
essentially to all experiments that aim at suspending the function of a
given organ or tissue, even though the measures employed are less drastic.

The comparative and the experimental methods are evidently of
equal value to the physiological anatomist as instruments of research.
Each of these methods has its strong and its weak points, while neither
can claim to give more reliable results than the other in all circumstances. The investigator must test the value of *both* modes of enquiry
in every individual case, and he must as far as possible arrange his
work in such a manner that the two methods will mutually supplement
and control one another.

This introductory chapter may fittingly conclude with a few remarks
concerning the history of physiological anatomy. Even the founders
of vegetable anatomy and physiology could not altogether fail to observe
that the various members of the plant body are endowed with special
functions. The protective function of epidermis and cork, the conducting capacity of vascular bundles, the absorptive activity of root-hairs, and
many similar relations, have long been matters of common knowledge.
The earlier work in this field, however, produced little more than a
mass of disconnected observations, which admitted only of the vaguest
generalisation; no methodical and exhaustive account of the connection between the structure and the functions of any tissue-system had
been written previous to the year 1874, when Schwendener published
his classical treatise on "The Mechanical Principles underlying the
Anatomical Structure of Monocotyledonous Plants" (*Das mechanische
Prinzip im anatomischen Bau der Monokotylen*), in which the definition
and diagnosis of a tissue-system was for the first time carried out consistently in accordance with the principles now generally accepted by
physiological anatomists. The tissue-system selected by Schwendener
was that which constitutes the skeleton of the plant, and the structure
and arrangement of this "mechanical" system were by him brought
into relation with its function in the most convincing manner.
Schwendener's treatise not only raised the anatomico-physiological
tendency of plant-anatomy to the higher status of an independent
branch of research, but also opened the way for a novel demarcation
and classification of the various tissues with reference to physiological
conceptions.

In 1881 the author published a paper entitled, "The Anatomy of the Photosynthetic Tissue-system of Plants," in which he adopted the Schwendenerian attitude; subsequently other pupils of Schwendener, such as Westermaier, Ambronn, Volkens, Tschirch, Zimmermann, etc., assisted in developing this new branch of vegetable anatomy. As early as 1884 the author was able—in the original edition of the present work—to attempt, for the first time, a connected account of the results that had so far been obtained along this line of investigation, and, by means of his physiological classification of tissue-systems, to outline a scheme into which all the various detailed observations might in due course be fitted. Since that time a host of investigators have been engaged in contributing to the advancement of physiological anatomy.

# CHAPTER I.

## THE CELLS AND TISSUES OF PLANTS.

### A. THE NATURE AND SIGNIFICANCE OF CELLULAR STRUCTURE.

EVERY discussion of the internal structure of plants, whether it be morphological or physiological in tendency, must be based upon a clear conception of the nature of vegetable cells. On this account the present chapter must be prefaced by a statement of certain elementary facts, involving no theoretical assumptions, concerning cellular structure.

If any filamentous Alga such as *Oedogonium* or *Spirogyra* be examined microscopically, it will be found that the liquid and semi-liquid constituents of the plant-body are enclosed within a thin-walled but relatively firm tube. This tube is however not traversed by a continuous cavity, but is at certain intervals divided into separate chambers or compartments by transverse septa consisting of the same substance as the outer wall of the tube. Each of these compartments encloses a separate portion of the above-mentioned liquid and semi-liquid contents. The filament is thus made up of a series of well-defined compartments, which may be regarded as the structural units or elements of the plant-body. Each unit is termed a **cell,** and the thallus of such an Alga is described as a cellular filament.

If a simple type of Moss-leaf be next examined from a similar point of view, it is at once evident that in this case the structural units are disposed in two dimensions of space, or, in other words, that such a leaf represents a cell-plate. Viewed from above, the cell-walls present the appearance of a delicate network, the different meshes or compartments of which vary greatly in shape and size. About the middle of the leaf the meshes are elongated, on either side thereof the cells are more polygonal in outline, while the leaf margin again is mainly occupied by elongated cells with tooth-like projections directed towards the outside. This variety of shape exhibited by the structural units, in the case of a cell-plate, becomes much more pronounced where cells are arranged in three dimensions of space so as to form a cellular mass or cell-body. All the vegetative organs of Higher Plants are constructed on this last-mentioned plan; they consist of structural elements of the most diverse shapes and sizes, comprising in addition to rounded or polyhedral cells of the isodiametric type, also elements which are tabular or prismatic, attenuated or sharply pointed, stellate or irregularly branched, and so forth. Moreover, not only do the cells of Higher Plants display almost endless diversity as regards their shape, but the character of their walls and contents is also extraordinarily diversified. In the case of a Higher Plant—more especially among Phanerogams—it is not always possible to demonstrate the cellular structure of an organ in its adult condition. Even as late as the nineteenth century it remained a matter of doubt whether the theory of the general occurrence of cellular structure could be extended so as to include water-conducting vessels and latex-tubes. But the researches of Von Mohl and Unger among others have rendered it quite certain that all Higher Plants—formerly often designated by De Candolle's term of "Vascular Plants," and thus to some extent contrasted with "Cellular Plants"—are cellular throughout, at any rate at certain stages of their development. The wood-vessels, though apparently non-cellular in the adult plant, always originate from rows of cells in which the transverse walls degenerate at an early stage; the same statement applies also to articulated latex-tubes and to certain other histological elements.

Hitherto in this discussion the cell has been treated merely as the ultimate *structural* element of the plant-body; but it must not be forgotten that every cell represents a unit not only in the morphological but also in the physiological sense. It is in fact a *functional* as well as a structural element.[2]

If the term "organ" be employed in general to denote the instrument wherewith a definite physiological function is performed, then the cell must be regarded as an elementary physiological instrument or "elementary organ." Every cell, namely, performs a definite physiological service either for the whole term of its life or at any rate at some period of its existence, while the sum total of the physiological functions of the various cells constitutes the vital activity of the entire plant. A cell may serve the interests of the organism as a whole, either indirectly, because it possesses a firm membrane or cell-wall, or directly, by virtue of the living cell-body—the protoplasmic cell-contents or **protoplast**—enclosed by the cell-wall. This physiological conception of the cell as a functional unit or elementary organ does not imply that every cell is necessarily possessed of independent vital activity. A cell may itself be quite lifeless, and yet play an important part in the life of the organism as a whole. This statement applies for example to the elements of which cork, fibrous strands and several other important tissues are composed. In such cases indeed it is the cell membrane alone that is entrusted with a definite task. When once this membrane has been manufactured by the living protoplast, the continued existence of the latter becomes a matter of indifference so far as the welfare of the whole organism is concerned; in such cases, as a matter of fact, the protoplast usually dies, and often disappears altogether, as soon as it is no longer required. The necessity of distinguishing between dead and living cells evidently in no way conflicts with the physiological conception which regards the cell as the functional unit of the plant-body.

The majority of cells represent not only elementary organs, but also elementary organisms; in other words, a cell, as a rule, does not merely work in the service of a higher living entity, namely, the entire plant, but also itself behaves as a living entity, though indeed as one of a lower order of magnitude. Thus each chlorophyll-containing palisade-cell in the leaf of a Phanerogam represents an elementary photosynthetic organ, but at the same time constitutes in itself a complete living organism: if proper precautions are observed, such a cell may be removed from the cell-community to which it belongs without incurring immediate destruction. The author has even succeeded in keeping such isolated elements alive for several weeks in suitable nutrient solutions; the cells continued their photosynthetic activity, and actually underwent an appreciable amount of growth after isolation.[3] Naturally this power of independent existence is least developed in the case of cells which are highly specialised, and which therefore represent very efficient elementary organs.

In the preceding discussion it has been shown that cells must be regarded as component parts or elementary organs of the whole plant; at the same time attention has been drawn to the fact that every cell also constitutes an independent entity possessing a distinct individuality of its own. Clearly every unprejudiced conception of the nature and significance of cellular structure must make equal allowance for these two different aspects of the character of a cell. About the middle of the nineteenth century, when the investigation of the cellular structure of plants was leading to quite unforeseen results, Schleiden, Nägeli and other prominent workers were not unnaturally inclined to lay stress upon the individuality of the structural units, and to emphasise the fact of their independent existence as elementary organisms. Schleiden, in particular, adopted a most uncompromising—one might justly say an extreme—attitude with regard to the matter. According to this view the growth and differentiation of each organ and of the plant as a whole is purely the result of formative processes which take place within the individual cells. A reaction against this one-sided conception was instituted by Hofmeister, who was the first to maintain that the growth of individual cells is dependent upon that of the entire organ of which they are the component parts.

In accordance with its twofold nature—as explained above—a cell plays now an active and now a passive part in the multifarious vital activities of the plant. As a part of the whole it is subject to the prevailing tendency of morphogenetic activity and growth; as an elementary organism it exerts a certain amount of control over these very processes, the extent of its influence depending upon its location and upon its special characteristics. Consider, for example, a single one among the innumerable green cells contained in an ordinary foliage leaf. By virtue of its photosynthetic activity such a cell influences the inception of new organs, and consequently the growth of the plant as ·a whole. On the other hand, this green cell would be unable to perform its characteristic functions, were not the leaf of which it is a part first of all enabled by definite formative processes of growth to assume its flatly expanded shape, and secondly induced by its specific mode of reaction towards photic stimuli to take up the position in which it is exposed to illumination in the most favourable manner. Since, namely, every green cell requires a certain intensity of illumination in connection with its photosynthetic activity, proper provision must be made to meet this demand, before the cell is able to perform its normal functions efficiently. On the one hand, therefore, the growth of the plant as a whole depends upon the activity of the individual green cells and upon the nature of the materials which they manufacture, while, on the other, this photosynthetic function of the green cells is itself influenced and controlled by the general growth of the plant and by the character of the architectural plan embodied in its external morphology.

This interaction between the cells in their character of elementary organisms on the one hand, and the plant-body as a whole on the other, may naturally assume a variety of aspects; the more marked the individuality of one agent becomes, the more does the other become subject to its control. Generally speaking, the independence of the individual cells is most pronounced and effective in the case of plants which stand at a relatively low level of evolution. Among the lowest of the Green Algae, the Protococcales, the individuality of the plant-body as a whole is so feebly developed, that the latter is not even regarded as an individual in the sense in which this term is applied to one of the Higher Plants; it is on the contrary customary in these cases to speak of "cell-colonies," "cell-families," or "coenobia," and thus to attach the notion of individuality exclusively to the component cells. The delicate cell-plates of *Pediastrum*, the sac-like net of *Hydro dictyon* and the cell-chains of *Scenedesmus* are all familiar examples of such cell-colonies or coenobia. With every rise in the scale of evolu tion, on the other hand, and with every increase in the perfection of organisation, the independence of the component cells becomes more and more subordinated to the individuality of the plant as, a whole. Certain elements may indeed lose their individuality altogether. Thus cells or cell-masses frequently coalesce with one another, or become disorganised in various ways. The "syncytes" (*Zell-fusionen*) produced in this way are employed for very diverse physiological pur poses, being utilised as water-conducting or as laticiferous vessels, as sieve-tubes, as secretory cavities, and the like. That complete loss of individuality on the part of the single elements is by no means incom patible with the performance of a physiological function, is further demonstrated by the extensive use which is made of dead cells in the construction of the plant-body. The living organism, whether vegetable or animal in nature, has often been described as a "society of cells." After what has been stated above it will be admitted that there is much to be said in favour of this comparison. The simile is undoubtedly justifiable so far as the main principle is concerned; for in the case of human society it is likewise necessary to discriminate between the individuality of the state, on the one hand, and that of the citizen who represents the social unit, on the other.

### B. THE STRUCTURE AND FUNCTIONS OF THE TYPICAL VEGETABLE CELL.

The use of the term "cell" to denote the structural and functional elements of which the bodies of plants and animals are composed, recalls the historical fact that Robert Hooke, the discoverer of the cellular structure of plants, was cognisant only of the relatively firm cell-membrane or wall.

The presence of a hard case of this kind enclosing the "living contents," or protoplasmic cell-body, is a characteristic feature of the typical vegetable cell, and distinguishes it from the cells of animals. In the present chapter, however, the living protoplasm and the various protoplasmic organs will be dealt with first; subsequently the cell-wall will be discussed as far as its general features are concerned.[4]

### 1. *General Characteristics of Protoplasm.*

In all cases, without exception, it is the **protoplasm** (v. Mohl), or **protoplast** (Hanstein), which must solely and exclusively be regarded as the vehicle of the physiological activity of the cell. The process of nutrition and the other chemical changes included in the general category of metabolism are carried on within the protoplasm, or at any rate are initiated by the living substance. It is in the protoplasm also that the interchange of energy occurs which becomes necessary when translocation of material takes place, when movements are carried out in connection with growth and in response to stimulation, when external resistance is overcome, in fact, whenever work is performed internally or externally by the plant or by one of its organs. Further, that remarkable property of living organisms which the physiologist calls irritability, likewise resides in the protoplasm. It is the living substance that controls all the morphogenetic processes which impress a specific character upon each cell as well as upon the entire plant. Finally, it is the protoplasm which initiates and carries out all the processes connected both with asexual and with sexual reproduction, and which itself acts as the vehicle of the heritable characters of the organism.

Every protoplast must evidently be a highly complicated organism, since an activity which is so excessively varied, but which is nevertheless governed by definite laws and characterised by the frequent repetition of similar actions, demands a correspondingly complex apparatus for its performance.[5] At first sight the normal condition of the protoplasm, as revealed by its outwardly perceptible physical characters, seems to be incompatible with the foregoing conclusion. Generally speaking, protoplasm has a semi-liquid, slimy consistency; this fact has again and again given rise to an erroneous doctrine, which refuses to admit the existence of the internal organisation that should properly belong to a living substance and regards the protoplasm as a complex liquid mixture or emulsion—in the physical sense,—within which vital activity proceeds in a manner analogous to the purely physical changes that may be observed in lifeless artificial solutions and mixtures.

A simile may help to explain how a substance, which superficially appears to possess the properties of a semi-liquid material, and the constituent particles of which are therefore readily displaceable with reference to one another, may nevertheless be endowed with a complicated internal structure and organisation. A great army marching into the field is by virtue of its elaborate organisation capable of executing the most varied and purposeful concerted movements and actions; its component divisions and subdivisions do not, however, form a rigid structure, but are on the contrary in the highest degree mobile relatively to one another. This internal mobility may indeed apparently develop into complete disintegration in the case of skirmishes and reconnaissances. Mobility of the component parts is in fact an indispensable condition of efficiency on the part of the entire army, which to the eye of an uninitiated onlooker seems to flow like a mighty stream over the countryside. Similarly protoplasm presents the appearance of a semi-liquid, structureless substance, whereas actually this outward semblance of liquidity merely expresses that very mobile condition of the constituent particles which is essential to the continuance of the highly intricate vital activity. This internal mobility entails, it is true, certain physical consequences; it is, in other words, accompanied by phenomena which appear also in connection with lifeless liquids and emulsions. But these purely physical phenomena have no immediate connection with what may be termed the vital aspect of protoplasmic structure and activity.[6]

It may be remarked that the consistency of the protoplasm is not necessarily the same in every part of a given cell, while one and the same portion of a protoplast may vary in consistency at different times; certain components of the protoplast may temporarily or permanently display a relatively solid character, doubtless in relation to the special functions with which these components are entrusted.

Naturally many attempts have been made to demonstrate that protoplasm is possessed of a *visible* structure and differentiation. Among the various theories which have been propounded in this connection Bütschli's hypothesis of the universal "foam structure" of protoplasm perhaps rests upon the widest basis of fact. There can indeed be no doubt that a delicate alveolar structure of the protoplasm is widely distributed as regards both animal and vegetable cells, and that this structure plays an important part in connection with various functions of the living substance. On the other hand, it cannot be denied that under certain conditions fibrillar structures also appear within the protoplasm. In many cases the fibrillar appearance in reality corresponds to the outlines of a reticulum which has been produced by the partial collapse of the walls of greatly elongated alveoli; in other instances, however, delicate fibrils actually seem to be differentiated within the protoplasm composing the alveolar walls. Strasburger accordingly discriminates between "alveolar" and "filar" plasma, at the same time attributing different functions to these two varieties of protoplasm. The former is termed **trophoplasm** by Strasburger, because it seems to be chiefly concerned with the process of nutrition; the latter is supposed to be active mainly in connection with developmental changes, and is therefore termed **kinoplasm**. Further investigation is required in order to decide how far this distinction is justified.

The chemical constitution of the living substance is quite in keeping with its complex morphological character; a great number of different substances enter into the composition of protoplasm, prominent among them being the *proteins*, which are themselves, of course, the most complex carbon-compounds known to the chemist.[7]

Since the correlation of structure with function is always evident, as regards both the gross anatomy and the histology of the plant-body, a similar connection between the morphological features of protoplasm and the functions of the individual protoplast would doubtless be demonstrable, were it not that most of the structures concerned are of ultramicroscopic dimensions. It is, nevertheless, possible to recognise within the living substance a number of well differentiated cell-organs which are entrusted with definite functions.

The arrangement of the protoplasm within the cell and the disposition of the plasmatic organs are not always the same, but depend at any given moment partly upon the stage of development to which the protoplast has attained and partly upon the nature of its special activities. Moreover, external stimuli may affect both the general disposition of the protoplasm and the arrangement of the cell-organs. In the case of the embryonic cells which compose the growing-points of a Higher Plant, it is usual for the cell-cavity to be completely filled with protoplasm. At the centre of such a cell will be found the most important of all the cell-organs, namely, the **nucleus**. The **centrosome** is a minute organ, which is of general occurrence in animal cells, but which among plants has been identified with certainty only in some of the Lower Cryptogams; where it is present, it is always closely associated with the nucleus. Grouped around the nucleus are several colourless or pale-green corpuscles of moderate size, the **chromatophores**. All the aforesaid organs lie embedded in the general cell-plasma or **cytoplasm**, which is bounded externally, that is to say towards the cell-wall, by the **ectoplast** (*Hautschicht*). In the adult cell the conditions are altered, owing to the appearance within the cytoplasm of cavities filled with **cell-sap**, the so-called **vacuoles**. The vacuoles may ultimately be replaced by a single large **sap-cavity**; in this case the protoplast, with its attendant organs, assumes the form of a **peripheral layer** of varying thickness (the "primordial utricle" of v. Mohl), which adheres to the inner surface of the cell-wall. Not invariably, however, does the whole of the cytoplasm shift on to the walls after the formation of a sap-cavity; very often the latter may be seen to be traversed in various directions by cytoplasmic strands and filaments which sometimes connect a central mass of plasma, in which the nucleus is suspended, with the peripheral layer (Fig. 1).

The cytoplasm—apart from the ectoplast, which is always stationary —very frequently displays a more or less energetic streaming movement which is termed **protoplasmic rotation or circulation**.[8]   This is not the place to consider in detail either the external features of the different forms of protoplasmic streaming or the internal causes of this phenomenon, since these matters pertain to the field of pure physiology. Brief reference must, however, be made to a circumstance which was

first clearly pointed out by de Vries, namely, the fact that these movements of the protoplasm greatly facilitate intracellular translocation and metabolism in general. Rotation and circulation of the protoplasm enable substances which enter the cell to distribute themselves and to intermingle with comparative rapidity, whereas these processes would take place very slowly if they were entirely dependent upon diffusion. The rapid intermixture effected in this manner in its turn appreciably accelerates the transference of material from cell to cell. According to Bierberg, potassium nitrate, lithium carbonate, and thallium sulphate all travel, in the leaves of *Elodea* and *Vallisneria*, by means of protoplasmic streaming, at three or four times the rate that could be maintained by diffusion alone. The importance of protoplasmic movement in this respect must not be underestimated upon the ground that mechanical intermixture can be affected also in various other ways. On the other hand, protoplasmic streaming is not so general a phenomenon as de Vries is inclined to suppose; in many instances, indeed, it takes place only as a result of injury, in which case the cells immediately adjoining the wound display the most active movement. The streaming which is thus induced or accelerated by traumatic stimulation undoubtedly serves to expedite those metabolic processes which tend to bring about the healing of the wound with the smallest possible loss of time.

Fig 1.

Single cell of a hair from a staminal filament of *Tradescantia virginica*. The cytoplasm consists of a peripheral layer, together with a number of protoplasmic strands traversing the sap-cavity. The nucleus lies near the distal end of the cell. ×400. After Kühne.

The course followed by protoplasmic currents of rotation or circulation on the whole bears out the notion that the movement serves to accelerate translocation. In an elongated cell, for example, the current usually flows in a direction parallel to the longitudinal axis of the element.

When it is therefore considered that in the case of a sporangiophore of *Phycomyces* the streaming protoplasm can travel all the way from the base of the organ to its apex in the course of a few hours, it will be realised how the plastic materials and reserve-substances required for the development of a sporangium can be concentrated at the proper point in a comparatively short space of time. When a cell-membrane is locally undergoing growth in thickness or in surface, it is quite usual for protoplasmic currents to move along those regions of the wall which are engaged in growth. Crüger and Dippel have observed this phenomenon in connection with the development of spiral and reticulate thickenings. In other cases streaming may result in a local accumulation of protoplasm near the region of growth; this condition is found in connection with such processes as the thickening of the outer wall of epidermal elements, the strengthening of the ventral walls of guard-cells, the protrusion and apical extension of root-hairs, and so forth.

While in all the instances so far considered the course of the protoplasmic currents is controlled by internal factors, cases are also known in which external influences determine their direction. Thus several investigators have asserted that photic or gravitational stimulation is followed by local accumulation of the protoplasm in root hairs, in the statocysts of root-caps, in sporangiophores, etc.; some of these observations, however, require further confirmation. There can also be no doubt that those changes in the position of chloroplasts, which depend upon variations in the intensity and direction of the incident illumination, are effected by definite though imperceptible movements of the general cytoplasm or of special cytoplasmic fibrillae. It is quite inconceivable that the chloroplasts should move about actively with the aid of pseudopodium-like processes, as was suggested many years ago by Schaarschmidt, and more recently again by Senn. Attention may finally be drawn to Tangl's discovery of the fact, that if a bulb-scale of *Allium Cepa* is subjected to mechanical injury, the protoplasm in the cells bordering upon the wounded surface accumulates on the walls which are nearest to the wound, and in so doing brings the nuclei into juxtaposition with these walls. Here again it is a traumatic stimulus that calls forth a movement of protoplasm in a definite direction. Nestler has since shown that movements of this character are of widespread occurrence. All these facts indicate that it is necessary to distinguish between several distinct forms of protoplasmic movement. The readily visible protoplasmic currents which not infrequently carry chromatophores and nuclei along with them, differ in nature from the imperceptible streaming that brings a nucleus into an *appropriate* position, or that effects a transposition of chloroplasts in response to photic or other stimuli.

## 2. *The Protoplasmic or Plasmatic Membranes.*

Every protoplast is surrounded by a special dermal layer, the external plasmatic membrane or ectoplast. Where this membrane is visible at all, it appears to consist of hyaline non-granular protoplasm, the so-called hyaloplasm; the granular protoplasm or polioplasm enclosed within the ectoplast, on the contrary, presents a more or less turbid appearance owing to the presence of numerous granular or vesicular inclusions, the microsomes. The protoplast is likewise separated from the various vacuoles which it contains by hyaloplasmic membranes, the vacuolar or internal plasmatic membranes.[9]

The ectoplast consists of relatively solid protoplasm, and is possessed of a special structure, by virtue of which it is able to control many of the relations which the protoplast maintains with the outer world. In the case of naked protoplasmic bodies, such as the plasmodia of Myxomycetes or the swarm-spores of Algae, the ectoplast serves principally for protection, a function elsewhere performed by the cell-wall. Further, whether the protoplast is clothed in a cell-membrane or not, the ectoplast invariably regulates the osmotic interchange with the external medium or with adjacent cells. Pfeffer's researches in particular have made it clear that this membrane controls both the entrance of dissolved substances required for the nutrition, and in fact for the general metabolism of the cell, and the egress of bodies manufactured by the protoplast. The osmotic interchange of material does not proceed entirely in accordance with physical laws, but is also influenced to a great extent by the physiological requirements of the cell. The specific structure of the plasmatic membranes is adapted to these requirements; it is for this reason that a substance of relatively high molecular weight may be able to penetrate freely through the ectoplast, while passage is denied to the far smaller molecules of another compound. These statements are equally applicable to the internal plasmatic or vacuolar membranes, the physiological behaviour of which has been particularly studied by de Vries.

The ectoplast is also closely concerned with the production of the cell-wall; indeed, in view of the fact that it lies in close contact with the cell-wall while the latter is undergoing growth and differentiation, there can be no doubt that this plasmatic membrane is directly involved in both these processes. Again, when a limited region of a cell-wall is increasing or diminishing in thickness, the underlying portion of the ectoplast must take an active part in the change, since all such processes of growth or disorganisation are under the immediate control of the living substance. As Noll points out, it is advantageous in this respect that the ectoplast is relatively solid and stationary, and hence capable of exerting a definite influence continuously at the same point.

The attribution of this controlling influence to the ectoplast must of course by no means be held to imply that the latter has a monopoly of morphogenetic activity. It is on the contrary highly probable, as will be shown later on, that the formative processes of the cell in general, and particularly those which relate to the cell-wall, take place under the directive influence of a totally different cell-organ, namely, the nucleus, and that the latter as it were employs the ectoplast as an executive agent. On account of its superficial position the external plasmatic membrane is that part of the protoplast which is first affected by external stimuli, and it is accordingly not improbable that the ectoplast represents the general perceptive organ of the cell; this supposed special function of the ectoplast cannot, however, be considered in detail until a later occasion (cf. Chap. XII.).

It is maintained by de Vries that both the external and the internal or vacuolar plasmatic membranes are independent, autonomous protoplasmic organs, invariably derived from pre-existing structures of the same kind, and never arising *de novo* by differentiation from ordinary polioplasm. This assumption is based upon the following observations, among others. If the ectoplast is subjected to mechanical injury—as, for instance, when a filament of *Vaucheria* is cut across— its severed margins always fuse together again immediately; further the vacuolar membrane can be artificially isolated from the surrounding polioplasm without at once losing its vitality or the organisation associated therewith. De Vries also cites certain observations made by Went, according to which vacuoles are distributed throughout every protoplast and multiply by division only, while those which appear to be in the nascent condition are in reality developing from pre-existing vacuole-forming organs or **tonoplasts.** (Went and de Vries apply this term also to the adult vacuolar membranes.)

This view is opposed by Pfeffer, who holds that the ectoplast and the vacuolar membranes are not autonomous protoplasmic organs comparable to nuclei or chromatophores, but that on the contrary both may be produced anew, as required, by differentiation out of ordinary cytoplasm; the latter, in fact, possesses an inherent capacity of separating itself, when necessary, by the formation of a special limiting layer, both from the surrounding medium and from included sap-cavities. Apart from theoretical considerations, Pfeffer bases his opinion principally upon the fact that formation of vacuoles—in no respect different from normal vacuoles—may be artificially effected by introducing minute crystals of asparagine or calcium sulphate into the interior of a Myxomycete-plasmodium. This is not the place to deal exhaustively with these divergent opinions. It may, however, be remarked that a compromise can be effected between the two attitudes, if it be admitted that plasmatic membranes may exhibit varying degrees of specialisation. When such a membrane merely represents a protective layer and an instrument for regulating interchange of material, it can scarcely be regarded as an autonomous organ. As its structure becomes more complex and as the principle of division of labour—which extends even to the individual protoplast—makes itself felt in connection with its activities, the plasmatic membrane acquires greater independence, and it may thus in certain circumstances even attain to a state of complete autonomy. The autonomous nuclei and chromatophores, which are discussed below (cf. Sections 3 and 4), must after all have been evolved in a similar fashion; at any rate, it is difficult to imagine how they can have originated otherwise.

### 3. *The Nucleus.*

Every typical vegetable protoplast is provided at any rate with one highly individualised cell-organ, namely, the nucleus. The nucleus is characterised as an organ of the first importance, not only by the constancy of its occurrence, but also by the uniformity of its structure.

The principal component of the nucleus is a meshwork of delicate threads, which are intertwined and connected to one another by numerous anastomoses. In properly fixed and stained preparations this **nuclear reticulum** can be investigated in detail. It is then found that the threads themselves are at most very slightly stained, but that they have embedded in them granules of a deeply staining material, the so-called **chromatin**. It is not improbable that these granules contain the *nucleins*, which are the characteristic chemical components of nuclear substance. Somewhere among the convolutions of the nuclear reticulum are situated one or more highly refractive, usually spherical **nucleoli**, which are readily distinguishable from chromatin granules by their staining properties. The meshes of the nuclear reticulum are filled with a homogeneous **nuclear sap**. From the surrounding cytoplasm the nucleus appears to be separated by a protoplasmic **nuclear membrane**.[10]

The shape of the nucleus is to some extent correlated with that of the cell to which it belongs. Thus it is usually rounded, and either spherical or disc-shaped, in isodiametric cells, while in elongated elements it often assumes a spindle- or rod-like form. There are, however, not a few exceptions to this rule. Many fibres (for instance, those of *Linum*) contain small spherical nuclei. The nuclei of the guard-cells of *Ornithogalum umbellatum* are crescentic in outline, while those of the corresponding cells of various grasses are dumbbell-shaped in accordance with the form of the cell-cavity.   More rarely the nucleus is lobed, as in the pollen-grains of various Angiosperms, or provided with slender pointed prolongations, as in the foliar epidermis of *Ornithogalum umbellatum*, in the petiolar hairs of *Pelargonium roseum* and *P. zonale*, and in the digestive cells of mycorrhizal organs, such as the roots of *Neottia Nidus avis* or the rhizomes of *Psilotum*.   In many of these instances the unusual shape of the nucleus is obviously correlated with the special nature of its functions; but further discussion of this point must be postponed.

The size of the nucleus is partly a specific character, which remains constant within certain limits of affinity; but it is also to some extent dependent upon the particular physiological function of the cell or tissue.   The first point may be illustrated by the following examples. Relatively large nuclei occur generally among Conifers and Monocotyledons, while these structures are remarkably small throughout the Fungi.   Among Dicotyledons, the RANUNCULACEAE and LORANTHACEAE are distinguished by the possession of comparatively large nuclei.   As regards the variation in the size of the nuclei in different tissues, prominence may be given to the fact that large nuclei are especially characteristic of meristematic tissues, in which, according to Strasburger, the diameter of the nucleus amounts on an average to two-thirds of the width of the entire cell.   As a matter of fact, Schwarz has shown by careful measurement that the nuclei of meristematic cells become at first even larger, when the latter are transformed into adult elements, but that their size gradually diminishes again as the growth of the cell comes to a standstill.[11]   Sachs was long ago led to attribute an administrative function to the nucleus, on account of its relatively large size in the case of meristematic cells.   Glandular cells also as a rule possess comparatively large nuclei, a circumstance which again suggests that it is the nucleus which controls the metabolic activity of the cell.

The typical vegetable cell is uninucleate.   Treub has shown that in Higher Plants several nuclei, or even a large number of these structures, occur in the elongated fibres and latex-tubes of various species of EUPHORBIACEAE, ASCLEPIADACEAE, APOCYNACEAE, and URTICACEAE, doubtless on account of the unusual size of the cells in question.   For whatever the special functions performed by the nucleus may be, it is clearly an advantage for a protoplast of more than ordinary dimensions to possess, in place of a single large nucleus, a number of equivalent organs of smaller size distributed evenly throughout the cell.   The multinucleate "cells" characteristic of certain Algae and of a great many Fungi must be regarded in a similar light.   On the other hand, the multinucleate condition of aged parenchymatous cells observed in certain Monocotyledons (*e.g. Tradescantia*) by Johow is the result of the fragmentation of a single primary nucleus; this disintegration is a senile phenomenon, and is probably devoid of any special physiological significance.[12]

A protoplasmic organ which is so conspicuous and so universally distributed as the nucleus must undoubtedly play a highly important part in the life of the cell; as a matter of fact, numerous investigators have been striving, during the past three decades, to gain further insight from various points of view into the functions of the nucleus.[13]

The majority of biologists at the present day regard the nucleus as the vehicle of the hereditary characters of the organism; in other words, the nucleus is supposed to contain the hereditary substance or idioplasm.

It was Nägeli, who, in the year 1884, first enunciated a theory, supported by ingenious arguments, to the effect that the protoplasm of every germ-cell must contain a comparatively solid and excessively complex germ-plasm or idioplasm; this substance initiates and regulates the "predetermined and specific developmental activity that results in the formation of cell-complexes of varying size, of a specific plant, of a root or of a hair pertaining to a specific plant." Every recognisable property of the completed organism is present in a rudimentary form in the idioplasm; the latter substance is, of course, not confined to the germ-cells, but occurs also in every somatic cell of the developing organism, initiating and directing its specific activities.

Nägeli pictured the idioplasm to himself as existing in the shape of a system of delicate strands, which form a continuous network penetrating into every part of the plant-body; but he hazarded the suggestion that this substance "is more especially aggregated within the nucleus." Soon afterwards this tentative hypothesis gave place to the theory which regards the idioplasm as located exclusively in the nucleus. This conception again may be traced back to Nägeli, who pointed out that the offspring in general derives its hereditary morphological features to an equal extent from both parents, in spite of the fact that the bulk of the female gamete, or egg, is usually many thousand times greater than that of the male gamete, or sperm. If, however, the capacity for inheritance possessed by the idioplasm is equally great in the case of both male and female gametes—and there is no reason to suppose the contrary,—it follows that equal quantities of idioplasm must be present in the two kinds of gametes, however greatly the latter may differ in respect of the total bulk of the protoplast. The question then arises, as to what part of the sperm contains the idioplasm. Since the cilia and the other *cytoplasmic* portions of the male gamete are adapted for the performance of special physiological functions, it follows that the idioplasm can only be located in the nucleus. This conclusion is supported by the fundamentally important observation that, in the process of fertilisation, the nuclei of the two gametes fuse with one another in a characteristic manner. Further, when the fertilised egg undergoes its first segmentation, the component parts of the nuclear reticulum, the so-called **chromosomes**, split longitudinally into halves, which distribute themselves between the two daughter-cells in such a manner that each of the latter receives equal numbers of " male " and " female " half-chromosomes [that is to say, equal chromosome-halves, derived respectively from the male and from the female parent]. The daughter-nuclei are formed by the combination of these half-chromosomes. Precisely the same apportionment of the chromosome-halves takes place at every subsequent cell-division.

This characteristic behaviour on the part of the two sexual nuclei and on that of the product of their union (the " fusion-nucleus " or " cleavage-nucleus ") was first discovered by van Beneden in the Nematode *Ascaris megalocephala*. Soon after, Strasburger, Hertwig, Kölliker and Weismann, all came to the conclusion, that the nucleus, or more precisely the nuclear reticulum which breaks up into the chromosomes during nuclear division, must be regarded as the vehicle of the hereditary substance.

This conception of the nucleus as a structure which embodies the hereditary properties of the organism, and as an organ which controls the growth and differentiation of the cell, was promptly subjected to the test of experiment both from the zoological and from the botanical side. While the general relations of the nucleus to the processes of regeneration were studied, particularly in the case of unicellular organisms, special attention was directed to the differences in behaviour exhibited by nucleated and non-nucleated fragments of protoplasts. Thus Klebs showed that if a protoplast of a filamentous Alga, such as *Zygnema*, *Spirogyra*, or *Oedogonium* be divided (by plasmolysis with 16-25 per cent. solution of cane-sugar) into nucleated and non-nucleated halves, it is the nucleated portion alone which is able to clothe itself with a new ·cell-wall, to grow in length and generally to resume the condition of a normal cell. The author has made a similar observation with regard to the cells of the hairs of certain CUCURBITACEAE. In these the protoplast not infrequently becomes separated into two parts owing to the very uneven deposition of thickening layers upon the lateral walls; if cell-wall formation continues at all after this separation has taken place, it is only the nucleated half of the protoplast that produces new layers of cellulose (Fig. 2). Schmitz and the author have further demonstrated, in the case of various SIPHONALES, such as *Valonia*, *Siphonocladus* and *Vaucheria*, that an isolated fragment of protoplasm cannot acquire a cell-wall or survive as an independent cell unless it contains at least one nucleus. The author has also observed, that when a filament of *Vaucheria* is cut in two, the chloroplasts are withdrawn from the vicinity of wound, whereas the numerous small nuclei, on the contrary, remain in close proximity to the new cell-wall which proceeds to develop over the wounded surface.

Closely related to the last-mentioned phenomenon are the facts recorded by the author concerning the position of the nucleus in growing cells. In such cases, namely, the nucleus usually lies in more or less close proximity to the region of most active, or of most protracted growth; this statement applies to the growth of the cell as a whole, as well as to the special processes of thickening or surface-extension of the cell-wall. Occasionally a direct communication is maintained between the nucleus and the region of active growth by means of protoplasmic strands. In those epidermal cells, for example, in which the outer walls become much thicker than the lateral and inner walls, the nucleus generally lies in contact with the outer wall to begin with, whereas it may shift on to one of the other walls, when the growth of the cell comes to an end. In *Aloe verrucosa* the external wall of each foliar epidermal cell is provided with a minute solid papilla, which first makes its appearance in the young cell in the form of a sharply defined cushion-like thickening of the membrane (Fig. 3 B). The nucleus places itself in close contact with this cushion, and it remains there until the thickening of the outer wall is completed (Fig. 3 c); subsequently it often assumes a different position. The epidermal cells of pericarps and seed-coats not infrequently have specially thickened inner walls; here again the nucleus is often (*e.g.* in *Carex* and *Scopolina*, Fig. 3 D) found lying in contact with the wall which is undergoing growth in thickness.

Fig. 2.

Cell from a hair of *Sicyos angulatus*, in which the protoplast has become divided into nucleated and non-nucleated halves; of these, only the former has secreted a new cell-wall.

Where growth in surface is restricted to a definite portion of the cell-wall, the nucleus likewise often lies in close proximity to this growing area. In *Pisum sativum* and in many other plants the protrusion of the outer wall of a rhizodermal cell, which represents the first visible stage in the development of a root-hair, always takes place exactly opposite the position of the nucleus (Fig. 3 A). Later the developing root-hairs exhibit pronounced apical growth, and the nucleus accordingly almost invariably takes up its position close to the apex.

In *Brassica Napus* some of the root-hairs branch, while remaining unicellular; in these the nucleus always lies in the branch which displays the most active growth in length.   On the basis of the facts detailed above, and in consideration of a number of similar observations, the author has come to the conclusion that the nucleus exerts a direct influence upon the growth and differentiation of every cell.   In view of what has been stated above, one is tempted to assume that it is the presence of the idioplasm which endows the nucleus with this power of organisation and administration.

The author's observations upon the connection between the position of the nucleus and its administrative functions have since been confirmed and supplemented by several other investigators.   With regard to the recent work upon this subject, it will only be necessary to discuss some of the facts discovered by Magnus, by Shibata, and by Guttenberg; all these observers attach a definite physiological significance not only to the location of the nucleus, but also to the shape assumed by this organ.

The roots of *Neottia Nidus avis* contain a fungus-mycelium which, in certain cells, becomes aggregated into dense coils of hyphae containing an abundance of protoplasm; sooner or later these coils die, whereupon their contents are rapidly digested and absorbed by

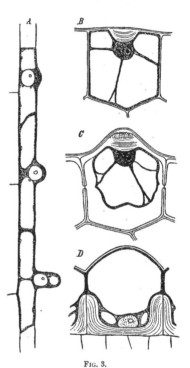

Fig. 3.

Location of the nucleus in growing cells. *A*. Origin of root-hairs in *Pisum sativum*.  *B, C.* Formation of a papilla in the outer wall of an epidermal cell of *Aloe verrucosa*.  *D*. Thickening of the inner epidermal walls in the testa of *Scopolina atropoides.*

the "host-cells."   The undigested remnant of each coil becomes compacted into a "clump" and surrounded by a sheath of protoplasm, which subsequently becomes transformed into a substance resembling cellulose.   The latter part of the process suggests an act of excretion, whereby the indigestible portions of the mycelium are rendered innocuous.   Now Magnus has noted that the nucleus of the host-cell always adheres closely to the clump; moreover, on the side nearest the clump the nucleus loses its definite outline, and may even put forth delicate processes, whereas on the other side its contour remains sharply defined. From the position taken up by the nucleus, and from the local increase of surface which it exhibits, Magnus quite properly concludes that this organ enters into active communication with the protoplasmic sheath, while the latter is undergoing transformation into cellulose. A very similar state of things has been observed by Shibata in the rhizome of *Psilotum triquetrum*. Here, again, the nuclei of the digestive cells are provided with amoeboid processes, and are situated close to the hyphal clumps, or actually adhere to the latter. In *Psilotum* a substance akin to *amyloid* is produced in place of cellulose for the purpose of cementing the clump together. In the galls produced on plants of *Zea Mays* by *Ustilago Maydis*, the hyphae of the parasite, on penetrating into the cells of the host, often become enclosed in a fairly thick sheath of cellulose secreted by the protoplasm of the host. Guttenberg describes the behaviour of an infected cell as follows: The nucleus moves towards the invading hypha, which thereupon rapidly becomes enveloped in cellulose; if the hypha continues to grow, the nucleus, which frequently assumes a lobed form, remains near the advancing apex, while the cellulose sheath is gradually extended.

If the nucleus really is the vehicle of the idioplasm, its controlling influence obviously cannot be confined to those ontogenetic processes which are concerned with the secretion and growth of the cell-wall. On the contrary, the nucleus must exercise some control over all the various components of the developing cell. But the nature of the subject is such, that very few data are as yet available, apart from the above-mentioned observations with regard to the formation of cellulose membranes. For the present, therefore, it must suffice to remark, that there are a number of circumstances which indicate that the nucleus exerts a certain amount of influence upon the growth and division of chromatophores. In the young cells of Higher Plants, the chromatophores either take the shape of pale-green chloroplasts, or else are present in the embryonic form of leucoplasts; in either case they are, to begin with, usually aggregated around the nucleus, and for a time grow and divide in this position. In species of *Selaginella* the author finds, in each cell of the fundamental meristem of the stem-apex, a single pale-green chloroplast, which almost invariably lies in close contact with the slightly larger nucleus. By repeated division this original chloroplast gives rise to a chain of chlorophyll-corpuscles, which at one point still remains in close relation with the nucleus. It occasionally happens that the original chloroplast is at first not actually in contact with the nucleus, but is only indirectly connected with it by means of a protoplasmic strand;

as soon, however, as the chloroplast begins to divide, the nucleus assumes its normal position in the immediate vicinity of the developing chlorophyll apparatus. These facts seem to indicate that growth and division of chloroplasts are, to some extent at any rate, controlled by the nucleus.

A still more convincing argument in favour of this view is furnished by Gerassimoff's observations upon *Spirogyra*. Gerassimoff has devised a method—which will be described below—for obtaining cells of *Spirogyra* furnished, either with two nuclei, or else with a single nucleus which is nearly twice as large as the nucleus of a normal cell. In such cells the ribbon-shaped chloroplasts are broader, more twisted and provided with more numerous marginal lobes in the neighbourhood of the nuclei than towards the ends of the cells. Moreover, these chloroplasts produce lateral out-growths, which detach themselves from the parent organ and develop into independent chloroplasts; for this reason, cells with the double allowance of nuclear substance usually contain from eleven to fourteen chloroplasts—in extreme cases as many as sixteen—as com-pared with the seven to nine which are present in normal cells. Gerassimoff is undoubtedly justified in attributing the difference in the development of the chlorophyll apparatus in the two cases to nuclear influences.

It is only reasonable to suppose, that the nucleus can exert such localised influences as have been illustrated by the preceding examples with the greatest certainty and efficiency, when it lies in close proximity to the site of the process affected; at the same time, there is no reason why a similar controlling influence should not in other cases be trans-mitted over considerable distances. The author has already pointed this out in his treatise on "The Relation between the Position of the Nucleus and its Functions," where he describes a number of cases, in which protoplasmic strands, connecting the nucleus with the growing region of the cell-wall, may serve to transmit stimuli from the con-trolling organ to the site of growth. The cystolith cells of *Ficus elastica* provide a case in point. Here the nucleus rests against the *inner* wall of the young cell, exactly opposite the peg of cellulose which forms the basis of the cystolith; almost invariably, however, the nucleus is connected with the peg by means of a special strand of pro-toplasm. Another aspect of the same phenomenon has been described by Miehe and Strasburger, among others; these observers state that the ectoplast is frequently connected to the nucleus by "kinoplasmic filaments," particularly in the case of embryonic cells. The proto-plasmic threads connecting nuclei with chromatophores (especially with chloroplasts), which have been noted by Lidforss as occurring in a

c

number of plants, are evidently of a similar nature. Townsend, finally, has not only observed the occurrence of a transmission of nuclear stimuli by means of protoplasmic threads, but has actually shown by experiment that such threads may serve to convey the influence of a nucleus to non-nucleated masses of cytoplasm. Similarly nuclear influences may be transferred from one cell to another through the protoplasmic connecting-threads which traverse the intervening wall. None of these facts, however, in any way detract from the importance that should be attributed to the controlling action which the nucleus can exert at close range. Pfeffer is no doubt justified in comparing the transmission of nuclear influences to the action of a telephone "which enables communications and orders to be repeated at a great distance from their original source"; nevertheless it cannot be denied, that commands and messages will be transmitted from master to servant with the greatest ease and with the smallest risk of confusion, when the two parties to the transaction are standing face to face.

The dominating influence exerted by the nucleus during the development of the cell has led many physiologists to assume, on insufficient grounds, that this organ similarly controls the entire vital activity of the *adult* cell. The behaviour of protoplasts which have been deprived of their nuclei shows that this inference is not justified. For it is found that such enucleate masses of protoplasm continue to respire, while protoplasmic streaming, the action of cilia and other movements also persist for some time ; the photosynthetic activity of the chloroplasts is not at first affected, and among CONJUGATAE the formation of starch is not inhibited, though in the case of *Funaria* this process does seem to be dependent upon the presence of the nucleus. Very instructive are Gerassimoff's observations upon enucleate cells of *Spirogyra* and other CONJUGATAE. In some of Gerassimoff's experiments the enucleate condition arose naturally, owing to the circumstance that on division of a nucleus both daughter-nuclei became enclosed in the same daughter-cell. A similar result can be artificially induced if nuclear division be interrupted, after cell-division has begun, by suddenly cooling the cell. Such enucleate cells always die after a few weeks, although the vital processes enumerated above are carried on in the meantime; they are also more susceptible to the influence of hostile environmental conditions, and seem in particular more liable to be attacked by certain parasitic Fungi. Observations of this kind prove that, while a number of vital processes may in the adult cell be independent of nuclear control, the nucleus nevertheless exerts an important though as yet undefined influence, at this stage also, upon the activities of the cell.

In what manner stimuli are transferred from the nucleus to the ectoplast and to the other cell-organs, is at present quite unknown.

The process may be purely physical and consist in the transmission of some particular form of molecular motion ; or it may involve chemical changes, such as the secretion of an enzyme or other active substance. Very likely one or the other type of action is employed alternatively, according to the kind of metabolic change or growth-process which has to be induced.

It has already been stated, that it is not the whole of the nucleus which should be regarded as the vehicle of the idioplasm, but only that portion thereof which gives rise to the chromosomes, namely, the nuclear reticulum. The physiological importance of the other components of the nucleus is still obscure. The nuclear membrane in all probability represents a plasmatic membrane (comparable to a tonoplast), which regulates the interchange of material between nucleus and cytoplasm. The nuclear sap presumably serves in the first instance merely as a medium in which the reticulum can be suitably suspended ; it may also to some extent act as a repository of reserve-materials. Opinions differ as to the significance of the nucleoli, which are ordinarily the most conspicuous inclusions of the nucleus. Some biologists hold that the nucleolar substance is a reserve-material which is absorbed by the chromosomes during mitosis, and secreted afresh when the daughter-nuclei are reconstituted. Strasburger likewise regards the nucleolar substance as a reserve-material, but believes that it serves to replenish the " kinoplasm," especially in connection with the formation of spindle-fibres. Häcker finally maintains that nucleolar material is a by-product of the vegetative metabolism of the nucleus, and that it is excreted into the cytoplasm, during mitosis, either in the solid form or in a state of solution.[14] The function of the nucleolus is likely to remain a matter of uncertainty, until our knowledge of all the chemical changes attendant upon mitosis is more exact than it is at the present day.

Whereas it was formerly supposed that nuclei might become differentiated *de novo* from cytoplasm (by so-called " free nuclear formation"), it is now established beyond fear of contradiction that these bodies never arise otherwise than by division [or fusion] of pre-existing nuclei. All the nuclei, therefore, which are present in the adult organism are the lineal descendants of the nucleus contained in the germ-cell. It is quite exceptional for a germ-cell to contain more than a single nucleus (instances are furnished by uredospores and by the ascospores of *Pertusaria*). The detailed discussion of nuclear division must be postponed until cell-division comes to be considered, because in all ordinary uninucleate cells the two processes are correlated with one another in a peculiar manner.

The structure of the nucleus is extraordinarily uniform not only among Algae and Fungi and in all the higher groups of plants, but also throughout the animal kingdom; it is therefore generally believed that the nuclei are homologous in all these various phyla. It is only in the very lowly organised Schizophyta (Cyanophyceae and Bacteria) that nuclei of the normal type have not as yet been identified. Whether the so-called chromatin-granules which occur in Bacteria partake of the nature of nuclei or not, is quite uncertain. The "central body" of the Cyanophyceae, which has been studied in detail by Zacharias, Bütschli and Palla among others, also differs in so many points of structure and behaviour from a typical nucleus, that the two organs cannot be homologized.[15] It is nevertheless conceivable that this central body may perform at any rate some of the functions of a nucleus, and that it may even stand in a certain remote phylogenetic relation to the typical nucleus of higher organisms.

### 4. *Chromatophores.*

The **chromatophores** are characteristic protoplasmic organs, which are peculiar to vegetable cells. They occur in all the great groups of plants, with the exception of the Fungi, Bacteria, and Myxomycetes. Chromatophores are not, however, necessarily present in all the cells and tissues of the plant-body, and they are abundantly developed only in tissues which are concerned with certain special physiological or ecological functions. Chromatophores are distinguished from other protoplasmic organs by the fact that they contain characteristic pigments, or, at any rate, possess the capacity of assuming a pigmented condition in certain circumstances. Since A. F. W. Schimper's classical researches on the subject, it has been customary to recognise three varieties of chromatophores, namely, **chloroplasts**, **chromoplasts**, and **leucoplasts**.[16]

**Chloroplasts** owe their characteristic colour to the presence of a green pigment, *chlorophyll*, which is accompanied by small quantities of yellow colouring matters, or *xanthophylls*. In those Algae which are not green in colour, chlorophyll is nevertheless present, but is masked by accessory pigments. Among Algae the shape of the chloroplasts varies greatly; the Higher Plants, on the other hand, almost invariably possess small discoid "chlorophyll-corpuscles." Chloroplasts are structures of the highest physiological importance. They are the actual photosynthetic organs of the cell, and hence attain their most characteristic development in specialised photosynthetic tissues, although they occur in smaller numbers in many other tissues as well. Since, however, they are not essential components of every unspecialised plant cell, they will not be discussed in detail until the photosynthetic system is under consideration (cf. Chap. VI.).

**Chromoplasts** are coloured yellow, orange, or red by various pigments, which are probably all closely related as regards their chemical constitution.   In some cases the characteristic pigment is suspended in a colourless protoplasmic matrix [or **stroma**] in an amorphous state, or rather in the shape of minute globules or vesicles (**grana**); in others it takes the form of crystals, which may be tabular or rod-shaped, but which most often are exceedingly slender and irregular spindle- or needle-shaped bodies scattered through the matrix or collected into bundles or sheaves (Fig. 4 b).   Occasionally more than one kind of colouring matter is present in the same chromoplast.   The chromoplasts of the mesocarp of *Lycopersicum esculentum* (the Tomato) and of *Solanum dulcamara*, for example, contain both yellow grana and red crystals.   In certain cases the crystalline pigment is accompanied by

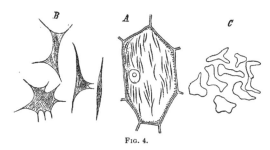

Fig. 4.

A. Cell from a perianth-segment of *Hemerocallis fulva*, showing spindle-shaped chromoplasts.  B. Chromoplasts from fruit of *Sorbus Aucuparia*, filled with a dense weft of thread-like crystals of colouring matter.  C. Lobate chromoplasts from the petals of *Genista tinctoria*.  A and B after Schimper.

protein-crystals (crystalloids).   Chromoplasts may be spherical, spindle-shaped, or quite irregular in outline (Fig. 4); their form is very often modified or altogether determined by the crystalline inclusions.   Unlike chloroplasts, chromoplasts are of little or no value from the point of view of nutritive metabolism; inasmuch, however, as they are often responsible for the bright attractive colours of petals and ripe fruits, their ecological importance is considerable.

**Leucoplasts** are typically colourless, but frequently possess the capacity of forming colouring matters under certain conditions, and of thus becoming transformed into chloroplasts or chromoplasts.   They are commonly minute spherical bodies, but their shape may also be greatly modified by inclusions, consisting of protein-crystals or starch-grains.   Leucoplasts are employed for a variety of purposes.   In embryonic cells they represent future chloroplasts or chromoplasts, which are still in a non-pigmented juvenile condition.   In storage-cells they are responsible for the formation of starch grains at the expense of soluble carbohydrate material, and are hence termed by

Schimper amyloplasts (*Starkebildner*). In epidermal cells, and especially in hairs, they often correspond to chloroplasts which have become abortive, and sometimes seem to be functionless throughout their existence. In certain plants (some ORCHIDACEAE and COMMELYNACEAE), on the other hand, they are so large and numerous as to arouse the suspicion that they perform some special but hitherto undiscovered function.

The three different types of chromatophore enumerated above are phylogenetically homologous structures, which in many cases also display direct genetic relationships. Thus a large proportion of the leucoplasts which occur in the embryonic cells of Phanerogams subsequently become converted into chloroplasts, while many of the latter in their turn change into chromoplasts at a later stage of development. A similar transformation of leucoplasts into chloroplasts may also take place within permanent tissues, for instance in the peripheral parenchyma of a Potato-tuber which has been exposed to light for some time. As, already stated, chloroplasts may on the other hand revert to the colourless condition in certain cases, for example, in epidermal cells and especially in trichomes. Schimper is probably right in assuming that the chloroplast, or rather the photosynthetic chromatophore generally, is the most primitive of the three known types; from the phylogenetic point of view leucoplasts and chromoplasts would thus represent later developments, which arose concurrently with an increase in tissue-differentiation, in accordance with the principle of division of labour.

Previous to the researches of Schmitz, Schimper and Arthur Meyer, it was generally believed that chromatophores—and more especially chloroplasts, which were at that time the only chromatophores that had been at all adequately investigated—could originate by differentiation from ordinary cytoplasm, as well as by division of pre-existing chromatophores. It is now almost universally admitted that chromatophores never arise otherwise than by division. This view was first put forward by Schmitz with reference to the chloroplasts of Algae, but has since been extended by Schimper and by Meyer so as to include chromatophores of every sort. Schimper was able to show that in a number of cases (roots of *Azolla* and *Lemna*; aërial roots of epiphytic Orchids; apical cells of Mosses) the cells of the apical meristem actually contain bright-green chloroplasts. Where the apical meristem is devoid of chlorophyll, it is, at any rate, provided with leucoplasts, which become converted into chloroplasts later on; it is an easy matter to verify the presence of these leucoplasts in the apical meristems of such plants as *Impatiens parviflora*, *Tropaeolum majus*, and *Dahlia variabilis*. According to the author's own observations, the growing-point of the stem of *Selaginella Kraussiana* and *S. Martensii* contains minute spherical leucoplasts, which become transformed into small pale-green chloroplasts at a distance of ·15 to ·2 mm. from the apex (Fig. 5B). In this case, each embryonic cell is provided with a single chromatophore, which, to begin with, divides once for every division of the cell; at a later stage the chromatophores multiply more rapidly than the cells. In the case of the pendent and flaccid young leaves of many tropical trees, multiplication of the chromatophores ·is postponed until an unusually late stage of· development,

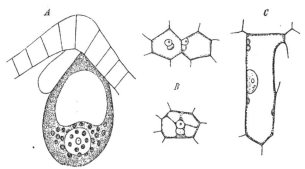

FIG. 5.

*A.* Egg-cell of *Hyacinthus nonscriptus*, showing leucoplasts with included starch.
*B.* Meristematic cells from the growing-point of the stem of *Selaginella Kraussiana* ; each contains a single chloroplast, which lies closely appressed to the nucleus.
*C.* Palisade-cell from a still colourless young leaf of *Humboldtia laurifolia* ; only a few pale-green chloroplasts are present. *A* after Schimper.

although the mesophyll may ultimately become very rich in chlorophyll. Each palisade-cell of *Humboldtia laurifolia* or *Maniltoa gemmipara*, for example, contains only one or two pale-green chloroplasts long after the leaves have emerged from the embryonic state, at a time when their internal differentiation is otherwise well advanced. This circumstance accounts for the white or pale-pink hue which is characteristic of young foliage of this type.

Not only can the chromatophores of permanent tissues be traced right back to the apical meristems, but the presence of these bodies in the germ-cells can also be clearly demonstrated. This fact was established first of all by Schmitz with reference to the asexual resting-spores, zoospores and female gametes of Algae. The spores of Bryophyta and Pteridophyta likewise contain chromatophores, which are particularly numerous and brightly coloured (green) in the case of *Equisetum*. Schimper discovered pale-green chloroplasts in the egg-cells of *Anthoceros laevis* and *Atrichum undulatum*, and among other Bryophyta ; the same observer was able to demonstrate the occurrence of more or less numerous leucoplasts within the egg-cells of several Phanerogams (*Daphne Blagayana, Hyacinthus non-scriptus, Torenia asiatica.* Cf. Fig. 5 A). Nothing whatever takes place, either during the ripening of the seed or at its germination, to justify the supposition that these leucoplasts are dissolved and replaced by newly differentiated chromatophores. The structures in question are not, indeed, always readily visible during this stage of the life history; but the difficulty in this respect arises from the presence of large numbers of more conspicuous inclusions, such as starch-grains, protein-granules, or oil-drops, and all the available evidence argues strongly in favour of the persistence of the chromatophores. In any case, there are no valid reasons for assuming that chromatophores originate *de novo* in the developing embryo.

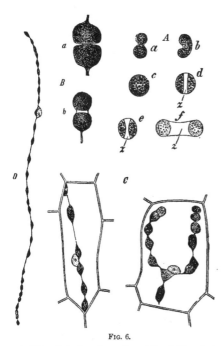

FIG. 6.

A. Chloroplasts from the cortex of the aërial root of *Chlorophytum comosum*, illustrating two modes of division; a and b, simple fission; d-f, division with formation of a colourless equatorial zone. B. Dividing chloroplasts from the cortical parenchyma of the stem of *Selaginella Kraussiana*. C. Chains of chloroplasts from the innermost cortical layer of *S. Kraussiana*. D. Long chain of chloroplasts from the outer cortex. A after Mikosch.

It may therefore be regarded as an established fact that chromatophores—like nuclei—invariably arise by division of pre-existing organs of the same kind. The act of division itself has been studied in detail mainly in the case of chloroplasts. Typically, chloroplasts divide by a process of simple constriction, which closely resembles the amitotic division of nuclei; so far as is known the structure of the chloroplast undergoes no alteration during this process. Fission of chloroplasts was first observed by Nägeli in certain Algae (*Bryopsis, Valonia*), in *Nitella* and in Fern-prothalli. Mikosch has observed a somewhat more complicated mode of division in the cortical cells of the aërial roots of *Chlorophytum comosum* (Fig. 6A). Here the chloroplast exhibits a uniform green coloration in the resting condition. Division is initiated by an aggregation of the pigment at two opposite poles, and by a concomitant disappearance of colour in the equatorial zone. Later, the chloroplast becomes almost dumbbell-shaped, the two polar areas becoming more and more sharply delimited from the colourless equatorial zone, and gradually rounding themselves off to form the daughter chloroplasts; as the latter move further and further apart, the colourless zone becomes more and more indistinct and finally disappears altogether. The author has found that the chloroplasts in the cortical cells of the stem of *Selaginella* illustrate every stage of transition between simple constriction and the more elaborate form of division which involves the early differentiation of a colourless equatorial zone. In the former type of division the substance of the chloroplast long remains green in the constricted region, which indeed does not lose its colour until it has been drawn out into a fine thread. In the other type a colourless zone is already visible at a time when the constriction is very slightly marked (Fig. 6 B). It is an interesting fact, that in these cortical cells of *Selaginella* the fission of the chloroplasts is never quite complete, inasmuch as the daughter-chloroplasts formed at each division remain joined together by a delicate colourless thread (Fig. 6 c, D); the chain of chloroplasts, which is found in every old cell of the cortex, is thus the product of the division of a single original chloroplast.

### 5. Cell-Sap.

It has been pointed out above that when an embryonic vegetable cell begins to grow the protoplast soon develops cavities of varying size, which become filled with a watery liquid, the **cell-sap**. As the cell increases in size, these vacuoles usually coalesce to form a single principal vacuole or sap-cavity.[17] The cytoplasm is separated from each vacuole, or from the single sap-cavity, by the internal plasmatic membrane or vacuolar membrane, which by de Vries is regarded as an autonomous cell-organ [cf. p. 25].

The cell-sap usually consists of a clear, watery liquid, exhibiting an acid reaction, and containing various inorganic and organic substances in solution, especially the latter. It is the nature of the soluble contents which determines the part played by the cell-sap of a given protoplast in the economy of the particular cell, and in that of the entire plant.

The most widely distributed constituents of cell-sap are certain organic acids (*malic acid, oxalic acid,* etc.), and their salts; these, together with other crystalloid substances, are responsible for the more or less considerable **osmotic pressure** exerted by the cell-sap upon the peripheral protoplasm and thus indirectly upon the cell-wall. The **turgor** of the cell, which is the resultant of the osmotic pressure of the protoplast and the elastic tension of the cell-wall, is of vital importance to the plant in a variety of ways. Thus the comparative rigidity of turgescent cells and tissues contributes very appreciably to the mechanical strength of the plant-body ; turgor may even be the sole source of mechanical strength, for instance in the case of small organs which are devoid of special strengthening cells, or where the entire plant is minute. It is by means of turgor-pressure that a growing plant overcomes internal and external resistances, while variations of turgor play an essential part in the execution of many paratonic movements.

Cell-sap also frequently contains reserve-materials. In the case of specialised water-tissues the water that forms the bulk of the cell-sap itself represents a store which is drawn upon by other living tissues, and in particular by the photosynthetic cells, in time of drought. Among the plastic materials—that is to say, such substances as are subsequently used up in connection with growth-processes—which are commonly contained in cell-sap, the carbohydrates (*sugars* and *inuline*) are the most important, but proteins and the amide *asparagine* are also of frequent occurrence. The so-called aleurone-grains, for example, which are often found in the storage tissues of seeds, are produced by the drying up of protein-containing vacuoles. Varying amounts of inorganic nutrient salts in the form of *nitrates, sulphates* and *phosphates* may also be stored in the cell-sap On the other hand, vacuoles not infrequently contain various by-products of metabolism—commonly termed excretory substances—such as *alkaloids*, as well as certain compounds, the physiological value of which is as yet uncertain, namely, *tannins, glucosides*, and so forth. Both these bodies of obscure significance, and the definitely excretory waste-products, may find an ecological application as instruments of protection against the attacks of animals. Substances dissolved in the cell-sap may also be utilised for other ecological purposes. In many flowers and fruits the presence of the pigment *anthocyanin*, forming a red solution in acid and a blue one in alkaline cell-sap, helps to attract pollinating or seed-dispersing animals ; in the case of fruits the sugar contained in the cell-sap performs a similar attractive function. Anthocyanin is also widely distributed as a constituent of the cell-sap in vegetative organs, especially in leaves, where it in many cases probably acts as a light-screen which prevents excessive illumination.

Sometimes the vacuoles contained in a protoplast are not all of the same nature, but differ from one another in the chemical composition of their contents or in other particulars. Thus one may find vacuoles containing coloured cell-sap side by side with others of which the contents are colourless ; similarly, vacuoles containing tannin occur in company with others which are devoid of this substance, and so forth. In this connection reference must be made to the **physodes** described by Crato, peculiar vacuoles—attaining their greatest development in certain Algae—which have the power of actively changing their shape. The **elaioplasts** discovered by Wakker are possibly also vacuolar in character; these structures, which are particularly prevalent in the epidermal cells of ORCHIDACEAE and LILIACEAE, appear to consist of a protoplasmic stroma containing numerous drops of oil in suspension.

. The **pulsating** or **contractile vacuoles**, finally, which occur in certain PROTOCOCCALES, and also in the zoospores of all the Myxomycetes, of many Algae and of a few Fungi, seem to be organs *sui generis*. The mechanism of their pulsation, that is, of their rhythmic disappearance and reappearance, is not properly understood; it is certain, however, that these structures perform very special functions, which are connected in some cases with metabolism, but in others with the process of locomotion.

### 6. *The Cell-Wall.*

Plants exist in the form of naked protoplasts only during the earliest stages of their ontogenetic development. Thus the zoospores of Algae and Fungi, ascospores and all gametes, are at first devoid of a cell-wall. Again, the synergidae and antipodal cells of the Angiospermic embryosac, and the pro-embryo-cells produced (by free cell-formation) from the oospore in many Gymnosperms are naked for a time, or even throughout their existence. Apart from such exceptional cases, however, it appears to be a general rule among the Higher Plants that the fertilised egg-cell and all the succeeding generations of cells are from the very first provided with a cell-membrane.

In the growing-points the cell-walls are always thin and delicate. The conversion of embryonic cells into permanent tissue is almost invariably attended by a more or less pronounced growth in thickness of the cell-walls, usually in relation to increased mechanical requirements. Comparatively rarely the thickening serves for a different purpose, such as the storage of water—in which case it is accompanied by mucilaginous modification—or the deposition of nutritive reserve-cellulose, as in the endosperm of the Date and in certain other storage organs. The cell-wall probably never undergoes uniform growth in thickness over the whole of its surface; in typical cases a number of more or less well defined areas remain entirely unthickened, while the intervening portions grow in thickness either in a **centrifugal** or in a **centripetal** direction.

**Centrifugal** growth in thickness is necessarily confined to such cells as expose a free surface all round, or at least on one side. The best examples of this mode of thickening are to be found among pollen-grains and spores. The warts, spines, or flanges, often elaborately sculptured or woven into complicated patterns, with which these structures are adorned, subserve a mechanical function by producing a rough surface and thus facilitating adhesion to the substratum—which in the case of pollen-grains is, of course, represented by the stigmatic surface. Centrifugally produced thickenings, of uncertain or entirely obscure significance, occur on the outer surface of epidermal cells and hairs, where they usually consist of slender ridges or minute nodules, less frequently taking the form of large prominent knobs.

**Centripetal** thickenings display far greater diversity as regards both their qualitative and their quantitative development. For various reasons a portion of the cell-wall almost always remains thin, in connection with this type of thickening. Sometimes partial thickening suffices to provide the required degree of mechanical strength and rigidity; thus, in the case of a velamen-cell or in that of a reticulate wood-vessel, the absence of thickening over a considerable area merely represents an economy of material. In other instances it is important that the growth in thickness of the membrane, which must take place in order to satisfy mechanical requirements, should not render the cell or vessel incapable of altering its shape, of increasing or diminishing its volume, and, if need be, of growing in length. A compromise is therefore effected, whereby certain regions of the wall are maintained in an unthickened, pliant and extensible condition. In the case of the water-tissue of *Aeschynanthus*, for example, the presence of annular thickening fibres in the walls does not in the least hinder the cells from contracting when water is withdrawn from them. Similarly the differentiation of annular and spiral wood-vessels does not prevent subsequent elongation of a stem. The same principle is illustrated by the unicellular hydathodes of *Gonocaryum* (Ch. X. II. A); here the funnel-shaped upper portion has a very thick wall, but the unthickened condition of the lower portion allows the volume of the entire hydathode to change in accordance with variations of turgor.

Most frequently, however, the retention of permanently unthickened areas in a cell-wall serves primarily to facilitate diffusion between adjacent cells. For although thickened walls are by no means impervious to dissolved substances, yet the rate of diffusion through a membrane varies inversely as the thickness of the latter; consequently the presence of unthickened areas in cell-walls must be highly advantageous from the point of view of translocation. These readily permeable spots generally take the shape of sharply defined areas of approximately circular cross-section, known as **pits**. In accordance with their function, the pits on the two sides of a cell-wall always correspond exactly, so that the adjoining **pit-cavities** are separated only by the thin primary **closing-membrane**. The ectoplast lining the pit-cavity adheres closely to the closing-membrane, and in this region probably exhibits peculiarities of structure that render it specially permeable. The mere fact that the plasmatic membrane does not induce growth in thickness of the cell-wall over certain areas proves that it is in some respects different, at these points, as compared with the remainder of the ectoplast. In most plants the closing-membranes of simple pits are perforated by minute pores, which are traversed by **protoplasmic connecting threads** or **plasmodesma**.[18] Not infrequently plasmodesma are developed also in connection with the thickened regions of the wall. These plasmodesma provide the only means of direct communication between adjacent protoplasts. In certain cases they help to expedite diffusion, but their principal function undoubtedly consists in the transmission of stimuli from cell to cell, for which reason the detailed discussion of these structures must be postponed until a later chapter. (Cf. intercellular transmission of stimuli, Ch. XIII. II. A.) The **bordered pit**, a peculiar modification of the ordinary or simple pit, is a characteristic and physiologically important feature of certain water-conducting elements, and for this reason is most conveniently discussed in connection with the conducting system (Ch. VII. II. A, 1 and 2). Generally speaking, in fact, the characteristic mode of thickening of the cell-wall is so closely correlated with the special functions of the cell or tissue to which it belongs, that this subject cannot properly be dealt with in a general account of the cell. It may, however, at once be stated that the great diversity which prevails with regard to the details of cell-wall thickening is partly a matter of variation of design [cf. p. 6].

The internal differentiation of secondarily thickened cell-walls [19] occasionally possesses a definite physiological significance, but more often merely represents a consequence of the special mode of development of the membrane.

**Stratification,** that is the differentiation of a number of concentric layers differing among one another in respect of refractive index, is an almost universal feature of thickened cell-walls. If such a stratified wall is caused to swell by treatment with suitable reagents, each individual stratum is often found to be composed, in its turn, of a number of delicate lamellae. Very frequently a thickened wall may be seen to consist of several stratified layers or complex strata, which differ from one another in chemical composition as well as in their optical behaviour; the so-called **primary, secondary** and **tertiary thickening layers** represent strata of this nature. As a rule the secondary layers form a massive stratum, while the tertiary layers are represented by a very thin and often highly refractive, pellicle [sometimes termed the " internal " or " limiting pellicle " (*Grenzhäutchen*)].

According to the view first put forward by Nägeli, stratification, and the optical differences which are responsible for the stratified appearance, are always due to differences in the water-content of successive layers of the wall. This theory has been severely criticised; but Correns has shown that it certainly applies in the case of certain bast-fibres, the walls of which lose their stratified appearance entirely, or to a large extent, when they are thoroughly dried. There are, however, undoubtedly also walls—for example in the pith of *Podocarpus*—the stratification of which is not affected by removal of the water of imbibition. In such cases the stratified appearance must depend upon differences of a chemical nature. It is quite conceivable that in yet other instances stratification is produced by a combination of physical and chemical factors.

Sometimes the thickening layers of a cell-wall, when examined in surface view, exhibit a system of delicate **striations,** which are generally directed obliquely with reference to the longitudinal axis of the cell. Occasionally two intersecting sets of striations are visible, in which case each belongs to a different stratum of the wall. Like stratification, striation may depend upon differences in water-content, upon chemical differences, or, finally, upon a combination of the two factors. The first-mentioned cause is responsible, as Correns has shown, for the transverse striation of the epidermal walls of *Hyacinthus* and *Ornithogalum*, and for the oblique striation of fibres. According to Krieg, the striation of Coniferous tracheides is similarly due to the alternation of relatively watery lamellae with layers which contain less water. It is at present unknown whether striation is ever actually produced by the other causes which have been specified.

Various investigators have attempted to demonstrate the presence, in cell-walls, of an organisation which is even more minute than that to which the phenomena of striation and stratification are partly attributable. Nägeli himself supposed that the walls of bast-fibres and other prosenchymatous cells are composed of exceedingly minute fibrils, and that the latter in their turn are built up by the concrescence in rows of the ultimate particles of cell-wall substance, or **micellae.** More recently Wiesner has devised a special method whereby the cell-wall can be broken up into minute particles, which he terms **dermatosomes.** This "atomisation" or "carbonisation" can be effected by soaking the fibres or other tissues in dilute hydrochloric acid for some hours and thereafter heating them to a temperature of about 50° to 60° C.; prolonged digestion with chlorine water produces similar results. Treatment according to either of these methods, in the majority of cases, causes the cell-wall to break up into a dust-like mass of extremely minute particles or dermatosomes. It is of course possible that these dermatosomes are mere artefacts; on the other hand, it seems not altogether improbable that the intact membrane may in certain cases consist of similar particles. Whether, however, the dermatosomes are formed from correspondingly minute protoplasmic particles or "plasomes," as Wiesner maintains, or whether, on the contrary, they arise by differentiation within a primarily amorphous membrane must remain undecided for the present. Similarly nothing definite can be predicated concerning the material which is dissolved away during the process of carbonisation, and which presumably serves as a cement for holding the dermatosomes together.

The principal chemical constituents [20] of typical vegetable cell-membranes are carbohydrates of the *cellulose* group. Cuprammonia (Schweizer's reagent) is the only liquid in which cellulose is soluble without decomposition. Cellulose is rapidly dissolved, and at the same time decomposed, by concentrated sulphuric acid. The classical colour reactions of **cellulose** walls are, first, the bright blue coloration produced by iodine in presence of concentrated sulphuric acid and, secondly, the appearance of a colour varying between blue and pinkish violet on treatment with chlor-zinc-iodine (Schulze's solution). The so-called *hemi-celluloses* form a large proportion of the thickening layers in thick-walled storage tissues, especially in the case of endosperms; these compounds are more easily hydrolysed than the genuine celluloses, and are hence readily mobilised and absorbed during germination. A somewhat similar "reserve-cellulose" is *amyloid*, a body which is coloured blue by iodine alone [*i.e.* without previous treatment with an acid]. A number of other organic compounds are widely distributed in cell-walls. The so-called *pectic substances*, which are closely allied to, if not identical with, the *hemi-celluloses*, are remarkable on account of their mucilaginous or gelatinous consistency; by treatment with dilute acids they become readily soluble in alkalies. *Chitin* is an important constituent of the cell-walls of Fungi. **Lignified** cell-walls contain several characteristic substances, such as *xylan* and the *lignic acids*, which latter compounds are perhaps chemically related to the still undetermined principal component of lignified membranes; most conspicuous of all, though present in relatively small quantities, are those bodies which are responsible for such characteristic colour reactions of lignified walls as the yellow coloration with aniline sulphate or chloride and the cherry-red tint produced by phloroglucin in presence of hydrochloric acid. Singer among others ascribes the reactions in question to the presence of *vanillin*, while Czapek regards the aromatic aldehyde *hadromal*, isolated by him from lignified tissues, as the cause of the aforesaid colorations. **Suberised** and **cutinised** (or **cuticularised**) cell-walls are impregnated with substances of a fatty nature; walls of this kind are insoluble in sulphuric acid, and are turned yellow by caustic potash and yellowish brown by chlor-zinc-iodine. **Mucilaginous** walls are formed by a special modification of cellulose or pectic membranes; their characteristic chemical features are, however, often apparent when the walls are first laid down. While all cell-walls contain a certain amount of mineral matter, the proportion of inorganic substance present is greatest in the case of old membranes. *Silica* and salts of *calcium* (*calcium carbonate* and *calcium oxalate*) are among the commonest mineral constituents of the cell-wall, and are sometimes present in large quantities.

Both the chemical changes which cell-walls undergo, and the impregnation with certain organic or inorganic substances to which they are subjected, are frequently correlated with special physiological or ecological relations. Suberisation, for instance, renders a membrane impervious, or at any rate much less pervious, to water and gases; hence the frequent occurrence of suberised walls in dermal tissues. Certain cell-walls, however, present the reactions of cutinised membranes, and are nevertheless readily permeable by water; the outer walls of epidermal hydathodes illustrate this latter condition. The physiological importance of lignification still remains unexplained. This chemical modification cannot have any mechanical significance, since fibres may be quite unlignified and yet display great tensile strength. Sachs believed that lignification endows a membrane with a special aptitude for conducting water; but apart from other difficulties, Sachs' view is inadmissible, because it is founded on the assumption that the transpiration-current travels in the walls, and not, as is now generally held to be the case, in the lumina of the conducting elements. Mucilaginous walls often serve as a means of water storage, especially when they occur in foliage leaves; in the case of seeds and fruits, on the other hand, mucilaginous membranes frequently assist in fixing the seed to the substratum during the first stages of germination. In a number of LEGUMINOSAE, the mucilaginous walls of the endosperm-cells represent a store of plastic material. The mucilage which so frequently occurs in submerged plants, finally, should perhaps be regarded as a means of protection against animal foes. The function of silicified and calcified walls is frequently mechanical, especially in the case of epidermal cells and hairs; impregnation with silica or lime renders the membrane harder and at the same time more brittle, and may thus become a very important feature in the case of organs which are specialised as instruments of protection against animals. The brittle condition which ensures the breaking off of the point of a stinging hair at the right moment, results from the combined effects of calcification and silicification.

Both the first formation of the cell-membrane and its subsequent growth are to a large extent dependent upon the activity of the living protoplasm. Generally speaking, in fact, the cell-wall may be regarded as a secretory product of the protoplast. The mode of origin of cell-membranes is, however, very imperfectly understood, and can only be discussed in the most general manner in the present treatise.

Theoretically there are three distinct methods of growth whereby a cell-wall may add to its thickness. Where growth takes place by **apposition** (in the strict sense) new particles of cell-wall substance—or perhaps new molecules of cellulose—are deposited separately and successively upon the pre-existent wall-surface, in a manner resembling the precipitation of metallic particles in the electrotype process. In the second type of growth—termed growth by **intussusception** by Nägeli—the new particles are deposited *in the interior* of the old membrane, where they become incorporated in the molecular or micellar framework of the wall under the influence of forces which reside in the latter. Finally, the increase in the thickness of a cell-wall may be due not to the deposition of separate particles but to the addition of successive *entire lamellae* of new cell-wall substance, which are superimposed one upon the other like the leaves of a book. This last-mentioned process is included under the traditional comprehensive definition of growth by apposition. [It might be termed growth by **superposition.**] Of these three conceivable modes of growth in thickness, the last-mentioned is the only one that lends itself to direct observation; as a matter of fact a whole series of investigators, including Schmitz, Strasburger, Klebs, Noll, Krabbe, etc., have succeeded in demonstrating the deposition of successive entire lamellae of cell-wall substance in a number of cases. A particularly clear and convincing illustration of the process is furnished by the growth in thickness of the walls of bast-fibres. It is, of course, possible, though by no means certain, that lamellae primarily laid down by apposition may subsequently add to their thickness by intussusception. In any case nothing but intussusception can explain either the occurrence of secondary differentiation within a cell-wall or the development of centrifugal thickenings ; it is conceivable that these processes are in certain cases facilitated by the penetration of living protoplasm into the substance of the wall.

Growth in surface on the part of a cell-wall often depends—especially among Algae—upon a passive extension of the pre-existent lamellae and a subsequent deposition of new layers, the area of which is naturally adapted to the increased surface available for deposition. In this case the increase in surface is, strictly speaking, not effected by growth at all, but is due purely to passive extension. Genuine growth

D

in surface can only take place by means of intussusception; there are a number of facts which suggest, that where membranes grow in this way, they do so more or less independently of the protoplasm.[21]

### C. THE DIMENSIONS OF CELLS.[22]

Among both plants and animals "the size of the individual is a quantity which varies within very wide limits, whereas the dimensions of the structural units or cells are restricted within a much narrower range of variation." The quotation is from Sachs, who was the first botanist to draw attention to this apparently unimportant but in reality highly significant circumstance. The tallest trees, such as the two species of *Sequoia* and certain *Eucalypti*, may attain a height of over 100 metres (330 feet), while the smallest Bacteria are scarcely one thousandth of a millimetre ($1\mu = \frac{1}{25000}$ inch) in length. The linear dimensions of the *individual* thus vary roughly between $1\mu$ and $100,000,000\mu$. In the case of the *cells* of which the individual is composed, the extreme limits of size lie much closer together. It appears from Amelung's measurements that the diameter of the more or less isodiametric parenchymatous cells which have the largest share in the composition of the plant-body, generally amounts to between ·015 and ·066 mm. Sachs' statement to the effect that "the transverse diameter of an adult parenchymatous cell is always measurable in hundredths of a millimetre" (or in other words always lies between ·01 and ·10 mm.) is a less exact but perhaps a more convenient generalisation. The cells composing the pulp of fleshy fruits and the pith of certain stems (*e.g. Sambucus nigra* and *Impatiens glandulifera*), may attain a diameter varying between ·13 mm. and 1 mm.; and may thus be readily distinguishable with the naked eye; but it is quite exceptional for parenchymatous cells to exceed the above-mentioned average dimensions. Elongated prosenchymatous elements often reach a much greater size. Thus bast-fibres are commonly between 1 and 2 mm. in length, but occasionally attain a length of 10 mm. (Flax and Hemp) or even of 50-200 mm. (certain *Urticaceae*). The diameter of the smallest cells, on the other hand, rarely falls below ·01 mm. in multicellular plants, which may in fact be said to consist of structural units, the linear dimensions of which lie, generally speaking, between $10\mu$ and $10,000\mu$. It is obvious, however, that such numerical data only give a very rough idea of the actual facts.

The principal outcome of Amelung's measurements is the demonstration of the fact, that when homologous organs of one and the same plant—for instance, different foliage leaves—vary in size, the dimensions of the component cells exhibit little or no variation    In other words, there is no fixed ratio between the size of an organ and the dimensions of its structural elements; the cells composing a small organ are *fewer in number* than those which make up a larger organ of the same kind, not smaller in size. It thus becomes evident why small multicellular plants generally display less histological complexity than larger forms, a circumstance upon which Sachs lays considerable stress. For on the one hand complex internal differentiation demands the presence of a large number of structural units, while there are on the other hand definite reasons why the individual cells cannot be reduced below a certain size. It should, however, be remarked that, although plants of microscopic size do not as a rule display a great amount of tissue-differentiation, they may nevertheless achieve a relatively high degree of internal elaboration in a different manner; the component cells, namely, may increase their cytological complexity by developing cell organs which are absent in larger multicellular plants. A unicellular Volvocine, for example, is provided with vibrating flagella, with an eye-spot and with a pulsating vacuole, in addition to the usual cytoplasm, nucleus, chromatophores and cell-wall.

Sachs regards the comparative constancy of the size of the cell among Higher Plants very much as the chemist regards the definite value which he ascribes to the atomic weight of an element, namely, as a well-established but at present inexplicable fact. The physiological anatomist cannot, however, resign himself to this attitude; in his continual endeavour to discover the relation between structure and function, he cannot ignore the problem presented by the relative constancy of cell-dimensions.

It has been stated above that the size of parenchymatous cells varies within comparatively narrow limits, however the plants to which they pertain may differ in size and other particulars, and however diverse the functions of the cells under consideration may be. This circumstance proves that the prevalence of certain average dimensions must involves some physiological advantage, which is not connected with the special performances of the individual elements, but which relates to properties that are common to all living cells. It is impossible at the present time to indicate the precise nature of this advantage. One is forced to assume that the actually observed average cell-dimensions are those which have proved most suitable with reference to rapidity of translocation, control of turgor, administrative activity of the nucleus, in short, all the aspects of metabolism that are in any degree affected by considerations of space.

In endeavouring to explain the relative sizes of the cells that compose *different* tissues, one is treading on somewhat firmer ground.

In the case of every tissue, namely, the size of the cells, no less than their form and structure, must be adapted to their physiological activity. Every cell accordingly exhibits specialisation for a definite purpose, not only with regard to its morphological features, but also in respect of its dimensions.

It is readily comprehensible, for example, why storage cells are, in general, much larger than photosynthetic elements. Practically the entire lumen of a cell can be utilised for the deposition of reserve-materials; chloroplasts, on the contrary, are necessarily restricted to the immediate vicinity of the cell-wall, so that in photosynthetic cells the available cell-wall surface should be as large as possible. Consequently, the cell-walls are multiplied in photosynthetic tissues and as a result the size of the individual cells is reduced as far as is compatible with their general vital activity on the one hand, and with a proper degree of illumination on the other. The pronounced elongation, which is in general characteristic of bast-fibres, must also be regarded as a factor in the specialisation for mechanical purposes exhibited by these cells, since elongation of the individual elements of a fibrous strand undoubtedly increases the strength of the whole bundle. Differences in the width of water-conducting vessels are similarly correlated with the varying requirements of the plant in respect of the amount and velocity of the transpiration current. Wind-borne pollen-grains, again, are generally smaller than those which are adapted for transportation by insect agency. In all the preceding instances, and in many others of a like nature, the adaptive character of the prevailing cell-dimensions can be readily demonstrated.

### D. ORIGIN AND ADVANTAGE OF CELLULAR STRUCTURE.

There are quite a number of Thallophyta, comprising both Algae and Fungi, which do not possess the typical cellular structure that alone has been under discussion so far; in these cases the plant-body consists of a single undivided protoplast enclosed within a continuous membrane. This unsegmented condition of the protoplast is characteristic of many Schizophyta and PROTOCOCCALES and of all the DESMIDIACEAE, DIATOMACEAE, SIPHONEAE and PHYCOMYCETES. Such plants are often termed **unicellular** in contrast to the **multicellular** forms which exhibit the typical cellular structure. This terminology implies that all plants without exception can be included in the " cellular " scheme. The term unicellular, however, is justifiable only when the plant to which it is applied is really homologous with a single cell of the most nearly related multicellular species, or where in other words the multicellular condition owes its phylogenetic origin to the association of a number of unicellular individuals. Ontogenetically this union often takes place at a very early stage, the asexual germs becoming associated together to form a multicellular body. The multicellular coenobia of certain PROTOCOCCALES, such as *Pediastrum* and *Hydrodictyon*, arise in this way; since those members of this group of Algae which lead a solitary existence ("eremobic" forms) are undoubtedly homologous with single cells, say of a *Pediastrum* colony, they may properly be designated unicellular organisms. In other cases the asexual reproductive cells may never display any tendency to disperse, but remain united from the moment of their origin (by division) onwards. Thus Nägeli assumes that the NOSTOCACEAE, in which the plant-body consists of multicellular filaments, are phylogenetically derived from unicellular CHROÖCOCCACEAE; he supposes that in the case of some of the Chroöcoccoid ancestors of the NOSTOCACEAE the several individuals produced by division of a single cell ceased to separate from one another and remained permanently united, thus giving rise to a single multicellular organism. Nägeli, in fact, goes so far as to suggest that this case illustrates a "principle which underlies the formation of tissues throughout the vegetable kingdom." [23]

There can be no doubt, however, that the multicellular condition has also arisen in other ways. In the genus *Caulerpa*, for example, the plant-body, though containing a single undivided protoplast—as in all SIPHONEAE,—is nevertheless differentiated into organs resembling rhizomes, roots and leaves, and altogether closely approximates to the habit of a (terrestrial) Higher Plant. Such a plant cannot properly be termed "unicellular"; on the contrary the *Caulerpa* individual is clearly comparable to an entire multicellular plant, and its continuous but multinucleate protoplast corresponds to the totality of the separate uninucleate protoplasts of the multicellular organism. If this view is the correct one, it follows, as was first recognised by Sachs,[24] that not only *Caulerpa*, but also the rest of the SIPHONEAE and all the PHYCO MYCETES, instead of being included among unicellular organisms, ought rather to be classed as **non-cellular** (or coenocytic[24a]) plants, and thus contrasted with the ordinary **cellular** forms.

Since very few non-cellular plants are highly differentiated, the acquisition of cellular structure must have been a powerful factor in the evolution of the higher forms of plant-life. The multicellular condition does in fact involve such undoubted advantages, that it is improbable that it owed its origin in every case to the modification of a reproductive process, as conjectured by Nägeli: in certain instances it must have arisen by a repeated septation and consequent subdivision of the primarily continuous protoplast of a non-cellular plant-body. The advantages attendant upon cellular structure are of course quite independent of the mode of origin of that structure; in other words, the nature and extent of these advantages is the same, whether the many-celled condition of a large and highly organised plant owes its origin to the (phylogenetic) aggregation of one-celled individuals or germ cells, or whether it is on the contrary due to the secondary sub-division of a non-cellular plant-body into a number of protoplasts enclosed in separate compartments.

It is mainly in connection with the **mechanical requirements** of the plant-body that cellular structure is highly advantageous. If a plant is to maintain a constant shape, and if it is to provide itself with permanent members, it must develop a comparatively solid skeletal framework, within which the shapeless living substance can arrange itself in a definite and orderly manner. In the absence of such a framework or skeleton of solid material, it would be impossible for a large plant to attain to any considerable degree of external organisation. The truth of this assertion is very clearly illustrated by the familiar case of the Myxomycete-plasmodium; in the living condition. the plasmodium, which is, of course, merely a naked, unsupported protoplast of unusual size, is continually changing its outline, or, in other words, represents a protean structure devoid of permanent shape. Even among plants of microscopic size, such as the Schizophyta and PROTO-COCCALES, the maintenance of a definite shape would be impossible if the protoplast were not invested by a cellulose membrane. The evolution of such non-cellular forms as the SIPHONEAE and PHYCOMYCETES may be figuratively described as an experiment on the part of Nature, which was intended to test how far the mere possession of a cellulose "exoskeleton" would enable a large plant to achieve a certain degree of external differentiation. As a matter of fact even many SIPHONEAE find this simple type of skeleton insufficient for their needs. In the genus *Caulerpa*, for instance, to which reference has already been made more than once, numerous trabeculae of cellulose are developed in all parts of the plant-body; these structures constitute an efficient buttressing system, which, as Janse has shown,[25] seems particularly designed to preserve the plant from deformation by the force of its own turgescence. Arrangements of this kind would, however, be of little value in the case of terrestrial plants, which require the much more effectual support provided by complete partitions; as it is further desirable that the walls should be disposed in several intersecting planes, the plant-body has become divided into numerous compartments, or, in other words, has acquired a cellular structure. It is by no means necessary to turn to the land-plants, with their exacting mechanical requirements, in order to illustrate the principle under discussion. Any filamentous Alga will afford a simple instance of its application. Here the transverse walls, that are developed at regular intervals, represent so many struts, which prevent the collapse of the delicate tube of cellulose; incidentally, they divide the tube into a series of compartments and the protoplast into an equal number of segments. The simple tubular plant-body has, in fact, become transformed into a cellular filament.

A second factor which determines the cellular structure of plants, and, indeed, of organisms in general, is the fundamental physiological principle of **division of labour**. From the mechanical standpoint the interpolation of solid walls is the main object of septation, the subdivision of the protoplast being merely an inevitable consequence of the septate condition. From the point of view of division of labour, on the contrary, subdivision of the continuous protoplast is the primary purpose of cellular structure, while the formation of cellulose septa merely represents the means employed to gain this end. The partition of a large protoplast gives rise to a number of separate structural units, each capable of independent physiological activity; such an association of functional units is far better adapted than a single large unsegmented protoplast for dealing effectively with a number of diverse functions.

The beginnings of this devolutionary process of septation are descernible even among the SIPHONEAE and PHYCOMYCETES. The plant-body of *Vaucheria* may, as far as its vegetative condition is concerned, be described as a variously-branched, non-septate tube. In connection with the reproductive processes, however, the living substance which is destined to give rise to asexual or sexual germs, usually becomes segregated from the somatic protoplasm. In *Vaucheria sessilis*, *V. tuberosa*, *V. gemmata*, etc., asexual propagation is initiated by a swelling up of the distal ends of certain branches; the protoplasmic contents of each swelling become completely separated from the rest of the protoplasm by a transverse wall before giving rise to a zoospore. The male and female sexual organs (antheridia and oogonia) likewise become shut off from the vegetative region of the thallus. Among the PHYCOMYCETES, also, the formation of sexual or asexual reproductive organs is always initiated by the appearance of a number of irregularly disposed transverse walls, which divide the previously continuous mycelial cavity into several compartments. In such cases the formation of septa cannot be explained on mechanical grounds; for in neither of the instances described does the awakening of reproductive activity necessitate any addition to the mechanical strength of that part of the plant-body in which septa appear. On the contrary, the septation here undoubtedly represents a secondary consequence of a physiological differentiation of the protoplasm. The dissociation of the reproductive and vegetative functions of the living substance, and the accompanying separation, by means of cell-walls, of the organs or protoplasmic regions concerned with the two aspects of physiological activity, probably represents one of the most fundamental applications of the principle of division of labour.

It follows without further explanation that division of labour within the limits of purely vegetative activity must also be facilitated by the acquisition of cellular structure; the distribution of the various nutritive functions among distinct tissue-systems, in particular, is scarcely conceivable in the absence of a cellular arrangement.

## II. TISSUES.

In all multicellular plants the individual structural units combine to form complex homogeneous units of a higher order of magnitude, which are known as **tissues**. In order that a cell-complex may rank as a tissue, in the anatomico-physiological sense of the term, it is necessary that its component elements should exhibit a certain degree of uniformity not only in respect of the functions which they perform, but also with regard to the structural features correlated with these functions. A great many-tissues contain a certain proportion of "foreign" cells, which differ very markedly as regards both structure and function from the "native" elements of the tissue. A typical illustration is afforded by the frequent occurrence in thin-walled green photosynthetic tissues of colourless thick-walled fibrous elements, which take no part in photosynthesis or in any other metabolic process, but are utilised for purely mechanical purposes. The term **idioblast**—first introduced by Sachs—is applied to elements of this nature. Where all the idioblasts contained in a given tissue are similar in structure and subserve the same purpose, they may in a sense be regarded as components of a special "diffuse" tissue.

### A. MODE OF ORIGIN OF TISSUES.

In all the Higher Plants, from the Bryophyta upwards, the formation of tissues depends, with few exceptions, upon **repeated cell-division**. One or more primordial mother-cells (such as a spore, an oospore, a single apical cell, or several initials) give rise by successive divisions not only to the diverse constituents of an individual tissue, but also to all the various tissues and tissue-systems in every organ of the plant-body. In many Thallophyta, also, tissues arise exclusively through cell-division; but among these lower plants a process of **secondary concrescence**, whereby cells or even cellular filaments or cell-masses, produced by division and originally separate, subsequently become united, is also of frequent occurrence.

Typical **cell-division**,[26] as illustrated by a uninucleate vegetative cell of a Higher Plant, must be considered in some detail. The process of cell-division comprises two separate series of change, which must be carefully distinguished, namely, first, the partition of the protoplast—which generally involves the formation of a cell-wall between the two daughter-cells—and, secondly, the division of the nucleus. These two processes are correlated with one another in a peculiar and characteristic manner, a circumstance which complicates the problem at issue. Nevertheless, thanks to the laborious researches of Flemming, Strasburger and numerous other investigators, the leading features of the process of cell-division are comparatively well understood.

When a cell is about to divide, the nucleus may be seen to undergo certain preparatory changes. The threads of the nuclear reticulum become thicker and shorter (Fig. 7 A), while at the same time the chromatin granules enlarge and assume the form of discs which are placed athwart the threads. The tangled thread or skein next resolves itself into a definite number of segments or **chromosomes** (Fig. 7 c), which usually take the form of U- or L-shaped rods. The chromosomes then arrange themselves in a single plane, the **equatorial plane**, collectively forming the so-called **equatorial plate**; the bend of each U or L always faces the equatorial plane. At this point it becomes obvious that each chromosome is divided longitudinally into two equal halves; this fission, however, is initiated at an earlier stage.

While the nuclear network is contracting and breaking up into chromosomes, other changes are also going on. The nuclear membrane becomes invested by delicate plasmatic strands which appear first at the two poles of the "mitotic figure" in the shape of the so-called **polar caps**; similar protoplasmic fibrillae subsequently become differentiated in the interior of the nucleus, where they take the form of two bundles tapering towards the poles (Fig. 7 B, c). Later the nucleolus and the nuclear membrane disappear, while the fibrils of the polar caps become prolonged inwards and attach themselves to the backs of the chromosomes; other fibrils derived from the two caps meet end to end and fuse to form threads running continuously from pole to pole. All these protoplasmic threads are termed **spindle fibres**; collectively they constitute the **nuclear spindle**. As stated above, the chromosomes, after forming the equatorial plate, undergo longitudinal fission; the two halves of each chromosome then travel along the spindle-fibres to opposite poles (Fig. 7 D), where they proceed to form the daughter-nuclei. For this purpose the chromosomes withdraw their freely extended ends and give rise by involution, branching and anastomosis to a typical nuclear reticulum. In the meantime a new nuclear membrane has developed, and so the reorganisation of the daughter-nuclei is complete. The nucleoli either disappear altogether before the formation of the equatorial plate, or else are extruded into the cytoplasm.

The process which has just been described is termed **indirect or mitotic division** of the nucleus, also **karyokinesis** (Schleicher) or **mitosis** (Flemming). In the opinion of Roux, among others, the complicated nature of mitosis indicates that this process serves to ensure an equal partition of the nuclear substance in general, and of the nuclear reticulum in particular, between the two daughter-nuclei. The necessity

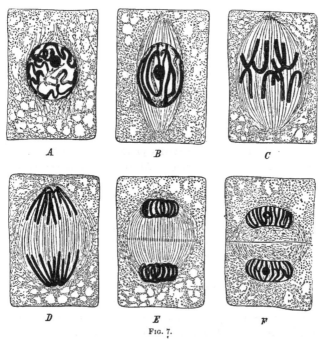

Fig. 7.

Various stages of nuclear and cell-division, as seen in the growing point of the root of *Allium Cep*a. Semi-diagrammatic. Explanation in the text.

for such an equal partition becomes self-evident, if the nucleus, or rather the chromatin, be regarded as the idioplasm or physical basis of inheritance. Hence the longitudinal fission of the chromosomes, by means of which the exact halving of the idioplasm is effected, represents by far the most important stage of the mitotic process.

The part played by the spindle-fibres during mitosis is not yet clearly understood. Many cytologists regard these " fibres " as elastic threads, which by their active contraction convey the daughter-chromosomes from the equatorial plate to the poles of the mitotic figure. Strasburger, in fact, discriminates between " traction-fibres " which attach themselves to the chromosomes and " supporting-fibres " which connect the poles with one another and, by furnishing the necessary resistance, enable the traction-fibres to pull the chromosome-halves apart. Another group of investigators regards the appearance of spindle-fibres merely as a consequence of special physical and chemical changes, originating mainly in the centrosomes which are of such regular occurrence in animal cells; according to this view the spindle-fibres are, as it were, comparable to "lines of magnetic force," and at most represent the paths along which the chromosomes travel. It is reserved for future workers in this field to decide between these divergent opinions.

The " supporting " spindle-fibres persist for some time after the reorganisation of the daughter-nuclei is completed in the form of the so-called " interzonal fibres" or " connecting fibres" ( *Verbindungsfaden*). Later these become reinforced by the interpolation of a number of additional fibrils, and as a consequence the whole mitotic figure expands and becomes more barrel-shaped (Fig. 7 E). If the dividing cell is rich in protoplasm, and not too wide, the expanding spindle finally comes into contact with the lateral walls in the plane in which the new transverse septum will later be laid down. The fibres then become thickened in the same plane, and the rod-like bodies thus produced fuse to form the so-called **cell-plate** (Fig. 7 E); this structure is at first a homogeneous plasmatic membrane similar in character to the ectoplast, but subsequently splits into two layers, between which the new cell-wall is secreted (Fig. 7 F). In this way the mother-cell forms two daughter-cells by **simultaneous** segmentation. If the cell has a large sap-cavity, and is consequently so wide that the expanding mitotic figure does not touch the lateral walls all round, the spindle moves across gradually from one side of the cell to the other, and in so doing lays down the new transverse wall **successively.** As soon as the division of the mother-cell is completed by the formation of the new cell-wall the interzonal fibres gradually lose their distinctness, and finally become merged in the cytoplasm of the daughter-cells.

Among Thallophyta the connection between nuclear and cell-division is not so intimate as in the case of Higher Plants; the new cell-walls are not formed within bundles of connecting fibres. Where the cells are uninucleate, the nucleus first divides independently. A new wall then arises, either simultaneously at the expense of a previously differentiated cell-plate, or else successively. In the latter case it begins as an annular ridge, laid down all round upon the lateral wall; the ridge extends further and further inwards until it finally cuts the protoplast in two and closes up to form a complete septum. Even in such cases, however, cell-division appears to stand in a definite relation to the division of the nucleus, or at any rate to the position taken up by that organ; for the new cell-wall is laid down midway between the two daughter-nuclei, while protoplasmic strands and fibrils may often be seen to extend from the two nuclei towards the advancing margin of the annular ridge. In the case of multi-nucleate Thallophyte-cells, however, septation seems—according to the data at present available—to take place quite independently of nuclear division.

The simultaneous segregation of numerous cells which takes place in connection with the development of endosperm in the Angiospermous embryo-sac—a process investigated more particularly by Strasburger—is a modification of the typical form of cell-division. In the case of large and rapidly-growing embryo-sacs the peripheral protoplasm contains a very large number (often several thousands) of nuclei, all of which owe their origin to the repeated division of a single nucleus, the so-called secondary nucleus of the embryo-sac (after the fusion of the latter with one of the generative cells of a pollen-tube). A transitory cell-plate is formed in connection with each of the free nuclear divisions which give rise to these numerous nuclei; this circumstance justifies

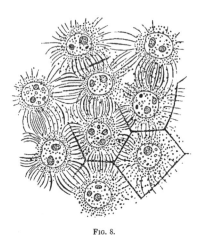

FIG. 8.

Formation of endosperm in the embryo-sac of *Agrimonia Eupatorium*; commencing development of partition-walls between the nuclei. After Strasburger (from Sachs, *Lectures on the Physiology of Plants*).

the inference that the mitoses in question were at some previous stage of evolution accompanied by cell-divisions, a conclusion which is further borne out by the fact that in small narrow embryo-sacs the endosperm is produced by ordinary cell-division. When the embryo-sac has finally completed its growth, but not before, each of the numerous peripheral nuclei becomes connected with all its neighbours by groups of spindle-fibres, within which cell-plates now once more appear (Fig. 8); but on this occasion the cell-plates give rise to septa which form a meshwork of cell-walls projecting inwards from the wall of the embryo-sac. The endosperm cells produced in this way are thus at first open on their inner side; ultimately, however, they become shut off by cell-walls from the liquid contents of the embryo-sac. During the further development of the endosperm, its cells multiply solely by ordinary division.

The new cell-walls formed in connection with cell-divisions are at first thin, unstratified and entirely homogeneous membranes; at this stage, in fact, they should be regarded as partitions separating adjacent protoplasts rather than as integral portions of individual cells. At a somewhat later period each wall usually consists of at least three obviously distinct layers, namely, a central **middle lamella,** overlain by cellulose layers belonging to and laid down by the protoplasts on either side of the wall. The middle lamella therefore constitutes a septum separating the cellulose membranes of two adjacent cells; it was at one time in fact regarded as an intercellular substance. Very often the structure described as the middle lamella merely represents the primary partition-wall, which has undergone a certain amount of chemical modification since its formation; more often, however, it also comprises the **primary thickening layers.** In all cases the middle lamella differs chemically from the **secondary and tertiary thickening layers.** In soft tissues it consists, according to Mangin, of a pectic compound of calcium. In ripe fruits this pectic substance often undergoes mucilaginous degeneration, with the result that the cells become separated from one another; a similar gelatinisation may be artificially effected by boiling the tissues. In the case of lignified and suberised tissues the middle lamellae are often more or less strongly lignified, and further characterised by their insolubility in sulphuric acid.

A striking contrast to the production of tissues by means of repeated cell-division is furnished by the **secondary concrescence** of originally distinct cells. This mode of tissue-formation is, however, comparatively rare. A familiar instance is provided by the development of the disc-shaped coenobium of *Pediastrum.* Here any cell of the disc may divide to form a number of zoospores, which emerge from their mother-cell still enclosed in a thin vesicle representing the innermost layer of the mother-cell membrane. The zoospores at first exhibit active swarming movements, but soon come to rest, and arrange themselves in a single plane, whereupon they acquire cell-walls, become attached to one another, and thus give rise to a small daughter-disc. While it is customary to describe the disc of *Pediastrum* as a coenobium or colony of unicellular individuals, the fact remains that these individuals are firmly knit together to form an entity of a higher order of magnitude; from this point of view a *Pediastrum*-disc must be regarded as a *tissue* produced by concrescence.

The above-described two extreme types of tissue-formation—namely, repeated cell-division and concrescence of originally separate cells—are connected by a number of intermediate forms, some of the most important of which may next be briefly discussed.

If a number of filaments, each the product of cell-division, arrange themselves side by side, or become closely intertwined or interwoven, a complex structure results which will resemble an ordinary tissue formed by cell-division more or less closely, according as the filaments are firmly united or only loosely combined; where complete con-crescence takes place between the component filaments, the resulting tissue is often hardly or not at all distinguishable from a tissue pro-duced by division. An instance in point is provided by the genus *Cutleria* among the PHAEOSPOREAE. Here the thallus is a flat structure composed of several superimposed layers, each of which breaks up at the margin into a number of narrow segments. If a single layer be examined separately, it is .found that each marginal segment consists of a cellular filament, which grows by means of an intercalary meris-tem; behind this meristematic zone lateral branches arise, and still further back adjacent filaments unite so completely that the fully developed tissue of the thallus shows no trace of its peculiar mode of development. Various other PHAEOSPOREAE, such as *Desmarestia* and *Stilophora*, form their tissues by the same ["trichothallic"] method, or by means of a closely similar process. In the CERAMIACEAE again the main axis becomes corticated through the concrescence of closely appressed lateral branches; in *Chara* cortication of the internodes is effected in a somewhat similar fashion.

Excellent illustrations of tissue-formation by concrescence and coalescence of simple filaments are furnished by the "tissues" of the Higher Fungi; these invariably consist of slender and often richly branched cellular filaments or **hyphae**, which may be regarded as histological elements of a higher order, as compared with the segments of which they are composed. Hyphae may become associated to form tissues in very various ways. They may be closely interwoven into felted membranous structures, or bound together to form strands of approximately parallel filaments; in other cases the arrangement that eventuates strongly resembles the parenchymatous tissue of a Higher Plant.

It is customary to describe those tissues which arise entirely by typical cell-division as "genuine" tissues, and thereby to contrast them with the "false" tissues produced by coalescence and concrescence of separate cells, cellular filaments or cell-masses; in the same way the various types of false tissue are distinguished by such names as "pseudo-parenchyma," etc., their development being described as a process of "apparent" tissue-formation. From the purely morpho-logical point of view this classification is thoroughly justified; the physiological anatomist, however, cannot recognise any such distinc-tions, since the functional activity of a tissue is in no way determined by its mode of origin. The importance of each tissue in the economy of the entire plant depends altogether upon its physiological value in the adult condition, and is not in the least affected by its ontogenetic development.

## B. THE PHYSIOLOGICAL FUNCTIONS OF TISSUES.

The **principal function** of a tissue is that form of physiological activity with which its most obvious and important anatomical features are correlated. In this connection the term " anatomical features " includes the topographical arrangement of each tissue in the various organs of the plant-body, as well as the histological characters of its component cells. The principal function of a tissue or tissue-system is often compounded of several **partial functions**. In the case of the epidermis, for example, the principal function consists in the protection which this layer affords the plant against a variety of hostile environmental factors. This general function of protection includes several partial functions, such as prevention of excessive transpiration, restriction of nocturnal radiation, protection against intense insolation, diminution of the risk of mechanical injury, etc., all of which are embodied in the anatomical structure of the epidermis. Every epidermal cell betrays in its histological peculiarities the influence of some of these partial functions, if not of all.[27] In other cases the partial functions of a complex tissue-system may be distributed among different component elements, the combined activities of which constitute the principal function of the system. In the case of the conducting system, for instance, the principal function is conduction in general. The various partial functions, such as transportation of water, conduction of carbo hydrates or translocation of protein-substances, are allocated to distinct component tissues. The histological complexity of a tissue or tissue-system is in fact largely determined by the number and variety of the partial functions which it has to perform.

There is rarely any fundamental difference of opinion with regard to the principal function of a tissue, or any serious doubt as to the anatomical characters which should serve to determine the nature of this principal function. The conditions are quite different, however, in the case of **subsidiary functions** and their anatomical criteria. It is quite certain that such subsidiary functions must be of very widespread occurrence. No one, at any rate, would venture to assert that a given tissue could never be confronted with physiological demands other than those normally connected with its principal function, even among the most highly-organised plants, where internal differentiation has been pushed to the furthest extent. For one thing an extreme degree of differentiation in itself entails certain disadvantages in the case of the vegetable organism, just as in human society; moreover, even the most intense specialisation in relation to one particular function scarcely ever renders a tissue utterly unfit for any other form of physiological activity. It is therefore hardly surprising to find that a tissue may be capable of undertaking one or more subsidiary functions. Thus the fact that both collenchymatous cells and bast-fibres are primarily mechanical elements does not prevent them from developing photosynthetic chlorophyll-corpuscles (collenchyma) or from serving as repositories of starch (bast-fibres).

The anatomical characters connected with subsidiary functions often in no wise interfere with those which pertain to the principal function. In the case of bast-fibres, for example, the elongated spindle-shaped form, and the thickened condition of the walls, are quite unaffected by the deposition of starch in the cavity. In other cases, indeed, the leading anatomical features may undergo a certain amount of modification in the interests of a subsidiary function, although the alteration produced in this way must always be restricted within comparatively narrow limits. An example will serve to elucidate this point. The specialised photosynthetic elements of terrestrial green plants are more or less cylindrical in shape, and are usually located immediately beneath the upper epidermis of the leaf, where they constitute the so-called palisade-tissue. The walls of these cells are soft and unthickened throughout, a feature which is the anatomical expression of the very active interchange of material that is continually going on between the palisade-cells and the adjacent tissues. Now there are certain Ferns, such as *Didymochlaena sinuosa*, in which the leaf is devoid of an upper epidermis in the physiological sense. Here the palisade-tissue, which in other plants is sub-epidermal, has become superficial, and in fact forms a part of the outer covering of the leaf. In these circumstances, it must evidently act as a dermal tissue, besides performing its primary photosynthetic function; as a matter of fact, the outer walls of the palisade-cells are in such cases always slightly thickened, and also cuticularised. Here a leading histological feature of the palisade-cell, namely, the unthickened condition of its cell-wall, is modified in the interests of the secondarily acquired subsidiary function of protection; but the change of structure in no way interferes with the principal function of the palisade-tissue.

The preceding remarks will have made it sufficiently clear that the anatomical and topographical principles underlying the construction of all tissues and tissue-systems are determined by the functions which the tissues are destined to perform. In discussing these principles more fully, it would obviously be inadvisable to consider the topography of the tissues separately from their anatomy and histology;

for the anatomical and topographical features are always referable to a common determining influence, to wit, the nature of the functions performed by the tissue. Although the detailed consideration of the principles embodied in the structure and arrangement of the several tissues and tissue-systems forms the subject-matter of the succeeding chapters of the present work, it is nevertheless desirable to devote a few words at this stage to four outstanding general principles which control the entire architectural plan of the plant-body, or which, at any rate, determine the leading features of a number of tissue-systems.

Reference must first of all be made once more to the principle of **division of labour**, which has already been mentioned on more than one occasion. It is mainly through the consistent application of this principle that plants and animals are enabled to perform their manifold functions in an efficient and reliable manner. Where every important function is assigned to a specialised organ or tissue, the morphological structure of each separate apparatus, and the function allocated thereto, can be mutually adjusted with the highest degree of precision. Physiological division of labour thus naturally leads to the morphological differentiation exhibited to a greater or less extent by the organs and tissues of every plant. The plants which stand at the highest level of organisation are those in which this principle of division of labour is carried out in the most thoroughgoing and consistent manner.

A second factor of fundamental importance, the influence of which is clearly perceptible in the structure of almost all the organs and tissues of the more highly organised plants, is the so-called **mechanical principle**. It is obvious that a certain degree of mechanical strength constitutes not only an essential feature of the general architectural plan of every plant, but also an indispensable condition for the undisturbed normal activity of each individual tissue and organ; similarly, in the case of a house, not only the arrangement of the foundations and roof, but also every detail of construction must satisfy a variety of mechanical conditions. Consequently, every highly-organised plant, besides developing a principal mechanical system or "skeleton," possesses in addition a number of minor mechanical arrangements, the majority of which are of purely local importance. Many of the histological details of the dermal, the photosynthetic and the conducting tissues, in particular, reveal the dominating influence of this mechanical principle in the clearest possible manner.

The principle of **economy of material** is most important from the ecological point of view. Owing to the intensity of the struggle

E

for existence, every plant is forced to obtain the greatest possible result with the smallest possible expenditure of material. This "economic" principle may be said, figuratively speaking, to lay down certain regulations for the construction and disposition of the various tissues. Plastic materials, in particular, are often economised through the selection of a suitable plan of construction. A notable illustration is afforded by the skeletal system; for the main advantage entailed by the construction of this system in accordance with approved mechanical principles consists in the resulting economy of material.

Finally, there is a principle of **exposure of maximum surface** which finds a very extensive and varied application. Any expansion of the superficial area of the partition between two cells must necessarily tend to increase the cohesion between the cells concerned; this circumstance is responsible, for example, for the very prevalent "interlocking" of the walls of adjacent epidermal cells. In the case of nutritive tissues, on the other hand, the development of large septa facilitates diosmotic interchange between different cells. The exposure of the largest possible amount of surface is, of course, likewise an advantage in the case of absorbing organs. The characteristic structure of the absorbing tissues of roots—in particular the presence of root-hairs—and of the haustoria of parasitic plants must be interpreted mainly from this point of view. The same principle is embodied in the construction of the photosynthetic system, where it is responsible for the interpolation of the partitions or flanges which provide a greater superficial area for the display of chloroplasts.

It is self-evident that none of the principles above enumerated—which may all be regarded as particular aspects of a common underlying **principle of efficiency**—represents an absolute "natural law." There are a great many anatomical features which appear to be incompatible with one or more of these governing principles. Thus, quite apart from other disabilities which are attendant upon extreme specialisation, the principle of division of labour occasionally comes into conflict with that of economy of material. It may even happen that different functions which have already been allocated to separate tissues become associated again in a later ontogenetic phase, being thereafter assigned to a single tissue. The trunks of *Dracaena* and other arborescent Liliiflorae, for example, are provided, before the commencement of secondary growth in thickness, both with fibres specialised for mechanical support and with typical vessels serving for water-conduction; in the bundles produced later by the secondary meristem, on the contrary, both functions are performed by one and the same form of tissue-element.

Sometimes it is the economic principle that is ignored; a good illustration is provided by the wasteful expenditure of spermatozoids or pollen-grains that takes place in connection with the process of fertilisation. In other cases, again, the mechanical principle is partially invalidated. The foliage of *Musa paradisaica* exhibits a peculiarity which never fails to impress the visitor to the tropics; the gigantic leaf-blades of this plant are torn, by the action of wind and rain, into a series of strips which are held together solely by the stout midrib. Anatomical investigation shows that the leaf of *Musa* displays none of those arrangements for preventing laceration of the leaf-margin which are otherwise of such general occurrence in leaves. Exceptional cases of this kind are, however, by no means necessarily irreconcilable with the general principle of efficiency; on the contrary, some special advantage usually accrues in such instances which must be set off against the loss entailed by the exceptional feature. The lacerated leaves of *Musa* are instructive from this point of view. For not only do the torn edges of the strips heal rapidly, so that no permanent damage results, but the fact that the strips subsequently maintain a pendent position is actually advantageous, because the leaf-surface is in this way withdrawn from the impact of the violent tropical rains and from the scorching rays of the sun.[28]

Attention must next be directed to the so-called **correlation of tissues**, a phenomenon which recurs in a great many different forms, but which is often of somewhat obscure significance; the term is intended to comprise all those cases in which different tissues—either forming part of the same organ or else located in widely separated regions of the plant-body—exert a definite and constant influence upon one another's activities. In the course of evolution those interrelations of different tissues which react favourably upon their physiological activity have gradually become fixed; in other words, tissues have become adapted to one another, and have, as it were, learnt to act in harmonious co-operation.[29] In the case of foliage leaves, for example, the structure and arrangement of the photo-synthetic tissue are intimately related to the course of the vascular bundles, which are instrumental in conveying water with mineral salts in solution to the photosynthetic cells, and in removing the plastic materials manufactured by the latter. Water-tissues are definitely orientated with reference to the photosynthetic cells, for the use of which the store of water is in the first instance designed. Even more remarkable is that type of correlation in which one tissue or organ directly controls the activity of another. Certain hydathodes, for instance, start their otherwise autonomous water-excreting activity only when the wood-vessels are water-logged and the hydrostatic

pressure within the latter rises to such a height that the intercellular spaces are in danger of becoming injected.   The insectivorous
leaf of *Dionaea muscipula* is provided with a special motor-tissue—
located in the midrib of the distal half of the lamina—which is
responsible for the characteristic snapping together of the two halves
of the trap; the motor-cells can, however, only be set in action
through the mechanical stimulation—by shock or contact—of one
of the six sensitive bristles situated on the upper surface of the leaf.
In these cases the direct control exerted by one organ upon some
physiological function of another takes the form of an act of stimulation.   The special condition of the controlling organ which results in
the stimulation of the executive structure may itself be induced by an
external stimulus; witness the case of *Dionaea*.   In all such instances of
direct correlation between separate organs, the requisite stimulus must
be transmitted through the intervening tissues; transmission of stimuli
may take place in a variety of ways.

Correlation of tissues does not always involve either the adaptive
regulation of one function by another or the co-operation of distinct
functions.   Quite apart from pathological conditions, it is certain that
in a perfectly normal plant the various tissues inevitably compete with
one another to a certain extent; usually it is a question of competition
for space or for a supply of plastic materials.   The success of a
particular tissue is naturally dependent upon a variety of circumstances.   In stems and other inflexible organs a certain amount of
competition not infrequently takes place between the mechanical and
the photosynthetic tissues; for the general structure of both these
systems necessitates their being situated as near the surface as possible.
Schwendener, however, has shown that in the haulms of Grasses the
issue of the struggle between the systems is determined by an
external factor, namely, illumination.   In the exposed portions of the
haulm—that is to say, in the region in which photosynthesis is favoured
by strong illumination—the photosynthetic tissue gains the upper
hand and occupies the periphery to the entire exclusion of the
mechanical cells.   In those parts of the haulm, on the other hand,
which are enclosed by the sheathing leaf-bases, the photosynthetic
system is placed at a disadvantage, owing to the relatively feeble
illumination, and accordingly gives way to the mechanical tissue.

In conclusion the influence of external conditions in general upon
the physiology and morphology of vegetable tissues must be briefly
considered.   Every tissue displays, in its hereditary morphological
characteristics and physiological properties, a very evident adaptation,
both to the general conditions of existence common to all plants, and
to the special features of the edaphic and climatic environment peculiar

to the individual species. Serious difficulties would however inevitably arise if the qualitative and quantitative development of the several tissues and mechanisms were absolutely constant for a given species, and if, consequently, the scope and intensity of the various physiological functions were precisely alike in every individual; since, namely, the environmental conditions are subject to considerable variation within the geographical range of any species, rigid adaptation to a particular set of conditions would necessarily expose certain individuals to grave risks, and might in certain circumstances be absolutely fatal. The hereditary characters of a species must therefore be not only *adaptive* but also *adaptable*, within certain limits. In other words, they must not be rigidly fixed, but must—to quote Nägeli—possess a certain plasticity, in order that the individual organism may be able to adjust itself directly to the environmental conditions associated with its particular habitat. Without some such property of self-regulation or accommodation no organism could survive for any considerable length of time. The senile decay and consequent gradual disappearance of certain species seems to be due mainly—as it is in the case of individuals—to decrease of plasticity on the part of the various adaptive characters, or, in other words, to a diminished capacity for physiological accommodation.

The plastic quality of the vegetable organism finds expression in the circumstance that the qualitative and quantitative development of the several tissues are both to a certain extent dependent upon the prevailing external conditions. In many cases, for example, the intensity of illumination determines the degree of development of the palisade-tissue. Similarly, differences in the humidity of the atmosphere may call forth an adaptive modification of the epidermal walls and of the ventilating system. Increased mechanical requirements lead to an increased production of skeletal elements. In the present state of physiological knowledge it is impossible to make any definite statements concerning the chain of causation which, in such cases, connects the external influence with the resultant internal modification; it is safer to emphasise the complex nature of the correlation by regarding the various environmental factors as stimuli, which may influence the development of tissues to the advantage of the individual organism.

## C. THE ANATOMICO-PHYSIOLOGICAL CLASSIFICATION OF TISSUES.

The term " tissue " ordinarily refers to some particular form or sort of tissue. In its widest sense, it denotes any aggregation of cells which have one or more characters in common ; these characters may be morphological, topographical, ontogenetic, or physiological in nature, and the aims of the particular investigator will largely determine the class of feature selected in a given case for purposes of definition and classification.

The physiological anatomist must of logical necessity define and arrange the various tissues of the plant-body in accordance with their anatomico-physiological characters. In so doing he starts with the assumption that the differentiation of a variety of tissues within the plant-body is mainly the outcome of division of labour, and that consequently the most characteristic features of each tissue are those which are most intimately connected with its physiological activity. The recognition of a physiological function does, in fact, presuppose the existence of a definite anatomical structure which is adapted for the performance of that function. Hence, every morphological feature of a tissue which can be shown to be thus adapted may be regarded as an anatomico-physiological character. Evidently, therefore, the anatomico-physiological definition and arrangement of tissues provides the broadest and most natural of all systems of tissue-classification, since from this point of view the plant-body is regarded not merely as a more or less complex aggregate of formal elements, but also as a living organism, composed of a number of functional units and engaged in a corresponding number of physiological activities, which all contribute to the safety and welfare of the whole.[30] It does not follow, however, that no other classification of tissues and tissue-systems is admissible in any circumstances. All that is required of a scheme of classification is that it should be carried out consistently from a single point of view. Those arrangements, on the other hand, which are merely intended to provide a simple and convenient list of the various forms of tissues are devoid of scientific value, however useful they may be for didactic purposes.

In defining and classifying the various tissues of the plant-body the physiological anatomist takes no account of their ontogenetic or phylogenetic relations. The homologies of tissues are of no interest to him in this connection ; his concern is solely with analogy. A single anatomico-physiological system—such as the mechanical system—may, therefore, comprise tissues of very diverse ontogenetic and phylogenetic origin ; it nevertheless represents a homogeneous structure from the present point of view, inasmuch as the various tissues included in the system are all concerned with similar functions. Conversely structures which have a common ontogenetic or phylogenetic origin may be assigned to systems that are widely separated in the anatomico-physiological scheme. Thus epidermal fibrous strands—such as are found in certain CYPERACEAE, stomata, epidermal hydathodes, root-hairs, etc., all pertain to the epidermis in the ontogenetic sense, whereas in the present work they are properly relegated to totally distinct tissue-systems.

Before proceeding to apply the principles outlined above to a preliminary review of the various kinds of tissues and organs, it is necessary to define the conception of an **anatomico-physiological system**. This term is employed, in the present work, to designate the totality of the tissues and mechanisms engaged. in the performance of a particular physiological function. The number of structures comprised in a given system will, of course, vary with the complexity of the function concerned and with the extent to which the latter is subdivided into partial functions. The great majority of anatomico-physiological systems are also tissue-systems, that is to say, they are mainly or entirely composed of actual tissues. That this is not necessarily the case is shown, for instance, by a consideration of the ventilating system, which is largely composed of intercellular air-spaces ; the secretory and excretory system, again, can only be regarded as a tissue-system in a metaphorical sense.

PRELIMINARY REVIEW OF THE VARIOUS ANATOMICO-PHYSIOLOGICAL SYSTEMS.

I. **Meristematic** tissues (primordial meristem ; primary and secondary meristems).

II. The **tegumentary** or **dermal** system (epidermis, cork, bark).

III. The **mechanical** system (bast-fibres, wood-fibres [libriform cells] ; collenchyma ; sclerenchyma).

IV. The **absorbing** system (absorbing tissues of terrestrial roots, especially root-hairs ; rhizoids ; the velamina of aërial roots ; water-absorbing trichomes of leaves : haustorial absorbing tissues of embryos and parasites).

V. The **photosynthetic** system (palisade-cells and spongy parenchyma ; chlorophyll-containing parenchyma generally).

VI The **conducting** system (" conducting parenchyma," including xylem-parenchyma, the parenchymatous elements of the pith, cortex and medullary rays, and the parenchymatous bundle-sheaths [border parenchyma] ; vascular bundles, comprising hadrome and leptome ; latex-tubes).

VII. The **storage** system (water-tissues ; storage-tissues of seeds, tubers and bulbs).

VIII. The **ventilating** system (the intercellular air-spaces together with their external openings or pneumathodes, especially the stomata and lenticels).

IX. **Secretory** organs and **excretory** reservoirs (hydathodes, nectaries, digestive glands ; mucilage-, oil-, and resin-glands ; mucilage-, oil-, resin-, and gum-passages ; raphide-saes, etc).

X. **Motor** tissues (flying-hairs and -tissues ; floating-tissues ; hygroscopic and other non-vital motor tissues : vital motor tissues).

XI. **Sense-organs** (tactile pits, papillae, hairs and bristles; statolith organs, such as starch-sheaths or the columellae of root-tips · ocelli and other light-perceiving organs).

XII. **Stimulus-transmitting** organs and tissues (protoplasmic connecting threads: the stimulus-transmitting tissue of *Mimosa pudica*).

The system of **meristematic** tissues is properly placed at the head of the above list, because, on the one hand, it comprises the embryonic conditions of all the permanent tissues, while from another point of view it may be regarded as an independent tissue-system which performs the special function of providing the raw material for the construction of these permanent tissues.   In view of their principal functions the dermal and mechanical systems may be grouped together under the head of **protective** systems.   The six following systems—namely, the absorbing, photosynthetic, conducting, storage and ventilating systems, together with the organs of secretion and excretion—are concerned with the various aspects of **metabolic activity.**   The last three systems, comprising the structures that serve for the **execution of movements** and for the **perception or transmission of stimuli,** are correlated with functions which were at one time considered to be peculiar to, or at any rate distinctly characteristic of, the animal organism; it is now, however, well known that the supposed distinction between animals and plants based upon the presence or absence of these functions, is valueless, except in special cases.

For the sake of convenience, the consideration of the anatomy and histology of the **reproductive organs** of plants is usually relegated to the text-books which deal with special morphology and taxonomy. The detailed study of this subject does not in any case come within the scope of the present work.   Hence, as regards reproductive organs in general, and the Angiospermous flower and fruit in particular, only those anatomical features will be taken into account which can be matched in the vegetative region of the plant-body; in other words, the flower and the fruit will be ignored, except in so far as their structure may help to illustrate the characteristics of one or other of the above-mentioned vegetative tissue-systems.

# CHAPTER II.

## MERISTEMATIC TISSUES.

### I. GENERAL CONSIDERATIONS.

THE gradual increase of histological differentiation, which is so striking a feature of the evolutionary history of the vegetable kingdom, is equally characteristic of the ontogenetic development of the shoot or root of any highly organised plant. A transverse or longitudinal section taken from the distal end of such an organ reveals the fact, that this region displays no trace of the complex structure which characterises the older portions of the organ, but is entirely made up of thin-walled cells with abundant protoplasmic contents. There is of course no sharp boundary dividing the homogeneous cell-mass of the **growing-point** [31] from the fully differentiated tissue-systems of the adult organ. Close behind the actual growing-point the uniform cell-mass becomes differentiated into several distinct layers, the cells of which, however, still retain certain common features, such as abundance of protoplasm, thinness of cell-walls and the power of active division. As the distance from the growing-point increases, the distinctive characteristics of the several tissues and tissue-systems come more and more into prominence, and the organ thus gradually attains to a degree of differentiation commensurate with the extent to which it embodies the principle of division of labour.

The above-described tissues, located in the growing-points and in the immediately adjacent distal regions of the shoot and root, represent the embryonic stages of the bulk of the tissue-systems which are discussed in the subsequent chapters of this work: the meristematic tissues may, in other words, be termed formative tissues in contrast to the executive permanent tissues. The ontogenetic development of every permanent tissue may, in fact, be divided into three stages. In the first of these the tissue becomes marked off from the general embryonic tissue of the growing-point, and undergoes a certain amount of preliminary differentiation. Then follows a period during which both growth and differentiation reach their greatest intensity. In the course of the third and final stage the tissue acquires its full functional activity.[32] It must not, however, be supposed that further histological changes never take place after a tissue has attained to the "permanent" condition; all that this term implies is, that subsequent modifications do not, strictly speaking, constitute a part of the onto-genetic development of the tissue. As a matter of fact, such secondary alterations are usually associated with a change or loss of function. Consider, for example, the case of the green cortical cells of a Beech-twig, which, after acting for a long time as conducting and storing elements, ultimately become thick-walled and lignified, and lose their living contents; here the nature of the secondary histogenetic changes shows that these cells in the end merely serve to increase the mechanical strength of the bark, or, in other words, that they pass out of the service of the nutritive into that of the protective systems.

A general account of the meristematic tissues would be incomplete if it treated them as nothing more than embryonic representatives of the permanent tissues. For it must be noted that, so long as an organ is undergoing growth, its embryonic tissues never become wholly converted into permanent elements; they are, on the contrary, constantly engaged in a process of self-regeneration, by virtue of which they maintain an existence which is in a sense independent of their genetic relation to the permanent tissues. It is from this point of view that the meristematic tissues may collectively be regarded as a tissue-system endowed with a definite physiological function, namely, that of providing the raw material employed in the construction of the permanent tissues. The highly differentiated adult organs are built up entirely out of the cells produced in the meristematic tissues; the most obvious indication of this relation is the frequency with which cell-division takes place at the growing-points. The term "meristematic tissue" or "meristem" was, in fact, selected by Nägeli with reference to this very feature.

Hitherto only a single kind of meristematic tissue has been mentioned, namely, that which is exemplified by the apical growing-point of a stem or root; while this **apical meristem** is indeed the most widely distributed type of meristematic tissue, meristems are also found in less outlying regions of the plant-body. Thus, in Grass-haulms, a zone of meristematic tissue occurs just above each node; each of these zones stands in the same genetic relation to the internode above it as does an apical meristem to the sub-apical region of the stem. Such **intercalary meristems**, interpolated between two regions of permanent tissue, are not at all uncommon; they are, moreover, not necessarily located immediately above the node—as is the case in Grasses, COMMELYNACEAE and other Monocotyledons—but may occupy the distal end (*Galeopsis Tetrahit* and other LABIATAE) or even the middle (*Pilea oreophila*) of the internode. The principal advantage of this interpolation of zones of meristematic activity between adult portions of the stem consists in the fact, that the apical growing-point is thereby enabled to make an early beginning with the formation of the floral region of the shoot; valuable time is thus saved which would otherwise have to be devoted to the production of cells sufficient for the ultimate development of the entire shoot. Such a division of labour between apical and intercalary meristems is, of course, particularly well suited to the requirements of short-lived annuals, such as many of the Grasses, which have to flower and fruit in the shortest possible time.

In the preceding discussion reference has repeatedly been made to the existence of more than one kind of meristematic tissue. The original meristematic tissue, from which all the cells of a growing-point are derived, may be termed the **primordial meristem**, because it in a sense contains within itself all the future tissues of the organ in an undeveloped or primordial condition. The primordial meristem sooner or later becomes differentiated into the three **primary meristems** of the apical region, namely, the **protoderm**, the **procambium** and the **fundamental meristem**. Here are already foreshadowed both the topographical arrangement and the leading anatomical characteristics of the principal permanent tissues; peripheral and central tissues are marked off from one another, while the segregation of prosenchymatous strands from masses of parenchyma, and the histological contrast between these two classes of tissue, are clearly indicated. Not infrequently, also, the disposition of the primary meristems throws some light upon the phylogenetic origin of the permanent tissues to which they subsequently give rise; it does not, however, afford any information as to the future physiological functions of these permanent tissues.

Quite distinct from the primary meristematic layers are the **secondary meristems**. Typically these originate from living permanent tissues, certain layers or masses of which undergo repeated cell-division and thus resume meristematic activity. In general terms, therefore, the development of secondary meristems may be referred to the class of phenomena comprised in the introductory chapter under the head of "change of function." A typical illustration of the formation of a secondary meristem is afforded by the conversion of a layer of green cortical cells—by tangential division—into a cork-cambium or phellogen.

The general properties of meristematic tissues may next be considered; these are, of course, intimately connected with the special activities of meristems, just as the histological peculiarities of permanent tissues are correlated with their physiological functions. Meristematic cells are invariably thin-walled; this property not only permits of any degree and form of secondary growth in thickness, but also facilitates that rapid influx of nutritive materials in the absence of which a meristematic tissue would soon be rendered inactive by exhaustion.

Another distinguishing feature of meristematic cells is the massive development of their protoplasts; usually the cell-cavities are completely filled with homogeneous protoplasm, vacuoles and sap-cavities of ordinary dimensions as well as larger inclusions, such as coarse-grained starch or drops of oil, being entirely absent. In other words, no storage of plastic materials takes place in meristematic cells, on account of their intense metabolic activity. The relatively large size of the nuclei in meristematic cells has already been mentioned and explained in terms of the probable functions of the nucleus (p. 27). Chromatophores, when present, are generally represented by leucoplasts, less frequently by small pale-green chloroplasts (Fig. 5 B, C). Finally, the small size of the cells and their great capacity for division may be included in this list of the general characteristics of meristematic tissues, although as a matter of fact, both these features are tacitly comprehended in the accepted definition of a meristematic tissue.

When a meristem is converted into permanent tissue, the relative positions of the constituent cells become altered in a variety of ways. For one thing, the cells of adult tissues vary greatly in size and shape: furthermore, the physiological requirements of the different cells and cell-masses often necessitate more or less extensive modifications of the spatial arrangement of the meristematic elements from which they are derived. The developing cells thus often undergo a very considerable amount of displacement owing to idiosyncrasies of growth; such individual or active displacements must not be confused with the passive or mechanical displacements imposed upon the cells as a result of the growth of the entire organ. Active displacement occurs, for example, when the branches of non-articulated latex-tubes push their way into adjacent tissues, when the pointed ends of bast-fibres grow past one another, when the hypodermal crystal-cells in the leaves of species of *Citrus* penetrate between the over-lying epidermal elements, or, finally, when expanding tracheae or sieve-tubes force apart and disarrange the cells which surround them.

Krabbe was the first to demonstrate the widespread occurrence of this phenomenon of displacement of cells, and to recognise its importance as a factor influencing the internal structure of plants. Krabbe refers the displacement in question to a process of **sliding growth**.[33] He supposes that large areas of a cell-wall may simultaneously undergo growth in surface: such extensive surface-growth, however, cannot take place unless the walls of adjacent cells become displaced relatively to one another, or in other words, unless these walls glide or slide past one another.  This hypothesis necessitates the further assumption that every cell which takes part in sliding growth is provided with a separate cell-wall; even where the partition between two adjacent cells may appear to be quite homogeneous, it must in such cases actually consist of at least two distinct layers.  There can be no protoplasmic connection between cells which are engaged in sliding growth, since any connecting threads would inevitably be ruptured in the process.  The plasmodesms which are found in adult tissues must therefore come into existence after sliding growth has ceased.[34]

This general discussion may conclude with a short account of the arrangements which serve for the **protection of meristems**.  Since meristems are among the most delicate of vegetable tissues, they evidently stand in need of effective protection against injurious mechanical and climatic influences.  The plant has a variety of means to this end at its disposal.  In the case of the outlying apical meristems, it is usual for the delicate meristematic tissue to be over-arched or completely enveloped by older and more resistant tissues or organs.  This type of protection is exceedingly widespread in its occurrence, although the details of construction vary considerably in different cases.  It is found even among Thallophyta; in the FUCACEAE, for example, the apical cell lies on the floor of a cavity which communicates with the outside only by means of a narrow cleft.  In many Liverworts again, and also in the prothallia of Ferns, the growing-point is situated at the base of a depression, which is produced by the more active growth of two projecting lateral lobes.  In the case of the vegetative and reproductive shoots of Phanerogams the task of protecting the more or less conical apical region usually devolves upon the young leaves, which converge and become folded together over the apex.  Very often these young leaves, while serving as a covering for the growing-point, nevertheless themselves also require a certain amount of protection.  This complication is more especially characteristic of buds which have to live through the winter, or which are developed underground and subsequently have to break through the soil.  In the case of most winter-buds all the delicate organs are encased in tough leathery scale-leaves (bud-scales).  In the case of subterranean structures on the other hand the entire bud often assumes a drooping or reflexed position, owing to nutation of the stem : the curved region of the axis thus forms the actual boring point, while the tender bud follows after in comparative safety.

A parallel development, which likewise serves to bring the

primordial meristem into a safe position, is the involution of the apex
which is seen in certain Seaweeds (spp. of *Polysiphonia* and *Helico
thamnion*).  An effective protection of the growing-point is most
urgently required in the case of structures which pass the whole of
their life in the soil and thus carry out their entire development
within a solid medium.   A typical illustration is provided by ordinary
roots.  ˙Here, moreover, owing to the absence of appendages which
might lend themselves to the formation of a bud, the growing-point
is, as it were, thrown back upon its own resources for the means of
protection.   As a matter of fact the meristem solves this problem
very successfully by producing a resistant root-cap; it consequently
appears to occupy an internal position in comparison with the obviously
superficial primordial meristem of the shoot.

## II. THE PRIMORDIAL MERISTEM.

Inasmuch as the primordial meristem provides all the structural
elements out of which stem, leaf and root are built. up, it represents
in the most literal sense the original or primordial formative tissue of
the plant.   In accordance with the universal nature of its relations,
the primordial meristem exhibits no differentiation which could be
regarded in the light of a preparation for the development of particular
anatomico-physiological tissue-systems.   Any inequalities of size, shape
or arrangement which its component cells actually display, depend
entirely upon the manner in which successive tissue-elements are
segregated from the meristematic cells.   While the sequence of the
cell-divisions accompanying this process of **segmentation** is sometimes
very regular and obvious, this is by no means always the case.   Since
Nägeli's classical researches upon the subject,[35] much time and labour
have been expended by a number of botanists upon its further
elucidation.

### A. MARGINAL AND APICAL [MERISTEMATIC] CELLS.[36]

The distribution and arrangement of the meristematic tissues in
the plant-body, or in a particular organ, depend upon the shape and
mode of growth of the structure under consideration.   Thus in the
case of a more or less circular plate of cells, the marginal elements,
which are all homologous and endowed with an equal capacity for
growth, collectively represent the primordial meristem.   These **marginal
cells** divide by walls parallel to the margin of the plate, and thus
gradually advance in a centrifugal direction; in this way they soon
become the terminal members of cellular filaments, which radiate
outwards like the leaves of a fan.   The thallus of *Melobcsia* (Fig. 9)

admirably illustrates this mode of growth. If such a cell-plate be supposed to rotate around either a transverse or a longitudinal axis, a spherical cell-body will eventuate; here the superficial cells effect a uniform expansion of the cell-body on all sides, and in their totality constitute the primordial meristem. In this case also, of course, the final result is the production of cellular filaments radiating outwards in every direction.

A more definite localisation of the primordial meristem prevails where the cell-plate or cell-mass grows most rapidly in one particular direction, where, in other words, a distinction can be drawn between

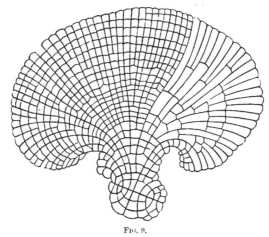

FIG. 9.

*Melobesia Lejolisii* (Florideae).   After Rosanoff and Sachs (from Sachs, *Lectures*).

longitudinal and transverse growth. In this event the cells have unequal powers of growth at different points in the margin. Those which occupy the anterior margin grow and divide most rapidly; they constitute the apical region, and are responsible for the **apical growth** of the elongating organ. In some cases such an apical region appears to consist of a number of divergent curved cellular filaments, terminating at the apical surface in marginal cells which are equivalent among one another (Fig. 10). It is more usual, however, for the cell-walls in an apical region to be arranged in such a manner that a single one among the marginal cells of the growing-point acts as the definite **initial cell**, from which all the remaining cells of the primordial meristem are genetically derivable. This primordial mother-cell occupies the apex of the growing-point, and is consequently termed the **apical cell**. In other cases, however, the arrangement of the cell-walls in the primordial meristem renders it necessary to assume the existence of several apical initials. The presence of such an apical cell-group comprising two or more initials naturally involves a scheme of cell-division which is more complicated than any that is found in connection with a single apical cell. But stress must at once be laid—if only on phylogenetic grounds—upon the fact that growth by means of a solitary apical cell cannot differ fundamentally from the process in which several initials are involved. It must of course be left to ontogenetic investigators to determine the features which are common to the different forms of apical growth, and to demonstrate the existence of intermediate stages between them.

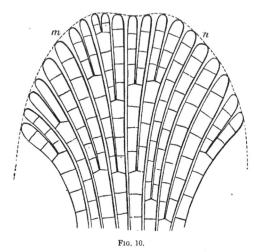

FIG. 10.

Apex of the thallus of *Stypopodium atomarium*; semi-diagrammatic, the cell-rows, which in reality form a compact tissue, being delineated as separate filaments. After Nägeli and Schwendener (from Nägeli and Schwendener, *Das Mikroskop*).

### B. GROWTH BY MEANS OF A SINGLE APICAL CELL.

In a number of Algae, among the Bryophyta and in the majority of Pteridophyta, apical growth is carried out by means of a single apical cell, which at definite intervals cuts off daughter-cells termed **segments**. The segments by their further division give rise to the many-celled primordial meristematic tissue; in the meantime the apical cell, the volume of which is of course diminished whenever a segment is cut off, grows vigorously until it regains its original size, when it is once more ready to undergo segmentation.[37]

In the simplest cases—such as that represented in Fig. 11—the apical cell becomes divided by successive transverse walls (I.-IV.), thereby giving rise to a simple longitudinal row of segments. Each segment soon becomes bisected by another transverse wall (1-4). The appearance of longitudinal walls in the segment-halves leads, first, to the formation of quadrants, and, later, to the differentiation of peripheral and central cells, which in their turn become subdivided by further transverse walls. In this way each segment becomes converted, without undergoing any appreciable growth, into a multicellular structure, which soon passes out of the meristematic condition and becomes a part of the permanent tissue. The branches originate at an early stage —arising indeed as lateral outgrowths of the apical cell—and behave in precisely the same manner as the main axis. Apical cells of this kind, which carry out all their divisions in a single plane, are found in certain genera of Algae— such as *Sphacelaria, Chaetopteris, Clado stephus, Stypocaulon*, etc.

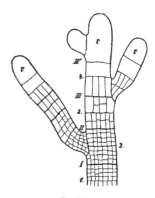

Fig. 11.

Branch of the thallus of *Sphacelaria scoparia*, bearing lateral branches; *v*, apical cell.

More usually the apical cell divides by oblique walls and in several planes; as a result it comes to be more or less deeply embedded in the rest of the primordial meristematic tissue. This scheme of segmentation natur-ally admits of considerable variation in detail. In some Liverworts

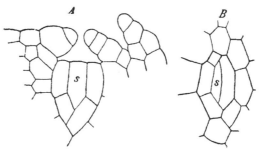

Fig. 12.

*A*. Apex of the stem of *Symphyogyna rhizoloba*, as seen in longitudinal section; *s*, apical cell. *B*. Apex of a young stem of *S. sinuata*, as seen from above. After Leitgeb.

(*Metzgeria, Ancura, Symphyogyna*), in the stems of *Salvinia* and *Azolla*, and in certain other instances, the segment-walls arise alternately right and left of the median plane, so that two rows of segments result. The accompanying figures (Fig. 12 A and B) illustrate the behaviour of such a "two-sided" apical cell, as seen in longitudinal section and in surface view respectively. Among the

F

Musci and Equisetaceae, as well as in the bulk of Filicales, the apical cell divides by three sets of oblique walls; its shape is therefore approximately that of a three-sided pyramid, with a convex base which faces outwards.    Division takes place in the three planes parallel to the sides of the pyramid in strict rotation, and the resulting segments consequently form a continuous spiral series.    Since the outer wall of each segment occupies about one-third of the circumference of the growing-point [as measured in a horizontal plane, of course] the segments appear to be arranged in three vertical rows.    These segments soon undergo further divisions.    In *Equisetum* (Fig. 13) the first of

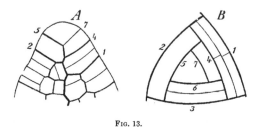

FIG. 13.

Stem-apex of an *Equisetum*.    *A*. Longitudinal section.    *B*. Apical cell, together with the most recently formed segments, as seen from above; 1-7, the successive primary segment-walls.    After Nageli and Schwendener (from Nageli and Sehwendener, *Das Mikroskop*).

the secondary walls is parallel to the primary segment-wall; the next wall to appear is radial and divides the segment into two homologous halves.    After this stage the tracing out of the successive divisions already becomes a matter of difficulty.    The roots of *Equisetum*, as well as those of Ferns and certain Monocotyledons, also grow by means of a three-sided apical cell.    Here, however, the segmentation is complicated—as Nägeli and Leitgeb first showed—by the appearance of transverse divisions.    The ordinary oblique walls cut off segments which give rise to the bulk of the organ, just as in the case of the stem.    The transverse divisions are responsible for the formation of the peculiar protective structure known as the root-cap; each segment cut off by a transverse wall forms a sort of hood over the apical cell, and the successive hoods piled one upon the other constitute the root-cap (Fig. 14).    The order of segmentation is usually such that a single transverse division alternates with three oblique divisions, each of the latter being in a different plane.

A four-sided (pyramidal) apical cell occurs in the seedling axis and in the young rhizophores of certain species of *Selaginella* (*e.g. S. Martensii*).

Even in one and the same growing-point the apical cell need not adhere permanently to a single mode of division.    Thus Treub has shown that the apical cell in the lateral shoots of *Selaginella Martensii* is at first four-sided, but subsequently becomes three- or even two sided; the two latter forms of apical cells may further replace one other quite arbitrarily in older apices, the appearance of an oblique segment-wall sufficing to convert a two-sided into a three-sided cell.   Goebel states, with reference to the origin of stems from the meristematic leaf-apices of *Adiantum Edgeworthii*, that the two-sided apical cell of the leaf becomes directly converted, by an appropriate division, into the three-sided apical cell of a stem. Similarly, the transformation of a root-apex into the growing-point of a

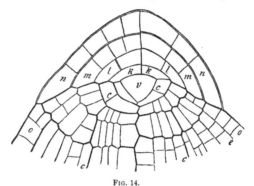

Fig. 14.

Longitudinal section through the end of the root of *Pteris hastata.*   *v*, apical cell; *k-n*, the successive hood-layers of the root-cap; *o-o*, protoderm.   After Nägeli and Leitgeb (from Sachs, *Text-Book of Botany*).

stem which takes place in certain Ferns (*e.g. Asplenium esculentum* and *Platycerium alcicorne*) is initiated, according to Rostowzew, by the omission of the outer or root-cap-forming segments from the segmentation-cycle of the three-sided apical cell.   Even among the Pteridophyta, therefore, traces may be found of a condition which is much more prevalent among Phanerogams, that is, the comparative inconstancy exhibited by one and the same plant, or even by a single organ, in respect of its mode of growth and as regards the form and arrangement of its apical cells.

### C. GROWTH BY MEANS OF SEVERAL APICAL CELLS.[38]

In view of the accepted definition of an apical cell, it is self-evident that if any given cell of the primordial meristem is to rank indisputably as an apical initial it must fulfil the following two conditions: it must retain its original position with reference to the axis of the apical cone, and it must divide actively, so long as apical growth continues.   There are three conceivable modes in which several initials may be associated, so that all conform to the aforesaid conditions.

In the first mode of arrangement the apical cells are all located in the same horizontal plane; they are, in fact, **juxtaposed** or **horizontally seriated**. In this case they must all abut directly against the median plane in a longitudinal section; as seen in horizontal projection they must be grouped around the axis, or, in other words, around the centre of the apical dome. Here the primordial meristematic tissues take the form of longitudinal strips derived from the several initials.

In the second case the apical cells are **superimposed** or **vertically seriated**, that is to say, they are situated in successive horizontal planes. In a radial section they are all seen to lie in the median plane, while in relation to the whole apical dome they are, of course, central. In this case the lowermost initial gives rise to an axile strand of primordial meristem, which represents, as it were, the core of the growing-point: the others, on the contrary, produce mantle-like meristematic layers, which form so many concentric shells around the axile strand. Each of these shells grows by means of an independent initial.

In the third and most complex type of arrangement the apical cells are **both juxtaposed and superimposed**. In other words, each of the successive horizontal planes includes a group of initials; every group must of course comply with the conditions laid down in the case of the first type for the single layer of initials. Since this plan obviously represents a combination of the first and second types of structure, it is hardly necessary to consider in detail the arrangement of the meristematic layers that arise from the several initials. It may, however, be noted that a stratification of the meristem due to the vertical seriation of the initials is sometimes evident for some distance behind the apex, whereas the horizontal seriation produces no such effect.

Each of the above three principal types of seriation includes a number of sub-types, which differ from one another in points of detail. Some of these varieties of arrangement may next be illustrated by examples. The simplest illustration of horizontal seriation of initials is furnished by the stem-apex of *Selaginella Wallichii*, which has been described by Strasburger. Here there are two wedge-shaped apical cells, of which one produces the right and the other the left half of the dorsiventral stem. The apical faces of these cells are narrow and rectangular, while their lateral surfaces are broadly triangular and extended at right angles to the plane of dorsiventrality. The segments are cut off in four vertical rows.

According to Schwendener four initial cells are grouped around the centre of the apical dome in the roots of MARATTIACEAE. A radial longitudinal section reveals two initials, one on either side of the middle line; each of these gives rise by periclinal divisions to segments, which on the one hand go to make up the root-cap, while on the other they become incorporated in the body of the root. In addition lateral segments are cut off by appropriately disposed longitudinal walls. The products of the four apical cells arrange themselves in quadrants, marked off from one another by walls which are thicker and more continuous than the rest.[39]

Schwendener states that four initials meeting at the centre of the apical dome also occur in the stems of Conifers (vegetative shoots of *Juniperus communis*, seedling axis of *Pinus inops, P. Laricio, P. sylvestris, Abies alba*). The occurrence of a single three-sided apical cell, previously described by Dingler as the normal condition among Conifers, appears to be quite an exceptional arrangement in this group of plants.

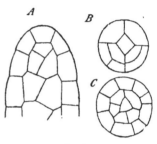

The simplest form—involving only two initials—of the second or vertical type of seriation is exemplified by the young leaf-segments of *Ceratophyllum demersum* (Fig. 15). According to the author's observations the upper of the two initials in this case takes the form of a three- or four-sided pyramid, with a convex outer face and a truncated inner end; if it is four-sided, the walls are usually so arranged that

FIG. 15.

Apex of a young leaf-segment of *Ceratophyllum demersum*. *A.* Longitudinal section. *B.* Surface view. *C.* Optical transverse section through the lower of the two apical initials.

two adaxial and two abaxial rows of segments result (Fig. 15 B). These segments divide solely by anticlinal walls and undergo no periclinal divisions whatsoever. Hence the shell of tissue derived from them comprises a single layer of cells, and in fact constitutes the outermost meristematic layer, which corresponds to the primary dermal layer of the adult organ, and which on that account has been termed the dermatogen by Hanstein. The second or lower of the two initials, which is shaped like a complete three- or four-sided pyramid, divides by oblique walls, in precisely the same manner as the solitary initial in the stem of a Fern or Horsetail (Fig. 15 A, C). Each of its segments becomes subdivided, by an approximately radial wall, into two slightly unequal daughter-cells; further division of the latter leads to the formation of a central mass of meristematic tissue—enveloped in the dermatogen—which gives rise to the whole of the permanent tissue of the leaf, with the single exception of the epidermis.

The author has further shown that the apical growth of young axillary branches of *Ceratophyllum demersum* is more complex, owing to the presence of three superimposed initials (Fig. 16). The upper-most of these gives rise, just as in the leaf, to a single superficial layer of meristem. The second or middle cell at first behaves in a similar fashion; but later, as the apical dome enlarges, periclinal as well as anticlinal walls appear in this second shell, so that the final product of the middle initial is a many-layered meristem. The third or lowermost initial, finally, divides by oblique walls and pro-duces an axile mass or core of meristematic tissue.

In the case of the third type of arrangement of apical cells [where the initials are both juxtaposed and superimposed] the difficulties

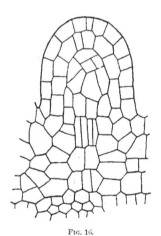

of observation are so great, that no instance has as yet been thoroughly worked out; it is highly probable, how-ever, that this complex form of apical growth is widely distributed among Angiosperms.[40]

It may, in any case, be regarded as certain that, among Phanerogams, the mode of apical growth is liable to considerable variation not only within a species but even in one and the same shoot. Thus the primary axis of the seedling often differs in this respect from a subsequently formed lateral shoot, while the latter in its turn may exemplify one type of arrangement of the apical initials when young, and a different type later on. The fact that different investigators so

FIG. 16.

Median longitudinal section through a young axillary branch of *Ceratophyllum demersum.*

frequently disagree concerning the apical growth of Phanerogams is no doubt to a large extent attributable to this variability. As has already been pointed out, the apical cells of certain Pteridophyta display a similar inconstancy of behaviour.

The discussion of apical growth by means of several initials naturally leads to a consideration of Hanstein's classification of the apical tissues of Angiosperms.[41] Hanstein assumes that the primordial meristem is always sharply separable into three distinct strata or " histogens." He distinguishes a central strand of **plerome**, composed mainly of cells extended in the longitudinal direction; this plerome is surrounded by a shell of **periblem** which usually consists of several strictly concentric layers of isodiametric cells. The periblem in turn is succeeded by a second shell of meristematic tissue, comprising a single layer of cells; this is the **dermatogen**, to which reference has already been made.  Each of these three sharply differentiated histogens is derived either from a single apical cell or from several apical initials, the latter arrangement being the more usual one.

As a matter of fact, such a differentiation of the primordial meristem is quite evident in the case of certain Phanerogamic stem-apices.  The most frequently quoted instance is that of *Hippuris vulgaris* (Fig. 17).  Here the dermatogen is clearly demarcated from the five-layered periblem, and the latter is also obviously marked off from the plerome, which can often be traced back to a single initial cell.  Other cases undoubtedly exist, to which Hanstein's

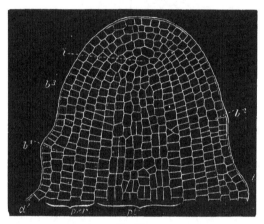

Fig. 17.

Longitudinal section through the growing-point of *Hippuris vulgaris*. *d*, dermatogen ; *per*, periblem ; *pl*, plerome ; *i*, plerome-initial ; *b-b3*, primordia of successive leaves.  After Warming (from Schenck's *Handbook*).

classification is equally applicable ; but in relation to the general theory of apical growth by means of several initials outlined above, Hanstein's classification amounts to nothing more than a summary of the features common to certain special cases of such apical growth, all of which can be quite naturally included in the second or third of the principal types of arrangement.  As a matter of fact, the structure of the primordial meristem must conform to Hanstein's scheme, wherever the following two conditions are fulfilled : in the first place, there must be at least three vertically superimposed initials or groups of initials present, and, secondly, the uppermost initial or group of initials must not give rise to more than a single layer of cells.  The above-described apex of the young axillary shoot of *Ceratophyllum demersum* with its three superimposed initials may accordingly be regarded as the simplest instance of a growing-point differentiated into plerome, periblem and dermatogen.

Hanstein's classification is, of course, by no means universally
applicable; for, as a matter of fact, the structure of growing-points
with several initials varies within comparatively wide limits.    Moreover, quite a number of cases are now known among Angiospermous
shoots and roots, in which Hanstein's three primary histogens cannot
be recognised.    In particular, it is not uncommon for the apical dome

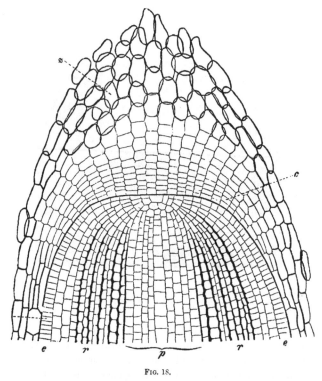

Fig. 18.

Longitudinal section through the root-apex of *Eriophorum vaginatum* [CYPERACEAE].
-, the oldest cells of the root-cap, which are sloughing off ; c, calyptrogen ; e, protoderm ; r, cortex (periblem) ; p, central cylinder (plerome). *Protoderm and cortex
arise from the same initial.*

of a shoot to be invested by a perfectly distinct dermatogen, while
at the same time the plerome and periblem arise from a common initial,
and hence can in no sense be regarded as two separate histogenetic
layers.    In many roots, again, the dermatogen and periblem obviously
have a common genetic origin.    The adjoining figure, for example
(Fig.  18), which represents a radial longitudinal section through
the root-apex of *Eriophorum vaginatum*, shows quite clearly that, in
this case, periblem and dermatogen are derived from one and the

same initial cell, and only become secondarily differentiated owing to the appearance of periclinal walls in the common mother-segments.

In conclusion, attention must be specially directed to the various complications that arise in connection with the **origin of the root-cap**, in the case of roots which grow by means of several apical initials.[42] As has been stated above, the roots of the MARATTIACEAE grow by means of four juxtaposed initials, and as regards the origin of the root-cap resemble those of *Equisetum* and most Filicales; the apical initials in fact undertake the construction of the root-cap as a subsidiary function, and to that end undergo occasional periclinal divisions. The formation of a root-cap is not such a simple matter, where both vertically and horizontally seriated initials occur together—the normal condition in the roots of Phanerogams. In order to obtain a clear understanding of the great variety that obtains with regard to the relations between primordial meristem and root-cap, it is necessary to keep in mind the circumstance that the root-cap is a secondarily evolved adaptive structure, which in the first instance constitutes a mechanically and generally protective appendage of the root-tip, and in addition usually contains the geotropic sense-organ of the root. The existence of a genetic relation between the primordial meristem and the developing root-cap must accordingly be regarded as a secondarily acquired feature. It is impossible to give a coherent account of the variety of structure exhibited by the growing-points of Phanerogamic roots, except from this phylogenetic point of view.

The various modes of origin of the root-cap in Phanerogams may all be referred to one or other of the following six principal types of development.

*Type* 1. The root-cap arises from a special meristematic layer or **calyptrogen**, which has no genetic connection with the main body of the root. The apical meristem of the root proper thus appears sharply delimited from the root-cap and calyptrogen; if these could be lifted bodily from the tip of the root, the surfaces of separation would be quite smooth and continuous. This mutual independence of root-cap and root proper is exemplified by the GRAMINEAE, CYPERACEAE (Fig. 18), JUNCACEAE and CANNACEAE.

*Type* 2. The histogenetic layer of the root-cap is continued backwards into the protoderm (dermatogen) of the root proper; or, in other words, the single protodermal layer separates, as it approaches the apex, at first into two and subsequently into three or more layers, of which the innermost continues to add to the protoderm, while the others form the successive shells of the root-cap. Eriksson has designated this type of histogen a "dermocalyptrogen," because it is genetically related both to the root-cap and to the protoderm of the root. From another point of view, however, the root-cap may, in a case of this kind, be regarded as a mere proliferation of the protoderm; this latter interpretation, which was long ago adopted by Hanstein, is undoubtedly more in accordance with the probable phylogenetic origin of the root-cap than the conception of a primitive dermocalyptrogen. The second type of structure is exemplified by the majority of Dicotyledonous roots; the following instances, among others, have been examined in detail: *Heli-*

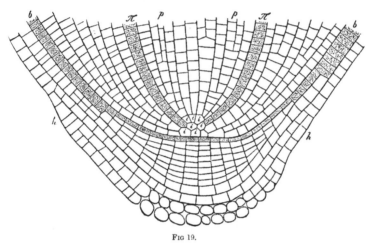

Fig 19.

Longitudinal section through the growing-point of the radicle of *Helianthus annuus.* *h-h*, root-cap: *b-b*, protoderm; *π-π*, pericycle (pericambium); *p-p*, plerome. After Reinke (from Sachs, *Text-book*).

*anthus annuus* (Fig. 19), *Fagopyrum, Brassica, Sinapis, Salix, Linum, Cyclamen, Lysimachia.*

*Type 3.* In addition to the protoderm, the adjacent cortical layers also take part in the production of the root-cap. Hence, on following up the protoderm towards the apex one does not, as in the preceding case, arrive at the inner limit of the root-cap, but, on the contrary, penetrates more or less deeply into the substance of the cap. It must, however, be noted that it is only the outer layers of the cortex or periblem that contribute to the formation of the root-cap. According to Flahault the roots of *Cercis Siliquastrum, Gymnocladus canadensis,* and *Juglans regia* belong to this type.

*Type 4.* The point of origin of the root-cap is more deeply seated even than in the preceding type; the *entire* cortical meristem, namely, gives rise to the root-cap, while the protoderm remains undivided. This type is illustrated by *Acacia, Mimosa, Tamarindus, Caesalpinia* and *Lupinus,* and by the Gymnosperms.

*Type 5.* The primordial meristem of the root-cap is fused with that of the root proper to form a common initial zone; the rows of cells which make up this common zone are, on the one hand, continuous with the several component shells of the root-cap, and on the other penetrate more or less deeply into the root proper. In extreme cases this joint

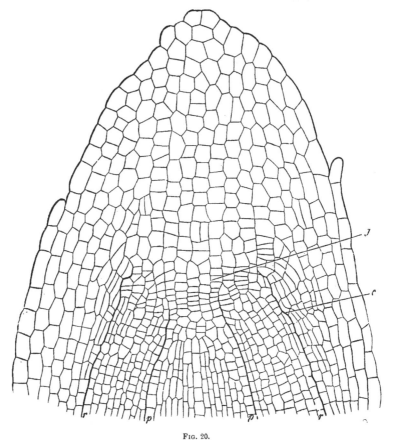

Fig. 20.

Longitudinal section through the root-apex of *Pisum sativum*. *J*. Common initial zone; *c*, its protodermal continuation; *p-p*, central cylinder (plerome); *r-r*, cortex (periblem). From de Bary, *Comparative Anatomy*.

meristem produces, not only the cortex, but also the central core (comprising both periblem and plerome) of the root. The protoderm, which extends forward as far as the common zone of meristem, also continually cuts off layers towards the outside which help to regenerate the root-cap (Fig. 20). The roots of *Pisum, Vicia, Cytisus, Acer, Lavatera, Foeniculum*, etc., belong to this category.

*Type* 6. Here, as in the preceding type, the meristems of root-cap and root proper coalesce to form a common primordial meristem. The protoderm, however, behaves differently, inasmuch as it takes no part in the formation of the root-cap—if, indeed, it is differentiated at all. The sixth type thus bears the same relation to the fifth, as the fourth does to the third. It is exemplified, according to Treub and Flahault, by the LILIACEAE, ASPIDISTREAE, OPHIOPOGONEAE, etc.; examples which have been carefully studied are *Calla palustris*, *Anthericum ramosum* and *Allium*.

If the first of these six types of apical growth be ignored for the moment, and the remaining five be considered in succession, a sort of panoramic view may be obtained of the changes which the root-tip of Phanerogams has undergone during the gradual evolution of a cap or hood for the protection of the primordial meristem. In the simplest case, the desired result is achieved by means of a proliferation of the peripheral layer or protoderm (Type 2). A more serious modification of structure is produced, where this hypertrophy extends to the cortical layers (Types 3 and 4). The influence of the root-cap thus tends to affect more and more deeply situated tissues of the root. The organic and genetic connection between parent organ and appendage is most intimate, where an initial zone common to the two is developed (Types 5 and 6).

There is a certain analogy, as will be shown later, between the behaviour of the primordial meristem in different roots and the variations in the mode of formation of cork. For cork, which is likewise a protective tissue, in certain cases originates in the epidermis, but in others arises from hypodermal parenchyma or even from more deeply seated layers.

Somewhat fuller consideration has still to be given to the first-mentioned type of arrangement, which is characterised by the fact that the root-cap is entirely independent of the main body of the root. The answer to the question, as to how the root has come to possess a root-cap at all in these circumstances, is provided by the phenomena attendant upon the first differentiation of lateral roots. The early history of the developing rootlets was thoroughly investigated long ago from this point of view by Nägeli and Lietgeb in the case of *Oryza sativa* [Fig. 21 A, B]. It appears that, in the first stage of development of such a root, the root-cap originates from two cells of the endodermis or innermost cortical layer of the parent organ; these initial cells, which are steadily pushed outwards owing to the radial extension of the developing root, divide at first by radial and later also by tangential walls, and so give rise to the primary root-cap. The structure subsequently becomes more complex, owing to the fact that the terminal element of the median cell-row in the body of the young root expands laterally and then divides by a transverse wall situated close to its anterior (or outer) face. The outer of the two daughter-cells gives rise by further divisions and by appropriate displacements of the resulting segments to a "hood-layer," which reinforces the primary root-cap. The formation of this solitary hood-layer appears to mark the termination of the development of the root-cap at the expense of the root proper; the subsequent growth and regeneration of the root-cap depend upon the activity of its own innermost layer, which thus constitutes the calyptrogen.

The fact that the root-cap is made up, at any rate during the early stages of its development, of two portions of totally distinct origin, has subsequently been confirmed for quite a number of different families by van Tieghem and Douliot. In all such cases the endodermis—with or without the assistance of the adjacent cortical layers—produces the outer apical envelope of the lateral root, termed by van Tieghem the "digestive pouch" (*poche* or *poche digestive*); the name is intended to suggest that the pouch secretes a digestive enzyme, which dissolves the cortical tissue of the parent organ and thus opens a path for the exit of the young rootlet. The inner apical envelope of the rootlet, or "calyptra," is produced by the rootlet itself, which, it may be remarked, always originates in the pericycle (pericambium). Typically, therefore, the apex of the rootlet is, to begin with, covered by a double envelope, consisting of a calyptra and a digestive pouch; at a later stage the pouch is cast off, and the rootlet is thenceforward covered only by the calyptra. In the CRUCIFERAE, CRASSULACEAE and CHENOPODIACEAE, in many CARYOPHYLLACEAE, among the Filicales and in many other cases, there is no digestive pouch. In the HYDROCHARITACEAE [Fig. 21 c], in *Lemna, Pistia, Pontederia,* and various other water-plants, no calyptra is formed, and the entire root-cap therefore corresponds to the pouch.

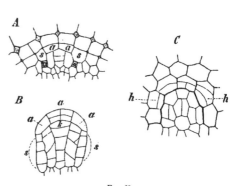

FIG. 21.

*A.* Transverse section through a root of *Oryza sativa,* showing the origin of a lateral root; *s, a, a, s,* cells of the innermost cortical layer, which enclose the primordium of the rootlet. The cells *a, a* give rise to the primary root-cap. *B.* Longitudinal section through a rootlet at a more advanced stage of development (the surrounding cambial and cortical cells are not shown); *s, s,* basal sheath ; *a, a,* primary root-cap ; *k,* the primary hood-cell cut off from the terminal cell of the median cell-row. *C.* Longitudinal section through the primordium of a cauline adventitious root of *Elodea canadensis*; *h-h,* cells of the stem-cortex, enveloping the root proper, which by periclinal divisions give rise to the entire root-cap. *A* and *B* after Nageli and Leitgeb.

While the root-cap in the first instance serves to protect the delicate primordial meristem of the root, it also acts as a boring-point which helps the root to force its way through the soil. The root-cap is able to perform this active function partly by virtue of its conical shape, but even more on account of the fact that its cell-membranes— or at any rate the middle lamellae of the old superficial cells—readily become gelatinous; the slimy surface which the root-cap consequently acquires materially assists the root in overcoming the frictional resistance of the soil-particles.[43] In addition, the axile region or "columella" of the root-cap harbours the statocysts or cells furnished with movable starch grains, which serve for the perception of gravitational stimuli.

## III. THE PRIMARY MERISTEMS.

In all the Higher Plants, the homogeneous primordial meristem becomes differentiated, at a varying distance from the growing-point, into several distinct meristematic layers, or **primary meristems**. To begin with, this differentiation amounts to little more than a topographical separation of a dermal layer from the more central tissues, accompanied by a segregation of "strands" or "bundles" from parenchymatous ground-tissue; it affords little or no indication of the nature of the functions that will be performed by the corresponding permanent tissues.

Such a differentiation of the primordial meristem into primary meristems is not a peculiar condition associated with any one mode of apical growth or dependent upon the presence of more than one initial, but takes place in every organ that attains to a certain degree of anatomical complexity in its adult condition. The following classification of primary meristems—devised by the author—is therefore equally applicable to the shoot of a Moss and to the stem, leaf, or root of a Dicotyledon.

In almost every case, three distinct primary meristems are present [namely, **protoderm, procambium** and **fundamental meristem**].[44] The **protoderm** is the outermost meristematic layer (Fig. 22, $p$), and represents the primitive dermal tissue of the young organ, though only in a topographical sense; it exhibits none of the anatomico-physiological characters of typical dermal tissues, and may not even necessarily acquire these features when it becomes converted into permanent tissue, since it not only gives rise to the epidermis but also regularly produces the principal absorbing organs, and in certain cases also takes part in the formation of mechanical or photosynthetic tissues.

The precise genetic relation of the protoderm to the primordial meristem depends upon the structure of the growing-point. Where a single apical cell is present, or where the initials are juxtaposed, the protodermal cells become cut off—sooner or later, according to circumstances—from the primary segments by periclinal walls. In the root-apex of Ferns, for example, the protoderm is differentiated, in this manner, at a very early stage, immediately after the appearance of "sextant-walls" in the root-segments cut off from the apical cells. Each sextant divides by a periclinal wall into an inner and an outer cell; the latter at once again divides by another periclinal wall, and the outer of the two resulting daughter-cells becomes an element of the protoderm. In the stem-apex of *Equisetum*, on the contrary, the protoderm separates at a much later stage, after the primary segments have undergone numerous radial, periclinal and anticlinal divisions. Where the apex contains several superimposed initials it is, as has already been explained, very usual indeed for the product of the uppermost initial, or group of initials, to remain permanently single-layered, in which case it becomes directly converted into the protoderm. In these special circumstances the protoderm constitutes an independent superficial histogenetic layer or dermatogen. It is, however, hardly necessary to insist that the author's "protoderm" embodies a much wider conception than Hanstein's "dermatogen"; the former term was introduced without any reference to the structure of the growing-point, whereas the latter connotes a special type of apical structure.

The boundary between the superficial protodermal layer and the more deeply situated meristematic layers is not equally well marked in all cases. Where the protoderm is late in appearing, the radial diameter of its component elements is variable, and its inner boundary is consequently somewhat ill-defined. Where, on the contrary, the protodermal layer is continued right over the apex as the dermatogen, its inner (periclinal) walls, when seen in transverse or longitudinal section, usually present the appearance of a very regular zigzag line.

Once the protoderm is differentiated, its subsequent divisions are usually all radial (anticlinal). Less frequently, divisions also take place in the tangential (periclinal) direction, with the result that several protodermal layers are formed; this latter condition is found where the adult epidermis is many-layered, as well as in connection with the development of certain special tissues or organs.

2. The second type of primary meristem, or **procambium**, consists of narrow, prosenchymatous meristematic cells (Fig. 22 c); it is con cerned with the production of the bulk of the vascular and fibrous "strands" of the adult plant. But just as the protoderm may form other structures besides the dermal tissues, so the procambium need not always give rise to bundles alone; it may, on the contrary, after

undergoing appropriate divisions, also produce photosynthetic cells or other parenchymatous elements. The disposition of procambial groups, as seen in a transverse section through a young organ, naturally corresponds to the primary arrangement of bundles in the adult organ. The procambium is thus often disposed in isolated longitudinal strands, for instance, in the stems of a great many Monocotyledons. Most roots,

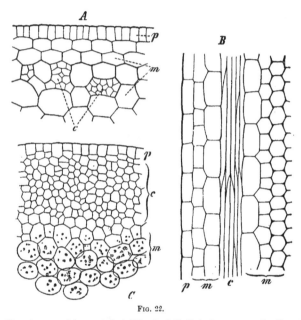

FIG. 22.

The primary meristems. *A*. Part of the adaxial half of a transverse section through a very young leaf of *Pandanus utilis*. *B*. Part of a radial longitudinal section through the same leaf, ×400. *C*. Part of a transverse section through a young petiole of *Asplenium foecundum*, ×125; *p*, protoderm; *c*, procambium; *m*, fundamental meristem. In *C* the cells of the protoderm are undergoing tangential division, while the fundamental meristem is beginning to be converted into chlorophyll-containing parenchyma.

on the other hand, are provided with a single central procambial strand. In other cases again, the procambium forms a continuous hollow cylinder, which is not infrequently supplemented by variously-arranged isolated strands.

The procambium always arises from longitudinal series of primordial mother-cells, in which further division takes place only by longitudinal walls, situated in various planes. To begin with, therefore, it consists of small prismatic cells with ordinary transverse walls. Very soon, however, the individual cells of the procambium undergo active apical growth, as a result of which their terminal walls become very oblique; the cells thus acquire the gable- or awl-shaped ends characteristic of prosenchymatous elements. The time of appearance of longitudinal division in the mother-cell of a procambial strand—or in other words the proximity of properly differentiated strands to the growing-point—depends as a rule upon the bulk of the strand, small procambial strands often making their appearance much later than large ones. Longitudinal division continues in a procambial strand for some time after it has become differentiated, in the manner described, from the primordial meristem ; secondary transverse divisions are comparatively infrequent, at any rate in typical cases. The adjacent cells of the fundamental meristem, on the other hand, divide equally in all directions, so that the procambial cells soon become much longer than the elements of the surrounding tissues ; the difference is accentuated by the active apical growth exhibited by the individual cells of the procambium.

Of the several primary meristems it is the procambium which usually remains longest in the meristematic condition. Among Gymnosperms and Dicotyledons, in fact, and exceptionally also among Monocotyledons, the majority of the procambial strands that give rise to vascular bundles do not become wholly converted into permanent tissue. A portion of each strand remains permanently in the condition of [procambial] primary meristem ; this so-called **fascicular** or **intrafascicular cambium** forms a strip which extends tangentially right across the bundle, separating from one another the two principal conducting tissues of the strand, the hadrome and leptome. The cells of the fascicular cambium are at first irregularly disposed, but later arrange themselves in rows which run parallel to the plane of symmetry of the bundle, thus forming a **serial cambium**. It is in this way that a regular secondary increase of hadrome and leptome is rendered possible.

The complete discussion of the manner in which these isolated cambial strips subsequently unite to form a closed cambial cylinder, which effects the secondary growth in thickness of the stem [and root], must be postponed until a later occasion (Chap. XIV.). It may be remarked in passing that the vascular bundles of Gymnosperms and Dicotyledons, with their persistent cambial strips, are often described as **open bundles**, in contrast to the **closed bundles** devoid of cambium which are characteristic of Monocotyledons and Ferns.

The procambium primarily represents a homogeneous meristem, although the permanent tissues to which it gives rise may be exceedingly diverse. The uniformity thus attributed to the procambium, however, applies only to its visible histological characteristics. How far this tissue is uniform in respect of its inherent morphogenetic tendencies is quite another question. In other words, it is a matter for special enquiry—and one which in any case falls outside the scope of the present discussion—to what extent and for what length of time an individual cell of the procambium retains the capacity for giving rise, according to circumstances, to one or other of a whole series of different permanent elements, such as stereides, segments of wood-vessels or sieve-tubes, tracheides, phloem-parenchyma cells and so forth. A similar uncertainty prevails with regard to this point in the case of all meristematic tissues.

**3. Fundamental meristem** or **ground-meristem** is the name given to the whole of the primary meristematic tissue that remains un-differentiated, after the protoderm and all the primary procambial strands have been segregated (Fig. 22, *m*). In contrast to the pro-cambium, the fundamental meristem is parenchymatous and relatively large-celled; it is, moreover, usually pervaded by conspicuous inter-cellular air-spaces. The last-mentioned characteristic provides a somewhat obvious point of distinction from the primordial meristem, the cells of which are generally in uninterrupted contact with one another at every point. As a rule, both the protoderm and the procambium agree with the primordial meristem in being entirely devoid of intercellular spaces.

The fundamental meristem gives rise to the bulk of the paren-chymatous permanent tissues such as the photosynthetic elements, a part of the conducting parenchyma, the pith, etc.; from the purely anatomical point of view, these tissues may be collectively termed the "fundamental parenchyma" or "parenchymatous ground-tissue" in contradistinction to the "fascicular tissue" which comprises the vascular and mechanical bundles or strands. But just as the pro-cambium exceptionally produces parenchymatous elements, so the fundamental meristem occasionally gives rise to fascicular tissue. An interesting illustration is provided by the young scape of *Allium*, in which, as the author has shown, the mechanical cylinder is derived from the fundamental meristem, at any rate in the majority of species.

## IV. SECONDARY MERISTEMS.

Secondary meristems typically originate within living permanent tissues which have previously been concerned for some time with a specific vegetative function. The physiological properties of the new permanent tissue or tissue-system produced by a secondary meristem are, as a rule, quite different from those of the parent permanent tissue. The activity of a secondary meristem thus frequently leads to a far-reaching change of function.

Among the various permanent tissues, it is the different kinds of thin-walled parenchyma which are best fitted to give rise to secondary meristems. Histologically considered, namely, parenchyma has made less progress than any other type of permanent tissue beyond the condition of the primary meristem from which it is derived. As a matter of fact, the majority of secondary meristems do arise within parenchymatous mother-tissues.

Ordinarily, secondary meristems are developed either for the purpose of effecting a secondary growth in thickness of the stem or root, or in connection with the production of cork. The normal secondary meristems therefore are the **interfascicular cambium** and the **phellogen**; both these tissues are discussed more fully in subsequent chapters.

In the case of wholly adventitious organs, even the "primordial" meristem may be secondary as regards its origin. Thus Hansen has shown, that the adventitious shoots which develop upon severed leaves of *Begonia* arise from adult, functional epidermal cells, through the mediation of a secondary meristem. Those adventitious buds on the other hand which appear on the upper or lower sides of the fronds in certain Ferns (*Asplenium bulbiferum, A. viviparum, Ceratopteris thalictroides*, etc.) originate from the protodermal layer, while the leaf is still quite young. In this case, according to Heinricher, a single element of the protoderm gives rise, by appropriate divisions, to a three-sided initial, which becomes the apical cell of the adventitious shoot.[45]

In conclusion, attention may be drawn to the fact—which is not without significance—that the difference between primary and secondary meristems is by no means so well-marked as might be inferred from the above-accepted definition of a typical secondary meristem. As a matter of fact, these two conditions of the meristem are connected by a complete series of intermediate stages; in other words, a primary meristem can become secondarily meristematic at any stage of its transformation into permanent tissue. This point is illustrated in the most instructive manner by the varying relations which exist between the (primary) procambium and the (secondary) cambial layer.[46]

# CHAPTER III.

## THE DERMAL SYSTEM.

### *I. GENERAL CONSIDERATIONS.*

Even in the case of the simplest of unicellular Algae, the living substance has found it necessary to shut out the surrounding medium—and thus to withdraw from the immediate influence of hostile environmental conditions—by secreting a protective membrane or cell-wall. Highly organised multicellular plants stand in still greater need of a **tegumentary** or **dermal system** which will serve to protect their internal tissues. Among the Lower Plants the cell-wall serves both as a limiting membrane and as an exoskeleton. In the case of Higher Plants, on the contrary, the dermal and the mechanical functions are assigned to distinct tissue-systems in accordance with the principle of division of labour.

The physiological anatomy of the dermal system will be more readily comprehended after some account has been given of the principal agencies against which this system is intended to afford protection; in this way it will at once become evident what conditions a tissue must fulfil, in order to rank as part of the dermal system.

The discussion may at first be confined to the aërial organs of ordinary terrestrial plants. Since such organs are liable—like other freely exposed moist bodies—to lose water by evaporation, the chief danger by which they are threatened is that of desiccation. Next in importance is the risk of mechanical injury, such as may be caused by the impact of violent rain, blown sand or dust, by the attacks of snails, insects or other animals, and so forth. Certain tissues also need protection against excessive illumination or against loss of heat by nocturnal radiation. During winter, finally, the living cells in perennial twigs and branches and in other hibernating organs have to be preserved from violent fluctuations of temperature and especially from the effects of sudden thawing.

In the case of submerged water-plants the dermal system does not need to cope with so many requirements, mainly because such plants run no risk of drying up under normal conditions. But a new factor is introduced owing to the fact that water-plants possess a highly developed ventilating system, which comprises numerous large air-passages in addition to the usual labyrinth of small intercellular spaces; evidently a continuous dermal tissue is required if these large spaces are to be kept water-tight. In the case of plants which grow in running water an additional element of danger arises from the shearing effect of the current, which might readily bring about a gradual disintegration of the tissues, if these were not enveloped in a continuous skin.

The demands which are made upon the dermal system of subterranean structures vary in accordance with the water-content of the soil; such organs, therefore, in this respect hold a position which is intermediate between that of aerial and aquatic organs, approaching more nearly to one or to the other according to the amount of water that is available.

The detailed discussion of the different tissues included in the dermal system will afford some indication as to the extent to which these tissues are actually able to protect the various parts of the plant-body against hostile external conditions.

## II. THE EPIDERMIS.

The **epidermis** represents the first stage in the evolution of a dermal system. Consisting in the majority of cases of a single layer of cells, it separates the underlying tissues from the external medium and preserves them from the dangers which have been enumerated above. This delimitation and protection jointly constitute the principal function of the epidermis. From the anatomico-physiological point of view, therefore, a superficial layer can only be regarded as epidermal if its histological features are really correlated with the performance of this [protective] function. It is by no means the case that every superficial layer is intended to serve as a protective epidermis. The outermost layer of a young root, for example—the so-called piliferous layer or rhizodermis [46a]—is mainly concerned with the absorption of water and nutrient salts, and is hence not an epidermis in the anatomico-physiological sense. Similarly it would not be consistent with anatomico-physiological principles to regard the guard-cells of stomata as part of the epidermis; these structures must on the contrary be assigned to the system which is concerned with ventilation. In the same way the "epidermal" hydathodes which occur on certain leaves,

and in fact all glandular organs that are ontogenetically or phylogenetically epidermal in origin, must be relegated to the secretory
system. Epidermal sense-organs again do not belong to the dermal
system. Lastly, specialised strengthening cells, even if superficial in
position and derived from the protoderm, must unquestionably be
referred to the mechanical system.

To sum up—the epidermis, in the anatomico-physiological sense,
comprises only those superficial cells or cell-layers, the histological
features of which clearly indicate that their principal function is that
of a primary tegumentary or dermal tissue.

## A. THE SIMPLE EPIDERMIS.[47]

### 1. *The shape of epidermal cells.*

Epidermal cells are usually more or less tabular in shape; in
the case of leaves they are not infrequently lenticular (biconvex).
They are always in uninterrupted contact with one another laterally.
In typical cases their radial diameter is small, though it may become
considerable where the water-storing capacity of the epidermis is pronounced. In organs like the majority of Dicotyledonous leaves,
which grow slowly and which are not conspicuously elongated in any
one direction, the epidermal cells are approximately isodiametric. In
various Monocotyledonous leaves, on the contrary, in most petioles and
stems and, in fact, in all distinctly elongated organs, they are as a rule
obviously elongated in the same sense as the whole organ. Exceptionally (as in Cycads, BROMELIACEAE, *Tradescantia, Crassula, Silene
fruticosa*, etc.) they are elongated at right angles to the long axis of
the leaf. Not uncommonly the epidermis is made up of cells of more
than one kind; in most Grasses, for instance, curious "dwarf cells" of
unknown significance are interpolated at regular intervals in the
longitudinal rows of typical elongated epidermal elements [Fig. 26].

### 2. *The outer epidermal wall. Wax.*

From the physiological point of view the outer **wall** must be
considered the most important part of an epidermal cell. This wall
usually differs from the others, first, in its greater thickness, and,
secondly, in certain physical and chemical peculiarities which are due
to the presence of fatty substances collectively termed *cutin*; it may
in addition be impregnated with compounds of a waxy nature.
Typically, the outer wall consists of three distinct zones. The innermost of these, bordering upon the cell-cavity, comprises the **cellulose
layers** (Fig. 23 A. *b.*), which, as their name implies, consist of unaltered
cellulose. Next follow the **cutinised layers** (Fig. 23 A, *cs*), which

contain a varying proportion of cutin. The outermost lamella of the wall contains the greatest amount of cutin and constitutes the cuticle (Fig. 23 c), which forms a continuous pellicle over the whole of the epidermis. The cuticle is always present, and cellulose layers are probably also of universal occurrence. The cutinised layers, on the other hand, are often absent when the outer wall is thin, and may be lacking even when it is quite thick (Fig. 23 B). When present, these cutinised layers are as a rule sharply marked off from the underlying cellulose layers; the surface of contact between the two may be smooth, but is sometimes uneven, owing to the fact that the cutinised layers project at a number of points in the form of minute teeth or

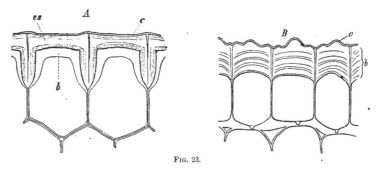

Fig. 23.

A. Cells from the foliar epidermis of *Aloe acinacifolia*. B. Cells from the foliar epidermis of *Allium Cepa*: c, cuticle ; cs, cutinised layers ; b, cellulose layers.

ridges of various shapes. This "interlocking" of the cellulose and cutinised zones, which is particularly well illustrated by the foliar epidermis of certain species of *Aloe*, undoubtedly helps to bind the two layers firmly together. The thickness of the outer wall has a two-fold significance. In the first place it serves to diminish the rate of transpiration, because the cutinised layers resemble the cuticle in being relatively impervious to water; secondly, it has the effect of increasing the mechanical strength of the epidermis. It will be convenient to discuss these two effects separately.

It is an easy matter to show, by comparative investigation, that there is a close correspondence between the thickness and degree of cutinisation of the outer epidermal wall, on the one hand, and the extent to which the underlying tissues require to be protected against excessive transpiration on the other. Thus the outer epidermal walls of submerged plants are generally delicate and hardly thicker than the radial or inner walls. Such plants accordingly wither and dry up very quickly if they are removed from their native element. In their case the epidermis is evidently incapable of restricting transpiration to any great extent.

As a matter of fact the delicate cuticle is quite pervious to water under normal conditions its functions are principally mechanical, while it may also serve to prevent Bacteria and other parasitic micro-organisms from gaining access to the interior of the plant. Among water-plants the epidermis is often thin-walled even in the case of organs which are not submerged; in the floating species of *Lemna*, for example, the outer epidermal wall is no thicker on the upper side of the fronds, which is exposed to air, than it is on the lower surface, which is constantly wetted.

The opposite extreme, as regards the structure of the outer epidermal wall, is illustrated by the plants which grow in arid regions; here the outer wall is greatly thickened and strongly cutinised, and the plants are thus enabled to reduce their cuticular transpiration enormously. This type of epidermis prevails, for example, among the plants inhabiting the deserts of Asia and Africa, in the Australian Grass-trees (*Xanthorrhoea*), PROTEACEAE and EPACRIDACEAE, and in many Arctic xerophytes. Even in a humid climate edaphic conditions may render the restriction of transpiration imperative. This circumstance explains the occurrence of a thick-walled epidermis in many epiphytes of the tropical rain-forest, and in halophytes. Many alpine plants similarly rely upon a heavily cutinised epidermis to counteract the tendency towards excessive transpiration induced by low atmospheric pressure and intense insolation.

The outer epidermal walls often display varying degrees of thickening and cutinisation in different organs of the same plant. It is, of course, quite evident that certain parts of a plant may stand in less urgent need of protection against loss of water than others. Floral organs, for example, are comparatively short-lived structures, and are, moreover, usually put forth at the time when there is least risk of desiccation. Hence they do not require to be protected against desiccation to the same extent as foliage-leaves which remain active throughout the vegetative season, and which are therefore more exposed to changes of weather. The epidermis is accordingly almost always thin-walled in the case of petals, stamens, stigmatic surfaces, and the like. Again, there is nothing surprising in the fact that the upper epidermis of a leaf is often thicker-walled than the lower. In *Daphne chrysantha* the outer wall is $4 \cdot 2\mu$ in thickness in the case of the lower epidermis of the leaf, while it is more than twice as thick ($8 \cdot 6\mu$) in the case of the upper epidermis. In *Vinca minor* the corresponding measurements are $3 \cdot 1\mu$ and $5 \cdot 4\mu$. Of the two leaf-surfaces it is obviously the upper that receives more light and heat, and that hence requires more effective protection against evaporation.

Experiments have repeatedly been performed for the purpose of testing the extent to which the epidermis as a whole, and its outer wall in particular, are capable of checking transpiration.[48] But the majority of the objects that have been employed for this purpose (such as apples, plums or pieces of *Cactus*-stem) are ill fitted to afford any exact information as to the restrictive effect exercised by the epidermal layers of foliage-leaves and other vegetative organs. A transpiration experiment of this type carried out by the author may be reproduced at length. Two similar pieces were cut from leaves of *Aloe vulgaris* and coated all over with tallow, except for an area of 10 sq. cm. in each case. The epidermis was then carefully removed, with the help of a scalpel, from the uncoated area of one of the pieces. A couple of apples were treated in precisely the same way. The loss of water by transpiration (estimated by weighing at stated intervals) worked out, in grams per sq. cm., as follows:

|  | *Leaf of Aloe vulgaris.* | | *Apple.* | |
| --- | --- | --- | --- | --- |
|  | Epidermis intact. | Epidermis removed. | Epidermis intact. | Epidermis removed. |
| After 3 hours, | ·022 | ·524 | ·015 | ·385 |
| After 24 hours, | ·160 | 2·502 | ·120 | 1·802 |

In *Aloe*, therefore, the loss by evaporation during the first three hours subsequent to the removal of the protective epidermis was 23·5 times as great as that suffered by the intact leaf; after twenty-four hours it was still 15·6 times as great. In the case of the apple the corresponding ratios were very similar, namely, 25·6 : 1 and 15 : 1. Evidently the epidermal layer of *Aloe vulgaris* and that of the particular sort of apple selected happen to possess a similar restrictive power.

In experiments like that which has just been described the controlling action of the thickened and cutinised outer walls cannot be fully effective, owing to the presence of more or less numerous stomata. As a matter of fact, the stomata probably close very soon under the conditions of the experiment; nevertheless entirely satisfactory measurements of cuticular transpiration can only be obtained by employing epidermal layers which are completely devoid of stomata and thereby eliminating the effects of stomatic transpiration altogether. There is yet another source of error which must be taken into account. It was known to Nägeli that potatoes and apples suffer a greater loss by evaporation, under otherwise similar conditions, if they are first killed by freezing, than if they are observed in the living state. The difference is probably due to a restrictive influence exerted upon evaporation by the plasmatic membranes of the living cells, an effect which, of course, disappears when the protoplasts are killed. This factor must also be eliminated before the restrictive effect of the cell-wall can be accurately determined.

In view of these facts the author has carried out the following experiment with leaves of *Ficus elastica*, *Hedera Helix* and *Aesculus Hippocastanum*. In each case one of a pair of leaves (or leaflets) was killed with chloroform vapour, while its fellow was used in the living condition. The lower surface—to which the stomata are restricted in all three species—was then coated with a layer of cacao-wax (a mixture of one part by weight of beeswax with three parts of cacao-butter, recommended for this purpose by Stahl), so that evaporation was thenceforward confined to the upper or astomatic surface; the whole of the petiole, including the cut end, was similarly waxed in every case. The experimental leaves were then allowed to transpire in a shaded portion of the laboratory, in an atmosphere with a relative humidity of 68-75 per cent., at temperatures ranging from 19° C. to 23° C. The loss of water due to evaporation was estimated by weighing the leaves daily at the same hour. For the sake of comparison a similar record was kept of the evaporation which took place under similar conditions from the freely exposed surface of water contained in a shallow vessel. After three days the average loss by evaporation in grams per sq. dcm. per twenty-four hours, was found to be as follows ·

|  | LIVING LEAF. | DEAD LEAF. |
|---|---|---|
| *Ficus elastica*, - - - - - - - - | ·032 | ·056 |
| *Hedera Helix*, - - - - - - - - | ·031 | ·044 |
| *Aesculus Hippocastanum*, - - - - - - | ·126 | ·156 |
| Water-surface, - - - - - - - - | — | 6·922 |

All the leaves were still quite fresh at this stage, those which had been killed being discoloured, as at the beginning of the experiment.

The above experiment shows in the first place that the smallness of the evaporation from the leaves, as compared with a freely exposed water-surface, is very largely attributable to the influence of the cuticle and cutinised layers, and not to that of the living protoplasm, although water vapour does escape somewhat more rapidly from dead leaves than it does from live ones. In the case of *Ficus elastica*, for example, the quantity of water retained mainly through the action of the outer epidermal wall amounts to 6·922 *minus* ·056, or 6·866 grams, and is thus 286 times as great as the small additional quantity (·024 grams) which is prevented from escaping by the plasmatic membranes of the living leaf. In every instance the amount of water lost by the dead leaves is only a small fraction of the quantity that evaporates from an equal area of the water-surface, namely $\frac{1}{44}$th in the case of *Aesculus*, $\frac{1}{123}$rd in that of *Ficus* and $\frac{1}{157}$th in that of *Hedera*. These figures afford a most striking proof of the high efficiency of the epidermis as an instrument for restricting evaporation. From observations extending over a month Unger deduced a much smaller value [49] for the ratio between evaporation from a freely exposed water-surface and the foliar transpiration of *Digitalis purpurea*; but the discrepancy is explained by the fact that Unger's method does not eliminate stomatic transpiration.

The effectiveness of the control exercised over transpiration by the epidermis is often enhanced by the addition of a coating of **wax**.[50] The characteristic "bloom" on a grape or a plum, and the glaucous appearance of many leaves [and stems] are caused by the presence of such waxy coatings. There are three principal types of waxy covering. In the great majority of cases the wax is deposited as a layer of minute, closely-crowded **granules**, with an average diameter of ·001 mm.; this condition is exemplified by the leaves and stems of many GRAMINEAE, LILIACEAE and IRIDACEAE. Less frequently the covering takes the form of a layer of **rods**, which stand up vertically above the cuticle;

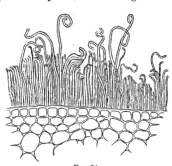

FIG. 24.

Coating of wax-rods on a node of *Saccharum officinarum*. ×142. After De Bary (from De Bary, *Comp. Anat.*).

these rods are often much taller than the epidermal cells, and may be hooked or rolled up at their distal ends. This type of coating is found in a number of GRAMINEAE and SCITAMINEAE (e.g. *Musa*, *Strelitzia*, *Canna*). Unusually tall rods—reaching a length of ·1 to ·15 mm.—occur on the nodes of the Sugar-cane (*Saccharum officinarum*, Fig. 24). A **wax-crust or -stratum**, finally, such as covers the leaves of *Thuja* and *Sempervivum*, consists of a brittle, translucent amorphous glaze, which is generally intersected by numerous cracks and flaws. As a rule the depth of such a crust scarcely exceeds $1\mu$; in special cases, however, it may be much greater. Exceptionally thick crusts occur, for instance, on the leaves and stems of *Panicum turgidum* (·03 mm.) and on the young leaves of *Copernicia cerifera* (·015-·019 mm.). In the Wax Palms (spp. of *Ceroxylon* and *Klopstockia*) the waxy incrustation may even reach a thickness of 5 mm. As regards the mode of formation of waxy coverings, De Bary has shown that the wax does not, as might be supposed, arise by chemical modification of the cuticle or of other layers of the outer wall, but that it is on the contrary a product of secretion.

F. Haberlandt was the first to investigate the effect of waxy coverings upon transpiration experimentally; he examined the leaves of the Swedish Turnip, which have a well-developed granular covering.

It was found that leaves from which the wax had been carefully wiped off transpired 4·03 grams of water per sq. dcm. in a day; control specimens with their waxy covering intact lost 3·6 grams under the same conditions. A second experiment gave as corresponding values 4·63 grams and 3·03 grams. The leaves thus lost on an average 30 per cent. more water when the wax was removed. Tschirch has obtained very similar results with the leaves of *Eucalyptus globulus*. These experimental data accord with the fact that waxy coverings are especially prevalent among xerophytes, such as the Mediterranean species of *Dianthus* and *Euphorbia*, the Australian *Acacias* and MYRTACEAE and numerous steppe- and desert-plants (e.g. *Capparis spinosa*, various CRUCIFERAE and RUBIACEAE, etc.) Tittmann has shown that certain spp. of *Sedum* and *Echeveria* produce less wax if transpiration is artificially diminished by growing the plants in a damp atmosphere. Waxy coverings may influence the transpiratory activity of the plant in a different way by preventing its surface from being wetted by drops of rain or dew; the capillary occlusion of stomata is often avoided in this way. In certain cases (*Euphorbia Tirucalli*, spp. of *Strelitzia*) each stoma is surrounded by a ring of wax, which thus forms an external air-chamber (cf. Chap. IX.).

According to the observations of Kerner and Delpino, a waxy covering or bloom sometimes (*Salix daphnoides* and *S. pruinosa*, *Fritillaria imperialis*, various UMBELLIFERAE, etc.) serves to create a slippery surface, and thus to prevent the access of ants to floral nectaries. The author has himself watched the behaviour of small ants which he transferred from the extrafloral nectaries of *Vicia sepium* to the bloom-covered scape of *Hyacinthus sylvestris*; on the treacherous waxed surface the insects moved very slowly, and only by dint of much exertion and at the cost of many slips, whereas a zone, several centimetres in width, from which the wax had been removed, was traversed with the greatest ease. Waxy coverings may thus secondarily acquire an ecological significance in connection with the relations between plants and animals; this statement in all probability applies more especially to stems in which the wax is confined to—or at any rate is more strongly developed on—certain comparatively circumscribed areas (*e.g.* the nodes in the Sugar-cane and in certain Bamboos).

Since wax, in the form of granules or rods, is easily rubbed off, it is an advantage if the plant is able to regenerate a waxy coating that has been removed. As a matter of fact, a considerable number of plants do possess this capacity; in the case of *Rubus biflorus* and *Macleya cordata* the act of regeneration may even—according to Tittmann—be repeated several times in succession.

Attention must next be directed to the mechanical significance of the thickened condition of the outer epidermal wall.[51]   It is obviously essential that a dermal tissue should be endowed with a certain degree of mechanical strength.   As a matter of fact, cutinised epidermal walls can actually withstand very considerable tensile stresses, the ultimate resistance of the average epidermis ranging, according to Damm, from 5 to 10 kg. per sq. mm.   Thus isolated strips of the epidermal layer of *Aristolochia Sipho* break under a load amounting on the average to 10·1 kg. per sq. mm.   In three other species examined by Damm, namely *Acacia dealbata, Aucuba japonica* and *Ilex aquifolium,* the breaking strengths are 9·2 kg., 7·7 kg. and 5·3 kg. respectively; in the case of ordinary cellulose walls the corresponding value, according to Schwendener, never exceeds 1 kg.   It would evidently be a mistake to suppose that the thickness and the cutinised condition of the outer epidermal wall are valueless except as means of restricting transpiration; for both these features not infrequently serve in the first instance to secure the requisite degree of mechanical strength.   Thus, in the case of many leathery-leaved tropical plants the epidermal wall is much thicker than is necessary for the adequate restriction of transpiration in a humid climate; here the thickness of the wall must be regarded as a mechanical adaptation, which helps to preserve the leaf from being injured by the violent rain that falls daily during the wet season.

The rigidity of the thickened epidermal wall entails a further important advantage; it minimises the distortion and compression of the underlying tissues that are liable to result from the contraction consequent upon any loss of water, and at the same time enables the collapsed tissues to return to their former dimensions and relative positions with ease and precision when water is reabsorbed.   In this respect the thickened outer walls of the epidermal cells may collectively be said to form a rigid case in which the less resistant tissues are safely stowed away.

The rigidity of the outer wall is often increased by a network of internally projecting flanges; each flange represents a prolongation of the cutinised layers, which extends more or less deeply into the substance of a radial wall (Fig. 23 A).   Every mesh of the network of flanges is thus coextensive with the lateral outline of a single epidermal cell.   In transverse section each flange presents a wedge-shaped, or more rarely a lanceolate, outline.

The firm texture of the outer epidermal wall is important in yet another way, because it affords more or less effective protection against the attacks of animals.   This interpretation applies more particularly to those cases in which the hardness of the thickened

walls is increased by impregnation with *silica* or *lime*.   Measurements
of the relative hardness of **silicified and calcified epidermal walls** have
been made by Emma Ott on the lines of the scratching method
customary in mineralogy; it appears that the stems of various species
of *Equisetum* (e.g. *E. ramosum, E. litorale, E. silvaticum, E. pratense,* etc.)
are capable of scratching calcite, which means that the silicified
epidermis has a relative hardness of 3 on the Mohs scale.   The more
highly silicified epidermal walls of *E. hiemale* and *E. Telmateia* scratches
fluorite (relative hardness = 4), while the spermoderm of *Coix Lacryma*
actually scratches opal (relative hardness = 7).   Calcified walls may
attain to a relative hardness of 3; the seed-coat of *Celtis*, for example,
scratches calcite readily.   It is not surprising, therefore, to find that
Grasses, Sedges and Horsetails, which all have highly silicified
epidermal walls, are to a great extent immune from the assaults of
snails.[52]   Stahl has remarked that snails will readily devour grass
which has been cultivated in the absence of silica, whereas they
invariably refuse normal silicified blades, or at most eat them with
obvious reluctance.   The inclusion of numerous small crystals of
*calcium oxalate* in the outer epidermal wall, a condition exemplified, for
instance, by *Welwitschia mirabilis*, by various species of *Mesembryanthemum* and by many NYCTAGINEAE, probably subserves a similar
ecological purpose.[53]

Certain peculiarities of the outer surface of epidermal layers
may be mentioned at this stage.   Many plants which grow in sunny
situations have leaves with smooth and shining upper surfaces.   Such
**polished or varnished leaves** are especially common in the tropics, where
the reflection of light by foliage constitutes a ubiquitous and characteristic physiognomical feature of the vegetation.[54]   It is probably
permissible to regard this reflective power of the cuticle as a protection
against excessive insolation, since it undoubtedly has the effect of
preventing a certain proportion of the incident light from penetrating
into the leaf.   An epidermis which is very smooth is often also easily
wetted; rain falling on such an epidermis runs off quickly, or else
spreads over the leaf as a thin film, which soon evaporates.   Obviously
this method of removing water from the leaf-surface is unsuitable
where the upper epidermis bears stomata, since these structures would
be constantly exposed to the danger of capillary occlusion.   The
stomatic epidermis is therefore usually provided with a waxy covering,
from which raindrops roll off without wetting the surface at all.

The **velvety leaf-surfaces** of many tropical plants (*Cyanophyllum
magnificum,* spp. of *Begonia,* various MARANTACEAE and ORCHIDACEAE,
etc.) are also very easily wetted, as Stahl has shown;[55] here the outer
wall of each epidermal cell assumes the form of a prominent papilla,

and as a consequence every drop of water that falls upon the velvety surface spreads out at once, owing to the action of capillarity, and is thus caused to evaporate rapidly. Possibly the network of delicate external ridges which is so frequently developed on epidermal surfaces —usually owing to corrugation of the cuticle—produces a similar effect.

This account of the structure and properties of the outer epidermal wall may conclude with a reference to the changes which the epidermal surface may undergo with age, partly owing to the natural weathering of the surface, and partly as a result of the growth in thickness of the organ to which the epidermis belongs. In this connection it will only be necessary to consider the so-called **persistent epidermal layers**, which act as the sole dermal covering for a number of years. Such a persistent epidermis is by no means rare; it occurs, for instance, in a number of LAURACEAE (*Cinnamomum officinarum, Laurus nobilis*), ROSACEAE (*Rosa alpina, R. canina, R. multiflora, Kerria japonica*), LEGU-MINOSAE (spp. of *Acacia, Sophora japonica*), AQUI-FOLIACEAE (spp. of *Ilex*), ACERACEAE (*Acer striatum, A. palmatum, A. Negundo*), CORNACEAE (*Cornus alternifolia, Aucuba japonica*), OLEACEAE (spp. of *Jasminum*), etc. There is nothing surprising in the fact that the outer wall attains a relatively enormous thickness in epidermal layers of this type. But it is interesting to note that this wall

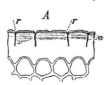

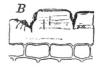

FIG. 25.

*A.* Epidermis of a yearling twig of *Acer striatum*; *r, r,* cracks in the cuticle and cutinised layers. × 370. *B.* Epidermis of a six-year-old branch of the same tree.

undergoes a process of continual regeneration; its outermost layers peel off or crumble away, while new layers are deposited on its inner side. *Acer striatum* affords a very instructive illustration of the crumbling process. On yearling branches of this tree the epidermis is quite smooth (Fig. 25 A). The cutinised layers, which are thick and impregnated with wax, are sharply delimited from the thin sinuous cellulose layer. The cuticle is very delicate, and is covered with a thin crust of wax. Late in the autumn of the first year numerous radial fissures develop in the outer wall; these appear over the radial walls of the epidermal cells, and extend inwards about half-way through the cutinised layers. This cracking is the first visible consequence of the passive tangential extension to which the outer epidermal wall is subjected, owing to the growth in thickness of the twig. The epidermis of a six-year-old branch (Fig. 25 B) shows a whole series of characteristic changes. The thickness of the outer wall—or rather that of the cutinised layers—has increased considerably. The cellulose layers are no longer sinuous, but have become flattened out. This fact, as well

as the greatly increased breadth of the cells, shows that the epidermis
has undergone considerable tangential extension.   The narrow radial
fissures have widened out into large gaps, while tangential cracks have
also made their appearance; in this way a more or less extensive
weathering and crumbling of the outer cutinised layers has gradually
taken place.[56]

### 3. *The radial and inner epidermal walls.*

The various properties and functions of the outer epidermal wall
having been discussed in detail, attention may now be directed to the
other walls.   The **radial or lateral walls**, like the outer wall, exhibit
adaptive features which are correlated with more than one physiological
function.   According to Westermaier, not the least important among
the functions of the epidermis is that of water-storage.   The detailed
consideration of this matter must be postponed until the contents of
epidermal cells come to be discussed.   For the present it must suffice
to note that the lateral walls must be thin, so as to allow water to pass
with ease from cell to cell, and also in order that the cells may be able
to contract and expand freely in the radial direction.   As a matter of
fact, it is the rule for epidermal cells to have very thin radial walls;
if the latter are thickened for mechanical reasons, then they are usually
provided with numerous circular or oval pits, which facilitate interchange of materials, and especially of water, between adjacent cells.
Burns has even found typical bordered pits—which otherwise are
confined to the walls of water-conducting elements—in the lateral
epidermal walls of certain species of *Stylidium* (especially *S. strepto-
carpum*).[57]

It is, of course, necessary, for mechanical reasons, that every
epidermal cell should be firmly attached to its neighbours; for the
epidermis is constantly liable to be subjected to a variety of tensile
stresses.   In the case of the stem these stresses arise mainly as a
result of transverse and longitudinal tissue-tensions, while the bending
action of the wind produces tangential stresses in the foliar epidermis.
The firm attachment of the epidermal cells to one another is effected by
the lateral walls, which are accordingly often thrown into folds in order
that the surface of contact between adjacent elements may be increased
(Fig. 26 A, B).   The individual cells thus acquire a lobed or toothed
outline, and their adjacent faces become interlocked in a manner
recalling the sutures of the Vertebrate skull.   In the case of elongated
organs, such as a stem or the leaf of a Grass, where the epidermal cells
are likewise elongated, it is only the longitudinal radial walls that are
conspicuously folded.   The GRAMINEAE in particular—which in every
respect exhibit a high degree of anatomical specialisation—exemplify

this lateral interlocking of the epidermal cells in a very characteristic manner (Fig. 26 B).    In most Dicotyledonous leaves the folding of the lateral walls is restricted to the lower epidermis, which usually possesses a relatively thin outer wall.    Where the waviness affects the whole extent of the lateral wall, it adds to the rigidity of the entire cell, and so helps to prevent collapse when water is withdrawn.    In delicate perianth-leaves the folds are often supplemented by radial flanges for

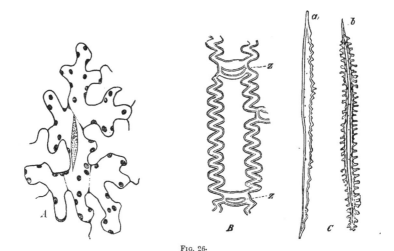

FIG. 26.

*A, B.* Epidermal cells with folded lateral walls. *A.* From the abaxial foliar epidermis of *Impatiens parviflora.* ×420. *B.* From the abaxial foliar epidermis of *Festuca ovina*; *z*, a semi-lunar dwarf-cell. ×400. *C.* Subepidermal mechanical cells from a pale of *Oryza sativa*, with tooth-shaped protuberances (after Von Höhnel); *a*, cell with one, *b*, cell with two sets of teeth.

similar reasons ; these flanges may even become hollowed out and thus transformed into tubular buttresses, which support the tangential walls and thus increase the rigidity of the whole epidermis.

The **inner epidermal walls** are responsible for the firm attachment of the epidermis to the underlying tissues, a duty which they are generally able to perform without the aid of any special contrivances. The average epidermis, indeed, adheres so firmly to the mesophyll that one cannot easily strip it off without tearing away one or more subepidermal layers as well.    A unique and exceedingly effective mode of attachment of the epidermis to the hypodermal layer is described by Von Höhnel [58] as occurring in the pales of certain Grasses (*Oryza sativa, Sorghum vulgare, Setaria germanica, Triticum Spelta, Hordeum vulgare*) ; the fibrous cells situated beneath the outer epidermis of the pale present a serrated appearance owing to the presence of numerous lateral protuberances of their walls, which fit into corresponding

H

depressions in the inner epidermal walls (Fig. 26 c).　In this way the cells of the epidermis and the hypodermal fibres are firmly dove-tailed together.

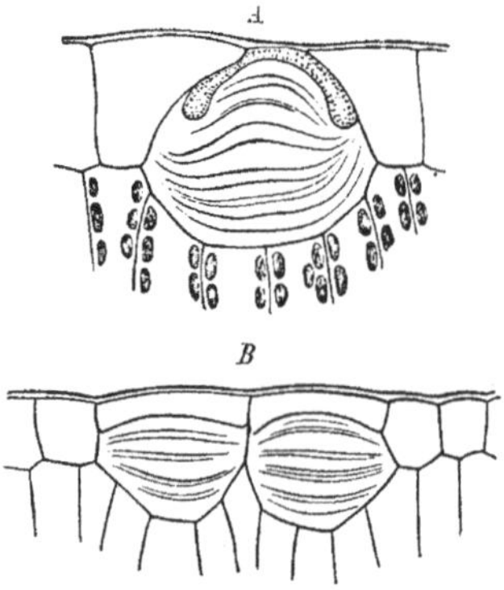

Fig. 27.

Epidermal cells with thick mucilaginous inner walls　*A*. From the adaxial foliar epidermis of *Theobroma Cacao*:　*B*. From the adaxial foliar epidermis of *Pterocarpus saxatilis*.

Like the lateral walls, the inner walls of epidermal cells are generally thinner and softer than the outer walls, so that diosmotic interchange with the underlying tissues can proceed unhindered.　In a number of plants (*Betula alba, Quercus pedunculata, Erica carnea,* spp. of *Salix, Daphne, Prunus* and *Cytisus,* certain SAPINDACEAE and DIOSMEAE, etc.) certain cells of the foliar epidermis are provided, according to Radlkofer,[59] with greatly thickened, **mucilaginous inner walls**, a condition which probably represents a water storing device (Fig. 27); in these cases the mucilaginous membrane undertakes a function which in the typical epidermis is assigned to the cell-sap.

### 4. *Cell-contents.*

Every typical epidermal cell is provided with a living protoplast in the shape of a thin peripheral layer of cytoplasm, containing a nucleus, and enclosing a sap-cavity which is filled with transparent and usually colourless cell-sap.　Very frequently leucoplasts (especially in Monocotyledons) or chloroplasts are also present.　Chloroplasts are, however, not conspicuously developed in the case of typical brightly illuminated epidermal layers.[60]　Nevertheless, the comparatively small chlorophyll-content of the epidermis must not be regarded as a consequence of the destructive action of intense light upon chloroplasts; the presence of a well-developed chlorophyll-apparatus in the guard-cells of stomata proves beyond doubt that the sharp contrast between colourless epidermis and green palisade-tissue cannot be altogether attributed to differences in the intensity of illumination.　The feeble development or complete absence of chloroplasts should rather be looked upon as the result of a division of labour, whereby the epidermis is entrusted with the task of storing water, and in return is relieved of photosynthetic duties.

Westermaier has, in fact, shown that the epidermal layers of vegetative organs, and especially those of foliage-leaves, play the part of superficial **water-jackets**, which protect the underlying tissues against damage through loss of water.[61]　A whole series of anatomical and physiological peculiarities of epidermal cells are connected with this partial function of water-storage.   In the first place, every epidermis possesses the power of alternately accumulating and delivering up water.   When an organ is transpiring too actively, the epidermis parts with a large portion of its store of water to the green photosynthetic cells, which exert a stronger "osmotic suction."   Hence when an organ begins to wilt, the first signs of collapse appear in the epidermis; the thin lateral walls of the epidermal cells are thrown into folds, and the cells gradually contract in the radial direction, as water is withdrawn from them.   When the cells are once more supplied with abundance of water, their lateral walls straighten themselves again.   That this "bellows-action" of the epidermal cells, which can be repeated whenever required, constitutes a very characteristic property of the water-storing epidermis, may be inferred, as has already been hinted, from the thin and flexible condition of the radial walls.   The following circumstance, upon which Westermaier quite properly lays stress, points to the same conclusion.   Photosynthetic organs are often provided with radial buttresses, consisting of specially thick cell-walls (*Kingia australis*), or, more usually, of entire mechanical cells (*Olea europea*), which serve to prevent the green tissues from collapsing when water is withdrawn from them; wherever these buttresses are present they never extend outwards beyond the inner epidermal walls.   Evidently it is not intended that the partial collapse of the epidermal cells themselves should be prevented.

As it is improbable that the same amount of water will be simultaneously withdrawn from the epidermis in every part of an organ, it would seem to be desirable that water should be able to circulate freely through the epidermis, so that it can readily flow towards the points at which it is most in demand at any given moment.   The requisite continuity of the epidermal water-storing system is in general sufficiently ensured by the circumstance that the radial walls are thin, or, if thickened, at any rate abundantly provided with pits.   There are in addition a number of special features, designed to produce the same result, which occur especially in the cells overlying subepidermal fibrous strands.   In the CYPERACEAE, for example, the epidermal cells which are situated above fibrous strands, in stem and leaf, are often very much flattened.   The inner walls of these cells are provided with peculiar, strongly silicified conical processes, which serve, according to Westermaier, to arrest the downward movement of the outer wall, and thus to prevent the interruption of the epidermal water-circulation which would result from the total collapse of these cells.

The water-storing capacity of the epidermis, and consequently its importance as a water-tissue, varies according to the height of its component cells; exceptionally tall epidermal cells occur in certain COMMELYNACEAE, ORCHIDACEAE, BEGONIACEAE, etc.    In xerophytes, and especially among desert-plants, the volume, and hence the water-storing capacity, of the epidermal cells is often increased by a papillose protrusion of their outer walls.    Occasionally individual cells become enormously distended; the familiar Ice-Plant (*Mesembryanthemum crystallinum*), the stems and leaves of which appear as if studded with beads of ice, owes its peculiar appearance to the presence of large numbers of such epidermal water-vesicles (Fig. 28).    According to Volkens similar structures occur in quite a number of the plants of

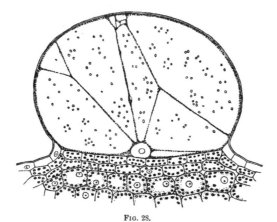

FIG. 28.

Enlarged water-storing epidermal cell (water-vesicle) of *Mesembryanthemum crystallinum*.   (L. S.)

the Egypto-Arabian desert-region (e.g. *Reseda pruinosa, R. arabica, Aizoon canariense*, etc).[62]    The water-storing capacity of the ordinary tabular epidermal cells is of course small, as compared with that of specialised structures like the above-mentioned water-vesicles; it is, nevertheless, by no means a negligible factor in the normal water economy of the plant.    In order to obtain at least an approximate quantitative idea of the water-storing capacity of an ordinary epidermis, the author has attempted to estimate what fraction of the daily foliar transpiration can be made good at the expense of the two epidermal layers of the leaf, in the case of certain woody plants; the calculation is based on the supposition that the epidermal cells contract to the extent of one half of their volume owing to abstraction of water by the photosynthetic tissue.    (Only the bare figures are stated, the details of the calculation being set forth in a note.)[63]    It appears that the foliar epidermis of *Aesculus Hippocastanum* would be able to cope with the average transpiration of the mesophyll for a period of two hours;

in the case of *Corylus Avellana* the corresponding period would be
barely three-quarters of an hour, and in that of *Pyrus communis* only
forty minutes. It follows from these measurements that the ordinary
epidermis would be hopelessly ineffective as a water-reservoir, if
stomatic transpiration were always going on. Under normal conditions, however, the epidermis is never called upon, in the case of
leaves that can close their stomata, to make good the large loss of
water which the photosynthetic tissues suffer on account of stomatic
transpiration; for Von Mohl, Leitgeb and others have shown that, as
soon as a leaf begins to wilt, the guard-cells at once begin to close
the stomatic orifices and thus to put a stop to stomatic transpiration.
Cuticular transpiration on the other hand is usually so small in
amount, even in bright sunshine, that the resulting loss of water can
be borne by the epidermis alone for a very considerable length of time.
Slightly withered leaves of Horse-Chestnut, Hazel and Pear lose so
little water, at any rate during the first twenty-four hours, that the
half-collapsed epidermal cells can make good the loss for a period
amounting in the three instances to $16\frac{1}{2}$, 6 (nearly) and 7 hours
respectively. During these periods, therefore, the photosynthetic
tissue need not suffer any actual loss of water, always provided that
the cell-sap does not become so concentrated in the transpiring epidermal
cells as to abstract water from the adjacent elements of the mesophyll.
There are, therefore, good reasons for regarding even the ordinary flat
epidermal cells as reservoirs of water which are able to cope effectively
with purely cuticular transpiration.

The epidermal cells of vegetative organs occasionally contain **pigments** dissolved in the cell-sap; *anthocyanin* (*erythrophyll*), which
varies in hue from bright red to bluish-violet, is perhaps the most
widely distributed of these colouring matters. Where this coloration
has any physiological significance at all, it is probably always connected,
in some way, with the relations of the plant to light.[64] In many cases
the coloured epidermis appears to act as a screen which protects the
underlying tissues against intense illumination and its secondary consequences, such as rapid destruction of chlorophyll (according to
Wiesner) or excessive respiration (according to Pringsheim). This
view is supported by a variety of circumstances. Thus Von Mohl long
ago drew attention to the frequent red coloration of young shoots and
seedlings, the nascent chlorophyll of which is particularly liable to
destruction by light. A great many evergreen leaves acquire a reddish
colour in winter time owing to the formation of anthocyanin; in this
case the chloroplasts require special protection during winter against
the injurious action of light, because no appreciable regeneration of
chlorophyll takes place at the low temperatures which prevail at that

season.  A different explanation must of course be sought for the fact
that the lower sides of the leaves are coloured red or violet in so many
shade-plants.  Here Kerner suggests that radiations which would
otherwise be reflected from the lower surfaces of the leaves are
absorbed by the coloured cell-sap and transformed into heat, which can
then be utilised by the plant for various purposes.

Kny was the first to investigate this subject experimentally.  He
found that an alcoholic chlorophyll-solution became discoloured more
slowly when shaded by a double bell-jar containing a solution of anthocyanin, than it did when the screen consisted of a colourless decoction
of sugar-beet.  Kny was also able to establish the fact that insolation
causes an unequal rise of temperature in equal quantities of red and
green leaves of the same species, the heating effect being almost
always greater in the case of the red leaves; the largest difference
observed amounted to 4° C.  Kny's observations were subsequently
confirmed and extended by Stahl with the aid of thermo-electric
methods.  The source of heat in Stahl's experiments was an ordinary
bat's-wing gas flame placed at a distance of 30 cm. from the leaf and
parallel to its surface.  Under these conditions red leaves of *Sarcanthus
rostratus* became 1·5-1·82° C. warmer than green leaves of the same
plant; in the case of *Sempervivum tectorum* the corresponding difference
was 1·67° C., in that of *Begonia heracleifolia, var. nigricans* 1·35° C., in
that of *Pelargonium peltatum* ·22° C.  A method based upon the
melting of a coating of cacao-butter gave concordant results.  In
Stahl's opinion the increased heat-absorption which results from the
red coloration entails a two-fold advantage.  In the first place, it
leads to an acceleration of metabolism in general and of translocation in
particular, and, secondly, it causes an increase in the rate of transpiration.  With reference to the latter point, Stahl draws attention to the
fact that leaves which are ornamented with red spots or blotches, or
which possess a uniformly red under-surface are particularly prevalent
among plants that grow in very moist surroundings, and especially in
tropical rain-forests.  In view of the fact, however, that the plants in
question for the most part grow in the deepest shade, it seems very
doubtful whether, in such cases, the presence of anthocyanin actually
produces heating effects sufficient to cause any appreciable increase of
transpiratory activity.  In order to settle this question it will be
necessary to carry out transpiration experiments upon suitable plants
in their natural habitats.  A further criticism applies to the experiments of Kny and Stahl; in many of the leaves examined by these
observers, the red cell-sap is not confined to the epidermal cells, but
occurs also in the photosynthetic tissues.  It must in short be
admitted that, in spite of numerous interesting detailed observations,

the general physiological and ecological significance of the presence of anthocyanin in vegetative organs is still very obscure.

Very often considerable quantities of *tannin*[65] are present in epidermal cells, especially in the case of leaves which persist through the winter.   Warming inclines to the opinion that compounds of this nature serve to diminish the risk of desiccation, a danger by which Arctic and alpine plants in particular are often threatened, especially in the absence of snow.   Stahl, on the other hand, regards tannin in the first instance as a means of defence against the attacks of snails · it might also quite conceivably assist in preventing parasitic Fungi from gaining access to the epidermal cells.

### 5. *Subsidiary functions of the epidermis.*

On account of its superficial position, the epidermis is exceptionally well fitted to undertake a variety of subsidiary functions ; this layer accordingly controls, or at any rate plays a prominent part in connection with, many ecological relations which have nothing to do with the protection of the plant.   Secondary relations of this kind may modify the character of epidermal cells to such an extent that they lose all or most of the features which are characteristic of protective dermal elements and become transformed into mechanical cells, photosynthetic elements, and so forth.   Any cells which have undergone such a complete change of function can no longer be regarded as epidermal elements in the anatomico-physiological sense, but must be assigned to other tissue-systems.   The present discussion is confined to those subsidiary relations of the epidermis which do not effectually mask its tegumentary character.

The **mechanical strength** of the thickened outer wall **of the epidermis** is sometimes greater than is required by this layer in its purely dermal capacity.   The most obvious illustration is provided by the epidermal cells which occupy the **margins of leaves** ; these are very often called upon to perform the specifically mechanical function of protecting the edge of the leaf against the shearing or tearing action of the wind, and are accordingly provided with much thicker outer walls than the rest of the epidermal elements.   Quite a different mechanical significance attaches, according to Raciborski, to the epidermal cells which form the **sutures** between adjacent perianth-leaves in certain flower-buds.[66]   There are two ways in which the union between neighbouring sepals or petals may be effected in such cases.   The epidermal cells may grow out bodily, along the suture, to form a series of interlocking teeth : such " cellular " sutures are exemplified by the calyces of ONAGRACEAE and by the valvate corollas of RUBIACEAE, ASCLEPIADACEAE, CAMPANULACEAE, LORANTHACEAE, UMBELLIFERAE, etc. (Fig 29 A).   In other instances the interlocking structures are ridge- or peg-like outgrowths of the outer epidermal walls or of the cuticle; this "cuticular" type of suture is well illustrated by *Hedera Helix* (Fig. 29 B). In the

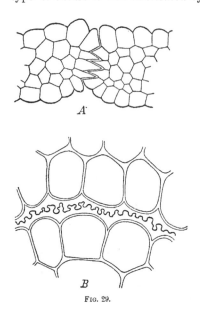

*A*

*B*

FIG. 29.

*A*. A cellular suture from the calyx of *Boisduvallia concinna* (after Raciborski). *B*. A cuticular suture from the perianth of *Hedera Helix*.

RHIZOPHORACEAE, and in certain other plants, both modes of union occur side by side. As a rule, perigonial sutures ultimately give way, owing to the pressure exerted by the growing internal organs. Occasionally, however, the lateral union is so strong that the periauth is either thrown off as a whole (*e.g.* the corolla of *Vitis*) or else remains closed altogether.

Very frequently the outer wall of each epidermal cell protrudes in a slightly papillose manner ; in many shade-plants this tendency is accentuated, so that large conical **epidermal papillae** eventuate. The papillose condition is correlated with certain **optical functions** which devolve upon the epidermis in these cases. The author [67] long ago suggested that such epidermal papillae might act as **condensing lenses,** concentrating the available light upon the chloroplasts which line the lateral walls of the palisade-cells—provided the latter were suitably arranged—and thereby producing an increase of photosynthetic activity ; at the present time he inclines to the opinion that this " lens-action " of epidermal papillae is primarily connected with the perception of photic stimuli (cf. Chap. XII.). Stahl [68] believes that the conical epidermal papillae of velvety leaves, besides facilitating the removal of water from the leaf-surface, also " enable the leaf to absorb very oblique rays of light, which would be entirely lost if the outer walls of the epidermal cells were of the ordinary flat type." The accompanying diagram (Fig. 30) illustrates the manner in which such epidermal papillae act as **light-traps.** The oblique ray *ab*, which would for the most part be reflected from a flat leaf-surface, here meets the sloping wall of a papilla at right angles, and thus penetrates without appreciable deflection through the adjoining epidermal cell into the underlying photosynthetic tissue. The ray *a'b'* suffers total internal reflection at *c,* and thereafter passes vertically downwards into the leaf. The ray $a''b''$, finally, which impinges vertically upon the leaf-surface is indeed very largely reflected at $b''$; the reflected portion is, however, not lost, but is merely diverted into a neighbouring cell. A leaf with a papillate epidermis can thus absorb a greater proportion of the available light than one in which the outer epidermal wall is flat, with the result that both photosynthesis and transpiration are accelerated. According to Stahl, this con-clusion is borne out, so far as transpiration is concerned, by the fact that velvety-leaved plants are almost entirely con-fined to very humid tropical districts.

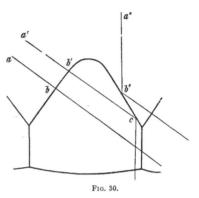

FIG. 30.

Diagram of a papillose epidermal cell, showing how the papilla acts as a light-trap (after Stahl). For explanation see text.

The pits, which sometimes occur in the outer wall of the epidermis, are probably con-cerned with certain subsidiary functions of this layer. Where the lateral walls of an epidermal cell are very sinuous, the outer wall is often furnished with **marginal pits**, which occupy the troughs of the undulations (*e.g.* GRAMINEAE, JUNCACEAE, CYPERACEAE and various Dicotyledons). Ambronn [69] con-siders that these marginal pits are, with few exceptions, function-less, and that they arise as a mechanical consequence of the corrugated structure of the lateral walls; he supposes that the corrugation results in the development of various tensile and com pressive stresses, which in their turn affect the growth in thickness of the outer walls in such a manner that pit-like structures eventuate. It seems more probable, however, that such marginal pits perform some definite function; in a subsequent chapter (Chap. XII.) it will be explained how these structures may be of use in connection with the perception of photic stimuli. Sometimes (e.g. *Coffea, Cocculus laurifolius, Cinnamomum aromaticum, Camellia japonica*) the whole of the outer wall is provided with scattered or even with numerous pits; here there is a strong presumption that the pits serve to facilitate some process of metabolic interchange between the plant and the surrounding medium. In the case of the. *lianes* of the genus *Salacia*, the nature of the process can be indicated more precisely. These plants emit un-usually large quantities of water, in the liquid form, from both sides of their leaves at night; since specialised hydathodes are absent, the water presumably exudes through the pits which are present in large numbers in the outer walls of the epidermal cells. Reference may lastly be made to the **tactile pits** of the outer epidermal wall which occur in the tendrils of CUCURBITACEAE, and in certain other organs that are sensitive to contact. The protoplasm which lines such pits is concerned with the perception of contact-stimuli; the detailed discussion of these structures must therefore be deferred until a later occasion (Chap. XII.).

### B. THE MULTIPLE EPIDERMIS.

Sometimes the primary dermal covering of a plant consists of more than a single layer of cells. This exceptional condition is in every case attributable to an increase in the demands that are made upon the protective capacity of the epidermis. The increased demand may

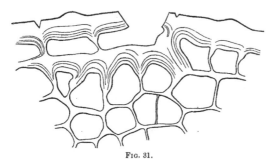

Fig. 31.

Development of cuticular epithelium in an old branch of *Viscum album*. (T. S.)

relate either to the control which the epidermis exercises over transpiration, or to the mechanical rôle of that layer, or, finally, to its water-storing function. These various possibilities may be illustrated by a few typical cases.

Damm [70] has shown that the epidermis of the stem of *Viscum album* and other VISCOIDEAE not only persists for a number of years, but is gradually reinforced by cutinised layers, which develop in the outer walls of the underlying cortical cells. To begin with, the guard-cells of the stomata and certain of the ordinary epidermal elements lose the power of adding further cutinised layers to their outer walls. As the neighbouring cells continue to thicken their walls, those which cease to do so soon sink below the general level of the epidermis; at the same time cutinised layers begin to appear in the outer walls of the hypodermal cells. In this way the inactive epidermal elements become conpletely enveloped in cutinised layers and consequently die. This process gradually extends inwards until it involves a number of cortical layers; in this way the older branches become invested by a multiple dermal tissue consisting essentially of numerous superimposed cutinised layers, which here and there enclose the remains of dead protoplasts. In old stems of *Viscum* the total thickness of this "cuticular epithelium," as Damm calls it, may amount to more than half a millimetre.  As the branches grow in thickness the surface of the epithelium becomes traversed by radial fissures (Fig. 31); these gradually enlarge to such an extent that the outermost layers of the cutinised mass ultimately crumble away or detach themselves in the form of scales.  The VISCOIDEAE form no periderm, so that here the multiple epidermis is the only dermal tissue produced.  Some MENISPERMACEAE, on the other hand, develop a transient cuticular epithelium, which after a certain number of years is replaced by cork.

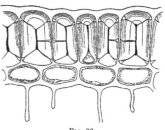

FIG. 32.

Double epidermis of the leaf of *Pinus montana*. (T. S.)

In *Pinus sylvestris*, and in certain other species of *Pinus*, the walls of the foliar epidermal cells are so greatly thickened on all sides that the cavities are reduced to narrow clefts, from which fine canals or pores radiate in various directions, but especially towards the corners of the cells (Fig. 32); in this way the mechanical strength of the epidermis is considerably increased.  The actual cutinised layers are only moderately thick; but all the other regions of the epidermal cell-membrane are also either cutinised or lignified.  This superficial cell-layer therefore, which corresponds morphologically to the epidermis, evidently functionates as a waterproof covering and as a hard exo-skeleton.  The hypodermis, as seen in transverse section, resembles a normal epidermis in the shape of its cells, in the colourless nature of the cell contents, and in the moderate thickness of its lateral and inner walls; this layer has apparently undertaken that partial function of the ordinary epidermis which is associated with these histological characteristics, that is to say, it represents a rather feebly developed peripheral water-tissue.  The foliar dermal system of *Pinus* therefore displays a relatively high degree of division of labour, inasmuch as the partial functions performed by an ordinary epidermis are, in this case, distributed between two cell-layers,

In the majority of BROMELIACEAE the leaf is likewise furnished with a double epidermis.  The cells of the outer layer have slightly thickened but highly cutinised external walls; their inner walls are, however, as a rule so thick that the cell-cavities are reduced to narrow clefts.  In certain species (*Tillandsia aloifolia, T. goyacensis, T. violacea*, etc.), according to Baumert, the thickened inner walls are shaped like concave mirrors; they accordingly reflect a portion of the light that falls upon the leaf and thus afford some protection against intense illumination.    In the case of the inner of the two cell-layers, all the walls may be equally thickened (Fig. 33 A), or else the outer walls may be somewhat thicker than the rest.    In the latter event the outer walls of the inner layer and the inner walls of the outer layer collectively form a remarkably thick and rigid composite membrane (Fig. 33 B).

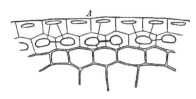

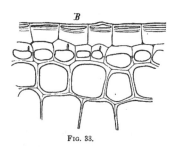

FIG. 33.

Double epidermis of the BROMELIACEAE.
A. Adaxial foliar epidermis of *Bulbergia viridiflora*.
B. Adaxial foliar epidermis of *Vriesea* sp.

As this membrane is quite free from cutin, it merely serves to increase the mechanical strength of the dermal tissue; the duty of controlling transpiration, on the other hand, is almost entirely relegated to the cutinised outer walls of the superficial layer.    Thus two partial functions which are both performed by the thickened, and at the same time cutinised, outer wall of an ordinary epidermis are, in the leaves of the BROMELIACEAE, dissociated and distributed between two different cell-layers.    The water-storing function is dropped altogether as far as the epidermis is concerned, because the leaves of Bromeliaceae are already provided with a well-developed *internal* water-tissue.    As a matter of fact, however, it is the necessity for the storage of water which most frequently leads to the development of a multiple epidermis.    Those cases in which a few scattered epidermal cells (*Tradescantia zebrina*) or a larger number (*e.g.* lower side of the leaf in *Passerina ericoides,* stem of *Ephedra* spp.) undergo a single tangential division, may be regarded as examples of the first stages in the multiplication of the epidermis. By a similar process of division the whole upper epidermis of the leaf becomes two-layered in *Peperomia arifolia,* two- to four-layered in *Begonia manicata, Peperomia blanda,* etc.    The further development of this tendency results in the formation of a typical many-layered peripheral water-tissue.    In these extreme cases, water-storage has become the principal or even the sole function of the multiple epidermis, except for its outermost layer; in order to avoid needless repetition these tissues will be discussed in connection with the storage-system (Chap. VIII.).

### C. APPENDAGES OF THE EPIDERMIS.[71]

In most plants a certain proportion of the young epidermal cells grow out either singly, or less often in groups, to form uni- or multicellular appendages of diverse shapes, which are included under the general categories of **hairs** or **trichomes**. While these epidermal outgrowths most frequently take the form of ordinary hairs, they may also develop into scales, prickles, warts or vesicles (bladders).

One can obtain a rough idea of the extraordinary variety of shapes which trichomes may assume by reflecting, first, that there are only a very few groups of plants (*e.g.* the POTAMOGETONACEAE and LEMNACEAE among Angiosperms) in which these structures are entirely or almost entirely absent ; secondly, that there is hardly any physiological function in connection with which hairs cannot be utilised ; and, finally, that one and the same organ often bears several different kinds of epidermal appendages.

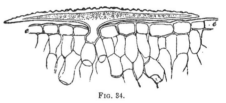

FIG. 84.

T-shaped unicellular hair [Malpighian hair] from the abaxial leaf-surface of *Cheiranthus Cheiri*. After De Bary (*Comp. Anat.*).

In its simplest form a trichome merely represents a tubular outward prolongation of an epidermal cell. This condition is illustrated by the CAMPANULACEAE and CRUCIFERAE ; in the latter family the hairs often fork or branch repeatedly, but without undergoing septation (Fig. 34). In most cases, however, the formation of a hair is accompanied by cell-division. The first wall laid down usually separates a **basal portion** or **foot**, which remains embedded in the epidermis, from the **body** of the hair or **hair proper**. Sometimes no further divisions take place ; more often, however, the hairs ultimately come to consist of several, or even of numerous cells, and assume the form of branched or unbranched cell-filaments, of cell-plates (scales), or even of bulky cell-masses (villi or shaggy hairs, warts and prickles). The foot of a trichome may lie at the same level as the rest of the epidermis, or it may be sunk in a small depression, or, on the contrary, raised up on an emergence composed of subepidermal tissues. Hairs which are directed obliquely to the surface are often provided with a little pad or cushion, situated on the convex or on the concave side of the base ; this pad is usually made up of some of the adjoining epidermal elements, but may in certain cases also include a few subepidermal cells. Renner asserts that such basal cushions serve to effect the erection of hairs which at first lie closely appressed to the leaf, stem [or other organ]. It is, however, quite conceivable that they act primarily as buttresses which add to the stiffness of the hairs; for such oblique hairs, especially if they are stiff and bristly, often assist in preventing small animals from crawling over the surface of the organs on which they occur. The cells which immediately surround the base of a hair generally differ in form, and often also in the character of their walls, from the ordinary epidermal elements; these are the **subsidiary cells,** which are so frequently described as forming a ring or rosette around the base of a hair.

According to the special nature of their functions hairs contain living protoplasts or consist of dead cells; the character of their walls is also naturally altogether dependent on their physiological properties.

As has already been pointed out, hairs are utilised for a great variety of purposes, many of which have nothing in common with the proper functions of the epidermis. In support of this statement reference may be made to water-absorbing and water-secreting hairs, to the extraordinarily varied types of glandular hairs, to the feathery hairs of which the "parachutes" of many seeds and fruits are composed and, lastly, to the remarkable trichomes that assist in the perception of certain stimuli. Obviously only those trichomes which assist the epidermis in its protective capacity can be regarded as epidermal appendages in an anatomico-physiological sense; the following discussion is therefore confined to hairs which fulfil this condition.

The value of a **hairy covering** [in relation to the dermal system] depends mainly upon the fact that it **causes a reduction of trans piratory activity,** and hence diminishes the risk of desiccation. A coat of densely packed dry hairs must obviously act as a screen, which protects the hairy organ from direct insolation and from the increased rate of transpiration that results therefrom;[72] it must also hinder the renewal of the stratum of air overlying the transpiring surface and thus further diminish transpiration. Care must always be taken to distinguish between these two factors in estimating the reduction of transpiratory activity which is effected by a dense covering of hairs. Thus the dense coat of hairs characteristic of numerous steppe- and desert-plants, as well as of many representatives of the Mediterranean and alpine floras, which extends even to the directly insolated faces of the leaves, must be primarily regarded as a light-screen. Where on the other hand, in dorsi-ventral leaves, stomata are confined to the shaded side—which usually corresponds to the abaxial surface—the hairy covering is evidently in the first instance designed to impede surface-ventilation by producing a labyrinth of spaces filled with stationary air. A similar purpose is served by the hairy coats which occur in many Arctic species and in a number of the plants that inhabit the *paramos* (alpine steppes) of the Venezuelan Andes ; here the risk of desiccation arises not from the heat of the sun, but from violent winds, which in the Arctic regions are in addition extraordinarily dry.

There are at present no exact experimental data available with regard to the effectiveness of a woolly or felted hairy covering as a means of reducing transpiration. In order to obtain the requisite quantitative information, at any rate in a single case, the author carried out an appropriate experiment with *Stachys lanata*. The leaves of this plant are thickly covered with woolly hairs on both sides ; the density of the covering will be appreciated when it is stated that there are no less than 120 hairs per sq. mm. on the upper side, or 156,000 on an entire leaf with a superficial area of 13 sq. cm. For the experiment two opposite leaves were selected, which were fully grown and approximately equal in size. After both had been painted on the under surface with cacao-wax, the hairs were carefully removed from the upper side of one leaf with the aid of curved scissors. The leaves were then allowed to transpire for twenty-four hours in the shade at a temperature of 20°-25° C., their short stalks being sealed into the necks of small flasks filled with water. Under these conditions the intact leaf lost ·646 grams by transpiration, the shaved leaf ·915 grams ; the ratio of these values is as 1 : 1·42. The leaves were next exposed to sunlight for an hour (during twenty minutes of which period, however, the sun was slightly obscured by clouds) ; the losses of weight were now found to be ·08 g. for the intact and ·167 for the shaved leaf, giving a ratio of 1 : 2·09.[73] In the second case, therefore, the reduction of transpiratory activity effected by the hairy covering amounted to more than 50 per cent. ; as in this plant the upper surface of the leaf is provided with numerous stomata, the measurements quoted refer mainly to stomatic transpiration. It is quite evident that a dense woolly coat of hairs reduces transpiration very considerably, when the organ concerned is exposed to direct insolation ; its influence in this respect is less powerful, though by no means negligible, in diffuse light, where its action consists mainly in the retardation of surface ventilation.

No doubt there are also other ways in which a dense coat of hairs may afford protection to the plant. The hairy coverings of plants which inhabit deserts or steppes may serve to diminish the loss of heat produced by nocturnal radiation ; in the case of young leaves which are just unfolding the hairs may, as Wiesner suggests, protect the developing chlorophyll-apparatus against the injurious action of sunlight. These juvenile or transitory hairy coverings, which are usually confined to the upper side of the leaf, often disappear completely at a later stage; the Colts-foot (*Tussilago Farfara*) affords a striking illustration of this temporary condition of hairiness.

The greatest possible diversity prevails with regard to the detailed structure of the hairy coverings by means of which plants protect themselves against these different climatic influences. The hairs employed for this purpose may be either branched or unbranched, but in either case are generally multicellular. Where they are closely appressed to the surface of the organ, and all point in the same direction, they often produce a **silky** glistening effect (*Convolvulus cneorum, C. nitidus*, etc.); if on the other hand they are irregularly twisted or curled, or coiled in corkscrew fashion, they form a **woolly** or **felted** coat (*Gnaphalium Leontopodium, Banksia stellata*, etc. etc., Fig. 35). A very dense and compact felt results if such twisted hairs become mutually entwined, entangled or interwoven. A particularly effective type of felted hairy covering has been described by Goebel[74] as occurring in the genus *Espeletia* (COMPOSITAE), which inhabits the *paramos* of Venezuela. Here both sides of the leaf are covered by unbranched hairs which, after rising vertically from the surface for some distance, describe a very flat spiral; they then once more continue in a more or less vertical direction, but soon undergo a second spiral twist,

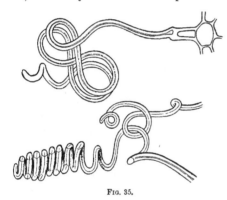

FIG. 35.

Coiled and twisted woolly hairs from the abaxial leaf-surface of *Banksia stellata*.

and so on alternately several times in succession. The most interesting point, however, is that the corresponding twisted regions lie approximately at the same level in all the hairs, so that the whole felted covering acquires a regularly stratified structure. This alternation of loose and compact zones still further impedes surface-ventilation, thus affording additional protection against the desiccating action of the violent winds which, as already stated, prevail in the *paramos*. More or less dense coatings are also formed by the so-called **stellate** and **tufted hairs**. Such hairs arise from an initial cell which divides by anticlinal walls into a number of segments; each one of the latter then develops independently, so that the hair ultimately consists of several branches or rays all inserted upon a common basal portion. Good examples of stellate hairs are found in the MALVACEAE and CISTINEAE, in spp. of *Croton*, etc.

In many plants, finally, the protective covering is made up of **peltate hairs** or **scales**; excellent instances of these are found among OLEACEAE and ELEAGNACEAE (Fig. 36). Perhaps the most remarkable and effective arrangement of peltate hairs occurs, according to Bobisut in the Palm *Arenga saccharifera.* The margins of the scales which cover the lower side of the leaf, in this Palm, are produced into numerous tubular branches; the latter are so closely interwoven that it is quite impossible to distinguish the individual scales. In addition, similar tubular branches grow downwards in dense masses from the lower side of each scale; these in their turn are abundantly branched, and the branches of adjoining tubes interlock so that the final result is an extraordinarily compact felted covering, which completely masks the actual epidermal surface.

FIG. 36.

Peltate scale from the adaxial leaf-surface of *Hippophae rhamnoides.*

We must next pass on to consider those hairs which serve for protection against the assaults of animals; here again one meets with great variety of detail, and in many cases also with remarkable specialisation. Protective hairs, other than those of the glandular type, may be grouped under two headings. The first group includes all hairs the action of which is purely **mechanical**; the second form of protective hair is in addition possessed of poisonous properties, and its effectiveness is indeed largely dependent upon **chemical** effects.

The **mechanical** type of protective trichome is exemplified by the short stiff bristles which are characteristic of certain families of Angiosperms; the walls of such bristles are often calcified or silicified and in addition covered with protuberances in the shape of knobs or bosses, which act as barbs. The BORAGINACEAE and many CRUCIFERAE are excellently protected by such rasping hairs—to use Stahl's term—against snails, caterpillars and other creeping animals, and also against browsing mammals. Even more effective are the fish-hook-like bristles of *Opuntia,* with their numerous barbs.

The **stinging-hairs**[75] which occur in the URTICACEAE and LOASACEAE, in *Jatropha* (EUPHORBIACEAE) and in *Wigandia* (HYDROLEACEAE) are among the most elaborate of the structures employed by plants for the purpose of protecting themselves against their animal foes. A typical stinging-hair consists essentially of a single large cell with abundant protoplasmic contents; the lower end of this cell is expanded to form a vesicle or bulb, embedded in a cup-shaped emergence. The walls of the hair are thick and brittle, except in the region of the bulb. In the genus *Urtica* the stiff and brittle condition is mainly due to silicification, so far as the distal part of the hair is concerned; the proximal portion, on the other hand, is calcified. In the LOASACEAE the walls of the stinging-hairs are calcified, while in the genus *Jatropha* they are strongly lignified. In all cases the structure of the tip displays several interesting adaptive features; this subject has received special attention at the hands of the author. Typically a stinging-hair ends in a minute swollen head; this terminal swelling breaks off at the slightest touch, leaving an open point, which can readily enter the body of the aggressor (Fig. 38 A, B). The head is more or less spherical or oval, and is attached obliquely; in the LOASACEAE it is usually so small as to be hardly distinguishable from the rest of the distal region (Fig. 30 G). On looking at the tip of a stinging-hair from the side, one is at once struck by the fact that its walls are of very unequal thickness in different parts. In the case of *Urtica dioica* a narrow strip of the convex side, situated just above the slight constriction or neck, is remarkably thin and often very sharply separated from the neighbouring thickened portions of the wall (Fig. 38 A). On the concave side of the tip the wall is also relatively thin in one place; but in this case the

FIG. 37.

Stinging-hair of *Urtica dioica.*

difference is not so marked, and, moreover, the thin area is much more extensive and passes over gradually into the thicker portion. The tip of the stinging-hair of *Loasa papaverifolia* is constructed on a very similar plan; here the extensive unthickened region of the concave side is extraordinarily thin (Fig. 38 G).

An important feature of the mechanism of stinging-hairs depends upon the fact that the tip normally breaks away along a plane which includes the above-mentioned thin areas; in other words, the precise line of separation is predetermined by the structure of the wall. This arrangement not only facilitates the detachment of the tip, but also ensures that the resulting point has the shape which is best suited for the purpose of piercing the object that it encounters. Owing to the fact that fracture takes place along an oblique plane, the resulting point is lancet-shaped and very sharp, and the aperture from which the poison issues is situated laterally some distance behind the point (Fig. 38 B). The open point of a stinging-hair is thus seen to be constructed after the pattern of the cannula of a hypodermic syringe or like the grooved poison-fang of a snake. In *Loasa papaverifolia* and in *Jatropha stimulata* the portion of the cell-wall which forms the actual sharp point—after the tip is broken off—is considerably thicker than the rest of the distal region (Fig. 38 G, H). In *Jatropha stimulata* the thin place is characteristically developed on the concave side of the tip, but is altogether absent on the convex face; on this side fracture always occurs at the constriction just behind the tip, where the thickening layers of the wall are sharply bent. It is interesting,

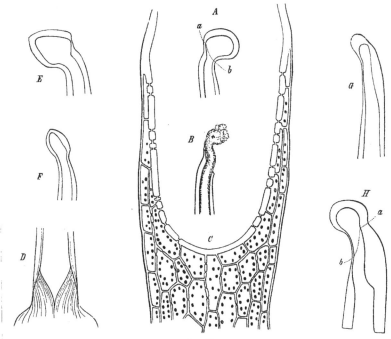

FIG. 3S.

Structure of various stinging-hairs. *A-D. Urtica dioica. A.* Apex of hair with tip attached. *B.* Apex after the tip has been broken off. *C.* Bulb of the hair [and the cup-shaped emergence in which the bulb is embedded] in L.S. *D.* Base of the hair-cell after treatment with sulphuric acid; the silicified region of the wall has not swollen. *E-H.* Apical portions of various stinging-hairs. *E. Urtica pilulifera. F. Laportea gigas. G. Loasa papaverifolia. H. Jatropha stimulata. a-b* (in *A* and *H*), line of separation of the tip.

from the phylogenetic point of view, to note that a series of transitions may be observed, within the limits of a single family, between an imperfectly constructed tip and the highly specialised structure which has just been described.

The author has shown that the substance which is injected from stinging-hairs into the wounds that they cause, is not, as was at one time generally assumed, *formic acid*. As a matter of fact, the very grave toxic effects (namely violent and persistent pain, tetanus-like symptoms and, in extreme cases, death) produced by the sting of certain tropical Nettles (such as *Urtica stimulans* in Java and *U. urentissima* in Timor) suffice to disprove the old theory. According to the author, the poison of *Urtica dioica* is an albuminoid substance, held in solution in the cell-sap, and in many respects resembling an enzyme, especially as regards its solubility.

Concerning the mode of liberation of the poisonous cell-contents it may be remarked that the elastic tension of the cell-wall of the bulb —which is neither calcified nor silicified—is sufficient to cause active ejection. Duval-Jouve long ago observed that, when the tip of a stinging-hair of the Nettle is broken off by a touch with a needle, a little drop of cell-sap at once exudes, or is even forcibly ejected. There can, however, be no doubt that the pressure exerted upon the bulb by the impinging object helps to produce ejection of the cell-sap. Among the reasons for this conclusion is the circumstance that one may be perceptibly stung twice in succession by the same hair; turgor obviously cannot have any share in inflicting the second sting.

In *Urtica dioica* the cells forming the basal cup are remarkably well provided with chloroplasts; the cup should therefore probably be regarded as a local photosynthetic apparatus pertaining to the hair. This view is supported by the fact that numerous pits are present in the thickened lateral walls of the bulb, a circumstance which points to the existence of an active interchange of material between the cup-cells and the principal cell, or body, of the hair (Fig. 38 c).

In many plants it is only the young leaves that possess a protective coat of hairs, the adult organs being entirely devoid of such a covering; this statement applies more especially to the upper surfaces of foliage leaves (*Tussilago Farfara, Petasites niveus*, etc.). Whether the hairs are cast in such cases, merely because they become superfluous when the cuticle and cutinised layers are fully developed and when the chlorophyll apparatus no longer requires protection, or because the persistence of a hairy covering would deprive the leaf of too large a proportion of the available light, or would otherwise interfere with its functions, is a question which must be left undecided. Detailed researches carried out by Keller[76] upon the phenomena attendant upon normal depilation have shown that this process follows one of two different methods. In the case of unicellular trichomes—and sometimes also in the multicellular types—the hair breaks off immediately above its point of insertion; the base of the hair is therefore laid open. In such cases the persistent basal walls usually become thickened and cutinised before separation takes place; in certain cases they may even acquire all the characteristics of typical outer epidermal walls. According to Keller special arrangements which facilitate the breaking off of a hair, or which predetermine the point of rupture, are but rarely found. The author has noticed that the deciduous unicellular hairs on the upper side of the leaf of *Coscinium Blumeanum* are noticeably constricted towards their lower ends.  A thin strip in the wall indicates the point at which the **T**-shaped unicellular hairs of *Banisteria* break across.  Multicellular hairs are usually cast or disarticulated; that is

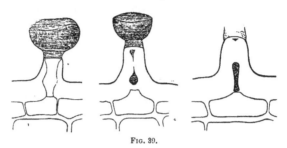

FIG. 39.

Occlusion of atrophying glandular hairs on the adaxial leaf-surface of *Rosmarinus officinalis.* The lateral walls of the stalk-cell become enormously thickened, and the cavity is largely obliterated. When the rest of the hair breaks off, a thick-walled almost solid papilla is left.

to say, the transverse wall separating the deciduous part of the hair from the persistent basal cell or foot splits across.  Those layers of the wall which ultimately become exposed undergo cutinisation before separation takes place.  The various modifications of this process of occlusion by means of cutinisation or thickening of the cell-wall, which may be seen in certain PAPILIONACEAE, PROTEACEAE, COMPOSITAE, etc., cannot be discussed here.  A similar occlusion not infrequently takes place in connection with the atrophy of glandular hairs; in this case it probably serves to maintain the continuity of the dermal tissue, and thus not only to prevent the occurrence of any uncontrolled transpiration, but also more particularly to hinder the entrance of parasitic Fungi.

### III.  PERIDERM.

When an organ has reached a certain age the epidermis as a rule no longer affords sufficient protection—for two reasons.  In the first place, this layer is after all a comparatively delicate structure, and is consequently not well fitted for the protection of bulky organs such as the branches of woody plants.  Secondly, apart from a few exceptional cases, the epidermis is incapable of keeping pace with the active growth in thickness which most axial organs exhibit, and therefore soon becomes ruptured and torn into shreds.  If, therefore, an organ which is growing in thickness is to remain enveloped in a protective covering, the epidermis must be replaced by a dermal tissue which is tougher and altogether more resistant, besides being capable of continued self-regeneration.  The epidermis fulfils these two conditions to a limited extent, inasmuch as its cells on the one hand possess thickened and cutinised outer walls, and on the other contain living protoplasts; but the physiological requirements referred to have become so much greater, when the more highly-developed type of dermal system is brought into action, that a single kind of tissue is no longer able to satisfy them all.  Hence the principle of division of labour once more comes into play.  The **periderm**, as the secondary dermal tissue is termed, accordingly comprises two different tissues.  One of these, the **cork**, is a permanent tissue which is concerned with the protection of the underlying layers against excessive evaporation, mechanical injury and deleterious influences in general; the other, the **phellogen**, is a meristematic layer, responsible for the regeneration of the dead cork, which is constantly undergoing disruption owing to the growth in thickness of the organ that it envelops.

A. CORK.[77]

The cells of the **cork** are usually prismatic or tabular in shape, with four-to six-sided end-walls.  Their walls may be thin, or more or less thickened, but in either case are usually of uniform thickness all over.  In a few instances the outer (*Salix, Xanthoxylon fraxineum, Cytisus Laburnum*) or the inner (*Mespilus germanica, Viburnum Opulus*) walls are thicker than the rest.  According to De Bary,—whose observations have in many respects been extended by Von Höhnel,—the wall separating two cork-cells generally comprises five distinct layers.  The inner surface on either side consists of a **cellulose layer**, which occasionally undergoes lignification.  Next follows, also on either side, a suberised layer, the so-called **suberin-lamella**, to which the physiologically important characteristics of the entire wall are due.  The central portion of the whole system is formed by the limiting membrane or **middle lamella,** which is either lignified or else appears to consist of unaltered cellulose.  Where the wall is thin, the innermost (or cellulose) layers may be absent, so that the whole membrane is suberised,—except for the middle lamella.

It was at one time generally assumed that a corky wall represents an ordinary cellulose membrane, the secondary thickening layers of which are impregnated with a fatty substance termed *suberin* (hence the term suberin-lamella).  This view has been strongly opposed by Gilson and by Van Wisselingh, who both maintain that suberised layers contain no true cellulose at all.  With regard to the chemical nature of *suberin*, it may be remarked that Kügler, Gilson and Van Wisselingh have all isolated *phellonic acid* from the cork of *Quercus*

*Suber*; according to Gilson this substance is accompanied by two other acids, namely *suberic acid* and *phloionic acid*. It is still uncertain whether these acids occur as glycerine-esters in suberised walls, as Kügler and Van Wisselingh assume, or whether, as Gilson asserts, *suberin* is made up of compound esters or of other products of the condensation or polymerisation of the three acids. The suberin-lamella may in addition be silicified, as Von Höhnel first demonstrated; it is noteworthy that silicified cork as a rule only occurs in plants which also contain a large amount of silica in their epidermis.

Among other physiologically interesting features of corky walls which are deserving of special mention, is the circumstance that pits are generally absent. Where they do occur, they are, according to Von Höhnel, confined to the inner cellulose layer, and never penetrate into the suberin-lamella. These pits are most conspicuous where the cellulose layer is secondarily thickened, and are evidently functional only so long as the cork cells are alive and

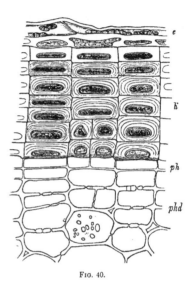

Fig. 40.

T.S. through the periderm of a branch of *Cytisus Laburnum* (in winter). *e*, dead epidermal cells containing Fungus-spores; *k*, layers of cork; *ph*, phellogen; *phd*, phelloderm.

undergoing differentiation; they serve to facilitate the access of plastic materials to the developing suberin-lamella. The fact that the suberin-lamella of the outer wall of a cork-cell is sometimes thicker than the corresponding layer of the inner wall is evidently connected with the functions of the fully-developed corky tissue; where this condition is found, the inner wall usually has a thicker cellulose layer than the outer wall.

The cells comprising corky tissues are always dead, and hence as a rule contain little besides air; this statement applies more especially to thin-walled cork. Whether remnants of the former protoplasts persist in the form of thin films on the inner surface of the cell-wall or not, is of little moment. Greater interest attaches to the fact that the cavities of thick-walled cork-cells are often filled with a yellow or reddish-brown substance, which consists in part at any rate of *tannins* and their decomposition-products (*phlobaphenes*). In special cases other compounds may be present, such as *betuline* in the cork of the

Birch, or acicular crystals of *cerine* in bottle-cork (*Quercus Suber*); *calcium oxalate* crystals also occasionally occur, either as dense masses (*Quercus Suber*) or in the form of raphides (*Testudinaria elephantipes*).

Cork, like the epidermis, always forms a continuous envelope, which is only interrupted by special intercellular passages that serve for ventilation [cf. Chap. IX. B]. Its component cells are usually arranged in radial rows. The total thickness of the corky envelope is very variable; for while some **cork-films**[78] comprise no more than two or three layers of cells, a **cork-crust** may be thicker than the branch which it envelops. Cork-films are smooth integuments, and as a rule consist entirely of tabular cells. Cork-crusts, on the contrary, are mainly composed of wide, soft-walled cork cells, narrow strips of tabular cells being intercalated towards the ends of successive vegetative periods. These massive crusts never form evenly distributed smooth integuments, but are arranged in the shape of projecting flanges or wings, separated from one another by deep longitudinal furrows. This condition arises partly owing to asymmetrical production of cork, and partly owing to the development of cracks in consequence of the growth in thickness of the whole organ. The majority of woody plants form cork-films; cork-crusts are comparatively rare, but occur, for instance, in the well-known Cork Oaks (*Quercus Suber, Q. occidentalis, Q. Pseudosuber*), on the young branches of *Acer campestre, Liquidambar styraciflua, Ulmus suberosa, Euonymus europaea*, and of spp. of *Banksia* and *Hakea*, and on the stems of climbing spp. of *Aristolochia*. It has been demonstrated by Von Höhnel in the case of *Ulmus suberosa, Euonymus europaea, Liquidambar styraciflua, Passiflora limbata*, etc., that the layers composing the crusts are to a large extent unsuberised, and consequently cannot be regarded as genuine corky tissues. For such tissues Von Höhnel proposes the name **phelloid**: their massive development is supposed to compensate to some extent for their feeble suberisation, the defect in quality being as it were counterbalanced by an increase in quantity. This interpretation cannot be applied universally; Von Höhnel himself states that in certain cases (*Quercus Suber, Acer campestre, Aristolochia cymbifera, A. Prixotoa*, etc.) the cork-wings are entirely made up of genuine cork.

Some account must next be given of the physical properties of cork. From the physiological point of view the most interesting property of cork is its very slight perviousness to water. This point has been quite clearly demonstrated by a number of experiments carried out in order to test the influence of cork upon the activity of transpiration. Both Nägeli and, more recently, Eder[79] have compared the transpiration from peeled potatoes with that of the intact tubers. In one of Eder's experiments the readings indicated that while an unpeeled potato lost ·0397 grams per cent. of water during twenty-four hours, the corresponding amount in the case of a very thinly peeled tuber was 2·5548 grams, or about 64 times as much.  In the course of a week the peeled potato lost 44 times as much water as the intact tuber.

In order to compare the permeability of a thin cork-film with that of an average epidermal layer, the author performed the following experiment upon two-year-old shoots of certain woody plants at the beginning of December.  Lengths of 20 cm. were cut from twigs of *Aesculus Hippocastanum, Syringa vulgaris, Sambucus nigra* and *Pirus communis.*  In every case the two severed ends, the leaf-scars, and the lenticels were carefully sealed with melted wax-mixture, and the experimental twigs were then allowed to transpire for three days in the laboratory, at a temperature of 17-19° C., and in an atmosphere with a relative humidity of 50-60 per cent.  The daily loss of weight was determined by weighing.  Simultaneously the evaporation from a freely exposed water-surface was measured under the same conditions, just as in the corresponding experiments carried out upon the epidermis (cf. p. 106).  In calculating the transpiring surface, the wax-coated areas were, of course, deducted.  Under these conditions the average amounts of water evaporated (in grams per sq. dcm. per 24 hrs.) were as follows ·

| | |
|---|---|
| *Aesculus Hippocastanum.* - - - - - | ·154 |
| *Syringa vulgaris,* - - - - - - | ·189 |
| *Sambucus nigra,* - - - - - - | ·317 |
| *Pirus communis,* - - - - - - | ·430 |
| Water-surface, - - - - - - - | 6·920 |

It happened that in this particular experiment the evaporation from the water-surface was approximately the same as in the previously described experiment, in which the transpiration from an astomatic leaf-surface was compared with the evaporation from a freely exposed water-surface.  Hence the rates of transpiration of the leaves and twigs can be directly compared.  It appears that in the Horse-chestnut the transpiratory activity of a two-year-old branch (·154 g.) is very nearly equal to that of a leaf (·156 g.); in other words, the outer epidermal walls and the two-year-old cork-film exert the same restrictive effect upon transpiration.  In the remaining instances, however, the transpiratory activity of the branches is, on the whole, considerably greater than that of the astomatic leaf-surfaces of the same plant.  Nevertheless, the presence of periderm diminishes the loss of water very appreciably ; for evaporation from a freely exposed water-surface was found to exceed transpiration from the branches sixteen-fold in the Pear, twenty-five-fold in the Elder, thirty-six-fold in the Lilac, and forty-five-fold in the Horse-chestnut.

The fact that a young cork-film in general restricts transpiration less effectually than a well-developed epidermis is very easily accounted for ; an astomatic epidermis is absolutely continuous, whereas, owing to the secondary growth in thickness of the branch, a two- or three-year-old cork-film is traversed by fine radial cracks, the effect of which in augmenting transpiration cannot be eliminated in experiments such as those described above. This conclusion is confirmed by the fact— established by Wiesner and Pacher,[80]—that two- and three-year-old twigs of Horse-chestnut transpire more vigorously than yearling branches ; from the fourth year onwards, however, a marked improvement is noticeable as regards the control of transpiration, a circumstance which must be attributed to the steady increase in the thickness of the corky integument.

In any case, corky tissues constitute, even in thin layers, a very effective instrument for the reduction of transpiration. Those branches, twigs, and buds of woody plants which have to live through the winter certainly require this protection, if only because transpiration goes on slowly even at temperatures below zero. Mention may be made of certain experiments, performed by Wiesner and Pacher, which bear upon this point. A two-year-old Horse-chestnut twig set up in a special transpiration apparatus lost ·323 per cent. of its weight after transpiring for twenty-four hours at temperatures ranging from $-3·5°$ C. to $-10·5°$ C.; between $-5·5°$ C. and $-13°$ C. the same branch lost ·199 per cent. of water in the same time. Similarly, a three-year-old Oak twig lost by transpiration in twenty-four hours ·251 per cent. by weight between $-3°$ C. and $-8·5°$ C., and ·192 per cent. between $-5·5°$ C. and $-13°$ C. The transpiratory activity of branches is thus by no means negligible, even at such low temperatures, and it is probable that this statement is equally applicable to winter buds, especially as these structures are known to suffer very considerable losses by transpiration at somewhat higher temperatures. Thus Wiesner and Pacher estimated the daily loss by transpiration at about $1·523 - 1·6$ per cent. in the case of a winter bud of Horse-chestnut. If, therefore, twigs and branches require to be protected against desiccation, the same necessity will arise in the case of winter-buds ; as a matter of fact, bud-scales very often are provided with a layer of cork, which is situated beneath the outer epidermis [81] (in the case of *Aesculus* the epidermis is cast when the cork appears).

Cork is very impervious to gases. Careful experiments carried out by Wiesner [82] have shown that quite thin plates of cork obtained from the Cork-Oak or from Potato tubers are practically impervious to ordinary air, even under considerable pressure. Wiesner's apparatus consists essentially of a glass T-tube. The plate of cork which is to be tested is cemented over the end of one transverse arm; mercury is sucked into the vertical arm through a rubber tube (provided with a screw-clamp), which is attached to the other transverse arm. Small plates prepared from the corky tissue of *Quercus Suber*, not exceeding ·05 to ·07 mm. in thickness, and hence comprising only two or three layers of cells, were in this way exposed to a (negative) pressure of 20 cm. mercury for several weeks without any leakage of the outer air taking place. With a suitable arrangement of the apparatus a plate of cork consisting only of three layers was found to be quite impervious to air, even under a difference of pressure amounting to a whole atmosphere.

Schwendener[33] has shown that corky tissues are in general both inextensible and inelastic. Isolated strips of the periderm of *Castanea vesca*, *Fagus sylvatica* and *Cytisus Laburnum* break before their longitudinal extension has reached 2 per cent. The thin-walled cork-layers of the Beech, which consist mainly of suberin, display a similar behaviour; the thick-walled layers on the contrary, in which cellulose predominates, are very extensible. The author has found that tangentially stretched strips of the periderm of arm-thick branches of *Tilia grandifolia* break when their elongation amounts to 4 per cent., whereas similar strips taken from two-year-old twigs can be stretched to the extent of 7 per cent. or 8 per cent.; the periderm thus seems to be somewhat more extensible while it is young. According to Schwendener the peridermal covering of *Prunus* is exceptional in this respect since it can undergo an extension of 10-12 per cent. without breaking. An apparent exception has been noted by the author in the case of ordinary bottle-cork; strips of this material cut in a tangential plane can be stretched to the extent of 25 per cent., and remain perfectly elastic up to elongations of 6-7 per cent. It is in all probability this remarkable stretching power of bottle-cork which is responsible for the old notion that corky tissues in general are highly extensible; the peculiar behaviour of bottle-cork, however, depends upon the *arrangement* of the component cells and not upon the physical properties of their walls. In the case of bottle-cork, namely, the cells of adjacent radial series alternate with one another; hence when a strip of cork is stretched in a tangential direction, the radial cell-walls are thrown into zig-zag folds. The elongation produced in this way is large simply because the radial diameter of each cork-cell is considerable. This statement can be easily verified by observing a strip of cork under the microscope while it is being stretched; it is further borne out by the fact that radial strips cannot be stretched beyond 4 per cent.; it is probable that the true extensibility of the tangential walls is no greater.

In considering the functions of corky tissues it must be borne in mind that cork is a very poor conductor of heat, mainly because its cells are filled with air. On account of this property cork would seem to be peculiarly well-fitted to act as a protective covering in the case of perennial aerial organs. Perennial branches contain, in addition to their various permanent tissues, two peripherally situated meristematic layers which require to be protected against sudden fluctuations of temperature, namely the phellogen and the cambium. It may, of course, in many cases be a matter of indifference whether the thawing of a frozen organ takes place rapidly or slowly; on the other hand there can be no doubt that rapid thawing is in general dangerous, especially where the same organ freezes and thaws several times in rapid succession. Plants therefore find it advantageous to guard against sudden variations of their internal temperature, and thus to minimise the deleterious effect of such violent changes. The gardener is, therefore, merely copying nature when he protects the more tender of his woody plants against frost by means of wrappings of straw or tow; only the materials employed by the plants themselves—namely cork and bark—are much better suited to their purpose.

The barest reference must be made to the fact that cork also constitutes an excellent means of protection against the attacks of parasitic Fungi and against the assaults of animals of every kind, a function in which it is very frequently assisted by various *tannins*, *alkaloids* and other poisonous or bitter substances which are so often deposited in bark.

Cork is also particularly well fitted to serve as a tissue of cicatrisation. As a matter of fact, wounds in the parenchymatous tissues of stems, roots or leaves are generally occluded by so-called **wound-cork**. The uninjured cells adjoining the wounded surface give rise by tangential division to a phellogen, which in its turn produces corky layers. It is in this way, for example, that the numerous wounds produced on woody twigs by the autumnal leaf-fall are occluded. Dead and diseased tissues also usually become cut off from the healthy parts of the organ in which they occur by layers of cork.

### B. THE PHELLOGEN.

The increase in thickness and the growth in surface of the outer epidermal wall are dependent upon the activity of the epidermal protoplasts; similarly the initiation and the continuous regeneration of cork are due to the activity of the meristematic tissue known as the **phellogen**. Typically the phellogen consists of a single layer of thin-

walled, tabular, meristematic cells with abundant protoplasmic
contents, which divide tangentially.   Most often the outer of the pair
of daughter-cells produced at each division becomes a cork-cell, while
the inner remains an element of the phellogen.   Occasionally, however,
the outer of the two daughter-cells remains a part of the phellogen, in

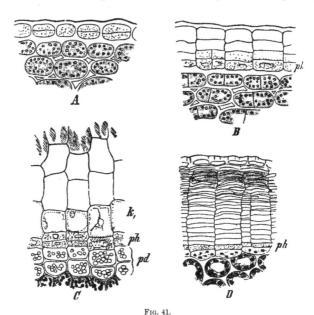

FIG. 41.

Cork.  *A, B.* T.S. of stems of *Scutellaria splendens* showing epidermal origin of
phellogen.   *C.* T.S. of twig of *Ulmus suberosa.*   The cork is very wide-celled.   In its
innermost layer (*k'*) living protoplasts are still present.   The phellogen (*ph*) is in process of division.   The phelloderm (*pd*) contains abundance of starch.   *D.*  Periderm of
a yearling twig of *Prunus Padus.*   (In all figures *ph* denotes phellogen.)

which case the inner becomes an element of the secondary cortical
parenchyma or phelloderm.   Whereas, namely, the phellogen is mainly
concerned with the production of cork, it also frequently—though not
invariably—performs one or more subsidiary functions, by adding
secondary elements in the form of phelloderm cells to the cortical
parenchyma (Fig. 40, *phd*;   Fig. 41 C, *pd*), or by contributing sclerenchymatous or collenchymatous cells to the mechanical system.

The phellogen very often originates [84] either in the epidermis or
else in the outermost layer of cortical parenchyma.   In such cases each
primary mother-cell usually undergoes two successive tangential
divisions giving rise to three superimposed daughter-cells of which
the middle one becomes an element of the phellogen.   The so-called
**superficial periderms** produced in this way are epidermal in origin in all

POMACEAE, in the genus *Salix*, in certain species of *Viburnum*, in *Nerium Oleander*, in *Staphylea pinnata* and in certain other cases (Fig. 42 A, B). Usually, however, it is the outermost cortical layer or hypodermis that gives rise to the superficial phellogen; this case is exemplified by most of our native trees and shrubs.

In many woody plants the phellogen arises at a greater distance from the surface, either in a deep-seated layer of cortical parenchyma or—as in most Dicotyledonous and Gymnospermous roots—in a layer belonging to the vascular cylinder. In the case of roots it is usually the pericycle (pericambium) which provides the primary mother-cells of the phellogen. It is evident that when a phellogen of this deep-seated type begins to form cork, all the tissues outside it must soon be entirely cut off from their water supply, and hence doomed to desiccation, unless, indeed, they are served by special water-paths (as in the case of *Salicornia*). The dried-up cortical tissues, together with the corky layers, constitute the compound structure known as **bark**.

The phellogen may sooner or later cease to divide, and itself become converted into cork. This condition is characteristic of organs with a limited growth in thickness, such as bud-scales and fruits (*e.g.* the Apple). In certain other cases the primary phellogen becomes inactive, but is replaced by a secondary phellogen of more profound origin; this process may be repeated again and again, so that a succession of periderms is formed, each more deep-seated than the last, and each, as it were, condemning a fresh slice of the cortex to speedy desiccation and death.

## IV. BARK.

**Bark** may be regarded as a third and final stage in the development of the dermal system; it merits separate consideration not so much on account of its physiological functions—which do not differ essentially from those of ordinary cork—as because of its anatomical complexity. A typical bark, namely, comprises, in addition to corky layers, also a variety of tissues which originally belonged to other systems. Superficial dead and dried-up tissues may indeed be utilised by the plant for purposes which are not connected with protection. That, on the other hand, any dried-up tissue—even if it originally formed part of a mass of succulent parenchyma—can restrict transpiration most effectively, may be demonstrated in the simplest way by noting the rapid fall in the daily rate of transpiration exhibited by an apple from which a slice has been removed. Masses of dried-up cells also provide a fairly satisfactory means of protection against mechanical injury.

In those woody plants which at first produce a superficial periderm, each of the successive peridermal layers abuts with its margins against its immediate predecessor, so as to cut off a portion of the bark in the form of a scale. It is in this way that the so-called **scale-bark** originates. But where the first peridermal layer is already deep-seated, the subsequent periderms are more or less concentric therewith; thus a **ring-bark** composed of a succession of annular zones is produced.

The anatomical structure of bark, of course, depends not only on the character of the peridermal layers themselves, but also upon the nature and variety of the tissues cut off by the periderm. Sometimes these tissues consist entirely of cortical parenchyma, but in other cases they may in addition comprise masses of collenchyma, fibrous strands, crystal-sacs and resin-ducts.

Special mention must further be made of those anatomical arrangements which serve to increase the mechanical strength of bark. It is self-evident that any strands of mechanical tissue which originally helped to give strength to a stem or root, must likewise increase the tenacity of the bark in which they become incorporated. A similar result, accompanied by an increase in the average hardness and solidity of the tissues, is produced by the conversion of primarily thin-walled parenchymatous elements into sclerenchyma. This transformation may be confined to individual elements, or may extend to whole groups of cells, and the elements affected may either retain their original form or else undergo a considerable change as regards both shape and size; an instance of the latter condition is provided by the multi-radiate stone-cells which are largely responsible for the very firm texture of certain kinds of bark.

It is characteristic of bark that it is forever undergoing exfoliation, the outermost scales continually peeling off from the underlying mass. That this process does not merely represent a passive desquamation, or, in other words, is not a necessary mechanical consequence of the increased girth of the organ, may be inferred from the fact that the plant produces special **absciss-layers** which facilitate the separation of the scales. The arrangements resemble those which occur in leaf-bases in connection with the autumnal leaf-fall. The peridermal absciss-tissues consist of special zones, which are interpolated between the layers of ordinary cork-cells; they agree with the latter as regards the shape of the constituent cells, but are distinguished by the entire absence of suberisation. For this reason they have been termed " absciss-phelloids " by Von Höhnel, who has devoted special attention to the anatomical features and, in some degree also, to the physiological behaviour of these layers. Von Höhnel discriminates between active and passive types of absciss-phelloid. It is characteristic of a passive absciss-phelloid that the cells of the cork are thick-walled and hard, while the intervening phelloid-layers are, on the contrary, thin-walled and in addition inclined to split along a predetermined plane; in this case it is hygroscopic deformation of the corky layers and dead cortical tissues that leads to disruption within the phelloid layer or along one of its margins. In the case of the active type the relations are reversed, the corky cells being thin-walled and the phelloid layers thick-walled and lignified; here it is the phelloid—with or, more often, without the co-operation of other tissues of the bark—which brings about separation within a zone of cork-cells. This complex subject, however, urgently requires further investigation, especially as regards the mechanics of the actual process of separation. Passive absciss-phelloids have been found by Von Höhnel in *Boswellia papyrifera, Philadelphus coronarius, Myrtus communis,* spp. of *Callistemon* and *Fuchsia,* etc. In *Boswellia papyrifera* each corky zone consists of some ten to fifteen layers of thick-walled flattened elements. The phelloid-cells, on the contrary, are arranged in single layers, and have thin outer and lateral walls; their inner walls are very peculiar, being greatly thickened, strongly lignified and, at the same time, silicified, and provided with inwardly projecting flanges, which run principally in the longitudinal direction. The lateral walls break very easily, so that the separation of a layer of bark always exposes the inner wall of a phelloid layer; owing to their resemblance to the thickened outer walls of an epidermis, these walls are particularly well fitted to form the outer boundary of the thick subjacent layers of cork.

Active absciss-phelloids are described by Von Höhnel as occurring in *Picca excelsa, Araucaria excelsa, Pinus sylvestris, Taxus baccata,* and *Salix europaea.* They almost invariably consist of several layers of very thick-walled cells, which alternate with strips of thin-walled cork.

The age at which our native trees begin to form bark varies greatly according to the species. Scale-bark appears on the trunk of *Pinus sylvestris* and *P. nigricans* between the 8th and 10th seasons (v. Mohl); the corresponding age for our native species of Oak is 25 to 36 (Hartig), for the Alder 15 to 20, for the Lime 10 to 12, and for the genus *Salix* 8 to 10, or even less. Bark formation is long postponed—until the 50th season, or even beyond that—in *Abies pectinata, Carpinus,* the Cork-Oaks, etc. The trunk of the Beech (*Fagus sylvatica*) never forms anything but superficial periderm. Where the formation of bark is tardy or altogether absent, the mechanical strength of the periderm is enhanced by the production of large quantities of sclerenchyma; such periderm is termed "stone bark" by Hartig.

## V. THE DERMAL SYSTEM AMONG THALLOPHYTA.

In those Algae which lead a submerged existence, the superficial layer of the plant-body will obviously possess only those features of a typical epidermis that are connected with its mechanical function or with its action as a light-screen. The mechanical factor finds expression in the presence of thickened outer walls and in the development of numerous vertical partitions in the superficial layer, which increase its power of resisting radial pressure. A typical epidermis is scarcely ever present, chiefly because submerged plants have no need to guard against the risk of drying up, so that the most important of the external agencies which affect the development of an epidermal layer is absent, but also because, in these circumstances, no reason exists why the photosynthetic system should not extend its sway over the outermost layer of the plant-body.

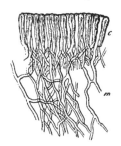

FIG. 42.

L.S. through the superficial region of a pileus of *Polyporus lucidus*. ×190. After De Bary.

In certain Sea-weeds, especially among the RHODOPHYCEAE (spp. of *Chylocladia*), peculiar iridescent plates are found in the superficial cells; these normally occupy the outer walls, but shift on to the lateral walls when illumination is feeble. On account of their behaviour, Berthold[85] believes that these bodies act as screens or reflectors, which intercept part of the light that falls upon the plant. It must be left undecided whether this is the correct interpretation of their significance or not.

According to Berthold, the hair-like appendages which occur in various Sea-weeds also serve for protection against excessive illumination. This view is supported by a variety of facts. In the case of many species which have a tufted habit, hairs are entirely absent from the branches that are hidden in the interior of the tufts; other species, again, are devoid of hairs in winter-time, or where the water is somewhat turbid, or when they grow in the shade of rocks, but are densely hairy if they are brightly illuminated. It is probable, however, that the hairs of Algae are in many instances concerned with a totally different function, namely, the absorption of food-materials.

Among Fungi the degree of specialisation of the dermal tissue varies according to the special functions, and more especially according to the longevity, of the organ concerned. The short-lived fleshy or juicy sporophores which are characteristic of so many Basidiomycetes may be compared with the fugacious floral structures of Phanerogams, as regards the development of their dermal system; the relatively moderate demands which are made upon this system in their case are satisfied by

K

a rather more compact arrangement of the peripheral hyphae, which often also have coloured and more or less mucilaginous walls. The long-lived woody fruit-bodies of various *Polypori* (*P. lucidus*, *P. fomentarius*), on the other hand, are provided with a comparatively tough dermal tissue composed of the thickened ends of hyphae, which are all placed

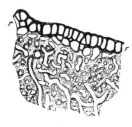

at right angles to the surface, so as to form a palisade-like layer (Fig. 42 c). In a number of cases the surface of the fruit-body also bears trichomes, which assume a variety of shapes and, presumably, perform a corresponding variety of functions. The so-called sclerotia of Fungi also usually possess a very firm thick-walled epidermis; the latter appears to consist of one or more layers of "cells," but is in reality made up of the peripheral branches of hyphae which further inwards constitute the storage-tissue or "medulla." In *Sclerotinia Fuckeliana*, for example, the epidermis ("cortex") of the sclerotium consists, according to De Bary,[86] of one or two layers of isodiametric "cells," which have tough dark-brown membranes and are very firmly united (Fig. 43, *r*). In *S. sclerotiorum*

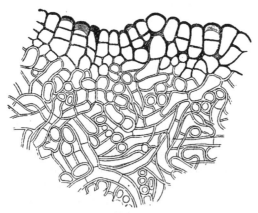

Fig. 44.

Thin T.S. through a ripe sclerotium of *Sclerotinia Sclerotiorum*, showing multiple dermal tissue.

a multiple epidermis is produced by tangential division (Fig 44). In certain species of *Typhula* (*T. phacorrhiza*, *T. gyrans*, *T. Euphorbiae*, *T. graminum*, etc.) the epidermal cells of the sclerotia are tabular or prismatic in form. The smooth or tuberculate outer walls are very thick, whilst the inner and lateral walls remain unthickened; the side walls are also frequently wavy. De Bary long ago remarked upon the close resemblance which exists between such a superficial layer and the thick astomatic epidermis of a Vascular Plant.

It is among Lichens, which so often grow in situations where they are exposed to sudden fluctuations of the climatic conditions, that an elaborately constructed and correspondingly efficient dermal tissue most frequently occurs. In *Usnea barbata* and other fruticose Lichens the dermal tissue consists of a fairly thick layer of closely interwoven hyphae, the walls of which are so greatly thickened as almost to obliterate their cavities. In certain foliose Lichens, again (*Parmelia, Physcia, Sticta, Peltigera,* etc.), it is composed of isodiametric cells with more or less thickened walls, arranged in several perfectly continuous layers. The nearest approach, however, to the condition of a typical epidermis is found in certain gelatinous Lichens (*Leptogium, Obryzum, Mallotium*). Here the dermal tissue usually consists of a single layer of polyhedral or tabular cells, which are in uninterrupted contact with one another; the outer walls of these cells are often somewhat thickened. In *Mallotium Hildebrandii,* which has been examined in some detail by the author, the superficial

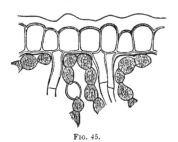

FIG. 45.

Upper surface of the thallus of *Mallotium Hildebrandii,* in vertical section, showing the epidermal layer.

tissue, on the upper side of the thallus, consists of a single layer of isodiametric cells with thin, colourless lateral and inner walls and greatly thickened outer walls. The outer wall is further differentiated into two sharply contrasted layers. The outer and thicker of these is colourless and homogeneous; it is bounded towards the outside by a relatively dense pellicle, which, however, in no way corresponds to a cuticle. The inner stratum is thin and reddish-brown in colour; like the cellulose layer of a typical outer epidermal wall it is continuous with the substance of the lateral walls. Treatment with concentrated sulphuric acid leads to the complete solution of the thick outer layer (which in this respect agrees with the walls of the medullary hyphae), while the thin, brown inner layer and the lateral and inner walls are not dissolved, and do not even swell up. Whereas, therefore, in the ordinary epidermis of a Higher Plant, it is the outermost layer of the outer wall which is chemically specialised (cutinised), the conditions are, in this respect, exactly reversed in *Mallotium.* Whether it is permissible to regard the innermost layer as cutinised, in this case, is another matter. The thickened outer wall is certainly quite pervious to water; for if a drop of water is placed upon a thallus, which has been allowed to dry up until it is quite brittle, the moistened spot at onces becomes soft and pliant. It should be stated that there are no openings in the epidermis through which water could enter. The epidermis of the lower side of the thallus, which is plentifully provided with hairs, has an entirely—or almost entirely—unthickened outer wall; the latter is insoluble in concentrated sulphuric acid, whereas the lateral and inner walls dissolve in this reagent.

## VI. ONTOGENY OF THE DERMAL SYSTEM.

In considering the relations of the various dermal tissues to the three primary apical meristems, namely, the protoderm, procambium and fundamental meristem, discussion must be confined to the epidermis, since cork—as has been shown above—always arises from the phellogen, which is a secondary meristem. In the case of aërial organs the epidermis almost always develops from the protoderm, which was indeed for this reason termed the "dermatogen" by Hanstein. In

Fig. 46.

Epidermal cell from the margin of one of the perforations in the leaf of *Monstera deliciosa*.

certain cases, however, a layer which is fully qualified, both anatomically and physiologically, to rank as an epidermis is ontogenetically derived from the fundamental meristem.[87] This abnormal relation is excellently illustrated by the epidermal cells surrounding the perforations and indentations which regularly occur in the fully-grown leaves of many Aroids. F. Schwarz has described the origin of the perforations in the case of *Monstera deliciosa* as follows. When the young leaves have reached a length of about 8 mm., the meristematic tissue dies and turns brown within certain sharply circumscribed, but not otherwise differentiated, areas situated between the secondary veins. The cells which immediately abut upon these patches of dead tissue repeatedly undergo division in a direction which is tangential with reference to the edge of the patch, and give rise to a tissue resembling periderm. The outermost cells of this marginal tissue then develop into a **secondary epidermis**, which thus forms a narrow strip interpolated between the upper and lower primary epidermal layers. Schwarz gives no details as to the histological features of this secondary epidermis derived from the fundamental meristem of the leaf. The author has therefore carefully compared the epidermis that bounds the edges of the marginal indentations with the primary epidermis of the leaf-margin, especially with regard to the detailed structure of the outer walls. He finds that the agreement between primary and secondary epidermal cells is complete in every particular. The outer walls of the secondary cells are covered by a thin cuticle. The underlying cutinised layer is distinctly stratified, and provided with a reticulate system of flanges which project into the lateral walls. Finally, there is a thin innermost layer of cellulose, so that the outer walls have a perfectly typical structure (Fig. 46). Even the relative thicknesses of the different layers correspond exactly in the primary and in the secondary cells of the margin. The total thickness of the outer wall is about $6\mu$, which is a little more than the thickness of the outer wall of an epidermal cell of the leaf-surface.

If young leaves suffer mechanical injury at an early stage of development, the margin of the wound may be occluded by a secondary epidermal layer, which again arises from the fundamental meristem. A typical case has been described in detail by Pfitzer. If a young leaf of *Peperomia pereskiaefolia* is injured by insect agency or otherwise, it first of all proceeds to form ordinary " wound-cork "; the outermost layer, however, that remains alive beneath the corky covering acquires thickened outer walls, resembling those of a typical epidermis. This process may properly be termed **regeneration of the epidermis.** More recently Vöchting has observed a similar phenomenon in the case of wounded tubers of Kohl Rabi (*Brassica oleracea,* var. *gongylodes*), where the regenerated epidermis reproduces all the essential features of the normal layer; it is a remarkable fact that, in this case, even the stomata are regenerated.

# CHAPTER IV

## THE MECHANICAL SYSTEM.

### I. GENERAL CONSIDERATIONS.

No plant can thrive, or even maintain its existence for long, unless it is provided with arrangements which ensure that the plant-body as a whole is firmly knit together, and that each of its organs is possessed of the requisite degree of mechanical strength.

Those plants which are highly differentiated and hence made up of very numerous and varied parts, are particularly subject to all kinds of mechanical injury. An insufficiently strengthened organ is liable, according to circumstances, to be broken across, to be torn asunder, to be bruised or crushed, and so forth. Every plant must safeguard itself against all the possible forms of injury by which its different organs are threatened; the latter must therefore be constructed so as to withstand, in some cases a transverse or bending stress, in others a longitudinal pull, in others again, longitudinal compressions, or radial pressures, or shearing stresses.

Such "mechanical" arrangements are indispensable to the welfare of all plants, from the most insignificant of Algae to the tallest or bulkiest of trees. The thallus of every filamentous Alga requires protection against the dangers of being torn asunder, of buckling and of breaking across; and it is evident that the mechanical requirements which have to be satisfied by the delicate tufted Algae that grow on stones in swiftly flowing mountain streams are far from negligible. To take a more obvious case, a little consideration will bring to mind the various demands that are likely to be made upon the strength of the different organs of an ordinary forest tree. The trunk has to support the weight of the massive crown, with its extensive array of branches and foliage; it must therefore be built like a pillar or column so as to be capable of resisting longitudinal compression. The branches also have to bear a heavy load; these, however, on account of their oblique or horizontal position, are chiefly exposed to bending stresses. If the fruits of the tree are attached to long pendent stalks—as in the case of the Plane—the latter are subjected to longitudinal tension. In a high wind, the mechanical demands are not only greater in intensity, but also more varied in quality. Trunk as well as branches must then withstand transverse stresses. In order, further, that the leaves may not be quickly torn to shreds, the cells and tissues of which they are composed must be firmly united; in addition, the leaf-margins are, as a rule, specially strengthened so as to prevent laceration. Since, finally, wind tends to uproot the tree bodily, the mechanical strength of the whole root-system is also tested, each portion in turn being stretched taut, like the cables of a ship riding at anchor in a gale.

It is primarily the presence of firm cellulose walls which makes it possible for a plant to preserve a constant shape and to attain to outward differentiation; for the morphogenetic capacities of naked shapeless protoplasts could not possibly become effective without the aid of some resistant material such as cellulose. Similarly, it is through its power of forming cell-walls, and thus acquiring the requisite mechanical strength, that the plant-body and its component organs are enabled to maintain a permanent specific shape. Both unthickened and thickened walls are useful in this respect. Thin walls do not in themselves produce any considerable degree of rigidity. Their mechanical significance depends wholly upon the circumstance that they can be distended by the osmotic pressure of the cell-sap and so become comparatively rigid if the turgor is sufficiently great. A parallel case is that of a soft thin-walled india-rubber tube, which becomes comparatively hard and inflexible if air or water is forced into it under pressure. The fact that herbaceous plants droop on wilting, proves that thin-walled cells are capable, when turgid, of sustaining the weight of the various organs which hang down in the withered plant, and thus of keeping them in the positions proper to their respective functions.

It is obvious, however, that the rigidity due to the turgescence of thin-walled cells is far too dependent upon external conditions, such as the humidity of soil and atmosphere, to ensure the permanent mechanical stability which is a vital necessity in the case of plants of large size. Those conditions—such as protracted high winds—which subject aërial organs to the severest stresses, are the very agencies that also tend to diminish turgor (owing to the increase of transpiratory activity which they cause), and consequently also the rigidity dependent upon turgor. Hence, in order to maintain the mechanical stability necessary to their welfare, large plants are forced to provide themselves with more reliable mechanical arrangements; this they can only do by once more applying the principle of division of labour, or, in other words, by assigning the task of maintaining stability to special tissues. Such **mechanical tissues** must of course be more or less perfectly adapted to their special function; this adaptation finds expression not only (quantitatively) in the very considerable thickness of the walls of mechanical elements, but also (qualitatively) in the tenacity of the material employed in the construction of those walls. In this way one can account for the differentiation of specialised mechanical cells or **stereides**; collectively, these stereides constitute the mechanical system or **stereome**, the construction and arrangement of which must next be considered in detail.

## II. MECHANICAL CELLS.

### A. FORM AND STRUCTURE OF MECHANICAL ELEMENTS.

#### 1. Bast-fibres (Bast-cells).

The term **bast** [ss] is considerably older than the science of vegetable anatomy. From time immemorial this term has been applied to the

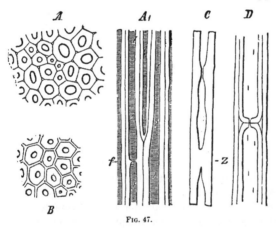

FIG. 47.

Bast-fibres. *A.* Group of bast-fibres from the (fruiting) spadix of *Phoenix dactyli-fera* in T.S. *A₁.* Group of bast-fibres from the (fruiting) spadix of *Phoenix dactyli-fera* in L.S. (cell-cavities shaded); the pointed end of a fibre is seen in the middle, a thin septum at *f. B.* Group of bast-fibres from a twig of *Buxus sempervirens* in T.S. *C.* Single bast-fibre of *Urena sinuata* with irregularly thickened walls; at *Z* the cavity is entirely obliterated (after Wiesner). *D.* Blunt-tipped bast-fibres from the scape of *Allium multibulbosum.*

flexible constituents of the rind of certain trees, which can be used as a binding material; in its original sense at any rate it referred to the conspicuous mechanical properties of the tissues in question. Bast must in fact be regarded as the most widely distributed and important of all mechanical tissues.

Bast-fibres [89] belong to the class of prosenchymatous cells; they are elongated and spindle-shaped, with sharp awl-shaped ends. Their characteristic form is largely produced by independent apical growth on the part of the individual cells; each fibre consequently wedges itself firmly between its neighbours. It is obvious that the cells of the bast can become very firmly attached to one another owing to the large surface of contact which is produced by their peculiar mode of growth; the firm consolidation of the separate elements of a bundle of bast-fibres, in fact, constitutes the physiological explanation of their prosenchymatous form.

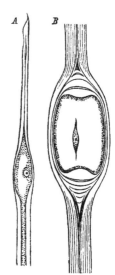

A morphological peculiarity of obscure significance is the local dilatation which is characteristic of the bast-fibres of certain plants (ASCLEPIADACEAE, APOCYNACEAE, *Linum usitatissimum*, etc.); as Krabbe has shown, the swellings are often occupied by the protoplasts, which are encased in secondary cellulose layers (Fig. 48.)

The stereides of the floral scapes of most species of *Allium* are among the few exceptions to the usual prosenchymatous shape; here the mechanical cells, which are about 4 to 5 mm. long, have obtuse ends because their cross-walls are not oblique but strictly transverse, as in the case of parenchymatous cells (Fig. 47 D.)

FIG. 48.

*A.* Local dilatation of a bast-fibre of *Linum perenne. B.* Bast-fibre of *Nerium Oleander*, showing a swelling in which the protoplast has become encysted through the deposition of secondary layers of cellulose.

The length of bast-fibres varies within wide limits; in general, however, it approximates to a definite value in each species. Most frequently it lies between 1 and 2 mm. The linear dimensions of some typical bast-fibres are given in the following table:

| | Length. | Greatest Width. |
|---|---|---|
| *Tilia* sp., - - - - | ·99–2·65 mm. | ·015 mm. (on an average). |
| *Corchorus* sp., | ·8–4·1 mm. | ·016 mm. |
| *Linum usitatissimum*, - | ·20–40 mm. | ·15–·17 mm. |
| *Cannabis sativa*, | 10 mm. and over. | ·15–·28 mm. |
| *Urtica dioica*, | up to 77 mm. | |
| *Boehmeria nivea*, | up to 220 mm. | ·04–·08 mm. |

It will be seen that certain bast-fibres represent the longest known vegetable cells.

The walls of bast-fibres are always more or less thickened; apart from a few exceptional cases the thickening takes place uniformly, and the lumen is very often greatly diminished or even partially obliterated (as in spp. of *Corchorus*). The middle lamella is generally very thin, though in a few cases (e.g. *Buxus*) it may be somewhat thickened. The secondary thickening layers are usually very massive and often exhibit distinct stratification. Striation depending upon inequalities of water-content occurs in the bast-fibres of the APOCYNACEAE and ASCLEPIADACEAE. Very characteristic of typical bast-fibres is the presence of narrow slit-shaped pits, the greatest (transverse) diameter of which is placed longitudinally or obliquely (Fig. 47 A₁, D). The oblique arrangement is the most frequent, and in this case the pits are almost always disposed in left-handed spiral series. The interest of this spiral arrangement of the pits consists in the fact that it indicates a corresponding arrangement of micellar rows. Various circumstances support the view that in prosenchymatous cells the micellae—*i.e.* the ultimate particles of the cell-wall—are united in rows which form ultramicroscopic fibrils.[90] An obliquely pitted bast-fibre may therefore be regarded as an aggregate of exceedingly numerous and delicate fibrillae twisted together into a spiral coil of many turns which surrounds a longitudinal canal consisting of the cell-cavity. It is in fact comparable to a rope, in which the several fibres also follow a spiral course; and, just as a rope owes some of its tenacity to the twisted condition of its strands, so it is probable that the spiral course of the micellar rows in the wall of a bast-fibre is advantageous from a mechanical point of view.

With regard to the chemical nature of the wall it may be stated that the thickening layers generally consist of practically unmodified cellulose. Lignified walls are, indeed, not uncommon, ·and every degree of lignification can be illustrated by various types of bast-fibres. There appears, however, to be no direct connection between the chemical character of hast-fibre walls and their mechanical properties.

It is self-evident that hast-fibres must retain their living contents as long as they are growing and thickening their walls. Trenb has shown that the fibres of ASCLEPIADACEAE, APOCYNACEAE and URTICACEAE are typically multinucleate; the same condition prevails, according to the author's own observations, in *Linum usitatissimum* and in certain LEGUMINOSAE. The presence of several nuclei appears advantageous when the very considerable length of many bast-cells and their active growth in length and thickness are taken into account. The septation which frequently takes place should very possibly be regarded in a similar light; it consists in the appearance of several thin transverse septa which divide the fibre into separate chambers, each corresponding ontogenetically to a cell. These delicate partitions are of no mechanical value, so that the mere formation of walls cannot be the ultimate object of the septation; it is possible, however, as has already been suggested, that the development of septa is determined by the same causes as the above-mentioned multiplication of nuclei. In the adult condition of a bast-fibre the living protoplast is superfluous, and hence usually undergoes degeneration; the cavity then becomes filled with a watery liquid or with air.

Stereides agreeing in every particular with typical hast-fibres are found even among Mosses. There the prosenchymatous attenuation of the ends is sometimes very marked indeed (*e.g.* midrib of the leaf of *Atrichum undulatum*), more so even than in most Monocotyledons; these stereides also sometimes possess slit-shaped pits (*Climacium dendroides*) of longitudinal or oblique (sinistrally inclined) cross-section. Bast-fibres, however, undoubtedly attain to their highest development among Monocotyledons, especially in the GRAMINEAE and CYPERACEAE.

### 2. *Wood-fibres or libriform cells.*

The distinction between bast-fibres and **wood-fibres** (or libriform cells) is not based to any great extent upon tangible morphological differences; so far as it has any justification at all, it depends upon topographical considerations. It has been a very widespread custom, since Sanio first introduced the conception of libriform or " hast-fibre-like " cells, to apply this term to the mechanical components of Dicotyledonous wood. More generally, and perhaps more correctly, one may designate all *intracambial* mechanical cells libriform cells or wood-fibres in contradistinction to the *extracambial* or genuine bast-fibres. Obviously, however, it is quite an arbitrary proceeding to employ location with reference to the cambium as a criterion for distinguishing between different types of cells. Any such distinction rests upon a purely topographical basis; from the point of view of the physiological value—in the present instance the mechanical significance—of a given type of cell, it is a matter of complete indifference whether the elements that conform to this type are situated within the cambial cylinder or outside it. Nevertheless for reasons of convenience the detailed discussion of wood-fibres will be deferred until the final chapter.

### 3. *Collenchyma.*[91]

While hast-fibres and wood-fibres perform the task of strengthening fully grown organs in a perfectly satisfactory manner, neither of these types of mechanical element is suited to the needs of young organs which are still growing in length. For it must be borne in mind

that fully developed fibrous tissues consist of nothing but a framework of dead cell-walls which are no longer capable of growth.    One
need only try to form a mental image of a young elongating organ
strengthened with strands of bast-fibres, in order at once to realise

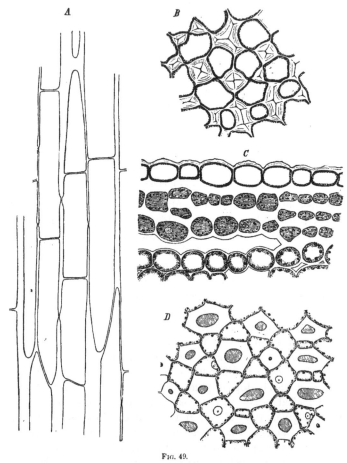

Fig. 49.

Collenchyma.   *A.* Collenchyma from the petiole of *Salvia Sclarea* in L.S.   *B.* Collenchyma from the petiole of *Salvia Sclarea* in T.S.   *C.* Lamellar collenchyma from the
petiole of *Astrantia major* in T.S.   *D.* Tubular collenchyma from the petiole of *Petasites
niveus* in T.S.   (Intercellular spaces shaded in *D.*)

the absurdity of such a combination.   One of two things would
inevitably happen in the circumstances suggested.   Either the bast
would offer such resistance as to prevent any appreciable elongation
of the organ; or else the fibres would give way and so be permanently
deprived of their mechanical usefulness.   Organs which are still

undergoing elongation must therefore make use of a mechanical tissue which is itself capable of active extension by means of growth. The tissue which satisfies these requirements is **collenchyma**. The fact that collenchyma regularly forms the skeletal system of growing organs does not preclude it from serving as the permanent mechanical tissue in many fully grown herbaceous structures (petioles, etc.). On account of its flexibility and power of growth collenchyma is further utilised in motor organs, such as the foliar pulvini of LEGUMINOSAE, the nodes of Grass-haulms, etc.

With reference to the shape of the individual elements, one may distinguish between prosenchymatous and parenchymatous collenchyma cells. The former sometimes attain to a considerable length (up to 2 mm.); like bast-fibres, they are often chambered and provided with slit-shaped, vertically elongated pits.

The walls of typical collenchymatous cells are thickened in a highly characteristic manner. The deposition of thickening layers, namely, instead of being uniformly distributed, is restricted to the edges of the cells, or is, at any rate, much more pronounced along the edges than in any other part of the wall (Fig. 49 B). This peculiarity is closely connected with the fact that collenchyma serves as the mechanical tissue of *growing* organs. The presence of unthickened or at most slightly thickened strips between the thick portions of the walls entails a twofold advantage. In the first place this mode of construction endows the whole tissue with greater plasticity, since the individual cells are not rigidly joined together, a circumstance which is particularly advantageous in connection with the growth in thickness of the organ concerned. The absence of thickening layers from certain regions of the wall also facilitates diffusion between neighbour ing cells, and hence greatly accelerates interchange of nutritive material.

The typical form of collenchyma which has just been described is termed "angular" collenchyma (*Eckencollenchym*) by Carl Müller. Where the thickened edges of the cells contain intercellular passages (*e.g.* spp. of *Petasites* and other COMPOSITAE, Fig. 49 D), the same author speaks of "tubular" collenchyma (*Lückencollenchym*). If the thickening is confined to the tangential walls, and is uniformly distributed over the latter, a "lamellar" collenchyma (*Plattencollenchym*) results (*Astrantia*, *Sanguisorba*, etc., Fig 49 c). If finally the walls are thickened on all sides, and if the innermost pellicle is sharply differentiated, while the middle lamella, on the contrary, is indistinguishable, one may use the term "cartilaginous" collenchyma (*Knorpelcollenchym*). Collenchyma in general may indeed not inaptly be compared to the cartilage of the animal body.

The thickened walls of collenchymatous cells have a high refractive index, and are hence particularly conspicuous in a transverse section on account of their brilliant appearance. When treated with chlor-zinc-iodine, or with iodine and sulphuric acid, they assume a bright blue colour, and are thereby shown to be more closely related even than the walls of bast-fibres to so-called pure cellulose, in respect of chemical composition. According to J. Cohn, collenchymatous walls contain the remarkably high proportion of 60–70 per cent. of water, whereas lignified bast- and wood-elements contain only 20–40 per cent. The water-content rises in the case of the individual cell-wall, as one passes from its inner surface to the middle of each thickened edge. Judging from the mode of contraction under the influence of absolute alcohol and other dehydrating agents, this water of imbibition is chiefly interpolated in the radial direction, less in the tangential and least of all longitudinally. Thus J. Cohn observed, in the case of *Eupatorium cannabinum*, contractions in the radial direction of 22–33 per cent., in the tangential of 7–12 per cent., in the longitudinal of not more than ·5–·75 per cent. The smallness of the longitudinal contraction had previously been noted by Ambronn; it proves that the ultimate particles of the membrane are most firmly knit together in the longitudinal direction, that is to say in the direction in which the mechanical strength of the membrane is most severely tested.

In contrast to bast-fibres, collenchymatous cells always retain their living contents, even where they serve as the permanent mechanical elements of fully developed organs. They also frequently contain chlorophyll corpuscles, though these are generally but few in number.

### 4. *Sclerenchymatous cells [or Sclereides].*[92]

All stereides which are not prosenchymatous will in the present work be termed **sclerenchymatous cells** or **sclereides** (Tschirch). Such cells are largely employed by plants for a number of local mechanical purposes, and accordingly exhibit considerable diversity as regards their morphological characters. More or less isodiametric scleren-chymatous cells (stone-cells or brachysclereides) occur particularly in the cortex of Dicotyledonous woody plants, where they usually owe their origin to the secondary sclerosis of thin-walled parenchyma. They generally serve to increase the incompressibility of the bark; their action may be compared to that of the sand which a mason uses to increase the tenacity of his mortar, or to that of the powdered glass which is added to gutta-percha in order to render it less com-pressible. The cortex of the young twigs of many deciduous trees (*Quercus, Juglans, Carpinus, Betula, Fraxinus,* etc., etc.) contains

numerous separate strands of bast-fibres which are linked together
by tangential arcs of brachysclereides to form a continuous cylinder
—the composite cylinder (*gemischte Ring*) of Tschirch—which is partly
responsible for the inflexibility of the organ.   It often happens that
this composite cylinder—or a mechanical cylinder which is primarily
homogeneous—later becomes interrupted by radial gaps, in consequence
of the tangential tensions that result from the growth in thickness of

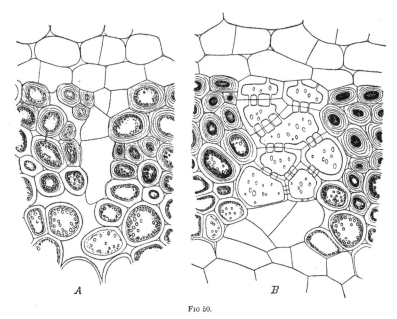

Fig 50.

Development of the composite mechanical cylinder in the stem of *Aristolochia Sipho*.
*A*. Early stage.  On the left a cell of the cortical parenchyma is beginning to penetrate
into a narrow radial cleft ; the larger gap on the right is already partly filled by four
intrusive cells, which are still thin walled.  *B*. A later stage.  A broad gap almost
filled by intrusive sclerenchyma.

the twig; in this event elements of the cortical parenchyma penetrate
into these gaps, the adjacent regions of their walls, protruding into the
clefts while these are still quite narrow.   The growth of the intrusive
cells keeps pace with the widening of the gap ; they divide, thicken
their walls and ultimately become transformed into sclereides, and
thus completely repair the broken cylinder (Fig. 50).   Another well-
known example of isodiametric sclereides is provided by the well-
known gritty particles in the flesh of the Pear and other Pomaceous
fruits (cf. p. 3).   Potonié regards these particles as relics of a stony
shell which was fully developed in some ancestral form ; in their
present condition they may play the same part in the soft flesh of the

fruit as is performed by stone-cells in the case of bark.    A similar
explanation possibly accounts for the presence of stone-cells in the
fleshy tuberous roots of *Paeonia, Dahlia*, etc.    Cells of the same nature
are also found in the pith of certain woody plants.

Rod-shaped sclerenchymatous cells (macrosclereides) with truncated
ends (Fig. 51 B) are also of frequent occurrence in the bark of trees (e.g.
*Cinchona*), and are very prevalent in pericarps and seedcoats, where they
are often arranged at right angles to the surface to form so-called
palisade-sclerenchyma (LEGUMINOSAE, *Cannabis*, etc., Fig. 51 C).    Closely

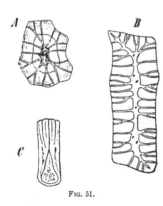

FIG. 51.

*A.* Sclerenchymatous cell from a walnut-
shell.  *B.* Rod-shaped sclereide from the stalk
of a pear.    *C.* Element of the palisade-
sclerenchyma from the testa of *Phaseolus
vulgaris.*

FIG. 52.

Columnar sclereide from
the leaf of *Hakea suaveoleus*
together with part of the
surrounding palisade-tis-
sue.

related, as regards the shape of the cells, are the bone-shaped osteo-
sclereides with dilated and sometimes branched ends, and the **I**- or
**T**-shaped columnar sclereides or buttress-cells which occur, for instance,
in the leaf of *Hakea* (Fig. 52).    Astrosclereides (ophiuroid cells,
spicular cells) are multi-radiate sclereides with their branches attenuated in a prosenchymatous manner.    Cells of this type are found both
in bark (*Abies pectinata, Larix europaea*) and in the green mesophyll of
certain leaves (*Camellia, Olea, Fagraea, Dammara, Sciadopitys, Gnetum,*
etc., Fig. 54); in the latter case they help to stiffen the leaf, being
indeed partly responsible for the tough leathery nature of that organ.
Mention must finally be made of the unbranched sclereides, resembling
bast-fibres in shape, but differing from them in the circular [cross-
sectional] outline of their pits, which occupy the central axis of the
glandular hairs borne by the leaf-blades and petioles of *Begonia
imperialis* (Fig. 53).    The coarse villi or shaggy hairs of the MELA-
STOMACEAE often contain whole bundles of sclereides, which penetrate

below into the mesophyll, and thus form a sort of root for the firmer attachment of the emergence.

Typical sclerenchymatous cells, such as occur, for instance, in bark, in the flesh of the Pear, or in the endocarp of a drupe, usually possess greatly thickened walls, which are also con-
spicuously stratified; frequently several concentric compound strata can be clearly distinguished. The secondary layers are traversed by numerous pits, which are often branched, and which are always circular in cross-section (Fig. 51 A). The walls are further commonly yellow or reddish-brown in colour, and strongly lignified. The protoplasts usually degenerate; the cell-cavities, which are often greatly reduced, contain a watery liquid or a reddish-brown granular substance.

### B. STRENGTH AND ELASTICITY OF MECHANICAL CELLS.

In the preceding general discussion reference has repeatedly been made to the harmony which exists between the histological structure of mechanical cells and their functions; in the present section it will be demonstrated at some length, that the chief physical properties of the cell-walls of mechanical elements are such that the cells in question are in this respect also fully qualified to act as the skeletal elements of the plant-body. Schwendener [93] was the first to undertake an experimental investigation of the elasticity and tensile strength of mechanical cells; his experiments, which deal principally with bast-fibres, are fully described in his classical treatise on the subject. Subsequently Ambronn made a detailed study of the strength and

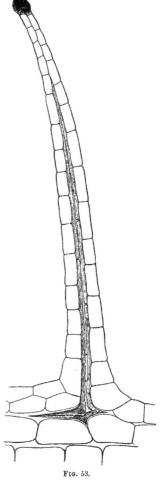

FIG. 53.

A shaggy glandular hair from the petiole of *Begonia imperialis*. The stalk is supported by a single large L-shaped sclereide.

elasticity of collenchyma. Data bearing on these questions have also been obtained by von Weinzierl, F. Haberlandt (father of the author), and others.

The experimental method employed by Schwendener is quite straightforward. Strips of tissue, 150-400 mm. in length and 2-5 mm. in breadth, are prepared from leaves, or from stems containing a large proportion of bast; one end of each strip is firmly fixed in the jaws of a vice, while to the other extremity a clamp is fastened to which weights can be attached. Simple accessory contrivances serve to measure the extension produced by a given load, and the subsequent contraction which takes place when the weight is removed. From each strip which has been tested in this way transverse sections are prepared; these are drawn to scale under suitable magnification with the help of a camera lucida, and the cross-sectional area of the resistant elements is calculated from this drawing. Ambronn employed a similar method to determine the tenacity and elasticity of collenchyma; he took special precautions with a view to estimating the mechanically effective cross-sectional area as accurately as possible. The transverse section of each collenchymatous strand was drawn upon paper of uniform thickness with the aid of a camera lucida; the drawing was then cut out and weighed. Next the areas corresponding to the cell-cavities were carefully removed; from the weight of the remaining paper net-work the cross-sectional area of the walls could then be readily calculated.

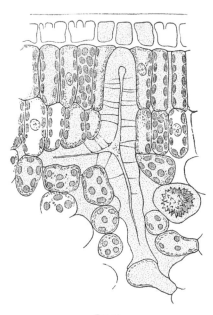

FIG. 54.

Part of a T.S. through the leaf of *Camellia japonica*, showing one of the multiradiate sclereides.

The subjoined table comprises a series of data selected from the measurements of Schwendener, Weinzierl, and Firtsch, to which are appended, for the sake of comparison, the corresponding values for certain metals; some of the latter (which are quoted from the fifth edition of Weisbach's *Ingenieur-und Maschinen mechanik*) are included in Schwendener's treatise on the "mechanical principle."

| PLANT OR METAL. | Modulus of Elasticity[94] in kg. per sq. mm. | Breaking Strength in kg. per sq. mm. | Elongation at the Elastic Limit per 1000 linear units. |
|---|---|---|---|
| Dianthus capitatus, - | 14·3 | | 7·5 |
| Dasylirion longifolium, - | 17·8 | 21·6 | 13·3 |
| Dracaena indivisa, | 17 | 21·8 | 17 |
| Phormium tenax, | 20 | 25 | 13 |
| Hyacinthus orientalis, | 12·3 | 16·3 | 50 |
| Allium Porrum, | 14·7 | 17·6 | 38 |
| Lilium auratum, | 19 | | 7·6 |
| Nolina (Pincenectia) recurvata - | 25 | | 14·5 |
| Papyrus antiquorum, - | 20 | | 15·2 |
| Molinia coerulea, - | 22 | | 11 |
| Secale cereale, - - | 15-20 | | 4·4 |
| Cibotium Schiedei, - | 18-20 | | 10 |
| Polytrichum juniperinum, Stem, | | 7·5 | |
| Do., Seta, | | 11·5 | |
| Silver, - - - - - | 11 | 29 | |
| Copper (Wire), - - | 12·1 | | 1 |
| Brass (Wire), - - - - | 13·3 | | 1·35 |
| Wrought Iron, - | 13·13 | 40·9 | ·67 |
| Steel (German), | 24·6 | 82 | 1·2 |

A consideration of the above figures leads to conclusions which are no less interesting than surprising. The tensile strength of bast below the limit of elasticity is quite remarkable. It lies in general between 15 and 20 kg. per sq. mm., and is thus equal to that of wrought iron; the bast of *Nolina recurvata* vies even with steel in this respect. Bast, however, differs from metals in two important particulars. For one thing it is far more extensible than any metal; the extension at the limit of elasticity varies from 1 per cent.-1·5 per cent. for bast, while for metals the corresponding value is on an average less than ·1 per cent. Further, there is in the case of bast only a very small difference between the limits of elasticity and cohesion; in other words, a very slight increase of tension over that which just suffices to produce a permanent elongation will cause the fibres to break. In the case of metals the difference is very much larger; the breaking strength of a rod of wrought-iron, for example, is very nearly three times as great as its modulus of elasticity. Schwendener remarks in this connection that "Nature has evidently concentrated her attention upon the tensile strength [of mechanical cells], and quite properly so; for great breaking strength would be of no value in the case of structures which cannot be stretched beyond the elastic limit without suffering injury."

There appears to be no definite correlation between the strength of bast-fibres and the degree of lignification of their walls. According to Sonntag, however, there does seem to be some connection between the extensibility of bast-fibres and the angle of inclination of the rows of micellae of which the walls are composed. In all fibres which are very extensible, the angle between the direction of the oblique micellar rows (fibrillae) and the longitudinal axis of the cell is very large.

Ambronn's experiments have shown that the absolute strength of collenchyma is little inferior to that of bast; on an average collenchymatous walls break under a strain of not less than 10-12 kg. per sq. mm. The essential difference between collenchyma and bast consists in the fact that the limit of elasticity is much lower in the case of the former tissue, a load of 1·5-2 kg. per sq. mm. sufficing to produce a permanent elongation. This physical peculiarity of collenchymatous walls is of the utmost importance in connection with the special function of this tissue. It is only by virtue of this property that collenchyma is able to " furnish mechanical support during intercalary growth, without preventing longitudinal extension."

In order fully to appreciate the excellent quality of the materials used in the construction of the skeletal system of plants, it is necessary to compare their mechanical constants with those of ordinary cellulose walls. According to Schwendener the tensile strength of the membranes of thin-walled parenchymatous medullary and cortical cells of various young Dicotyledonous stems amounts on an average to about 1 kg. per sq. mm. There is, therefore, no great risk of error in assuming that the breaking strain never exceeds 1 kg. per sq. mm., in the case of fully imbibed cell-walls composed of relatively pure cellulose and not specialised for mechanical purposes. For the walls of stereides, the breaking strain is from ten to twenty-five times as great.

The statements which have been made so far apply only to mechanical strands that have been freshly removed from the living plant, and that hence contain their normal proportion of water of imbibition. The elasticity and strength of tissues are naturally affected by the removal of water. Schwendener remarked that the absolute strength of bast increases with diminishing water-content, while its elasticity is correspondingly diminished. More recently Weinzierl has come to similar conclusions on the basis of elaborate researches. The point may be illustrated by an extract from one of Weinzierl's tables, referring to *Dasylirion longifolium* :

|  | Percentage Water-content of the Leaf. | Modulus of Elasticity in kg. per sq. mm. | Breaking Strength in kg. per sq. mm. | Elongation at the Elastic Limit per 1000 linear units. |
|---|---|---|---|---|
| I. | 45 | 17·8 | 21·6 | 13·3 |
| II. | 11 | 23·2 | 26·7 | 10·4 |

The physiological interest of these alterations depends entirely upon the circumstance that completely dried-up fibrous strands occasionally occur as constituents of bark.

### III. PRINCIPLES GOVERNING THE CONSTRUCTION OF THE MECHANICAL SYSTEM.[95]

One of the most powerful of the considerations that influence an engineer (or architect), in deciding upon the plan of construction of a bridge (or roof), is the desire to economise material. In other words, a design is not satisfactory unless it ensures the maximum of strength and solidity with the minimum expenditure of material. To this end the resistant elements of the structure must be arranged in a particular manner in accordance with approved mechanical principles. An organism is confronted with a precisely similar problem when it is called upon to provide itself with a sufficient degree of mechanical strength. In this case also the desired result must be obtained with the smallest possible expenditure of material; hence the principles of construction to which the human builder adheres are precisely those which control the morphogenetic activities of organisms. The very same mechanical principles that are embodied in a modern railway bridge of bold and elegant design, were expressed perhaps with even greater perfection hundreds and thousands of years ago in the skeletal systems of the plants of former geological periods.

In the succeeding sections of this chapter we shall become more closely acquainted with the most important of the principles that govern the construction of the mechanical system.

#### A. INFLEXIBILITY.[96]

If a straight girder be supported at both ends and weighted in the middle, it will bend to a greater or less extent according to the magnitude of the load; as a result of this curvature the upper side of the girder must be slightly shortened, while the lower side is correspondingly lengthened. The shortening, of course, produces a state of compression, the lengthening a state of tension, in the corresponding halves of the girder; the effect, in either case, is most pronounced at the upper and lower surfaces, while on approaching the centre of the cross-section from either side the corresponding tension gradually diminishes, then falls to zero and finally passes over into the tension of opposite sign. The layer of zero tension is known as the neutral surface.[96a] In order, therefore, that a girder may possess maximum inflexibility, the available material must be concentrated in the regions of greatest tension, that is to say, near the upper and lower faces. A typical girder thus comes to consist of upper and lower **flanges**, which are firmly joined together by a connecting piece, or **web**. The cross-section of such a girder usually resembles an I or a combination of two T's ( $\mathbf{I}$ ), the horizontal strokes representing the flanges and the vertical line the web (Fig. 55 A).

The strength of a girder depends, *ceteris paribus*, upon the strength of its flanges. It also increases as the distance between the two flanges becomes greater, because the tension due to the load varies inversely as the distance between the flanges. Since the web has to bear only a small proportion of the total strain, it may be considerably lighter in construction than the flanges which it links together. In the girders of bridges the web usually takes the form of a lattice-work or honey-comb structure; if a girder is composed of more than one kind of substance, the inferior material is used in the construction of the web. Similarly in the plant the flanges of a girder are always composed of mechanical cells, whereas the web may consist of vascular tissue or of parenchyma.

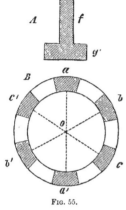

Fig. 55.

*A*. Cross-section of an I-girder; *g*, upper flange; *g'*, lower flange. *B*. Cross-section of a compound girder, comprising three I-girders, *aa'*, *bb'*, *cc'*, with their common neutral axis at *o*.

A simple **I**-girder resists bending only in a single plane. In order to obtain a structure which is inflexible in several planes—*i.e.* in several directions at right angles to the long axis of the organ—several **I**-girders must be combined in such a way that they have a common neutral axis [96a] The accompanying diagram (Fig. 55 B) shows such a combination comprising three **I**-girders, *aa*, *bb*, and *cc* being the respective pairs of flanges. The broken lines represent imaginary webs, which are superfluous here because adjoining flanges are firmly joined together by means of tangential connections. In this way the two flanges pertaining to each girder are connected just as securely as if the web were actually present. Compound girders of this kind are exceedingly widespread in vegetable organs. By supposing a number of girders to become so closely crowded in a ring, that their flanges undergo lateral fusion, one arrives at the conception of a hollow cylinder or tube, likewise a very prevalent type of mechanical construction among plants.

Each of the numerous flanges of a radially inflexible compound girder may be exposed alternately to compression and tension. In order that the individual flanges may not warp or buckle, when compressed, before the elastic limit of the tissue is reached, each is also constructed so as to resist bending, its cross-section resembling that of an entire girder, with the outline of an **I**, a cross or ring, or other appropriate figure. Such an adaptive modification, whereby the individual flanges of the main girders are themselves converted into secondary girders, is also frequently found in the plant-body.

In the case of radially inflexible compound girders or hollow cylinders, the continuous flange, or wall of the tube, must not be too thin, as otherwise it is liable to collapse under the transverse tensions which result when the structure is subjected to bending stresses ; such a collapse is always preceded by an alteration of the cross-sectional outline of the girder, and even this minor deformation must obviously be avoided as far as possible.   The thickness of the wall or continuous flange must therefore not fall below a certain fraction of the total diameter of the cylinder or compound girder.   Schwendener estimates the minimum thickness of the wall which is admissible in such a radial compound girder at about one-seventh or one-eighth of the total diameter.   A thinner wall is only permissible where special stiffening arrangements are provided, which help to preserve the cross-sectional outline of the girder.   Later on we shall become more intimately acquainted with such special reinforcing devices.

### B. INEXTENSIBILITY.

The degree of inextensibility of a structure depends entirely upon the cross-sectional area of its resistant elements ; the disposition of these elements is therefore theoretically a matter of indifference.   The theory, however, assumes that the stretching force is uniformly distributed, a condition which is not likely to be fulfilled unless the area over which the resistant elements are distributed is very small.   If, therefore, the resistant elements were scattered, an unequal distribution of tensions would be very liable to occur, and individual strands might break, with the result that the strength of the whole structure would be seriously impaired.   The more closely aggregated on the other hand the resistant elements are, the more uniform the distribution of tensions is likely to be ; hence the aggregation of the mechanical cells into a single compact and solid mass is the most advantageous arrangement for an organ which has to withstand longitudinal tension.

### C. INCOMPRESSIBILITY.

#### 1. *Resistance to longitudinal compression.*

If an erect prismatic or cylindrical body, with a longitudinal axis greatly exceeding its transverse diameter, be fixed at its lower extremity and weighted at its upper end, in such a manner that the load acts in the direction of the longitudinal axis, it will suffer longitudinal compression.   The middle line of such a body can, however, only remain vertical, if the line of action of the load coincides *exactly* with the longitudinal axis ; in this event the load will give rise to the same amount of compression at every point within the body, and thus produce

a definite pressure per unit cross-section. In nature this ideal condition can scarcely ever be realised; for almost inevitably some accessory
circumstance, such as the effect of a lateral pressure, or a slight
asymmetry of construction, will produce a small deflection, which is
thereupon at once accentuated by the action of the load. Thus any
columnar structures which occur in the plant-body must be constructed
so as to withstand *bending*, and the mechanical principles discussed in
the section dealing with inflexibility must apply equally in their case.

### 2. *Resistance to radial pressure.*

If a cylindrical body is to withstand radial or crushing pressures,
it must evidently have its resistant elements arranged in the form of
a strong peripheral shell. It will be seen later on that all subterranean
and submerged organs have to be protected in this way against the
radial pressure of the surrounding soil or water. The pressure-resisting
hollow cylinder or tube must not of course be confused with the similar
structure which affords protection against bending strains. As already
explained, one may imagine the latter to be produced—from a purely
mechanical point of view—by the lateral fusion of a number of isolated
girders; this explanation does not, however, account for the origin of
the pressure-proof hollow cylinder, which cannot in any circumstances
be replaced by a ring of isolated girders.

### IV. ARRANGEMENT OF MECHANICAL TISSUES.

A detailed examination of the mechanical system in the various
organs of the plant-body, and particularly in the stem, leaf and root,
makes it quite evident that the disposition of the mechanical tissues
conforms exactly to the structural principles which have been discussed in the preceding sections. There is indeed a remarkably close
agreement in this respect between theoretical expectation and observed
fact. In no department of physiological anatomy are more striking
and obvious adaptations to be met with, mainly because there is no
section of physiology in which the physical laws that underlie the
physiological adaptations have been determined with greater complete-
ness and precision.

Before the mechanical system can be considered in detail, a few
words must be devoted to the general structure and arrangement of
mechanical tissues.

The specialised mechanical elements, namely bast-fibres and collenchymatous cells, are generally united to form more or less massive
strands or bundles, which are as a rule elongated in the same direction
as their component cells. The disposition of mechanical strands within

the organ which they serve to strengthen almost invariably reveals the dominating influence of the mechanical principle. Incidentally, of course, the mechanical strands also display relations to other anatomico-physiological systems, and these relations must be taken into account, before one can comprehend every detail in the structure and arrangement of the mechanical system. These secondary relations will be referred to again later on. At the present stage it must suffice to draw attention to the frequent association of fibrous strands with vascular bundles, a phenomenon the true significance of which was long misunderstood. It is very usual, namely, for a fibrous strand to apply itself closely to a vascular bundle, the resulting complex morphological entity being termed a fibro-vascular bundle. In this case the fibrous strands, when seen in transverse section, appear as semi-lunar or sickle-shaped borders or partial sheaths embracing one or both halves of the vascular strands, or as complete sheaths which entirely surround the bundles. Schwendener was the first to point out that the association of bast with vascular tissue in these fibro-vascular bundles is not the effect of some obscure morphological law, but really represents a case of physiological opportunism.

Since a vascular bundle is in part composed of very delicate cells, it requires to be protected by some coarser and more resistant tissue, and hence tends to associate itself with mechanical strands. A gutter-shaped fibrous sheath acts like a splint or greave, behind which the delicate parts of the vascular bundle find the necessary shelter. Vascular bundles are, on the other hand, particularly well fitted to form the web between two flanges of bast, more especially in the case of small girders, so that the association of fibrous and vascular tissues appears advantageous from this point of view as well. The term **mestome**—signifying " filling tissue "—was applied by Schwendener to unmixed " vascular " bundles,—that is, to conducting strands which are entirely free from any admixture of mechanical elements,—" in order to give expression to the fact that vascular bundles are so frequently interpolated between the flanges of girders, or else embedded in a complete cylinder of bast, or finally scattered throughout a continuous mass of wood-fibres, as in woody Dicotyledonous stems, where the conducting strands seem to occupy the gaps and interstices of a mechanical framework." In the present work it will be regularly used in referring to these relations between the vascular and mechanical systems.

### A. THE STRUCTURE OF INFLEXIBLE ORGANS.

Most aërial plant organs are constantly exposed to bending stresses. In cylindrical structures such as Grass-haulms, floral axes, etc., the bending forces may act in any plane at right angles to the longitudinal axis; in such cases, therefore, a system of compound girders, such as has been described above, is required to produce the desired radial inflexibility. In these circumstances the fibrous or collenchymatous strands may in general be expected to arrange themselves in one or more circles near the periphery. The centrifugal tendency of the mechanical system will be most pronounced, where the dominance of the mechanical principle is least affected by the claims of other tissue-systems.

Flat organs, such as foliage-leaves, are chiefly exposed to bending stresses in the plane which is perpendicular to their greatest surface; inflexibility is therefore principally required in that plane. The desired result is attained by arranging the girders in transverse series.

Schwendener has shown that the details of the mechanical construction of inflexible organs vary in an extraordinary degree, especially among Monocotyledons. With the object of reducing the numerous varieties of arrangement to some sort of order, Schwendener undertook a systematic classification of the various types of construction, collecting them into a number of comprehensive classes called mechanical systems. In the present treatise no attempt will be made to deal with all the twenty-eight types of inflexible organ, which Schwendener recognised and illustrated by instructive examples among Monocotyledons alone. It will only be necessary to demonstrate the correctness of Schwendener's attitude by selecting a few typical instances for fuller consideration. Convenience will decide whether any particular example is chosen from among Monocotyledons or Dicotyledons; it will, further, be regarded as a matter of indifference whether the mechanical tissues selected for consideration consist of genuine bast, of wood-fibres or of collenchyma.

### 1. *Cylindrical organs.*

We shall follow Schwendener in including among cylindrical organs all structures which are liable to bend in any direction at right angles to their longitudinal axis, irrespectively of the question whether they are truly cylindrical—in a geometrical sense—or polyhedral in shape.

### (a) *First system. Subepidermal girders.*

The two types of arrangement belonging to the first mechanical system are illustrated by the petioles and inflorescence-axes of certain Aroids—as regards fibrous strands,—and by the stems of LABIATAE and

the petioles of some UMBELLIFERAE—as regards collenchyma.   In all cases
the mechanical strands are disposed either in a single zig-zag series or in
two concentric circles.   The adjoin-
ing diagram (Fig. 56 B) depicts
the arrangement of the fibrous
strands in the petiole of *Colocasia
antiquorum*:  here the strands
vary in size and are regularly
associated with mestome-bundles.
The four-sided stem which is
characteristic of the LABIATAE
(Fig. 56 A) is strengthened by
two diagonally placed collenchymatous girders;  this is the
simplest possible type of mechanical construction in the
case of a cylindrical inflexible
organ.

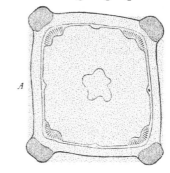

In this connection mention
may also be made of the
occurrence of a circle of fibrous
strands in the yearling twigs
of many woody Dicotyledons
(*Fagus, Betula, Acer campestre,
Cornus sanguinea, Nerium Oleander, Paulownia*, etc.);  at a
later stage, when the output
of secondary wood suffices for

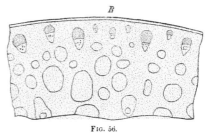

FIG. 56.

Subepidermal girders.   *A*. T.S. through an internode
of *Lamium album*; the collenchymatous strands in the
angles of the stem form a pair of diagonally placed
girders.   ×12.   *B*. Part of a T.S. through the petiole of
*Colocasia antiquorum*; the fibrous strands are accompanied by mestome-bundles.   ×30.   (After Schwendener.)   Both figures diagrammatic.   (In all the diagrams
dealing with the mechanical system, the fibrous and collenchymatous strands are distinguished by shading.)

mechanical purposes as well as for conduction, this primary mechanical
system is cast off as a result of cork-formation.

(*b*)  *Second system.   Composite peripheral girders.*

The second form of mechanical system is especially prevalent among
the CYPERACEAE and JUNCAGINACEAE.   The six types included in this
system by Schwendener comprise a great variety of structural forms ·
only three of these need be described here.

The haulm of *Trichophorum germanicum* is supported by a sub-
epidermal circle of **I**-girders—most often five or six in number—
which are placed at regular intervals from one another.   The outer
flange of each girder abuts directly against the epidermis, while the
inner flange forms a semi-lunar sheath around the mestome-strand
which represents the web of the girder (Fig. 57 A).   There is in addition
a series of more deeply seated subsidiary girders which alternate with

the main girders.   In *Cyperus vegetus* (Fig. 57 B) the disposition of the
mechanical strands is very similar.   Here again the haulm contains
a number of subepidermal **I**-girders arranged in a fairly regular
circular series.   In this case, however, the two flanges of each girder
are so widely separated, that the mestome-strand no longer suffices
to fill the whole of the intervening space, but has to be supplemented
by parenchymatous cells interpolated between the outer flange and

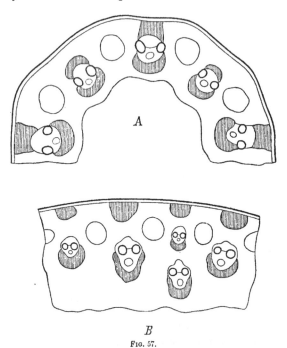

FIG. 57.

Composite peripheral girders.  *A.* Half of a T.S. through the haulm of *Tricho-*
*phorum germanicum*;  the web of each girder consists of a mestome-strand.   × 100.
*B.* Part of a T.S. through the haulm of *Cyperus vegetus*;  the web of each girder
consists of a mestome-strand (half-encircled by the semi-lunar inner flange) together
with some parenchyma (which separates the mestome-bundle from the outer flange).
× 66.

the mestome-strand, which itself adheres closely to the inner flange.
The parenchymatous cells employed for this purpose are specialised
photosynthetic elements (palisade-cells); the whole girder thus assumes
the form shown in the adjoining figure (Fig. 58).

A much more complicated and elegant type of construction is met
with in the haulms of *Juncus glaucus*.   Here again there is a ring of
regularly spaced main girders, the outer flanges of which consist of
stout fibrous strands (Fig. 59 A), while the inner flanges are composed
of sickle-shaped fibrous sheaths embracing the large mestome-bundles

which lie on the same radii.  The space between the two flanges of each girder is occupied by the following tissues—in centrifugal order ; first, a large mestome-strand, next the small outer fibrous sheath, then thin-walled parenchyma, then a wide air-passage, beyond which parenchymatous tissue once more appears.  The web here exhibits two unexpected features, the more remarkable of these being the occurrence of an air-passage, which must weaken the radial connection between the two flanges, notwithstanding the presence of a loose "stellate tissue" in its interior.  It must not be forgotten, however (cf. above, Section III., A.), that there is no need for a continuous web, provided that the tangential connection between the adjacent girders, or groups of girders, is sufficiently strong. From this point of view it must be admitted that the inclusion of the large air-passages in the web of the girders is not only free from objection, but actually advantageous, inasmuch as the passages interfere in this way as little as possible with the tangential con-tinuity of the mechanical system.  The second outstanding peculiarity of the

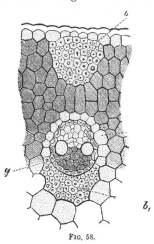

FIG. 58.

A single composite peripheral girder from the haulm of *Cyperus vegetus*.  b, outer flange of bast ; $b_1$, inner flange ; g, mestome-bundle ; s, parenchymatous bundle-sheath.

girders in *Juncus glaucus* is the presence of small fibrous sheaths on the *outer* sides of the mestome-strands.  It is not at once obvious why plastic materials are expended in the production of these outer sheaths, instead of being utilised to strengthen the inner flanges.  It is necessary, however, to recollect that fibrous strands may have a strictly local mechanical value ; in the present instance, for example, the outer sheath serves to protect the delicate protein-conducting portion of the mestome-strand.

The somewhat wide intervals between the successive main girders of *Juncus glaucus* are occupied by symmetrically arranged subsidiary girders, with webs consisting of mestome-strands and nothing else. Each group of subsidiary girders, which " forms a veritable phalanx of four to six members," serves to give the necessary rigidity to the comparatively wide partitions that intervene between the main girders.

*(c) Third system.   Subcortical fibro-vascular strands.*

In the types belonging to the third system the fibrous bundles are, as it were, repelled from the surface, and thus forced into a more central

position, by the intrusion of sub-epidermal parenchyma.  The general
disposition of the mechanical strands nevertheless remains decidedly
peripheral.  In contrast to the preceding systems, the success of which
depends more upon the advantageous arrangement of the mechanical
tissue than upon its massive development, the present system is chiefly
remarkable for the large size of the fibrous strands, which can thus
readily provide the requisite degree of inflexibility, provided they can

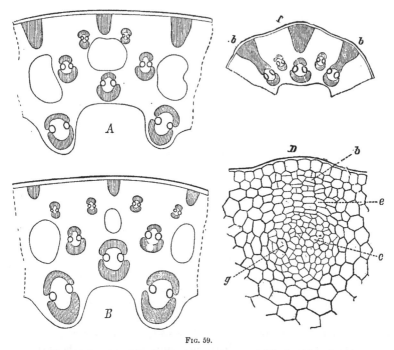

Fig. 59.

Mechanical system of the haulm of *Juncus glaucus*.  *A*. T.S. of aerial portion of
haulm.  *B*. T.S. of subterranean portion.  *C*. T.S. 1·5 cm. below the apex of the
haulm; at *b* the subepidermal fibrous strands have fused with the outer fibrous
sheaths of the subtending mestome-bundles.  Development of a composite peripheral
I-girder from a common procambial strand; *b*, bast-procambium; *c*, mestome-
procambium; *g*, protoxylem vessel; *e*, the portion of the procambial strand which
will give rise to parenchyma.

find a place anywhere near the periphery.  The haulms of Bamboos,
and the stems and inflorescence-axes of Palms, are cases in point.  In
Palms the fibrous sheaths of the mestome-bundles frequently become
fused tangentially into an irregular ribbon-shaped strip; in this case
they collectively form a hollow cylinder (Fig. 60), with a wall which
is thin in comparison with the diameter of the whole organ.  The
more deeply-seated mestome-bundles have relatively slight fibrous
sheaths, which serve solely for purposes of local protection, without

affecting the inflexibility of the stem.  Those fibrous strands, on the other hand, which lie on the outside of the main stereome-cylinder, probably help to prevent the cortex from splitting or being stripped off when the stem is bent.

A very remarkable variety of the third mechanical system is the "corrugated" type of stereome—to use Potonié's term—which is found in the stems of Tree Ferns (*Cyathea, Alsophila*), belonging to the tropical family CYATHEACEAE.[97]  Here the mestome-bundles take the form of ribbons or sheets enclosed in thick stereome-sheaths.  Each ribbon is curved in a horizontal plane, being as a rule roughly **V**-shaped in cross-section, with the sharp edge facing the centre of the stem.  Since such a curved sheet opposes a much greater resistance to bending strains than a flat sheet of equal thickness, the advantages of this corrugated mode of construction—which is, of course, widely employed in engineering and building practice—are sufficiently obvious.

FIG. 60.

Sub-cortical fibro-vascular strands.  Peripheral portion of a T.S. through the spadix of *Phoenix dactylifera*.  Each of the small peripheral fibrous strands contains an excentrically placed mestome-bundle.  The sub-cortical mestome-strands are provided with stout external fibrous sheaths which in places unite to form irregular ribbons or sheets.  Towards the centre of the section the fibrous sheaths of the mestome-bundles become gradually weaker.

(*d*) *Fourth system.   Simple hollow cylinder or tube of stereome with embedded or apposed mestome-strands.*

The fourth system is so largely employed among Monocotyledons that it may be regarded as the dominant type in this group of plants. It is particularly prevalent in the LILIACEAE (Fig. 61 A, B), IRIDACEAE, ORCHIDACEAE and DIOSCOREACEAE; but many Dicotyledons also (*e.g.* the CHENOPODIACEAE, CARYOPHYLLACEAE, GERANIACEAE, PRIMULACEAE and CUCURBITACEAE, and the genera *Phytolacca, Aristolochia, Papaver, Geum, Saxifraga, Armeria, Plantago, Lonicera*), conform to this scheme, so that there are obvious points of contact between the two great divisions of Angiosperms, in respect of their skeletal arrangements.

In Monocotyledons the mestome-strands may be apposed to either side of the fibrous cylinder, or they may be completely embedded in the latter.  Among Dicotyledons a series of transitions may be observed leading from the completely extra-cambial to the completely intra-cambial form of mechanical cylinder. It is customary to discriminate between the extreme types, and to restrict the term " bast-cylinder or -ring " to the extra-cambial variety; it has, however, already been pointed out that this distinction rests upon a purely topographical

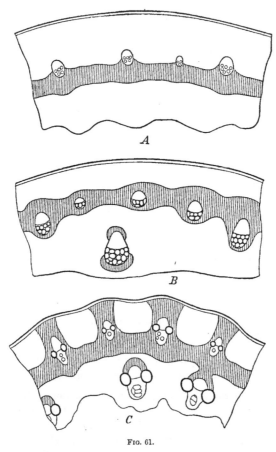

Fig. 61.

Mechanical cylinders of various Monocotyledons. *A*. Sub-cortical fibrous cylinder of *Allium odorum*, with externally opposed mestome-bundles. *B*. Sub-cortical fibrous cylinder of *Convallaria verticillata*, with embedded mestome-bundles. *C*. Sub-cortical fibrous cylinder, reinforced by sub-epidermal girders, of *Molinia coerulea*.

basis. An interesting intermediate condition is found in the stem of *Astrantia major*, where the leptome-portions of the vascular bundles are embedded in the stereome-cylinder, whereas the hadrome-portions are merely apposed to its inner face. The leptome and hadrome components of the same bundle are thus separated by a strip of mechanical tissue (Fig. 62). Examples of typical intra-cambial mechanical cylinders are furnished by the stems of *Tropaeolum majus, Impatiens nolitangere, Centranthus ruber, Sedum reflexum,* etc.

In a purely mechanical sense it is permissible to regard the continuous bast- or wood-fibre-cylinder as the product of the lateral coalescence of a circle of closely crowded fibrous flanges; but it would be a mistake to assume that the ontogenetic development of the bast-cylinder actually involves any such process of fusion. There is no reason to suppose that any stereome-cylinder has arisen in this way, either in the ontogeny of the individual or during the evolution of a species. As Schwendener has pointed out, the bast-cylinder is "an independent structure, the shape and position of which are not

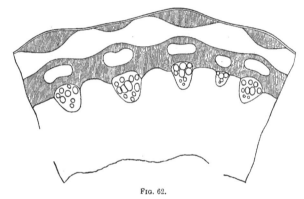

FIG. 62.

Mechanical system of the stem of *Astrantia major.* A subcortical fibrous cylinder, with embedded leptome- and internally apposed hadrome-strands is reinforced by subepidermal girders of collenchyma.

appreciably affected by other tissues. Its structure and arrangement are entirely determined by the mechanical principle. It is not in the least subject to the influence of the peripheral vascular bundles; on the contrary, it is the latter that adapt themselves to the peculiarities of the mechanical cylinder, apposing themselves to its outer or inner face, or becoming completely embedded in its interior, for the sake of the shelter or support that it affords" (Fig. 61 A, B).

The distance of the fibrous cylinder from the surface of the stem rarely exceeds one-twentieth, and may be as little as one-fiftieth, of the total diameter of the organ. On its cortical side it is almost always very thick-walled and hence as a rule sharply differentiated. On its inner side, on the contrary, it often passes insensibly into the medullary parenchyma, the cells becoming gradually shorter, wider and less thick-walled, while the cross-walls at the same time approximate more and more to the strictly transverse position.

M

The disposition of the mechanically effective elements in the form of a hollow cylinder or tube is the simplest, and hence also the most primitive, arrangement of the skeletal system in inflexible organs. A consideration of the mechanical arrangements of the Lower Plants (Bryophyta and Thallophyta) brings this fact out very clearly. In the erect stems and setae of Mosses the mechanical system takes the form of a typical subepidermal hollow cylinder; on its inner side the latter as a rule passes over quite gradually into the conducting parenchyma, but occasionally the two regions are sharply differentiated (stem of *Meesia longiseta*).

(e) *Fifth system. Peripheral hollow cylinder or tube of stereome, reinforced by isolated subepidermal girders.*

The haulms of ordinary Grasses are provided with a tubular stereome, supplemented by isolated subepidermal girders, the flanges of which project from the outer surface of the tube at more or less regular intervals. This is evidently a particularly effective arrangement, and as a matter of fact analogous types of construction find a wide application in architectural and engineering practice. The mestome-strands are mostly apposed to the inner side of the fibrous cylinder, but may also run freely in the medullary tissue. Usually a small peripheral mestome-bundle is embedded in the fibrous tissue just below the insertion of each supplementary girder upon the cylinder. One of the adjoining diagrams (Fig. 61 c) illustrates the type of bast-cylinder reinforced by well-developed subepidermal flanges which is characteristic of the GRAMINEAE.

## 2. *Bilateral organs.*

The foliage-leaf is the most important of the bilaterally symmetrical organs of the plant-body; in accordance with the nature of the mechanical requirements of leaves, their girders are all arranged parallel to one another, and at right angles to the surface. Schwendener distinguishes between "subepidermal" [superficial], "internal" and "mixed" girders, which differ in their relations to the epidermis, and which serve to define the three principal systems of mechanical construction employed in bilateral organs.

In the simplest cases subepidermal I-girders extend from one face of the leaf to the other. The leaves of *Typha, Cordyline, Phormium* (Fig. 63 A), *Pandanus* and *Musa*, and the blades of Grasses and Sedges exemplify this type of structure. The two flanges of each girder are similar in shape and nearly equal in size; the web consists either of mestome in conjunction with parenchyma, or of mestome alone.

A more advanced stage of specialisation is illustrated by those leaves (and midribs) in which the mechanical system is unequally developed on the two faces, in accordance with an unequal distribution of mechanical requirements.    In all long, curved, ribbon-shaped leaves, the upper side is particularly exposed to tension, while the lower side is more liable to be subjected to compression.  Evidently, broad sub-epidermal plates of bast are best fitted to ensure a uniform distribution of tensions on the upper side, whereas the lower side can be satisfactorily protected against compression by means of ordinary girders.  This form of mechanical system is exemplified by the leaf-midribs of *Erianthus, Saccharum; Zea* (Fig. 63 c), *Gynerium,* etc.    Another illustration is provided by the leaf-blades of many *Carices*; here the leaf is **V**- or **U**-shaped in section, so that the need for inextensibility is greatest along the two margins, each of which is accordingly provided with a special fibrous strand.

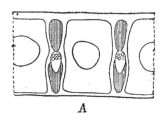

*A*

*B*

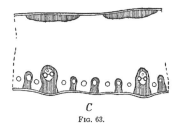

*C*

Fig. 63.

Girders of the internal type, characterised by the fact that the outer fibrous sheaths of the mestome-bundles are separated from the epidermis by photosynthetic tissues, are found in certain species of *Allium,* in the genera *Tritonia, Aspidistra,* etc.

Girders in various Monocotyledonous leaves, seen in T.S.  *A. Phormium tenax,* Upper and loWer flanges of approximately equal size; web composed of mestome.  *B. Carex sylvatica.*  Upper flange small and tangentially flattened ; loWer flange large ; web comprises a mestome-bundle With a fibrous sheath on the hadrome-side, and colourless parenchyma.  *C. Zea Mays* (mid-rib).  Upper flanges developed as tension-flanges.  The loWer or compression-flanges enclose mestome-bundles.

Mixed girders occur in the leaf-blades of certain Palms.  The larger of the bundles that lie deeply embedded in the mesophyll are embraced both internally and externally by fibrous sheaths ; the resulting fibro-vascular strands constitute internal girders, the location of which midway between the two surfaces of the leaf appears at first sight to be unsuitable from the mechanical point of view.  In many cases (*Livistona chinensis, Chamaerops excelsa*), indeed, where the fibrous sheaths of the mestome-strands are feebly developed, their mechanical significance must be mainly local, consisting in the protection of the individual conducting strands.  In other instances, on

the contrary (e.g. *Phoenix dactylifera, Sabal Adansoni*), the two flanges
are so massive that they undoubtedly serve to strengthen the organ as
a whole.    These internal girders are accompanied by a large number
of comparatively stout subepidermal fibrous strands.    In the Date
Palm the latter are developed approximately to the same extent on
both sides of the leaf; they are, however, placed at irregular intervals,
and it only occasionally happens that two strands correspond on
opposite sides, and so combine to form a typical girder.    The author
has observed precisely the same lack of correspondence between the
fibrous strands on the opposite sides of the leaf in many other Palms
(*Livistona chinensis, Sabal Adansoni, Chamaerops excelsa*).

This peculiar arrangement of the mechanical tissues in many
Palm leaves seems at first sight irrational, and its true significance is
not realised until one has had an opportunity of watching the behaviour
of a Palm when exposed to a high wind under natural conditions.
The inflexibly constructed stem displays a remarkable degree of rigidity,
and oscillates without undergoing any appreciable flexure.    The leaves
and leaflets behave quite differently; they flutter rapidly to and fro,
and in so doing are at one moment violently bent, and at the next,
recoil with equal violence, while ultimately they return to their "resting" position without having sustained any permanent deformation.
The leaves are, therefore, not inflexible at all, in the sense in which
the term applies to the stem; it would indeed entail far too great an
expenditure of material to render an organ like a Palm leaf, which
exposes a very large surface to the wind, actually inflexible.    Such
leaves are accordingly constructed so as to combine strength with
pliancy; the withdrawal of the mechanical strands from the periphery of
the cross-section to the centre thus fully accords with the behaviour of
the leaves in the wind, as Stahl was the first to note.[98]    The centralisation of the mechanical tissues has the further advantage of increasing
the resistance of the pinnae to the longitudinal tensions produced by
wind action.

Another example of the bilateral or dorsiventral type of organ
which has to be inflexible in a single plane is provided, according to
Worgitzky,[99] by the spirally coiled portions of certain tendrils, which
provide a firm and elastic connection between the climber and its
support.    When such a "watch-spring tendril" is stretched, it is only
the straight portion that is subjected to longitudinal tension; in the
coiled region the pull is resolved into bending and twisting components.    As the coils are pulled apart, every portion of the spiral
region tends to straighten itself, with the result that its concave face
is stretched, and its convex face compressed.    The curvature always
takes place in the same plane and in the same sense.    Hence, the

mechanical tissue must be strongly developed as a " tension-flange " on the concave side of the coiled region, whereas turgescent parenchyma often provides a sufficiently effective " compression-flange " on the convex side. In *Cyclanthera explodens*, for example (Fig. 70), the tension-flange consists of a broad ribbon-shaped mass of fibrous tissue with slightly incurved edges ; this is supplemented by three vascular bundles, which abut against its inner side.  On the convex side there are two smaller vascular bundles, one near either edge, with large fibrous sheaths ; these represent the compression-flanges of the tendril, which is thus dorsiventral from the first, as in all CUCURBITACEAE. The tendrils of *Passiflora* and *Bignonia* are at first radially symmetrical, but subsequently become dorsiventral, owing to the fact that the secondary woody cylinder develops much more strongly on the concave side.

### 3. *Mechanical arrangements for the prevention of deformation.*

When a cylindrical organ is bent, its normally circular cross-section tends to become elliptical.  This fact can be easily verified by bending a piece of india-rubber tubing.  Similarly, when a foliage-leaf or other flattened organ bends, the two faces tend to approach one another.  Bending thus invariably subjects the cross-sectional outline of an organ to a certain amount of deformation ; it is incumbent upon every plant to guard as far as possible against such distortions, which are decidedly injurious.  Any alteration of the cross-sectional outline of an organ entails corresponding changes in the relative positions of its mechanical constituents ; but a mechanical system is reliable only so long as its peripheral girders and other components preserve certain fixed relative positions.  It must also be kept in mind that every deformation of the cross-section involves radial distensions and compressions, which may react so severely upon particular tissues as to disturb their normal activities, or which may appreciably diminish the cohesion of the entire organ. It is evident, for example, that undue compression must seriously interfere with the functions of the conducting tissues, while excessive distension may easily lead to the disorganisation of loosely constructed tissues, such as those comprised in the photosynthetic system.

Special stiffening arrangements are particularly necessary in the case of organs which are pervaded by wide air-passages.  The richly branched cellular trabeculae that intersect the air-passages of some CYPERACEAE—especially in the transverse direction—collectively forming a very elegant system of internal buttresses, are of this nature ; they attain their most remarkable development in certain species of *Scirpus*.  A similar mechanical effect is produced by the subdivision of the large air-passages in the stems of *Papyrus antiquorum, Juncus glaucus, Scirpus lacustris* and many other water- and marsh-plants, into compartments, by means of transverse plates or diaphragms. In *Juncus glaucus* and *Scirpus lacustris* these diaphragms succeed one another at intervals of between 5 and 10 mm.; as a rule they contain small transverse vascular strands or mestome-anastomoses, which link up the longitudinal bundles with one another, thus representing a further stiffening device.

According to Mágócsy-Dietz [100], the septate pith which occurs in many woody plants also belongs to the category of special stiffening arrangements, especially when the diaphragms are largely composed of sclerotic cells (*Liriodendron tulipifera*). The diaphragms lose their mechanical significance more and more as the twigs grow older, and at the same time acquire increased importance as repositories of storage material.

### 4. *Mechanical arrangements in connection with intercalary meristems.*

There are comparatively few Phanerogams in which cell-formation and -extension are strictly confined to the apical region of the axis and to the youngest internodes. Such a " direct superposition of new structural elements upon an adequately strengthened foundation," as Schwendener terms this mode of growth, occurs for instance among Palms, in the genus *Dracaena*, in the PANDANACEAE, in many LILIACEAE, etc. Much more usually, however, the longitudinal extension of axial organs is largely dependent upon intercalary growth, a condition which necessitates special adaptations on the part of the mechanical system.

In the axial organs of most Dicotyledons and many Monocotyledons the growing region is rather extensive; it generally comprises a number of internodes, while its total length may amount to anything between two and fifty centimetres. As such growing regions have to be inflexible in structure, the general construction of their mechanical system does not differ appreciably from that which prevails in the fully developed portion of the organ. The difference between the two stages of development consists rather in the quality of the material employed for the construction of the skeletal system; for reasons which have been explained in detail in a previous section (Section A. 3), bast is here replaced by collenchyma; this tissue, which is usually developed in the form of subepidermal plates or ribs, thus represents a temporary skeleton, analogous to the scaffolding that surrounds a half-built house. In a number of Monocotyledons (*Tradescantia, Dioscorea, Tamus*) and herbaceous Dicotyledons (CUCURBITACEAE, UMBELLIFERAE), it supplements the true stereome throughout the first season; in woody plants, on the other hand, it is generally cast off owing to the formation of cork, when the twig has reached a sufficient thickness.

Organs which are engaged in intercalary growth may be strengthened in a totally different manner, namely by local enlargement of the internode in the region of most active growth. In *Tradescantia erecta*, where an intercalary growing-zone is situated at the base of each internode, the diameter of the stem is nearly twice as great at these points as it is at the upper ends of the internodes. The converse relation is illustrated by the peduncles of certain COMPOSITAE, in which intercalary growth is most active towards the distal end of the organ. It was Westermaier who pointed out that the unusual shape of these peduncles is correlated with their peculiar type of intercalary growth.[101] In such cases the local enlargement is sometimes very marked. The peduncle of *Arnoseris minima*, for example, may be six times as thick immediately below the capitulum as it is near its base. A zone of intercalary growth may, finally, be located near the middle of an internode. This somewhat unusual condition is exemplified, according to Westermaier, by *Pilea oreophila*; here, accordingly, it is the middle of the internode that is enlarged.

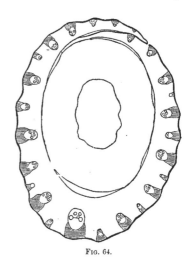

FIG. 64.

T.S. through the leaf-sheath of *Brachypodium sylvaticum*, ·5 cm. above the node. Subepidermal fibrous strands with apposed mestome-bundles.

The most perfect arrangements for the mechanical protection of intercalary growing-zones are found in the haulms of GRAMINEAE and CYPERACEAE. In these organs there is, at the base of each internode, a relatively short zone which remains permanently meristematic, thus constituting an intercalary growing-point. The zone in question is of course soft and mechanically weak. In order, therefore, that the base of the internode may not be broken by the first puff of wind, or by the mere weight of the overlying portion of the haulm, it must be strengthened by means of some special mechanical device. The requisite protection is provided by the sheathing leaf-base, which completely encloses the whole meristematic and growing region of the internode. The mechanical system of the leaf-sheath takes the place of the still undeveloped skeletal system of the ensheathed portion of the stem. Since the leaf-sheath is tubular, it must be inflexible in structure, for this reason its mechanical system is not like that of an ordinary bilateral leaf-blade, but on the contrary resembles that of a hollow stem.    The fibrous strands display a centrifugal tendency ; in other words they are well developed beneath the outer epidermis, whereas the inner half of the sheath is altogether devoid of fibrous flanges, or at most provided with very feeble strands of bast (Fig. 64).    This instance once more clearly demonstrates the fact that the disposition of fibrous strands is solely determined by mechanical considerations.

The same office which is performed by the leaf-sheaths of the GRAMINEAE and CYPERACEAE pertains to the peculiar tubular spathe that surrounds the apical portion of the scape of *Armeria*.    Here also, as Westermaier has convincingly demonstrated, everything points to the conclusion that the spathe provides the necessary mechanical support for an intercalary growing-zone.

### B. THE STRUCTURE OF INEXTENSIBLE ORGANS.

All the organs that serve to attach a plant to its substratum, whether the latter be the soil, the face of a rock, a tree trunk, a branch, or other object, are frequently subjected to longitudinal tension.    This statement applies more particularly to **roots**, which may in fact be regarded in general as " models of inextensibility."    It has already been explained that the most advantageous disposition of the mechanically effective tissues, in organs which have to be inextensible, consists in the concentration of the resistant elements into a single compact, cable-like central strand.    In all ordinary roots, accordingly, the mechanical tissue combines with the conducting elements to form an axile bundle.    A number of different types of inextensible construction may be distinguished, according to the varying disposition of the actual mechanical tissue in the composite central strand.    In its simplest form the axile strand consists of a central stereome-bundle surrounded by several hadrome- and leptome-groups (lateral roots of *Stachys sylvatica*, *Mentha aquatica* and *Hedera Helix*).    The axile strand may on the other hand be produced externally into ridges of mechanical tissue, which separate the leptome- and hadrome-bundles from one another (lateral roots of most Grasses and Sedges).    Sometimes the stereome is restricted to strips corresponding to the ridges of the preceding type, while the central area is occupied by a parenchymatous pith (*Smilax*), which may contain additional isolated fibrous strands (many **P**alms).    The stereome may finally take the form of semi-lunar sheaths embracing the outer faces of leptome-bundles (**P***isum*, **P***haseolus* and other LEGUMINOSAE).

Those **rhizomes** which serve to fix the plant in the ground agree with roots in having their mechanical tissues united to form a stout axile tube or a solid central strand; this centralisation of the mechanical system is very marked in the rhizomes of Grasses, Sedges and Rushes, which accordingly—when regarded from an anatomico-physiological standpoint—approximate more closely to roots than to the aërial stems of which they are the morphological equivalents (Fig. 65 A).

The anatomical contrast between inflexible aërial and inextensible subterranean structures is sometimes even exemplified by different portions of the same stem. This point is excellently illustrated by certain CYPERACEAE and JUNCACEAE. The lower portion of the haulm of

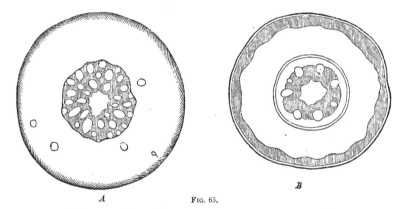

A    Fig. 65.    B

A. T.S. through a rhizome of *Carex glauca*. The mechanical system is in the form of a thick-walled axile tube of stereome with numerous mestome-bundles embedded therein. B. Lateral root of *Zea Mays* in T.S. The mechanical system comprises (1) a thick-walled axile tube of stereome (to ensure inextensibility), (2) a subepidermal fibrous cylinder (to ensure longitudinal incompressibility).

*Juncus glaucus* extends several centimetres below the surface of the soil; like a tap-root it is exposed solely to longitudinal tensions, when the aërial portion is bent. It is *a priori* improbable that this subterranean prolongation of the haulm should possess the structure of a rhizome. Anatomical investigation does in fact show that the disposition of the fibrous strands and mestome-bundles is almost identical with that which prevails in the aërial region of the haulm. But the two portions of the stem differ appreciably as regards the quantitative development of the bast in the individual strands. A glance at the two transverse sections (Fig. 59 A and B) at once reveals the centripetal tendency of the mechanical tissue in the subterranean, and its predominantly centrifugal disposition in the aërial, region of the haulm. The most obvious point of contrast is provided by the development of the subepidermal girders; these are much broader, and also project inwards to a far greater extent, in the inflexible region. In the inextensible underground portion the average cross-section of a girder only amounts to between one-third and one-fourth of its area in the upper inflexible region; the inner flanges of the main girders, on the other hand, show a distinct increase in thickness, while the external fibrous sheaths of the large mestome-bundles are so stout that they can scarcely possess a purely local significance.

With the aforesaid subterranean organ may be associated the **stems of twining plants and of climbers in general,** since these are also continually subjected to longitudinal tensions. In the case of twining stems, the tensions in question arise from the growth in thickness of the supporting structure, and from the flexures produced in the latter by the action of the wind. In other types of climbers allowance must also be made for the fact that the different points at which a support is grasped may subsequently move further apart; further, more or less extensive regions of a climbing shoot may temporarily or permanently assume a pendent position.[102] In accordance with these peculiar mechanical conditions, the mechanical system of climbing stems displays a centripetal tendency, which is sometimes merely indicated by a sclerotic development of the pith (*Tamus* and *Dioscorea*), while in other cases it may lead to the centralisation of the mechanical strands themselves. Thus in certain Monocotyledonous climbers—such as *Calamus Rotang* and species of *Carludovica*—even the innermost vascular bundles of the stem are furnished with stout fibrous sheaths; again, in the stems of certain PIPERACEAE (*Piper* and *Macropiper*) a mechanical cylinder is interpolated between the pith and the peripheral circle of vascular bundles.

All **submerged plants,** whether they inhabit running or stagnant water, must have more or less inextensible stems. A special interest attaches in this connection to the genus *Potamogeton,* because of the various habitats of the different species. Schwendener has demonstrated, with the help of elaborate arguments, that the species which are restricted to stagnant or slowly-flowing water (**P.** *crispus,* **P.** *densus,* **P.** *pectinatus*) develop no specialised mechanical cells, either in the cortex, or in the central mestome-cylinder, the latter being itself able to cope successfully with the very slight demands that are made upon the tensile strength of the stem. Those species, on the other hand, which are adapted for life in rapidly streaming water (**P.** *lanceolatus,* **P.** *longifolius,* **P.** *compressus,* etc.), not only exhibit a corresponding reinforcement of the central cylinder by mechanical elements, but are in addition provided with a number of scattered fibrous strands embedded in the parenchymatous mesh-work of the cortex; these isolated strands are evidently intended to prevent the loose cortical tissue, with its numerous air-passages, from becoming torn, or from being stripped off bodily.

At this point attention may also be drawn to the inextensible construction of **pendulous fruit-stalks** and **inflorescence-axes.** Here the requisite inextensibility is provided for, either by contraction of the fibrous cylinder, or by strengthening of the fibrous sheaths that accompany the internal mestome-strands (*Platanus, Stanhopea insignis*). Mention may finally be made of the fact that the pendulous pinnae of the Palm *Martinezia Lindeniana* are, according to Koop, inextensible in structure, the central region of the mesophyll being traversed by a very large number of stout fibrous strands. There can be no doubt, however, that these deep-seated strands do not, in the first instance, serve to support the comparatively insignificant weight of the pinna; they come into action, principally, when the pinnae are agitated by the wind and thus subjected to much severer tensions. A similar structure has already been described as characteristic of the majority of pinnate Palm-leaves.

### C. STRUCTURE OF INCOMPRESSIBLE ORGANS.

#### 1. *Resistance to radial pressure.*

Every subterranean or submerged organ is subjected to radial compression by the surrounding medium. The cortical parenchyma of such organs usually contains air-passages, and is otherwise loose in texture; in this event the structure has to be protected against compression by means of a firm peripheral shell. Sometimes the epidermis, in conjunction with one or two of the subjacent layers, is able to provide the requisite protection (for example, in the genera *Naias* and *Sagittaria*, in *Juncus supinus, Alisma natans* and other plants which grow in stagnant or slowly running water). In such organs, on the other hand, as grow in mud or in any saturated soil, the large size of the cortical air-passages necessitates the presence of thick-walled parenchyma or genuine bast-fibres. The roots of certain *Carices* (*C. stricta, C. caespitosa, C. vulgaris*), and of many GRAMINEAE, are accordingly provided with a tubular sheath of fibres with partially suberised walls.

#### 2. *Resistance to longitudinal compression.*

Every upright axial organ which has to bear the weight of branches and foliage must possess the same kind of rigidity as a pillar or column. It has already been explained, at the beginning of the present chapter, that general mechanical principles render it necessary for the disposition of resistant material in such longitudinally incompressible

organs to be the same as in inflexible structures, and the subject need
not be pursued any further.

Greater interest attaches to those organs which behave as inextensible or as longitudinally incompressible structures according to
circumstances.  This complex condition is exemplified by the **stilt-** or
**buttress-roots** which occur in the genus *Pandanus*, in *Iriartea exorrhiza*,
and in *Rhizophora Mangle* and other RHIZOPHORACEAE; further, by the
whorled adventitious roots which issue from the lower nodes of the
stem of the Indian Corn, and which are physiologically equivalent to
the aforesaid buttress-roots.

Warming [103] has shown that the stilt-roots of *Rhizophora Mangle*
are very stem-like in structure; the centre of each root is occupied by
an extensive pith, which is surrounded by a number of alternating
groups of leptome and hadrome that collectively form a hollow polyarch vascular cylinder or stele.  The water-conducting portions of the
vascular cylinder are accompanied, on the medullary side, by thick-
walled mechanical cells, which are clearly arranged so as to render the
root inflexible [and hence longitudinally incompressible].  At a later
stage the organ is still further strengthened by the development of
a secondary woody cylinder containing numerous fibrous elements.  The
stilt roots of *Rhizophora mucronata* and *Bruguiera eriopetala* are similarly
constructed.  Those of *Zea Mays*, however, require more detailed consideration.  The subterranean root-system of the Indian Corn does not
provide adequate support for the rapidly growing shoot.  Hence,
adventitious roots are produced from the lower part of the aërial stem,
where they form a very regular whorl at each node.  These roots do
not, however, grow straight downwards, but follow a more or less
oblique course, all the members of a whorl forming about the same
specific angle with the vertical (Fig. 66, V, V'); they enter the soil
at a distance from the stem which varies according to the height of
their point of origin, and thereupon produce numerous rootlets, which
anchor them firmly in the ground.

Evidently when a Maize-stem bends before the wind, the stilt-
roots on the windward side are stretched, while those which extend to
the leeward are longitudinally compressed.  Each root therefore has to
behave, at different moments of its existence, now as an inextensible,
and now as an inflexible organ; the structure of the root corresponds
exactly to this alternation of mechanical requirements.  The fibro-
vascular cylinder, instead of being solid as in most roots, encloses
a wide core of pith, so that the mestome and the accompanying stereides
together form a hollow cylinder (Fig. 65 B).  The latter has its
mestome elements arranged in the manner characteristic of normal
roots.  The most conspicuous components are certain very wide vessels

which are disposed in a ring. The intervals between the mestome-groups are occupied by relatively thin-walled bast-fibres, which add to the mechanical strength of the fibro-vascular cylinder. A thick-walled endodermis of the **C**-type (cf. Chap. VII., III. D.) surrounds the stele. The cortex consists of thick-walled parenchyma, which towards the periphery passes gradually into a subepidermal zone of stereome of varying width, composed of fibres with very thick and abundantly pitted walls.

The mechanical system of the stilt-roots of *Zea Mays* thus consists of two concentric hollow cylinders. The outer of these is entirely

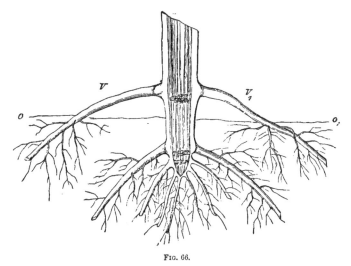

Fig. 66.

Basal part of a stem of *Zea Mays* in L.S. *V*, *V*₁. Adventitious roots—issuing from a node above the level of the soil (*O*, *O*₁)—which act as buttresses, besides assisting the subterranean roots to fix the stem in the ground. For the sake of clearness only a few of the rootlets are figured.

made up of bast-fibres and is principally responsible for the inflexibility of the organ; the inner includes the mestome elements and serves chiefly to secure inextensibility. On the whole, however, the anatomy of these stilt-roots seems to indicate that they are constructed as inflexible rather than as inextensible organs; otherwise it is not clear why the solid fibro-vascular cylinder which is characteristic of the roots of Grasses should here be dilated into a tubular structure. Stilt-roots of this kind are not peculiar to the Indian Corn: very similar organs recur in the genus *Sorghum* and in other tall and robust GRAMINEAE.

The stilt-roots of *Pandanus* are adapted to their twofold mechanical duty in quite a different manner. Here a transverse section of the central cylinder reveals a large number of scattered vascular bundles, each accompanied by a strong fibrous sheath, while the cortex is also traversed by numerous longitudinal fibrous strands of various sizes. The mechanical system is thus distributed very evenly over the entire cross-section, an arrangement which can only be justified on the score that both tension and compression have to be provided for.

Straight **spines** and **thorns** must also be capable of resisting longitudinal compression. According to Schwendener, this condition is satisfied, at any rate among MONOCOTYLEDONS (*e.g.* *Agave americana*, *Chamaerops humilis*, *Livistona sinensis*).

### D. ARRANGEMENTS FOR RESISTING SHEARING STRESSES.

When the action of a mechanical force upon a body is such that the constituent particles of the latter tend to become displaced with reference to one another, the force is said to be a **shearing** one; the property inherent in a body which enables it to resist such shearing action, or which in other words prevents the component particles from sliding over one another, may be termed **shear-resistance**. Every flexure of a cylindrical or flat organ is accompanied by shearing stresses; in the case of a simple flexure, these stresses are comparatively insignificant and do not necessitate the provision of any special shear-resisting arrangements. When such adaptations are nevertheless found to occur, especially in flattened organs, the reason must be sought in the increased shearing action consequent upon violent movements of the surrounding air or water. A leaf fluttering in the wind is exposed to shearing stresses acting at right angles to the surface and tending to cause laceration. In order to cope satisfactorily with this danger, the girders that serve to ensure inflexibility must be firmly anchored together by the greatest possible number of cross-ties. This office is performed by the vascular anastomoses to · which reference has already been made. These anastomoses form a dense reticulum in the leaves of Monocotyledons and Dicotyledons; they are seen with the greatest clearness in "nature-printed" figures of leaves. In these anastomoses the vascular tissues are often accompanied by bast-fibres; in the leaf of *Maranta arundinacea* indeed the majority of the anastomoses consist of stereides alone. In this connection mention may also be made of the "false veins"—described long ago by Mettenius—which traverse the delicate fronds of certain species of *Trichomanes*; these are also entirely composed of mechanical elements.

It is, of course, the **leaf-margin** which is particularly exposed to the risk of tearing, and which therefore most frequently requires special mechanical protection.[104] The simplest, and at the same time the most frequent, method of protecting the leaf-margin consists in an increase in the thickness of the outer epidermal walls; the marginal cells often differ very noticeably from those of the leaf-surface in this respect. Frequently subepidermal layers, consisting either of more or less thick-walled collenchyma (AROIDEAE), or of well-differentiated bast-fibres, are also employed to strengthen the leaf-margin. The cross sectional outline of these subepidermal collenchymatous or fibrous

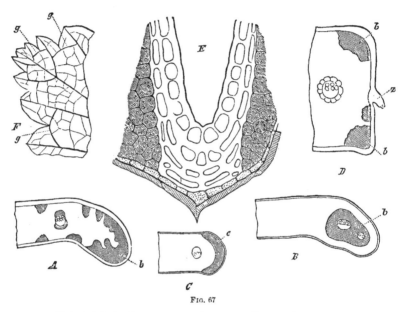

FIG. 67

Mechanical devices for strengthening the margins of leaves. *A-D.* Cross-sections of various leaf-margins. *A. Theophrasta Jussieui;* *b*, a subepidermal marginal fibrous strand of irregular outline. *B. T. imperalis.* The marginal fibrous strand, *b*, in which two mestome-bundles are embedded, is separated from the epidermis by a few layers of parenchyma. *C. Eucalyptus* sp.; *c*, semi-lunar subepidermal plate of collenchyma. *D. Phoenix dactylifera* (pinna); *b*, fibrous strands beneath the upper and lower angles of the margin; *z*, remains of the strip of tissue by means of which the pinna was formerly connected to its neighbour. *E. Sanguisorba carnea.* Tangential section through an indentation of the leaf-margin. Tearing is prevented by a sickle-shaped mass of collenchyma, which is further supported by a tangential mestome-anastomosis. (All that is shown of this anastomosis is a single row of tracheides together with part of the parenchymatous bundle-sheath.) *F.* Part of the leaf-margin of *Ribes rubrum* in surface-view; each indentation is strengthened by a mestome-anastomosis (*g*), which all but intersects its base.

groups varies greatly according to circumstances. Where the leaf has a sharp edge, the subepidermal fibrous strand is often sagittate in cross-section (e.g. in *Iris, Eryngium,* etc.). If the edge is rounded, the marginal strip of stereome usually has a sickle-shaped or semi-lunar cross-section (*Eucalyptus* sp., Fig. 67 c); or it may be broken up into several small strands, which are irregular in shape and unequal in size. In the Date Palm the pinnae appear to be cut off short along the margins, which are of the same thickness as the rest of leaf. Each pinna thus has two straight edges, the height of which exactly equals the thickness of the leaf; from the middle of either edge there projects a narrow strip of dried-up tissue (Fig. 67 D, z). This strip, which is approximately conical in cross-section, is a relic of the tissue that formerly connected two adjacent leaflets, the pinnate condition being here—as is well known—due to the splitting up of a primarily continuous lamina. In the angle contained by the edge of the pinna on the one hand, and by its upper or lower surface on the other, there is situated a single large bundle of fibres. In this way the strands which form the two most powerful subepidermal girders of the whole pinna serve at the same time for the special protection of the margin.

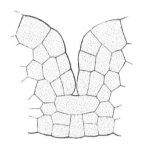

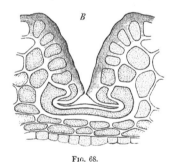

Fig. 68.

T.S. across one of the commissures of a leaf-sheath of *Equisetum hiemale*. *A.* Meristematic condition. *B.* Adult state.

The marginal fibrous strands of leaves are, however, not always strictly subepidermal; they may be separated from the epidermis by one or two layers of green parenchyma, in which case it is usual for the marginal mestome-bundle to abut against one side of the fibrous strand or to be embedded within the latter (*Theophrasta imperialis*, Fig. 67 B). Finally, it is not at all uncommon for other subepidermal strands to unite with the previously strengthened marginal veins, in which case, of course, a great variety of cross-sectional outlines may result.

The strengthening of the leaf-margin is effected in a very remarkable manner in the genus *Aloe*, where the requisite special protection is afforded by a subepidermal layer of palisade-shaped sclerenchymatous cells. Towards the flat surface of the leaf this layer passes over gradually into ordinary photosynthetic palisade-tissue. There seems, therefore, no doubt, that in *Aloe* the need for special protection of the leaf-margin has forced the marginal photosynthetic cells to give up their original function, and to become converted into mechanical elements.

Where the leaf-margin is entire, the arrangements for its special protection are very much the same at every point. In the case of

sinuate, crenate or serrate leaves, on the contrary, it is often necessary
for the indentations to be specially protected, since these are, of course,

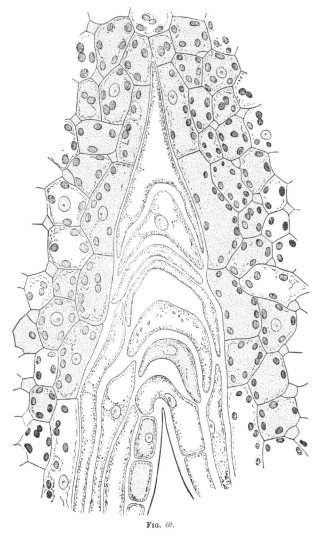

FIG. 69.

Base of an indentation of the leaf-margin of *Asplenium diversifolium*, showing a series of
crescentic stereides which prevent tearing.  After Bennesch.

particularly exposed to the danger of tearing.   This strictly local protection is often effected by short mestome-anastomoses, which extend
tangentially so as almost to intersect the base of each indentation.

N

Cross-ties of this kind are particularly well-developed in certain species of *Ribes* (e.g. *R. rubrum*, Fig. 67 F, *gg*). In other instances, the base of each indentation is embraced by a sickle-shaped strand composed of strongly thickened—usually collenchymatous—cells. Such strands, composed of thick-walled colourless cells elongated in a direction parallel to the margin, are well seen in species of *Poterium* and *Sanguisorba* (Fig. 67 E), where they contrast very sharply with the adjoining photosynthetic parenchyma. These local mechanical strands often abut internally against one of the above-described tangential mestome-anastomoses.[105]

In the leaves of many Ferns, each indentation of the margin is protected by a whole series of very thick-walled crescentic stereides; *Asplenium diversifolium* affords an excellent illustration of this remarkable device for the local strengthening of the leaf-margin (Fig. 69). In this connection mention must also be made of the peculiar "anchor-cells" of *Equisetum*, which were first described by C. Müller (Fig. 68); these help to prevent the leaf-sheath from tearing along the relatively weak "commissures" (*i.e.* between the successive teeth). In view of their function these peculiar mechanical elements might more aptly be termed "clamp-cells," since they serve as clamps for holding together adjacent segments of the sheath.[106]

## V. THE INFLUENCE OF EXTERNAL FACTORS UPON THE DEVELOPMENT OF THE MECHANICAL SYSTEM.

Generally speaking, both the qualitative and the quantitative development of the mechanical system are included among the hereditary characters of the species. Nevertheless a certain amount of direct accommodation to external conditions on the part of the mechanical system may take place during the life of the individual, or in the course of development of some of its organs.

In this connection special stress must be laid upon the relations between the quantitative development of the mechanical system and the action of those mechanical forces which are ultimately responsible for the presence of the entire skeletal system.[107] It is probable that the forces in question act as stimuli that tend to accentuate the qualitative and quantitative development of the stereome, at any rate so long as they do not exceed certain limits of intensity. Certain facts bearing on this point were recorded over a hundred years ago by Knight. A number of young [seedling] Apple-trees, with stems from six to seven feet in height [between the roots and the first branches], were tied to stakes in such a manner that the lower portions of the stems were nearly deprived of all motion to the height of three feet from the ground, the upper parts of the stems and branches being left in their natural state.   The trees were planted so as to be freely exposed to the action of the wind.   " In the succeeding summer, much new wood accumulated in the parts which were kept in motion by the wind; but the lower parts of the stems and roots increased very little in size."   In the following winter one of these trees was fixed so that it could only move towards the north and south.   " Thus circumstanced, the diameter of the tree, from north to south, in that part of its stem which was most exercised by the wind, exceeded that in the opposite direction [*i.e.* from east to west], in the following autumn, in the proportion of thirteen to eleven."   There can be no doubt that an unequal production of wood-fibres was partly responsible for the greater increase in thickness on the two sides that were subjected to the more severe mechanical stress.

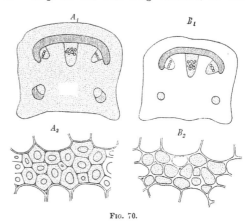

FIG. 70.

Tendrils of *Cyclanthera explodens* in T.S.   For explanation see text.

Recently, observations of a similar nature have for . the first time been carried out upon **tendrils**.   It has long been known, that tendrils which fail to grasp a support remain much thinner than those which become attached in accordance with the function for which they are designed; as a rule, also, unattached tendrils soon die, dry up and fall off.   The adjoining. figure (Fig. 70) depicts transverse sections across the coiled regions of two equally old tendrils of *Cyclanthera explodens,* one of which ($A_1$) has grasped a support, while the other ($B_1$) has remained unattached.   The former shows, on its inner side, a well-developed ribbon-shaped tension-flange composed of two or three layers of very thick-walled bast-fibres ($A_2$); further, each of the two vascular bundles which run near the margins of the outer side is provided with a strong sheath of bast, comprising from seven to ten cells in transverse section.   In the unattached tendril, on the other hand, the development of the fibres in the stereome-ribbon has been arrested at an early stage; except in isolated cases the cells are only slightly thickened in a collenchymatous manner, and are soft and consequently often distorted ($B_2$).   The two vascular strands on the outer side are entirely devoid of fibrous sheaths.

At this point reference may also be made to Treub's observations upon the active increase in thickness which takes place in the irritable **climbing hooks** of *Artabotrys, Ancistrocladus, Uncaria*, etc., when these organs take hold of a support. In such cases the primary object of the secondary thickening seems to be the production of an increased number of mechanical elements. As regards the stimulus which starts the secondary growth, it is unknown whether the pressure of the support upon the hook, or the longitudinal tension experienced by the latter after fixation, plays the greater part. The same uncertainty prevails in the case of ordinary tendrils.

Interesting experimental results have been obtained by Worgitzky with regard to the tensile strength of certain tendrils; the subjoined figures refer to two spirally coiled tendrils of *Passiflora quadrangularis*.

| | Load at the elastic limit in grams. | Load which just suffices to straighten all the coils, in grams. | Load at the breaking-point in grams. |
|---|---|---|---|
| Attached Tendril, - - - | 40 | 500 | 600 |
| Unattached Tendril, | 8 | 250 | 350 |

In the first [German] edition of the present work the author drew attention to the lack of any detailed experimental investigation of the effects of artificial tension or compression upon the development of the mechanical system. Since then quite a number of physiologists have carried out researches of this nature, but for the most part without arriving at any positive or unequivocal conclusions. The most remarkable results are those obtained by Vöchting in the course of certain experiments upon potted plants of Curly Kale (*Brassica oleracea*, var. *bullata*), which were placed in a horizontal position, with weights attached to the distal end of the stem. He found that in these circumstances the secondary woody cylinder grew more actively, and contained more mechanical elements, along the upper and lower sides of the stem—that is to say, in the regions of greatest tension—than it did on its flanks. In other experiments of Vöchting's, pumpkins were allowed to ripen, not, as is usually the case, while resting upon the ground, but while hanging freely in the air by their stalks, so that the latter were continuously subjected to considerable longitudinal tension. Here again a general increase in the thickness of the cell-walls was observed, not only in the case of the mechanical tissues but also in that of the parenchymatous ground-tissue; the secondary wood was likewise found to contain an unusually large proportion of mechanical elements.

In subsequent researches upon this subject, allowance will have to be made for the fact that a plant cannot be expected to accommodate itself to artificial mechanical influences, unless the latter accentuate, or at any rate imitate, the natural mechanical forces to the action of which the organ under investigation is normally exposed.  In other words, only such structures as roots, tendrils and pendulous fruit-stalks, should be subjected to continuous longitudinal tension.  Stems and petioles should be bent to and fro; for under natural conditions inflexibly constructed organs are never continuously exposed to longitudinal tension.

The development of the mechanical system may be affected by influences which are themselves not of a mechanical nature at all.[108] We are indebted to F. Haberlandt for some observations concerning the influence of the humidity of the soil, which furnish incontrovertible proof of the favourable effect of watering upon the strength of the bast-fibres of *Cannabis sativa*.  The average load required to break a strip of bast 2 mm. in width was 4·12 kg., if the fibre was taken from unwatered plants; in the case of plants growing in well-watered soil, the corresponding load was as much as 5·48 kg.  In this experiment, however, it was left undecided, whether the increase of strength caused by watering was due to an increase in cross-sectional area of the fibrous strands, or to increased resistance on the part of the cell-walls. Kohl, on the other hand, has observed that certain plants (*Mentha aquatica, Thalictrum galioides, Menyanthes trifoliata*, etc.), develop more collenchyma and bast if grown in dry air—*i.e.* under conditions favourable to transpiration,—than they would produce in a moist atmosphere —*i.e.* with their transpiration reduced.  Here, again, it is impossible to state with certainty whether the process is adaptive and self-regulatory or not.  It should, however, be noted that in the case of herbaceous plants growing in a dry atmosphere, or in fact under xerophytic conditions in general, turgor has a smaller mechanical value than usual, because the risk of temporary wilting is so great; in these circumstances, any increase in the development of mechanical tissue must be advantageous.

## VI. THE MECHANICAL SYSTEM AMONG THE THALLOPHYTA.

Fresh-water Algae that grow in running water, and Sea-weeds that are exposed to the action of the waves, must evidently be possessed of inextensibility and shear-resistance; otherwise they could not long withstand the stresses to which they are constantly subjected. According to Wille [109] the FUCACEAE are actually provided with special thick-walled mechanical cells, which are highly extensible and elastic : these occur mainly in the stalks and midribs, where they are disposed

more or less centrally in accordance with the tension-resisting character
of those parts of the thallus.

Among Fungi thick-walled mechanical hyphae are on the whole rare.
The subterranean rhizomorphs found in the PHALLOIDEAE and LYCOPER-
DACEAE, and in some species of *Agaricus*, consist of a relatively loose
"cortex" composed of thin-walled hyphae and enclosing a core of thick-
walled gelatinous filaments, which are evidently responsible for the inextensibility of the whole structure ; it is doubtful whether this core may
not in addition serve for translocation, and perhaps also for other purposes.
Experiments performed by the author indicate that the "medullary
strand" in the branches of the fruticose lichen *Usnea barbata* likewise
represents a tension-resisting device (Fig. 71 A).   In this instance the

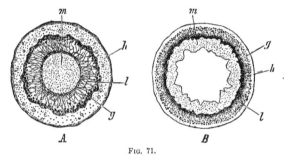

FIG. 71.

*A.* T.S. through a pendulous branch of *Usnea barbata* (explanation in text).  *B.* T.S.
through the hollow inflexible podetium of *Cladonia filiformis.*  ×40.  *h*, dermal tissue ;
*g*, gonidial layer ; *l*, loose plectenchyma representing the ventilating system ; *m*,
mechanical system.

central bundle of thick-walled hyphae resembles, as regards its strength
and elasticity, such a material as india-rubber, rather than bast or collenchyma.   For one thing it is extraordinarily extensible.   In the fresh
and fully imbibed condition, the medullary strand in a branch ·5 to
1 mm. in diameter, pertaining to a pendulous variety of this lichen,
could be stretched to more than twice its original length (an elongation
of 100-110 per cent.) without breaking.   In the case of a more
shrubby and thick-set variety, branches of the same thickness were
found to break when stretched to the extent of 60-70 per cent.   In
both instances the medullary strands remained perfectly elastic after
undergoing an extension of 20 per cent.   The much smaller extensibility of the cortical and dermal tissues betrayed itself, in these
experiments, by the appearance of numerous wide transverse fissures in
the outer part of the branch.   The breaking-weight of the medullary
strands amounts on an average to 1·7 kg. per sq. mm., and is thus
approximately the same as that of the foliar epidermis of *Allium
Porrum*, which is estimated by Weinzierl at 1·5-1·8 kg. per sq. mm.

The hollow stipes of Agarics and certain other Fungi must be regarded as inflexible structures, although they are devoid of special mechanical hyphae. The tubular sporophores (podetia) of *Cladonia* are provided with a sharply differentiated mechanical cylinder composed of special thick-walled hyphae (Fig. 71 B).

## VII. ONTOGENY OF THE MECHANICAL SYSTEM.

The skeletal system of a plant may originate from any or all of the three primary meristematic layers of the growing-point; this statement applies equally to bast and to collenchyma, so that the two principal types of mechanical tissue are closely connected ontogenetically as well as in other respects.[110]

### 1. Development of stereome from procambium.

As the author has shown, isolated groups of bast or collenchyma generally arise from similarly disposed procambial strands (Fig. 72 A, B). This rule holds good, not only for subepidermal skeletal strands, but also for bundles which are more deeply situated (*e.g.* the fibrous strands of *Scirpus Holoschoenus, Typha latifolia, Phoenix dactylifera*, etc.; the collenchyma of *Salvia officinalis, Cucurbita Pepo*, etc.). Where the mechanical system consists of a hollow cylinder which is entirely separated from the mestome-bundles, the antecedent procambial cylinder as a rule likewise originates independently; this mode of origin is exemplified by the peripheral fibrous cylinder of the rhizomes of certain *Carices*, and by the subcortical bast cylinder of the stem of *Cucurbita Pepo* (cf. also Fig. 72 c). Not infrequently, however, the mechanical system consists of independent strands or of a separate cylinder in the adult state, but nevertheless has a common origin with mestome-bundles in the growing-point. This peculiar method of development is best explained with the aid of one or two concrete instances. In *Juncus glaucus*, each of the principal girders, as already stated, comprises a subepidermal fibrous strand—forming the outer flange,—and a large semilunar bast-sheath which constitutes the inner flange; the interval between the two flanges is occupied by the following tissues, considered in centripetal order: a strip of parenchyma, an air-passage, a second parenchymatous zone, a fibrous sheath which serves for local protection, and finally a large mestome-bundle. The whole of this complex array of tissues originates from a single subepidermal procambial strand, which projects far into the fundamental meristem and appears hour-glass-shaped in cross-section (Fig. 59 D). Whereas, however, longitudinal division continues for a long time in the innermost and outermost regions of this composite strand, this process ceases entirely at an early stage of development in the constricted middle zone; here the cells expand, undergo repeated transverse divisions, and so become gradually converted into parenchymatous tissue. In consequence of this peculiar behaviour, the primarily continuous procambial strand becomes separated into two parallel bundles which complete their development quite independently of one another, the outer one becoming a subepidermal fibrous strand, while the inner one gives rise

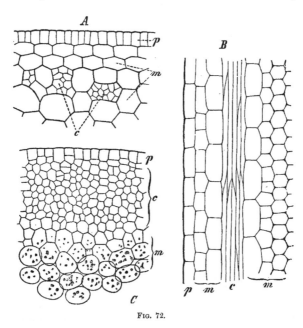

Fig. 72.

Development of stereome from procambium. *A*. Part of the adaxial half of a T.S. through a young leaf of *Pandanus utilis*; *c*, procambial strands which will give rise to small fibrous strands. *B* Part of a radial longitudinal section through the same leaf; *c*, as in *A*. ×400. *C*. Part of a T.S. through a young petiole of *Asplenium foecundum*, ×125; *c*, procambial cylinder which will give rise to the subepidermal fibrous cylinder. (In all figs. *p*=protoderm, and *m*=fundamental meristem.)

to the mestome-bundle together with its two semi-lunar fibrous sheaths. In some CARYOPHYLLACEAE (*Lychnis, Melandryum, Dianthus, Saponaria*) the fibrous cylinder and the mestome-bundles likewise arise from a common procambial cylinder, which becomes secondarily separated into an external cylinder of bast-procambium, an internal circle of ribbons and strands representing the mestome-procambium and an intermediate zone of parenchyma.

According to Ambronn, collenchyma and mestome may similarly originate from a common primary meristem and only become secondarily separated; this arrangement is of very widespread occurrence in the AROIDEAE, UMBELLIFERAE and PIPERACEAE.

Where, as so frequently occurs, the strands of mechanical tissue and the mestome-bundles are associated to form "fibro-vascular" bundles, they almost always originate from a common meristem, consisting either of a homogeneous procambial cylinder or of a series of homogeneous procambial strands, within which mechanical and conducting elements become differentiated in accordance with the varying requirements of the plant. Very often longitudinal division continues for

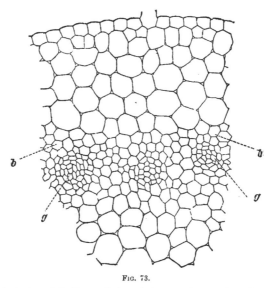

FIG. 73.

Mode of origin of the fibrous cylinder and the internally apposed vascular bundles in the scape of *Primula sinensis* (T.S.); *b*, bast-procambium ; *g*, mestome-procambium (primordia of the vascular bundles).

a longer period in those portions of the common procambium which are destined to give rise to vascular tissue; this condition prevails, for instance, in the scape of *Primula sinensis*, where the "primary" procambial cylinder gives rise to a fibrous cylinder, while the mestome-bundles, which abut against the inner side of the latter, appear to arise from "secondary" procambial strands (Fig. 73).

Attention may finally be drawn to an important ontogenetic distinction between stereome and mestome. The conversion of a strand or strip of bast-procambium into permanent tissue takes place simultaneously, or nearly so, over the whole of the cross-section. In the case of mestome-procambium, on the contrary, the transformation into permanent tissue normally begins first of all at two or more definite points in the cross-section and extends gradually from these to the other portions. In other words differentiation is simultaneous in a fibrous strand, but successive in a vascular bundle.

### 2. *Development of stereome from protoderm.*

As was pointed out in a previous section of this chapter (IV. A. 1), the centrifugal tendency of the mechanical system in inflexible organs often brings the fibrous or collenchymatous tissue into immediate contact with the epidermis. It might be anticipated that this close

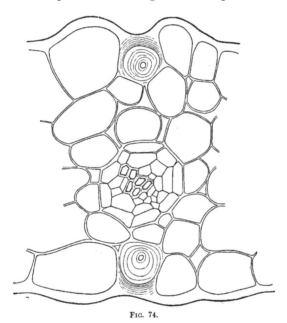

FIG. 74.

Small portion of a T.S. through the lamina of *Pteris serrata*, passing through one of the I-girders; each flange consists—in cross-section—of a single bast-fibre of protodermal origin. After Kraus.

proximity of the mechanical system would not be without its effect upon the structure of the epidermis itself. As a matter of fact the epidermis is, in these circumstances, employed to supplement the sub-epidermal bast or collenchyma in a degree which varies within wide limits. In extreme cases the superficial layer entirely relinquishes its dermal character and becomes completely converted into mechanical tissue. Even where there is no such close approximation of the skeletal and dermal systems, the epidermis may be pressed into the service of the former under the dominating influence of the mechanical principle. For it must be remembered that in inflexible organs the advantage gained by the conversion of any layer into mechanical tissue increases with the distance of that layer from the centre, provided of course that the transformation is not rendered inadvisable by opposing physiological considerations of even greater moment. The collenchymatous thickening of the epidermal walls which is seen in the leaves of many LILIACEAE (especially in the genus *Allium*), the thick-walled and lignified condition of the epidermis in the colourless bracts of *Papyrus antiquorum* and in the glumes of some species of *Cyperus*, and many similar phenomena, can be most readily explained from this point of view.

Reference has been made on a previous occasion to the conversion of epidermal cells into typical stereides; this so-called transformation of course merely consists in the fact that the meristematic layer, which in some ancestor of the plant under consideration was entirely engaged in the production of epidermal cells, at the present time gives rise in the same region to stereides instead of typical dermal elements. The meristematic layer in question is, of course, the protoderm. In the simplest type of transformation the protodermal cells are converted straightway—*i.e.* without any preliminary tangential division—into thick-walled prosenchymatous stereides. This case is illustrated by the leaves of many Ferns, very excellently, for example, by those of *Pteris serrata*, where, in the smaller veins, each flange of the **I**-girder is seen in cross-section to consist of a single "bast-fibre" of protodermal origin (Fig. 74). More often, however, such protodermal stereides are merely the outermost members of a typical strand or cylinder of bast; thus, for example, in the haulms of many Grasses (*Bromus*, *Melica*, etc.). In other cases, again, the protodermal cells first undergo tangential divisions; of the resulting daughter-cells only the internal ones give rise to fibrous tissue, whereas the outermost become secondarily incorporated in the epidermis. The small but complete protodermal procambial strands that give rise to perfectly normal fibrous bundles in the leaves of certain CYPERACEAE (e.g. *Cyperus vegetus*, *C. longus*, *C. glaber*, *C. glomeratus* and other species of *Cyperus*) originate in this manner. In cross-section such a procambial strand may consist either of a single protodermal cell or of a group of protodermal elements. [For further details the reader may consult Fig. 75 A-E, and the author's dissertation on "The Ontogeny of the Mechanical System."] It is worth noting that in many cases— *e.g.* in the haulms of *Papyrus antiquorum*—the peripheral fibrous strands originate at the boundary between protoderm and fundamental meristem, so that both these meristematic layers may take part in the formation of the same procambial strand (Fig. 75 F, G).

Collenchyma, like bast, may originate from protodermal cells without any preliminary tangential division taking place; thus, for instance, in the scape of *Allium ursinum.* In *Peperomia latifolia,* on the other hand, the outer portion of the subepidermal cylinder of collenchyma owes its origin to the repeated tangential division of the protoderm.

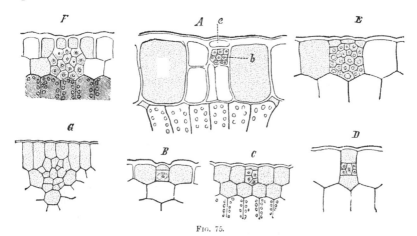

FIG. 75.

Protodermal fibrous strands in T.S. *A.* Adaxial side of a foliage-leaf of *Cyperus glaber*; *b,* fibrous strand; *e,* "secondary" epidermal element. *B.* Abaxial side of a small bract; a protodermal bast-fibre is shown, overlain by a "secondary" element of the epidermis. *C-G. Papyrus antiquorum. C.* Fibrous strand from the abaxial side of a bract. *D.* Fibrous strand from the adaxial side; the mechanical tissue has evidently arisen from a mother-cell cut out in the middle of a narrow protodermal element. *E.* Fibrous strand from the adaxial side (a slightly different case). The greater portion of the protodermal cell has been converted into mechanical tissue; the smaller portion has only produced three flattened "secondary" epidermal elements. *F.* Peripheral fibrous strand from the haulm; its outer portion is protodermal in origin, while its inner portion is derived from fundamental meristem. *G.* A similar strand in the procambial stage.

The sclerenchymatous layers of seed-coats and pericarps are also in many cases (*e.g.* among the LEGUMINOSAE) protodermal in origin.

### 3. Development of stereome from fundamental meristem.

In some Monocotyledons, and particularly among the LILIACEAE, the fibrous cylinder originates in the fundamental meristem, certain cells of which undergo repeated division and so give rise to a secondary meristem; the latter cannot be regarded as a *typical* procambium, because its cells are often not prosenchymatous in character, and also because it contains very well developed intercellular spaces.

The fundamental meristem is still more frequently responsible for the production of collenchyma; in this case also the intervention of a secondary meristem is the rule.

## VIII. SPECIAL MECHANICAL DEVICES.

In its widest sense the mechanical system may be held to include a variety of structures, which have indeed no immediate connection with the strengthening of the plant-body as a whole, but which nevertheless perform definite mechanical functions. As a rule the importance of such structures is purely ecological. Many of them find their natural place in the chapter which treats of the motor

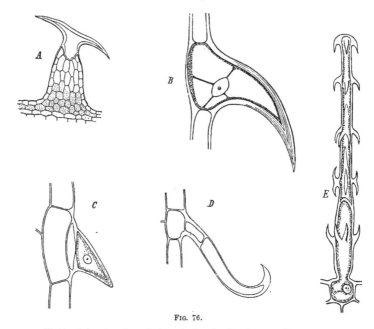

Fig. 76.

Climbing-hairs of various plants. *A. Humulus Lupulus. B. Galium aparine. C. Apios tuberosa. D. Phaseolus multiflorus. E. Loasa hispida*; in this case the nucleated proximal portion of the protoplast is shut off from the distal portion by a septum.

system. The present section will be chiefly devoted to structures which serve solely for fixation or attachment.

Attention may first be directed to the so-called **climbing-hairs** of twining plants and climbers in general; though varying greatly in detail, these hairs are generally arranged so as to prevent the stem from slipping downwards along its support, while offering no obstacle to the upward movement necessitated by longitudinal growth. If the stem of such a plant as *Galium aparine, Apios tuberosa*, or *Humulus Lupulus* be stroked between the fingers in the downward direction, no serious opposition is encountered; when, on the other hand, one attempts to draw the fingers upwards along the stem, the friction experienced is often sufficient to arrest the movement altogether. The explanation is very simple, consisting in the fact that the barbs of the climbing-hairs all point downwards. The climbing-hairs of *Phaseolus vulgaris* and **P.** *multiflorus* are exceptions to the rule, since they are not arranged in any definite manner. They do indeed often point downwards (*i.e.* towards the base of the stem), but just as often have their tips directed upwards or even sideways. This aberrant behaviour is probably connected with the fact that the hairs are somewhat flexible,

FIG. 77.

Anchoring-hairs from the hypocotyl of *Avicennia officinalis.*

instead of being rigidly attached to the stem. The cell which forms the body of the hair is thin-walled towards the base, where it is inserted upon a stiff thick-walled stalk-cell; the hooked body-cell can therefore twist and turn in various directions around the rigid stalk-cell (Fig. 76 D).

The simplest type of climbing-hair consists of a single cell, which ends in a strongly recurved sharply pointed tip; the thickness of the cell-wall increases towards the point (*Galium aparine*, Fig. 76 B). In *Humulus Lupulus* the two-armed grappling-hair is obliquely inserted upon a multicellular pedestal or emergence, in such a way that its distance from the surface is greatest on the basiscopic side (Fig. 76 A). A very effective arrangement is found in the LOASACEAE, where the unicellular climbing-hairs bear several whorls of minute but very sharp barbs; these originate as tiny protuberances of the cell-wall, but when fully developed take the form of massive calcified projections (Fig. 76 E).

Closely related to these trichomes of climbing plants are the remarkable **anchoring-hairs** of *Avicennia officinalis*, first discovered by Karsten.[111] These structures, which serve to fix the seedlings in the mud when they first fall from the tree, occur in great numbers on the hypocotyl. Each consists of a row of elongated cells, the side walls of which become constantly thicker towards the apex of the hair. The terminal cell is strongly recurved, and ends in a sharp point (Fig. 77). The hairs are freely movable in all directions, the thin-walled basal cells acting like the cable of a ship's anchor; in the long run, this arrangement greatly increases the chances of fixation.

A very similar anchoring device has been observed by Fritz Müller[112] in the seeds of a Bromeliaceous epiphyte, *Catopsis nutans*, which almost always attaches itself to the very outermost and thinnest

branches of trees.   Here each member of the tuft of silky hairs which
acts as the parachute-organ of the seed terminates in a pointed, thick-
walled and sharply recurved hook.   By means of these grappling-hooks
the seed is enabled to adhere to the thin and often slippery twigs which
the plant inhabits.   In the case of an epiphytic Orchid, also (*Phygma-
tidium*), the same observer found both ends of the minute cylindrical
seeds to be furnished with a number of
hooked processes.

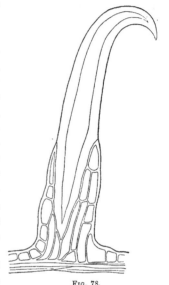

Some account must next be given
of the exceedingly diversified **hook- or
claw-shaped attachment-organs** that assist
in the dispersal **of seeds and fruits**, by
fixing these structures to the fur or
plumage of animals.   The morphological
value of such hooks cannot be considered here; the simplest of them represent trichomes.   Anatomically they
all agree in the thick-walled character
of their constituent cells, which sometimes indeed consist of typical stereides.
A few examples must suffice to illustrate
the interesting and varied forms assumed
by these organs.

In the case of *Circaea lutetiana*
hooked unicellular thick-walled hairs
assist in the dispersal of the fruits.
Each hair has an expanded foot or base

Fig. 78.

Hooked hair from a pale of *Lappago
racemosa*.

which is surrounded by a rosette of   narrow, radially elongated
epidermal elements.   Unicellular hooked hairs also occur on the pales
of *Lappago racemosa* (GRAMINEAE); here the body-cell of the hair has
its more or less attenuated base deeply embedded in a cup-shaped
multicellular pedestal (Fig. 78).

In *Caucalis daucoides* (UMBELLIFERAE) the mericarps are covered
with hooks about 2·5 mm. in length; each hook consists of a slender
tapering bundle of thick-walled prosenchymatous stereides, the whole
being surmounted by a single very large transversely directed cell bent
in the shape of a **f** (Fig. 79).   The distal end of this terminal cell,
which acts as a barb, is sharply pointed, and so thick-walled that its
cavity is reduced to a narrow slit.   The wide proximal portion of the
lumen contains the remnants of a massive protoplast.   The wall abutting
against the bundle of stereides is moderately thickened and provided with
numerous pits.   In the Burdock (*Lappa major*) each of the numerous
hooks of the involucre consists essentially of a tapering, distally recurved

bundle of stereides; in this case there is no specialised terminal barb-
cell, corresponding to that of *Caucalis*.

Somewhat complicated appendages, bearing numerous barbs, are
found on the fruits of *Cynoglossum cheirifolium* (Fig. 80). Each of
these structures takes the form of a slender cone, about ·5 to ·8 mm. in
height; it consists of a central core of thin-walled parenchyma, surrounded by a sheath of thicker-walled epidermal cells, the long axes of
which extend obliquely outwards. The radial walls of each epidermal
cell are thrown into folds; a small two- or many-toothed central
protuberance adds to the roughness of the outer wall. The appendage
terminates in a group of from four to six thick-walled epidermal
elements, with short re-
curved distal ends and long
embedded basal portions.
The radial walls of the terminal and sub-terminal cells
are not folded. The whole
appendage is a striking example of inflexible construction, and altogether con-
stitutes a highly effective
organ of attachment.

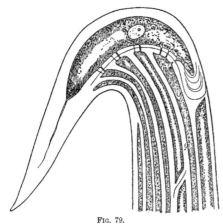

FIG. 79.

Distal portion of a hook from the fruit of *Caucalis daucoides*.

The mechanism of
another type of fixing organ
depends upon adhesion to,
or fusion with, the support
or substratum, instead of
upon the presence of hooks
or barbs; an illustration is provided by the **adhesive discs** or pads
peculiar to the tendrils of certain BIGNONIACEAE and of species
of *Ampelopsis*. In the latter genus the rudiments of the adhesive
apparatus are present even in the unstimulated condition of the
tendril. The first result of contact-stimulation is the abundant
secretion of a mucilaginous substance, which owes its origin [in part
at any rate] to the gelatinisation of the entire outer epidermal wall,
with the exception of the cuticle and the innermost pellicle. According to Lengerken [113], mucilage is also secreted within the cell-cavity,
whence it passes to the outside. Whatever its origin, the mucilage
ultimately exudes, after rupturing the cuticle, and glues the disc to
the substratum. The contact-stimulus further induces a considerable
hypertrophy of the epidermis, with the result that hair-like or
papillose protuberances penetrate into every nook and cranny in the
substratum, fitting themselves exactly to all the inequalities of the

surface and thus attaching the tendril very firmly to its support. Many parasites fasten themselves in a similar manner to the surface of the host, before producing their actual haustoria.

The fixation of many seeds and fruits in the soil is effected by means of special **mucilaginous layers**; these are also of some importance in connection with water-storage, a point which will be more fully discussed in a subsequent chapter (Chap. VIII.). A comparatively specialised type of structure is exemplified by the seed-coat of *Ruellia*, where peculiar **mucilage-hairs** are present. In *R. strepens* the walls of each hair consist of a number of layers which swell greatly in contact with water, and of a thin cutinised stratum which is provided with numerous internal annular thickenings. When swelling takes place, the hairs, which in the dry state are closely appressed to the surface of the testa, gradually erect themselves and at the same time twist in various directions; in so doing they elongate very considerably, while the in-

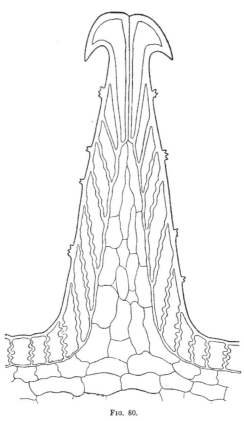

FIG. 80.

L.S. through one of the barbed appendages of the fruit of *Cynoglossum cheirifolium*.

ternal layers of the wall swell so as to obliterate the cell-cavity almost entirely. The cutinised layer is consequently torn across into a number of strips, which, on account of their annular thickenings, act as so many hoops binding the mucilaginous layers together. The most remarkable mucilage-hairs occur in the seed-coats of certain LYTHRACEAE; they were first observed by Kiärksou, and have been described in greater detail by Köhne, Klebs and Correns.[114] In this instance the so-called hairs are not trichomes in a morphological sense at all. In

o

*Cuphea viscosissima* each epidermal cell of the dry seed-coat contains an irregularly coiled thread-like protuberance, projecting inwards from its outer wall and almost filling its cavity (Fig. 81).  This coiled filament is covered with grooves arranged in very gently ascending spiral series. On the access of water, a circular patch of the outer wall immediately overlying the insertion of the filament breaks away on one side like a lid, and the filament begins to turn inside out.   In this way a "hair" is produced, the wall of which consists of the everted outer membrane of the filament, while the swollen contents of the latter now form a coating of mucilage on the outside of the hair.   As the filament

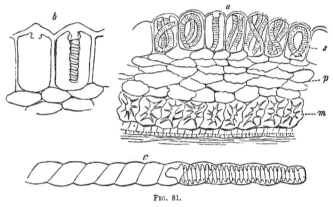

Fig. 81.

Mucilage-hairs of the seed of *Cuphea petiolata*. *a*, T.S. through the testa ; *s*, epidermis ; *p*, parenchyma ; *m*, stereides ; *b*, two epidermal cells more highly magnified ; *c*, part of an everted hair.   After Klebs.

turns inside out, its spiral folds straighten themselves more or less completely.   Correns has shown that the energy required for the eversion of the filament is derived from the swelling of a substance contained in the cell-cavity ;  the same agency may even produce a certain amount of " turgescence " in the completely everted hair.

A totally different mechanical function is performed by the so-called **eel-trap hairs** (*Reusenhaare*), which are developed in connection with certain ecological relations between plants and crawling insects or other minute animals.   We have already had occasion, in discussing trichomes in general, to mention the fact that stiff hairs are often directed obliquely downwards, especially on stems and petioles, so as to act as obstacles to insects which may attempt to crawl upwards.   In such cases the aim in view is of course the repulse of "unbidden guests" ; in the insectivorous genus *Sarracenia*, on the other hand, the eel-trap hairs serve to prevent the escape of captured animals.   The pitchers of *Sarracenia purpurea* are furnished with two different kinds of these hairs. The flap overhanging the mouth of the pitcher has its adaxial surface—which faces the opening—covered with numerous unicellular hairs that point obliquely downwards; the cuticle of each hair is provided with conspicuous longitudinal ridges, which extend, at regular intervals, from the broad base or foot right up to the apex of the hair. The other eel-trap hairs occur on the inner surface of the narrow lower portion of the pitcher. These have thick walls, like the first-mentioned hairs, but are much smaller and devoid of cuticular ridges, while the foot is not broad but elongated.

The more elaborate types of eel-trap hair are furnished with special locking devices, which prevent the hair from being pushed back so far as to allow the insect to climb over the obstacle. According to the author's observations [115], the stem of *Biophytum proliferum* bears, just below the insertion of each whorl of leaves, a ring of densely crowded, stiff hairs, arranged in a somewhat irregular manner, but all pointing obliquely downwards. Each hair is about 3 mm. in length and consists of a single cell, with very thick lignified walls showing obvious stratification; it has a short sunken foot which is likewise thick-walled and provided with numerous pits. Just above the foot, the hair is produced, on its upper side, into a blunt process of varying length; this protuberance is also

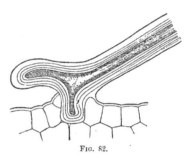

FIG. 82.

Basal portion of an eel-trap hair from the stem of *Biophytum proliferum*.

very thick-walled, and is often bent backwards towards the surface of the stem (Fig. 82). When an insect which is trying to ascend the stem pushes one of these hairs upwards, the blunt end of the aforesaid process soon comes into contact with the stem, and so arrests the movement of the hair; the protuberance thus acts as a catch or stop.

The eel-trap hairs in the perianth-tube of *Aristolochia Clematitis*, which have been investigated by Hildebrand and by Correns, are provided with a locking arrangement that acts on a similar principle.[116] As is well known, these hairs serve to prevent the premature egress of small flies which find their way into the perianth-tube; the flower is protogynous, and the flies are only allowed to escape, by the collapse of the hairs, when they have been covered with pollen owing to the dehiscence of the anthers. Each hair (Fig. 83) consists of three parts. The foot is composed of two cells, the upper of which has very thick outer walls. The solitary hinge-cell is narrow below, but expands towards its distal extremity; its outer wall is thin and very extensible. The body of the hair, finally, consists of a long row of rather thin-walled and highly turgescent cells; at its base it is expanded on its upper side so as to form a hump, which acts in the same manner as the basal protuberance in the case of *Biophytum*. As each hair is sunk in a shallow depression of the inner surface of the perianth-tube, the hump soon comes into contact with the margin

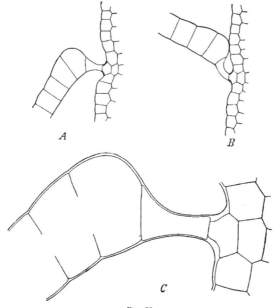

FIG. 83.

Basal portion of an eel-trap hair from the perianth-tube of *Aristolochia Clematitis*.
*A.* Hair in the "normal" position.  *C.* Hair in the "normal" position, more highly
magnified.  *B.* Hair in the "locked" position.

of the depression, if the hair is pushed upwards, and thus arrests the movement very quickly (Fig. 83 B).

In the "locked" position, the body of the hair stands out at right angles from the surface of the perianth-tube, or is inclined slightly upwards and outwards. The different hairs project on every side and interlock with one another, so that the flies are effectually prevented from escaping. The exit remains closed in this manner until the hairs die and shrivel up; in the course of this process of collapse the thin outer walls are thrown into folds, and the cross walls approach one another. The highly adaptive character of these remarkable eel-trap hairs is shown, not only by the differentiation of a hinge-cell and a locking device, but also by the circumstance that the inflexibility of the hair-body is ensured, not by the presence of a thick outer wall—which would have been the simplest arrangement—but by the turgescence of a thin-walled tube, stiffened at intervals by transverse buttresses in the shape of cross-walls. In this way the hairs are enabled to collapse, after they have performed their temporary function, and thus to leave the way open for the escape of the pollen-covered insects.

# CHAPTER V.

## THE ABSORBING SYSTEM.

### I. GENERAL CONSIDERATIONS.

ALL the materials that are taken into the plant-body from the surrounding medium must enter in the liquid condition or in a state of solution. It is true that plants can render solid bodies available for absorption with the aid of enzymes and acid substances which they secrete for that purpose; such solvent actions come into prominence during the germination of albuminous seeds, and also play an important part in the so-called digestive processes of carnivorous plants. But in these cases one is after all dealing with changes which go on *outside* the plant-body, and which merely prepare the solid substances for absorption. In fact, if the phenomena of fertilisation, and the ingestion of solid matter which takes place in the Myxomycetes, be left out of consideration, it may be stated quite generally that foreign material cannot penetrate into the interior of a living cell except by diosmosis.

Among the substances which are absorbed by all plants, water occupies a unique position. A comparatively small amount of this substance is needed for nutritive purposes, in order, that is, to provide the hydrogen and oxygen which enter into the composition of most organic compounds. A somewhat larger quantity is required to serve as the so-called water of imbibition which saturates protoplasm, cell-walls, starch grains, and so forth, and to provide the bulk of the cell-sap. By far the greatest proportion, however, of the water absorbed, by land-plants at any rate, is needed to make good the loss due to transpiration.

All the other materials which are normally absorbed by a plant represent, from the nutritive point of view, either indispensable food constituents, or else so much useless matter, which is taken in merely because the organism, though possessed of a certain amount of selective capacity, is not altogether able to exclude substances that are devoid of nutritive value.

It will be convenient, in considering the conditions under which absorption is carried on, to begin with the case of a **submerged green plant** such as a unicellular or multicellular Green Alga. Plants of this kind, surrounded as they are on all sides by a medium which contains all the necessary food-constituents in the proper degree of concentration, can absorb nutrient materials over the whole of their outer surface. In these circumstances special absorbing tissues can, as a rule, be dispensed with; the root-like organs with which many Algae are provided probably never exactly correspond to the roots of land plants, but serve solely as organs of attachment, and thus properly belong to the mechanical system

**Terrestrial green plants** are placed under very different conditions. They obtain water and food materials, partly from the soil in which they are rooted, and partly from the atmosphere that envelops their aërial organs. As a rule the air only provides a single one among the food-constituents, namely, carbon dioxide; but this substance is of the very greatest importance, since it represents the raw material from which the green cells extract carbon by the process of photosynthesis. The relation of a terrestrial plant to atmospheric carbon dioxide is thus comparable to that of a submerged plant⁻ to the whole of its food-materials. The photosynthetic organs may absorb carbon dioxide over their entire surface, as is the case among Mosses; where on the contrary the external surface is insufficient or unsuitable, the defect is remedied by the development of a system of internal air-spaces. The intake of carbon dioxide thus forms part of the more general process of gaseous interchange—just like the absorption of respiratory oxygen. No special tissue is known to exist for the purpose of absorbing carbon dioxide or of conveying it in a state of solution to the places where it is utilised.

From the soil, terrestrial green plants obtain water and the nutrient salts dissolved therein; for this purpose they develop a system of special absorbing organs or **roots**; these structures are further provided with a specialised absorbing tissue, which in a purely topographical and ontogenetic sense corresponds to the epidermis of the aërial organs.

Modifications of the typical absorbing arrangements are far from uncommon among land plants; they generally arise in connection with the special mode of life of the plants in question, or, in other words, in relation to certain climatic or edaphic peculiarities of their habitat. The most frequent of these modifications depends upon the water-absorbing capacity possessed by many aërial organs, and particularly by leaves. This mode of absorption plays a prominent part in the water economy of the Bryophyta, and is also widely distributed among Phanerogams. It is necessary, however, to distinguish between two degrees of specialisation in this respect. The leaves of a great many plants are able to absorb water when partly withered, simply because the cutinisation of the outer epidermal walls does not render them absolutely impervious. In such cases one is dealing with a purely incidental property, which may occasionally redound to the advantage of the plant, but which cannot in any way affect the normal balance of its water-interchange. Many epiphytes on the other hand—especially BROMELIACEAE—and also certain steppe- and desert-plants, depend for their water-supply entirely upon the ability of their leaves to absorb the moisture that becomes available in the form of rain or dew. In these cases, accordingly, the aërial portions of the plant are provided with special absorbing organs, which not infrequently exhibit a complicated and highly adaptive structure.

Among **epiphytic** ORCHIDACEAE and AROIDEAE, again, the aërial roots possess, in the so-called " velamen," a special apparatus which serves for the absorption of dew and rain, and perhaps also for the condensation of water-vapour.

The case of Fungi and other **non-chlorophyllous plants** must next be considered. Owing to the absence of chloroplasts, such plants must lead a saprophytic or parasitic existence, or, in other words, are entirely dependent upon organic food-materials. Here also, absorption is carried on over the whole external surface in the simplest types, such as Yeasts and Bacteria. At a higher stage of evolution, division of labour leads to the segregation of the propagative organs from a purely vegetative region of the plant-body, the task of absorption being then entrusted to the latter, which in the most specialised types becomes entirely converted into a ramified absorbing organ. The filamentous mycelium, which is so characteristic of the Fungi, serves mainly as such an organ of absorption ; in the case of certain Phanerogamic parasites the vegetative body becomes likewise reduced to a system of haustorial organs.

In the matter of absorption, **embryos**—and also **seedlings**, so long as they are dependent upon the mother-plant for their supplies of plastic material—occupy a position which is in a sense intermediate between the condition of green plants and that of non-chlorophyllous forms. In these instances, also, the office of absorption is, among the more specialised types, entrusted to special haustorial organs ; if the latter are bulky multicellular structures, they are generally furnished with a definite absorbing tissue.

The preceding general remarks will have made it evident that the structure of the absorbing system gives a clear indication of the general mode of nutrition of a plant, while it also throws some light upon the nature of its habitat and upon its other ecological peculiarities.

In conclusion, the general features which are common to all the different types of absorbing system may be briefly indicated. The principle which, above all, dominates the construction of this system, is that of the **maximum exposure of surface**; for, clearly, the quantity of nutritive material that can be absorbed by osmosis, is *ceteris paribus* directly proportional to the area of the absorbing surface. With regard to the disposition of absorbing tissues, it is obvious that the latter must be situated near the periphery, and in fact must consist of superficial cell-layers. A third feature of absorbing cells, and one which greatly facilitates diosmosis, is the unthickened character of their walls; this condition is, however, not incompatible with the development of local thickenings for mechanical purposes. Mention must finally be made of a purely physiological but very characteristic and important feature of living absorbing tissues, namely, their capacity for producing and excreting acid substances and enzymes; were it not for this property the solution of solid food materials would be very slow, and in many cases quite impracticable.

## II. THE ABSORPTION OF INORGANIC COMPOUNDS.

### A. THE ABSORBING SYSTEM OF SUBTERRANEAN ROOTS.

The subterranean root-system of terrestrial green plants, the axial and foliar organs which, in certain cases, perform the functions of roots and the roots of water plants, carry out the absorption of water and soluble salts with the aid of tissues which are essentially of the same type in every case. In discussing this form of absorbing tissue, we shall throughout regard the ordinary subterranean root as the typical absorbing organ.

Absorption is not carried on over the whole surface of a subterranean root-system. It can, on the contrary, be easily shown that the functional absorbing tissue only occurs on the youngest rootlets, and that it is, moreover, confined to a definite zone, which begins close behind the subapical growing region, and ends at a varying distance from the tip owing to the degeneration and exfoliation of its cells on the older parts of the root. In the case of the commonest type of absorbing system, which is characterised by the presence of " root-hairs " (see below), the restricted distribution of the actual absorbing tissue may be demonstrated by means of a very simple experiment. All that is necessary is to dig up a plant carefully with its roots intact, and to attempt to remove the soil-particles adhering to the latter by shaking and washing. It will be found that the soil cannot be completely removed, except from the actual root-tips,—which hence have a smooth and whitish appearance —and from the portions which betray their age by their brown coloration and in other ways. The zones bearing the absorbing tissue, on the contrary, tenaciously retain a large proportion of the soil-particles,— which are indeed firmly glued to the root-hairs—and thus appear clothed in a coating of earth.

The single superficial layer which represents the entire absorbing tissue of the root, corresponds morphologically to the epidermis. The cells of this layer are extended parallel to the long axis of the root. They contain a moderately thick peripheral layer of protoplasm. Their cellulose membranes are generally thin and soft, particularly in the case of the outer walls, which constitute the actual absorbing surface. These walls are further devoid of a cuticle, but instead usually possess a coating of mucilage. The rest of the outer wall consists of cellulose layers, which, according to Kroemer, sometimes undergo lignification or become impregnated with protective substances of unknown composition.

Among roots, one may recognise two stages in the specialisation of the absorbing tissues, which differ from one another in respect of the amount of surface that is made available for purposes of diffusion. Some plants content themselves with the increase of surface which can be obtained by augmenting the output of roots; here the outer walls of the absorbing cells are flat or slightly convex, so that the absorbing region of the root is smooth. In other cases the enlargement of the absorbing surface depends chiefly upon the fact that the absorbing cells are prolonged externally into so-called **root-hairs**. This more advanced condition represents the normal type of absorbing tissue in the case of roots.

The simple form of absorbing tissue, in which root-hairs are absent, suffices for the needs of such plants as grow under conditions that are so favourable to the absorption of water and nutrient salts as to render the exposure of a large absorbing surface unnecessary. Most marsh- and water-plants are thus fortunately circumstanced. Hence the roots of such species as *Butomus umbellatus, Caltha palustris, Hippuris vulgaris, Lemna minor, Nymphaea alba, Pistia Stratiotes,* are entirely devoid of root-hairs. Certain forms which grow in similar localities (e.g. *Scirpus sylvaticus* and *Carex paludosa*) are nevertheless provided with root-hairs; as a rule the plants in question can grow temporarily in drier surroundings.

The two stages of specialisation are, as a matter of fact, connected by numerous intermediate steps. In this connection the varying ecological relations of the plant play a great part, as appears from the circumstance that either type of absorbing tissue may be developed within a single species, or even in one and the same individual, according to the nature of its requirements for the time being. There are quite a number of water-plants, such as *Elodea canadensis, Nuphar luteum, Acorus calamus* and *Cicuta virosa*, the roots of which produce no root-hairs so long as they are immersed in water, while they develop these structures in abundance on entering the soil.

Fuller consideration must be given to the type of absorbing tissue which is characterised by the constant presence of root-hairs.[117]  The greatest interest attaches to the hairs themselves, since these structures are almost entirely responsible for the functions which are ascribed to the whole absorbing tissue. Reference has already been made, on more than one occasion, to the fact that root-hairs owe their existence to the principle which aims at the maximum exposure of surface. The increase of absorbing surface which is effected by the development of these structures is indeed very considerable. Schwarz has estimated its numerical value in certain cases. His method consists in determining the number of root-hairs borne on 1 sq. mm. of root surface and in measuring their average length and diameter ; the total surface exposed per unit length can then be readily calculated. In this way it was found, for example, that the surface of a Maize root growing in a moist chamber, and consequently well provided with root-hairs, exceeds the surface of an [imaginary] completely hairless root in the ratio of 5·5 to 1. For roots of *Pisum sativum*, likewise grown in a moist chamber, the corresponding ratio is 12·4 to 1, for the aërial roots of *Scindapsus pinnatus* 18·7 to 1, and for the lateral roots of *Trianea bogotensis*—grown in water—6·63 to 1. It must not, however, be supposed that these figures indicate any fixed ratios peculiar to the different species ; on the contrary the increase of surface effected by the formation of root-hairs corresponds very closely to the relative humidity of the soil, inasmuch as both the number and the average length of the hairs are reduced in very wet soil, where water and nutrient salts are available in abundance. This reduction is carried to its logical conclusion, when the root-hairs are entirely suppressed in contact with liquid water ; this condition, which occurs, according to Schwarz, in *Allium Cepa, Cicer arietinum, Cucurbita Pepo, Helianthus annuus, Phaseolus communis, P. multiflorus, Ricinus communis, Zea Mays*, etc., may be regarded as a reversion to the simpler type of absorbing tissue. A low degree of humidity in the soil, on the other hand, renders the absorption of water and nutrient materials more difficult, and thus necessitates an increased production of root-hairs. A very dry soil, finally, retards the appearance of the hairs and may even inhibit their formation altogether; this effect must be regarded as a pathological condition of arrested development, induced by the unfavourable environment. Under normal conditions, most plants are undoubtedly able to accommodate their output of root-hairs very closely to the prevailing external conditions.

An abundant production of root-hairs is only possible if all the absorbing cells are capable of giving rise to these structures. Where this is the case, the hairs may be exceedingly numerous. Schwarz estimates that a piece of Maize root (grown in a moist chamber), 1 mm. in length, bears on an average 1925 root-hairs, a number which in a root 1·44 mm. in diameter represents a density of 425 hairs per sq. mm. A similar piece, taken from a root of *Pisum sativum*, bore (under the same conditions) 1094 hairs altogether, or 232 per sq. mm. It should be noted that the absorbing cells only retain the capacity for producing root-hairs for a limited time. Hence it is probable that fresh hairs are never interpolated among pre-existing ones; in other words, root-hairs always arise, so far as is known, strictly in acropetal succession.

In certain cases, especially among water-plants, the power of forming root-hairs is restricted to particular absorbing cells, which usually differ markedly in appearance from the surrounding hairless cells. In *Nuphar luteum*, *Sagittaria sagittaefolia*, *Elodea canadensis*, etc., these piliferous cells are much shorter than the rest, being cut off by trans-verse walls at an early stage of development. In *Hydromistria stolonifera* and *Hydrocharis Morsus-Ranae*, on the contrary, Kny found that the piliferous cells are distinguishable, while still covered by the root-cap, owing to their greater width and depth. In certain ERIOCAU-LACEAE (*Paepalanthus* spp.) and JUNCACEAE, according to van Tieghem, the short piliferous cells generally become divided by a longitudinal wall, each of the two daughter-cells thereupon growing out to form a root-hair, so that these structures come to be associated in pairs. In *Distichia* these "twin-hairs" are fused for half their length, but diverge above so as to resemble a single forked hair. In *Lycopodium*, according to Nägeli and Leitgeb, a whole group of root-hairs arises from a small common initial cell; the latter divides into two to four daughter-cells, each of which gives rise to a separate root-hair.

Attention must next be directed to the morphology and physiology of the individual root-hairs. Almost invariably it is only a limited and rather sharply defined region of the outer wall of a piliferous cell that grows out to form the root-hair. Frequently this portion is situated at the acroscopic end of the cell; there are, indeed, a number of plants in which this relation is quite constant. Now, as a rule, it is only the main root that grows straight downwards, while the smaller rootlets may form the most diverse angles with the vertical; it thus seems *a priori* probable, that the aforesaid localised origin of the root-hairs has nothing to do with the stimulus of gravity, but rather represents a case of correlation between the mother-organ and its appendages.

Among *Phanerogams* every root-hair is a mere diverticulum of one of the absorbing cells. It is obvious that the continuity of the cell-cavity favours the rapid translocation of absorbed materials. The walls of root-hairs are covered internally by a thin peripheral layer of proto-plasm, which is usually somewhat thicker towards the apex of the hair; as a rule the nucleus lies either within this apical protoplasm or close behind it.

To begin with, every root-hair consists of an unbranched cylindrical tube with a rounded tip; this primitive form is, of course, only retained if the root-hair develops in water or in damp air. It is under these conditions, also, that root-hairs generally attain the greatest length. In the various species examined by Schwarz the maximum length was found to vary between ·15 and 8 mm. To quote a few specific instances, the maximum length in spp. of *Potamogeton* is 5 mm. (in water), in *Elodea canadensis* 4 mm. (in mud), in *Brassica Napus* 3 mm. (in damp air), in *Pisum sativum* and *Avena sativa* 2·5 mm., in *Vicia Faba* ·8 mm., in *Muscari botryoides* ·5 mm., etc. When growing in soil, root-hairs usually suffer a very considerable reduction in length; the changes of shape which these structures undergo in contact with soil-particles are of even greater interest, because they indicate the physiological significance of the root-hairs in the clearest possible manner.

Every root-hair has an inherent tendency to grow out at right angles to the surface of the root; on its way through the soil, however, it is certain before long to encounter a solid particle, which will force it to turn aside. It then grows on in contact with the particle, until it meets with a crevice filled with air or water, and is thus enabled to resume its original line of advance. In this way a root-hair may come into contact with the same solid object at several different points; at each contact the hair describes a knee-shaped curve, and at the same time opposes the greatest possible surface to the obstacle. It consequently often expands into a disc-like structure, or produces lobed outgrowths which grasp the soil-particle like the fingers of a hand: the surface of the hair, in fact, faithfully retains the imprint of all the irregularities in the outlines of the objects to which it adheres (Fig. 84). These changes in the shape of the root-hair are accompanied by a retardation of its general growth, so that the hair often reaches a mere fraction of the length which it could attain if it were allowed to develop in damp air. The significance of this diminution in length is quite obvious.

The great elongation of a root-hair in moist air, after all, only represents a means to an end; it is a response, which enables the hair to traverse considerable interstices and cavities in the soil, and so ultimately to come into contact with fresh solid particles. Absence of light, as is well known, causes an excessive elongation of aërial organs; consequently, even if a shoot begins its development far below the surface of the ground, it is usually able to reach the light sooner or later. Similarly, absence of contact with solid bodies causes an excessive elongation of root-hairs, whereby these structures are ultimately enabled to penetrate into solid ground. Evidently these two phenomena, though so widely different in their physiological significance, are nevertheless closely comparable from an ecological point of view. The above-mentioned modifications of the root-hairs must undoubtedly be regarded, not as cases of mere arrested development, but rather as responses to the stimulus of contact with solid particles.

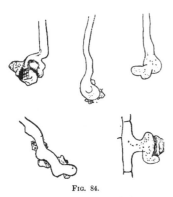

FIG. 84.

Root-hairs of *Linaria Cymbalaria.*

The connection between root-hairs and the soil-particles to which they adhere, is usually rendered more intimate and permanent by a process of agglutination; the existence of this process was assumed by Charles Darwin, and the phenomenon has since been more closely studied by Schwarz, who attributes it to gelatinisation of the outermost layers of the root-hair membrane.

The type of longitudinal growth exhibited by root-hairs entirely accords with the manner in which they penetrate the soil and glue themselves to solid particles. According to observations carried out by the author[118] upon root-hairs of *Cucurbita Pepo, Pisum sativum, Polygonum Fagopyrum* and *Helianthus annuus,* which had been marked by dusting them with rice-powder, longitudinal growth is strictly confined to the dome-shaped tip of the hair, and ceases altogether immediately behind the apex. Root-hairs thus exemplify an extreme type of apical growth, and in this respect also show themselves to be exceedingly well fitted for the task of pushing their way between the solid particles of the soil. The significance of the very close apposition of the root-hairs to solid particles only becomes fully evident, when one considers how the water and nutrient salts are distributed within the soil. As a matter of fact only a small part of the most important food-constituents exist in a state of solution in the soil-water; by far the larger proportion are **adsorbed**, that is, retained in the undissolved condition by virtue of the molecular forces resident in the minute soil-particles. The adsorbed nutrient materials include, for example, salts of *potassium* and *ammonium*, as well as *phosphates* and *iron*-compounds. We need not concern ourselves here with the chemical and physical forces that underlie this phenomenon of adsorption. It will suffice if we imagine the adsorbed food-materials to exist in the form of a very fine precipitate which envelopes each tiny soil-particle, being in its turn surrounded by films of water of varying thickness; for the water in the soil is also, in part at least, attached to the soil-particles by molecular attraction.

In the light of these facts we can now fully comprehend the significance of the close connection—amounting almost to fusion—which exists between root-hairs and soil-particles. It is only through this intimate contact that the root-hairs are enabled to withdraw water and adsorbed nutrient material from the soil by which they are so firmly retained.

The absorption of mineral salts is further greatly facilitated by the circumstance that roots secrete acid substances and thus exert a solvent action upon the surrounding medium. This fact is easily demonstrated by a simple experiment, which was first performed by Sachs.[119] A slab of polished marble is covered with a moderate thickness of soil, in which a plant is allowed to develop its root-system; the roots that come in contact with the slab corrode its surface, and after a time produce a very distinct etched pattern. The lines of such **root-prints** are very sharply defined, a fact which indicates that the acids secreted by the absorbing tissue do not distribute themselves throughout the soil-water, but are on the contrary largely retained in the water of imbibition of the cell-walls and in the innermost of the water-films that adhere to the outside of the roots. This circumstance furnishes an additional reason for the exposure of the greatest possible surface on the part of the roots.

Czapek has shown that root-prints are produced, mainly if not entirely, owing to the excretion of carbon dioxide. According to Stoklasa and Ernest, organic acids (*acetic acid* and *formic acid*) are only excreted when the roots are insufficiently supplied with oxygen. As soon as any solids have been brought into solution, they can readily penetrate as far as the ectoplast—which regulates the diosmotic interchanges of the root-hair—since the wall of the hair is, with few exceptions, not more than ·0006 to ·001 mm. in thickness. At the rounded tip of the hair the wall is often twice or even thrice as thick as in the other portions, an arrangement which evidently assists the hair in penetrating the soil.   This thickened apical membrane, in fact, acts like a miniature root-cap.

Owing to the thinness of their cell-walls, root-hairs are naturally somewhat delicate and short-lived structures; at a certain distance from the root-tip they collapse, turn brown and die.   The whole absorbing tissue indeed likewise soon dies and disappears; the root is thenceforth covered by the suberised "exodermis," which represents the outermost layer of the cortex.   Rapid alternations of dryness and humidity, transplantation and similar causes also usually lead to the destruction of the root-hairs; under these conditions death may be preceded by pathological changes of form, such as irregular branching (e.g. *Brassica* spp.), etc.

According to the author's observations, two epiphytic Ferns which are exceedingly common in Java and the rest of the Malay Archipelago, namely, *Drymoglossum nummularifolium* and *D. piloselloides*, are distinguished by the possession of exceptionally resistant and long-lived root-hairs.   Here, namely, when a root-hair dries up in consequence of long-continued shortage of water, the cytoplasm and nucleus withdraw to a basal portion of the hair which becomes marked off from the rest by a more or less regular transverse constriction.   At the level of this constriction a cap-shaped septum is laid down, cutting off the encysted protoplast from the withered distal region of the hair, which is thereupon cast off.   In this way a dormant piliferous cell is produced, which only requires the access of water, in order to resume its activity and give rise to a new root-hair.

Hitherto discussion has been confined to the principal or **nutritive function** of root-hairs.   Some account must next be given of the **mechanical function** which these structures perform in connection with the fixation and distribution of the root-system in the soil.   It is self-evident that the above-described adhesion of the root-hairs to solid particles must play an important part in anchoring the plant in the ground.   But this adhesion also produces a mechanical effect of a different kind, in connection with the passage of the root through the soil.   By fixing the sub-apical region of the root firmly in the ground, the root-hairs provide the resistance which that organ requires in order to overcome the friction of the soil.   Schwarz, however, has pointed out, how essential it is that the actual growing apex should not be deprived of its flexibility.   As he says: "If the tip of the root were fixed to the soil-particles by hairs, the stimuli produced by contact, moisture and gravity would be largely ineffective, while the growth of the root would also be retarded.   For this reason the root-hairs almost invariably first appear at some little distance from the apex.   It is interesting to note, that they approach more closely to the tip in roots which grow slowly or in very stiff soil, than they do in roots which meet with little resistance. A slender root, again, requires a better purchase than a stout one, because it is more easily pushed aside; in thin roots we accordingly find that the distance of the youngest hairs from the apex is relatively small." Schwarz estimates the average distance of the anterior limit of the piliferous zone from the apex at ·75 to 1 mm. for slowly-growing roots, and at 2 or 3 mm. for those that grow rapidly. In very thin roots, such as those of *Poa pratensis*, the youngest hairs may be situated only ·11 mm. from the tip. Certain CRASSULACEAE (e.g. *Sedum Andersoni* and *Sempervivum Funkii*) behave in an exceptional manner, inasmuch as their roots bear hairs right up to the apex; this aberrant condition is probably connected with the fact that the plants in question always grow in very hard and dry soils.

Certain seedlings display very instructive adaptations in connection with the mechanical function of their root-hairs. Ecologically considered, one of the most important conditions for successful germination is the early establishment of a satisfactory connection with the soil; the urgent nature of this necessity becomes evident, when we consider that, under natural conditions, seeds usually germinate *upon the surface* of the soil. For this reason, numerous long root-hairs are generally produced during the very first stages of germination. Thus they have been observed to arise in great numbers at the junction of root and hypocotyl in water-plants by Warming,[120] and in *Eucalyptus* and other MYRTACEAE by Briosi. In the seedlings of *Panicum miliaceum* and *Setaria italica*, the first-formed root-hairs appear, according to Schwarz, on the so-called coleorhiza or sheath of the radicle; the grain is therefore actually fastened to the ground before the root emerges from the coleorhiza. A similar arrangement has been observed by the author in *Briza minor*, and is probably somewhat prevalent among Grasses in general. In all such cases, of course, the hairs serve for absorption as well as for fixation.

As the preceding examples show, root-hairs may occur on organs other than those from which they derive their name. This aberrant location of the absorbing tissue is still more noticeable, where the place of the root-system is taken by modified axial or foliar organs. In the case of *Psilotum triquetrum*, for instance, where genuine roots are altogether absent, the rhizome is furnished with root-hairs. Mention must also be made of the HYMENOPHYLLACEAE, in which root-hairs appear on modified axial organs and sometimes even on the leaves. Goebel records the occurrence of tufts of root-hairs on the bases of the outer members of the leaf-rosette in certain Orchids (*Microstylis*, *Malaxis*, *Sturmia*). The submerged leaves of *Salvinia natans*, finally, which in habit closely resemble genuine roots, are also covered with absorbing hairs. Both in the HYMENOPHYLLACEAE and in *Salvinia*, however, the root-hairs no longer represent mere prolongations of the superficial cells, but are separated by cross-walls from the piliferous elements. In *Salvinia*, in fact, each hair is divided into from seven to nine compartments by a corresponding number of transverse walls; the terminal segment is short and conical, and ends in a sharp point.

It has already been stated that the absorbing tissue of the root inclusive of the root-hairs—is short-lived and deciduous. After its disappearance the root is bounded on the outside by the outermost cortical layer, which acquires suberised walls and thereafter constitutes a continuous secondary epidermis, or, as it is usually termed, an **exodermis.** If this suberisation takes place while the absorbing tissue is still active (*Coleus, Lamium, Hedera*; most Monocotyledons), special short cells with unsuberised walls and abundant protoplasmic contents are interpolated at regular intervals between the ordinary long exodermal cells; the short elements serve as **passage-cells** for the transmission of absorbed materials from the absorbing tissue to the living parenchyma of the cortex. The most specialised types of exodermis are found in aërial roots.

<h3 style="text-align:center">B. RHIZOIDS.[121]</h3>

The possession of genuine roots is characteristic of the most advanced stages in the differentiation of the plant-body. It is, in fact, only the Phanerogams and the sporophytes of the Pteridophyta that are furnished with true roots; and even in these most highly organised vegetable forms the root-system is not infrequently suppressed in connection with some special adaptation. In addition to the Thallophyta, which may be put on one side for the present, all the Bryophyta—Mosses as well as Liverworts—and the gametophytes of the Vascular Cryptogams are devoid of roots. Among the Bryophyta, and in those Pteridophyte prothallia that are autotrophic, the functions which among more highly organised plants are entrusted to the root-system and its absorbing tissue, devolve entirely upon special trichomes known as **rhizoids.** The simplest type of rhizoid is hardly distinguishable from a root-hair. The term "root-hair" is indeed often applied to these characteristic gametophytic trichomes. There are, however, grave objections to this extended definition of a root-hair. Physiologically considered, the trichomes in question represent both roots and root-hairs; that is to say, they are endowed with the properties of typical root-hairs, but in addition possess many of the capacities of roots; in particular, rhizoids agree with roots in being sensitive to the influence of light, gravity and moisture, whereas genuine root-hairs are quite free from these forms of irritability.

The rhizoids of Liverworts and Fern-prothallia scarcely differ from typical root-hairs, except in the presence of a basal wall. Each rhizoid consists of a single thin-walled tubular cell, which, in contact with solid particles, displays the same tendency towards expansion and adhesion that has already been discussed in connection with root-hairs (Fig. 85). Rhizoids further agree with root-hairs in exhibiting very pronounced apical growth.

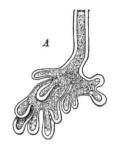

The members of the most specialised series of Liverworts, the MARCHANTIACEAE, possess two sorts of rhizoids, namely the ordinary or "simple" kind, and, in addition, peculiar "pegged" rhizoids, distinguished by the presence of numerous internal thickenings of the cell-wall, in the shape of simple or branched pegs, which project into the cavity of the rhizoid on every side; the pegs are closely crowded and often distinctly arranged in spiral series. The simple rhizoids are principally attached to the midrib of the thallus, where they extend close up to the apical region. The pegged rhizoids, on the contrary, occur chiefly on the wings of the thallus. It is these pegged rhizoids which are mainly responsible for the supply of water and mineral salts to the plant.

FIG. 85.

Distal ends of two rhizoids of *Mastigobryum trilobatum. A.* With numerous lobes. *B.* Flattened into a disc-like structure.

The author is therefore inclined to suspect that the pegs have no special mechanical significance, but that they serve in the first instance to augment the absorbing surface of the rhizoid by throwing the ectoplast into folds. According to this view one is dealing with a case of internal increase of surface; and the pegged rhizoids are to be regarded as better fitted for the task of absorption than their smooth-walled prototypes. Kamerling has put forward a different theory concerning the purpose of the pegs. · He believes that they serve to keep the gas-bubbles, which appear in the rhizoid when the influx of water diminishes, "suspended in the centre of the cavity" and so to enable water to pass between the bubbles and the wall. Further investigation is certainly required, before one can come to a final decision with regard to the physiological significance of pegged rhizoids.

The rhizoids of Mosses are among the most remarkable of trichomes; in every respect they display a very far-reaching adaptation to their special functions. As a rule each rhizoid represents a system of

abundantly branched filaments.   The main branches are often five or
six times as thick as the ultimate ramifications; the former take the

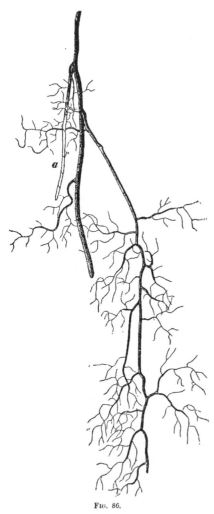

place of the roots of a Vascular Plant, while the latter
are comparable to root-hairs.
There is thus a close physiological resemblance between a
fully developed rhizoid and an
entire root-system (Fig. 86).
Except for the slender ultimate
branches, all the ramifications of
the rhizoidal system consist of
cellular filaments; the individual cells are usually long in
comparison with their breadth
and are separated from one
another by oblique transverse
walls.   Both the thin-walled
character of the septa and their
oblique position—which is an
embodiment of the principle of
maximum exposure of surface
—facilitate diosmosis between
the successive cells, and thus
increase the conducting capacity of the whole filament.   In
the   fine   ultimate   branches,
which are responsible for a much
smaller amount of conduction,
the   septa   are   often   of   the
ordinary transverse type.   The
lateral walls of Moss-rhizoids
are generally brown in colour,
in which case they exhibit an
evident differentiation into two
layers.   The cell-contents com-

FIG. 86.

Fully developed rhizoid of *Tortula muralis*.  *a*, a young
lateral branch.

prise a peripheral protoplasmic
layer, which sometimes contains
leucoplasts or small pale-green chloroplasts, and a central cavity filled
with colourless sap.   All the branches, as well as the main axis, display
very pronounced apical growth.

The above general account may be supplemented by a detailed
description of the rhizoidal system of *Tortula muralis*, a plant which

has furnished the author with material for the investigation of several interesting ecological peculiarities of Moss-rhizoids. On examining the growing main axis of a rhizoid of *Tortula,* one is at once struck by the fact that all the transverse walls are laid down straightway in the oblique position; their obliquity is thus not brought about by secondary displacement. The lateral branches make their first appearance at a short distance from the growing apex, and invariably arise at the acroscopic end (*i.e.* at the end nearest the apex of the branch) of a cell. Opposite the oblique septum of such a cell the lateral wall develops a papillate protuberance, the basal diameter of which is roughly equal to the thickness of the main axis (Fig. 87 A). The protuberance, which is filled with dense protoplasm, becomes cut off from the mother-cell by a thin "watch-glass-shaped" wall, and now represents the initial cell of a lateral branch. These initial cells do not, however, necessarily at once develop into branches, but in a certain number of cases—though not in many—pass first through a period of rest, during which they may be quite fairly compared to the dormant buds of a Dicotyledonous tree-trunk. The majority of initial cells, however, develop without passing through this resting stage. If they arise from young rhizoids, they grow out into thick lateral branches, which turn downwards in response to the stimulus of gravity, just like the main axis. Initial cells produced by older rhizoids, on the contrary, give rise only to comparatively slender branches, which slope downwards quite gently, or which may even follow a horizontal course. These thin branches are cut off by means of transverse septa from the initial cells, which maintain their individuality unchanged and, as their subsequent behaviour shows, become for the time being converted into dormant initials.

When the main axis has ceased to grow, and the principal branches have reached a certain age, some of the dormant initials give rise to a new series of thick lateral branches. If such an initial had at a previous stage produced a slender branch, the new member invariably arises on the physically lower side of the old one (Fig. 87 B). Those initials, however, which previously at once entered the resting state, on resuming their activity as a rule straightway produce thick branches; in exceptional cases they first produce thin branches, and give rise to thicker members later on. Every thick lateral branch behaves just like the main axis in respect of its method of ramification. The above-described arrangements evidently provide a very effective means of regulating the output of absorbing organs. The plant is thus enabled to take advantage of favourable conditions without loss of time, while with the aid of its resting initials it can survive unfavourable periods with comparatively slight risk of damage.

The ecological importance of the dormant initials becomes even more evident, when one considers that lateral branches never arise adventitiously from other parts of the rhizoids and that, under normal conditions, the green protonemal branches also originate from the aforesaid initials.

The resemblance of the members of such a rhizoidal system to root-hairs is most obvious in the case of the thin-walled slender branches. In contact with soil-particles, these ultimate branches undergo the characteristic modifications of form which have already been repeatedly mentioned; they flatten themselves, embrace the particles and adhere to them precisely like root-hairs. Even the main axis occasionally undergoes similar changes of shape. One of our figures (Fig. 87 c) depicts a rhizoid which in its downward course has evidently encountered a solid object of considerable size. As a result it has first flattened itself, and has then on one side given rise to a stout lateral branch which has clung closely to the obstacle, while on the other side it has continued its growth in the original direction.

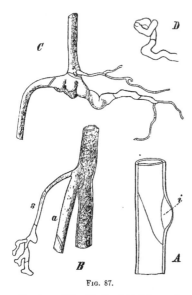

Fig. 87.

Structure of the rhizoids of *Tortula muralis.*
*A.* Small portion of the main axis of a rhizoid; *i*, initial cell of a lateral branch. ×450. *B.* Portion of a rhizoid in a more advanced stage of development; an initial cell has given rise first to a slender branch (*s*), and subsequently to a stout one (*a*). *C.* Part of a rhizoid showing the effect of contact with a large soil-particle upon the main axis (see text). *D.* Coiled extremity of a slender lateral branch.

Like roots and root-hairs, rhizoids serve as organs of attachment, besides performing their principal function of absorption. In those Mosses which inhabit bark, and which consequently derive their sustenance almost entirely from the atmospheric precipitation and from the dust that accumulates among their tufted shoots, the mechanical significance of the rhizoids comes especially into prominence. This statement applies likewise to many of the species that grow on rocks. Among such Mosses again as live in running water (*Fontinalis, Cinclidotus*), the rhizoids are also purely mechanical in function, and are, according to Paul, remarkable for the thickness of their walls. The cable-like rhizoidal strands of certain POLYTRICHACEAE, finally, which consist of a number of slender rhizoids twisted together around a stouter central one, are also probably constructed with a view to inextensibility.

The stems of certain Mosses (*Paludella squarrosa*, spp. of *Dicranum*, *Meesea*, and *Mnium*) are covered for the greater part of their length with a dense felted mass of rhizoids. Oltmanns has shown that this rhizoidal jacket represents a device for the capillary retention and conduction of water.

### C. THE ABSORBING TISSUES OF AERIAL ROOTS.[122]

The aërial roots of tropical Orchids and of many epiphytic Aroids are distinguished by the presence of a peculiar covering, which has

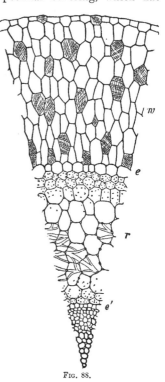

long been known by the names of "root-sheath" (*Wurzel-hülle*), *velamen radicum*, or shortly **velamen**; the anatomy and development of this tissue have been most closely and thoroughly studied by Leitgeb, upon whose observations the following account is largely based. The velamen usually consists of a silvery parchment-like sheath of variable thickness. A study of its development shows it to be a product of the protoderm, which close behind the root-tip becomes many-layered by tangential division. In a few exceptional cases (*Vanilla planifolia, V. aphylla, Dendrocolla teres*, etc.) the protoderm divides by anticlinal walls only, and the velamen consequently remains single-layered. The number of layers comprised in the velamen ranges from a single one to as many as eighteen (*Cyrtopodium* spp.), but remains at a fairly constant figure in each species.

The cells of the velamen are in uninterrupted contact with one another. Their shape varies to a considerable extent. Sometimes they

FIG. 88.

Sector of a T.S. through the aërial root of *Stanhopea oculata. w*, velamen; *e*, exodermis; *r*, cortical parenchyma; *e'*, endodermis. ×150.

are approximately isodiametric but somewhat elongated radially, as seen in transverse section; in other cases their long axes are parallel to that of the root itself. The walls are strengthened in a great variety of ways, but most often by means of spiral thickening fibres (Figs. 88 and 89); these fibres may run parallel to one another, or they may enclose meshes of varying width, or, finally, they may be associated to form broad ribbons or belts.  Reticulate thickenings are less common than those of the spiral type, but occur for example in *Dendrocolla teres* and *Vanda furva*.  Rarely the walls are uniformly thickened and provided with more or less numerous pits (*Angraecum subulatum*); in a few instances, finally, the cell-wall remains thin throughout (*Trichotosia ferox*).  All these various forms of thickening sometimes occur side by side in the same velamen, while the types of thickening may differ even on different sides of one and the same cell (*Renanthera matutina*).  Between the thickening fibres the cell-walls are often perforated by holes ; these perforations are not confined to the septa between adjoining cells, but occur also in the outer walls of the superficial layer.  The presence of the holes can be simply demonstrated by dipping a [dry] aërial root in water, when the liquid is found to be rapidly absorbed.  The frequent occurrence of minute Algae (*Protococcus, Raphidium*, etc.) in the interior of velamen-cells likewise indicates the presence of perforations.  Leitgeb, finally, has proved the existence of holes in the cell-walls by injection of the velamen with an

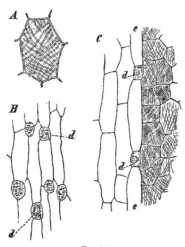

FIG. 89.

*A.* Velamen-cell from an aërial root of *Stanhopea oculata*, showing the thickening fibres of the Wall. × 480.  *B.* Small portion of the exodermis in surface view (from a tangential L.S. through the root); *d, d*, passage-cells.  *C.* Part of a radial L.S. through the root; *e*, exodermis ; *d, d*, passage-cells.

emulsion of vermilion and water.  In their adult condition the velamen cells almost invariably contain nothing but air ; to this fact is due the silvery appearance of such aërial roots.

From a physiological stand-point, the velamen must be regarded as the absorbing tissue of the aërial root, a fact recognised long ago by Schleiden and Unger.  When it is recollected that the walls of the velamen-cells are perforated at a number of points, and that their air-containing cavities communicate freely with the outer atmosphere, one can readily comprehend why this spongy tissue greedily absorbs rain-water and dew.  This notion of the capillary action of the velamen was brought into prominence first of all by Duchartre, and subsequently by Schimper and Goebel.  While probably correct, it is by no means incompatible with the opinion of the earlier observers (such as Schleiden, Unger, Chatin, and Leitgeb), who believed that the spongy texture of the velamen enables it to condense water-vapour and other gaseous constituents of the atmosphere (*e.g.* ammonia), and thus to render them available for the benefit of the shoot.  The absorption of ammonia by the velamen has indeed been experimentally demonstrated by Goebel in the case of *Odontoglossum Barkeri*.  The more important part of the problem, however, is that which refers to the condensation of water-vapour.  The experiments which have been carried out with regard to this matter by various investigators in greenhouses or in the laboratory have indeed mostly led to negative results.  It is, nevertheless, quite possible that the water-condensing powers of aërial roots, *under natural conditions*, are sufficient to confer an appreciable ecological advantage upon plants that possess such organs.  In any case the experiments in question were mostly carried out upon *severed* roots ; but in this way a serious error is introduced, owing to the stoppage of the current of water which normally flows out of the absorbing organ.  The problem can only be solved by means of experiments performed upon intact plants growing in their natural surroundings.

The histological structure of the velamen certainly supports the condensation-theory.  The number and delicacy of the fibrous thickenings hardly seem consistent with the idea that they merely represent the resistant framework of a capillary apparatus.  Both the aforesaid features, however, greatly increase the surface available for condensation, especially in the case of the so-called "fibrous bodies," which in certain instances are associated with the exodermal passage cells ; these fibrous bodies will be more fully discussed later on.

We have still to consider how the water absorbed—or possibly condensed—by the velamen is transferred (together with the substances dissolved therein) to the cortex of the root.  The cortical parenchyma is separated from the velamen by a very characteristic intermediate layer, the **exodermis** (or outer endodermis), which in all essential features resembles a typical endodermis (cf. Chap. VII.).  The exodermis comprises two different sorts of cells.  Those of the first type are more or less elongated ; their outer walls are very often thickened, but never pitted.  Among these long cells, shorter, rounded, and almost invariably thin-walled elements, with abundant protoplasmic contents, are interspersed, usually in vertical series (Fig. 89, *d*).  The long cells are the exodermal elements in the strict sense ; they have suberised walls, and are consequently relatively impervious to water. They accordingly serve to prevent the aërial roots from drying up during periods of prolonged drought, the velamen itself being of course incapable of undertaking the protective functions usually performed by a dermal tissue.  The small thin-walled cells, on the other hand, act as channels for the inward passage of the water collected by the velamen.

In the aërial roots of certain Orchids the elements of the velamen which immediately overlie the passage-cells—the "cover-cells" of Leitgeb—are distinguished by the presence of very remarkable disc-shaped or spherical local thickenings on their inner walls. The structures in question were first observed in the genus *Sobralia* by Leitgeb, who also demonstrated their fibrous texture. More recently Meinecke has made a detailed study of these peculiar thickenings, which he terms **fibrous bodies**. A typical fibrous body develops as follows. First a number of delicate fibrous ridges make their appearance upon the inside of the cell-wall; from these arise numerous exceedingly slender rodlets, which at first project at right angles. Soon the fibres and rods all become interwoven into a felted mass,

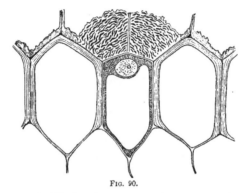

Fig. 90.

Small portion of a T.S. through an aerial root of *Sobralia macrantha*, showing a passage-cell with a fibrous body attached to its outer wall.

which may reach a considerable size. Leitgeb supposes that the highly porous body produced in this way first absorbs and retains any water that reaches it, and subsequently transfers the liquid *gradually* to the underlying passage cell. The author, however, considers it more probable that these fibrous bodies represent minute but highly efficient condensing organs. Their location in contact with the outer walls of the passage-cells agrees quite well with this interpretation. It must, however, be admitted that those who deny the condensing capacity of the velamen are entitled to regard the fibrous bodies as protective plugs serving to retard evaporation from the thin-walled passage-cells, a possibility which was indeed recognised by Leitgeb. In any case the controversy as to the function of these interesting structures can only be settled by experimental investigation.

In certain cases (*e.g.* in *Angraecum subulatum*) the velamen is cast off, when the root has reached a certain age, the exodermis thenceforth filling the place of a normal epidermis. A similar exfoliation of the velamen takes place, if a root which is at first aërial subsequently enters the ground.

At this stage reference may be made to certain anatomical specialisations among Mosses,[123] which may be regarded as the physiological counterparts of the velamen of aërial roots, inasmuch as in both cases water is absorbed through aërial organs by means of capillary action.

In the genus *Leucobryum*, and in the LEUCOBRYACEAE in general, the leaf consists of a single median layer of green cells enveloped on either side by one or more layers of large colourless cells devoid of protoplasmic contents, which communicate with one another and with the outer atmosphere by means of large circular holes in their walls. The photosynthetic tissue is thus enveloped in a system of capillary tubes. So long as these capillaries are full of air, the leaf has a whitish appearance—hence the name *Leucobryum*; when the leaf is wetted, however, the capillaries instantly become filled with water, and the green colour of the photosynthetic cells becomes apparent.

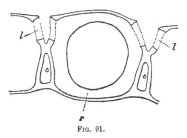

FIG. 91.

Part of a leaf (from a lateral branch) of *Sphagnum cymbifolium*, in T.S. : *r*, annular thickening fibre of a capillary cell ; *l*, perforations in the outer walls of the capillary cells ; *c*, photosynthetic cells. After Russow.

In the genus *Sphagnum* the tubular photosynthetic cells of the leaf are arranged in the form of a network, the meshes being occupied by colourless " capillary cells " (Fig. 91). The walls of the latter are stiffened by means of annular or spiral fibrous thickenings, and in this respect display a higher stage of adaptation than the thin-walled capillary cells of *Leucobryum*. In the distal portions of some of the leaves the annular and spiral thickenings expand, according to Russow, into broad plates or ribbons, which constitute a particularly effective stiffening apparatus. The unthickened tracts of the walls are perforated by large openings, which are usually circular in outline ; very frequently the margin of each hole is surrounded by an annular thickening fibre. The stem of *Sphagnum* is likewise provided with a capillary apparatus comprising from two to four of the outermost cortical layers. The capillary cells resemble the analogous elements of the leaf ; both their transverse and their longitudinal walls are perforated, so that the whole tissue can become saturated with water in a very short time. As the individual cells are of considerable size, relatively large amounts of water can be stored in this cauline capillary tissue, while a certain amount of upward movement of the water may also be effected by capillary action. We are therefore, in

this case, concerned with a tissue which simultaneously performs several
different functions (viz. absorption, storage and conduction of water).
It might, accordingly, have been dealt with in a later chapter; on
account of its histological features, however, it is most naturally
associated with the foliar absorbing tissue of the LEUCOBRYACEAE.

### D. THE WATER-ABSORBING HAIRS OF FOLIAGE LEAVES.[124]

Not only the Bryophyta and certain Ferns (HYMENOPHYLLACEAE),
but also many Phanerogams, possess the capacity of absorbing water
through the surface of their foliage-leaves.    If withered leaves or leafy
shoots are immersed in water—the cut surfaces being kept dry—they,
as a rule, regain their turgescence completely after a certain time,
and sometimes even in the course of a few hours.    While the water-
absorbing capacity, which many leaves reveal under these conditions,
may acquire some ecological importance in certain circumstances, it
does not necessarily constitute a specially evolved useful adaptation;
for every article that can be wetted at all is in some degree pervious
to water, and the living cells of a withering leaf will thus in any case
absorb water deposited upon the outside of the epidermal wall to a
certain extent, by virtue of their osmotic activity.    Like transpiration,
this superficial absorption is primarily a physical process, whereas its
regulation in the interests of the uninterrupted normal activity of the
foliage leaf is effected by means of special adaptations, which come
into being only in relation to particular ecological conditions.    It is
only in dry climates, or more generally under xerophytic conditions,
that plants find it necessary to make immediate use—by superficial
absorption—of every drop of rain or dew that comes in contact with
their foliage.

The first stage in the regulation of foliar water-absorption, whereby
this phenomenon ceases to be purely incidental and becomes an integral
part of the vital activity of the plant, consists in the fact that absorption no longer takes place over the entire surface of the leaf, but is
restricted to definite points of entrance, which, as a rule, exhibit some
evident histological specialisation.    It sometimes happens that the
outer epidermal walls are more pervious to water over the veins; this
condition may be looked upon as the first step towards the localisation
of foliar absorption.    Almost always, however, it is some specialised
form of trichome that is entrusted with the task of absorbing water;
such structures will henceforth be referred to simply under the name
of **absorbing hairs** (or scales).    The absorption of water may also be
assigned to organs of a somewhat different type, which likewise
correspond morphologically to hairs or to modified epidermal cells;
these are the so-called **hydathodes,** which are fully dealt with in a

subsequent chapter (Chap. X., II., A.). The principal function of hydathodes is the *secretion* of water, but many of them are also capable of *absorbing* liquids. It is, in fact, impossible to draw any sharp distinction between secretory hydathodes and water-absorbing trichomes. After a certain stage in the life of a leaf the hydathodes not infrequently devote themselves exclusively to absorption. Again, in closely related species which differ greatly in their mode of life, homologous trichomes may in one case serve as hydathodes, and in another act as absorbing hairs.

In spite of a number of statements and suggestions to the contrary, it may be regarded as an established fact that leaves never absorb water through their stomata, at any rate under normal conditions. Apart from the strictly localised "water-pores," stomata serve solely as the external openings of the ventilating system, and for this reason are actually furnished with numerous devices for preventing capillary occlusion of the stomatic pores and access of water to the intercellular spaces. Leaves that have been immersed in water for purposes of experiment may become translucent in places, owing to the partial injection of the intercellular spaces with water which has entered by the stomata; but this phenomenon must be regarded as a result of the unnatural conditions to which the leaf is exposed and not as an illustration of the normal action of the stomata.

Absorbing hairs are found on the leaves of many plants of hot, dry climates, but especially among species inhabiting desert regions; it is, in fact, the possession of such hairs that enables desert-plants to make use of the nocturnal dews which in the Egypto-Arabian desert, for instance, occur almost nightly from November to April, and often collectively represent a very considerable amount of precipitation. Many members of the Mediterranean flora are also provided with absorbing hairs, and similar organs doubtless occur in a number of our native plants which inhabit dry, sunny situations. As regards epiphytes, water-absorbing trichomes have hitherto only been recorded in the BROMELIACEAE, where, however, they attain a very high degree of specialisation.

In respect of their external form, water-absorbing trichomes may be developed as ordinary simple hairs or bristles, as capitate or clavate hairs, or, finally, as peltate hairs or scales; both unicellular and multicellular forms are known.

**Unicellular absorbing hairs** occur, according to Volkens, in *Diplotaxis Harra* and in certain species of *Heliotropium* inhabiting the Egypto-Arabian deserts. The author's observations upon material collected by himself confirm Volkens' statements in all essentials. The leaves of *Diplotaxis Harra* (Fig. 92 A) bear a number of stiff, spreading bristles, with a rough surface and a bulbous base; the hairs are covered throughout by a delicate cuticle, which rests immediately upon the cellulose layers of the wall. The moderately thick waxy covering of the epidermis stops short at the base of each bristle; hence the hairs are readily wetted, although water will not adhere to the general surface of the epidermis. Except for the bulbous base, the cavity of the hair is occluded by layers of cellulose, which afford a splendid illustration of stratification. The basal portion contains a massive peripheral protoplasmic layer with a nucleus; its thickened lateral and basal walls are abundantly pitted. The adjoining subepidermal mesophyll cells are radially elongated and colourless, and constitute a localised water-storing tissue. If a severed leaf of this plant is allowed to wither and is then repeatedly moistened with water, it fully regains its normal turgidity in a very short space of time. Since the wax-covered epidermal cells cannot be wetted at all, the absorption of water must be entirely effected by the hairs. Under natural conditions, the process probably takes place somewhat as follows. Drops of dew run down the outside of a hair and are absorbed through the outer wall of the swollen base; after absorption the water diffuses through the pitted walls of the bulb into the adjacent water-storing cells. In *Diplotaxis* the annular absorbing zone at the base of the hair is not differentiated histologically; in the somewhat less bulky, but otherwise very similar, absorbing hairs of *Heliotropium luteum, H. undulatum* and *H. arbaïnense,* however, the corresponding region is specially thin-walled. The contrast between the thick outer epidermal wall and the thin-walled base of the hair is particularly striking in the first mentioned species (Fig. 92 B).

**Multicellular absorbing hairs** often form a felted covering on both sides of the leaf. It is, of course, not every felted hairy coat that is capable of absorbing water. If such a covering is easily wetted and rapidly absorbs drops of water; if further the hairy leaf rapidly recovers its turgidity when immersed or besprinkled in the withered condition; if, finally, places for the entrance of water through the hairs are indicated by the presence of thin-walled basal cells with abundant protoplasmic contents, it may be safely assumed that the hairs in question serve to some extent for the absorption of water. According to Volkens and E. Gregory, hairs of this type are usually constructed in such a way that one or more thin-walled living basal absorbing elements are surmounted by an elongated cellular filament, which is filled with air, or which has its cell-cavities occluded by massive thickening layers. These dead distal portions of the hair may become interwoven into a felted mass (*Petasites albus,* **P.** *niveus, Helichrysum graveolens, Salvia argentea, Alfredia nivea, Inula Helenium, Atractylis flava, Ifloga spicata,* etc.), or they may all point in the same direction and so produce a smooth glistening coat (*Convolvulus Cneorum*, *Plantago cylindrica*, etc.). Besides helping to reduce the amount of transpiration (cf. Chap. III., II., C.), these distal parts of the hairs, by virtue of their capillary action, retain drops of rain or dew and conduct them to the basal absorbing portions.

Two examples will serve further to explain the structure of this type of absorbing hair. The leaf of *Centaurea argentea* bears, on both surfaces, a very dense felted covering composed of unbranched absorbing hairs (Fig. 92 c). Each hair comprises an embedded basal cell, one

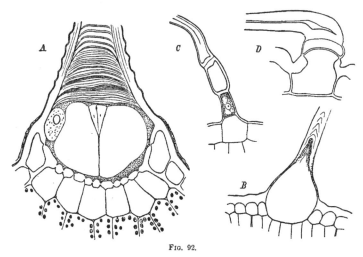

Fig. 92.

Water-absorbing hairs of various xerophytes.  *A. Diplotaxis Harra.*  *B. Heliotropium luteum.*  *C. Centaurea argentea.*  *D. Convolvulus Cneorum.*

to three thin-walled, slightly elongated absorbing cells, a broader and somewhat thick-walled subterminal cell and a greatly elongated, air-containing terminal cell, which is separated from the subterminal cell by a very thick transverse wall. A wilted leaf which was immersed in water—except for the cut surface of the petiole—was found to resume its normal turgidity in the course of twenty-four hours, during which period it absorbed water to the extent of 13 per cent. of its weight. Both sides of the leaf of *Convolvulus Cneorum* are covered with a shining coat of densely crowded hairs; each hair has a very thick-walled, elongated terminal cell, while the outer wall of the short disc-shaped absorbing cell—which intervenes between the terminal cells and the embedded basal cell—is exceedingly thin (Fig. 92 D). The thin-walled character of the absorbing cell is rendered more striking by the great thickness of the outer epidermal walls. A wilted leaf—

the hairy covering of which was found to absorb water with avidity—
showed an increase of weight of 10 per cent., after being immersed for
twenty-four hours (the same precautions being observed as in the previous
case). It may be noted in passing that particular care must be taken, in
experiments of this kind, not to carry out the second determination of
weight, until the water which is retained by capillarity in the interstices
between the hairs has evaporated.

De Bary long ago remarked upon the frequent occurrence of
embedded, sessile or shortly stalked, capitate or clavate hairs of a non-
glandular character. These hairs appear in many cases to represent
absorbing organs of a special type, the actual absorption being carried
out by the one- or more-celled capitate or clavate distal end of the
hair, which is provided with abundant protoplasmic contents. The
outer wall of the stalk cell is often furnished with a strongly cutinised
annular thickened zone, serving to keep the diameter of the channel
through which water passes from the hair into the mesophyll constant,
even in the wilted condition of the leaf. The one- or several-celled
basal portion, finally, is often enlarged, in order that the greatest
possible number of photosynthetic cells may be brought into direct
communication with this part of the absorbing organ. In *Peperomia
scandens* the basal cell is prolonged inwards, and often projects far down
into the water-tissue.

Where organs of this kind occur on leaves which are found by
experiment to be capable of absorbing large quantities of water in a
short space of time, where further the quick penetration of coloured
solutions into the cavity of the hair and the rapidity with which
plasmolysis takes place indicate that the outer wall, though cutinised,
is nevertheless comparatively pervious, it is fairly safe to conclude that
the hairs in question represent water-absorbing organs. The absorption
of thin films of water is further facilitated by the circumstance that
such hairs are generally inserted below the general level of the
epidermis, or else are brought distally into close contact with the
epidermal surface, owing to curvature of the stalk-cell. As examples
we may mention the capitate hairs of *Syringa vulgaris* and the clavate
hairs of *Vaccinium Vitis Idaea* (the latter acc. to Lundström). Both
these instances, however, require further investigation. As a matter of
fact, our knowledge of absorbing hairs of this type is altogether
insufficient; in many cases they certainly also serve as water-secreting
organs (cf. Chap. X.).

The most highly organised foliar water-absorbing organs are the
absorbing scale-hairs of the BROMELIACEAE, our knowledge of which we
owe chiefly to the researches of A. F. W. Schimper. Among the forms
that have their leaves arranged in rosettes, the absorbing scales are

located mainly on the leaf-bases, which collectively form a water-reservoir; in the matted and long-stemmed species (e.g. *Tillandsia*), on the other hand, both leaves and stems are covered all over with peltate scales.

The histological structure of such an absorbing scale is in general as follows (Fig. 93). A more or less expanded one- to several-celled basal portion, or **foot**, is continued into a funnel-shaped **stalk**, which is sunk below the level of the epidermis. This stalk consists of three or four flat, thin-walled cells containing abundant protoplasm, which

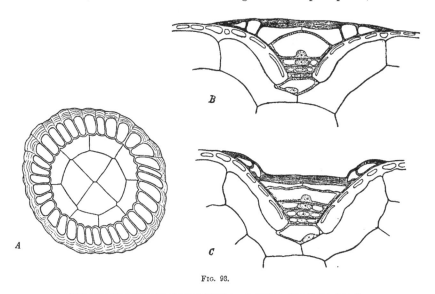

Fig. 93.

Water-absorbing scales of *Vriesea psittacina.* *A.* Surface view of the disc of a scale. *B.* Vertical section of a scale in the turgescent condition. *C.* Vertical section of a scale in the collapsed state.

represent the actual absorbing apparatus; it is surmounted by an approximately circular, or, in some cases, irregularly-shaped **disc**. The marginal cells of the disc are often radially elongated; in the genus *Tillandsia* the membrane of these cells is produced into a scarious radially folded wing. In contrast to the stalk-cells, both the marginal and the central cells of the disc are devoid of living contents. The upper (*i.e.* outer) cell-walls of the disc are only slightly thickened in the case of species which grow in moist shady situations, or which have their scales completely sunk below the surface of the epidermis; where, on the other hand, the scales project, the corresponding walls are of very considerable thickness. In all circumstances, however, these outer walls of the disc are entirely free from cutin, and are completely

Q

dissolved by treatment with concentrated sulphuric acid. According to Mez, they consist of a cellulose matrix impregnated with a high percentage of pectic compounds. The cuticle itself may also be soluble in sulphuric acid (*Vriesea psittacina*), or it may persist after treatment with this reagent in the form of an extremely delicate pellicle (*Tillandsia usneoides*).

When a Bromeliaceous leaf which is provided with a dense covering of peltate scales is moistened, water is rapidly absorbed, and the leaf, which was whitish in the dry state, assumes a greenish coloration. Microscopic examination shows that the cells of the scales have become filled with water. If a drop of caustic potash be placed upon the epidermis and wiped off after a few seconds, it will be found that, around each scale, the previously colourless contents of the epidermal cells have assumed a golden-yellow tint. After treatment with salt solution, the plasmolytic contraction of the protoplasts first appears in the parenchymatous cells surrounding the base of each scale. All these facts, taken in conjunction with the circumstance that the epiphytic BROMELIACEAE are either altogether rootless or at most possess only grasping roots, and nevertheless remain fresh for weeks at a time, provided that their leaves are occasionally moistened, indicate that the scales are extremely well fitted for the task of absorbing water and aqueous nutrient solutions.

A few words must finally be devoted to the function of the thickened outer walls of the disc. In the dry condition of the hair the thin lateral walls of the disc-cells are crumpled up, while the thick outer walls almost touch the living stalk-cells, which are sheltered by them as by a lid. On the access of water, the lateral walls expand once more to their former size, and the lid is raised. The disc thus represents a protective device, which prevents the escape of water-vapour through the thin-walled stalk-cells during periods of prolonged drought.[125] The action of the lid is very clearly seen in the case of *Vriesea psittacina*. Here the circular disc (Fig. 93 A) comprises, first, a number of radially elongated marginal cells, secondly, a circle of eight tangentially elongated intermediate cells, and, thirdly, a group of four central elements placed cross-wise. It is only the central cells that have thick outer walls, and thus form the actual lid (Fig. 93 B). When transpiration has been going on actively for some time, the lid sinks down, until its margin rests upon the upper edge of the funnel-shaped stalk, which thus becomes completely shut off from the outer atmosphere (Fig. 93 c).

Water-absorbing trichomes in general are, with few exceptions, covered with a cuticle insoluble in sulphuric acid, even at the points through which water actually enters. Whether the undoubted permeability of this cuticle depends upon aberrant chemical composition, or upon special structural peculiarities—such as the presence of ultra-microscopic pores—is a question which cannot be answered in the present state of our knowledge.

In conclusion, it is necessary to consider whether the living proto-plasts of water-absorbing trichomes take any *direct* part in the process of absorption—by acting, as it were, like so many tiny pumps—or whether, on the other hand, the entrance of water is merely a physical consequence of the osmotic suction produced in the mesophyll cells by transpiration. In the latter event the living elements of the hairs would merely serve as readily permeable passage-cells; although this hypothesis cannot be regarded as an impossible one, and in fact does in all likelihood apply to the less specialised types of absorbing hairs, it seems more probable that, in other cases, those elements of the hairs which are distinguished by the presence of well-developed proto-plasts (and also the passage-cells of aërial roots) do actually take an active part in absorption, by providing the requisite energy for " pump-ing" water into the mesophyll. The question is, however, really a physiological one, and can therefore· only be settled by experiment.


## III. THE ABSORPTION OF ORGANIC FOOD-MATERIALS.

### A. THE ABSORBING SYSTEM OF EMBRYOS AND SEEDLINGS.

Every young sporophyte may be regarded as a parasitic organism, at any rate during the early stages of its development, when it is entirely dependent upon the reserve materials stored in the maternal tissues. This period of parasitic existence begins with the first forma-tion of the embryo, and continues until the end of germination. The arrangements which are connected with the absorption of plastic materials, in such cases, may be grouped under two heads. The first of these comprises the haustorial organs that transfer plastic materials to the **embryo**, while it is developing during the ripening of the fruit, and is therefore still in connection with the mother-plant. The second class includes all the structures which assist the **young seedling** to absorb reserve materials, when the seed begins to germinate, after it has separated from the mother-plant and has passed through a more or less prolonged period of rest in the ground.

Haustorial organs of the first type may be developed from struc-tures of the most varied morphological value, such as the embryo-sac, the endosperm, the suspensor, the antipodal cells (in certain COMPOSITAE and RUBIACEAE), one of the synergidae (*Calendula*), or even the pollen-tube (*Cucurbita*, according to Longo). From among these various possibilities the three first-mentioned may be selected for further illustration.[126]

In several species of *Linum* the lower region of the **embryo-sac** (or megaspore) serves as a haustorium; this portion, which ultimately becomes marked off from the upper part by a constriction, contains protoplasm and receives a few of the endosperm-nuclei, but no cell-walls are ever formed between the latter. In such a case as this the haustorial portion of the embryo-sac remains unbranched. In other instances an increased exposure of surface is effected by the more or less extensive branching of the haustorium. This more advanced condition is exemplified, according to Billings, by *Globularia cordifolia*. Here the upper extremity of the embryo-sac grows out as a tubular process through the micropyle, spreading all over the upper end of the ovule and coming into contact with the wall of the ovary; this process puts forth a number of filamentous outgrowths, some of which push their way downwards between the growing seed and the pericarp, while others extend upwards along the funicle towards the placenta.

The formation of **endospermic haustoria** was first observed by Treub in *Avicennia officinalis* (VERBENACEAE). In this plant the endosperm—together with the enclosed embryo—passes bodily through the micropyle and into the cavity of the ovary, with the exception of a single huge and abundantly branched cell (the "*cellule cotyloide*"), which acts as a haustorial organ; the branches of this haustorial cell ramify throughout the nucellus, and finally even invade the placenta. In the above-mentioned *Globularia cordifolia* the lowermost cells of the endosperm grow out in tubular fashion into the adjoining integument; the latter thus becomes riddled by a system of haustorial tubes, which ultimately absorb the whole of its substance.

The author has elsewhere fully described the very remarkable endospermic haustoria discovered by him in certain viviparous Mangroves. In *Bruguiera eriopetala* (RHIZOPHORACEAE) the endosperm, which forms a thin layer surrounding the four basally connate cotyledons, grows out into numerous haustorial lobes resembling root-hairs. These arise in the following manner. During the early stages of development of the fruit, the primary endosperm is entirely obliterated by the growing cotyledons, except for isolated approximately hemispherical cells with abundant protoplasmic contents, which adhere closely to the cotyledonary surface. These persistent endospermic cells later develop into multicellular discs or "islands" of endosperm, which ultimately become more or less fused with one another. This peculiar endospermic tissue puts forth numerous one- to many-celled haustorial processes, which penetrate the very loose parenchyma of the integument (Fig. 94 B). It is noteworthy that, at a number

of points, individual cells of the endosperm give rise to tubular
processes on their *inner* side, which insert themselves between
the radially elongated palisade-like epithelial cells of the cotyledons
(Fig. 94 A).  These latter processes are of course not actually haustorial organs, but evidently serve to produce the closest possible
connection between the absorbing endosperm and the seedling which is
destined to receive the absorbed material.
Here the principle of maximum exposure
of surface affects the structure, not only
of the receptive part of the endosperm,
but also of its discharging surface.

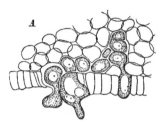

In *Aegiceras majus* (MYRSINACEAE),
which is likewise viviparous, endospermic
haustoria occur in the form of wedge-
shaped lobes and folds or villi, which are
not infrequently branched; but these
developments are strictly confined to the
placental region, where the influx of
plastic and reserve-materials takes place.
That the presence of these endospermic
haustoria must facilitate the nutrition of
the large seedlings of viviparous plants
is sufficiently obvious.  One is involun-
tarily reminded of the richly branched
chorionic villi and lobes of the Mam-
malian placenta, which are themselves
indeed nothing more than elaborate
haustorial developments.

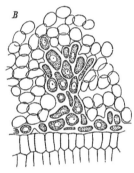

FIG. 94.

Endospermic haustoria of *Bruguiera
eriopetala*.  *A*. Portion of the integument with the adhering palisade-like
epithelium of a cotyledon; the latter is
traversed by several small branches of
the endospermic haustoria.  *B*. One of
the larger haustoria; the round celled
tissue belongs to the integument, the
palisade-like layers to one of the cotyledons.

Where haustoria are formed by the
**embryo** itself, they generally originate
from the **suspensor** ; here again the
dominating influence of the principle of
maximum exposure of surface is often
beautifully illustrated.  Thus Treub has
described how, in species of *Phalaenopsis* (*P. grandiflora*, *P. Schil-
leriana*, etc.), each of the four cells of the suspensor grows out into
a pair of long tubular structures; one set of these haustorial tubes
penetrates the micropyle, while the rest surround the embryo like a
bundle of Fungus-hyphae.  In another Orchid, *Stanhopea oculata*, the
fertilised egg-cell first of all divides to form a ten- to fifteen-celled
spherical structure, called the proembryo; a single cell of this proembryo gives rise to the true embryo, while all the rest grow out into
long tubes, which push their way among the cells of the nucellus right

up to its outermost layer, sometimes even extending into the micropyle. According to Koorders, the large absorbing vesicles that appear on the proximal portion of the suspensor of *Tectona grandis* represent modified cells of the endosperm, which have become secondarily fused with the suspensor. Thus structures which are by no means homologous may serve as haustorial organs in one and the same plant.

Attention must next be directed to the **absorption of reserve-materials during** the process of **germination**. Where the whole of the stored up plastic material provided by the mother-plant for the use of the offspring is contained within the confines of the embryo—in which case it is usually deposited in the cotyledons—the transitory parasitism of the young plant produces little or no effect either upon its external form or upon its internal organisation. Special arrangements for absorption only occur, where the reserve-materials are deposited in tissues, such as the endosperm—or more rarely the perisperm—which are not in organic connection with the embryo. In such circumstances, namely, the germinating embryo has to extract its plastic materials from foreign tissues, like a genuine parasite; for this purpose it develops absorbing tissues, and in cases of greater specialisation provides itself with definite organs of absorption.

Where the embryo is entirely surrounded by the storage tissue, it generally absorbs the soluble food materials through every part of the surface which is in contact with that tissue; it is more especially the protodermal tissue of the cotyledons that is at first actively engaged in absorption, whereas later the same layer becomes converted into a typical epidermis. This primitive arrangement provides an excellent illustration of change of function. In such cases there is really no specialised absorbing system. The method of absorption is usually very similar to that just described, when the embryo is attached to one side of the endosperm. In the seeds of *Agrostemma Githago*, for instance, the lower surface of one of the cotyledons is the only part of the embryo that is in contact with the endosperm; the superficial layer in this region at first acts as the absorbing tissue, but subsequently becomes converted into an epidermis with stomata, in every respect resembling the epidermal layer that arises directly from the protoderm of the outer cotyledon, which is all along in contact with the testa. The only difference between the absorbing epithelium and the non-absorbing protoderm consists in the fact that the cells of the former undergo distinct radial elongation, a modification which represents the first step towards the development of a specialised absorbing tissue.

The embryo agrees with the adult plant in the fact that, where definite absorbing tissues are present, these are located in special absorbing organs.  Structures of this nature, comparable to haustoria and often actually designated by this name, are especially characteristic of Monocotyledonous seedlings.  Most often it is the distal portion of the cotyledon that remains embedded in the endosperm during germination in order to serve as an organ of absorption.  A familiar instance is that of the Date Palm.

The absorbing tissues of germinating embryos resemble those of roots in the fact that they can be referred to two distinct types, of which one is more specialised than the other.  The simpler type is characterised by the possession of absorbing cells with flat, or at most slightly papillose, outer walls.  The absorbing surface exposed is thus relatively small, and the absorption of reserve-materials proceeds but slowly, weeks or even months elapsing before the endosperm is completely exhausted. This slow absorption is, however, necessitated by ecological conditions in the case of many plants, which accordingly find the simple form of absorbing tissue quite sufficient for their needs; the latter is, indeed, the only type that is known to occur among Palms, and in the LILIACEAE, IRIDACEAE, ZINGIBERACEAE, MARANTACEAE, CYPERACEAE, etc.

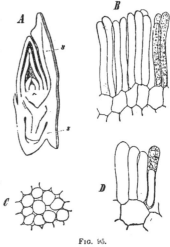

FIG. 95.

*A.* Embryo of Wheat in the resting condition (L.S.): *s, s,* scutellum (its absorbing epithelium is shaded).   ×16.   *B.* Part of the scutellar epithelium at an advanced stage of germination, showing the tubular absorbing cells.   ×230.   *C.* Part of the scutellar epithelium at an advanced stage of germination; absorbing cells in T.S.   *D.* Cells of the scutellar epithelium from a seedling of *Stipa Calamagrostis.*   ×320.

The embryos of Grasses are furnished with a more elaborate form of absorbing tissue,[127] which is located on the dorsal surface of the scutellum, where the latter abuts against the endosperm (Fig. 95 A). Even in the resting condition of the embryo, the absorbing cells are elongated at right angles to the surface of the scutellum [forming the so-called scutellar epithelium]; at this stage they are in uninterrupted contact with one another laterally.  During germination the scutellum increases considerably in size; since the epithelial cells do not expand laterally in the same degree, they become partially or completely separated from one another.  In this way numerous intercellular spaces open to the outside [*i.e.* towards the endosperm] are produced, while individual cells, or small groups of cells, become isolated all round (Fig. 95 B).  The increased surface which becomes available in this way is

still further augmented by active growth on the part of the epithelial
elements themselves.    The resulting elongation is often very considerable.    The average length of the epithelial cells of Wheat, for example,

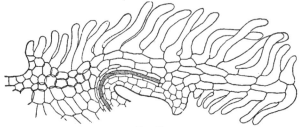

FIG. 96.

Scutellum (haustorial organ) of a seedling of *Briza minor* ; the absorbing tissue
resembles the piliferous layer of a root.

is ·023 mm. in the resting condition, but rises to ·09 mm. during the
period of most active absorption ; for Maize the corresponding figures
are ·025 mm. and ·071 mm.    When fully developed, the absorbing cells
of the scutellum thus generally appear as
blunt elongated sacs or tubes, which are from
four to twelve times longer than their
breadth.    As might be expected, the walls
of these cells are very thin ; another characteristic feature is the abundance of their
protoplasmic contents.    The nucleus usually
lies near the base of each cell.    When germination is completed and the endosperm
entirely exhausted, the absorbing cells col
lapse ; their lateral walls crumple up and the
protoplasmic contents disappear altogether.

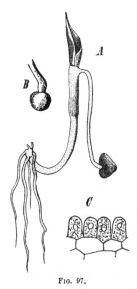

The author has shown that the scutellar
epithelium of *Briza minor* displays a remarkable resemblance to the absorbing tissue of
an ordinary root (Fig. 96).    Before germination begins, the absorbing cells are roughly
cubical (with an average diameter of ·017
mm.) ; when the plumule of the seedling has
reached a height of 3 mm., these cells may
be ·086 mm. long, while they finally average
as much as ·15 mm. in length.    Here,
moreover, the absorbing tubes arise by local

FIG. 97.

*A.* Seedling of *Tradescantia erecta* :
*s,* seed.   × 4.   *B.* Haustorial organ, dissected out and more highly magnified.
*C.* Absorbing cells of the haustorial
organ.

prolongation of the superficial cells of the absorbing organ, just as is
the case with typical root-hairs.    The fully grown scutellum presents a

peculiar appearance in longitudinal section, because its upper margin grows out as a sort of fringe or crest bearing absorbing cells on both sides.

Apart from the Grasses, a highly organised absorbing tissue has been described only in the seedlings of COMMELYNACEAE. In *Tradescantia erecta* the knob-shaped end of the filiform cotyledonary stalk remains embedded in the seed. This haustorial organ (Fig. 97 B), which is about the size of a pin's head, is covered on all sides by a layer of specialised absorbing elements; the latter, which are about ·07 mm. in height and ·03 mm. in width, contain a large amount of protoplasm, and are quite loosely connected together (Fig. 97 c).

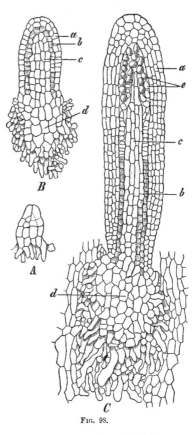

FIG. 98.

*A.* Very young sporogonium (embryo) of *Dendroceros cichoriaceus. B.* Slightly older sporogonium of *D. crispatus. C.* Still older sporogonium of *D. crispatus. d,* foot (haustorium) With its tubular absorbing cells. After Leitgeb (from Engler-Prantl, Natürliche Pflanzenfamilien).

In the case of starchy seeds the absorbing tissue of the germinating embryo secretes an amyloclastic enzyme, *diastase,* and thus helps to render the stored-up starch available for absorption. This process is analogous to the secretion of acid substances by root-hairs. If the embryo of a germinating grain of wheat is carefully separated from the endosperm, and a paste of wheatstarch and water spread thinly upon the back of the scutellum, the great majority of the starchgrains will become extensively corroded by the action of the exuded diastase in the course of twenty-four hours. A similar corrosion occurs — though more slowly—if the experiment is carried out upon the club-shaped haustorial organ of *Canna indica.* It should be noted that, in the case of Grasses, the secretion of diastase is not confined to the absorbing tissue of the scutellum ; this point will be more fully discussed in a subsequent chapter (Chap. X., II., B. 2.). The principal enzyme secreted by the haustorium of the Date and other

Palms is a cellulose-attacking *cytase*, in accordance with the fact that the non-nitrogenous reserve material is in this instance represented by the greatly thickened cell-walls of the endosperm.

It is interesting to compare the young **sporogonia** of Mosses and Liverworts with the seedlings of Phanerogams, in respect of their haustorial arrangements.[128] Where the sporogonium possesses a sharply differentiated basal region, the latter penetrates into the body of the parent plant—*i.e.* the gametophyte—from which it derives a large amount of nourishment. This haustorial **foot** of the sporogonium is, as might be expected, provided with a specialised absorbing tissue ; among Mosses the latter consists of thin-walled, more or less papillose cells, while in certain Liverworts (*Anthoceros, Dendroceros, Notothylas*) it is represented by short tubular outgrowths, which push their way actively into the tissues of the gametophyte (Fig. 98). Here again, therefore, one meets with the same two stages in the specialisation of the absorbing system which have already been distinguished on other occasions.

### B. THE ABSORBING SYSTEM AMONG SAPROPHYTES, PARASITES AND CARNIVOROUS PLANTS.

It will be found convenient to postpone the discussion of the numerous intermediate stages connecting green plants with the various heterotrophic forms for the moment, and to pay attention only to the general considerations that determine the structure of the absorbing system in plants which are devoid of chlorophyll, and hence largely or entirely dependent upon a supply of organic food-material. Evidently there are two opposing [morphogenetic] influences at work in such cases. On the one hand, namely, greater demands are made upon the absorbing system, because in the absence of photosynthetic organs it becomes responsible for the acquisition of the whole of the food-materials ; the lack of actively transpiring leaves, on the other hand, permits of reduction of the water-absorbing system. The other ecological peculiarities of the plant will, of course, decide as to which of these factors will exert the greater influence upon the structure of the absorbing system, in any particular instance.

**Phanerogamic saprophytes**[129] may be dealt with first. Among such humus-loving plants, those which retain their green leaves and thus manufacture at any rate a portion of their non-nitrogenous plastic materials, have to take in considerable quantities of water, and hence require a large absorbing surface. In these circumstances root-hairs are needed which can not only absorb water, but also withdraw organic compounds from the humus particles to which they attach themselves. It is highly probable that such root-hairs (*e.g.* those of most terrestrial

Orchids) secrete special enzymes, which are able to convert some of the insoluble organic constituents of the humus, such as the starch contained in the dead vegetable tissues, into readily diffusible compounds.

The members of a different class of green humus-plants—comprising the CUPULIFERAE and BETULACEAE besides many Conifers—instead of producing numerous root-hairs, provide themselves with a large absorbing surface by entering into symbiotic association with a Fungus-mycelium. At the beginning of the present chapter, it was pointed out that the filamentous mycelium which is the normal form of plant-body among Fungi primarily represents an organ of absorption. The aforesaid woody plants, and also many herbaceous saprophytes, make use of this highly efficient foreign absorbing structure for their own purposes. In such plants, the absorbing roots are completely enveloped—from the tip downwards—in a perfectly continuous mycelial sheath composed of densely interwoven hyphae; individual branches of this sheathing mycelium penetrate between the hairless absorbing cells of the root, which thus become likewise more or less enveloped in hyphae. The sheath, on the other hand, usually sends forth a number of hyphae from its outer surface, which pervade the surrounding humus in all directions. The symbiotic hyphae are more efficient than ordinary root-hairs, not only quantitatively, on account of the larger surface which they expose, but also qualitatively, because they have a greater inherent capacity for utilising the various constituents of the humus. Fungus-hyphae are, in other words, more thoroughly equipped than root-hairs for a saprophytic mode of life.

The water and the soluble inorganic and organic food constituents absorbed by the fungus-hyphae are taken over from them by the outermost cell-layer of the **mycorrhiza**[129a]—as this particular type of root is called—whence they pass to the ordinary conducting tissues of the root; this outermost layer, of course, corresponds to the piliferous layer (rhizodermis) of an ordinary root. Frank was able to show experimentally that the association with a Fungus-mycelium is actually beneficial to a mycorrhizal plant. Young plants of Beech were grown in pots containing forest-soil rich in humus. A certain number of the pots were previously sterilised by heating to 100° C. The plants growing in the unsterilised humus soon developed the characteristic mycorrhiza, and continued to thrive; those which were planted in the sterilised soil, on the contrary, formed no mycorrhiza and gradually sickened and died.[130]

Many Vascular Cryptogams (prothallia of *Lycopodium*, HYMENO-PHYLLACEAE) and Mosses likewise lead a more or less saprophytic existence. The author has shown that the rhizoids of several Mosses

are specially adapted for saprophytic nutrition.[131]  In the genus
*Buxbaumia* (*B. aphylla*, *B. indusiata*), the protonema is green and
capable of photosynthesis ; but the bulbous stem, and the leaves with
which it is clothed, are entirely devoid of chlorophyll.  There is, on
the other hand, an extraordinarily well-developed system of rhizoids,
which differ from typical Moss-rhizoids in having quite thin colourless
cell-walls.  Another remarkable feature is the not uncommon occurrence of fusions between separate rhizoid branches, which lead to the
production of H-shaped connections, and here and there even to the
formation of a regular netwoik of rhizoids.  Altogether the rhizoidal
system of *Buxbaumia* strikingly resembles an ordinary Fungus-
mycelium.

The rhizoids of *Rhynchostegium murale* often penetrate into the
dead stems and roots of Vascular Plants ; in so doing, they do not
undergo any great change of form, except that they are always more or
less constricted where they pass through the substance of a cell-wall.
In another member of the HYPNACEAE, *Eurhynchium praelongum*,
which generally grows among rotting leaves, the rhizoids perforate
the outer walls of the epidermis, and then pass from one epidermal
cell to another by boring through the lateral walls.  These rhizoids
develop very conspicuous lobed outgrowths, which accommodate themselves to the contours of the epidermal cells, and sometimes completely
fill their cavities (Fig. 99 A).  Where such rhizoids happen to penetrate
into elements of the mesophyll, they show the same tendency to occupy
the entire cell-cavity.

A very interesting specialisation has been observed by the author
in the case of a variety of *Webera nutans*, which grows upon moist,
rotten Fir wood.  In this instance the rhizoids on penetrating the
substance of the wood grow mainly *along* the walls of the tracheides ;
here and there, however, a rhizoid puts forth short processes which
perforate the thick walls of the tracheides just like the hyphae of a
parasitic Fungus (Fig. 99 B).  These processes usually arise from
irregular, rounded or lobed expansions of the rhizoids ; their thickness
is from one-third to one-sixth of the average diameter of the parent
rhizoid.  After a process has passed through the wall of a tracheide,
its tip once more expands, and thenceforward grows on as an ordinary
lateral branch.

**Phanerogamic parasites** [132] abstract nourishment from their hosts in
a variety of ways.  The development of their absorbing system is
naturally very largely dependent upon the extent of their parasitism.
Where the parasite retains possession of functional foliage leaves, the
haustorial absorbing tissue does not expose a very large surface, and is
altogether comparatively unspecialised.  Frequently, as in the case of

the Mistletoe and certain other LORANTHACEAE, the whole aim of the parasite is to effect a connection with the water-conducting system of the host. Those hemi-parasites which are rooted in the ground often develop typical root-hairs, in this respect resembling seedlings which have formed a root system, but which still draw upon the store of plastic material deposited in the seed. Where a parasite is entirely devoid of chlorophyll, its haustorial organs frequently develop pencil-shaped or filamentous processes, in accòrdance with the principle of maximum exposure of

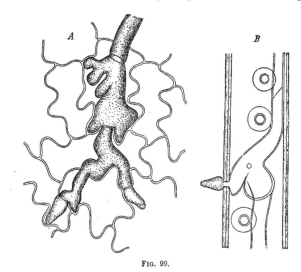

FIG. 99.

Rhizoids of saprophytic mosses. *A*. A rhizoid of *Eurhynchium praelongum*, growing in the epidermis of a decaying beech-leaf and producing lobed haustorial processes. *B*. Perforation of the wall of a Coniferous tracheide by a rhizoid of *Webera nutans*.

surface. In such cases the absorbing tissue takes the form of hypha-like strands of cells, which penetrate the tissues of the host in various directions (*Lathraea squamaria, Cuscuta, Orobanche*). This type of absorbing tissue not infrequently displays a very considerable division of labour. Tracheidal cells, which are continuous with the water-conducting system of the parasite, strive to place themselves in contact with the hadrome elements of the host; sieve-tube-like haustorial cells exploit the protein-conducting leptome, while the less specialised absorbing strands withdraw non-nitrogenous plastic materials from the medullary and cortical parenchyma (*Orobanche*). The most far-reaching modification in connection with a parasitic mode of life is seen in the RAFFLESIACEAE, where the vegetative body is reduced to a mycelium-like system of absorbing strands.

After these general remarks, a few of the cases which have been

studied  in  detail  by  Solms-Laubach,  L.  Koch,  Heinricher  and  others,
may  be  discussed  somewhat  more  fully.

*Thesium  pratense* (SANTALACEAE)  is  a  parasitic  herb,  the  roots  of
which  produce  peculiar  haustorial  organs  in  the  shape  of  ovoid  or
almost  campanulate  outgrowths  (Fig.  100 A).    According  to  Schwarz,
the  young  haustorium  attaches  itself  to  the  surface  of  a  host-root  by
means  of  root-hairs.    When  fully  developed,  the  "apex  of  the  haustorium  sits  upon  the  root,  just  like  a  saddle  on  a  horse's  back."    In  a
longitudinal  section  the  haustorium  is  seen  to  consist  of  an  axile
portion  surrounded  by  a  sheath  of  cortical  tissue.    The  former,  which

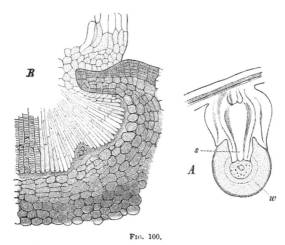

Fig. 100.

Haustorium of *Thesium pratense*.  *A.* Section through the root of the (Dicotyledonous) host and a haustorium of the parasite, slightly magnified ; root (*w*) in T.S.,
haustorium in L.S. ; *s*, haustorial process.  *B.* Portion of a L.S. through the root and
the haustorial process,  × 80 ; the elongated absorbing cells of the latter are arranged
in a fan-shaped manner.  After Solms-Laubach.

may  be  termed  the  "haustorial  core,"  is  more  or  less  flask-shaped.
"The  ovoid  belly  of  the  flask  is  completely  invested  by  the  cortical
tissue  of  the  haustorium ;  a  conical  process,  which  corresponds  to  the
neck,  projects  into  the  tissues  of  the  host-root,  far  beyond  the  surface
of  attachment."    This  **haustorial  process** (*Saugfortsatz*)  is  the  only  part
of  the  organ  that  interests  us  at  present,  since  it  is  there  that  the
whole  of  the  absorbing  tissue  of  the  parasite  is  located.    It  consists
mainly  of  elongated  parenchymatous  cells  rich  in  protoplasm,  but  also
contains  two  vascular  bundles  which  are  crescentic  in  transverse  section.
The  terminal  cells  of  this  process,  which  constitute  the  actual  absorbing
tissue,  are  distinguished  by  their  greater  length ;  their  tips  are  rounded
or  club-shaped,  and  are  closely  apposed,  either  to  extracambial  cells  of
the  host-root,  or  to  elements  of  the  woody  cylinder.    These  terminal

cells (Fig. 100 B) remind one forcibly of the scutellar epithelial cells of Grasses. Each of the vascular strands of the haustorial process likewise expands distally into a tuft of tracheidal elements, which place themselves in direct communication with the wood-vessels of the host.

The above description of the haustorium of *Thesium pratense* applies only to those cases in which the host-root pertains to a Dicotyledonous plant. When the host is a Grass or other Monocotyledon, both the haustorium as a whole, and the haustorial process in particular, undergo a considerable amount of modification. The haustoria of *Thesium* thus possess a certain power of accommodating themselves to different host-roots.

All the species of *Cuscuta* (CONVOLVULACEAE) are plants with filiform twining stems, which envelop the aërial organs of their hosts in an inextricable tangle of branches. They possess neither roots nor green leaves. The entire supply of food materials and water must therefore be taken in through the haustorial organs which are borne on the stems ; these haustoria are accordingly provided with a larger absorbing surface—or, in other words, are more highly specialised—than

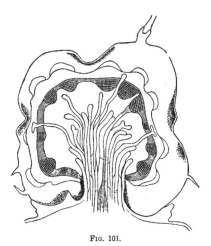

Fig. 101.

A haustorium of *Cuscuta europaea* penetrating the stem of *Urtica dioica* (latter in T.S.). The absorbing tubes are most abundantly developed in the pith ; a few penetrate as far as the hadrome strands, or even reach the leptome after breaking through the woody cylinder.

the corresponding organs of *Thesium*. Such a haustorium of *Cuscuta* consists of a circular or irregularly lobed adhesive disc (appressorium), which is firmly anchored in the cortex of the host-stem by means of a conical haustorial process. The tip of the process bears the actual absorbing tissue, which consists of a fan- or pencil-shaped tuft of elongated hair-like cells. In the case of *C. epilinum* some of these tubular absorbing elements penetrate between the cortical cells of the host, while others lay themselves against the woody cylinder, as a rule, however, without breaking through the latter. The haustorial process of *C. europaea*—which most often grows upon the Nettle—penetrates still more deeply into the host-stem (Figs. 101 and 102), actually breaking right through the woody cylinder, and developing its absorbing tissue most abundantly within the pith. Some of these absorbing tubes soon acquire the characteristics of tracheides and approach the oldest

vessels of the primary hadrome groups, to which they become closely
attached (Fig. 102 A). Others force their way between the primary
bundles [on the opposite side of the pith], and for the second time
break through the feebly developed mechanical tissue; in this way
they reach the primary and secondary leptome, where they produce
irregular lobed expansions (Fig. 102 B). The majority of the tubes,
however—which all contain abundant protoplasm and large nuclei—

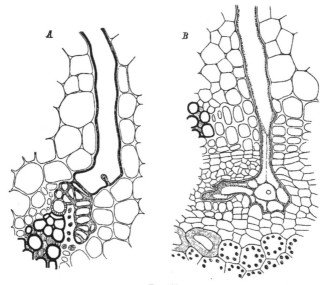

FIG. 102.

Haustorial tubes of *Cuscuta europaea*. *A*. Tracheidal haustorial tube Which has
attached itself to a hadrome-strand of the host-stem (*Urtica dioica*), after traversing
the pith. Its walls are lignified throughout, and (toWards the branched distal
end) provided With scalariform or reticulate thickenings. One of the terminal
branches is in immediate contact With a protohadrome element; the other is separated
from the Wood-vessels by a layer of hadrome-parenchyma cells, the Walls of Which are
sWelling. *B*. Haustorial tube Which has penetrated to the leptome of the host-stem
(*Urtica dioica*), after traversing the pith and breaking through the cylinder of Wood-
fibres; the expanded and irregularly lobed distal extremity of the tube contains
abundant cytoplasm and a nucleus.

remain in the pith, which they traverse in all directions. The
haustoria of *C. epithymum* behave in a very similar fashion.

Among the most remarkable of all Phanerogamic parasites are the
RAFFLESIACEAE, to which reference has already been made. While the
flowers of the genus *Rafflesia*, as is well known, reach gigantic
dimensions, there is in these plants not the slightest trace of stem,
leaf or root, or indeed of any vegetative organs of the ordinary type.
The plant-body is, in fact, represented by a *thallus*, composed of
numerous branching cellular strands, which is quite comparable to the

mycelium of a Fungus.   Occasionally the strands may expand to form single-layered plates of cells, or even larger cell-masses; the largest of these expansions are the so-called "floral cushions" which ultimately give rise to the flowers.

*Rafflesia* is parasitic on the stems and roots of spp. of *Cissus*.   According to Schaar, the thallus of *R. Rochussenii* consists mainly of hypha-like strands of cells which traverse the protein-conducting secondary leptome of the host (Fig. 103).   Other branches pass radially through the cambial zone and the secondary xylem, thus simulating medullary

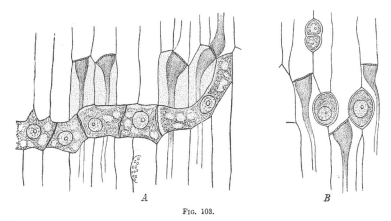

FIG. 103.

Filaments of the "thallus" of *Rafflesia Rochussenii* in the secondary leptome of a root of *Cissus* sp.   *A*. Radial L.S. through the root.   *B*. Tangential L.S.   (The leptome consists mainly of sieve-tubes and companion-cells.)

rays.   Others, again, run in the genuine secondary xylem- and phloem-rays, which contain starch.   These radial thallus-strands give off numerous longitudinally directed branches—especially in the region of the secondary phloem,—which follow a very sinuous course.   It is note-worthy that any radial thallus-strands that may have travelled through the phloem as far as the outermost sieve-tubes—which are entirely devoid of contents,—turn back and once more grow towards the centre. Evidently the parasite by no means pervades the tissues of the host at random, but, on the contrary, confines its attention to the tissues that are richest in nutrient materials.   In passing from phloem to xylem, the thallus filaments have of course to traverse the cambial zone; there must, therefore, be some arrangement which prevents the secondary growth of the root from breaking the connection between the strands in the phloem and those in the xylem.   This continuity is, in fact, ensured by the circumstance that the radial thallus strands themselves become meristematic at the level of the cambium.   In this region the cells of the strands are thin-walled, and remain permanently capable of division; they cut off segments on both sides at such a rate that the elongation of the strand keeps pace exactly with the radial expansion of the root.

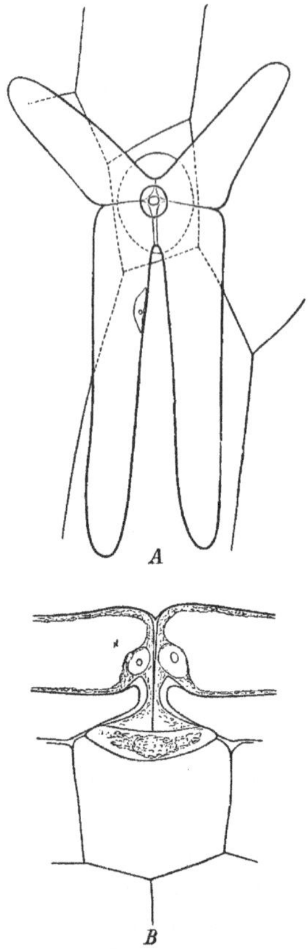

After thus dealing with the absorbing system of saprophytes and parasites, we may devote a few words to the mechanism of absorption in **carnivorous plants**.[133] In certain instances (*e.g. Sarracenia*, according to Balatin and Goebel) the nutrient solution produced by the digestion of the captured animals is taken in through the general surface of the organ which serves as a trap, just as in the case of hairless roots. In other instances it is probable that the digestive glands act secondarily as organs of absorption (*Drosophyllum, Dionaea, Pinguicula*). A few carnivorous plants, finally, are provided with special organs that serve, at any rate in the first instance, for the absorption of the products of digestion. According to Goebel, this advanced condition is exemplified by *Utricularia vulgaris*. The bladders of this plant bear on their inner surface peculiar four-armed absorbing hairs; the arms, which resemble root-hairs in form, are filled with oil-drops when the bladder has been recently "fed." The structure of such an absorbing hair is illustrated in the adjoining figure (Fig. 104). Two of the arms are considerably shorter than the other pair, and also diverge more strongly. Each arm consists of a single cell, continued below into a very narrow downwardly directed process with a relatively thick outer wall. These basal prolongations are fused to form a short and slender stalk, which probably serves as a hinge endowing the hair with a certain amount of flexibility and thus preventing it from being damaged by the struggles of the animals caught in the trap. The stalk expands below, where it rests upon a very thin-walled disc-shaped cell (the "intermediate cell" of Goebel), which in turn is seated upon the embedded basal cell or foot of the hair.

FIG. 104.

Absorbing hair of *Utricularia vulgaris. A.* Surface view. *B.* Vertical section.

## IV. THE ABSORBING SYSTEM AMONG THE THALLOPHYTA.

Most Algae absorb food-materials through the whole of their outer surface; these plants, accordingly, have in general no need of a special absorbing system. The multifarious root-like structures which nevertheless occur in this group serve almost without exception only as organs of attachment. There are a few species of Algae, such as *Botrydium granulatum*, which lead a terrestrial existence; the richly branched rhizoids by which *Botrydium* is fastened to the soil are no doubt utilised for absorption as well as for fixation. Further

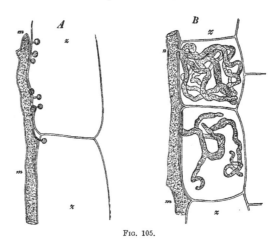

Fig. 105.

Haustoria of parasitic Fungi (Peronosporeae). *A. Cystopus candidus* in the pith of *Lepidium sativum*; *m*, intercellular hypha With knob-shaped haustoria. *B. Peronospora calotheca* in the pith of *Asperula odorata*; *m*, intercellular hypha with much-branched filiform haustoria; *z, z*, host-cells.

investigation is needed in order to determine how far the various trichomes which occur in the higher groups of Algae, such as the FLORIDEAE and PHAEOPHYCEAE, are employed for the absorption of nutrient materials. Reinke and Wille [134] are probably right in supposing that the tufted hairs of the LAMINARIACEAE and FUCACEAE, which occur on the surface of the thallus in the former group, but are sunk in the so-called sterile conceptacles [cryptostomata] in the latter, are organs of absorption comparable to root-hairs. In support of this contention Wille lays stress on the ease with which many coloured solutions pass into these trichomes, and from them into the conducting tissues of the thallus.

Among the Fungi [135] the great importance of the absorptive function exert a far-reaching influence upon the form and structure of the plant-body as a whole.    As was pointed out at the beginning of this chapter, the richly branched filamentous mycelium of the Fungi, in fact, represents an enormously developed system of absorbing organs. Where the mycelium takes the form of more or less bulky strands (rhizomorphs)—as in the PHALLOIDEAE and in many LYCOPERDACEAE, AGARICINEAE, etc.,—the latter put forth special hair-like hyphae, which project from the surface and perform the office of absorption, being, in fact, directly comparable to the root-hairs of Higher Plants.

Among Lichens the lower side of the thallus produces numbers of **rhizoidal hyphae**, which likewise act as root-hairs.    In certain species they are united to form stout strands, the so-called **rhizines**.

Many parasitic Fungi, such as the PERONOSPOREAE, UREDINEAE and ERYSIPHALES, absorb their food-materials with the aid of haustoria, which penetrate into the interior of the host-cells.    In its simplest form such a haustorium consists of a vesicle of varying size joined to the parent hyphae by a slender stalk, which traverses the wall of the host-cell (certain ERYSIPHALES, *Cystopus candidus*, Fig. 105 A).    The more elaborate types of haustoria are usually branched or lobed, and thus expose a far greater absorbing surface (*Erysiphe graminis, Peronospora parasitica, P. calotheca*, Fig. 105 B); mention may also be made of the tufts of exceedingly delicate hyphae, which represent the haustorial organs in *Piptocephalis freseniana* and other small MUCORINEAE that are parasitic upon the larger members of the same family.

# CHAPTER VI.

## THE PHOTOSYNTHETIC SYSTEM.

### I. GENERAL CONSIDERATIONS.

THE term assimilation is used by plant physiologists sometimes in a wider and sometimes in a narrower sense. It may be defined so as to comprehend *all* the metabolic processes whereby organic or inorganic food materials are incorporated in the living organism. Schleiden, Pfeffer and Wiesner, among others, employ the term in this wide sense, and the same usage prevails among animal physiologists. Sachs, on the other hand, limits its application to the most fundamental and remarkable of all assimilatory processes, namely, that in which organic materials are synthesised from carbon-dioxide and water, while oxygen is evolved as a by-product. Every vegetable organism, whether green or otherwise, "assimilates" in the wide sense of the term, whereas assimilation in the narrower sense is almost exclusively confined to plants that are provided with chloroplasts.

If the comprehensive definition of assimilation be maintained, then obviously no special assimilatory system exists; this statement holds good even if attention is strictly confined to the incorporation of those food-materials that are employed for the synthesis of carbohydrate and proteins (to the exclusion of the "ash-constituents," which are left as a residue when the other components of the plant-body are destroyed by combustion). Since it has been proved by experiment that the presence of chlorophyll is not an essential condition for the formation of protein compounds, we cannot *a priori* assume any living cell of a plant—no matter to what tissue-system it may belong—to be incapable of manufacturing protein substances. Whether all the living cells of a plant do participate to an equal extent in this process is another question. The principle of division of labour may quite conceivably hold good in this respect, as it does in the case of so many other functions; in other words, the synthesis of protein—or, in more general terms, the assimilation of nitrogen—may be mainly carried on by a special form of tissue. Treub [136] has indeed shown, in the case of *Pangium edule*, not only that the first demonstrable product of nitrogen-assimilation is *hydrocyanic acid*—a discovery which is of great importance for the general theory of nutritive metabolism—but also that this substance is in all probability manufactured only by particular foliar epidermal cells, and by certain idioblasts in the periderm and pith of the stem, which are at the same time concerned with protein-synthesis. Further investigation is required in order to determine how far the conditions in *Pangium*—at present an isolated case—exemplify a widespread phenomenon. In the meantime we have no choice but to focus our attention upon the synthesis of carbohydrates—that is, assimilation in the strict sense [or **photosynthesis**,[136a] as it will henceforth be termed here] —and to make a detailed study of the tissues that are concerned with this process. Other things being equal, the photosynthetic capacity of a cell can be expressed with considerable accuracy in terms of its chlorophyll-content; the latter feature, in fact, provides a definite criterion, which, whatever its imperfections may be, at any rate greatly facilitates the task of delimiting the photosynthetic system.

Just as a cell cannot be straightway identified as a mechanical element, simply because it is thick-walled, so the photosynthetic system does not necessarily include all the chlorophyll-containing cells of a plant. A green cell should not in fact be regarded as a photosynthetic element, unless photosynthesis is its principal function. This rule must be all the more strictly enforced because photosynthesis is, in accordance with its great physiological importance, one of the most widespread of subsidiary functions. Neither the external nor the internal conditions which determine the presence of chlorophyll are of a very exacting nature. Chloroplasts may consequently occur in a great variety of cells, such as ordinary epidermal elements or hairs, bast-fibres or collenchymatous cells, elements of the parenchymatous bundle-sheath and the like. While such "incidentally" photosynthetic cells undoubtedly represent additions to the chlorophyll-apparatus of the plant which are not altogether negligible ecologically, they are nevertheless all of secondary importance from the physiological point of view. Hence, in determining whether a given cell is a photosynthetic element or not, it is not safe to rely solely upon its relative chlorophyll-content. In order to rank as such an element, a cell must also fulfil certain histological and topographical conditions and generally conform to the principles of construction that are characteristic of the photo-synthetic system as a whole.

## II. PHOTOSYNTHETIC CELLS.

### A. SHAPE OF THE CELLS. CHARACTER OF THE CELL-WALLS.

The outward shape of the elements that compose the photosynthetic system varies within wide limits. In their simplest form photosynthetic cells are isodiametric, with a tendency to become rounded off at the edges. Much more frequently, however, they are elongated in a definite direction and approximately cylindrical or tubular in shape; the relation of this prevalent shape to the functions of the cells will be explained later on. The orientation of green cells with reference to the surface of the photosynthetic organ is also variable. Sometimes they lie parallel to the surface, in which case they may be extended either *along* the principal axis of the whole organ, as in the leaves of *Elodea canadensis, Galanthus nivalis* and *Leucojum vernum*, or *athwart* this axis, as in *Iris germanica, Erythronium Dens canis* and spp. of *Tritonia* and *Gladiolus*. Usually, however, they are placed at right angles to the surface (rarely they are obliquely inclined), and in this case receive the long-established name of **palisade-cells**. The ratio of length (or height) to width is very variable ; some palisade-cells are not much longer than their width, while in those of a more slender type the height may exceed the breadth ten- or twelve-fold. Closely related to the palisade-cell is the obconical type of photosynthetic element or **funnel-cell** ; this has an upper wide end, which nearly always abuts immediately against the epidermis, and a narrow lower end, which generally rests upon elements of the spongy parenchyma.

A remarkable modification, and one which is of great importance for the understanding of the palisade form, is the so-called **arm-palisade-cell** ; in this case the palisade, instead of consisting of entire cells, is made up of groups of cell-branches or -arms. It should be remarked that the palisade-cell is merely a particular variety—characterised by its special orientation—of the generalised tubular photosynthetic element; this fact must not be overlooked when one seeks to obtain a physiological explanation of the shape and arrangement of palisade-cells. From the arm-palisade cell the transition is easy to the **tabular photosynthetic elements** that are found in various Conifers (**P**inus*, Cedrus*) and GRAMINEAE ; these are polyhedral cells, with more or less pronounced flanges projecting inwards from the cell-wall. Finally, we may include in this list of cell-shapes the **spongy parenchyma** type, which has numerous radiating branches, and may even approach a stellate form, though as a matter of fact photosynthesis is normally a subsidiary function for cells of this character.

The walls of photosynthetic cells are in general thin and soft.

Simple pits are sometimes present; for example, in the rounded photosynthetic cells of succulents, in the mesophyll of the pinnae of Cycads, and in the green parenchyma of the cladodes of *Ruscus*. Local thickenings of the wall are rarely found in photosynthetic elements. Probably the best-known instance is furnished by the palisade-cells of *Cycas*, which are regularly provided with longitudinal thickening fibres; these evidently in the first instance enable the individual thin-walled palisade-cells to withstand longitudinal compression, while they also protect the whole tissue against radial pressure.

## B. CHLOROPLASTS.

### 1. *Form and structure of chloroplasts.*[137]

Among the Algae, and particularly in the CHLOROPHYCEAE, the chloroplasts assume a great variety of forms. The simplest condition is that in which each cell contains a single large chloroplast; the latter is often saucer- or bowl-shaped, or tabular in form, and may either be closely appressed to the wall (PALMELLACEAE, *Ulva*, *Enteromorpha*, *Coleochaete*) or else lie suspended in the middle of the cell-cavity (*Mougeotia*). At the next stage of specialisation the chloroplast is likewise solitary and flattened, but has developed marginal lobes or frills, or has become perforated after the fashion of a gridiron or lattice (*Oedogonium*, *Cladophora arcta*). Occasionally it takes the form of a narrow ribbon, which, again, may be straight, curved, or sinuous; the ribbon-shaped chloroplasts of *Spirogyra* are spirally twisted, and in addition provided with inwardly projecting ridges. *Zygnema cruciatum*, finally, has star-shaped chloroplasts: every cell has one such stellate body at each end, the space between the two members of a pair being bridged by a mass of cytoplasm enclosing the nucleus.

Sharply contrasted with the great diversity of the forms assumed by the chloroplasts in certain *Chlorophyceae*, is the uniformity of type which characterises these structures in most of the groups of Algae, and in almost all the Bryophyta, Pteridophyta and Phanerogams, where they are commonly developed as small rounded, ellipsoidal, or (owing to mutual pressure) polyhedral bodies, usually termed **chlorophyll corpuscles** (Fig. 107). In specialised photosynthetic cells these chlorophyll corpuscles are always present in large numbers, forming a green peripheral layer which covers the cell-walls more or less completely.

Among the Higher Plants (that is, the groups from the Bryophyta upwards) there are but few aberrations from the typical chorophyll-apparatus composed of numerous "corpuscles." The genus *Anthoceros* among Liverworts provides one of these exceptions. In this case every photosynthetic cell of the gametophyte contains a single bowl-shaped chloroplast. Schimper has, however, pointed out that the first indications of a fragmentation of the primitive single chloroplasts may be observed in the sporogonium of this genus; here most of the cells have two chloroplasts, while several may occur in each of the epidermal elements. The only other exception hitherto recorded among Higher Plants is furnished by the genus *Selaginella*. The author has himself shown that each of the funnel-shaped photosynthetic cells of *Selaginella Martensii* and *S. grandis* possesses but a single large bowl-shaped chloroplast, which completely covers the cell-wall in the lower half of the cell (Fig. 106 A). In *S. Kraussiana* there may be either a single chloroplast or a pair of these structures in each ·green cell. In *S. caesia*, finally, every green cell contains two chloroplasts, the arrangement of which leaves no doubt that they correspond to the two halves of a single bowl-shaped chromatophore (Fig. 106 B)

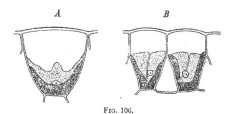

FIG. 106.

*A.* Photosynthetic cell of *Selaginella Martensii*, with a solitary bowl-shaped chloroplast. *B.* Two photosynthetic cells of *S. caesia*, each containing a pair of chloroplasts.

It should further be noted that, towards the base of the leaf, the chloroplasts are more or less deeply constricted, or even divided into several distinct portions. Here, as in *Anthoceros*, there is an unmistakable tendency for an originally solitary chloroplast to undergo fragmentation.

In reviewing the various forms of chloroplasts that have been described one receives the impression that Nature has, as it were, experimented among the lower green plants in order first to test the efficiency of the various modifications, and then to select the best pattern for further elaboration among the Higher Plants. It is, as a matter of fact, easy to show that the form of chlorophyll apparatus which is characterised by the presence of numerous small chloroplasts in each photosynthetic element has many advantages over other types. The immediate duty of every active chloroplast is the absorption of the carbon-dioxide that diffuses into the cell. Other things being equal, the rapidity and efficiency with which this task can be performed depends upon the surface available for purposes of absorption, and obviously the maximum exposure of surface will be obtained when the functions of the chloroplasts are distributed among a number of small corpuscles. This type of chlorophyll apparatus possesses the following additional advantages: first, enhanced mobility of the whole apparatus, a factor which reacts favourably upon photosynthetic activity in a way which will be explained in more detail below; secondly, acceleration of the efflux of synthetic products; and, lastly, a more even distribution of light throughout the chlorophyll apparatus, especially when illumination is feeble.

In many cases chloroplasts have the power of changing their shape in response to external influences,—particularly photic stimulation. Micheli, Moore, Senn and others have shown that, among Algae and Mosses, and also in Phanerogamic palisade-cells, the chloroplasts contract and become rounded or hemispherical in very intense sunlight (as well as after prolonged darkening).

As regards the **internal structure of chloroplasts**, it has come to be generally accepted, in consequence of the researches of Pringsheim, A. Meyer and Schimper, that the chlorophyll-corpuscle of the Higher Plants consists of a colourless spongy protoplasmic matrix or **stroma**, in which are embedded a number of green viscous drops or granules, the **grana**. Such a granular structure is very evident, for example, in the case of the chloroplasts of Orchids and Fern-prothallia; in Algal chromatophores, on the other hand, the grana are exceedingly minute or quite unrecognisable, and the same statement applies to *Anthoceros.* In certain cases treatment with water reveals a peculiar system of curved radial striae; this striation was observed first in *Bryopsis* by Rosanoff, and subsequently in *Anthoceros* by Schimper, while the author has found it to be particularly conspicuous in *Selaginella Martensii.* The striated appearance only becomes evident after swelling has taken place; what the corresponding internal structure may be, is a matter for further enquiry.

It is not yet definitely known—in spite of the researches of Von Mohl, Nägeli and others—whether chloroplasts are invested by special membranes or not. Most botanists, it is true, agree with Von Mohl that, in general, no *visible* plasmatic membrane is present. There are, however, undoubtedly some exceptions to this rule; in *Selaginella Martensii*, for example, the bowl-shaped chloroplast found in each photosynthetic cell (Fig. 106 A) possesses, on its concave side, a somewhat highly refractive, relatively thick (about $\cdot3\text{-}\cdot4\,\mu$ in thickness) and granular plasmatic membrane, which may possibly represent a special light-perceiving apparatus.

The characteristic **pigments** of chloroplasts [138] are soluble in alcohol. Crude alcoholic "chlorophyll-extract" contains both green and yellow colouring matters, which may be separated from one another by shaking up the extract with benzol (Kraus's method); the lighter benzol solution floats as a deep-green stratum above the yellow alcoholic layer. According to Schunck and Marchlewski, *chlorophyll* proper is frequently accompanied by a second green substance, which they call *allo-chlorophyll*; the yellow pigments are known as *xanthophylls*.

The spectrum of the crude extract is characterised by six absorption-bands, of which four, belonging to *chlorophyll* proper, are located in the "red" half of the spectrum, while the other two, situated in the more highly refrangible or "blue" region, pertain to the *xanthophylls*. Very soon after the discovery of these absorption-bands attempts were made to bring them into relation with the photosynthetic dissociation of carbon-dioxide, which, as had long been known, only takes place under the influence of light.[139] Lommel, Jamin and Becquerel have all assumed, on theoretical grounds, that it is precisely those radiations which are absorbed by a solution of chlorophyll—as shown by its spectrum,—that are also responsible for the photosynthetic decomposition of carbon-dioxide. Timiriazeff and Engelmann have tried to obtain experimental confirmation of this conclusion. As a matter of fact, the graphs corresponding respectively to the intensities of [optical] absorption and to the rates of photosynthesis in different parts of the spectrum do on the whole follow parallel courses; this statement applies more particularly to the "red" half of the spectrum, for which the photosynthetic maximum falls between the Frauenhofer lines B and C, or, in other words, just in the position of the most intense absorption-band (No. 1) of chlorophyll-extract. In the other half of the spectrum the curve of absorption rises steadily, whereas the curve of photosynthesis reaches a secondary maximum near the Frauenhofer line F, falling off again beyond this point. The discrepancy is probably due to the fact that the total energy of solar radiation diminishes somewhat rapidly in this region of the spectrum.

These views in favour of a close correspondence between [optical] absorption and photosynthetic activity have received strong confirmation from the discovery of the phenomenon of **complementary chromatic adaptation** by Engelmann and Gaidukov. Engelmann found, in the case of seaweeds growing at different depths, that the rays of light which are most effective in respect of photosynthesis are invariably those that are complementary to the coloration of the respective plants, the reason being that it is, of course, precisely these complementary rays that are absorbed by the pigments concerned. Even at a comparatively slight depth below the surface of the ocean green and blue-green light predominates; in these circumstances the FLORIDEAE derive an undoubted advantage from their red coloration. It has since been shown by Gaidukov that cultures of certain species of *Oscillatoria* change colour if they are exposed for long periods to coloured light, the alteration always consisting in a closer approximation to the tint which is complementary to that of the light employed. In red light, for example, the plants take on a green, in green light a red, and in blue light a brownish yellow hue.

Stahl has even extended the views based on these interesting discoveries to the Higher Plants, by suggesting that the chloroplast pigments of the latter have been evolved in adaptation to the mixed composition of ordinary white light. The green fraction of the crude chlorophyll extract—comprising *chlorophyll* proper—serves, in Stahl's opinion, for the absorption of the orange and red rays which preponderate in sunlight that has passed through the cloudy atmosphere of the earth. The yellow portion, on the other hand, composed of the *xanthophylls*, is concerned with the absorption of the blue and violet rays which form a great part of the diffuse light reflected from the blue sky.

Chloroplasts are in many cases provided with characteristic inclusions ; among the most widely distributed and important of these are the familiar **included starch-grains.**[140] These bodies usually belong to the class of compound grains. While the more or less numerous constituent granules are quite young, they are separated from one another by green protoplasmic material ; later, when the grains have reached a considerable size, the actual substance of the chloroplast may be reduced to a thin film surrounding the whole inclusion, or may apparently even be ruptured by the latter at certain points. These included starch-grains do not all arise in the same way. The material utilised for their production may be photosynthetically manufactured by the enveloping chloroplast itself; on the other hand, the latter may merely convert into starch carbohydrate material that has been imported from without in a soluble form. Hence the mere presence of included starch-grains cannot be accepted as conclusive proof of antecedent photosynthetic activity. Conversely even the most active photosynthesis may not lead to the accumulation of included starch, if the translocation of synthetic products is sufficiently rapid. The latter condition is frequently fulfilled in the case of palisade-cells and other specialised photosynthetic elements ; on the other hand, chloroplasts in spongy-parenchyma cells or in the cortex of stems, although exhibiting relatively slight photosynthetic activity, nevertheless retain their included starch-grains for long periods, when these have once been formed.

Chloroplasts almost always also contain inclusions of the nature of **oil-drops** ; these vary in size, and are soluble in alcohol. As a rule, they appear only in the older cells; in *Vaucheria*, however, oil-drops are found adhering even to the young chloroplasts, while in the genus

*Iris* the chromatophores present a peculiar granular appearance in cells of all ages, owing to the presence of innumerable minute drops of oil.    In general, such oily inclusions may be regarded as waste-products produced in consequence of a senile " fatty degeneration " of the chloroplast ; but it is possible that in certain cases (such as that of *Vaucheria*) they represent the normal product of photosynthesis.

**Protein-crystals** (crystalloids) have been observed in the chloroplasts of certain plants (*Phaius grandifolius, Zingiber officinale, Pellionia Daveauana,* etc.), where they seem to play the part of reserve-materials.    Mention must further be made of the **pyrenoids,** which form a characteristic feature of the chloroplasts among Algae, and also in the genus *Anthoceros.*    According to Schimper, a pyrenoid often consists of a single protein-crystal, which may or may not be enveloped in a layer of substance resembling protein.    In certain cases (*Zygnema, Cosmarium*) the pyrenoids are not crystalline.    Pyrenoids are frequently surrounded by numerous small starch-grains arranged in the form of concentric shells.

## 2. *Arrangement of chloroplasts in the cell.*[141]

As regards their arrangement in the individual cell, chloroplasts usually exhibit definite relations to the other components of the proto plast, and to the various regions of the cell-wall ; their disposition is, in fact, in the first instance dependent upon certain internal factors, which are more or less directly connected with the photosynthetic function. As a rule, however, the arrangement of the chloroplasts is by no means absolutely determined by these internal relations, but is, on the contrary, constantly subject to alteration under the influence of various external stimuli, among which illumination plays the leading part.    The resulting movements of the chloroplasts generally react in a favourable manner upon the activity of the chlorophyll apparatus.

Beginning with the **internal factors** that affect the arrangement of chloroplasts, we must note that these bodies almost invariably take up a peripheral position in the cell ; in this case they always form a single layer, which is closely appressed to the cell-wall, or rather to the ectoplast.    This peripheral location not only enables the chloroplasts to utilise the light that reaches them as fully as possible, but also facilitates the gaseous interchanges associated with the photosynthetic function.    The latter consideration further explains why, in the photosynthetic tissues of Higher Plants, the chloroplasts—if they are not too closely packed—adhere exclusively, or in great part, to those walls which abut upon air-spaces ; by this means they evidently obtain the most favourable conditions for the absorption of carbon-dioxide.[142] The last-mentioned point can be readily verified in the case of the radical leaves of species of *Sempervivum*.  Here the photosynthetic system is made up of radial plates of parenchyma, which extend parallel to the long axis of the leaf.  Each plate consists of a single layer of cells, and is separated from the adjoining plates by air-spaces. The chloroplasts are normally confined to the walls bounding these intercellular spaces, while the transverse septa in each plate are left entirely bare.  In the case of typical palisade-tissue, also, it is not unusual to find the chlorophyll corpuscles adhering chiefly to those strips of the anticlinal walls which are in contact with air-spaces (*Leucojum vernum, Echinops exaltatus, Centaurea macrophylla, Cirsium pannonicum, C. palustre,* Fig. 107 B).

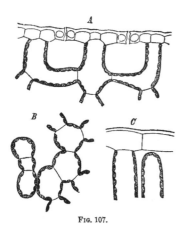

Fig. 107.

Arrangement of chlorophyll corpuscles in palisade-cells.  *A.* Curved palisade-cells of *Scilla bifolia.*  *B.* Palisade-cells of *Cirsium pannonicum* (in T.S).  *C.* Palisade-cells of *Brassica Rápa.*  For explanation see text.

Just as certain regions of the wall in photosynthetic cells, are, as it were, sought out by the chloroplasts, so others seem to be avoided by these structures.  In palisade-cells, according to the author's own observations, the walls that are generally devoid of chloroplasts are those which lie athwart the paths normally followed by streams of diffusing material.  It is, of course, the periclinal walls that are constantly traversed, on the one hand, by currents of water flowing into the green cells from the epidermis or the water-tissue, and, on the other hand, by outward currents of synthetic products travelling towards the spongy mesophyll or the conducting parenchyma.  It must not be supposed that the periclinal orientation of the walls is the determining factor in such cases.  Where, namely, the upper end of a palisade-cell projects into an air-space, the freely exposed tangential wall is occupied by chlorophyll-corpuscles (Fig. 107 c); in curved palisade cells, on the other hand, no chloroplasts are found on the inner tangential walls, even where the latter are obliquely inclined, or almost perpendicular to the surface (Fig. 107 A).

The influence of the nucleus upon the position of chloroplasts must be included among the internal factors that regulate the arrangement of these bodies.  A relation of this kind is often clearly demonstrable even among the Algae.  Thus in *Mougeotia* the nucleus always lies close up against the axile chloroplast, near the centre of one of its broad faces.  In some species of *Spirogyra* the nucleus likewise adheres closely to the ribbon-shaped chromatophore, while in others, where it is suspended in the centre of the cell, it is connected with the pyrenoids by protoplasmic strands. Among Pteridophyta the genus *Selaginella* affords striking illustrations of the same phenomenon. In the funnel-cells of *S. Martensii* (and *S. grandis*),—each of which, it will be recollected, contains but a single bowl-shaped chloroplast—the nucleus invariably lies at the bottom of the bowl in close contact with the substance of the chromatophore (Fig. 106 A); no less evident is the constant association of the nucleus with the chain of chloroplasts which is commonly found in each of the green parenchymatous cells of the stem (Fig. 6). In the Higher Plants, however, these special relations between nucleus and chloroplasts (or amyloplasts) are most conspicuous in the case of tissues which are concerned with the manufacture of starch at the expense of other plastic materials. Hence it is particularly in young growing organs and in storage tissues that one finds the chromatophores grouped closely around the nucleus; they generally disperse when their included starch-grains have grown very large, but may approach the nucleus again, if their starch is once more dissolved (*Orchis fusca, Adoxa moschatellina,* according to A. Meyer). The nucleus thus appears to exercise a certain amount of control over the formation of starch in chromatophores. This idea is supported by the author's discovery of the fact, that in *Selaginella Martensii* the included starch-grains are not evenly distributed throughout the bowl-shaped chloroplast, at any rate where the amount of starch is small; under these conditions they are massed together near the point of contact of the chloroplast with the nucleus, while the rest of the chromatophore is quite free from starch. Further confirmation is provided by Pringsheim's observations—which have already been referred to—upon the species of *Spirogyra* which have the nucleus suspended in the centre of the cell; here, namely, each of the protoplasmic strands that radiate from the nucleus to the chromatophore enters the latter near a pyrenoid. The author has noted something of the same kind in the case of a potato which was turning green through exposure to light; in this instance, also, most of the protoplasmic strands radiating from the nucleus were connected to the peripheral starch-forming chloroplasts (Fig. 108).

Attention must next be directed to the influence of **external factors** upon the arrangement of chloroplasts; we may begin with the action of light, which was first described by Boehm [in 1856], since which time this phenomenon has been studied by quite a number of investigators.

An orientating effect of light upon the chromatophores can be readily observed in certain Algae. The axile plate-like chloroplast of *Mougeotia* does not take up its position in the cell at random; on the contrary, Stahl's experiments have conclusively shown that in diffuse light (*i.e.* in relatively feeble illumination) the broad face of the chloroplast is always placed at right angles to the direction of the incident rays (surface or full-face **position**), whereas in direct sunlight (*i.e.* in intense illumination) the narrow edge is presented to the light (**profile position**). It is quite an easy matter to observe the rotation of these chloroplasts in response to changes of illumination. The advantage of the mobility is obvious; under favourable conditions of illumination the chloroplasts expose their full surface, and by so doing are enabled to absorb the maximum amount of light. In very bright light, which is injurious because of the rapid decomposition of chlorophyll induced thereby, and also perhaps for other reasons, they take up the profile position and thus escape the greater proportion of the light that falls upon the cell.

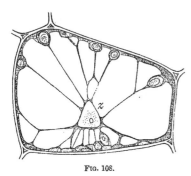

FIG. 108.

Parenchymatous cell from the peripheral tissue of a Potato-tuber, which has turned green through exposure to light. The nucleus is connected with the chloroplasts by protoplasmic strands.

Chloroplasts which adhere closely to the ectoplast obtain the suitable intensity of illumination by moving along the cell-wall. This arrangement is exemplified by the tubular thallus of *Vaucheria*, by the protenemal filaments of Mosses and by the basal filamentous region of a Fern-prothallus. A very similar state of things prevails in the leaves of Mosses, in the expanded portion of a Fern-prothallus, in the fronds of *Lemna* (Fig. 109), and in other organs composed entirely of one or more layers of green parenchyma. In all these cases it is always the difference in the *intensity of illumination* at different points of the cell that controls the movements of the chloroplasts. The *direction* of the incident rays is of no importance, except in so far as it is responsible for the aforesaid difference of intensity. Senn's detailed investigations have shown that chloroplasts in general contrive to take up their position in those parts of the cell-wall which are for the time being exposed to the most favourable intensity of illumination; in moderately intense diffuse light they move towards the illuminated tracts, while in direct sunlight they seek the shaded portions. Thus, in the case of a Moss-leaf or Fern-prothallus, the periclinal walls are covered with chloroplasts in diffuse light, while the anticlinal walls are bare; to this condition Frank has given the name **epistrophe**. In the brighter illumination produced by partially intercepted sunlight, the chlorophyll-corpuscles shift from the tangential to the radial walls; this arrangement Frank calls **apostrophe.**   Here we are once more dealing with the "surface" and "profile" portions of chloroplasts, which in this case, however, are not immediately determined by the *direction* of the incident light.   Very intense illumination, such as that produced by bright sunshine, causes an aggregation of the chlorophyll corpuscles, which has been termed **systrophe** by Schimper.   In this condition the chloroplasts leave the cell-walls altogether, and become massed together into a more or less compact clump in each cell; they consequently

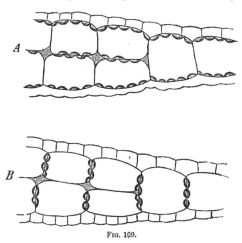

Fig. 109.

T.S.S. through fronds of *Lemna trisulca. A.* Chloroplasts in surface-position (epi-strophe). *B.* Chloroplasts in profile-position (apostrophe).   After Stahl.

shade one another, and thus obtain as much protection as possible against the injurious effects of very intense illumination.

In the palisade-tissues of Higher Plants, it is generally the radial walls that enjoy the conditions of illumination which are most congenial to the chloroplasts; the latter are in any case forced to occupy these walls to a large extent, since only a small number can. be accommodated on the tangential walls.   Stahl's description of this arrangement as a "profile" position of the chloroplasts with reference to the incident light is however inadmissible.   The error is obvious in the case of leaves which maintain a fixed position, since here the light meets the leaf at various angles at different times of the day, as the sun changes its position.   The author has further observed that in curved palisade-cells the chloroplasts are generally massed upon the radial walls; if, however, the ends of the cells project freely into airspaces, chlorophyll-corpuscles also occur upon the tangential walls, although those which do so must necessarily assume the "surface" position.

**External agents other than light** may induce far-reaching changes in the arrangement of the chloroplasts. Mechanical shock or injury, defective water-supply, fluctuations of temperature and chemical stimulation may all bring about a transition from the epistrophic to the apostrophic, or even to the systrophic condition. Thus Kraus has shown, in the case of evergreen leaves, that the chloroplasts congregate in a mass at the base of each palisade-cell during winter; the same statement applies to the green cortical parenchyma of many woody plants. Finally, it must be noted that darkness may also induce apostrophe, especially when its action is prolonged.

So far as is known, the various movements of chloroplasts are not due to any power of active locomotion on the part of these bodies, but depend upon passive transportation by the general cytoplasm or by specialised cytoplasmic structures. According to Knoll and Linsbauer, for example, the chlorophyll corpuscles in the leaf of *Funaria*, are drawn to and fro by the strands of a protoplasmic reticulum, which does not, however, represent a permanently differentiated structure.[143] Whether chloroplasts may not be possessed of some power of independent amoeboid movement in special cases, particularly among Algae, is nucertain. But even if the actual movements of the chloroplasts are always passive, it is at least highly probable that the perception of the stimuli which induce the movements is located in these bodies and that they are, in particular, very sensitive to photic stimulation. As to the manner in which the chloroplasts in their turn incite the cytoplasm or the special cytoplasmic motor organs to carry out the requisite movements, we are absolutely without information.

### 3. *Relations between chlorophyll-content and photosynthetic activity.*

The photosynthetic capacity of a cell must clearly be to a very large extent dependent upon the number of chlorophyll corpuscles that it contains. Palisade-cells take the first place in respect of chlorophyll-content, a circumstance which in itself suffices to characterise these cells as the special photosynthetic elements of the plant. Since the mesophyll of the majority of leaves is distinctly differentiated into palisade-layers and spongy parenchyma, it becomes a matter of special interest to compare these two types of tissue with reference to their chlorophyll-content, and so to obtain a criterion for a comparison of their respective photosynthetic activities. With this end in view, the author has made rough determinations—by actual counting—of the average number of chloroplasts present per cell, in these two tissues, in the case of various leaves. The results of some of these estimations may be stated as follows. In the leaf of *Ricinus communis* there are, on an average, 36 chlorophyll corpuscles in a palisade-cell, and only 20 in a cell of the spongy parenchyma. As the palisade-cells outnumber the spongy elements approximately in the ratio of two to one, the palisade-tissue is furnished with about 403,200 chloroplasts per sq. mm. of surface, whereas not more than 92,000 per sq. mm. are present in the spongy parenchyma. The former tissue thus contains 82 per cent. of the total number of chloroplasts present. In the following table the corresponding values are set forth for various other plants ·

| | Palisade Tissue. | Spongy Tissue. |
|---|---|---|
| *Fragaria elatior*, - - - - - - | 86 | 14 |
| *Pulmonaria officinalis*, - - - - - | 85 | 15 |
| *Ricinus communis*, - - - - - | 82 | 18 |
| *Brassica Rapa*, - - - - - - | 80 | 20 |
| *Galeopsis Tetrahit*, - - - - - | 79 | 21 |
| *Tropaeolum majus*, - - - - - | 77 | 23 |
| *Helianthus annuus*, - - - - - | 73 | 27 |
| *Phaseolus multiflorus*, - - - - - | 69 | 31 |
| *Bellis perennis*, - - - - - - | 67 | 33 |

The palisade-tissue therefore contains at least twice, very often three to five times, and in extreme cases six times as many chlorophyll corpuscles as the spongy parenchyma. These two tissues undoubtedly differ in a similar degree in regard to their photosynthetic activities; the contrast in this respect is in fact probably even greater, for a twofold reason. In the first place, the palisade-tissue, being situated next the upper side of the leaf, is much better illuminated than the spongy parenchyma which it overlies and shades; secondly, the removal of synthetic products proceeds—as is shown below in more detail—much more rapidly in the palisade-cells than it does in the spongy tissue.

The generalisation which may be founded upon all these facts, is to the effect that the palisade-layers constitute the special photosynthetic tissue of the ordinary dorsiventral leaf. The same conclusion can be drawn from a comparative anatomical study of the axial structures in those leafless plants that have transferred their photosynthetic system to the stem. *Equisetum, Ephedra, Asparagus, Casuarina, Spartium*, species of *Genista*, etc., belong to this class. In such "switch-plants" the cortical parenchyma of the younger branches assumes the character of a typical palisade-tissue, a circumstance which shows that this type of tissue does not merely represent a morphological constituent of the foliage-leaf, but is really the essential anatomical component of highly organised photosynthetic organs in general.

Pick[144] has demonstrated the **great photosynthetic activity of pali-sade-cells** by actual gasometric analysis. He carried out comparative estimations of the evolution of oxygen, in light, from the green branches of leafless switch-plants on the one hand, and from the twigs of leafy plants—the cortical cells of which contain comparatively little chlorophyll—on the other. The amount of oxygen evolved by the stems of the leafy plants was always small, and often negligible, whereas all the switch-plants exhaled considerable quantities of this gas. Thus the amount evolved after three hours insolation was 1·6 c.c. per sq. cm. in the case of *Casuarina excelsa* and 2·3 c.c. in that of *Spartium monospermum*.

In comparing the photosynthetic energy of foliage-leaves which conform to the same general plan of construction, the relative chlorophyll-content is a fairly safe criterion. Weber [145] has estimated the average amount of dry substance produced photosynthetically by unit area of leaf, in the course of a day, for various plants. Each species was found to possess a "specific photosynthetic capacity," which was not, however, further investigated by Weber. In these circumstances it seemed desirable to the author to determine the numbers of chloroplasts per unit area in Weber's plants, in order to ascertain if any definite relation exists between photosynthetic activity and chlorophyll content. If both quantities are reckoned as equal to 100 in the case of *Tropaeolum majus*, the following values are obtained:

|  | Specific Photosynthetic Capacity per unit area. | Number of Chlorophyll Corpuscles per unit area. |
|---|---|---|
| *Tropaeolum majus*, - - - - - | 100 | 100 |
| *Phaseolus multiflorus*, - | 72 | 64 |
| *Ricinus communis*, - - - - - | 118·5 | 120 |
| *Helianthus annuus*, - - - | 124·5 | 122 |

The close agreement between the two sets of values is obvious and striking. The slight discrepancies are no doubt due partly to the imperfect experimental methods employed, and partly to differences in the structure of the leaves and in the size of the chloroplasts.

Most probably there is a specific photosynthetic capacity characteristic of the individual chloroplast, which varies in different plants. In other words, even where the chloroplasts of two plants agree in size and shape, those of one species may display greater photosynthetic activity under the same external conditions. Such specific differences are, however, certainly of secondary importance in comparison with the varying quantitative development of the chlorophyll apparatus, at any rate as far as Phanerogams are concerned.

## III. ANATOMY OF THE PHOTOSYNTHETIC SYSTEM.[146]

### A. THE PRINCIPLE OF MAXIMUM EXPOSURE OF SURFACE.

**Palisade-parenchyma** has been recognised as the most important of specialised photosynthetic tissues, in the preceding section; this tissue must therefore form the starting-point of any attempt to interpret the anatomical structure of the entire photosynthetic system in terms of its physiological activity. It must, however, be kept in mind that the radially elongated palisade-cell merely represents *a particular type* of photosynthetic element, and that cases of tangential (longitudinal or transverse) extension also occur, which have to be taken into account in any anatomical exposition of the photosynthetic system that claims to be comprehensive.

It will be found convenient to begin with a consideration of the green tissue of the leaf of *Pinus*. If a transverse section of a Pine-needle is examined microscopically, the walls of the large polyhedral photosynthetic cells, which are in uninterrupted contact with one another laterally, are seen to possess peculiar folds or flanges projecting more or less deeply into the cell-cavity. The orientation of these flanges depends upon the position of the cell (Fig. 110 A); in the more deeply seated cells, namely, it shows no definite relation to the leaf-surface, whereas in the peripheral elements almost all the flanges are placed at right angles to the surface, so that the cells in question, though poly-hedral or tabular in form, come to consist of a number of branches, each of which resembles a palisade-cell in shape. Commonly there are two radial folds starting from opposite sides of the cell, so that in transverse section the latter appears like a capital **H**, with an abbreviated cross-stroke. It might reasonably be argued that there is no real analogy between this counterfeit palisade-tissue and that which consists of genuine palisade-cells, but merely a superficial resemblance from which no reliable conclusions can be drawn. This question cannot be settled by investigation of the *Pinus* needle alone, since anatomically the latter represents a very aberrant type of leaf. Arm-palisade-cells similar to those of *Pinus*, however, recur in the leaves of certain RANUNCULACEAE (e.g. *Trollius europaeus*, *Caltha palustris*, *Aconitum Napellus*, *A. dissectum*, spp. of *Paeonia* and *Anemone*, etc.), where they are located between upper epidermis and spongy parenchyma in precisely the same manner as the typical palisade-tissue that is found in other members of the same family (*e.g.* spp. of *Ranunculus* and *Helleborus*, *Eranthis hiemalis*, *Aquilegia vulgaris*, etc.). There can, therefore, be no doubt that these peculiar flanged photosynthetic cells actually represent a remarkable modification of ordinary palisade-tissue.

Before proceeding to examine the further consequences of this conclusion, we may pause to consider the structure of **arm-palisade-tissue** in a little more detail. Arm-palisade-cells occur sporadically in all the great divisions of vascular plants. Among Dicotyledons they are found more particularly, as already stated, in the RANUNCULACEAE, but also in the genera *Sambucus* (Fig. 110 B), *Viburnum*, *Saurauja*, *Meliosma*,

*Acanthopanax, Cussonia, Schwenkia, Chloranthus, Phyllanthus,* etc.   Instances of their occurrence in Monocotyledons are afforded by certain species of *Bambusa* and *Arundinaria,* further by the genera *Elymus* and *Calamagrostis,* and by *Alstroemeria psittacina.*   Gymnosperms are represented in this connection by the genera *Pinus* and *Cedrus,* and Ferns by *Aspidium aculeatum, A. Sieboldi, Lomaria gibba, Todea aspera, Didymochlaena sinuosa,* and various species of *Adiantum* (Fig. 110 c).

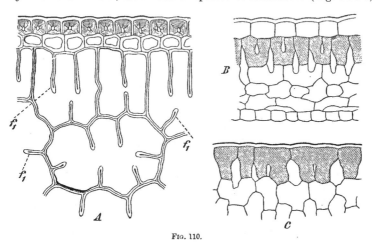

FIG. 110.

Arm-palisade-tissue in various leaves.  *A. Pinus Laricio* (T.S.).   In the cells immediately below the hypodermis the flanges (*f₁*) are all arranged at right angles to the leaf-surface ; in the more deeply-seated layers they are disposed in various planes. *B. Sambucus nigra* (T.S.).   *C. Adiantum trapeziforme* ; the arm-palisade layer is protodermal in origin.   Cell-contents omitted in all the figures.   Arm-palisade-cells shaded in *B* and *C.*

The characteristic flanges may develop only from the upper or epidermal end of the cell, in which case the cell-cavity is continuous as regards the lower half of the cell, while above it is divided into two or more branches or arms ; in other instances (e.g. *Anemone sylvestris* and *Clematis recta*) folds arise from both ends of the cell, thus giving rise to the H-shaped elements noted above in the case of *Pinus.*   In *Bambusa Simonii,* again, the flanges are only developed in connection with the lower ends of the tabular green cells ; the photosynthetic elements of Ferns also often conform to this comb-like or pectinate type, though in their case additional flanges are not infrequently present on the radial walls.   The number of "arms" developed by each cell is most often three or four, but ranges from two in *Caltha palustris* to eight or more in *Todea aspera.*   The length of the flanges usually amounts to between one-third and two-thirds of the radial diameter of the cell.

The foregoing interpretation of the arm-palisade-cell as a modification of the typical palisade-cell introduces the conception of a **palisade-**

unit, which may correspond either to an entire cell or else merely to a cell-branch, but which has in either event the same physiological value. The radial walls of ordinary palisade-cells may then be regarded as flanges which have extended right across the cell; conversely, flanges may be looked upon as incomplete partition-walls. This notion of a palisade-unit may be further illustrated by comparing an arm-palisade-cell to a group of normal palisade-cells which have become partially fused.

The case of *Pinus* may now be reconsidered in the light of the conclusions arrived at in the preceding argument. It is evident that the radial flanges in the peripheral arm-palisade-cells are of the same nature as the irregularly disposed folds in the more deeply situated cells, and that the same structural principle is responsible for the existence of both; in the one instance, however, it is only the *presence* of the flanges that demands an explanation, whereas in the other it is necessary to account for their definite *orientation* as well. An enquiry into the significance of palisade-tissue thus involves two questions, namely first, as to what physiological benefits result from the development of folds or partitions of any kind in photosynthetic cells, and, secondly, as to what special advantage is gained by arranging of these flanges or walls at right angles to the surface of the organ. The present section is only concerned with the former question; in these circumstances the physiological significance of the irregularly disposed flanges in the *Pinus* needle is the problem that lies nearest at hand.

Let us first consider how the photosynthetic efficiency of a cell is increased by the development of flanges. It is found not only that chloroplasts invariably take up a peripheral position, but also that they tend to spread themselves over any projections or fold of the wall that may be present. Hence the physiological advantage of the flanged type of photosynthetic cell consists principally in the fact that a larger internal surface is provided for the display of chloroplasts. The principle of maximum exposure of surface is thus not only responsible for the breaking up of the primitive centralised chlorophyll apparatus into separate corpuscles, but also governs the construction of the photosynthetic tissue as a whole.

The increase in the number of chloroplasts that is rendered possible by the development of flanges, or by the presence of their analogues, the radial walls of ordinary palisade-tissue, is very considerable. If the internal surface of a cell devoid of flanges be reckoned as equal to 100 superficial units, the corresponding value for flanged cells may be anything from 120 to 150. Typical values are: 145 for a three-flanged pectinate cell of *Bambusa Simonii*, 115-135 for tabular cells of *Pinus sylvestris* with irregularly disposed folds, 127 for an H-shaped arm-palisade-cell of *Anemone sylvestris*, and 148 for a four-armed cell of *Sambucus nigra*. The interpolation of complete partitions is of course even more effective.

Quite a unique method of surface-increase is employed in the leaves of certain HYMENOPHYLLACEAE (*Trichomanes auriculatum, Hymenophyllum Karstenianum, H. speciosum, H. plumosum, H. Malingii*). Mettenius long ago noted that the external walls of the photosynthetic cells—which are here superficial — project in the form of papillae provided with a dense peripheral layer of chloroplasts. In *Hymenophyllum Malingii* the papillae are two to four times longer than their width, and altogether closely resemble palisade-cells (Fig. 111). It might, in fact, be maintained that the palisade-layer of this plant is composed of trichomes.

FIG. 111.

T.S. through a pinna of *Hymenophyllum Malingii*. After Giesenhagen.

### B. THE PRINCIPLE OF EXPEDITIOUS TRANSLOCATION.

Attention must next be directed to the second principle that governs the construction of the photosynthetic system, namely, the **principle of expeditious translocation.** It is this principle that determines the *orientation* of the flanges or partitions which owe their existence to the influence of the first principle, while the great diversity that characterises the detailed structure of the photosynthetic system depends above all upon the fact that this second principle is carried out to a very varying extent in different plants.

The activity of photosynthetic cells obviously cannot long remain uninterrupted unless the products of synthesis are quickly removed from the place of their formation. Under ideal conditions each individual chlorophyll-corpuscle should be entirely engaged in the formation of starch from the products of its own photosynthetic activity, and should have little or nothing to do with imported carbohydrate material. It is a well-known fact that every chemical reaction proceeds most smoothly and completely, as well as with the greatest velocity, when the products of the reaction are removed as fast as they are formed; for this reason alone, it is a disadvantage if the products of synthesis are denied immediate egress from the photosynthetic element, while still greater harm results if they are permitted to circulate in the cell, or are actually deposited therein. It is, indeed, a general law of vegetable physiology, that a metabolic process can only be carried on continuously so long as the products are

promptly removed.   It is, therefore, to be expected that plants should
have evolved arrangements for the most direct and expeditious removal
of the products of photosynthesis from the green cells ; it follows, as a
corollary, that the utilisation of one and the same tissue for translocation and photosynthesis must be avoided as far as possible.   In an
ideal photosynthetic system every green cell would be in a position to
pass on its synthetic products at once to some member of a different
tissue specialised for translocation.

The course followed by the outgoing current of synthetic products in any particular case depends upon the extent to which this
principle of expeditious translocation is carried out.   When this
principle is ignored altogether, the photosynthetic cells themselves
serve for translocation, and the direction of the efferent stream is
longitudinal over the whole of its course from apex to base of the
leaf.   In more highly organised leaves, the stream at first either flows
sideways in the leaf, thus taking the shortest path to the longitudinal
vascular strands, or else passes vertically inwards towards a dense
network of larger and smaller vascular bundles which extends in every
direction beneath the photosynthetic tissue.

The flanges or partitions evolved in accordance with the principle of
maximum exposure of surface, arrange themselves in relation to the
lines of flow of the translocatory stream ; hence the elongated tubular
form affected by the majority of photosynthetic cells, and particularly
by the palisade-units.

### 1. *Various Modes of Construction of the Photosynthetic System.*

It has already been explained that the great diversity displayed by
the photosynthetic system in respect of the details of its construction
—a diversity which will presently be illustrated by a number of
concrete examples—is due to the fact that the principle of expeditious
translocation is itself carried out with a varying degree of thoroughness.
The author has accordingly classified the various modes of construction
of the photosynthetic system with reference to ten leading types, which
in their turn may all be relegated to one or other of three principal
systems of construction.

*First System.*—The simplest type of photosynthetic system is characterised by the fact that the **photosynthetic tissue** is itself **responsible
for the removal of synthetic products** from the entire organ.   As the
partitions are accordingly placed parallel to the long axis of the organ,
the photosynthetic cells are elongated in the same sense, and lie in a
plane parallel to the surface.   These cells form a single layer in the
leaves of most Mosses, while two layers are present in *Elodea canadensis* ;
in *Galanthus, Leucojum* and *Sempervivum* the corresponding tissue

consists of several layers, and is sometimes quite massive. In the case of *Galanthus nivalis* the adaxial half of the leaf comprises, in addition to the epidermis, three to four layers of cells provided with numerous chloroplasts. These elements are extended longitudinally, and form very regular series running along the leaf parallel to its surface; the individual cells vary greatly in length, but are, as a rule, from two to five times longer than their width. The lower half of the mesophyll likewise consists of rows of longitudinally extended cells, forming a photosynthetic layer which is nearly as thick as that in the adaxial half; this abaxial tissue is, however, less richly provided with chlorophyll, and is, on the other hand, traversed by more numerous intercellular spaces. Some of the plants exemplifying this first system exhibit the beginnings of a division of labour between photosynthetic and translocatory cells. In *Elodea canadensis*, for example, those cells in the lower of the two component layers of the leaf which contain least chlorophyll are longer and narrower, and are thus more especially adapted for purposes of translocation.

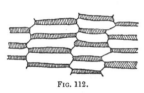

FIG. 112.

Surface section (Tangential L.S.) through the leaf of *Gladiolus floribundus*. Intercellular spaces shaded.

*Second System.*—At a somewhat higher level of organisation, **distinct tissues** are regularly set apart **for photosynthesis and for translocation,** the synthetic products being transferred directly from one to the other. The types in which this plan is carried out, display the greatest variety as regards minor features; there are, in fact, innumerable modifications differing in points of detail, but nevertheless all evidently conforming to the two aforesaid guiding principles.

The simplest variety of the second system is exemplified by the genera *Gladiolus* and *Tritonia,* and also by *Iris germanica*; it is characterised by the combination of longitudinal translocatory channels with transversely extended photosynthetic cells. *Gladiolus* illustrates this variety in its most highly developed form (Fig. 112). Here the transverse diameter of the photosynthetic cells exceeds the length of their (morphological) longitudinal axis from four- to seven-fold. As seen in a transverse section of the leaf, these cells appear to be in almost uninterrupted contact with one another; a surface (tangential) section, however, reveals the presence of fairly extensive intercellular spaces between their transverse walls. These spaces act as transverse barriers which prevent the synthetic products from travelling at right angles to the long axis of the cells. . The outgoing stream of plastic material is thus forced to travel across the leaf towards the proper channels of translocation, which are represented partly by the parenchymatous sheaths of the minor vascular bundles, and partly by the larger plates and strands of conducting parenchyma.

*Pinus* illustrates a type which is in some respects even simpler than that which has just been described, but which is, on the other hand, complicated by the introduction of arm-palisade-cells (Fig. 110 A). Full consideration has already been given, in the preceding section of this chapter, to the structure and arrangement of the flanged tabular cells of this genus; it may be noted in addition, with reference to the principle of expeditious translocation, that the photosynthetic tissue of *Pinus* is made up of sharply defined cell-plates, each composed of a single layer of cells which are in contact with one another only at a few points, being otherwise separated by air-spaces. In these circumstances, it is obvious that the synthetic material manufactured in each individual cell-plate must be transferred directly to the parenchymatous sheath and conducting parenchyma of the central cylinder. A similar transversely laminated photosynthetic system is found also in certain other Conifers, such as *Thuja plicata*, *Cryptomeria elegans* (Fig. 116 C), species of *Abies*, etc.

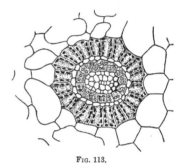

FIG. 113.

T.S. through one of the smaller vascular strands of the foliage-leaf of *Papyrus cicuta*. The bundle is enclosed within three concentric sheathing layers, viz. an internal and an external sheath of green cells, separated from one another by the prosenchymatous en odermis. The external green sheath, which consists of radially elongated cells, represents the specialised photosynthetic tissue.

A third modification of the second system, which is especially common among Monocotyledons, shows a further advance in the develop ment of typical palisade-layers, which are supported by an efferent tissue arranged without any definite relation to the vascular bundles; this condition is exemplified by the leaves and stems of the genera *Allium, Ornithogalum, Asphodelus*, etc. Here, again, the photosynthetic tissue is abundantly provided with transverse air-spaces, which prevent the outgoing stream of material from travelling for any appreciable distance within the palisade-tissue itself. The synthetic products cannot therefore, by any means, pass from the site of their formation in a direct line across the palisade-cells to the base of the leaf; they are on the contrary forced, in consequence of the arrangement of the mesophyll, to pursue a centripetal course, and thus, by following the long axis of the palisade-cells, to leave the photosynthetic elements as quickly as possible.

In the case of a very similar and closely related type of structure palisade-tissue is associated with a common parenchymatous sheath,

which envelops the entire circle of vascular bundles (or extends around
the fibrous cylinder); this arrangement is characteristic of cylindrical
or prismatic stems with a well-developed photosynthetic system, such
as those of *Spartium junceum, Genista bracteolata, Tunica Saxifraga* and
spp. of *Aparagus*.

Yet another type of structure, which is not uncommon among both
Monocotyledons and Dicotyledons, is characterised by the fact that the
tubular photosynthetic cells are arranged in more or less radial rows,
or at any rate in curved series, converging towards the parenchymatous
bundle-sheaths or the strands of conducting parenchyma; here also a
certain proportion of the photosynthetic cells belong to the palisade-type.

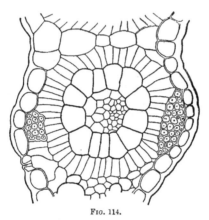

FIG. 114.

T.S. through the leaf of *Panicum turgidum.*  Photosynthetic system of the girdle-type.
After Volkens.

This **girdle-structure**, as it may be termed, is seen in its greatest perfection
in certain species of *Cyperus* (Fig. 113), in which each vascular bundle
of the leaf is surrounded by a layer of radially elongated photosynthetic
cells; translocation probably goes on in a second green parenchymatous
sheath, which is situated immediately within the prosenchymatous
endodermis.   This girdle-arrangement is likewise excellently illustrated
by many Grasses (e.g. *Saccharum officinarum* and *Spartina cynosuroides*),
more especially by species that inhabit desert-regions (*Pennisetum
dichotomum, Cynodon Dactylon, Panicum turgidum* [Fig. 114], *Andro-
pogon hirtus, A. foveolatus, Danthonia Forskalii*, according to Volkens).
Among the Grasses the sheath of conducting parenchyma, though in
the first place acting as the efferent tissue, also contains numerous
chloroplasts, which are moreover often large and very brightly coloured.
It is uncertain whether this green inner sheath merely represents an
unimportant addition to the chlorophyll-apparatus of the plant, or

whether there exists some as yet undiscovered division of labour
between the chloroplasts in the sheath and those in the girdle-cells.

*Third System.*—The third and most effective type of photosynthetic
system is characterised by the fact that the synthetic products are not
transferred directly from the photosynthetic elements to the efferent
channels, but first pass through a special **intermediary tissue** (*Zuleitungs-
gewebe*). This scheme of construction comprises two sub-types, which
differ from one another in one important feature; in one case, namely,
the efferent vascular strands all follow more or less parallel courses,
whereas in the other they form a complicated reticulum. The intermediary tissue is accordingly composed, in the former case of transversely
elongated cells, and in the latter of elements with several radiating

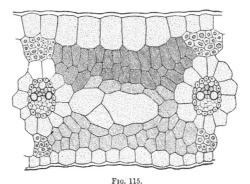

FIG. 115.

T.S. through a bract of *Cyperus alternifolius.*   × 300.

branches. The actual photosynthetic tissue again consists of palisade-
cells. The first-named type of structure prevails in the leaves of most
Grasses and *Carices*, and in many species of *Cyperus*; further, in certain
LILIACEAE, among Cycads and in *Taxus baccata*. The other type is
the normal arrangement among Dicotyledons, while it likewise occurs
in the majority of Ferns.

The anatomy of the bracts of *Cyperus alternifolius* (Fig. 115) will
serve to illustrate the type of photosynthetic system that is characteristic of the GLUMIFLORAE. Here the web of each **I**-girder consists of
a vascular bundle enclosed in a sheath of large colourless cells. The
photosynthetic tissue takes the form of cell-plates situated above and
below the air-passages that alternate with the girders. On the
adaxial side of the leaf it consists of palisade-cells; in transverse
sections these cells appear to be in uninterrupted contact with one
another, but a radial section shows that they are separated from one
another by intercellular spaces of various sizes. At the margins of each

plate the palisade-cells bend round towards the adjacent bundle-sheath, while above the air-passages they abut against the intermediary tissue; the latter consists of rows of transversely elongated cells poorly provided with chlorophyll, and effects a connection between the parenchymatous sheaths of adjoining vascular bundles. The photosynthetic tissue on the abaxial side of the leaf is very similar, except that the subepidermal cells are not of the palisade-type. On collating the anatomical data obtained

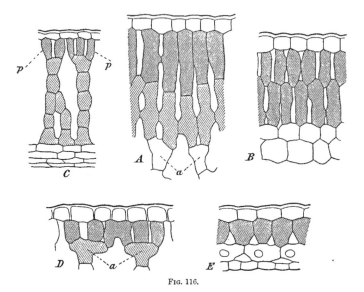

Fig. 116.

Palisade-tissue in various leaves. *A. Asplenium Ruta muraria*, adaxial side of leaf; *a*, funnel-shaped collecting-cells. *B. Asphodelus Villarsii*. T.S. ×106. *C. Cryptomeria elegans*. ×65. Photosynthetic system of the laminated type. *D. Aspidium Sieboldi*. T.S. *a*, collecting cells. *E. Asplenium Belangeri*, fertile portion of pinna. Mesophyll only comprises two layers, viz. an adaxial layer of funnel-cells and an abaxial layer of spongy-parenchyma.

from the transverse and longitudinal sections, one realises that the whole construction of the photosynthetic system is such as to ensure the transference of synthetic products to the main conducting channels by the shortest route, thus clearly revealing the dominating influence of the principle of expeditious translocation.

The general structure of the ordinary dorsiventral Dicotyledonous leaf has been repeatedly discussed, since its first thorough investigation at the hands of Brogniart and Treviranus. In this case the upper epidermis is always underlain by a one- to many-layered palisade-tissue, which is usually very loose in texture. The originally prismatic photosynthetic elements show a marked inclination to become cylindrical, so that the radial walls are particularly liable to separate from

one another. This feature is partly connected with the necessity for
the presence of air-spaces in immediate contact with the photosynthetic
cells ; it also indicates that each palisade element has a tendency to
become independent of its neighbours, neither receiving raw food-
materials from the latter nor supplying them with synthetic products.
Such a palisade-cell in fact only carries on interchange of materials
with the tissues that are in contact with its two extremities, that is to
say, with the epidermis or the water-tissue on the one hand, and with
the spongy parenchyma or the parenchymatous bundle-sheaths on the
other. There is another circumstance which proves that no translocation takes place from one palisade-cell to another. At the base of the
leaf, namely, the development of the palisade-tissue is in no way
different from that which prevails at the distal end, although the
quantities of translocated material passing through
the cross-section of the leaf are very different in the
two cases.

The stream of synthetic products in the palisade-cells thus undoubtedly seems to follow the long
axis of the elements in question, or, in other words,
to travel at right angles to the surface of the organ.
This interpretation also accounts for the existence
of a particular variety of palisade-tissue which is
otherwise inexplicable. Not infrequently a small
group of from two to ten palisade-cells converge at
their lower ends so as to form a little fan-shaped
group resting upon a single underlying cell, the
upper end of which is correspondingly dilated in a
funnel-shaped manner (Figs. 116 and 117); the
obvious inference is that these supporting elements

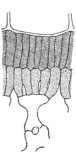

FIG. 117.

Small portion of a
T.S. through the leaf
of *Ficus elastica*, showing a collecting-cell
and its associated
group of palisade-cells.

are **collecting-cells**, which receive the synthetic products from all the
members of a group of palisade-cells, and transmit them more or less
directly to the main channels of translocation. The author has noticed
this peculiar feature of the photosynthetic system in *Ficus elastica*—
where it is particularly well-developed—further in *Pulmonaria officinalis*,
*Juglans regia, Eleagnus angustifolia, Eranthis hiemalis*, etc. Sometimes
the groups of palisade-cells abut directly upon elements of the intermediary tissue, especially when the leaf is thin and not very highly
differentiated.

The intermediary tissue, which constitutes the physiological link
connecting the photosynthetic tissue with the efferent channels, is
represented by the familiar **spongy parenchyma**. It lies beneath the
palisade-layers, and generally consists of elements provided with a
ally directed branches, constituting so many

supply-pipes, through which the synthetic products are conveyed from
the palisade-tissue to the ultimate branches of the vascular reticulum.
Incidentally it may be noted that the spongy parenchyma also contains
some chlorophyll, though not a great amount, and is thus capable of
photosynthetic activity in a minor degree, while it also acts as the
principal ventilating tissue of the leaf by virtue of the numerous air-
spaces with which it is provided.   The spongy parenchyma, therefore,
illustrates the unusual phenomenon of a tissue which is simultaneously
adapted in relation to several distinct functions [of approximately
equal importance].

The intermediary tissue occupies the larger and smaller meshes of
the network of efferent strands composed of the parenchymatous

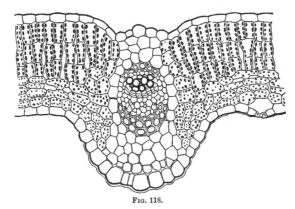

Fig. 118.

T.S. through the leaf of *Raphanus sativus*, including a secondary vein and the
adjoining mesophyll.

sheaths that surround the vascular bundles.   As far as the finer
branches of the vascular system are concerned, these conducting
parenchyma-sheaths [border parenchyma] [146a] consist of a single layer
of more or less elongated cells containing but little chlorophyll.   Not
infrequently the cells of the sheath are provided with lateral extensions,
in order to facilitate communication with the spongy tissue ; properly
speaking, these lateral branches belong to the efferent system.   In
passing from smaller to larger vascular strands, one finds that the
simple bundle-sheath becomes replaced by a many-layered structure,
which in its turn gradually gives way to the more massive and less
individualised " nerve-parenchyma " (Fig. 118) that forms the great
bulk of the mid-rib and the principal lateral veins.   This nerve-
parenchyma is continuous with the parenchymatous ground-tissue of
the petiole and stem.   It is thus quite evident that the enlargement
of the cross-sectional area of the conducting

hand with the increase in the amount of plastic material that has to be transported; a similar spectacle is afforded by a complicated river-system, with its main channel and countless tributaries.

The several types which conform to this third system of construction, differ from one another mainly as regards the relative degree of development of the component tissues. Thus palisade-tissue may preponderate over spongy parenchyma and *vice versa*; even different leaves of the same species may differ appreciably in this respect. The causes of such variation will be discussed in a subsequent section of the present chapter (Section IV.). An excellent illustration of complex organisation, combined with high physiological efficiency, is

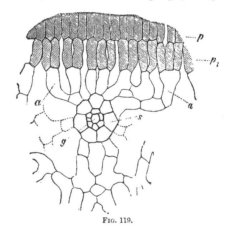

FIG. 119.

Part of a T.S. through the leaf of *Ficus elastica*. *p, p₁*, the two adaxial palisade-layers; *a*, collecting-cells; *g*, a small vascular strand composed entirely of tracheides; *s*, parenchymatous sheath of the strand. × 230.

provided by the leaf of *Ficus elastica* (described in detail in the author's *Vergleichende Anatomie des Assimilationssystems*). The mesophyll of this leaf comprises, apart from the vascular bundles and their sheaths (Fig. 119): (*a*) two palisade-layers; (*b*) from six to ten layers of spongy tissue; (*c*) a layer of funnel-shaped collecting cells interposed between (*a*) and (*b*); (*d*) a layer of cells, situated immediately within the lower epidermis, which, though shorter than the palisade-cells, approach the latter in general form. The presence of the last-mentioned layer illustrates a tendency on the part of the plant to extend the photo-synthetic system beyond its ordinary limits, wherever illumination provides an opportunity. Such an abaxial palisade-layer recurs in various other plants; it is generally composed of funnel- or hour-glass-shaped cells, and constitutes, as it were, a miniature copy of the principal adaxial system. The opposite extreme as regards the organisation of

T

the mesophyll is well illustrated by the fertile region of the leaf of
*Asplenium Belangeri,* where nothing but a single stratum of funnel-cells
and one layer of spongy parenchyma intervenes between the upper and
the lower epidermis (Fig. 116 E).

The above-mentioned funnel-cells require a little more notice.   In
many shade-plants they take the place of ordinary palisade-cells even
on the upper side of the leaf.   In some cases—e.g. *Oxalis Acetosella*—
the chlorophyll-corpuscles are confined to the lateral walls in these
funnel-cells ; here the advantage of this particular shape of cell seems
to consist in the fact that the chloroplasts lie, as it were, half-way
between the profile- and surface-positions, and thus receive more
light than they would if they were permanently fixed in the profile-
position, which is an unfavourable one where the *average* illumination
is feeble.   Noll has shown that chloroplasts obtain still better illumination, if they are located at the base of funnel-shaped cells, a condition
which is exemplified by certain species of *Selaginella* (Fig. 97).
Here the funnel-like shape causes the cells to act as condensing lenses.
Owing to the concave form of the outer wall, namely, the rays of
light that fall perpendicularly upon the leaf-surface are made to converge, by refraction and total reflection, towards the base of the cell,
so that the chloroplasts assembled at that end of the cell are more
brightly illuminated than they otherwise would be.   The same optical
device is found in a much more perfect form in the protonema of
*Schistostega osmundacea.*   Here the chlorophyll-apparatus in each
funnel-shaped cell is so brilliantly illuminated, that it acts against a
dark background as a self-luminous body, any rays that are not
absorbed by the chloroplasts being reflected and retraversing the
whole optical system in the sense contrary to their path of entry.
We are indebted to Noll[147] for the above explanation of the well-
known luminosity of this protonema.

The deviation of the funnel-cell from the normal palisade-form
is obviously connected with special conditions of illumination : but
there are modifications in respect of shape and orientation of another
sort which are related to the function of translocation, and yet others
that result from certain developmental processes, and the physio
logical value of which is obscure—if it exists at all.   Where the
palisade-tissue is loosely put together, individual cells thereof are
frequently found to assume an oblique position or to bend to one side
in order to reach a collecting-cell which is situated some distance away.
Similar displacements often take place in the case of palisade-cells that
underlie water-absorbing or water-secreting trichomes.   Sometimes *all*
the palisade-cells of a leaf (or green stem) are obliquely directed, a
circumstance first discovered by Pick ; this peculiarity probably does

not as a rule bear any relation either to translocatory arrangements or to the conditions of illumination. Pick himself, it is true, regards the obliquity—which he found to occur especially in vertically-extended photosynthetic organs—as an adaptation directly induced by oblique illumination; but Heinricher has shown that this view cannot possibly be maintained in face of the fact that the degree of obliquity may vary in different parts of the same leaf. In *Isolepis australis*, for example, the inclination of the palisade-cells to the vertical is 40° beneath the sub-epidermal fibrous strands, but only 14° where the palisade-tissue abuts immediately against the epidermis. The obliquity is also often more pronounced on the lower side of the leaf; in *Asperula longiflora*, for instance, the corresponding angles are 20° for the lower side, and not more than 5° for the upper. Heinricher has further shown in the case of *Isolepis australis* (and the author has confirmed his results for certain other plants) that the palisade-cells all slope in the same direction—namely, towards the leaf-apex—in spite of the fact that only the basal portion of the leaf is erect, while the apical region is reclinate; as far as the drooping part of the organ is concerned, the orientation of the palisade-cells is obviously incompatible with Pick's hypothesis. The author has, moreover, observed that the displacement of the palisade-cells may take place while the leaves in question are quite young and still enclosed within a bud or buried in the ground (*Dactylis glomerata, Poa annua, Ornithogalum nutans, Scilla bifolia, Allium ascalonicum, Narcissus poeticus*; cf. Fig. 120); in these circumstances the influence of light is, of course, entirely excluded. Further research is, however, needed, before the origin and significance of this by no means uncommon oblique orientation of the palisade-tissue can be satisfactorily explained.

FIG. 120.

Oblique palisade-cells in the young leaf of *Ornithogalum byzantinum*; the portion figured is still below ground at this stage.

### 2. *Experimental Data concerning the Emigration of Carbohydrates from the Photosynthetic System.*[148]

The author has succeeded in identifying the paths by which the carbohydrate material manufactured in the photosynthetic cells emigrates from the leaves, by the methods of comparative anatomy. In accordance with the principle of expeditious translocation, the synthetic products pass, in the case of a typical Dicotyledonous leaf, from the palisade-tissue through the collecting cells into the intermediary tissue or spongy parenchyma, and thence into the parenchymatous bundle-sheaths, or the conducting parenchyma of the principal veins, as the case may be; the reticulately-disposed **conducting parenchyma** thus represents

the actual efferent tissue of the leaf.  The evidence of comparative
anatomy suffices for the identification of the paths of translocation in
photosynthetic organs, since it proves that the migration of synthetic products is necessarily confined to certain definite tracts, on account of the
general anatomical construction of these organs ; nevertheless the fact
that other investigators have arrived at similar conclusions concerning
the conduction of carbohydrates in foliage leaves, on experimental
grounds, provides a highly satisfactory confirmation of the author's
views.

With regard, first of all, to the assumption that synthetic material
is removed as quickly as possible from the palisade-tissue, we may
quote an observation, made by Sachs, upon leaves which he tested with a
solution of iodine, after they had been killed and decolourised by boiling
in alcohol.  Sachs states, that if the leaves, before being tested, " had
not yet accumulated the full amount of starch, or had, on the other
hand, already lost a portion thereof, they appeared brown, or at most
brownish black, on the upper side, in cases where the lower half of the
mesophyll was coal-black, or even metallic black in colour."  The conclusion to be drawn is, of course, *not* that the dorsal palisade-tissue has
a smaller photosynthetic capacity than the ventral spongy parenchyma,
but rather that the synthetic products are removed more rapidly from
the former tissue.  In other words, the included starch-grains of
chloroplasts merely represent the temporary surplus which results,
when photosynthesis produces more carbohydrate material than can
be removed by translocation in a given time.  The author himself has
observed an extraordinarily rapid depletion of photosynthetic cells in
a representative of the girdle-type of leaf-structure.  A leaf of *Saccharum officinarum* picked just after sun-rise, at Buitenzorg, contained
no starch, either in the radially elongated photosynthetic girdle-cells,
or in the elements of the parenchymatous bundle-sheaths—which are
likewise well provided with chloroplasts ; in other words, the entire
synthetic products [of the previous day] had been removed during the
night.  A leaf of the same plant picked shortly before the usual daily
thunderstorm—about 3 p.m.,—after a sunny morning, also contained
no starch in the specialised photosynthetic cells ; in this case, however,
the chloroplasts of the bundle-sheaths enclosed large numbers of minute
starch-grains.  Evidently the migration of carbohydrates from the
photosynthetic cells had been so rapid, that no surplus remained in
the form of starch, although photosynthesis must have been proceeding
briskly all the time.

When, therefore, a leaf has become heavily charged with starch after
energetic photosynthesis, it is clearly the specialised photosynthetic
tissue that is relieved first of all ; the spongy parenchyma next follows

suit, and when translocation has been going on for some time in the absence of photosynthesis (in darkened leaves), transitory starch [148a] is no longer to be found anywhere, except in the conducting parenchyma. At this stage the iodine test produces a blue-black network upon a yellow ground. Finally, the bundle-sheaths also become emptied; in their case the process of depletion begins with the smallest bundles and gradually extends to the larger veins, while it simultaneously advances in the leaf as a whole in the basipetal direction. According to Schimper, the course of depletion can be very readily observed in the leaves of *Hydrocharis Morsus ranae.*

In many plants—perhaps in the majority—the migrating carbohydrate material may be temporarily converted into starch in any cell through which it happens to be passing. This so-called **transitory starch** is found to some extent even in the internal layers of the palisade-tissue, but is abundant in the spongy parenchyma, and accumulates in even larger quantities in the conducting parenchyma, where it is also most persistent. In certain cases, on the other hand, the migrating carbohydrates always occur in the soluble form as *glucose*, either from their first formation, that is to say, even in the palisade-cells (spp. of *Allium*), or at any rate from the bundle-sheaths onwards (*Impatiens parviflora* according to Schimper).

The solution of the surplus starch that accumulates in the photosynthetic system, is brought about, according to the detailed observations of Brown and Morris, by the action of *diastase*, just as in the case of the starch deposited in storage tissues (cf. Chap. VIII.).

## IV. INFLUENCE OF LIGHT UPON THE DISTRIBUTION AND ORGANISATION OF THE PHOTOSYNTHETIC SYSTEM.[149]

The synthesis of organic substance from carbon-dioxide and water is well known to be dependent upon the access of light. In this respect the requirements of different species vary considerably; many shade plants in particular are satisfied with a somewhat scanty supply of light. Nevertheless, every plant endeavours to arrange its photosynthetic tissues in such a manner that they will obtain the most favourable illumination. Hence light-intensity is the factor which primarily determines the location of the photosynthetic system, invariably causing it to take up its position as near the periphery as possible, no matter what special type or modification is in question. This fact is most evident in the case of those cylindrical or prismatic photosynthetic organs which at the same time contain voluminous water-storing tissues, and which are devoid of sharply differentiated "light" and "shade" surfaces. Even in flattened structures, however, such as ordinary leaves, phyllodes and cladodes, the shape of which ensures the most effective illumination of the internal tissues, the differences of light-intensity at different points in the interior are sufficient to exert an appreciable influence upon the disposition of the specialised photosynthetic parenchyma, more especially where the latter is developed as palisade-tissue. In the case of many more or less erect leaves which receive approximately the same amount of light on either side, the chances of effective illumination of the interior are so good, that the entire thickness of the mesophyll, from

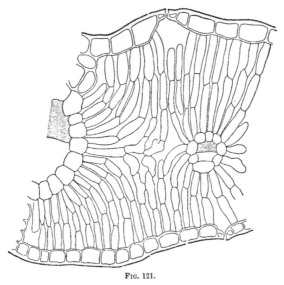

FIG. 121.

T.S. through the isobilateral leaf of *Scabiosa ucrainica*. After Heinricher.

one epidermis to the other, is made up of tubular photosynthetic cells or palisade-tissue, with at most a single layer of intermediary cells in the very middle. According to Heinricher, this type of structure is exemplified by the leaves of *Scabiosa ucrainica* (Fig. 121), *Moricandia arvensis*, *Chelone Torreyi*, *Turgenia latifolia*, etc. In a slightly different form of isobilateral leaf the spongy tissue is rather more extensive; it makes up the central portion of the mesophyll, while the peripheral palisade-tissue is equally developed on both sides of the leaf.

The connection between the disposition of the specialised photo-synthetic tissue on the one hand, and the intensity of illumination on the other, is still more obvious in the case of typical dorsiventral leaves, which are extended at right angles to the direction of the most intense diffuse light, and which consequently possess distinct " light " and "shade" faces.    As a rule, the "light" face coincides with the adaxial side of the organ, but cases are known in which the abaxial side becomes the "light" face, owing to torsion of the petiole or leaf-base (*Alstroemeria, Allium ursinum*; *Gynerium argenteum* and some other grasses).    In dorsiventral leaves the specialised photosynthetic tissue is generally confined to the "light" face.    In exceptional cases, however (*Corylus Avellana, Caccinia strigosa, Galium purpureum, Nigella damascena*), palisade-cells may be developed on the "shade" side as well, a circumstance which proves that the palisade form is not necessarily correlated with *intense* illumination.    In these instances the plant, as it were, prefers to develop a palisade-layer immediately within the lower epidermis, instead of adding a deep-seated layer to the adaxial palisade-tissue, the reason evidently being, that in the prevailing circumstances better illumination can be obtained close to the lower surface of the leaf.

We may next pass on to consider the influence of light upon the degree of differentiation of the photosynthetic system.    In the case of the individual plant, and *a fortiori* in that of the individual photosynthetic organ, the available light is in a high degree dependent upon the environment, no less than upon the position of the organ in the shoot system; as this external illumination may itself be a very variable quantity, one can readily understand why most plants display a greater or less degree of plasticity in respect of the development of their photosynthetic system, and, in fact, appear to be able to regulate that development in accordance with the prevailing light-intensity.    This power of accommodation depends on the fact, that light has a direct influence upon the differentiation of photosynthetic cells, in the sense that an increase in the intensity of illumination not only permits of increased photosynthetic activity, but also results in a more extensive development of photosynthetic tissue.    If, therefore, the specialised photosynthetic tissue takes the form of palisade-layers—as it generally does among Higher Plants,—then increased illumination favours both the quantitative and the qualitative development of that tissue; in other words, the individual palisade-cells become larger, and additional layers are also often produced.    In the case of the Beech (*Fagus sylvatica*), for example, a leaf which has developed under the influence of direct sunlight contains an abaxial layer of palisade-cells, in addition to the two or three adaxial layers, whereas a leaf grown entirely in the shade is only provided with a solitary adaxial palisade-layer, which moreover consists of relatively short cells.    A "sun" leaf may be thrice as thick as a "shade" leaf, solely on account of the more abundant development of its palisade-tissue, the spongy tissue comprising two or three layers in both cases.    A "sun" leaf of the Sycamore (*Acer*

*Pseudoplatanus*) contains a layer of very tall palisade-cells arranged in groups on the top of more or less funnel-shaped collecting-cells. In a " shade " leaf of the same plant the palisade-cells are shorter by one half, while collecting-cells are altogether absent. Both " sun " and " shade " leaves contain two layers of spongy parenchyma (Fig. 122).

The great majority of plants agree with the Sycamore and Beech in the fact that, while bright illumination leads to an increased development of palisade-tissue, the latter is never entirely lacking, even when the plant or organ is growing in the deepest shade. According to Stahl, however, there are instances (*Lactuca Scariola*, *Iris Pseudacorus*) in which the difference between sun and shade leaves is more fundamental ; in these cases palisade-tissue only develops under the influence of intense illumination, and is entirely absent from the shade leaves. No doubt, the plants in question possess the normal power of producing palisade-tissue ; but this capacity is only awakened by very powerful photic stimulation, whereas in the majority of plants relatively feeble illumination suffices to start the series of developmental processes which culminates in the differentiation of palisade-cells. In most cases, indeed, the inception of the palisade-tissue takes place quite independently of illumination, while the leaf is lying in complete darkness within a bud, and any effect that light may produce is purely quantitative, leading to the formation of larger or more numerous palisade-cells.[150]

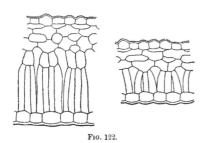

FIG. 122.

A shade leaf (right) and a sun leaf (left) of *Acer Pseudoplatanus*, in T.S.

The same circumstance also largely accounts for the fact, demonstrated by Bonnier and in more detail by Wagner, that the leaves of alpine plants develop a more massive palisade-tissue at high levels ; for it is well known that the intensity of light increases with altitude.[151]

Stahl has pointed out that the photosynthetic system of evergreen plants shows a remarkable want of plasticity. The shade leaves of *Vaccinium Vitis Idaea*, for instance, contain at least three layers of palisade-cells, while in the case of *Ilex aquifolium*, *Vinca minor* and spp. of *Pyrola* the palisade-tissue is comparatively well-developed even in the shadiest localities. The same statement applies to *Buxus sempervirens*, the shade leaves of which do not differ appreciably from the sun leaves, either as regards their total thickness or in respect of the development of the palisade-tissue. Since, moreover, many characteristic shade plants possess typical palisade-tissue, the palisade-cell must not be regarded as a form of photosynthetic element that has been evolved in adaptation to intense illumination, although the development of the tissue is undoubtedly favoured by bright light, as Stahl himself first discovered. Similarly, it cannot be argued that the shape and orientation of palisade-cells depends primarily upon the *direction* of the incident light.[152]

## V. THE PHOTOSYNTHETIC SYSTEM AMONG BRYO-PHYTA AND ALGAE.

The photosynthetic system reveals the dominating influence of the two principles of maximum exposure of surface and expeditious trans-location in various degrees among the Bryophyta and Algae, just as it does in the case of higher Plants; on account of certain peculiarities, how-ever, the lower green plants are most conveniently treated separately.

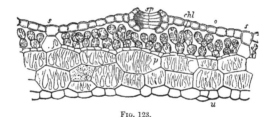

Fig. 123.

T.S. through the thallus of *Marchantia.* *sp*, air-pore ; *chl*, photosynthetic filaments.

Among thallose Liverworts it is more especially the MARCHANTIALES that are provided with a highly differentiated photosynthetic system (Fig. 123). Here the upper surface of the thallus is divided into a number of lozenge-shaped areas, each of which is underlain by a sub-epidermal air-chamber; these chambers are separated from one another by partitions, composed of a single layer of cells, and corresponding to the boundaries of the lozenge-shaped areas. The epidermis is perforated over the centre of each chamber by a wide circular air-pore. From the floor of the chamber arise numerous erect branching filaments, composed of cells containing abundant chloroplasts. These structures, which remind one of Algal filaments, constitute the actual photosynthetic tissue of the plant. The basal element of each filament is funnel-shaped, and represents a collecting cell, which serves to transmit the synthetic material manufactured in the upper cells of the filament to the under-lying colourless parenchyma. The entire layer of filaments thus agrees with a palisade-tissue as regards the course of the current of migrating plastic material.   According to Stahl, indeed, when *Marchantia poly-morpha* is grown in sunlight, its photosynthetic filaments actually come to resemble a palisade-tissue ; in these circumstances, namely, the individual cells of the filaments are elongated in palisade fashion, that is to say with their long axes at right angles to the surface of the thallus.

In foliose Liverworts, and in the majority of Mosses, the photosynthetic system has remained at a low level of organisation ; usually it comprises only a single layer of cells, and serves simultaneously as the efferent tissue.   The individual cells are accordingly often ' elongated so as to form curved series running obliquely from the apex to the base of the leaf and inwards towards the midrib ; in this case the removal of synthetic products is effected in part at least by the midrib. In *Polytrichum* the photosynthetic tissue takes the form of longitudinal plates or lamellae, which stand up perpendicularly from the upper side of the leaf ; each plate consists of a single layer of cells, and is separated from its neighbours by fairly wide air-spaces.   In its essential features, therefore, the photosynthetic system of *Polytrichum* agrees with that of *Sempervivum*.   In the genera *Aloina* and *Crossidium* the broad midrib bears, in the apical half of the leaf, a number of erect dichotomously branching cell-filaments, containing abundant chlorophyll.   In the former genus, the wings of the leaf consist of a single layer of cells poorly provided with chlorophyll, and arch over the photosynthetic filaments after the manner of a protective epidermis ; a cross section of the leaf recalls a side-view of one of the air-chambers of *Marchantia*.

A far more elaborate photosynthetic system is found in many Moss-capsules,[153] where it is located partly in the wall of the sporogonium proper, and partly in the so-called **apophysis**.   Even in the exaggerated form which it assumes in the SPLACHNACEAE, the apophysis represents nothing more than the distal end of the seta, which has become specialised for photosynthetic activity, just like an *Equisetum*-stem or a twig of *Spartium* [or other switch-plant].   According to the author's own observations, the wall of the capsule of *Funaria hygrometrica* (Fig. 124) contains two layers of photosynthetic elements.   The outer of these is largely composed of typical funnel-cells.   The inner layer, on the other hand, consists of elements resembling spongy parenchyma-cells.   It is connected to the outer spore-sac by cellular filaments (trabeculae), which serve to convey the synthetic products manufactured in the wall to the developing archesporium across the intervening air-space.   The photosynthetic system of the massive apophysis consists of typical palisade-tissue, and forms a three- to five-layered sheath around the clavate parenchymatous core which represents a prolongation of the vascular bundle of the seta ; just above the core, a little central mass of spongy parenchyma is interpolated, and this in its turn is connected with the columella by a bunch of trabeculae. These three structures, namely the parenchymatous core of the apophysis, the central mass of spongy parenchyma, and the bunch of trabeculae, presumably co-operate in transporting the synthetic materials manufactured in the palisade-tissue. The archesporium thus receives food-materials from two sides, namely first, from the capsule wall through the outer spore-sac, and, secondly, from the apophysis, through the inner spore-sac.

The photosynthetic system of the *Funaria* sporogonium is a highly efficient one, as may indeed be inferred from the large amount of chlorophyll with which it is provided. The chlorophyll-content of a

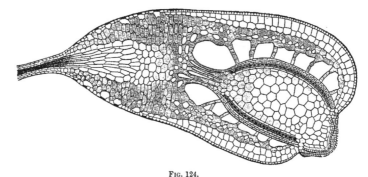

Fig. 124.

L.S. through an unripe sporogonium of *Funaria hygrometrica*. For explanation see text.

fully grown, but still green, capsule is equal to that of about fourteen foliage leaves, or, since a shoot of *Funaria* bears on an average from seven to ten leaves, to that of one and a half entire shoots. The author has further obtained experimental proof of the efficiency of the sporogonial photosynthetic tissue, by removing unripe capsules from the parent plants and thereafter cultivating them in purely inorganic nutrient solutions. In one experiment of this kind, ten sporogonia, in which the spore mother cells were just being differentiated, were cut off and grown in a culture solution for three weeks; at the end of this period, they were ripe and had produced a crop of perfectly normal spores, their dry weight meanwhile having increased by 150 per cent. Many Mosses besides *Funaria* are furnished with an equally active sporogonial photosynthetic system; in all such cases, the capsule is independent of the shoot on which it is borne, so far as the supply of plastic materials is concerned.

Among Algae it is—as might have been expected—only the more highly differentiated PHAEOPHYCEAE and RHODOPHYCEAE that develop a special photosynthetic system. Where such a system occurs, it always occupies a peripheral position, and, in fact, actually extends to the surface of the plant-body, the presence of a protective epidermal layer being generally unnecessary in the case of submerged plants. The influence of the principle of maximum exposure of surface may account . for the fact, that the photosynthetic cells are frequently very small, the total cell-wall-surface available for display of chromatophores being correspondingly increased. The principle of expeditious translocation also makes itself felt to a varying extent. Wille [154] indeed has succeeded in finding examples among Algae, not only of all the main systems of construction recognised by the author in the case of the Higher Plants, but also of many of the individual types of arrangement that have been described above. Thus in a number of types (*c.g.*, the *Ulva*,- *Polysiphonia*-

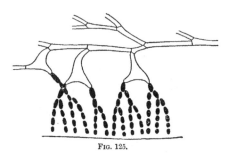

Fig. 125.

Photosynthetic system of *Gelidium corneum*. The photosynthetic elements (cavities shaded) are arranged in radial series converging toWards funnel-shaped collecting-cells, which in their turn communicate with the conducting system.

and *Lithoderma*-types) the photosynthetic and efferent tissues are identical. In another series of genera (*Chorda, Chordaria, Desmarestia, Dictyota, Ahnfeltia, Ceramium, Corallina, Furcellaria, Odonthalia, Rhodomela*) the synthetic products are transferred directly from the photosynthetic elements to the conducting tissue. Several forms, finally (*Nothogenia, Rhodophyllis, Crypto-siphonia*), conform to the third system, in which distinct intermediary elements are interpolated between the photosynthetic and translocatory tissues. Palisade-tissue appears first in the second of the above-mentioned groups (*e.g.* in *Odonthalia dentata*); but the resemblance to the Higher Plants is greatest, where the photosynthetic cells are elongated at right angles to the surface—or arranged in anticlinal series—and converge towards funnel-shaped collecting cells (*Gelidium corneum*, Fig. 125), or where the intermediary elements are disposed so as to form curved translocatory paths (*Cryptosiphonia Grayana, Calosiphonia Finisterrae*), just as in the leaves of many Phanerogams.

## VI. ONTOGENY OF THE PHOTOSYNTHETIC SYSTEM.

In the great majority of cases, the photosynthetic system develops, like other purely parenchymatous systems, from the fundamental meristem. In *Cyperus pannonicus*, however, the layer of palisade cells that is interpolated between each subepidermal fibrous strand and its associated mestome-bundle arises from procambial elements; here, namely, the fibres, the mestome elements and the intervening palisade-layer all become differentiated within a primarily homogeneous procambial strand.   In certain species of *Adiantum* and in *Didymochlaena sinuosa*, the highly characteristic arm-palisade-tissue originates from the protoderm (Fig. 110 c).   In these plants the photosynthetic tissue forms the superficial layer on the adaxial side of the leaf.   A genuine epidermis —in the anatomico-physiological sense—is thus absent from this face. The somewhat thickened external walls of these superficial palisade-cells do indeed possess the structural characteristics of outer epidermal walls ; the elements in question are nevertheless primarily specialised for photosynthesis, and only perform the functions of a dermal layer in a very minor degree.   The same remark applies to the superficial funnel-cells in the leaf of *Selaginella*.   The leaf of *Elodea canadensis*, which consists only of two cell-layers, both containing numerous chloroplasts, arises—except for the midrib—directly from the protoderm of the stem ; the midrib on the other hand is formed, as the author has shown,[155] by the repeated and very regular division of a hypodermal element of the periblem (fundamental meristem) and its derivative segments.

In conclusion, a few words may be devoted to the mode of differentiation of palisade-tissue, with special reference to the stage of development at which this process takes place.   The differentiation of palisade-cells is always initiated by the appearance of active anticlinal division in approximately isodiametric mother-cells.   The additional walls demanded by the principle of maximum exposure of surface are thus, as it were, interpolated under the eyes of the observer ; certainly palisade-cells never arise, as is often alleged, by the mere elongation of isodiametric meristem elements.   The stage of development at which these partitions appear varies in different plants.   In *Ficus elastica* the first differentiation of the palisade-tissue is contemporaneous with the commencement of differentiation within the protodermal layer, and with the origin of the smaller vascular strands, while at the same stage the development of the mechanical system is already half completed.   In *Caragana frutescens*, on the other hand, the palisade-tissue is initiated at a much earlier stage, making its appearance after the segregation of the principal veins and before that of the smaller vascular bundles ; the same statement applies to the green twigs of *Ephedra vulgaris*, in which the palisade-cells and the subepidermal fibrous strands appear almost simultaneously.

# CHAPTER VII.

## THE VASCULAR OR CONDUCTING SYSTEM.

### I. GENERAL CONSIDERATIONS.

IN every highly-organised plant, substances of various kinds are constantly being moved from place to place, at any rate during the periods of active vegetation; the nature and amount of the materials thus transported varies greatly at different times. Before dealing in detail with the paths along which this **translocation** or **conduction** of materials principally takes place, we must first consider the nature and physiological importance of the materials themselves.

In land plants the most energetic transportation of materials takes place in the so-called **transpiration current**; this current serves mainly to replace the moisture that escapes by transpiration, and also, to a smaller extent, to supply water to the several tissues for nutritive purposes and as "water of imbibition." The various inorganic and organic food-materials travel, either in this transpiration current or independently, from the absorbing organs to the photosynthetic tissues and to all the organs that are engaged in active growth. In green plants—which alone need be considered for the present—it is especially the mineral salts or "ash-constituents" that are transported in this way. Lastly, starch, sugar and other carbohydrates, fats, amides, proteins, in short, plastic or synthetic materials generally, are constantly migrating from photosynthetic tissues and from organs of temporary storage to growing regions and permanent storage-tissues—in fact, to every part of the plant-body where such materials are required.

The movements of these diverse substances are altogether regulated by the momentary requirements of the plant and its component organs. Thus materials travel from the root into the stems and leaves, from storage-tissues into all parts of the shoot or root, from foliage-leaves into buds, flowers, ripening seeds or fruits, growing roots and so forth. It is obviously inadmissible to speak simply of an ascending current of crude or nutritive sap and a descending flow of elaborated or formative sap; the old theory of the circulation of sap modelled on these lines was not only a far too extensive generalisation, but was also based upon views regarding individual translocatory processes which were in themselves inadequate, and which have since been greatly modified, mainly owing to the important discoveries of Sachs.

The conducting tissues form the most complicated system in the plant-body, and include the greatest variety of constituent elements, comprising as they do both syncytes and single cells, and dead as well as living elements. This circumstance is quite in harmony with the diversity of the materials that have to be transported; for, in accordance with the principle of division of labour, a regular and uninterrupted flow of material is most likely to be maintained, where separate paths of translocation are provided for the most important classes of substances. There is, besides, a mechanical reason for the prevailing segregation of conducting channels, inasmuch as different substances very often have to be transported simultaneously in opposite directions.

On account of their diversity, the several conducting tissues possess but few characteristics in common with one another; such as they are, these common characters are all referable to the influence of a fundamental principle, which has as its aim the modification of any structural features that would tend to interfere with the flow of material.

The translocation of a given substance may depend altogether upon molecular motion; it may, on the other hand, be a movement *en masse*, or, finally, the two modes of transport may be combined. The question as to the energy employed in translocation is one which concerns the pure physiologist, and which, therefore, need not be considered here. It may, however, be remarked that the presence of numerous transverse septa in the conducting channels must obviously retard translocation, whether this process takes place by means of diosmosis or filtration, or by any other mode of movement, at any rate where the material moves in the cell cavities and not in the walls. Hence it is an essential condition of the efficiency of a conducting tissue, that the number of transverse septa should be reduced and their permeability towards migrating materials increased as far as possible. One accordingly finds, in the first place, that all conducting elements are more or less markedly elongated; further, that their transverse walls are thin, and often provided with numerous pits or even with actual perforations; and, lastly, that the surface available for diffusion is often augmented by some special arrangement.

Little need be said here concerning the first of these general characteristics of conducting tissues. Wherever translocation takes place chiefly in one direction, the conducting elements become elongated in a like sense. This particular adaptation of structure to function is especially striking in a case such as that of the Dicotyledonous woody cylinder, where two systems of conducting channels cross one another at right angles. There is also evidently a certain correlation between the length of the individual conducting elements and the activity of translocation. It would, however, be going too far to assume that great elongation of the conducting elements always indicates a high rate of translocation, since translocation may also be accelerated by the development of more numerous pits on the septa, by an increase in the calibre of the conducting elements and a consequent diminution of frictional resistance, or in other ways. Nevertheless, the correlation in question can be readily demonstrated in a number of cases. The water-conducting channels, for example, generally consist of long vessels in stems, whereas much shorter elements prevail in the bundle-ends of leaves. In the same way the carbohydrate-conducting bundle-sheath-cells of leaves in general adjust their length very closely to the average rate at which synthetic material passes through them.

With regard to the second principal feature, namely, the provision of pits, it should first of all be stated that numerous intermediate stages may be found between the normal pitted condition of the transverse wall and complete elimination of the septum. Most often the septa bear simple pits which are circular or oval in outline, and which not infrequently have their limiting membranes perforated by protoplasmic connections. Such pitted septa are particularly characteristic of the various kinds of conducting parenchyma (including xylem-parenchyma, medullary rays and parenchymatous bundle-sheaths). If a pit is large, and its limiting membrane very thin, there is a risk of rupture, when the osmotic pressure differs considerably on the two sides. In order to guard against this danger, large pits are often subdivided by ridge-like thickenings of the wall, which afford mechanical support to the limiting membrane. Russow, among others, has observed such areolated pits in the secondary phloem of a great many woody plants.

In the case of the great majority of pits, the limiting membrane is perforated by more or less numerous, very fine pores, through which the neighbouring protoplasts communicate by means of exceedingly delicate protoplasmic connecting threads (cf. Chap. I. B. 6); many unpitted membranes display the same feature. It is uncertain how far these protoplasmic connections represent open channels of translocation. Considering the ease with which diffusion takes place through thin portions of the cell-wall and through the associated ectoplasts, and in view of the extreme tenuity of the protoplasmic filaments (their total cross-sectional area only amounts to a fraction of the area of unperforated membrane), it seems improbable that these connecting threads can play any considerable part in connection with intercellular translocation. It is nevertheless quite possible that they acquire greater importance as paths of conduction in special cases, for instance, in certain endosperms. Wherever, in particular, the canals are unusually wide they may have this significance; in the case of many sieve-tubes, for example, it is certain that large quantities of protein material pass through the comparatively wide pores of the sieve-plates.

The reduction of the transverse partitions in the conducting tubes is carried a step further, where the limiting membranes sooner or later become completely obliterated. This condition is exemplified by the perforated transverse walls of the latex-tubes of *Musa* and *Chelidonium*. A far more important illustration is provided by the tracheae or true vessels which, among Angiosperms, constitute the principal conducting elements, both in the primary hadrome and in the secondary wood. Here the transverse walls are sometimes only perforated by several large openings; more often, however, they disappear altogether except for a narrow marginal ring, which persists and marks the former position of the septum. A still more complete reduction of the transverse walls occurs in the latex-vessels of the majority of CICHORIACEAE, PAPAVERACEAE, PAPAYACEAE, etc., which in the adult condition show no trace whatever of septation.

The final stage in this progressive degeneration of transverse walls is exemplified by coenocytic latex-cells of the type found in the genus *Euphorbia*, which are devoid of septa from their first formation. These structures do not arise by the fusion of rows of cells; each, on the contrary, consists of a single cell that has become enormously elongated and abundantly ramified, so as to give rise to a complex system of tubes, which push their way among the other tissues like the hyphae of a parasitic Fungus.

In addition to their elongated form, and the pitted or perforated condition of the transverse walls, conducting elements often display yet another histological feature which tends to facilitate the passage of material, namely, an increase in the surface available for diffusion. This increase is effected by the enlargement of the area, either of the entire transverse wall or of the limiting membrane of the pits borne thereon. The total area of a septum may be augmented in two different ways. In the one case the transverse wall retains its primary orientation, but is enabled to expand by the dilatation of the cells on both sides of the septum; this condition is exemplified by the sieve-tubes of primary leptome strands, by the peculiar latex-cells which occur in the bulb-scales of *Allium*, and by the conducting elements in the stem of *Polytrichum* (both in the central strand and in the leaf-traces). Increased exposure of surface is attained in a different

way, when the transverse walls assume an oblique position, or where,
in other words, the conducting elements become prosenchymatous.
Examples of this latter condition are furnished by the sieve-tubes of
secondary leptome-strands, by phloem-parenchyma (cambiform cells),
and by ordinary conducting tracheides.   Where it is the pit-membranes
that are enlarged, the pit-cavities are dilated in a trumpet-shaped manner;
this condition is analogous to the first of the above-described modes of
expansion of the entire septum.    Dilated pits occur in many Monocotyledonous endosperms and in the bast of certain LILIACEAE.   They
are mentioned here principally because, physiologically speaking, they
represent a stage intermediate between the "simple" and the "bordered"
types of pit.    It must be noted, however, that the peculiarities of the
bordered pit do not merely serve to increase the area of the diffusion-
membrane, but are, on the contrary, determined by special considerations
connected with the conduction of water.

In the preceding paragraphs we have only dealt with the *histological*
features of the conducting system which facilitate and accelerate the
progress of translocation.    There are, of course, also purely *physiological*
factors, that tend to produce the same result.    Under this latter head
must be included, in the first place, those rotating and circulating movements of protoplasm which expedite osmosis by effecting the mechanical
mixture of the diffusible substances ; further, all arrangements that
serve to maintain a concentration-gradient in a series of cells by altering the chemical character of the diffusing substances, or to accentuate
this gradient, when it has once been established.    Here we become
involved in the still mysterious question as to the part played by the
living protoplasts in the process of translocation, a subject which does
not properly lie within the scope of the present work.[156]

## II. THE STRUCTURAL ELEMENTS OF CONDUCTING TISSUES.

### A. CONDUCTION OF WATER AND MINERAL SALTS.

Almost every one of the numerous researches that have been carried
out upon the conduction of water in plants deals mainly with the
woody cylinder of Dicotyledons and Gymnosperms.    The familiar
experiment in which it is shown that the leaves of a transpiring shoot
remain fresh and turgid, after a ring of bark has been removed at its
base, whereas the foliage soon withers if the wood is cut through, was
known to Stephen Hales.    Since similar results are obtained with the
stems of herbaceous Dicotyledons, it may be inferred that conduction of
water takes place solely in the xylem portions of the vascular bundles,
previous to the formation of a secondary cylinder in Dicotyledons, and

permanently in Monocotyledons, where secondary wood is ordinarily altogether absent. If the cut ends of such herbaceous stems are dipped into a coloured solution (say, of aniline blue or eosin), mere inspection will show, that the rapid rise of the coloured liquid which results if transpiration is sufficiently active, takes place only in the vascular bundles; microscopic examination, carried out with suitable precautions, proves conclusively that the solution travels exclusively in the xylem strands. The translucent stems of species of *Impatiens*, and flowers with white petals, furnish very suitable material for experiments of this nature. The same method was employed by the author, in order to demonstrate the water-conducting capacity of the central strand in the stems and setae of various Mosses.

The identification of the tissues within which conduction of water takes place is a comparatively easy matter; it is, however, much more difficult to point out the actual tissue-elements through which the current travels. The difficulty arises very largely from the fact that the secondary woody cylinder of Gymnosperms and Dicotyledons—which has formed the principal subject of decisive experiments—is, physiologically speaking, a heterogeneous structure, the anatomical characters of which represent a compromise between two conflicting influences. The woody cylinder is in fact responsible both for the mechanical strength of the stem and for the transport of water through this organ ; consequently it is not an easy matter to predicate with certainty regarding each organic element of the woody cylinder whether it subserves both of these functions or only one (cf. Ch. XII.).

The mineral salts absorbed from the soil and the water rising from the roots undoubtedly travel along identical paths. This conclusion may reasonably be drawn from the fact that, when transpiration is active, the ascending current of water (transpiration current) can be shown to carry with it a considerable quantity of the absorbed mineral salts ; it seems improbable that the remaining portion of the mineral matter, which travels upwards in immediate response to the varying demands of nutritive metabolism, should follow a different path.

## 1. *Histology of the Water-Conducting Tubes.*

The elements that are utilised for the conduction of water are the **wood-vessels** or **tracheae** and the **tracheides**; collectively these form a water-conducting system which extends into every part of the plant-body. We must first of all deal with the histological features of these elements.[157]

Vessels and tracheides are on the whole very similar in structure. The chief difference between the two consists in the fact, that a tracheide is surrounded by a perfectly continuous cell-membrane or, in other words,

represents a single cell, whereas a true vessel arises from a longitudinal
series of cells which become secondarily continuous with one another.
Even in a fully developed vessel, it is still quite easy to make out the
limits of the individual cells or segments of which it is composed.   If
the transverse septa between the segments are placed at right angles to
the long axis of the vessel, the perforation through which communication is established usually takes the form of a large circular hole,
encircled by an annular ridge representing the persistent marginal zone
of the septum ("simple perforation"); where, on the other hand, the
septa are more or less oblique, the perforation usually consists of
several parallel slits ("scalariform
perforation").

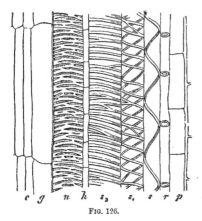

FIG. 126.

Radial longitudinal section through the hadrome
portion of a vascular bundle from the stem of
*Oenothera odorata.* *p*, conducting parenchyma; *r*,
annular (protohadrome) vessel; *s*, spiral vessel, with
the coils of its solitary thickening fibre very greatly
drawn out; $s_1$, younger spiral vessel, with two thickening fibres which anastomose at certain points; *h*,
row of wood-parenchyma cells; *n*, reticulate vessel;
*g*, vessel in process of development, with its septa
still intact; *c*, cambium.

Tracheides are generally elongated, prosenchymatous cells.
Their average length rarely exceeds 1 mm.; in the stems and
petioles of *Musa* and *Canna* they
may be more than 1 cm. in length
(with a diam. of ·08–·1 mm.),
while in the case of *Nelumbium
speciosum*, Caspary has measured
tracheides which were over 12
cm. long.   Vessels, of course,
attain to a far greater length
than tracheides, but do not on
an average exceed 10 cm. (according to Strasburger).   In the
wood of *Quercus pedunculata*,
tracheae 2 m. in length are
not uncommon, while in *Robinia
Pseudacacia*, a relatively large
proportion of the vessels are 1 m. long.   Among lianes even longer
vessels are not at all unusual.   It must be a comparatively isolated
occurrence for a single vessel to extend without interruption through
the whole extent of the plant-body, or even throughout a single organ.
The diameter of the tracheae is also very variable, the widest vessels
(·3–·7 mm.) being found among twining plants.

The walls of vessels and tracheides are always partially thickened.
According to the distribution of the thickening layers, it is customary
to discriminate between **annular, spiral, reticulate, scalariform** and **pitted
vessels** or **tracheides** (Fig. 126); the last-mentioned type includes two
varieties, namely, that with ordinary or simple pits, and that in which the
pits are bordered.   In the spiral type of vessel the number of thickening

fibres ranges from one to four or even more.   The individual fibres are often forked in places; or successive turns of the spiral may be connected by oblique anastomoses.   The first-formed vessels, which are differentiated while the young organ is still growing in length, are always annular or spiral, because it is only these two types of thickening that admit in any considerable degree of a subsequent longitudinal extension of the thickened walls.   The thickening rings or spirals do in fact move further and further apart in these young vessels, as the organ grows in length.   The reticulate and the pitted types of thickening are connected by insensible gradations.

A special feature of the thickening fibres has recently been brought to light by Rothert,[158] who has demonstrated that in most plants these fibres are more or less distinctly contracted immediately above their insertion upon the unthickened membrane (Fig. 127).   The contracted basal portion is often sharply delimited from the broader part, so that the whole fibre is ⊣-shaped in cross-section.   In other cases the two regions pass insensibly into one another.   The cross-sectional outline of thickening fibres is altogether very variable.   Rothert believes the physiological value of the narrow base

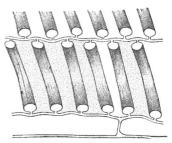

FIG. 127.

L.S. through two spiral vessels of *Cucurbita Pepo*. After Rothert.

to consist in the circumstance, that this mode of insertion of the thickening fibre causes the smallest possible loss of permeable surface in the shape of unthickened membrane.   Schwendener had long previously attributed a similar significance to the peculiar structure of bordered pits (see below).   Rothert himself indeed institutes a comparison between these pits and the annular, spiral and reticulate thickenings.   Clearly, however, the increase in the area of unthickened and hence readily permeable wall-surface, produced by the narrow insertion of the thickening fibres, is inconsiderable, unless the fibres are very closely crowded.   There are, moreover, quite a number of plants (*Equisetum*, CACTACEAE, etc.) in which the insertion of the fibres is not contracted at all.

All the above-mentioned modes of thickening (the bordered pits excepted) are designed to serve the same end, namely, to give sufficient rigidity to the conducting tubes without unduly interfering with diffusion between adjacent elements.   The necessity for the provision of special strengthening arrangements arises from the circumstance that tracheides and vessels, being dead structures and hence unable to

develop a turgor-pressure of their own, are subject to compression at
the hands of the surrounding parenchymatous tissues.

The structure of **bordered pits**[159] demands fuller consideration.
The cavity of a simple pit remains approximately of the same width
throughout.   In the case of a bordered pit, on the contrary, the cavity
is greatly enlarged in the neighbourhood of the limiting membrane;
since the pits correspond on the two sides of the wall, the result is the
formation of a lenticular pit-chamber divided into two equal halves by
the limiting membrane.   This membrane is not equally thick all over,
but bears at its centre a more or less thickened pad or **torus**, which
is slightly wider than the orifice of the pit.   Among Conifers the
torus is rod-shaped in optical section in the spring-wood, biconvex
(lenticular) in the autumn-wood (Fig. 128 A).   In contrast to the

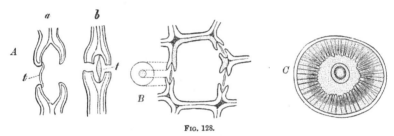

FIG. 128.

Structure of the bordered pits of *Pinus sylvestris*.  *A*. Bordered pits as seen in a
tangential section of the Wood ;  *a*, from spring Wood (air-dried) ;  *b*, from autumn
Wood ;  *t*, torus.   ×750.   *B*. T.S. through a tracheide ; the bordered pits are confined
to the radial Walls.   *C*. A bordered pit in surface vieW, showing the radial striation
of the thin marginal region of the limiting membrane.   ×1000.   *A* and *C* after
RussoW, *B* after Strasburger.

torus the rest of the closing membrane is exceedingly delicate; in
Conifers this marginal area exhibits very distinct radial striae, which
extend right up to the edge of the torus (Fig. 128 c).   According to
Russow, this striation probably indicates the presence of slightly
thickened strips separated by thinner areas.   It should be noted that
the limiting membrane has a very strong affinity for certain dyes, and
particularly for haematoxylin and eosin.

The appearance of a bordered pit in surface view varies considerably according to the shape of the border and the cavity.   Either of
these portions may be circular, elliptical or narrowly oblong in outline ;
moreover, the border and the orifice of the cavity need not be of the
same shape, so that a great number of different combinations may
occur.

As regards the disposition of bordered pits, it may be stated that
these structures are generally arranged in vertical or horizontal rows
or in spiral series : the spiral arrangement is more particularly charac-

teristic of narrow slit-like pits.   The pits are sometimes placed at wide intervals, while in other cases they are crowded closely together; in the latter event the border often acquires a polygonal outline.

Hitherto discussion has been confined to the typical " two-sided " bordered pits, such as occur on a wall which separates two vessels or tracheides from one another,   Wherever a tracheal—that is, a water-conducting—element abuts against a parenchymatous cell engaged in the storage or conduction of plastic materials, there one meets with " one-sided " bordered pits, that is to say, pits which are only bordered

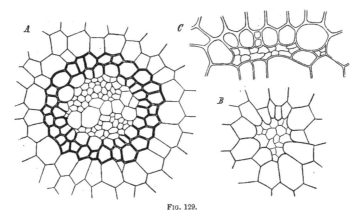

FIG. 129.

Water-conducting tissues of mosses.  *A.* Central strand of the seta of *Mecsea longiseta,* consisting of a bundle of tracheides enclosed in a protective sheath.  *B.* Central strand of *Rhynchostegium murale.*  *C.* A ribbon-shaped leaf-trace-bundle from the cortex of the stem of *Mnium rostratum.*  (All in T.S.)

on the side of the wall that pertains to the tracheal element.   In such cases the limiting membrane is never provided with a torus, but remains thin all over; the turgor-pressure in the parenchymatous cell causes it to bulge towards the border.   Under certain conditions the limiting membrane may even grow out into a tubular process, which penetrates right through the pit-cavity and protrudes into the lumen of the vessel; this phenomenon is considered in more detail below (II. A. 5).

The present section may conclude with a brief account of the **rudimentary water-conducting tissues** which are frequently found in the stems and setae **of Mosses.**[160]   The author has proved that the cells composing the simple central strand which is characteristic of the axial organs of Mosses (Fig. 129), are to be regarded as rudimentary tracheides.   The cells in question are elongated prosenchymatous elements, with narrow cavities, and usually with thin and delicate walls. The oblique transverse walls in particular are always thin, even where the lateral walls are more or less thickened. The walls are almost invariably smooth throughout, that is, devoid of any trace of pitting or other form of sculpture. In the female plants of *Mnium punctatum* and *Bryum leucothrix*, however, the walls are provided with transversely elongated pits in the distal enlarged portion of the central strand: these pits are particularly numerous in the last-mentioned species. In central strands which are mainly thin-walled, the edges of the cells are not infrequently thickened in such a manner, that the tissue presents a somewhat collenchymatous appearance in transverse section (*Dicranum scoparium*). In the genus *Polytrichum* the tracheides of the central strand are arranged in a number of isolated groups, separated from one another by greatly thickened membranes, which are usually reddish-brown in colour (Fig. 130 *t*); within each group the cells are separated from one another by thin walls. An examination of serial sections shows, that these apparently isolated tracheidal groups are in reality continuous with one another at certain points, and that some of the thin partitions within the groups represent the very oblique septa of the prosenchymatous tracheides, while the rest consist of the thin regions of longitudinal walls, which become gradually thicker as they are traced upwards or downwards. The thickened tracts of the tracheidal walls thus constitute a stiffening frame-work, which performs the same office as pertains to the various types of thickening exemplified by the tracheides and vessels of Vascular Plants.

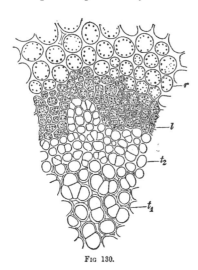

FIG 130.

Part of a T.S. through the vascular bundle [central strand] of the stem of *Polytrichum juniperinum*; *r*, cortical parenchyma; *l*, leptome; *t₁*, central tracheides with thick brown Walls; *t₂*, peripheral tracheides with colourless Walls.

## 2. *The Functions of Vessels and Tracheides.*[161]

Botanists formerly believed that vessels and tracheides served for purposes of ventilation; it is now established beyond any doubt that these structures represent the water-conducting organs of the plant. The modern view is based in the first instance upon observations respecting the nature of the **contents** of vessels and tracheides at different seasons and at various times of the day.

Direct inspection of the contents under the microscope was first undertaken by Volkens, who worked both with herbaceous plants and with the leaves, petioles and young twigs of various deciduous trees. Volkens's method was to cut a piece 2-3 cm. in length from the stem or twig by means of special double shears, so that the two ends were severed simultaneously. From this piece longitudinal sections were then prepared and examined as soon as possible as dry objects under the microscope ; in this way the presence of water in vessels and tracheides can be readily demonstrated. From observations carried out in the manner described Volkens concludes that " in herbaceous plants the vessels contain nothing but water during the early morning hours, when transpiration is slight or completely in abeyance ; in the course of the day, however, a portion of this water is replaced by air." In the afore-said organs of certain trees (*Prunus*, *Platanus*) the vessels were found to contain a certain amount of water on the hottest summer days, even at a height of more than 40 feet from the ground. In the case of leaves which had been transpiring actively for some time, the vessels and tracheides of the lamina were usually found to contain short water columns alternating with air-bubbles, though occasionally the contents consisted solely of air and in a few cases of water alone. The tracheides composing the foliar bundle-ends generally contain nothing but water.

Both Schwendener and Strasburger have similarly determined the nature of the tracheal contents by direct inspection, in the case of thicker branches and large trunks. Schwendener obtained cylindrical samples of wood with the aid of a Pressler growth-borer (an instrument used by foresters). In order to guard against the entrance of air into the severed vessels and tracheides, the cavity of the instrument was filled with boiled water or glycerine before insertion ; for a similar reason the sections cut from the cylinder were mounted in boiled water for microscopic examination. Samples were removed in this way during the month of May from trunks of Beech, Oak, Alder, Birch, Willow, and Pine ; in all cases the vessels (or tracheides) were found to contain more or less numerous bubbles of air in addition to water. In the case of the Beech, Schwendener further determined the average lengths of the alternating columns of water and air. According to measurements carried out between 30th May and 23rd June, the average length of the water-columns was found to lie between ·06 and ·182 mm., and that of the air-bubbles between ·22 and ·392 mm. It should be noted that, in these experiments, the tracheal air was measured at atmospheric pressure. Under normal conditions, however, as will be further expla'ned below, the air in the vessels is more or less rarefied, so that in the intact elements the air-bubbles are longer than they would appear to be from the above measurements. Strasburger likewise concludes that all vessels and tracheides contain air-bubbles as well

as water.    But he maintains, that the outermost water-channels—which
are known to convey the greater part of the transpiration-current—
always contain the smallest proportion of air.    If both vessels and
tracheides are present, the former, which have the wider cavities, contain the larger amount of air; where, however, true vessels alone are
present—as in the wood of *Ficus*, *Acacia* and *Salix*—they never contain
much air.    The protoxylem vessels of roots likewise contain very few
air-bubbles.

A number of investigators have devised experimental methods for
determining the water- and air-content of vessels and tracheides.    Boehm
was the first who endeavoured to prove the presence of water in vessels,
in opposition to the opinion prevalent at the time.    He attempted to
force air into branches (of Sycamore, Horse-chestnut, Birch, Lime, etc.)
from 1-2 cm. in diameter and about 50 cm. in length, through the cut
end, with the aid of a column of mercury.    He found all the vessels,
or at any rate those near the periphery, to be impervious to air under
an additional pressure of one atmosphere, though this pressure often
caused water to exude from the tracheal elements for a short time.
Boehm inferred from his experiments that the wood vessels contain
water.

Hartig was the first to suggest that the vessels of transpiring twigs
and branches contain **rarefied air.**    This fact was, however, definitely
established for the first time by Von Höhnel, who showed that when a
transpiring branch is severed under mercury, the latter rises in the
vessels to a considerable height (up to 70 cm.) in spite of the great
resistance opposed by capillary forces.    Von Höhnel's experiment clearly
proves that the tracheal air is in a highly rarefied condition, a phenomenon which can only be explained by assuming that during periods of
active transpiration water is removed from the vessels faster than it can
be replaced by the osmotic pumping action of the adjacent living cells.
Since the vessels and tracheides are separated by almost air-tight walls
from the ventilating spaces of the plant, air only gains access very
slowly to the partially depleted water-channels; hence the reduction of
pressure which Von Höhnel found to result from active transpiration.

R. Hartig endeavoured to estimate the proportions of water and
air in vessels and tracheides by yet another experimental method,
namely, by accurate determinations of the weights of samples of wood.
He came to the conclusion that the water-conducting elements of the
woody cylinder contain both sap and air at all seasons; he further
estimates that the proportion is sufficient, assuming the tracheal walls
to be completely saturated, to occupy at least one-third, but often as
much as two-thirds, of the lumen of each element in the case of
deciduous trees provided with true vessels.    In Coniferous wood, which

is devoid of true vessels, sap occupies at least two-thirds, and at most nine-tenths of the tracheal cavities.  Hartig's estimates, however, take no account of the fact that different vessels in one and the same woody cylinder may often contain very different amounts of water at the same moment.

The author has shown that the rudimentary tracheides in the central strands of Moss-stems likewise contain a watery liquid, which after active transpiration becomes replaced by rarefied air.  In isolated cases (*Mnium undulatum, Polytrichum*) these elements of the central strand contain starch-grains and drops of oil, even when they are fully developed.

The preceding discussion of the contents of vessels and tracheides naturally leads to a consideration of the experimental and anatomical data from which it may be inferred that the elements in question do not merely contain water, but are actually engaged in the **transportation** of this substance.  Attention may first be directed to the experimental observations that have been carried out upon the supply of water to vessels and tracheides, and upon its removal from these organs under normal conditions in the intact plant.  It is known, on the one hand, that the water-conducting tubes of the root become filled with water owing to the osmotic action of the living parenchyma, while, on the other hand, it has been shown, that, in leaves, a more or less rapid depletion of the vessels and tracheides is effected by the osmotic "suction" of the transpiring mesophyll cells.  The co-existence of these two processes necessarily implies a transportation of water in that portion of the vascular system which intervenes between the absorbing roots and the transpiring leaves.

From the earliest days of physiology onwards, attempts have been made to obtain an ocular demonstration of the conduction of water.  For this purpose it is customary to make use of coloured liquids, or of solutions which give rise to coloured precipitates on the addition of suitable reagents, the liquid being in either case applied to the severed ends of branches or to the roots of intact plants.  Experiments of this nature were carried out as early as the eighteenth century, and have been frequently repeated, in the original form and with various modifications, up to the present day.  In every instance the conclusion to be drawn is the same, namely, that the liquids rise most rapidly in the vessels and tracheides.  It should, however, be noted that the rate of ascent of the liquid is, in such cases, generally estimated by the coloration of the cell-walls in which the pigment becomes deposited, a circumstance which may easily lead to erroneous conclusions, inasmuch as the water-conducting capacity of a cell is, of course, in no way related to the staining properties of its membranes.  *A priori*, therefore, one would

feel disposed to place most reliance upon those experiments in which
the rise of the coloured solution itself is observed, rather than some
secondary phenomenon such as the coloration of the walls.   Elfving,
for example, has shown that when a watery solution of eosin is sucked
through the wood of *Taxus baccata* or other Conifer, the cavities of the
tracheides become filled with the deep-red liquid, while their thickened
walls remain unstained.   In any case the method of coloured solutions
is inadequate, where it is desired to determine the individual water-
paths, and not merely the general course of the transpiration-current.

Various physiologists have devised experiments for the special
purpose of proving that the transpiration-current moves in the **cavities of
the vessels and tracheides,** and not—as has been alleged especially by
Unger and by Sachs—within the lignified walls of these elements.   All
these experiments consist in observing the effect of artificial occlusion of
the tracheal cavities upon the transpiration-current.   Elfving injected
branches of Yew, Ash, Oak, etc., and also haulms of Indian Corn, with
melted cacao-butter (previously coloured red); after allowing the fat to
congeal, he cut a fresh surface and attempted to force water through
the stem, but invariably without success.   Wood is thus evidently
rendered impervious to water, when the cavities of the vessels and
tracheides are blocked with fat.   In order to meet the objection that
the injection of fat interferes with the permeability of the cell-walls,
both Scheit and Errera replace the cacao-butter by gelatine coloured
with eosin or Indian ink.   In Errera's experiments, a branch was
allowed to absorb the gelatine through its cut end by the force of its
transpiration, with the result that the leaves soon drooped.   In this
way it has been clearly demonstrated that the transpiration current
travels in the cavities of the conducting tubes.

In searching for arguments in favour of the water-conducting function
of vessels and tracheides, **anatomical and histological evidence** can only
be regarded as of secondary importance in comparison with experimental data.   Nevertheless, though the evidence of anatomy is not in
itself conclusive, it cannot be altogether ignored in a professedly
anatomical treatise.   In this connection we may first of all lay stress
upon the circumstance, that the vessels and tracheides invariably form a
continuous system in every part of the plant-body, from the finest
rootlets up to the extreme tips of the uppermost leaves.   If these
structures were merely reservoirs of water, as has been suggested in
certain quarters, this continuity would be unnecessary, not to say
incomprehensible.   A further argument may be derived from the
elongated tubular form of the vessels and tracheides, a fact which, considered in conjunction with the experimental evidence, clearly indicates
their conducting function.   Another noteworthy circumstance is the very

close correlation which exists between the requirements of particular organs in respect of water-conduction on the one hand, and the number and diameter of their tracheal elements on the other. This correlation will be referred to again on a subsequent occasion. As regards the histology of the walls of water-conducting elements, special reference may once more be made to the bordered pits ; the significance of these peculiar structures becomes partially comprehensible, if one assumes that water moves in the cavities of the cells to which they belong. This point will also receive more attention later on.

At this stage it will be desirable to summarise in half a dozen sentences those facts concerning the conduction of water which may be regarded as definitely established

1. Vessels and tracheides normally contain both air and water, the relative amounts of the two substances varying according to the season and the time of day.

2. The pressure of the air in the tracheal elements varies, but is frequently very low.

3. The injection of water into the vessels and tracheides is effected by osmotic forces, which reside chiefly in the root-system.

4. The partial removal of water from the vessels and tracheides is due to the osmotic " suction " developed in the mesophyll.

5. The rarefaction of tracheal air must be attributed to rapid removal of water from the vessels.

6. The vessels and tracheides serve as water-conducting tubes, and the water travels in their cavities.

So far as small herbaceous plants are concerned, these facts are quite sufficient to account for the upward movement of water in the vessels and tracheides of the stem. In such plants, root-pressure is well able to keep the entire tracheal system filled with water from the roots right up to the tips of the leaves ; frequently, indeed, this pressure is so great that special organs—the hydathodes—are brought into play in order to get rid of the excess of water, and so to prevent the ventilating spaces from becoming flooded. On the other hand, the entire conducting region lies within the sphere of influence of the osmotic suction developed in the transpiring mesophyll.

Very different conditions, however, prevail in the case of tall trees, which naturally form the centre of interest in every enquiry into the causes of the " ascent of sap." In these plants the vessels and tracheides of the woody cylinder rarely contain continuous water-columns during any part of the vegetative season ; the physiologist is thus at once confronted with a host of difficulties, if he attempts to discover the details of the mechanism underlying the ascent of sap in tall trees, or even tries to determine what forces are available for the purpose of raising large quantities of water to considerable heights.

Where the vessels and tracheides are associated in bundles in such a manner that their readily permeable walls are laterally contiguous, water can pass in a continuous stream past the relatively stationary air-bubbles, provided the latter are not too long. Here the bubbles are comparable—as Schwendener has pointed out—to islands which break up a river into a network of channels. The conditions are different, if the transpiration current flows through vessels which are surrounded on every side by non-conducting tissues; this arrangement is exemplified, according to Strasburger, by the genera *Salix* and *Ficus*, and also by certain LEGUMINOSAE. In these circumstances, there are several ways in which the ascent of sap might theoretically take place,—and the same statement doubtless applies to all cases in which continuous water-tubes are present. One may, in the first place, suppose the alternating columns of water and air—which constitute a so-called Jamin's chain—to move bodily under the influence of appropriate forces. It must not be forgotten, however, that it is very exceptional for vessels to be as long as the stem or trunk in which they are located. Hence, in estimating the force which opposes the movement of the Jamin's chain as a whole, one must take into account not only the resistance opposed to displacement of the chain itself, but also the resistance which the air-bubbles would have to overcome in passing through the successive transverse septa; since the walls are highly impervious to air, this latter resistance must be very considerable. This mode of ascent of sap would, of course, be even more difficult of execution in a system of tracheides, on account of the far greater number of transverse walls. In the second place, the ascent of sap might consist—as Vesque maintained and Strasburger has reasserted—not in a bodily displacement of the Jamin's chain, but in a movement of water past the stationary bubbles; if this is actually the case, there must be a thin film of water between the air-bubbles and the wall of the tube, linking up the apparently isolated water-columns. Finally, it is conceivable—as Westermaier first asserted—that the living xylem-parenchyma cells which are closely associated with all vessels and tracheides, and perhaps also the elements of the medullary rays, may act like so many minute suction- and compression-pumps, withdrawing water, by virtue of their osmotic activity, from the several liquid columns of the Jamin's chain, only to discharge it at a higher level, either into another liquid column in the same chain, or into a different section of the conducting tube. These different theoretical possibilities might, of course, also be combined in various ways. The whole question, however, has not as yet progressed beyond the stage of more or less well-founded hypotheses and suggestions.

Similarly, it is not yet possible to indicate the precise function performed by the **bordered pits** which constitute such a characteristic feature of the walls of water-conducting elements. As Schwendener has pointed out, the construction of a bordered pit tends to increase the surface available for diffusion without reducing the strength of the wall more than is absolutely unavoidable (cf. Ch. I.) The development of the border which overlaps the pit-cavity, thus in a sense represents a compromise, between the requirements of the mechanical principle on the one hand, and the demands of translocation on the other. This explanation, however, does not account for the presence of the most characteristic part of the limiting membrane, namely, the torus. Direct observation has shown, that the torus may become closely appressed to the orifice of the pit, so that the latter is tightly closed. The bordered pits are thus undoubtedly adapted for a very special purpose. Russow states that they represent minute valves, a view which is certainly borne out by every feature of their construction. The torus must be assumed to be relatively impervious to water, and, when in a state of imbibition, also to air ; if, therefore, the pressure becomes unequal on the two sides of the limiting membrane, the torus is forced against the mouth of the pit, which thus becomes more or less effectually closed. Where the border is thin, the margin of the aperture bulges slightly inwards, an arrangement which helps to ensure a good contact with the limiting membrane; this point was first noted by Russow. In the closed condition of the pit the thin marginal area of the limiting membrane stretches elastically so as to fit the border exactly, if it is not already wide enough for this purpose. When the pressures on the two sides of the wall are equal or not greatly different, the torus presses quite loosely, or not at all, against the orifice ; in this condition of the pit, diffusion can proceed freely through the thin marginal area of the limiting membrane, which is highly pervious, at any rate to water.

What difference of pressure—whether of water or of air—is required in order to induce the bordered pits to close, is not known with any certainty.[162] Russow, however, points out, that when fresh Coniferous wood is allowed to dry, the pits invariably close, all the limiting membranes shifting towards the low-pressure side. A difference of pressure of less than one atmosphere is thus certainly effective. In this particular experiment the limiting membranes become pressed so tightly against the border, that the tori assume the form of hemispherical protuberances projecting through the pit orifices. It has been calculated by Pappenheim, on the basis of certain filtration experiments, that most of the bordered pits in a piece of Coniferous wood will close under a difference of pressure of 5 cm. of mercury (about $\frac{1}{15}$th of an atmosphere). But for various reasons estimates of this kind cannot be regarded even as approximately accurate. The attraction of the tori cannot, of course, lead to an actually air- or water-tight sealing of the pits; Pappenheim and Strasburger have, in fact, shown experimentally, that it merely results in a great reduction of the permeability of these structures towards water and air.

Precisely what part the valve action of the bordered pits plays in connection with the ascent of sap, is not clear. Strasburger supposes, that these pits represent an automatic mechanism for preventing the low tension that arises in a partially depleted water-channel from being relieved by the entrance of *air*, before the deficiency of *water* can be made good; for this reason, depleted vessels do not necessarily lose their water-conducting capacity, whereas they undoubtedly would do so, if large quantities of air were to gain admittance.

In conclusion, a few sentences may be devoted to the hitherto like wise unsolved problem of the sources of the **energy employed in raising water** to the tops of tall trees. Since this question really belongs to the domain of pure physiology, it will, in the first instance, suffice to remark that neither capillarity, nor atmospheric pressure, nor root-pressure can supply the requisite energy, while it is quite inadmissible to suppose that the water moves either by osmosis alone, or in the form of vapour. A very considerable amount of energy may, however, become available in consequence of transpiration, since this process can, in certain circumstances, give rise to a very powerful osmotic suction, particularly when the removal of water is rapid enough to cause a large diminution of turgor in the mesophyll, which is the actual site both of the evaporation and of the osmotic action. Under ordinary conditions, however, no very great diminution of turgor actually takes place. Hence the suggestion recently put forward by Dixon and Joly, and independently by Askenasy, to the effect that the cohesive power of water probably enables the suction produced by transpiration to be transmitted to the root system, cannot seriously affect our estimate of the value of osmotic suction as a *source* of energy.

Elaborate calculations, and a careful consideration of the general behaviour of Jamin's chains, led Schwendener to conclude that the upward suction produced in these chains as a result of transpiration is usually restricted to the smaller twigs, and in all probability seldom extends below the base of the leafy crown. As, on the other hand, the action of root-pressure cannot as a rule raise water more than 1 or 2 metres above the ground, in the case of our native deciduous trees, "the movement of the Jamin's chains in that part of the trunk which lies between the aforesaid limits can only be due to forces resident in the trunk itself." Schwendener accordingly assumes, that the requisite energy is furnished in some as yet unexplained fashion by the living elements of the wood, thereby confessing his adherence to the views previously formulated by Westermaier, Godlewski and Janse, who all maintain that the ascent of sap is a vital and not a purely physical process.

Strasburger, on the contrary, has endeavoured to disprove the vitalistic theory experimentally. He has succeeded in showing that poisonous liquids, such as solutions of copper sulphate or picric acid, which rapidly kill the living cells of the wood, may nevertheless rise for days in succession, in trunks 20 metres (65 feet) in height, right up to the transpiring foliage. Strasburger's experiments are, however, open to many objections, and are certainly not conclusive. On the contrary, recent experiments carried out by Ursprung appear to support the opposite view, according to which the living cells of the wood do take an active part in raising water. Ursprung finds that, if a portion of a branch or trunk is boiled, etherised, or cooled down sufficiently, so that its living cells are rendered inactive, the transference of water is invariably retarded to such an extent that the experimental plants soon wither and die.

### 3. *The special uses of vessels and of tracheides.*

Between vessels and tracheides no very sharp distinction can be drawn in respect either of length·or of diameter; nevertheless, these two types of water-conducting element are so clearly differentiated in most Angiosperms, that one is forced to enquire whether both are employed for precisely the same purposes or not. Anatomical investigation supplies an unequivocal answer to this question: for—as Schwendener long ago remarked,—wherever this differentiation exists, the vessels serve mainly to transport water over considerable distances, whereas the tracheides are chiefly employed in satisfying local requirements. A comparison between a typical leaf and a typical stem brings out this difference of function very clearly. In the stem the principal channels of the transpiration-current consist mainly or exclusively of vessels, while the lateral offshoots of these main channels—forming in the leaves a dense network, which serves to effect a uniform distribution of water throughout the mesophyll—are largely or entirely composed of tracheides. Tracheides also predominate in the lateral anastomoses that connect adjacent vascular bundles, or different parts of the secondary woody cylinder (the latter case being exemplified by the Oak), with one another.

The fact that vessels represent the principal water-paths, is indicated not only by the great length of these structures, but also by their width, which is on an average greater than that of tracheides. This argument is in no sense vitiated by the circumstance that tracheides

x

may also be very long and wide in exceptional cases—for instance, in the stems and petioles of *Musa*, *Canna* and *Nelumbium* (cf. above, p. 308).

The division of labour which finds expression in the differentiation of vessels and tracheides, of course corresponds to a high level of general organisation. In accordance with this rule, one finds that the water-conducting tissue of Mosses consists entirely of rudimentary tracheides. In the Pteridophyta, also, tracheides alone are recorded, with few exceptions (*Pteris aquilina*, root of *Athyrium filix femina*). Among Conifers true vessels are unrepresented, even in the primary bundles, while the secondary woody cylinder likewise consists solely of " fibre tracheides." It is, in fact, only among Dicotyledons that true vessels predominate in the secondary xylem : in *Salix*, in *Ficus* and in certain LEGUMINOSAE they are the sole water-conducting elements.

### 4. *Variations in the development of the water-conducting system in relation to different requirements.*[163]

The demands that are made upon the water-conducting capacity of the vascular system, of course vary directly in accordance with the transpiratory activity of the plant. Consequently, the size and number of the foliage leaves, and the climatic and edaphic conditions to which the plant is exposed, must exert a marked influence upon the quantitative and qualitative development of the vessels and tracheides. Special ecological relations, finally, and peculiarities in the general scheme of construction of the plant-body, also affect the development of the paths of water-conduction.

The general correlation which exists between the number and width of the water-conducting tubes on the one hand, and the extent of the foliar transpiring surface on the other, stands in no need of explanation ; great interest, however, attaches to the fact—established experimentally by Jost—that this correlation sometimes affects the differentiation of the vascular system during the development of the individual in a very remarkable manner. If a seedling of *Phaseolus multiflorus* is deprived of one of its two primary leaves, or even of later-formed leaves together with one or more axillary buds,—in either case before the organs in question have unfolded—it is found that the vascular bundles supplying the amputated leaves or buds remain rudimentary, especially as regards their water-conducting components. Jost obtained confirmatory results with seedlings of *Helianthus annuus* and *Vicia Faba*. In these cases there is evidently a process of adaptive self-regulation at work, the elimination of the transpiring surfaces leading to a timely arrest of the development of the corresponding water-conducting strands.

The water-conducting system likewise undergoes reduction if a diminution or complete stoppage of transpiration is brought about, not by the removal of the transpiring organs, but by an increase in the humidity of the atmosphere, or by actual immersion in water. Kohl has demonstrated this point by comparing the structure of the stem in plants of *Lamium album, Isopyrum thalictroides, Aster chinensis* and *Lycopus europaeus*, grown in a damp and in a dry atmosphere respectively. A very considerable adaptability in this respect has also been observed, by Schenck, in the case of *Cardamine pratensis*; if individuals of this ordinarily terrestrial plant are grown in water, the water-conducting tissue of the stem becomes greatly reduced. As might be expected, plants which are naturally amphibious exhibit a similar plasticity.

The most far-reaching reduction of the water-conducting strands is naturally to be found among submerged water-plants. In the majority of such plants, spiral and annular vessels are indeed differentiated in the young axial organs, but soon degenerate completely over long distances, being replaced by a lysigenous intercellular canal. This condition is exemplified by certain species of *Potamogeton*, and by the genera *Zannichellia, Althenia* and *Cymodocea*; here the vessels only remain intact at the nodes. In *Elodea canadensis* one or two axile vessels are differentiated in the apical region, but disappear altogether, even at the nodes, when the stem begins to elongate. According to Sanio, *Ceratophyllum* never forms any vessels at all. It should be noted that this vascular reduction by no means involves the tissues that serve for the conduction of proteins; as a matter of fact, the translocation of these substances is not in the least affected by the nature of the medium in which the plant grows.

The fact that vessels are still differentiated in the young internodes of most submerged Angiosperms, might be regarded as the consequence of a hereditary tendency, since these plants are—like all aquatic Phanerogams—undoubtedly descended from terrestrial ancestors. It is, however, unlikely that the vascular tissue of the young stem is altogether functionless, even in such cases. The frequent occurrence of so-called water-pores or similar apertures in connection with the foliar bundle-ends, furnishes indirect evidence of an actual flow of sap along these transitory strands. The openings in question evidently serve for the exudation of sap, a process which may to some extent assist in the transportation of food materials, while it also facilitates the removal of metabolic by-products.

The development of the water-conducting system is noticeably affected, not only by a diminution, but also by any increase in the demands which are made upon it. Instructive illustrations of this fact are furnished by twining plants, and by climbers in general.   In such plants the construction of the conducting system is governed—as Westermaier and Ambronn have shown—by two factors, namely the great length of the conducting region and the relatively small cross-sectional area available for the disposition of the conducting elements.

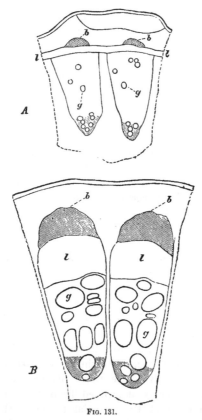

Fig. 131.

Diagram to shoW the leading anatomical differences betWeen a tendril and a leafy branch of the Vine.  *A.* Sector of a T.S. through a fully-developed tendril.  *B.* Similar sector of a T.S. through a yearling branch ; *l*, leptome ; *g*, hadrome-vessels ; *b*, fibrous strands.  Primary hadrome and fibrous strands shaded.  Both figures are draWn to scale With the aid of a camera lucida.  For further explanation see text.

These conditions render it highly necessary that all features which tend to lower the rate of conduction should be eliminated ; in the case of water-conducting vessels the most serious obstacle to rapid flow arises from the adhesion of the water to the walls of the tube.   This difficulty can be most readily overcome by an increase in the diameter of the tube, since in the case of a cylinder of given height the area of the wall is directly proportional to the radius, whereas the volume varies as the square of the radius. As a matter of fact, it is well known that the vessels in the stems of climbers are very wide in comparison with those of ordinary plants, being indeed often readily visible with the naked eye. The average width of the larger vessels—as measured by Westermaier and Ambronn—of certain climbers is set forth in the adjoining table; for the sake of comparison, the corresponding figures are appended for a few common deciduous trees:

| CLIMBERS. | Average Width of large vessels in $\mu$. | TREES. | Average Width of large vessels in $\mu$. |
|---|---|---|---|
| *Hypanthera Guapeva* - | 600-700 | Oak - - - - | - 200-300 |
| *Calamus Rotang*, - | 350 | Elm - | - 158 |
| *Anisosperma Passiflora*- | 300 | Ash - - - - | - 140 |
| *Passiflora laurifolia* - | 200 | Birch - - - - | 85 |
| *P. edulis* - | 200 | Alder - - - - | 76 |
| *Glycine sinensis* | 200 | Lime - - - - | 60 |
| *Aristolochia sp.* | 140 | Pear - - - - | 40 |
| *Serjania sp.* - - - - | 120 | Box - - - - | 28 |

The vessels of young branches are in general narrower than those of the older wood; the difference is probably connected with the fact that the distance over which water has to be transported is less in the former case.

The ecologically conditioned variation in the width of the vessels appears especially striking, when one compares transverse sections of the stems of climbing and erect species belonging to the same genus. Thus the total cross-sectional area of the larger vessels is six times as great in *Galium aparine* as it is in *G. verum*. Similar differences may even be observed in different organs of one and the same plant. It is very instructive, for instance, to compare the width and number of the vessels in the tendrils and in the leafy branches of *Vitis vinifera*. As a glance at the adjoining figure (Fig. 131) will show, no better illustration could be found of the manner in which different physiological requirements, in respect of water-conduction, are reflected in the anatomical structure of the organs concerned.

### 5. *Tyloses.*[164]

The cavities of water-conducting elements often become blocked, for varying distances, by more or less closely packed bladder-shaped intrusions derived from the adjoining parenchymatous cells. These intrusive vesicles, the development of which was first studied and explained by Hermine von Reichenbach, are known as **tyloses**. Most frequently they arise in connection with one-sided bordered pits, the limiting membranes of which undergo active surface growth and thus push their way into the cavities of the vessels; in annular and spiral vessels, tyloses similarly arise from circumscribed areas of the thin membrane separating two adjacent thickening rings or two successive turns of a spiral thickening fibre (Fig. 132). A single parenchymatous cell may give rise to several tyloses. As long as it remains alive, of course, a tylosis contains cytoplasm and cell-sap, while it may also be furnished with a nucleus which has passed over from the parent cell. The walls of tyloses are generally thin; where adjoining tyloses become concrescent, the common walls generally bear pits which correspond on both sides in the ordinary way. Hermine von Reichenbach did not fail to note the fact that tyloses are but rarely cut off from their parent cells by septa. Molisch was also unable to discover any septation, except in *Cuspidaria pterocarpa* and in *Robinia*. Multicellular tyloses, closely resembling hairs, are recorded by Winkler as occurring in *Jacquemontia violacea*. As a rule, however, these structures do not consist of independent cells, but represent mere diverticula of parenchymatous elements.

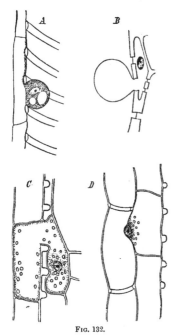

FIG. 132.

Tyloses. *A.* Young tylosis developing in a spiral vessel of the petiole of *Monstera deliciosa*. *B.* Tylosis in a pitted vessel of a branch of *Robinia Pseudacacia*. *D.* A tylosis just beginning to develop in an annular vessel of the petiole of *Cucurbita maxima*. *C.* A similar tylosis at a later stage of development.

The functions of tyloses seem to be somewhat variable. They occur with especial frequency beneath the scars of branches that have broken off, and also near the severed ends of cuttings, being in both cases developed in such numbers that they become flattened by mutual pressure, and completely fill the cavities of the vessels for a considerable distance; in these instances it can hardly be doubted that they serve, as Boehm first suggested, to plug up the cavities of the conducting tubes that have suffered mechanical injury. The tyloses which appear in heart-wood, or in the older portions of splint-wood, in the absence of any injury, apparently fulfil a similar purpose. Sclerotic tyloses, such as occur in *Piratinera guianensis* and *Mespilodaphne Sassafras*, must obviously add to the mechanical strength of the heart-wood. Those tyloses, finally, which contain starch grains,—like the cells from which they arise—should be regarded as reservoirs of carbohydrate material, a fact recognised by the discoverer of these structures. In the rhizomes of *Aristolochia Clematitis* and *A. Serpentaria*, for example, the tyloses contain so much starch in winter-time, that the vessels appear to be crammed with this reserve-material for great distances.

While these various hypotheses concerning the functions of tyloses are doubtless applicable to the special cases selected for mention above, they do not provide an explanation of the fact that these structures so frequently develop in uninjured vessels in herbaceous plants,—and also in many woody species—without reaching a size sufficient to bring them into contact with one another. Such tyloses obviously cannot act as plugs ; starch is also by no means always present in these structures. The author believes that tyloses of this last-mentioned type take some part in the process of conduction, by increasing the surface of contact between the vessels and the neighbouring parenchymatous cells. In this way they might, for instance, accelerate the development of exudation-pressure in the vessels. They might also conceivably excrete sugar into the cavity, or, on the other hand, withdraw certain constituents of the sap, thus acting like haustoria, which they indeed resemble in form. This view with regard to the function of tyloses is supported by the following circumstance, which was first noted by Reess : In vessels which live for several years, the formation of tyloses may continue for a long time, just as if the first-formed tyloses gradually became functionless, and hence had to be replaced by freshly-developed organs of the same kind. It is quite evident that much remains to be done towards a complete elucidation of the physiological anatomy of tyloses.

### B. TRANSLOCATION OF PLASTIC MATERIAL.

The first step towards internal differentation within the conducting system is taken, when separate channels are developed for the conduction of water and for the transportation of plastic substances. The second stage of specialisation involves a further division of labour, whereby the more diffusible compounds are segregated from the less diffusible substances, each group of bodies being relegated to a separate conducting tissue. The readily diffusible compounds are exemplified by the soluble carbohydrates, and by *asparagine*—a decomposition product of proteins—and other amides. These substances travel in the conducting parenchyma. Among the substances which diffuse with difficulty the various proteins are the most important. In the Higher Plants these compounds travel chiefly in the sieve-tubes, but are probably also transported to a certain extent in the phloem-parenchyma (cambiform cells). In *Pangium edule* the *hydrocyanic acid* which Treuh believes to be the first demonstrable nitrogenous product of synthesis, follows practically the same path as the protein-compounds.

Latex-tubes occupy a special position, since they take part in the conduction of both classes of plastic compounds, besides often acting as repositories of waste-products.

## 1. *Conducting parenchyma.*

The category of **conducting parenchyma** (*Leitparenchym*) comprises a variety of parenchymatous tissues, which all perform essentially similar functions; the parenchymatous bundle-sheaths and the nerve-parenchyma of leaf-blades, the bulk of the cortical parenchyma of the petiole and stem and the parenchymatous constituents of the primary leptome- and hadrome-strands all come under this heading. Physiologically considered, these various parenchymatous tissues collectively constitute a definite tissue, inasmuch as they all serve principally for the conduction of non-nitrogenous plastic substances in general, and of carbohydrates (hexoses, transitory starch) in particular.[165]

Conducting parenchyma is always composed of thin-walled, more or less elongated elements; where it forms part of a hadrome strand, the walls are generally lignified. The cells contain living protoplasts, and their transverse walls are usually furnished with simple pits for greater ease of diffusion. In Dicotyledonous leaves the cells of the parenchymatous bundle-sheaths not infrequently develop lateral branches, which extend toward the adjoining elements of the intermediary tissue or spongy parenchyma. In addition to protoplasm, conducting parenchyma cells usually contain various sugars dissolved in the cell-sap, or fine-grained transitory starch. Very often the cells in question are also provided with chloroplasts; their photosynthetic activity is however, in all probability, never very great, and is always subordinated to their principal function of conduction. A totally different subsidiary function—namely, storage—may assume greater importance; during the resting season, in fact, when translocation is largely in abeyance, storage often becomes the principal function of the conducting parenchyma.

## 2. *Leptome-parenchyma* (*cambiform cells*).

**Leptome-parenchyma** consists of elongated thin-walled cells, pointed at either end or attenuated in prosenchymatous fashion, and provided with more or less well-developed protoplasts (Fig. 136). In their typical form these cells closely resemble the cambial or procambial elements from which they are derived; it is on account of this resemblance that they have been termed cambiform cells. They are well provided with pits, which are especially numerous on the terminal walls; the fact that the lateral walls also bear a considerable number of pits suggests the existence of an active interchange of materials between the leptome-parenchyma on the one hand, and the contiguous conducting parenchyma and companion-cells on the other. Leptome parenchyma cells frequently become subdivided into two or more segments by secondary transverse walls; such septate leptome-parenchyma is connected by various intermediate stages with typical non-prosenchymatous conducting parenchyma.[166]

Little is known concerning the functions of leptome-parenchyma. A plausible theory, which, however, still requires confirmation, is that it serves for the conduction of the more readily diffusible among the protein compounds over considerable distances. Most probably this tissue is also to some extent responsible for transferring protein compounds to and from the sieve-tubes. Such leptome-parenchyma cells as approximate closely to ordinary conducting parenchyma in structure, may perhaps take part in the transportation of carbohydrates.

### 3. Sieve-tubes and companion-cells.[167]

**Sieve-tubes** are formed by the longitudinal fusion of rows of elongated cells. The length of the individual segments—which are clearly distinguishable even when the sieve-tube is fully developed—varies within wide limits; the largest known segments (exceeding 2 mm.) occur in climbing plants, which likewise possess the widest sieve-tubes ($\cdot02$-$\cdot08$ mm. in diam.). The partitions separating successive segments from one another are as a rule strictly transverse in primary bundles, whereas in the secondary leptome of Dicotyledons and Gymnosperms they are very oblique. The septa bear the **sieve-plates**, which, from an onto-genetic point of view, merely represent the peculiarly modified limiting membranes of unusually large pits. Where the septum is strictly transverse or nearly so, it becomes transformed bodily into a single sieve-plate, except for a very narrow marginal zone (Figs. 133 and 134); a distinctly oblique septum, on the contrary, always bears several transversely elongated sieve-plates, placed one above the other and separated by narrow strips of imperforate membrane (Fig. 135). Sieve-plates also occur on the lateral walls of the tubes; these are variously disposed, but are always restricted to the walls that abut against other sieve-tubes.

Every sieve-plate is partially thickened in a characteristic manner, the thick portions forming a delicate network. The meshes of this reticulum are known as **sieve-fields**. Among Gymnosperms they are perforated by protoplasmic threads, which later become converted into so-called **slime-strings**. In Angiosperms, on the other hand, the regions of the limiting membrane corresponding to the sieve-fields of Gymnosperms become completely absorbed; the sieve-plates are thus finally traversed by closely crowded equidistant circular or polygonal pores.

In a certain sense, therefore, the sieve-tubes of Gymnosperms are comparable to tracheides, while those of Angiosperms correspond to genuine vessels. The width of the smallest known sieve-pores hardly exceeds the transverse diameter of the exceedingly minute pores that traversed the limiting membranes of ordinary pits. At the opposite end of the scale must be placed the sieve-pores of such plants as *Cucurbita* or *Lagenaria*, which may reach a diameter of $5\mu$; this is another instance of the prevalence of large conducting channels in climbing plants.

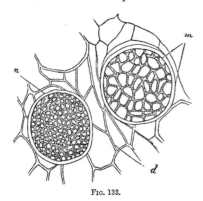

FIG. 133.

Small portion of a leptome-strand of *Lagenaria vulgaris*, in T.S., showing two sieve-plates, $m$ and $n$, each occupying the whole extent of a transverse septum of a sieve-tube. One sieve-plate ($m$) has wider meshes than the other ($n$). ×375. After De Bary.

Sooner or later every sieve-plate becomes invested by a peculiar highly refractive substance, to which Hanstein has given the name of **callus**. These callus-pads consist of a special carbohydrate, *callose*, which is characterised by the fact that it is insoluble in cuprammonia, but soluble in a cold 1 per cent. solution of caustic potash and in soda solution. It is further stained reddish-brown by chlor-zinc-iodine and bright blue by aniline blue. The callus does not merely cover both surfaces of the sieve-plate, but also lines the walls of the sieve-pores (Fig. 136, *ca*); the pores consequently become partially or entirely blocked. Since, however, *callose* is a readily soluble substance, the contracted or occluded pore may be reopened at a later stage. In many cases, therefore, as was noted long ago by Wilhelm, callus serves the purpose of regulating the width of the sieve-pores in accordance with variations in the intensity of translocation. The formation of callus at the beginning of winter, for example, often leads to the complete occlusion of the sieve-pores, and hence to an entire cessation of the flow of material in the sieve-tubes. In the spring the sieve-tubes may be reopened by the dissolution of the callus-pads. Such an alternate opening and closing of the tubes takes place, for example, in *Vitis* and in certain other woody Dicotyledons, as well as in many Monocotyledonous rhizomes. In the majority of plants, however, the deposition of callus results in the permanent closure of the sieve-pores, and for this reason only takes place in old sieve-tubes which are about to become inactive.

In its young state—before perforation of the septa takes place—
every sieve-tube segment is provided with a more or less well-developed
peripheral layer of protoplasm and with a large nucleus, while the sap-
cavity is filled with a watery non-coagulable liquid.    Later a number
of highly refractive homogeneous slimy globules of protein material
appear within the peripheral protoplasm.    In the CUCURBITACEAE these
slimy masses dissolve again before the sieve-tubes are fully developed,
but in other plants they are permanently retained.    In the sieve-tubes of
*Robinia Pseudacacia* and other LEGUMINOSAE the slimy protein substance
is generally collected, according to Strasburger, into a large ellipsoidal
mass suspended near the middle of the tube by two fine threads, one of
which passes to each of the terminal sieve-plates.    The peripheral
protoplasmic layer persists in the adult sieve-tube, and doubtless
extends over the walls of the sieve-pores, so that this peripheral protoplasm is continuous from segment to segment.    The nuclei, on the
other hand, become disorganised and disappear, while the sieve-tube is
still in process of development.    The peripheral protoplasm frequently
contains leucoplasts, with included starch-grains which assume a wine-
red coloration on treatment with iodine.    The sap-cavities of fully-
formed sieve-tubes are filled with a clear, more or less concentrated
solution of protein compounds, which in the CUCURBITACEAE has a slimy
consistency.    As A. Fischer has shown, it is impossible to obtain a
correct idea of the distribution of the protein material in the tube,
unless the organ that is to be examined is immersed bodily in boiling
water before being cut into sections.    When this precaution is observed,
it is found that the coagulated albuminoid slime is uniformly distributed
throughout each segment, and also extends through the sieve-pores.    If,
however, the sections are prepared from living stems—as had been
customary previous to Fischer's work—the sieve-tubes become partially
emptied ; what remains of the contents forms a mass of slime in contact with each sieve-plate in the shape either of a thin disc or a thick
plug (*Schlauchkopf*).    In the majority of plants, as a matter of fact,
the concentration of the protein solution in the sieve-tubes is too low to
furnish any appreciable coagulum on boiling.

Leaving aside the Thallophyta, which will be dealt with later on in a
separate section, it is found that sieve-tube-like rows of cells first
appear in the rudimentary leptome-groups of the central strand of the
stem in the most highly organised Mosses, namely, the POLYTRICHACEAE.
The terminal enlargement of these cells, and the presence of a large
amount of protein material in their protoplasts, are both features which
recall the sieve-tube segments in the primary bundles of Angiosperms
Typical sieve-plates are not found in the POLYTRICHACEAE, nor does any
formation of callus take place.    It is nevertheless probable that

numerous exceedingly minute sieve-pores are present; for if the cells in question are plasmolysed, the protoplasts hardly ever separate from the transverse walls, with which they are evidently in very close rela-tion. According to Poirault, the sieve-plates of the Pteridophyta are likewise only perforated by very minute pores, similar to those which traverse the limiting membranes of ordinary pits. Poirault was able to place the existence of these fine pores beyond doubt in the case of *Angiopteris* and *Ophioglossum.* Among Gymnosperms, also, the sieve-plates—though areolated—are, as already stated, traversed only by very fine pores; the sieve-fields never become obliterated, as they do in Angiosperms.

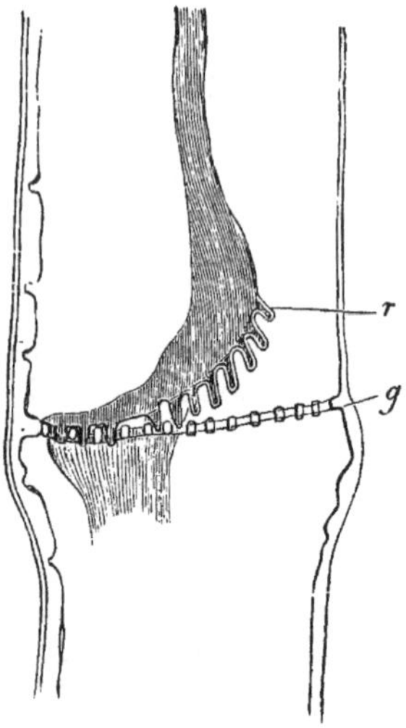

Not unnaturally, the physiological functions of such characteristic organs as the sieve-tubes have long been the sub-ject of speculation. The most widely accepted view is that first put forward by Nägeli, which regards the sieve-tubes as organs serving for the conduction of the less diffusible plastic materials, and especially of protein compounds.[168]

More recently, Czapek has endea-voured to prove by experiment that the sieve-tubes are also instrumental in trans-porting carbohydrates, and are thus re-sponsible for the conduction of the greater proportion of the non-nitrogenous plastic materials as well.[169] His results may, however, be interpreted in quite a differ-ent sense, and do not in any case con-trovert the hypothesis maintained by Schimper, the author and others, who believe that the translocation of carbohydrates takes place mainly in the conducting parenchyma. Certainly there seems no cogent reason for abandoning the view of Nägeli, according to which the sieve-tubes are principally concerned with the transportation of protein-compounds.

Fig. 134.

Large sieve-tube of *Lagenaria vulgaris,* in L.S., after treatment with alcohol and iodine ; *g,* sieve-plate ; *r,* contracted con-tents. ×375. After De Bary.

With regard to the anatomical evidence bearing upon this matter, it may first of all be pointed out that the form and structure of the sieve-tubes, and particularly the perforation of the sieve-plates, are most readily explained by means of Nägeli's theory. If it be objected that the pores traversing the sieve-plates are in many cases so narrow, that the pressures which prevail in the sieve-tubes could hardly suffice to force " protoplasm " through them, it may be replied that it is not protoplasm that passes through the sieve-pores, but a solution of protein compounds, which, moreover, is generally of a more or less watery consistency. It is just where the solution is slimy—as in the CUCURBITACEAE—that the sieve-pores are unusually wide. Another strong argument in favour of the protein-conducting function of the sieve-tubes is derived from the disposition of these structures in the plant-body. For, like the tracheal elements, the sieve-tubes are arranged in continuous strands, which traverse all the organs that have to be supplied with protein-compounds; in particular, they link up the foliar mesophyll, which is very probably the site of the most active protein-synthesis, with the regions where protein com-

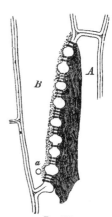

pounds are in greatest demand. It appears, further, that the total cross-sectional area of the leptome-strands in a particular organ—which is an approximate measure of the total number of sieve-tubes—in general corresponds very closely to the relative quantity of protein-material that passes through that organ. If, for example, the transverse section of a tendril of the Vine be compared with that of a leafy branch of the same plant, the leptome is always found to be much more massively developed in the last-mentioned organ (Fig. 131). On the other hand, a similar comparison between a vegetative stem or branch, and an inflorescence-axis or peduncle of the same plant, often demonstrates that the increased supply of protein-compounds required for the purpose of pollen- or seed-production may necessitate a corresponding enlargement of the leptome-strands. A single instance may be cited in support of this statement. If the cross-sectional area of the leptome in a yearling vegetative twig of the Hazel (*Corylus Avellana*) be denoted by the figure 100, the corresponding value for this tissue in a floral

FIG. 135.

Small portion of a tangential longitudinal section through the secondary leptome of the stem of *Vitis vinifera*, showing one of the oblique septa of a sieve-tube. *A, B*, the two adjacent segments of the sieve-tube. The living contents, contracted owing to the action of alcohol, are seen in *A*, adhering to the septum and sending forth blunt processes through all the sieve-pores towards *B*. *a*, a starch grain. × 600. After De Bary.

axis, bearing two staminate catkins, was in a given instance found to be equal to 146. It is also significant, that sieve-tubes of large calibre are present in large numbers in the walls of the insect-catching and -digesting pitchers of *Nepenthes*. Increased demands upon the conducting capacity of the leptome may, in certain circumstances, react not only upon the quantitative but also upon the qualitative development of the sieve-tubes. Among climbers, for example, the factors which have been referred to above, in connection with the organisation of the water-conducting system, exert an equally marked influence upon the devolopment of the sieve-tubes; attention has indeed already been drawn to the unusual width of the sieve-pores in the case of climbing plants. It is thus not a mere coincidence, that when we desire to demonstrate the structure of sieve-tubes, or to carry out researches upon these organs, we generally make use of climbers such as *Cucurbita Pepo, Lagenaria vulgaris, Vitis vinifera,* and *Calamus Rotang.* According to Westermaier and Ambronn, particularly well-developed sieve-tubes occur also in *Humulus Lupulus, Clematis Vitalba, Lonicera Caprifolium, Dioscorea Batatas, Tamus communis,* in the genus *Serjania,* in the PASSIFLORACEAE, etc.

The conclusions concerning the functions of leptome-tissue which have been arrived at above, on an anatomical basis, receive strong confirmation from the results of the so-called ringing experiments [170] which were first employed for scientific purposes by Hanstein. In the majority of Dicotyledonous stems and branches, the leptome-strands are all extracambial. If, now, an annular zone comprising all the extracambial tissues be removed from a severed twig—say of a Willow—near its basal end, only a few small adventitious roots subsequently arise from the short piece below the decorticated strip. The larger piece, on the other hand, swells out just above the incision, owing to the formation of a mass of callus-tissue, and gives rise to numerous long roots. In this experiment the continuity of the leptome is completely broken, whereas the conducting parenchyma of the wood is left intact; the result clearly indicates that the protein-compounds required for the production of roots travel exclusively in the leptome-strands, or, in other words, in the sieve-tubes, and perhaps to some extent also in the leptome-parenchyma (cambiform cells). Hanstein further found that ringing a severed branch does not seriously interfere with the access of protein-compounds to the shorter piece, in those Dicotyledons which possess additional leptome-strands in the pith (such as the PIPERACEAE, NYCTAGINACEAE, SOLANACEAE, CUCURBITACEAE, etc.). This result is entirely in agreement with the interpretation placed upon the preceding experiment; for, in the second case, only a portion of the strands that are believed to serve for protein-conduction are severed.

As regards the nature of the forces that control the movements of protein-compounds in the sieve-tubes, there seems, in the first place, no doubt that the liquid contents of the intact sieve-tubes are under a certain amount of pressure, which causes them to flow in the direction of least resistance. When, namely, a petiole or stem of *Cucurbita* is cut across, large quantities of slimy protein-material exude from the severed sieve-tubes. With reference to this point, A. Fischer has proved, that the effects of the partial depletion of the sieve-tubes extend backwards from the cut surface of a petiole through one or two internodes at the very least. This observation indicates that the pressure in the sieve-tubes is sufficient to overcome the resistance opposed by a very considerable number of sieve-plates. Thence we may infer that any differences of pressure which arise within the intact sieve-tube system, owing to the partial depletion of the tubes at certain points, are at once equalised by a more or less rapid displacement of the liquid contents in the corresponding direction.

Whether the hydrostatic pressure in the sieve-tubes owes its origin to the osmotic properties of the liquid contents, or whether it is due to compression of the sieve-tubes by the highly turgescent adjoining tissues (leptome-parenchyma and companion cells), is still uncertain. Most probably both factors have a share in producing the pressure observed. If sieve-tubes are able to develop a turgor-pressure on their own account, this capacity is entirely due to the presence of a peripheral layer of cytoplasm (or at any rate of a plasmatic membrane). It is, indeed, conceivable that the living contents of sieve-tubes are mainly concerned with the development and regulation of pressure. Strasburger, on the other hand, believes that the peripheral protoplasm serves to retain the substances which are undergoing translocation within the confines of the sieve-tube, and thus to prevent them from diffusing into the adjoining cells. Czapek finally supposes that the living protoplasm plays some *active* part in the process of translocation quite apart from its rôle as the producer of turgor-pressure. This matter evidently requires further investigation.

Translocation of the starch which is often contained in sieve-tubes has never been observed; it undoubtedly only takes place in very exceptional circumstances. According to Strasburger, the presence of starch is in some way connected with the formation of callus, growth of the callus-pads being correlated with disappearance of the starch. Besides proteins and starch, sieve-tubes often contain various sugars, and therefore probably take some part in the transportation of carbo-hydrates; the bulk of the carbohydrate material, however, certainly travels in the conducting parenchyma.

Among Angiosperms the primary mother-cells of sieve-tube segments as a rule undergo several divisions. This process was first described by De Bary, and was afterwards studied in greater detail by Wilhelm and others. The largest of the resulting daughter-cells becomes the actual sieve-tube segment, while the rest develop into the much narrower **companion-cells**, as these sister elements are termed by Wilhelm. (In the bundle-ends the relative sizes of sieve-tubes and companion-cells are reversed.) A companion-cell (Fig. 136, *g*) may be readily dis-tinguished from ordinary leptome-parenchyma (cambiform cells) by its narrow cavity, by its abundant protoplasm and large nucleus, and more particularly by the fact that the wall separating a companion-cell from a sieve-tube always bears numerous transversely elongated pits which correspond on the two sides, whereas the walls dividing sieve-tubes from leptome-parenchyma cells are as a rule altogether devoid of pits.

Among Monocotyledons, and also in the primary leptome-strands of certain Dicotyledons, the companion-cells are arranged in continuous series. In most Dicotyledons, however, this is not the case, at anyrate as regards the secondary phloem. In Gymnosperms and Pteridophytes these structures are absent. Their place is taken by rows of special parenchymatous elements with very abundant protoplasmic contents; in the secondary phloem of Gymnosperms these "albu minous cells" only form continuous series over short distances, in which respect they resemble the companion-cells of Dicotyledons.

The physiological significance of the com panion-cells (and of the analogous albuminous cells) is quite unknown. It has already been stated that they do not always form continuous series for long distances; assuming that they take part in the transportation of protein compounds at all, they may, as Strasburger supposes, be concerned with purely local requirements. In view of their close anatomical relation to the sieve-tubes, however, it seems possible that they may co-operate with the latter in some other way, although the precise nature of the assistance which they render cannot at present be indicated. They might, for example, conceivably play a leading part in connection with the movement of the liquid contents of the sieve-tubes. A. Fischer believes them to be organs engaged in the synthesis of proteins; but this view is no more deserving of acceptance than the hypothesis of Sachs, which attributes a similar function to the sieve-tubes themselves.

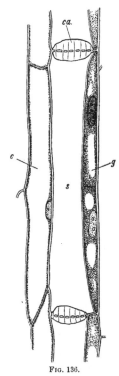

FIG. 136.

Small portion of a L.S. through the leptome of the stem of *Ecballium Elaterium*. *c*, leptome-parenchyma cell (cambiform cell); *s*, sieve-tube; *ca*, callus-pads deposited on the sieve-plates; *g*, companion cell.

#### 4. *Latex-tubes or laticiferous elements.*[171]

In the great majority of higher plants, the sieve-tubes, the leptome-parenchyma (cambiform tissue) and the conducting parenchyma suffice for the translocation of the whole supply of plastic materials. But certain families of Angiosperms are characterised by the possession of an additional conducting tissue in the shape of **laticiferous elements** or **latex-tubes**, structures which owe their name to the milky appearance of their contents.  These laticiferous elements are long, and as a rule abundantly branched tubes, which traverse the roots, stems and leaves in all directions; they are accordingly well fitted for transporting materials on a considerable scale.

The walls of latex-tubes are soft, and as a rule also thin, except in the genus *Euphorbia,* where they attain a thickness of from ·01-·016 cm., being at the same time unusually extensible and elastic.  Pits are scantily developed even in the more thick-walled types.  In *Euphorbia Lathyris* the author has noticed that exceedingly narrow pits sometimes occur where the wall of a latex-tube abuts against palisade-cells, while the regions of the wall bordering upon spongy parenchyma bear larger pits, with their closing-membranes perforated by very conspicuous protoplasmic threads.  Kienitz-Gerloff has similarly demonstrated the existence of protoplasmic connections between laticiferous tubes and adjacent parenchymatous cells in *Euphorbia Cyparissias* and in *Nerium.*

From the researches of Treub and Emil Schmidt it appears, that even fully-developed latex-tubes contain a peripheral layer of protoplasm, in which large numbers of nuclei may be suspended.  Various circumstances, such as the production of adventitious branches and the occlusion of severed latex-tubes by means of a callus-like formation (in the root of *Scorzonera* and the stem of *Euphorbia splendens*), lead one to conclude that the protoplasts in these structures retain their vitality for a long time.

**Latex,** the characteristic substance contained in latex-tubes, is usually a milk-white, more rarely a pale yellow (*Argemone*) or even orange-coloured (*Chelidonium*), liquid.  Ontogenetically considered, it corresponds to the cell-sap, at any rate according to Schmidt, Kallen and Molisch; Berthold, however, considers that it represents a peculiarly modified liquid protoplast.  Latex always takes the form of an emulsion composed of innumerable minute granules and drops, suspended in a watery liquor which contains other substances in solution.  The suspended particles are of very diverse kinds, and comprise granules of resin, gum or caoutchouc—which tend to form coherent masses when the latex exudes—oil-drops, tannin vesicles, protein crystalloids and (in the EUPHORBIACEAE) small starch-grains.  These starch-grains are remarkable for their peculiar shapes; in our native *Euphorbias* they are generally spindle-shaped, while in the tropical species they are flattened with expanded ends, thus resembling miniature thigh-bones or dumb-bells.  Potter states that they are formed in connection with amyloplasts, just like ordinary starch-grains.  Molisch has recorded the occurrence of large numbers of protein-granules, which are likewise formed by special plastids, in the latex of *Cecropia peltata* and *Brosimum macrocarpum.*  The watery matrix of the latex may contain a variety of substances in solution, such as mineral salts (*e.g. calcium malate* in species of *Euphorbia, magnesium* salts in *Ficus elastica* and many other plants), proteins, sugars, gums, tannins or alkaloids (such as the *morphine* contained in opium). Proteolytic enzymes are also known to occur in the latex of certain plants (*e.g. Carica Papaya* and *Ficus Carica*). Many additional facts concerning the qualitative and quantitative composition of latex are recorded in a recent dissertation on the subject by Molisch.

The general conclusion to be drawn from the preceding statements is, that latex must be regarded as a nutritive fluid, because of the large quantities of plastic substances present therein, but that it nevertheless also contains substances which are really of an excretory nature, or which, in other words, represent by-products of metabolism.

From a morphological point of view, it is necessary to distinguish between two different types of laticiferous organs, namely, articulated latex-tubes or **latex-vessels** and non-articulated latex-tubes or **latex-cells**. **Latex-vessels** are characteristic of the CHICORIACEAE, CAMPANULACEAE, PAPAVERACEAE, LOBELIACEAE and PAPAYACEAE, also of many ARACEAE and MUSACEAE, and of the genera *Manihot* and *Hevea* among the EUPHORBIACEAE. They originate from rows of meristematic cells, in which the transverse septa become absorbed at an early stage of development. In a few instances (*e.g. Musa* and *Chelidonium*), the transverse walls persist, but are perforated by one or more large pores. Similar perforations occur, in all cases, on the longitudinal walls, wherever two latex-tubes come into contact with one another. Typically, latex-vessels develop numerous branches, which anastomose freely, so that a dense network of tubes results (Fig. 137 B). Both the primary branches, and the anastomoses, may arise either by the fusion of rows of meristematic cells, or by the formation of lateral outgrowths which push their way between the surrounding tissues.

**Latex-cells** occur in the majority of EUPHORBIACEAE, and in the URTICACEAE (including the ARTOCARPACEAE and MORACEAE), APOCYNACEAE and ASCLEPIADACEAE. In these cases each tube arises from a *single* mother-cell, which becomes enormously elongated and abundantly ramified. Anastomoses are altogether absent, or at any rate infrequent (Fig. 137 A). According to Schwendener, the mother-cells of the latex-tubes are clearly recognisable even in the very young embryos of the EUPHORBIACEAE. Schwendener's observations have been confirmed by Chauveaud, and extended with essentially similar results to the rest of the above-mentioned families. The initial cells in question appear at the inner margin of the primary cortical tissue in the first or cotyledonary node, and by means of active apical growth soon develop into elongated branching tubes. Their branches behave like the hyphae of a parasitic

Fungus, forcing themselves between the neighbouring cells and penetrating into the growing points of shoot and root. Since latex-cells never arise *de novo* during the subsequent development of the plant, the entire laticiferous system of the plant-body is derived from the small number of initials which become differentiated in the embryo. If subsequent additions to the system are required—for instance, in connection with the formation of adventitious buds—the existing latex-tubes simply give rise to new branches at appropriate points according to

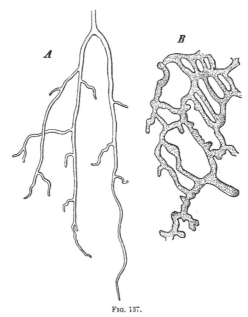

FIG. 137.

Portions of latex-tubes dissected out. *A*. Ultimate ramifications of a latex-cell from a young leaf of *Euphorbia Myrsinites*. *B*. Anastomosing latex-vessels from the cortex of the root of *Scorzonera hispanica*. ×120. *B*. after Unger.

Chauveaud, the "secondary" latex-tubes which appear among the products of cambial activity are produced in the same way. The differentiation of latex-tube initials thus appears to cease altogether at an early stage of embryonic development

In considering the **functions of latex-tubes**, we may once more begin with a discussion of the anatomical data from the physiological point of view. If the course of the latex-tubes be carefully traced through root or stem, petiole or mid-rib, it will be found that these structures everywhere maintain close relations both with the conducting parenchyma and with the leptome-strands, being either opposed to, or embedded in, these conducting tissues. This circumstance in itself strongly suggests that the functions of the two sets of organs are similar, an impression which is strengthened by a consideration of the chemical composition of latex. The anatomical relations, finally, which exist between the laticiferous and photosynthetic systems render it still more probable that latex-tubes play a considerable part in the translocation of carbohydrate materials at any rate. The author's observations with regard to this point—which have been confirmed and largely extended by Pirotta and Marcatili, and independently by Gaucher—may be summarised as follows. In foliage leaves the latex-tubes ramify most abundantly immediately beneath the palisade-layers which represent the special photosynthetic tissue. Not infrequently, individual branches of the laticiferous system depart from the main trunks that accompany the vascular bundles, turn upwards (Fig. 138) and insinuate themselves between the palisade-cells, where they are in a position to receive the synthetic products at first hand. The ultimate branches of latex-tubes often act as the points of attachment for converging groups of palisade-cells (*Euphorbia* spp., *Ficus nitida, Hypochaeris radicata*) ; where this direct contact is not feasible, funnel-shaped collecting cells are interpolated, which effect the transference of synthetic products from the palisade-cells to the latex-tubes (Fig. 138 B). Such cases at once recall the previously described arrangements for linking up the photosynthetic system with the efferent parenchymatous bundle-sheaths, which are found in the foliar organs of plants that do not possess latex-tubes.

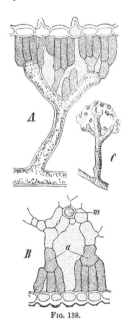

Fig. 138.

Relations of latex-tubes to the photosynthetic tissues in various leaves. *A. Euphorbia Myrsinites* (exceptional case). ×170. *B. E. biglandulosa; a,* collecting cells ; *m,* latex-tube. ×210. *C. Hypochaeris radicata.* ×170. (For explanation see text.)

Certain tissue-correlations also suggest that latex-tubes are concerned with the conduction of plastic materials. Thus the degree of development of the laticiferous network in a leaf is found to be inversely proportional to that of the conducting parenchyma. Where the laticiferous system is abundantly developed in the mesophyll, it relieves the foliar conducting parenchyma—and particularly the bundle-sheaths—to a large extent of the office of translocation ; in these circumstances, the normal efferent tissues exhibit a comparatively slight and imperfect organisation. This reduction of the conducting parenchyma in the presence of a laticiferous system is very strikingly illustrated, according to the author's observations, by *Euphorbia Myrsinites* and *E.*

*biglandulosa.* In these species the parenchymatous sheaths of the smaller vascular bundles are interrupted at numerous points, while the individual cells are not of the normal elongated type, but often appear as broad as they are long, and generally present an irregular outline. In the mid-ribs the ordinarily massive conducting parenchyma is almost entirely suppressed, its place apparently being taken by from six to eight wide latex-tubes, which in places lie in immediate contact with the leptome. To what extent the vicarious activity of the laticiferous system may affect the development of the leptome itself, is uncertain. De Bary goes so far as to maintain that the sieve-tubes of many laticiferous plants (*e.g.* the ASCLEPIADACEAE) are unusually narrow and inconspicuous; Kniep, however, disagrees with this statement, at any rate so far as stems are concerned.

Anatomical evidence alone cannot, of course, supply a convincing proof of the importance of the laticiferous system in connection with the translocation of plastic materials. As a matter of fact, all attempts to reach a definite conclusion regarding this point along experimental lines have also so far met with indifferent success. In most of the experiments in question, the method employed has been that of determining the composition of the latex, either at different points of the life-history of the plant, or else under normal and abnormal external conditions; from the data obtained in this manner, conclusions have been drawn as to the nutritive value of the latex.

The variations in the composition of latex which are correlated with different stages of development of one and the same species, have been investigated in detail in the case of *Euphorbia Lathyris* by Schullerus. In the developing embryo the latex contains a large amount of starch, oil, protein compounds and tannin. When the embryo enters upon its period of rest, in the ripe seed, the latex has become transparent, and contains a much smaller proportion of plastic materials, the starch-grains having disappeared altogether, while both oil and tannin are present only in small quantities. During the germination of the seed the latex again becomes viscid and full of suspended particles; the rod-shaped starch-grains reappear in large numbers, and the rest of the above-mentioned plastic substances once more show a decided increase. In the adult plant the latex retains this viscous consistency and nutritious character throughout the vegetative season. During the resting period—that is in winter time—its quality again deteriorates. At this time it is characterised—so far as the stem is concerned—by the small amount of plastic material which it contains, and by the appearance of a fresh constituent in the form of *calcium malate*; in the roots the amount of protein in the latex actually increases during this period, whereas non-nitrogenous plastic materials, such as oil and starch, disappear altogether.

The above results led Schullerus to conclude, that latex in the first instance represents a " plastic sap," and that the latex-tubes serve for its conduction.   Faivre had previously arrived at similar conclusions from his observations upon the composition of latex in plants grown under various abnormal conditions.   If, for example, seedlings of *Tragopogon porrifolius* are kept in the dark, the latex soon becomes watery, owing to a diminution in the amount of solid constituents.   The same effect is seen in plants grown in the light but deprived of carbon dioxide.   In either case, if the plant is subsequently permitted to resume its normal photosynthetic activity, the latex soon regains its ordinary consistency.   Faivre accordingly concludes that latex originates in the photosynthetic organs, and is transported thence in order to be utilised for nutritive purposes in other parts of the plant.

The views of Schullerus and Faivre have not escaped criticism. Kniep, in particular, has recently objected to the conclusions of these investigators on various grounds.   He insists more especially upon the fact that increased " wateriness " of latex does not necessarily indicate any decrease in the absolute amount of solid matter present.   The change of consistency may in such cases be due to mere dilution, that is to say, to the distribution of the same quantity of solid material in a larger bulk of liquid ; the continued growth of a latex-tube may be accompanied by a corresponding increase of liquid contents, even when photosynthesis is in abeyance [whereas no addition to the solid contents is to be expected under these conditions].   Kniep also states, that when *Euphorbia* seedlings are kept in the dark for lengthy periods, the starch-grains in the latex-tubes do not disappear, or at most undergo partial solution after a very long time.   The persistence of these starch-grains certainly does not support the contention that they represent a store of reserve material.   On the other hand, it must not be forgotten that the presence of starch-grains in the latex is a peculiarity of the EUPHOR-BIACEAE, so that the behaviour of these bodies need not affect our judgment as to the nutritive value of latex in general ; the significance of these remarkable starch-grains is, in fact, a matter for separate enquiry.   Kniep himself does not altogether deny the nutritive value of latex.

The preceding discussion sufficiently emphasises the need for further experimental research upon this subject.   It should not be forgotten that latex may not have the same value in different families of plants ; while it may be mainly nutritive in character in certain Natural Orders, its ecologically important protective action—which is further explained below—may come more into prominence in others.

Even in the present imperfect state of our knowledge, however, we may confidently assert that the latex of a great many plants contains a large proportion of plastic material, which is subsequently utilised for purposes of nutrition. The far-famed Cow Tree (*Brosimum Galactodendron*), for example, produces large quantities of a sweet and palatable latex, which is used in its native country in place of cow's milk; in this case no one would be disposed to regard the latex as a means of protection against animal foes, or to suppose that the laticiferous system is merely a repository for useless excretory products.

Various circumstances indicate that actual streaming of the latex takes place within the intact laticiferous system. The dilution of the latex which occurs in starved plants presupposes a dispersal of the solid particles in the liquid matrix, quite apart from the question whether some of the particles of oil, resin, etc., are absorbed or not; but, as Schwendener has pointed out, this dispersal cannot be effected on any considerable scale without displacement of the latex *en masse*. In *Euphorbia* the behaviour of the suspended starch grains also throws some light on this question. Schwendener states that, in the foliar latex-tubes of *Euphorbia splendens*, the starch-grains are all minute and rod-shaped, without expanded ends. In the stem, a few centimetres below the insertion of a leaf, every intermediate stage may be found between larger simple rods and the peculiar dumb-bell-shaped grains which never occur in the foliar tubes. The obvious inference is, that the rod-shaped grains migrate from the leaf to the stem, and there gradually assume the dumb-bell shape; such a migration, however, necessarily implies a bodily displacement of the latex. The latex-tubes of *Euphorbia* sometimes become obstructed at certain points; where this has happened, microscopic examination reveals a dense mass of starch-grains adhering to one side of the obstruction. Evidently the obstructing body acts as a filter, which holds back the starch-grains but allows the liquid portion of the latex to pass. Schwendener, finally, actually observed a jerky movement of the latex in transparent seedlings of *Chelidonium majus*, which were transpiring actively through their leaves or roots.

We are thus led to consider the sources of the energy that is employed in transporting latex. The liquid contained in the intact laticiferous system is exposed to very considerable pressure, as is shown by the fact that it often exudes in large quantities if the tubes are cut open. Where the latex-tubes are thin-walled, the pressure is most probably due to the turgor of the surrounding parenchymatous tissues. In various species of *Euphorbia*, Schwendener was able to prove the existence of an independent turgor-pressure within the latex-tubes themselves, which produces an elastic extension of their walls. When the pressure is relaxed, these latex-tubes undergo a contraction amounting to as much as 4 or 5 per cent. of their diameter. The elastic tension of the walls, of course, constitutes a reserve of potential energy that will cause the latex to flow towards any part of the system in which a diminution of pressure may happen to occur.

The conclusion that latex represents, in the first instance, a mixture of plastic materials, is not invalidated by the undoubted occurrence of useless by-products of metabolism among the contents of laticiferous tubes. It is difficult to say whether the particles of resin and caoutchouc, to which the milky appearance of typical latex is due, should be reckoned as plastic substances or as excretory compounds. The dilution of the latex in starved plants suggests that the particles in question are partially absorbed under special conditions; but, as Schwendener has pointed out, the latex-tubes in such cases often contain plug-like masses, enclosing large numbers of the solid particles, which are evidently produced by partial coagulation of the latex. At the same time, it is improbable that the dilution of the latex is entirely due to this cause. On the other hand, it seems unlikely that the large amounts of solid matter which are deposited in the latex-tubes of many tropical trees in the form of resin and caoutchouc particles merely represent a heterogeneous mass of useless (metabolic) by-products. Schwendener thinks that the particles in question may possibly perform the mechanical function of preventing the relatively light oil-drops from rising, and the relatively heavy starch-grains from sinking, in the watery matrix, so that a given distribution of the various plastic constituents in the emulsion may remain undisturbed, except in so far as it may be modified by streaming movements.

In conclusion, mention must be made of two subsidiary ecologically important functions that are frequently performed by latex. The first of these depends upon the circumstance, that latex readily coagulates on exposure to air, and is thus excellently fitted to serve as an occluding material in cases of mechanical injury. The other ecological function of latex consists in the protection which it affords against animal foes, especially when it is—as often happens—endowed with poisonous properties. In many cases this protective function assumes such importance, that it even affects the location and disposition of the ultimate branches of the laticiferous system. In the genus *Euphorbia*, the latex-tubes that traverse the mesophyll frequently push their way towards both leaf-surfaces as far as the inner side of the epidermis, where they extend tangentially in every direction. If small portions of the epidermis are removed from leaves of *Euphorbia Lathyris* or *E. Myrsinites*, the existence of this abundantly developed subepidermal laticiferous system is revealed by the adhesion of large numbers of its branches to the inner surface of the severed piece. In these circumstances the slightest mechanical injury to the leaf results in the exudation of abundance of protective latex. In certain succulent *Euphorbias* (*E. officinarum*, *E. canariensis*), some of the distal branches of the latex-tubes actually insinuate themselves between adjoining epidermal cells, so that their tips come to lie immediately beneath the cuticle.

Finally, Stahl, Piccioli and Kny have shown, that in the involucral bracts of species of *Lactuca* and other CICHORIACEAE, the ultimate branches of the laticiferous system develop into peculiar **latex-hairs**. These hairs are from three to five times longer than their width, and may have blunt or pointed ends ; as a rule their walls are heavily cutinised. According to Zander, each hair arises, together with the pediment-cells that half-surround its base, from a single epidermal mother-cell. Morphologically, therefore, these organs represent genuine hairs, which have become continuous with underlying latex-tubes, owing to the absorption of the intervening basal wall, or else by perforation of the pediment-cells. Usually each involucral bract bears from 60 to 100 latex-hairs. At the slightest touch these break and liberate drops of latex.

Among the various tubular structures which are more or less closely related, both phylogenetically and physiologically, to genuine latex-tubes, special mention may be made of the remarkable rows of vesicles that occur in the bulb-scales of species of *Allium*. The cell-rows in question traverse the scales in the longitudinal direction, lying close beneath their outer surface ; towards the base of each scale the different rows are connected by means of anastomoses. The individual sacs are somewhat elongated, and are separated from one another by densely pitted walls. They are filled with a turbid liquid resembling latex, which, at the time when the bulb begins to sprout, appears to contain a large amount of plastic material. In this connection we may also refer to the similar latex-cells of the CONVOLVULACEAE, which have been investigated in detail by Czapek. The cells in question are arranged in longitudinal series, and occur in stem, leaf and root, in close relation to the vascular strands. Except in *Dichondra*, the transverse septa do not undergo degeneration. The contents comprise a peripheral layer of protoplasm, and "latex" of unknown composition. As each internode completes its longitudinal growth, the latex-cells lose their contents and collapse.

## III. THE STRUCTURE OF VASCULAR BUNDLES.

Having made ourselves acquainted with the several structural elements that are employed for purposes of conduction, we must next proceed to review the various ways in which these elements may be combined so as to form **composite conducting strands** [or vascular bundles]. In the present chapter we shall confine our attention strictly to the *primary* conducting tissues.

### A. SIMPLE CONDUCTING STRANDS.

The simplest possible form of conducting strand is entirely made up of water-conducting elements. The delicate vascular anastomoses of leaves and stems, for example, often consist of nothing beyond a few tracheides, and thus represent minute channels serving exclusively for the conduction of water. Most frequently, indeed, these tracheides are enclosed within a glucose-conducting parenchymatous bundle-sheath, in which case the anastomoses do not, strictly speaking, pertain to the category of simple strands; in other instances, again, the sheath is fibrous, as, for example, in certain thick Monocotyledonous leaves (*Rhapis*, *Vanda furva*) and in the wing-like inflorescence-bract of the Lime. As a further example of the purely water-conducting type of bundle, we may once more mention the strand of rudimentary tracheides which occupies the centre of the stem in Mosses.

Of far more frequent occurrence are simple bundles consisting exclusively of protein-conducting elements, in the form of sieve-tubes accompanied by companion-cells and leptome-parenchyma (cambiform cells). Floral axes, and other organs in which the translocation of protein compounds assumes large proportions, often contain isolated leptome-strands in addition to the ordinary vascular bundles. Such **accessory leptome-strands** are found, for instance, in the perimedullary region of the stem in many CAMPANULACEAE and CICHORIACEAE, and in species of *Solanum*. In the scape of *Plantago lanceolata* they are arranged in tangential series between the larger and smaller vascular bundles. In many CUCURBITACEAE they occur in the extracambial parenchyma, both on the inside of the mechanical cylinder, and outside the latter, between the collenchyma and the ordinary cortical parenchyma; occasionally they may even run immediately beneath the epidermis. In *Cucurbita* all the accessory leptome-strands are linked up by cross-connections; they serve in all probability, as A. Fischer has suggested, to supply the developing mechanical tissues

—namely, the fibrous cylinder and the collenchyma—with protein compounds.

### B. COMPOSITE CONDUCTING STRANDS.[172]

#### 1. *General considerations.*

In the great majority of cases, the different conducting elements are associated to form composite strands. A typical conducting strand, in the anatomico-physiological sense, therefore comprises tissues of three different kinds. The protein-conducting elements, namely, the sieve-tubes and companion-cells, and, it may be, also leptome-parenchyma (cambiform cells), form—sometimes in conjunction with conducting parenchyma, but more frequently alone—the delicate **leptome** portion (*Sieb-teil, Cribral-teil*) of the strand. The water-conducting vessels and tracheides constitute—almost always in conjunction with conducting parenchyma —the resistant **hadrome** portion (*Gefäss-teil, Vasal-teil*). Both hadrome and leptome, therefore, but especially the former, comprise, in addition to the characteristic components that serve for the conduction of water and protein-compounds, elements of the nature of conducting parenchyma, which may in short be termed hadrome- and leptome-parenchyma respectively. The composite strands formed by the combination of one or more hadrome- and leptome-strand with one another are known as **mestome-strands** or **vascular bundles**. In most cases each vascular bundle is enclosed within a **bundle-sheath** of one kind or another. Sheaths of **conducting parenchyma** are characteristic of leaf-blades ; **starch-sheaths**, which really represent sense-organs, are found in stems and petioles ; **endodermal sheaths** (*Schutzscheiden*), finally, may occur in stem, leaf or root.

Before it was realised that the mechanical strands form an independent system, and that their frequent association with vascular bundles is merely the result of physiological opportunism, a vascular strand, the groups of wood fibres associated therewith and the surrounding bundle-sheath, were collectively held to constitute a morpho logical unit, termed a **fibro-vascular bundle**. The **phloem** portion of such a fibro-vascular bundle corresponds to the leptome together with its fibrous sheath ; the **xylem** includes the hadrome with its associated wood-fibres (libriform cells). If the leptome has no fibrous sheath, it of course becomes synonymous with phloem ; where, as in most Mono cotyledons, no wood-fibres are developed, xylem is the exact equivalent of hadrome.[173]

The adjoining table will help to explain the terminology employed in the present work in referring to the various conducting elements, as constituents of conducting strands and of fibro-vascular bundles respectively:

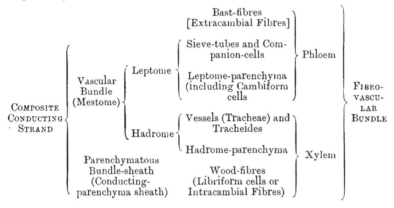

Since any one of the three forms of conducting tissue that take part in the formation of a composite strand may also follow a separate course and carry out its functions independently of the rest, it behoves us to enquire why plants in general show a distinct preference for the constant association of these tissues with one another. Although this question cannot be answered at all satisfactorily in the present state of our knowledge, we may nevertheless indicate what general considerations must regulate the further investigation of the problem.

In the first place, it is worthy of note that water-conducting elements, and particularly wood-vessels, nearly always have parenchymatous tissue (conducting parenchyma) associated with them. From this circumstance it may be inferred, that there is some connection between the functions of vessels and those of the accompanying hadrome-parenchyma. It has already been suggested on a previous occasion, that this connection may possibly depend upon the fact that the hadrome-parenchyma takes an active part in the conduction of water. But the almost constant association of parenchyma with vessels is capable of a totally different explanation, if it be assumed that the plastic materials utilised in the formation of the characteristic thickenings of the vessel-walls are conveyed to the several vessels through the mediation of special rows of hadrome-parenchyma cells.

The very constant association of two tissues, namely, the protein-conducting leptome and the mainly water-transporting hadrome, which functionally are quite independent of one another, to form composite bundles, constitutes another difficult problem. It is probable that in this case a variety of circumstances have combined to produce the final result. The soft and vulnerable nature of the thin-walled leptome no doubt in itself furnished a sufficient reason for its juxtaposition to the comparatively resistent hadrome; in any case, this association most probably first took place at a stage of evolution antecedent to the differentiation of a special mechanical system (which the author regards as a relatively recent development). According to this view, the now prevalent fibrous sheaths were added at a later period for the sake of additional protection. In the case of photosynthetic organs, a further advantage accrues from the association of hadrome and leptome, inasmuch as in these circumstances a single set of intermediary tissues suffices to connect the photosynthetic tissue with all the three types of conducting tissue. If hadrome, leptome and conducting parenchyma all formed separate strands, the efferent arrangements would have to be greatly amplified and elaborated.

## 2. *Various types of composite conducting strand.*

It is customary to recognise three principal types of vascular bundle, which are defined in accordance with the relative position of hadrome and leptome, namely, the **concentric**, the **radial** and the **collateral** types. It must, however, be clearly stated at the outset, that these different forms of vascular bundle are not all homologous or phylogenetically equivalent structures.

A vascular strand is described as **concentric**, when one of the two principal components occupies its centre and is surrounded by the other as by a sheath. If the hadrome is central and the leptome peripheral, the bundle may be termed **hadrocentric** (amphicribral). Bundles of this type prevail in the stems and leaves (the smaller bundles being excepted in the latter case) of Ferns, and also occur in certain Dicotyledons. The opposite or **leptocentric** (amphivasal) condition is exemplified by the leaf-trace bundles in many Monocotyledonous rhizomes (such as those of *Cyperus aureus, Papyrus, Acorus Calamus, Iris germanica,* etc.), and by the medullary strands of a number of Dicotyledons (PIPERACEAE, *Phytolacca dioica,* spp. of *Rheum, Rumex, Drosera, Geranium, Aralia, Begonia, Statice, Campanula, Scorzonera,* etc).

In their simplest form, the hadrocentric conducting strands of Ferns are circular or elliptical in cross-section (Fig. 139); more often, however, they assume a flattened or ribbon-like shape, while they may in addition be folded so as to resemble a V, U or X when seen in transverse section. The outline of the hadrome, which, as already stated, occupies the centre of the strand, either follows the outline of the entire bundle more or less closely, or else displays a symmetry which is quite independent of the latter. In the smaller bundles the hadrome consists almost exclusively of spindle-shaped scalariform tracheides, only the narrow first-formed elements being annular or spiral. Genuine vessels only occur in exceptional cases (*Pteris aquilina*). In the larger bundles parenchymatous cells are interspersed among the tracheides, while other parenchymatous elements separate the latter from the surrounding leptome. The leptome itself consists of sieve-tubes, and of leptome-parenchyma cells, containing abundant protoplasm, which take the place of companion-cells. The entire bundle is enveloped by two concentric sheaths. The inner of these consists of one or more

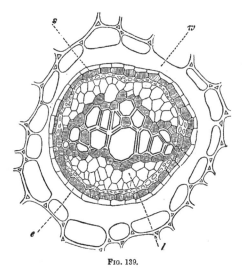

Fig. 139.

One of the smaller bundles (meristeles) from the rhizome of *Polypodium glauco-phyllum* in T.S. *l*, leptome ; *p*, parenchymatous sheath (pericycle); *e*, endodermis · *ic*, the thickened inner walls of the layer immediately external to the endodermis. (The conducting parenchyma is shaded.) ×320. After Potonié.

layers of starch-containing parenchymatous cells, and may be termed the "phloem-sheath" (*Phloem-scheide*); it is surrounded in its turn by a "protective sheath" (*Schutzscheide*) or **endodermis**, a structure which will be dealt with in more detail later on.

The concentric type of vascular structure is foreshadowed, as it were, by the arrangement of the conducting tissue in the [gametophytic] stems of the most highly organised Mosses, namely, the POLYTRICHACEAE. Here the single conducting strand comprises, in the first place, a central water-conducting tissue, which may be regarded as a primitive type of hadrome. In *Polytrichum juniperinum* and *P. commune* this hadrome is further differentiated into a harder internal region (Fig. 140, *t*), in which the thickened regions of the cell-wall are red-brown in colour, and a more delicate outer zone with colourless, or at most pale yellow, walls (Fig. 140, *t₂*); in withered stems this external region of the hadrome is greatly contracted. The central hadrome-strand is surrounded by three to five layers of rudimentary leptome-tissue, some of the elements of which resemble sieve-tubes, while others correspond to leptome-parenchyma (cambiform cells). When translocation is active, particularly during the development of the sporogonium, this leptome-tissue contains large quantities of protein-compounds. If translocation is retarded or altogether inhibited, starch also appears in varying quantities in the leptome tract.

In the genus *Polytrichum*, and also in *Pogonatum aloides*, the hadrome of the erect aërial stem is entirely composed of water-conducting tracheides ; but in *Atrichum un-dulatum* these are accompanied by starch-containing conducting-parenchyma cells, just as in the case of more highly organised plants. In the subterranean rhizomatous axis of *Polytrichum*, and in the aërial stem of *Dawsonia superba* (a Polytrichaceous Moss from New Zealand), thick-walled mechanical cells, which are directly comparable to wood-fibres (libriform cells), occur in association with the water-conducting elements ; in these cases, therefore, the rudimentary hadrome-strand is equivalent to a xylem-bundle. The concentric axial vascular bundle of the POLYTRICHACEAE is not sharply delimited from the cortex. The entire absence of bundle-sheaths comparable to those

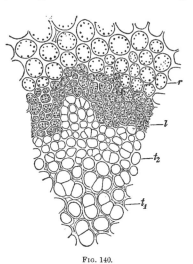

FIG. 140.

Part of a T.S. through the conducting strand of the stem of *Polytrichum juniperinum* ; *r*, cortical parenchyma ; *l*, leptome ; $t_1$, thick-walled central ; $t_2$, thin-walled peripheral portion of the hadrome.

of the Ferns is, in fact, one of the features that indicate the rudimentary character of the vascular bundles in Mosses.

The **radial bundles**[173a] of roots are characterised—as their name implies—by the radiate arrangement of the hadrome- and leptome-strands. The hadrome is disposed in the form of more or less numerous radial plates, which alternate with an equal number of isolated strips of leptome. The intervals between successive " rays " of vascular tissue are occupied by groups of conducting parenchyma, which are usually about two cells in width. Generally speaking, the rays are more numerous in stout roots than in slender ones. Among Dicotyledons there are typically two, three, four, six or eight rays, while in Mono-cotyledons their number may amount to twenty, fifty or even more. With reference to the mode of development of the hadrome-plates, which become differentiated centripetally from a corresponding number of peripheral points of origin, it is customary to discriminate between diarch, tetrarch, pentarch, hexarch and polyarch bundles. The radial hadrome-plates may meet at the centre of the bundle, in which case they often converge towards a single specially wide axile vessel; this condition is exemplified by the primary roots of a number of Grasses and of certain other Monocotyledons (Fig. 141). In other cases the central region of the bundle is occupied by a parenchymatous pith (Fig. 142) or by a mass of fibrous tissue.

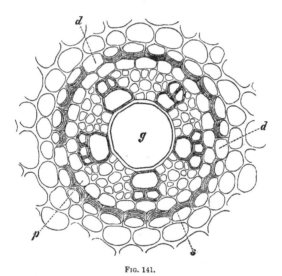

FIG. 141.

Radial bundle of the root of *Allium Ascalonicum*, in T.S.; *g*, large central vessel, towards which the five hadrome-plates converge; *p*, pericycle; *s*, endodermis; *d*, passage-cells, situated opposite the hadrome-plates.

Since roots, in which the radial type of bundle prevails, are normally constructed as inextensible organs, the mechanical elements are often distributed over the different vascular "rays"; there is, in fact, often no other way in which they can take up the central position prescribed by the mechanical principle. Semilunar fibrous sheaths on the outer side of the leptome-strands have only been observed in the roots of certain PAPILIONACEAE (*Pisum, Phaseolus*). The whole radial bundle is surrounded by a pericambium or **pericycle**, usually consisting of a single layer of thin-walled parenchyma, which is chiefly concerned with the production of lateral roots. In the GRAMINEAE, CYPERACEAE and JUNCACEAE, and in certain allied families, the pericycle is interrupted— in the majority of investigated species—opposite the hadrome-plates, which thus abut directly against the endodermis.

In roots, any departures from the typical radial structure of the vascular strands are generally correlated with special environmental conditions, or arise from the necessity of increasing the amount of available conducting tissue. Some of these modifications are connected with mechanical requirements, and have accordingly been already dealt with on a previous occasion (Ch. IV.). In most cases of aberrant vascular structure in roots, the essential feature consists in the dilatation of the central cylinder and the concomitant differentiation of a number of scattered hadrome-vessels and leptome-strands on the inner side of the ordinary circle of radial hadrome- and leptome-plates. This type of

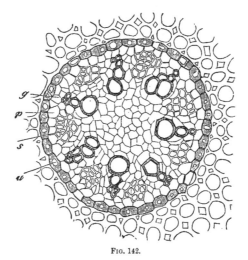

FIG. 142.

T.S. through the heptarch radial bundle [stele] of an adventitious root of *Primula Auricula*; *g*, hadrome-plate; *s*, leptome-strand; *p*, pericycle; *u*, endodermis. × 225. De Bary.

modification is illustrated by the aërial roots of many epiphytic Aroids, by the ordinary roots of MUSACEAE and certain Palms, and by the stilt-roots of *Pandanus*. In stout aërial roots the central cylinder is almost always greatly dilated; the consequent approximation of the water conducting elements to the periphery evidently facilitates the transference of water to the cortical tissues. The structural contrast between the grasping and absorbing roots of some of these Aroid epiphytes—a contrast first pointed out by Schimper [174]—affords a very clear demonstration of the way in which this particular modification of vascular structure is correlated with an increase in the demands made upon the conducting system. In *Monstera deliciosa* the width of the central cylinder amounts to one-third of the total diameter of the organ in the case of the grasping roots by which the plant is attached to its

z

support; here the pith is sclerotic and devoid of vascular elements, while both the vessels and the sieve-tubes of the characteristic vascular "rays" are of relatively small calibre.   In the absorbing roots, on the contrary, which grow vertically downwards and finally enter the soil, the width of the central cylinder is equal to one-half of the total diameter, while the ordinary hadrome- and leptome-strands are supplemented by numerous wide vessels and sieve-tubes scattered throughout the pith.

Radial bundles also occur in the stems of LYCOPODIACEAE and in the stolons of *Nephrolepis*.   In the upright species of *Lycopodium*, the hadrome takes the form of radiating plates, which meet at the centre of the stem.   In those forms, on the other hand, which have prostrate dorsiventral shoots, the hadrome is broken up into a number of separate horizontally extended ribbon-shaped tracts.   The more or less incurved margin of each tract consists of annular and spiral tracheides, whereas the central portion is entirely made up of scalariform elements.   Each tracheidal tract is enveloped in a layer of hadrome-parenchyma.   The intervals between adjacent hadrome-tracts are occupied by the leptome.

A **collateral vascular bundle** consists of a hadrome- and a leptome-strand, which run side by side in actual contact with one another.   If the hadrome is enclosed between *two* leptome-strands, the bundle is termed **bicollateral**.   Collateral bundles predominate in the leaves and stems of Angiosperms and Gymnosperms, and occur also in the stem of *Equisetum* ; in most Ferns the smaller foliar bundles conform to this type.   In stems the orientation of the collateral bundles is usually such that the leptome is external.   A corresponding arrangement prevails in dorsiventral leaves, the leptome facing the abaxial and the hadrome the adaxial side.

Collateral bundles exhibit less diversity of cross-sectional outline than concentric strands.   Most often they are approximately circular, elliptical or oval; but they may also be laterally compressed or ribbon-shaped, as in the leaves of *Hypochaeris, Pandanus, Hyacinthus,* the SCITAMINEAE, etc.   As a rule, the hadrome and leptome are combined in such a way that the composite bundle possesses only a single plane of symmetry.

As regards minor details, the construction of collateral bundles is exceedingly variable.   A very well-defined type, which occurs in a great many Monocotyledons, is characterised by the V- or Y-shaped cross-section of the hadrome-strand, the base of the V being occupied by a few spiral or annular vessels, or sometimes by an intercellular passage, while the distal end of each arm is marked by a single very wide vessel with spiral or closely reticulate thickenings (Fig. 143).   The space between the arms is either taken up by the leptome-strand or else filled by a group of narrow vessels, the latter alternative being exemplified by many Grasses. In certain other Monocotyledons (*e.g. Musa, Maranta, Calamus* [Fig. 144]), the vascular bundles are characterised by the presence of a single very wide vessel in the centre of each hadrome-strand. According to Scherer, the development of isolated vessels of

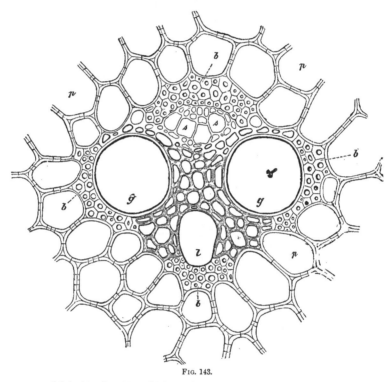

Fig. 143.

Collateral bundle, with partial fibrous sheaths, from the haulm of *Bambusa Simonii*. *g*, large pitted vessel ; *l*, intercellular passage [which has taken the place of the proto-hadrome]. The space between *g* and *l* is occupied by smaller vessels and tracheides. *s*, sieve-tubes, accompanied by companion-cells and phloem-parenchyma (cambiform cells); *b*, the four partial fibrous sheaths ; of these the outer shelters the leptome strand, while the two lateral ones protect the large pitted vessel ; *p*, thick-walled parenchymatous ground-tissue.

exceptional width which usually takes place immediately after the forma-tion of a few narrow [protoxylem] elements, is possibly connected with the general mode of life of the plants in question. For the most part the species provided with this type of bundle are bulbous or tuberous forms, which develop their vegetative organs very rapidly. Their transpiratory activity, and hence the demands made upon their water-conducting system, are thus liable to increase quite suddenly. It is this sudden demand that is provided for by the formation of the aforesaid large vessels. The leptome-tissue of Monocotyledonous bundles also exhibits considerable variety as regards the nature and arrangement of its con-

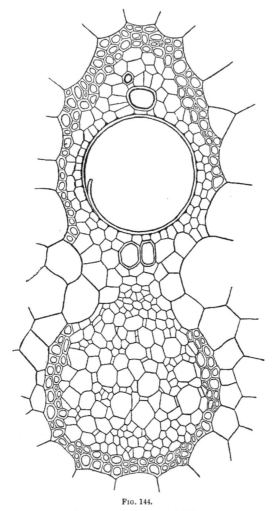

Fig. 144.

Vascular bundle from the sheathing leaf-base of *Musa paradisiaca* (T.S.).

stituent elements; it is, however, always principally composed of sieve-tubes and companion-cells, while a small amount of parenchymatous tissue (including cambiform cells) is also present, especially towards the flanks of the bundle. Kny has shown that, in the petioles of many Palms, the protective sheaths of fibrous tissue are produced inwards in the form of flanges which divide the leptome into two or more sections ; by this means the leptome is more effectively protected against compression. In the collateral bundles of Dicotyledons and Gymnosperms the hadrome elements are usually disposed in rows which run parallel to the plane of symmetry of the bundle (Fig. 145). This arrangement

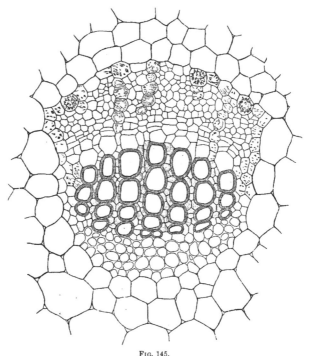

FIG. 145.
Collateral bundle from the leaf of *Malva sylvestris* (T.S.).

is due to the fact, that a serial cambium is formed, at an early stage of development, which remains active at the junction of hadrome and leptome for a long time, or even indefinitely ; in the latter event the persistent strips of fascicular cambium form the starting points of the secondary cambial cylinder. Monocotyledonous plants, on the other hand, never develop a serial fascicular cambium ; hence there is no persistent cambial zone at the junction of hadrome and leptome in their bundles. In other words, the entire primary procambium is straightway converted into permanent tissue.

As already explained, an intercellular passage of varying width occurs in each hadrome-strand in many Monocotyledons (*e.g. Acorus Calamus, Butomus, Sagittaria, Alisma*, CYPERACEAE, JUNCACEAE, etc.), and in *Equisetum*. Traces of the disintegrated annular and spiral vessels which are replaced by this intercellular passage, persist in the form of characteristic thickening fibres adhering to the walls of the canal. According to Westermaier, these passages contain water, at any rate at certain times;[175] it is uncertain, however, whether they merely serve as reservoirs of water or whether they take an active part in the work of conduction.

3. *The physiological significance of the different types of vascular bundle.*

We must preface our remarks concerning the physiological significance of the principal types of conducting strand with the statement, that this subject cannot be treated satisfactorily, unless the phylogenetic relations of the several forms of vascular structure are also taken into account. We must, accordingly, begin with that type of conducting system which may be presumed to be the oldest in the phylogenetic sense. The most primitive type of vascular strand is the hadrocentric bundle (concentric bundle with central hadrome), which makes its first appearance in the stems of the most highly organised Mosses—the POLYTRICHACEAE—and recurs in the basal region of the young stem in many Ferns. Both the radial and collateral types of vascular structure must be regarded as derivative forms.

In order to appreciate the physiological significance of the hadrocentric type of vascular structure, we may formulate the following question : Given a homogeneous non-vascular axial organ, how may a central concentric vascular bundle become evolved by progressive differentiation ? Let us imagine a small and simply constructed plant body, consisting of an upright leafy shoot, with absorbing organs attached to its lower end. The problem of conduction or translocation will assuredly be of the simplest nature in the region of the stem that lies below the insertion of the lowermost leaf. Here the water—containing mineral salts in solution—which is ascending to the leaves, meets the synthetic products on their way downwards to the growing organs of absorption ; the notion of ascending and descending sap-currents is, in fact, realised in its simplest form. In these circumstances, the separation of the two currents is obviously most essential to the uninterrupted progress of translocation. Hence the first step towards the evolution of a vascular system, will be the differentiation of a definite strand of water-conducting tissue, which in an upright radially symmetrical stem will naturally take up a central position. This first stage in our theoretical scheme of evolution is exemplified by those leafy Moss-stems which possess a simple type of central strand, consisting of a water-conducting bundle, surrounded by a sheath of parenchyma serving for the conduction both of protein-compounds and of carbohydrates. By a further division of labour, special channels are set apart for each of the last-mentioned groups of plastic materials, the cortical parenchyma being reserved for the translocation of carbohydrates, while the transport of protein substances is assigned to a new type of conducting tissue, namely, the leptome. This leptome, being a delicate tissue, and hence very liable to mechanical injury, will naturally tend—in an inflexibly constructed organ—to take up its position as close as possible to the neutral axis. Since, however, the centre of the stem is already occupied by the water-conducting strand, the leptome is forced to assume the form of an annular sheath enveloping the hadrome. By these successive steps a typical axile concentric composite bundle eventuates.

The preceding account of the evolution of the concentric vascular bundle implies, that this type of conducting strand has not been a simple histological unit throughout its evolutionary history; in other words, that it has not developed by a process of gradual differentiation from a homogeneous primitive conducting strand. On the contrary, the concentric bundle is here assumed to have arisen under the influence of the same tendency on the part of heterogeneous strands to combine to form histological units of a higher order, which was previously invoked in order to explain the association of fibrous strands with mestome-bundles. The primitive axile conducting strand of the stem, delimited from the surrounding cortical tissue by an endodermis— which ontogenetically, as a rule, pertains to the cortex—may be regarded as the first stage in the evolution of the structure known as the **central cylinder** or **stele**.[175a]

The central cylinder of the root, which consists of a radial vascular bundle enclosed in a pericyclic sheath (pericambium), is homologous with the primary central cylinder of the stem; in other words, the radial type of bundle is undoubtedly derived from the hadrocentric type. The transition from one type of structure to the other may conceivably have taken place as follows: The hadrome gradually extended outwards along two or three rays, and finally broke through the surrounding leptome-cylinder, which thus became divided up into a corresponding number of separate portions. It need hardly be pointed out, that if the vascular structure of the root were of the concentric type, the water taken in by the absorbing region of the root-system would have to traverse a sheath of protein-conducting leptome, before it could reach the proper conducting channels; consequently the extension of the hadrome towards the periphery at a number of points, which is characteristic of the radial type of vascular structure, not only facilitates the passage of water from the cortical parenchyma to the actual water-conducting channels, but also prevents any interruption of the longitudinal flow of protein-material by transverse currents of water.

The radial construction of the vascular bundle in roots is also advantageous in relation to the mode of origin of lateral roots. In order that the continuity of the water-conducting system may be maintained, it is necessary that the vessels of the lateral roots should be directly attached to those of the parent organ. It would be difficult to ensure this continuity, if the newly-formed vessels had to traverse a more or less extensive sheath of leptome. The connection could, in fact, only be effected with the aid of a secondary meristem; but the leptome, consisting as it does mainly of sieve-tubes, is exceedingly ill-fitted for producing the requisite meristematic tissue. In the radial type of central cylinder, on the other hand, where the outer ends of the hadrome-plates abut immediately against the pericycle, which is normally the root-producing layer, the problem of linking up the two sets of vessels presents no difficulty.

How far the occurrence of radial bundles in the stems of *Lycopodium* is capable of a physiological explanation, must be left undecided. It has been suggested, that in this case the central vascular strand has been produced by the phylogenetic aggregation of several previously distinct strands; if this view is accepted, the problem resolves itself into the question as to the advantages consequent upon such a fusion.

In dealing with the physiological significance of the collateral type of vascular bundle, stems and leaves must be treated under separate headings. As regards leaves, we may first of all consider the gradual transition from concentric to collateral structure of the bundles which, as the author has shown, regularly takes place in Ferns. Here the hadrome-portion of the concentric strand gradually shifts further and further from its original central position towards the adaxial side of the bundle, with the result that a large portion of the overlying leptome is pushed aside, and the entire bundle loses its concentric character. The displacement often begins in the petiole, and at latest appears in the principal veins of the lamina. In the smaller veins, the last remnants of the adaxial leptome disappear, and the bundle becomes truly collateral.

These facts—which may be quite easily verified—prove that, in the leaves of Ferns at any rate, the collateral structure of the vascular bundle, and its characteristic orientation, are consequences of the dorsiventral construction of the entire organ; the same argument most probably applies to the leaves of Phanerogams. It is natural further to suppose, that the collateral structure of foliar bundles is in some way correlated with the special physiological activities of the leaf. This assumption also finds support in the data of comparative anatomy; for it is found, that the collateral or excentric tendency of the vascular bundles is most pronounced, where the palisade-tissue—which represents the special photosynthetic apparatus—is most highly differentiated. The palisade-tissue may, in fact, be said—figuratively speaking— to exercise an attraction upon the hadrome-portions of the vascular bundle, and thus to cause that disturbance of the concentric arrangement which produces first the exceptric and finally the collateral type of structure.   This "attraction" is undoubtedly closely connected with the fact, that it is the hadrome which supplies the photosynthetic tissue with water and mineral salts.   The outward flow of these substances from the vascular bundles is mainly directed towards the upper surface of the leaf, and would thus be impeded if leptome were developed on the adaxial side of the bundles.   In order, therefore, to ensure direct communication between hadrome and palisade-tissue, the former shifts upwards, and in the end entirely displaces the leptome which originally occupied the adaxial margin of the concentric bundles.   The outward diffusion of water and nutrient salts, of course, takes place chiefly in the smaller bundles, a circumstance which explains why only the lesser veins become collateral in Fern-leaves. .

According to the view set forth above, the radial and collateral types of bundle have both been derived from the primitive concentric condition under the influence of the same principle.   In either case, the change of structure facilitates the diffusion of water and mineral salts, which in the root have to pass from the cortex to the hadrome, and in the leaf are transferred from the hadrome- to the palisade-tissue. The radial symmetry of the root entails a centrifugal displacement of the hadrome along several radii, whereas in the dorsiventral leaf the water-conducting tissue always shifts in the adaxial direction ; thus totally different derivative types of vascular structure result in the two cases.

The occurrence of collateral vascular bundles with adaxial hadrome in sepals, petals, bracts, bud-scales, etc., as well as in dorsiventral foliage-leaves, is probably to be regarded as the result of an inherited tendency, depending on the fact that these various structures are in all likelihood phylogenetically derived from foliage-leaves.

It has already been stated that the stems of Dicotyledons and Gymnosperms—and also those of *Equisetum*—contain a circle of collateral bundles with external leptome ; here phylogenetic considerations throw some light upon the probable origin of the collateral condition.   In all the above-mentioned cases, namely, the whole circle of vascular bundles, the pith enclosed thereby, the primary medullary rays which divide the bundles from one another and the pericycle which separates them from the endodermis (or starch-sheath) collectively represent a single central cylinder homologous with the central cylinder of a root or a young Fern-stem.   The problem thus resolves itself into an enquiry concerning the derivation of a central cylinder composed of several collateral bundles from the primitive type of central cylinder which consists essentially of a single hadrocentric strand.    Now it is by no means unusual, even in the case of roots, for the middle of the central cylinder to be taken up by a parenchymatous **pith**, so that the vascular tract is annular in transverse section.    Similarly, the middle of the primitive hadrocentric bundle of the stem has in the course of evolution been replaced by a core of medullary parenchyma of gradually increasing diameter ; the vascular tract of the central cylinder has accordingly expanded into a double ring—or rather a hollow cylinder—composed of hadrome and leptome.[175b]

It is impossible to decide, whether medullation, or the acquisition of centrifugal tendency by the vascular tissue, was the primary phenomenon ; in all probability both factors have contributed to the final result.    Before the differentiation of a mechanical system—which, as already stated, is phylogenetically a comparatively recent development— the duty of strengthening the various organs rested chiefly with the thick-walled hadrome, a state of things which still persists in many weakly constructed plants.    In these circumstances the hadrome tended to shift towards the periphery, in organs which had to resist bending strains, and at the same time to assume the form of a hollow cylinder ; as a secondary consequence of this centrifugal displacement of the hadrome, the leptome was also forced further outwards.    A demand for an improved water-supply to the transpiring cortex may have helped to " attract " the hadrome towards the surface, in the stem as well as in the leaf.    On the other hand, the formation of a core of medullary parenchyma also entails a number of special advantages, of which one deserves particular mention.    By the development of a pith, the stem acquired a tissue which could be employed for the temporary storage of various substances, with the result that the primary cortex could be more effectively utilised for purposes of translocation.

The breaking up of the primitive continuous leptome-hadrome cylinder into a circle of distinct collateral bundles was probably a secondary change, brought about by the necessity of ensuring free communication between the conducting tissue of the cortex and the storage-cells of the pith.    This necessity led to the formation of the **primary medullary rays**, which, to begin with, no doubt merely represented centrifugal extensions of the medullary parenchyma, that broke up the hadrome, without interrupting the continuity of the leptome-cylinder. The careful researches of Zanetti have shown, that this condition persists at the present day in the stem of *Osmunda regalis*.    In this respect, therefore, *Osmunda* may be regarded as a relic of that stage in the evolution of vascular plants, at which the primitive continuous concentric strand of the stem was about to be transformed into a circle of separate bundles. When the radial extensions of the pith broke through the leptome, the vascular cylinder became finally resolved into a ring of isolated collateral strands, with external leptome.

The disposal of the collateral vascular bundles in a single ring, which is characteristic of *Equisetum*, of the Gymnosperms and of most Dicotyledons, is evidently a more primitive condition, than the system of concentric circles that occurs in some Dicotyledons, or the "scattered" arrangement which prevails among Monocotyledons. The more detailed discussion of these matters must be reserved for a subsequent section (IV., A). For the present it will suffice to remark, that these aberrant types of arrangement were doubtless in the first instance evolved with reference to special physiological requirements; it would consequently be very unwise to regard all the bundles, in such abnormal types of vascular structure, as necessarily derived from a single ancestral central cylinder. The vascular system is, in fact—like other anatomico-physiological systems—by no means invariably homogeneous in a phylogenetic sense. When once the idioplasm has developed "determinants" corresponding to collateral vascular bundles, conducting strands of this type may be formed quite independently of the original vascular cylinder, at any point where a physiological demand for these structures happens to arise; such accessory bundles, which may take the form of medullary or cortical strands, or of anastomoses in the mesophyll or at the nodes of stems, must be regarded as entirely new departures from a phylogenetic point of view.

As has been explained above, the orientation of the collateral bundles in stems is such, that the leptome portions are next the periphery, while the hadrome-strands face the centre. This arrangement is determined by two factors, which both act in the same sense. In the case of [horizontally extended] leaves, physiological considerations render it necessary that the hadrome should be dorsal, and the leptome ventral, in position. If a leaf-trace bundle of this type enters the stem without torsion, the customary arrangement follows as a matter of course. There is, however, a phylogenetic factor, which tends to produce the same results. If, namely, the relative positions of hadrome and leptome remain unaltered after the disruption of the primitive hadrocentric central cylinder, the resulting collateral bundles will naturally have their leptome portions on the outer side. This purely morphological explanation does not, of course, imply that the characteristic orientation of the axial bundles is altogether devoid of physiological significance. Among Dicotyledons and Gymnosperms, at any rate, this orientation seems almost obligatory in view of the subsequent occurrence of secondary thickening; moreover, since stems are typically inflexible in construction, a leptome which is external, can readily become associated with protective fibrous strands, or with a continuous mechanical cylinder. In special circumstances, on the other hand, this necessity of protecting the leptome may lead to an inverted orientation of the vascular bundles. In the genus *Centaurea*, for example, the small [accessory] cortical strands often have their leptome portions directed towards the subcortical stereome, while the hadrome portions face the periphery; here the inversion results in a further advantage, inasmuch as the water-conducting hadrome is thereby brought into closer relation with the cortical photosynthetic system. Similar deviations from the normal arrangement of the collateral bundle are not at all uncommon.

We have still to consider two types of vascular bundle, which may both be readily derived from the simple collateral strand. It has already been stated, that the **bicollateral bundle** is only distinguished from the simple collateral type by the presence of an additional leptome group on the inner side of the hadrome. The duplication of the leptome should probably be attributed to an increase in the physiological demands made upon this section of the conducting system, a view which is supported by the fact that stems provided with bicollateral bundles frequently contain isolated accessory leptome-strands as well; this latter condition is exemplified by the CUCURBITACEAE and CICHORIACEAE, and by many SOLANACEAE. It remains to be explained, why a plant should prefer the addition of an inner strand of phloem to a mere enlargement of the existing outer strand. Once more the decisive factor appears to have been the necessity for ensuring effective protection of the vulnerable leptome-tissue. In the chapter dealing with secondary thickening (Ch. XIV.), a full description is given of the special arrangements for local protection—in the shape of intrusive strands and plates of fibrous tissue—which often become necessary where the peripheral leptome is bulky. In axial organs of annual duration, the same effect can be obtained more economically by shifting a portion of the leptome tissue to the inner side of the comparatively resistant hadrome strands, where, quite apart from the local protection afforded by the latter, the tensions due to curvature are, in any case, mitigated on account of the greater proximity to the neutral axis.

It is possible, however, that the duplication of the leptome-strands serves to provide separate conducting channels for two opposite currents of plastic material. A. Fischer states, that the dorsal leptome-strands in the bicollateral foliar bundles of *Cucurbita*, which are specially active in the young leaf, are empty when that organ is full-grown, whereas the sieve-tubes of the ventral strands always retain their contents even in adult leaves. From these facts Strasburger concludes, that the dorsal strand, in which the sieve-tubes are all comparatively narrow, supplies the developing tissues with plastic material, but becomes inactive when the leaf is fully grown; the ventral strand, on the contrary, serves—just like the single leptome-strand of a normal collateral bundle—to collect and carry away the nitrogenous compounds manufactured in the leaf.

The **leptocentric bundle**—that is, the concentric bundle with central leptome—is connected with the collateral type by a series of intermediate forms. The frequent occurrence of leptocentric strands in Monocotyledonous rhizomes, and in the medullary tissue of certain Dicotyledons, gives a clue to the direction in which one has to look for the physiological significance of this type of vascular structure. Such bundles, in short, occur principally in storage-organs and -tissues, and more particularly in those which are concerned with the storage of carbohydrates and of water, or of the former alone. The extension of the hadrome in the form of a sheath completely surrounding the leptome, is apparently designed, as Strasburger has pointed out, to bring the greatest possible number of vessels and tracheides into direct contact with the surrounding storage-tissue. As will be explained later on (Ch. XIV., II. C. 2), the water-conducting channels of many woody plants are utilised for transporting carbohydrates as well as water, when the buds are unfolding in spring ; it is therefore quite conceivable, that the tracheal elements of leptocentric bundles likewise perform this twofold conducting function, in which case their peripheral disposition—within the individual bundles—must undoubtedly facilitate the depletion of the adjacent storage tissues. In any case, this argument is applicable to the cases in which leptocentric strands are associated with water storing tissues. Certainly Strasburger's theory of the leptocentric (amphivasal) type of vascular structure, if not finally established, seems at any rate very plausible.

## C. THE PERIPHERAL BUNDLE-ENDS.

The numerous ultimate ramifications of the vascular system usually terminate in the photosynthetic mesophyll ; this arrangement is quite in accordance with expectation, since the latter tissue on the one hand absorbs the largest proportion of the transpiration current, and on the other hand supplies the bulk of the plastic material which has to be transported in the opposite direction. Both these circumstances are reflected in the construction of the bundle-ends, which in some respects show a progressive simplification of structure, but at the same time develop certain features that are absent from the rest of the vascular system.

The hadrome-portion of such a bundle-end is usually composed of one

or more rows of spiral or closely reticulate tracheides.   Each of these
small water-conducting strands is enclosed in a carbohydrate-transport-
ing parenchymatous sheath, which generally extends right to the end of
the tracheide-bundle (Fig. 146), forming a sort of cap or hood over its
slightly dilated extremity.   In exceptional cases the distal ends of the
[terminal] tracheides may project freely into an intercellular space;
this condition occurs very frequently in the thick-leaved species of
*Euphorbia* (e.g. *E. biglandulosa* and *E. Myrsinites*), where the protruding
tips of the tracheides often expand into more or less spherical vesicles.
Sometimes the terminal tracheides are altogether unusually wide, in
which case they undoubtedly also act as
organs of water-storage (for details cf. Ch.
VIII.).

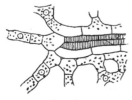

Fig. 146.

Foliar bundle-end of *Ficus elastica*,
consisting of two rows of spiral tracheides enveloped in a sheath of conducting parenchyma (parenchymatous
bundle-sheath) which is two cells thick
on one side.

As a rule the leptome does not extend
to the very tips of the bundle-ends, though
it may do so in exceptional cases.   In the
bundle-ends of Angiosperms, according to
A. Fischer,[176] the sieve-tubes gradually
decrease in width and the sieve-plates
become more and more imperfect.   The
companion-cells, on the other hand, retain
their ordinary dimensions, and may even
become somewhat enlarged (Fig. 147 B, C).   The relation between
sieve-tubes and companion-cells, in point of size, which prevails in
the bundle-trunks or larger conducting strands of the leaf and stem,
is thus reversed in the bundle-ends.   In the last resort, the common
mother-cells of sieve-tube segments and companion-cells cease to divide
altogether (Fig. 147 A).   Both these terminal undivided elements, which
possess the abundant protoplasmic contents and large nuclei of genuine
companion-cells, and the above-mentioned unusually wide, but otherwise normal companion-cells, are termed **transitional cells** by Fischer.
The physiological significance of these structures is obscure.   Possibly
they transfer to the sieve-tubes protein-compounds manufactured in
the photosynthetic tissues.   Fischer believes that they are themselves
the protein-manufacturing elements of the leaf.

The photosynthetic organs of Conifers and Cycads are entirely
devoid of fine vascular ramifications.   Here the uniform distribution of
water to the various parts of the transpiring mesophyll is ensured by the
development of a fringe of tracheal tissue along both sides of each
foliar hadrome-strand, the so-called **transfusion tissue**[177] (Von Mohl) or
" border of tracheides " (De Bary).   This tracheal fringe may take the
form of two lateral wings projecting into the adjacent photosynthetic
tissue; or it may embrace the hadrome or leptome portion of the vein;

in *Picea excelsa* and in the genus *Pinus*, finally, the whole vascular strand is surrounded by transfusion tissue. As De Bary first pointed out, this transfusion tissue consists of parenchymatous tracheides, with pits of the bordered type. In the CUPRESSINEAE the walls of the transfusion tracheides are provided with peculiar peg-like projections, while in other cases they are furnished with spiral or reticulate thickenings.

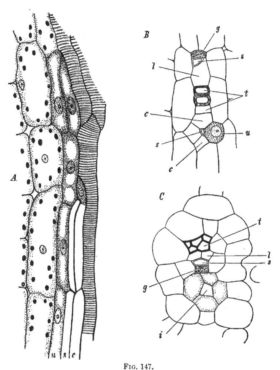

FIG. 147.

Bundle-ends. *A.* L.S. through a bundle-end from one of the leaf-teeth of *Fuchsia globosa*; *u*, transitional cell; *s*, sieve-tube; *c*, phloem-parenchyma cell (cambiform cell). *B.* T.S. through a bi-collateral bundle-end from the lamina of *Ecballium Elaterium*; *t*, tracheide; *l*, conducting parenchyma; *c*, phloem-parenchyma (cambiform cells); *s*, sieve-tubes; *g*, companion cell; *u*, transitional cell (exceptionally large companion cell). *C.* T.S. through a bundle-end from the lamina of *Aralia Sieboldi*; *t*, tracheide; *l*, conducting parenchyma; *s*, sieve-tube; *g*, companion cell (transitional cell); *i*, secretary intercellular passage.

According to Scheit, these thickenings are most strongly developed where the transfusion tissue abuts directly against photosynthetic parenchyma. If, on the other hand, the transfusion tissue is contained within a thick-walled parenchymatous sheath—as in *Pinus* and other ABIETINEAE—the walls of the tracheides do not require to be specially strengthened, and therefore only bear bordered pits. Scheit states that the quantitative development of the transfusion tissue depends upon the average transpiratory activity of the plant, which in its turn is determined by the nature of the environment. Thus the tissue in question is poorly represented in the Spruce (*Picea*) and the Fir (*Abies*), which both prefer a somewhat moist and shady habitat, but is strongly developed in the sun-loving Pine (*Pinus*).

According to Strasburger, special "albuminous cells" distinguished by the size of their nuclei and by their large content of protein-material are closely associated with the leptome-strands and form a counterpart to the border of tracheides. The elements in question, which evidently take the place of the transitional-cells of Dicotyledonous bundle-ends and doubtless perform similar functions, have been observed by Strasburger on the flanks of the foliar bundles in a number of Conifers (*Pinus sylvestris*, *Abies pectinata*, *Picea excelsa*, *Juniperus communis*, *Taxus baccata*) and in *Cycas circinalis*; they can only be clearly made out in spirit material.

### D. THE ENDODERMIS (*SCHUTZSCHEIDE*).[178]

In a great many plants the vascular tracts are separated from the adjoining tissues by special sheathing layers, which sometimes surround each individual conducting strand, and in other cases extend continuously around the entire vascular cylinder. This layer, which does not possess the same morphological value in every case, was termed the **protective sheath** (*Schutzscheide*) by Caspary, but in English text-books is more generally known as the **endodermis**. As a rule endodermal layers differ markedly in their anatomico-physiological characters from carbohydrate-conducting parenchymatous sheaths. Occasionally, however, a sheath of the latter type undertakes the duties of an endodermis as a subsidiary function, in which case it also acquires some of the correlative histological features; according to Schwendener, this condition is exemplified by the leaves of those Grasses which do not possess a typical endodermis.

Endodermal layers serve, in the first instance, to restrict **translocation** to certain definite paths, and to prevent the premature escape of substances which are travelling in the vascular bundles. In addition they often afford mechanical protection to the conducting strands. With these two physiological functions are correlated two outstanding structural features of the endodermis, namely its relative imperviousness and its mechanical strength.

The cells of the endodermis are always in uninterrupted contact with one another. They are usually of an elongated parenchymatous form, but in exceptional cases become distinctly prosenchymatous. Frequently all the walls are suberised, in which case, according to **Von Höhnel** and **Kroemer**, they resemble the membranes of cork cells, the primary layer being overlain by a suberin lamella, which in its turn is covered by a layer of cellulose; it is the last-mentioned layer that may become thickened for mechanical purposes, in which event it frequently exhibits the reactions of lignified membranes. This form of endodermal cell is especially prevalent among Monocotyledons. Another type of endodermis is characterised by the fact that its cells always remain thin-walled, the tangential walls being altogether unsuberised, while the radial septa are furnished with a so-called **Casparian strip**. This Casparian strip is a peculiarly modified narrow zone of the radial [and transverse] endodermal walls; more rarely a wider strip, or even the entire radial wall, is specialised in the manner about to be described. Microscopic examination of an endodermal cell of this second type shows that the region of the wall corresponding to the Casparian strip is thrown into folds; in a transverse section this corrugated area appears as a dark patch or shadow. In the strip the *primary* layer of the wall is chemically modified; according to a widely accepted view, it is suberised, like the suberin lamella in an endodermal cell of the first-mentioned type. The recent investigations of Kroemer, however, have rendered it very doubtful, whether the Casparian strip can be properly described as suberised. Nevertheless, as the effect of the chemical modification of the strip is such as to diminish its permeability, the current designation will be retained in the present treatise. This conservative attitude is further justified by the fact that the term " suberisation " is still used as a collective name for a number of distinct, though doubtless closely related, chemical modifications involving certain characteristic alterations of the physical properties of the cell-wall.

According to Schwendener, the Casparian strip, as seen in a section, is not a natural feature of the endodermal cell, but is directly due to the mode of preparation. As he points out, the living endodermal cells are in a state of tension, either as a result of their own turgescence, or owing to the pressure exerted by adjoining cells. On cutting a section the turgor falls, and the tension in the walls is relieved. But those parts of the membrane which consist of unaltered cellulose, contract much more strongly than the less extensible suberised strips ; the latter consequently become thrown into folds. Van Wisselingh accepts this explanation for a number of cases, but maintains that in other instances the corrugation of the Casparian strip has nothing to do with the mode of preparation and pre-exists in the intact organ. In the latter event the folding must arise in the course of development, owing to the fact that the suberised strips undergo more active growth in surface than the rest of the wall.

The suberised condition of the cell membrane renders the endodermis comparatively impervious. This physiologically important fact

2 A

was experimentally demonstrated by Schwendener with the aid of
staining solutions, and may also be inferred from certain observations
made upon plants growing under natural conditions.  In many
GRAMINEAE and CYPERACEAE, for example, the primary cortical tissues of
the root die and shrivel up at a comparatively early stage.  The central
cylinder is thenceforward enclosed only in the endodermis, but nevertheless remains quite fresh and continues to perform its functions in a
perfectly normal fashion.  In a number of roots, *e.g.* among GRAMINEAE,
CYPERACEAE, JUNCACEAE, in many Palms and in various other Monocotyledons, the endodermis only becomes impervious [at a late stage],

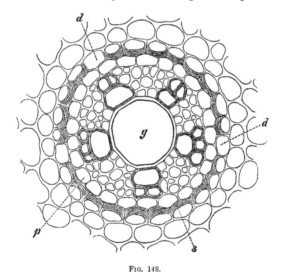

FIG. 148.

Radial bundle of the root of *Allium ascalonicum*, in T.S. ; *g*, large central vessel,
towards which the five hadrome-plates converge ; *p*, pericycle ; *s*, endodermis ·
d, passage-cells, situated opposite the hadrome-plates.

when there is no longer any need for freedom of interchange by diffusion
between central cylinder and cortex.  In such cases the endodermis
becomes equally impervious at every point.  Among Dicotyledons and
Ferns, on the other hand, as well as in the LILIACEAE and ORCHIDACEAE,
certain circumscribed and histologically differentiated areas of the endodermis often remain pervious, even when this layer is fully developed.
In such cases the endodermis comprises two different kinds of cells,
namely, typical endodermal elements, and interspersed among them the
so-called **passage-cells**, the tangential walls of which are not only unthickened, but also entirely unsuberised and hence readily permeable.
These passage-cells are always placed opposite the hadrome-plates,
and in fact abut directly against the first-formed [protoxylem] vessels

(Fig. 148). In tangential section they are seen to be scattered in certain cases, while in other instances they form longitudinal series.

Schwendener has shown by experiment, that the passage-cells do actually represent permeable spots in an otherwise impervious endodermis; he has also drawn certain conclusions from his experiments concerning the physiological significance of these structures, and especially as regards their topographical relation to the vessels. In view of the fact that the vessels are water-conducting tubes, the passage-cells must undoubtedly serve, in Schwendener's opinion, " to establish direct communication between the water-conducting channels and the living cortex, thus corresponding, as it were, to the open side-sluices of an extensive system of irrigating canals, the main channels of which are represented by the large vessels."

In stems and foliar organs the passage-cells are arranged on a similar plan. A deviation from the ordinary arrangement occurs in the leaves of many ORCHIDACEAE and BROMELIACEAE, where the passage-cells, instead of being placed on the inner face of the collateral bundle opposite the annular and spiral protoxylem vessels, are situated on the flanks of the bundle, either exactly along the line of junction of hadrome and leptome, or more often rather nearer the inner margin of the leptome. As Schwendener remarks, this unusual position clearly suggests that, in such cases, the passage-cells stand in close functional relation to the leptome as well as to the hadrome.

That an endodermis which has both its tangential and its radial walls suberised, must be relatively impervious, is self-evident. Different conditions, however, prevail, where the tangential walls are entirely unsuberised, while in the case of the radial septa suberisation is restricted to the above-mentioned narrow strip. Opinions differ considerably as to the physiological significance of the Casparian strips. Collectively they form a tubular meshwork, to which Schwendener ascribes a purely mechanical function; in this investigator's opinion, in fact, the endodermis may be compared to the protective net-work of ropes in which a balloon is encased. According to Strasburger, the suberised strips of the radial walls perform a mechanical function of a different nature, by increasing the lateral cohesion between adjacent cells of the endodermis, and thus preventing the formation of radial intercellular spaces within that layer. By this means the ventilating system of the cortex is permanently shut off from that of the central cylinder, with the result that considerable negative pressures can be maintained in the water-conducting channels. It should be noted, that De Vries had previously drawn attention to the importance of the endodermis as a pressure-boundary between cortex and central cylinder, in the case of roots, and had also shown that when water is injected under high pressure into the central cylinder of a severed root, it does not escape from incisions made in the cortical parenchyma, near the root-tip, so long as the endodermis remains intact. The living protoplasts of the endodermal cells allow no water to pass, because their turgor-pressure is much greater than the exudation-pressure in the vessels and tracheides, while the strips of suberised membrane form a continuous barrier which effectually prevents any movement of water through the substance of the radial walls of the endodermis. The same argument naturally applies in the case of the dissolved substances that travel in the vascular bundles. These are likewise confined to their proper channels, partly by the plasmatic membranes of the endodermal protoplasts, and partly by the suberised strips of the radial walls. An endodermal layer can therefore prevent the escape of all the substances that are transported in the vascular bundles, even though its walls are only partially suberised.

It should be remarked that mere suberisation of the walls does not increase the mechanical strength of the endodermis to any great extent. This result can only be achieved by pronounced thickening of the walls, such as takes place more particularly among Monocotyledons. The endodermal walls may be equally thickened all round (Russow's "O-type"), or the thickening may be confined to the radial and the *inner* tangential walls (Russow's "C-type"). As regards such points of detail, however,—which are in any case not of great mechanical importance,—differences may occur even within a single genus (cf. *Carex, Smilax, Ruscus, Potamogeton*).

Very frequently the mechanical efficiency of the endodermis is enhanced by the development of thickened walls in neighbouring layers. Among Ferns, indeed, the endodermal cells themselves remain permanently thin-walled, and the thickening is entirely confined to the adjoining layer of the cortex. In certain roots (*e.g.* those of *Taxus*, the CUPRESSINEAE, *Viburnum*, some POMACEAE, etc.), the cortical layers immediately outside the endodermis are furnished with thickening ridges forming a close and continuous meshwork, the mechanical effect of which is similar to that of the above-described suberised meshwork of the Casparian strips. These accessory mechanical sheaths constitute the "Φ-endodermis" of Russow, so-called—like his other types —on account of the appearance which the thickened walls present in transverse section. In the roots of LAURACEAE, the endodermis is locally strengthened by semilunar fibrous sheaths associated with the leptome-strands. The endodermis itself is not infrequently more strongly developed opposite the delicate leptome-groups, its cells tending to become radially elongated at these points (*e.g.* in the IRIDACEAE and in the aërial roots of Orchids).

While it is clear that such strengthening devices must tend to increase the resistant properties of the endodermis, it it nevertheless by no means easy to form a clear conception of the mechanical sigui- ficance of that layer.   We may, however, regard it as certain, that this mechanical function is in some way connected with the tissue-tensions that are set up between the parenchymatous ground-tissue and the central cylinder.   Injury to the leptome and other vulnerable vascular tissues by the resulting tensions and compressions, is prevented by the interpolation of an appropriately constructed protective sheath in the shape of the endodermis.   "Young *Iris*-roots, for instance, in which the tissue-tensions are never very great, possess an unthickened endodermis: in older roots, on the contrary, in which the endodermis is strongly thickened, very considerable tissue-tensions prevail."

Schwendener further remarks that "climatic and edaphic conditions also react upon the structure of the endodermis.   Thus the endo- dermis is always specially strengthened in the roots of lithophytes and steppe-plants.   In certain megathermic members of these ecological classes (*Dasylirion, Restio*), and also in some of our native Ferns, which grow on limestone rocks or on dry walls, the endodermis becomes thickened to quite an extraordinary extent, evidently in adaptation to the alternation of periods of abundant water supply with severe droughts.   Greatly thickened endodermal layers are also found in many marsh plants which inhabit places that are liable to dry up periodically.   Obviously, rapid variations of turgor are just as likely to occur under these conditions, as in the case of the above-mentioned xerophytes.   Where, on the contrary, the substratum is constantly moist and soft, as in deep bogs, backwaters, lakes, etc., the endodermal cells of the root do not as a rule develop any cellulose thickening layers (e.g. *Najas, Potamogeton, Sparganium natans, Sagittaria sagittae- folia, Alisma natans, Calla palustris*, etc.).   In the case of plants that grow in running water, however, the *stem* may be furnished with a thick-walled endodermis even if the corresponding layer in the root is unthickened."

Very little is known concerning the factors which determine the presence or absence of a typical endodermis; why, further, in some cases each vascular bundle is enveloped in a distinct protective sheath, while in other instances the entire central cylinder is surrounded by a common endodermis, is also quite obscure.   Schwendener has shown that the presence of a foliar endodermis in some tribes of the Gramineae, and its absence in others, cannot be ascribed to the influence of environ- ment.   Thus each bundle of the leaf is furnished with a typical endo- dermis in *all* the FESTUCEAE, in the hygrophilous genus *Glyceria*, as well as in the xerophilous *Festucas*.   The presence or absence of endodermal

layers in the leaves of Grasses cannot therefore depend upon ecological
factors, but must be entirely governed by physiological considerations,
such as differences in the methods of translocation, or possibly still unde-
tected tissue-correlations, which are not the same in the various tribes of
the family.  Similarly, the significance of the " individual " and the
"common" types of endodermis cannot be explained, until all the possible
determining factors—both physiological and phylogenetic—are properly
understood.  No reason, for instance, can be given at present for the
fact that a " common " endodermis occurs in the stems of *Ranunculus
aconitifolius*, *R. parnassifolius*, and *R. amplexicaulis*, whereas *R. Lingua*
and *R. Flammula*, though belonging to the same section of the genus, are
nevertheless provided with protective sheaths of the " individual " type.

## IV. THE ARRANGEMENT AND COURSE OF VASCULAR BUNDLES.[179]

The various conducting strands of the plant collectively form a
system which traverses every part of the shoot and root.  The vascular
system may even extend into epidermal appendages, if local translocation
within the latter is unusually active for one reason or another ; this
latter point is illustrated by the digestive glands of *Drosera rotundifolia*,
which will be dealt with in more detail later on (Ch. X. II. B. 1).  Where,
on the other hand, translocation is reduced to a minimum, the vascular
bundles may be entirely suppressed : this extreme reduction is exemplified by *Wolffia*, and by a few other small floating plants belonging to
the same family (LEMNACEAE).  Evidently physiological requirements have
a large share in determining the development of the conducting system.

The disposition of the vascular bundles in the different parts of the
plant-body, displays a variety which is scarcely inferior to the diversity
of arrangement with which we have already become acquainted in the
case of the mechanical system.  It is, however, much more difficult, in
the present instance, to correlate the arrangement in every case with the
underlying physiological principles.  As has already been explained, an
ordinary vascular bundle represents a combination of heterogeneous
tissues.  Hence that particular course of a bundle which is the most
advantageous, so far as the protein-conducting leptome is concerned, is
not necessarily best suited to the requirements of the water-conducting
tissue.  Moreover, the disposition of the vascular bundle is determined
by mechanical requirements as well as by considerations of nutrition.
Thus, delicate conducting strands tend to seek the shelter of skeletal
elements, while, on the other hand, stout resistant vascular bundles can
themselves extend protection to more vulnerable tissues.  The external
morphological features of the plant-body, finally, also influence the

arrangement of the conducting system to a very marked extent; thus, the course of the vascular bundles in the stem is in general closely related to the scheme of phyllotaxis, whereas the leaf-arrangement affects the disposition of the stereome only in a very minor degree. In short, correlation between different organs or tissues exerts a far-reaching influence upon the arrangement of the bundles.

### A. COURSE OF THE VASCULAR BUNDLES IN LEAFY STEMS.

In stems the vascular bundles generally follow a longitudinal course; they may run parallel to the longitudinal axis or else diverge obliquely either in a radial or in a tangential plane. Considerable variety prevails as to the manner in which bundles terminate above and below. A bundle which passes out above into a leaf, is termed a **common bundle,** the lower cauline portion being distinguished as the **leaf-trace.** **Cauline** bundles lie wholly within the stem; frequently they have leaf-traces inserted upon them at various points. At its lower extremity a bundle may either remain isolated ("separate" or "individual" bundles), or it may fuse with another bundle ("united" bundles). Sometimes each leaf sends but a single vascular strand into the stem. If the leaf-trace comprises several bundles, these often pectinate with strands belonging to other leaves; the whole vascular arrangement may be further complicated owing to the fact that some or all of the bundles fuse with one another at certain points.

After these preliminary remarks, we may pass on to describe the **principal types of arrangement of vascular bundles.**

1. The simplest and most primitive form of vascular system consists of an **axile strand,** from which single bundles pass out into the leaves. This axile strand may be cauline, or it may represent a sympodial structure, built up of the cauline portions of successive leaf-traces.

In the stems of POLYTRICHACEAE, in the adult stem of *Hymeno-phyllum, Gleichenia* and *Lygodium,* as well as in "seedling" Ferns generally, and also in certain species of *Selaginella,* this axile strand constitutes a primitive central cylinder [protostele [179a]]. In the POLYTRI-CHACEAE the leaf-traces join the concentric cauline strand, after traversing the cortex in an obliquely radial direction; those elements of the leaf-trace which correspond to sieve-tubes and conducting parenchyma, become merged in the leptome-sheath, while the tracheal tissue is continued across the sheath in the original direction of the leaf-trace, finally becoming inserted upon the central core of hadrome. In Mosses which have a simple central conducting strand in the stem, the leaf-traces may likewise be attached to this main trunk of the vascular system; in other cases, however, they do not penetrate so deeply into the stem, but end blindly in the cortical parenchyma. In the latter event the leaf-traces can only obtain their supply of water from the central strand indirectly, through the mediation of the intervening parenchymatous cells. The imperfect organisation which is exhibited by the conducting system, in such cases, is of special interest, because it proves that the vascular bundles of the leaf are not necessarily homologous with those of the stem ; hence it is quite possible, that the prevailing continuity of the

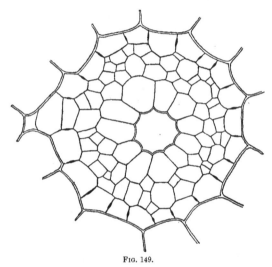

FIG. 149.

T.S. through the vascular bundle of the stem of *Najas major*. The centre of the strand is occupied by an inter-cellular passage surrounded by a circle of conducting parenchyma cells, which in turn are enclosed in a sheath of leptome tissue ; the entire bundle is enveloped in the endodermis.

vascular system in Higher Plants has been brought about by the fusion of phylogenetically distinct foliar and cauline strands.

In the genus *Lycopodium*, and in a number of aquatic Phanerogams (such as *Bulliardia aquatica, Potamogeton, Hippuris, Myriophyllum, Ceratophyllum, Najas, Elodea canadensis*, etc.), the axile strand almost certainly owes its origin to the [phylogenetic] fusion of several vascular bundles. As far as the Phanerogams are concerned, such a fusion would merely represent a reversion to the primitive state from the derivative decentralised condition. The internodes of different species of *Potamogeton* illustrate various stages in this process of fusion. In *Potamogeton natans* and *P. perfoliatus*, the several bundles of which the axile strand is composed are still clearly distinguishable. They are, however, already very closely associated with one another, being separated only by a few layers of parenchymatous cells, which represent the remains of the pith and primary medullary rays. Small fibrous strands are embedded in this [stelar] parenchyma. The fusion is considerably further advanced in *P. lucens* and *P. gramineus*. In *P. pectinatus* and *P. pusillus*, finally, the axile strand is reduced to a central intercellular passage which has taken the place of the water-conducting tissue, surrounded by a wide sheath of leptome. Here the vascular system no longer shows any signs of having been formed by the fusion of several distinct strands, and the same remark applies to the substantially similar axile strands of *Zannichellia, Elodea, Hydrilla, Ceratophyllum, Najas* (Fig. 149), etc. (Concerning the reduction of the water-conducting tissues in these cases cf. above, p. 323.)

Schwendener explains this tendency of the vascular system to revert to its simplest condition, in water-plants, on mechanical grounds. The stems of aquatic plants are subjected to continued tension in an upward direction, on account of the large amount of air which they contain, and very often have in addition to contend with the action of water-currents. As, however, most of the plants in question prefer stagnant or slowly running water, their mechanical strength need not be of a very high order, and hence their requirements in this respect are usually satisfied by a centralisation of the vascular tissue. The cortical parenchyma is never in danger of drying up, even if it is far removed from the vascular system, so that on that score also the central disposition of the conducting tissues is entirely free from objection.

2. The **simple tubular vascular system** [including the solenostele [179a] and the dictyostele] which occurs in a large number of Ferns, represents the intermediate link connecting the simple axile strand with the more complex types of vascular structure that prevail among Phanerogams. In most Ferns the young stem contains a solid axile vascular strand, which expands, as the stem increases in thickness, to form a tubular structure enclosing a parenchymatous pith and itself surrounded by a parenchymatous cortex. Since the pith contains both conducting parenchyma and storage tissue, complete isolation of this region must be avoided; free communication with the pith and conducting parenchyma of the petioles is secured by the development, at the insertion of each leaf, of a more or less extensive break or gap in the vascular cylinder, the so-called **leaf-gap**. If the stem is short and the leaves closely crowded, the tubular stele acquires the form of a perforated cylinder or cylindrical meshwork. From the margin of each mesh, or leaf-gap, the bundles of the leaf-trace ascend obliquely into the petiole (Fig. 150).

While the above-described arrangement of the vascular system is characteristic of the majority of the Filicales, a number of deviations from the type occur within the phylum; in some cases, for example, several concentric circles of bundles are developed (MARATTIACEAE, spp. of *Pteris*), while in others the main vascular cylinder is supplemented by small accessory medullary and cortical strands.  It is probable that these complications arise as a result of increased demands in respect of translocation ; but the detailed investigations which might furnish evidence in support of this suggestion, have yet to be carried out.

3.  Attention must next be directed to a third and very widely distributed type of vascular arrangement, which may be shortly entitled the **Dicotyledonous type.**  The characteristic features of this type, which occurs in the great majority of Dicotyledons, and in the Coniferales and Gnetales, as well as in a few Monocotyledons (DIOSCOREACEAE) and Vascular Cryptogams (*Equisetum* and the OSMUNDACEAE), are three in number. In the first place, all the primary vascular strands are common bundles, which enter the stem from the leaf along a curved path, and afterwards run down through several internodes.  Secondly, while a bundle is traversing the stem in a vertical direction, it always remains approximately at the same distance from the centre.  Lastly, each leaf-trace bundle sooner or later attaches itself laterally, either with or without previous division, to a neighbouring leaf-trace belonging to an older leaf; the leaf-traces thus all combine to form a continuous system, which is either unilaterally sympodial or reticulate in character.  In consequence of this behaviour of the vascular strands, a transverse section of the stem reveals a single circle of bundles, surrounding the pith and itself surrounded by the cortex, these two portions of the ground-tissue being connected by the primary medullary rays which occupy the radii between the bundles.  It may be remarked in passing, that these interfascicular strips should not be called medullary rays, in the anatomico-physiological sense of the term, unless they actually consist of parenchyma and thus constitute a histological link between cortex and pith ; when, on the contrary, the vascular bundles are embedded in an uninterrupted fibrous cylinder, the interfasicular strips of bast must not be called medullary rays, since a layer of mechanical tissue forms just as effective a barrier between pith and cortex as the vascular bundles themselves.

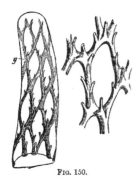

FIG. 150.

Tubular vascular network (dietyo-stele) in the distal portion of the stock of *Aspidium Filix mas*; on the right a single mesh of the network, more highly magnified, to show the insertion of the foliar strands.

We may next consider a few examples of the Dicotyledonous type of vascular arrangement in greater detail.  The adjoining figure (Fig. 151A) illustrates the course of the vascular bundles in a young twig of

*Juniperus nana.*  It should be noted at the start, that it is generally considered preferable, in making a diagram of this sort, not to attempt a perspective drawing, which is both troublesome to execute and difficult to interpret, but rather to imagine the vascular cylinder split open by a vertical cut and then laid out on a flat surface.

The leaves of *Juniperus nana* are arranged in alternating whorls of three ; the members of each whorl are not all inserted exactly at the same level, but are slightly displaced spirally.  A single vascular bundle enters the stem from each leaf ; this single leaf-trace descends without dividing through an entire internode, but towards the middle of the second internode forks into two shanks, which insert themselves right and left upon the adjoining leaf-traces.  A transverse section of a twig thus reveals either six or nine vascular strands, according to the region of the internode selected ; there will be six bundles present, if the section is taken where the leaf-traces are still undivided, whereas nine will be met with further down, where the leaf-traces have already forked (Fig. 151 B).  Every axillary bud receives a pair of small bundles.

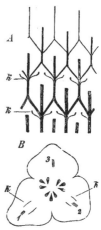

Fig. 151.

*A.* Diagram showing the course of the vascular bundles in a twig of *Juniperus nana* [the cylindrical surface being reduced to one plane]. *k*, the strands supplying the axillary buds.  *B.* T.S. through a young shoot. 1, 2, 3, the leaf-traces which pass out, in the order of the numerals, into the successive members of a whorl.  After Geyler (from De Bary, *Comp. Anat.*).

If it be asked whether this very regular reticulate arrangement of the leaf-traces serves any useful purpose, the answer will most decidedly be in the affirmative.  It must be remembered that each leaf-trace not only supplies the leaf to which it belongs with water and mineral salts, but also removes the surplus of plastic material manufactured in that leaf.  If, now, one of the uppermost bundles in the diagram be traced downwards, it will readily be seen, that by the time it has reached the third internode below the insertion of the leaf to which it belongs, it has become indirectly connected with every alternate member of the entire circle of bundles, that is with three leaf-traces in all.  Precisely the same statement, of course, applies to all the other leaf-traces.  As a result of the orderly arrangement of the vascular strands, therefore, a proper supply of water and mineral salts is secured to every leaf, while the effects of any inequality in the development of the different bundles are practically neutralised through being distributed over a considerable number of leaves.  Similarly the synthetic products of the leaves—or at any rate that portion of them which travels in the vascular tissues—soon become uniformly distributed over the entire cross-section of the stem; in this way, differences in the physiological activities of the several leaves are equalised. Nor is the mode of origin of the vascular bundles which enter the axillary buds altogether devoid of significance; for it is evident that the buds do not depend entirely for their nutrition upon their subtending leaves, since they receive plastic materials from members of the next (higher) whorls as well.

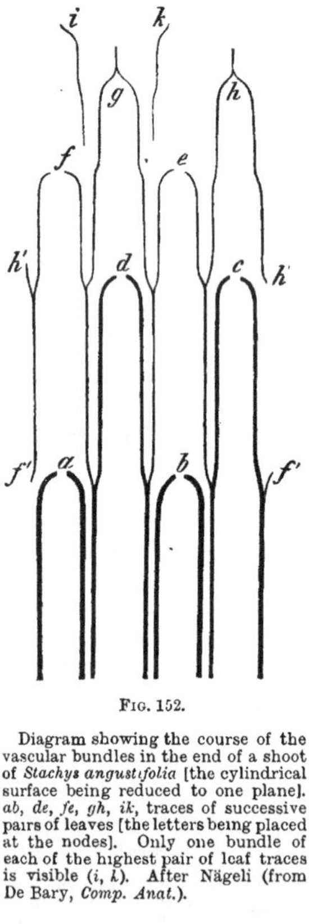

In order to illustrate the course of the vascular bundles in the stem of a Dicotyledonous plant, we may consider the case of *Stachys angustifolia*—described by Nägeli—with the aid of the accompanying diagram (Fig. 152). Here the leaves are decussately arranged, or in other words, placed in pairs with the median planes of alternate pairs at right angles to one another. The petiole of each leaf contains a single bundle, which forks as soon as it enters the stem; the two shanks of the leaf-trace diverge and, after passing down through two internodes, unite at the second lower node with the corresponding shanks of the next leaf-trace below. A transverse section across the stem, taken a short distance below the apex, thus shows eight bundles grouped in pairs in the corners. If one of the larger bundles is traced upwards, it will be found that the strands that branch off from it pass out into leaves belonging to two separate rows; if, therefore, we imagine the flow of material through such a bundle to be interrupted or retarded at any point, it is quite evident that the ill effects of this disturbance will be evenly distributed among a number of leaves belonging to two separate rows, and not concentrated upon a single row, much less upon an individual leaf.

FIG. 152.

Diagram showing the course of the vascular bundles in the end of a shoot of *Stachys angustifolia* [the cylindrical surface being reduced to one plane]. *ab, de, fe, gh, ik,* traces of successive pairs of leaves [the letters being placed at the nodes]. Only one bundle of each of the highest pair of leaf traces is visible (*i, k*). After Nägeli (from De Bary, *Comp. Anat.*).

In addition to such simple cases of bundle arrangement as the two just described, a large number of much more complicated types have been examined by Nägeli and others. In spite of the straightforward course of the individual bundles, the complications, in such cases, may be very elaborate; they depend for the most part upon the presence of multiple leaf-traces which become interwoven in diverse ways. All the manifold varieties of bundle arrangement, however, clearly embody the same governing principle, namely, that which aims at securing for every organ of the plant body a uniform and uninterrupted supply of the nutritive materials necessary to its growth and activity. As already explained, this co-operation among the various leaf-traces prevents inequalities in the incidence of external factors of nutrition from inducing an asymmetrical development of the plant-body, and thus ensures that the normal plan of construction is carried out, in spite of the frequent occurrence of such inequalities.

For the sake of greater clearness, we may consider an imaginary case in which a single bundle enters the stem from each of the numerous leaves, and then runs down perpendicularly *without* forming any connection with other leaf-traces. This arrangement would, in a sense, represent the most complete contrast to the type of vascular system that consists of a simple axile strand. If now the flow of water and nutrient salts were to be interrupted on one side of our imaginary stem, owing to some local injury, all the leaves and axillary buds on that side would inevitably wither and die, and the entire plant would thus suffer a serious loss. The scheme of vascular arrangement which actually prevails, enables the organism to react as a co-ordinated whole towards the various external agencies that tend to interfere with translocation ; as a result, the plant can deal more effectively with inequalities in the flow of food materials, and can even suffer local disorganisation of the vascular system without serious inconvenience.

Modifications of the normal Dicotyledonous type of vascular system arise in a great many different ways. A very frequent abnormality consists in the presence of numerous accessory **medullary bundles**, which may represent leaf-traces that have penetrated very far into the stem, or which may, on the contrary, be independent cauline strands. The former case is exemplified by most CUCURBITACEAE, by the PIPERACEAE, and by species of *Papaver, Thalictrum, Actaea,* etc., the latter by various species of *Begonia, Aralia, Orobanche,* by certain MELASTO-MACEAE, UMBELLIFERAE, etc. The presence of accessory vascular strands in the cortex, i.e. *outside* the circle of ordinary bundles, is on the whole less frequent. Such **cortical bundles** may merely represent loops or branches of ordinary leaf-traces (*Lathyrus Aphaca, L. Pseudaphaca, Casuarina, Salicornia,* the CACTACEAE, many species of *Begonia*), or they may constitute an independent cortical system of leaf-traces (CALYCANTHACEAE and many MELASTOMACEAE). According to Heinricher, finally, the cortical strands that occur in many species of *Centaurea* are in all probability genuine cauline bundles. The few detailed investigations which have hitherto been carried out upon the subject, clearly show that these " anomalies " of vascular structure form a most promising field for anatomico-physiological research. Westermaier has made a careful physiological study of the medullary vascular system in the BEGONIACEAE. In this Natural Order, the medullary strands are for the most part cauline bundles of secondary origin. At the nodes they curve outwards and unite at random with traces belonging to leaves which are inserted higher up. Their physiological significance is clearly indicated by the circumstance, that their occurrence is restricted to those BEGONIACEAE that hibernate by means of tubers or rhizomes (*Begonia boliviensis*, *B. tuberculata*, *B. hybrida*, *B. ignea*, *B. robusta*, *B. Rex*, *B. hydrocoty-lifolia*, etc.). In such plants the quantity of translocated materials passing through the stem is far greater than in the case of species with perennial woody shoots. The increased demands made upon the conducting system, in consequence of the alternate collection and redistribution of reserve-materials, are met by the development of accessory bundles in the pith.

Another interesting case that has been elucidated by Westermaier, is that of certain species of *Campanula* (such as *C. Trachelium* and *C. multiflora*), which are distinguished by the circumstance that their flowers are very numerous and at the same time often closely crowded together. The portion of the axis below the fascicular or capitate inflorescence has to satisfy the needs of a vast number of seeds, all ripening more or less simultaneously, and must hence be provided with a larger amount of conducting tissue, than is required by the stems of related species in which the flowers are fewer in number, or at any rate more widely scattered and put forth successively. In the former case, accordingly, the ordinary conducting system is augmented by medullary strands. In *Campanula Trachelium* these accessory strands lie in close proximity to the inner side of the normal woody cylinder; in *C. multiflora* they are arranged in two concentric circles, the outer lying close to the circle of ordinary bundles, while the much smaller inner circle encloses the central portion of the pith.

The presence of cortical bundles can also be referred to physiological causes in certain cases. Wherever, namely, the cortical parenchyma develops an unusually large amount of chlorophyll and generally approximates more or less to the condition of typical photosynthetic tissue, there cortical vascular bundles also tend to appear; these accessory strands evidently serve to supply this photosynthetic tissue with water and mineral salts, and to remove the whole or part of the plastic materials manufactured in its cells. De Bary was the first to put forward this explanation of the occurrence of a cortical vascular reticulum in the leafy marginal expansions of the so-called "winged" stems of species of *Lathyrus* and other plants; the same author has also remarked upon the close resemblance between the vascular reticulum in the cortex of many succulents (*Salicornia*, species of *Mesembryanthemum*, CACTACEAE) and the vascular system of an ordinary leaf. Heinricher, finally, has demonstrated the presence of a cortical vascular system in association with axial photosynthetic tissue in the genus *Centaurea*.

.4. The majority of Monocotyledons are provided with the form of vascular system termed the **Palm-type** by Von Mohl, which is characterised by the following features. In the first place, all the bundles are common, and as a rule a large number of bundles (up to several hundreds) enter the stem from the sheathing base of each leaf. Secondly, instead of all the strands of each leaf-trace running at the same depth in the stem, the median bundles always pass further inwards than the lateral ones. Thirdly, every bundle that penetrates to any considerable depth finally turns outwards again, thereafter pursuing an obliquely radial and downward course towards the periphery ; only the lateral bundles descend almost perpendicularly. The majority of the strands are not only radially inclined, but also deviate tangentially and at the same time become spirally twisted, both features being most pronounced in short internodes. Lastly, every bundle descends as an independent strand through a large number of internodes, finally uniting near the periphery with a strand which pertains to a leaf inserted much further down the stem. It is on account of this characteristic behaviour that the vascular bundles in the transverse section of a Monocotyledonous stem appear to be arranged in a "scattered" or "irregular" fashion, with a tendency to become aggregated near the periphery.

In attempting to explain the physiological significance of the Palm type of vascular system, we must first of all note that the equalisation of the supply of water and nutrient salts to the several leaves, and the uniform distribution of the synthetic products over the entire cross-section of the stem, are in this case sufficiently ensured by the inclusion of a very large number of strands in each leaf-trace, and by the fact that these strands enter the stem over a large portion of its circumference. The insertion of the lower ends of the leaf-trace strands upon bundles which enter the stem further down, is therefore a matter of less urgency here than it is where the Dicotyledonous type of arrangement prevails, and is consequently a feature of minor importance in stems with Palm-structure. The outward trend of the leaf-trace strands in the latter part of their course is, in all probability, mainly determined by mechanical considerations. The vascular bundles are generally provided with partial or complete fibrous sheaths; in accordance with the rules of inflexible construction, however, these mechanical strands tend to approach the periphery, and, in so doing, as

it were, carry the conducting strands with them.   The spiral torsion
of the bundles also seems to be a device for increasing the mechanical
strength of the stem (cf. Ch. XIV.).

Numerous modifications of the Palm-type are known, but only a
few of them can be mentioned here.   In ordinary, hollow-stemmed
Grasses, the true form of the vascular system is obscured owing to the
great elongation of the internodes.   Here the haulms appear to be
traversed by parallel bundles, which descend
perpendicularly and are connected at the nodes
by an abundantly branched network of transverse anastomoses.   The development of such
a nodal vascular network is partly correlated
with the production of axillary shoots, but
also, doubtless, serves to ensure the access of
a sufficiency of plastic material to the overlying
intercalary meristematic zone.   In addition the
nodal anastomoses have a mechanical significance, inasmuch as they act as cross-ties
between the main girders.

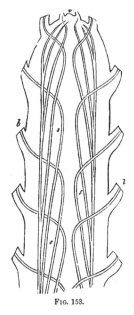

FIG. 153.

Simplified  diagram  of  the
course of the vascular bundles
in stems of the "Palm-type."
b, leaf-base ; v, growing-point.
After Falkenberg (from Sach's*
Lectures).

In the haulms of many CYPERACEAE and
JUNCAGINACEAE the longitudinal bundles are
likewise linked together by means of numerous
transverse connections ; these are embedded
in the parenchymatous diaphragms that divide
the large air passages into a series of compartments.   The mechanical function of such
" mestome-anastomoses " has already been discussed (Ch. IV.); their anatomical composition,
however, shows that they are also concerned
with the conduction of water ; for they consist
of narrow annular or spiral (rarely pitted)
tracheides, together with more or less elongated,
and in part thickwalled, parenchymatous elements.   In all probability
it is the photosynthetic activity of the haulms in the CYPERACEAE
and JUNCAGINACEAE, that accounts for the presence of numerous
small cross-connections between the longitudinal water-channels.   The
parenchyma which occupies the intervals between the peripheral
girders, contains abundance of chlorophyll, and is indeed often
developed as typical palisade-tissue, while in most species of Cyperus
the vascular bundles are surrounded by the green sheaths which
have already been referred to in a different connection (p. 284).
As the transpiratory activity of such haulms must be relatively
considerable, it is not surprising to find arrangements, which enable

one side of the haulm, when it is more brightly illuminated—and hence for the time being more actively engaged in transpiration and photosynthesis,—to enlist the vascular strands of the other side in its service, as well as its own bundles.   In order to test the water-conducting capacity of these vascular cross-connections, the author carried out the following experiment upon *Scirpus lacustris*.  Deep cuts were made in a number of haulms, each incision extending over about one-third of the circumference.  The. strips of tissue above the cuts remained perfectly fresh and turgid, and the wounded haulms were in other respects indistinguishable from uninjured specimens, even after the lapse of several weeks.

In certain Monocotyledons, the course of the vascular strands differs so greatly from the scheme of the Palm-stem, as to necessitate the recognition of distinct types cf vascular structure.   These aberrant plans of arrangement approximate in many respects to the ordinary Dicotyledonous condition.

### B.  COURSE OF THE VASCULAR BUNDLES IN ROOTS.

An ordinary subterranean root of a Vascular Cryptogam or Phanerogam typically contains an axile vascular strand, from which similar bundles pass into the lateral roots without any further complication.   The simplicity of the vascular arrangement is, in the first instance, correlated with the fact, that the medium in which roots grow is more or less constantly moist, so that the cortex is hardly ever exposed to the danger of drying up, even in those parts of the root which have already cast off their absorbing tissue.   In these circumstances peripheral bundles are not needed.   The central position of the vascular bundle also accords with the inextensible character of the organ ; in the absence of specialised mechanical elements, the vascular tissue can cope with moderate mechanical requirements, if it is massed around the longitudinal axis of the root.   From a physiological point of view, therefore, the central mestome-core of a root is in many respects comparable to the axile strands which so frequently occur in the stems of water-plants.

Some account has already been given, in an earlier part of the present chapter, of those anomalies in the vascular structure of roots which consist in the appearance of numerous isolated vessels and leptome-strands within a dilated central cylinder.   Reinhardt has carried out researches with regard to the relations of the accessory bundles to one another, and to the normal hadrome- and leptome-plates, in such cases. It appears that completely isolated hadrome-strands, extending from the stem to the apex of the root, occur in a number of Palms (*Caryota purpurascens, Phoenix dactylifera*, species of *Cocos* and *Chamaedorea*, etc.).

2 B

Isolated leptome-bundles are found in *Chamaedorea*, in some of the roots of *Musa rosacea*, and probably also in the genera *Monstera* and *Raphido-phora*. The accessory hadrome-strands anastomose more or less freely in the CYCLANTHACEAE, and also in *Chamaerops humilis* and *Areca rubra*. It is unusual for accessory leptome-strands to be connected by anastomoses; more often the outermost strands unite with the peripheral leptome-plates, or adjacent accessory strands may fuse in pairs. It appears, therefore, that conducting strands which occur as accessory structures within the primary central cylinder of a root, generally serve in the first instance for the transportation of materials through the whole length of the organ, or at any rate over considerable distances.

Where lateral roots are modified to form tuberous storage organs (*Dioscorea Batatas, Sedum Telephium*), the typical axile strand is replaced by, or resolved into, a number of distinct collateral bundles ; it can hardly be doubted that this exceptional arrangement tends to facilitate the deposition and removal of reserve-materials. The root-tubers of the OPHRYDEAE are characterised by the presence of several radial bundles, each enclosed within a typical endodermis. Van Tieghem believes that each of these "bundles" really represents an independent central cylinder, and that the tuber owes its origin to the (phylogenetic) fusion of several distinct lateral roots.

FIG. 154.

Venation of the leaf of *Convallaria latifolia*. After Von Ettingshausen (from Sachs, *Lectures*).

### C. COURSE OF THE VASCULAR BUNDLES IN LEAVES.

The arrangement of vascular bundles in petioles presents no features of special interest; attention may therefore forthwith be directed to flat leaf-blades and to the flattened axial organs which resemble leaves.

It is a familiar fact, that leaves differ greatly among one another in respect of what is generally termed their "venation"; this character has indeed been widely employed for taxonomic purposes, especially by palaeontologists, and a special terminology of venation has been invented to this end. The study of leaf-venation from the physiological point of view, on the other hand, is a comparatively recent development.

All the various forms of venation may be referred to one or other of two principal types. In one case the bundles are isolated, and end freely without undergoing fusion. This is the simpler, more un-specialised type, and mostly occurs in leaves which never require a very large supply of water and nutrient materials, either on account of their small size, or because their transpiratory activity is slight and their photosynthetic capacity (in the case of foliage-leaves) low ; it is exempli-fied by the leaves of Mosses, by the small scale-like foliar organs of *Equisetum, Casuarina,* and *Ephedra,* by the foliage leaves of all Conifers and of many Ferns, by the small Rataphylls and floral leaves of many Angiosperms and, finally, by the submerged leaves of certain water-plants (*Batrachium, Myriophyllum,* the HYDRILLEAE). In all these cases, the leaf is traversed, either by a single medium strand, or by a system of bundles which branch but do not anastomose.

The second principal scheme of vena-tion is characterised by the presence of numerous anastomoses, which link together the different branches of the vascular system. In the majority of Monocoty-ledons, the main strands all extend in straight lines or along slightly curved paths to the tip of the leaf (Fig. 154); here the anastomoses take the form of slender cross-connections between the prin-cipal veins. In almost all Dicotyledons, on the other hand, the bundles put forth branches in all directions in the plane of the leaf, which anastomose very frequently, the whole system assuming the character of a dense network, with its ultimate rami-fications [bundle-ends] terminating blindly

FIG. 155.

Venation of the leaf of *Salix grandi-folia.* After Von Ettingshausen (from Sachs, *Lectures*).

in its meshes (Fig. 155). According to Schuster, the disposition of the smaller branches is such that the meshes form "areas of minimum perimeter." This arrangement enables every part of the leaf to obtain its proper share of water and mineral salts with the smallest possible expenditure of vascular tissue, and, of course, similarly facilitates the removal of synthetic products from the mesophyll.

Generally speaking, the vascular network is most extensive and most closely woven in leaves which transpire very actively. The great difference in transpiratory activity between petals and foliage-leaves of the same plant is clearly reflected in the development of the vascular system in the two sets of organs (Fig. 156). A similar contrast prevails, with regard to the development of the foliar vascular system, between plants which grow in dry places and those which affect damp localities.   We are indebted to Zalenski for detailed observations upon this point.   Zalenski expressed the development of vascular tissue in terms of its linear dimensions per unit area of leaf-surface. Some of his measurements are set forth in the accompanying table:

| NAME. | HABITAT. | Linear dimensions of vascular tissue in mm. per sq. cm. of leaf-surface. |
| --- | --- | --- |
| Chrysosplenium alternifolium,- | River bank, | 170 |
| Maianthemum bifolium, | Wood, - - - - | 177 |
| Actaea spicata, - - | Do. - | 188 |
| Asplenium Filix femina, | Do. - | 213 |
| Geranium palustre, | Shady pasture, - | 454 |
| Trifolium repens, - | Sunny pasture, | 791 |
| Psoralea bituminosa, | Limestone rocks | 1130 |
| Reseda lutea, - - - - - | Dry sandy slope, - | 1160 |
| Astragalus austriacus, | Steppe, - | 1191 |
| Scutellaria alpina,- | Limestone rocks, | 1450 |

According to Schuster, a comparison between the vigorously transpiring "sun-leaves" and the feebly transpiring "shade-leaves" of one and the same plant reveals similar differences as regards the linear dimensions of the vascular system.

The numerous **anastomoses** of the vascular network are of great physiological importance.   Apart from their mechanical significance, which has already been discussed on a previous occasion (Ch. IV.), they play an indispensable part in equalising the water-supply at different points in the photosynthetic system, at any rate where the latter is continuous over a large extent of surface.   This last-mentioned function of the anastomoses was tested by the author with the aid of the following experiment.   A number of leaves of Sycamore (*Acer Pseudoplatanus*) were selected in the month of May, and either one or two of their principal veins severed close to the base, care being taken that where two veins were cut, these were not immediate neighbours.   In the absence of anastomoses, the regions of the leaf blades traversed by the severed veins would inevitably have dried up in a short time, owing to insufficient water-supply.   As a matter of fact, the leaves seemed to suffer no ill effects in consequence of the apparently serious injury inflicted upon their vascular system.   Whether shaded by other leaves or exposed to daily insolation, they continued, so far as could be judged, to functionate in a perfectly normal manner, and certainly did not undergo the usual autumnal discoloration any sooner than uninjured leaves on the same tree ; injured and uninjured leaves also fell at about the same date in the autumn.   When it is remembered, how often individual conducting channels in foliage-leaves must be rendered inactive, under natural conditions, by the attacks of insects or by hail, it will be readily admitted that the anastomoses are of the greatest importance, not only for the welfare of the individual leaf, but also as regards the economy of the whole plant. It is true that Vöchting and Simon have recently shown, that when vascular bundles are severed in certain stems and roots, the continuity of the interrupted water-conducting channels may

be re-established by the formation of new tracheidal connecting strands. Freundlich has observed a similar reconstruction of interrupted water-channels in some Dicotyledonous leaves, whereas this regeneration does not appear to take place in the cases of Monocotyledons or Pteridophytes.[180] Such observations do not, however, in any way detract from the importance which has been assigned to the anastomoses. The formation of fresh tracheidal connections involves considerable delay, during which any parts of the leaf that had been cut off from their direct water-supply would be exposed to the danger of drying up, if no vascular anastomoses were present. In any case, the injuries to which leaves are subject

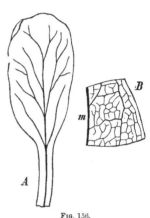

Fig. 156.

*A.* Course of the vascular bundles in a petal of *Barbarea vulgaris*. *B.* Small portion of a leaf of the same, drawn to the same scale, for comparison with *A.* ×10.

under natural conditions, are often so extensive that they could not possibly be repaired by the formation of new vascular connections.

In flattened, photosynthetically active axial organs, such as the cladodes of *Ruscus*, the arrangement of the vascular bundles is exactly the same as in flat foliage-leaves, a fact which requires no explanation from a physiological point of view.

## V. THE CONDUCTING SYSTEM IN THE THALLOPHYTA.[181]

In many Rhodophyceae and Phaeophyceae the central tissue of the thallus-branches contains certain rows of elongated cells, with transverse septa which display a considerable resemblance to sieve-plates. These cell-series are evidently the representatives of the conducting system of the Higher Plants, or rather of one of the most remarkable constituents of that system, namely the sieve-tubes.

Among the Rhodophyceae sieve-tubes were first discovered by J. Klein, and subsequently described by Wille (*Lophura tenuis, Rhytiphloea pinastroides, Helicothamium scorpoides, Cystoclonium pur-*

*purascens*). As a rule the entire wall separating two adjacent segments of such a sieve-tube is occupied—except for a narrow marginal zone— by a single pit of unusual size, the thin closing membrane (sieve-plate) of which is traversed by fine perforations. These sieve-tubes contain large quantities of "protoplasm" (? slimy protein material). The very characteristic sieve-tubes of the PHAEOPHYCEAE were first observed by H. Will in the older parts of the stipe of *Macrocystis luxurians* (Fig. 157 A). The structures in question occur in radial rows at the periphery of the central core of hyphal tissue. Their horizontal or slightly oblique transverse walls are converted bodily into sieve-plates, with wide perforations, which in the older sieve-tubes become occluded by callus-plates. Their longitudinal walls are moderately thick. The

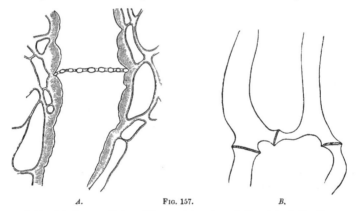

A.                    FIG. 157.                    B.

A. Sieve-tube from an old stipe of *Macrocystis luxurians*. B. Sieve-tubes in the stipe of *Fucus serratus*. After Wille.

segments are rather wide and are furnished with a peripheral layer of protoplasm, which is often covered with a slimy, vacuolated substance. Almost contemporaneous with Will's observations were N. Wille's researches upon the sieve-tubes of various species of *Laminaria* and *Fucus*. In the former genus, the sieve-tube segments are comparatively narrow, but expand at both ends just like the corresponding structures of Angiosperms. The transverse septa are finely perforated over their entire surface. These sieve-tubes, which contain large quantities of "protoplasm" (? slimy protein material) extend mainly in the longitudinal direction, but are connected at frequent intervals by anastomoses (Fig. 157 B). The sieve-tube system of the stipe is continued into the flattened leaf-like portion of the thallus, where it is located in the tissue that intervenes between the two photosynthetic layers. According to Wille and Hanosteen, the sieve-tubes of the FUCACEAE closely resemble those of the LAMINARIACEAE as regards both structure and arrangement.

Whether the "sieve-tubes" of RHODOPHYCEAE and PHAEOPHYCEAE serve to transport synthetic products other than protein-compounds, or whether special conducting elements are provided for that purpose, is a matter for further investigation.

In the relatively complex **rhizomorphs** of certain Fungi (PHALLACEAE, LYCOPERDACEAE, some AGARICINEAE), those longitudinal "medullary" hyphae which are not mechanical in character, probably act as organs of translocation. In some Hymenomycetes—especially *Agaricus praecox*, *A. olearius*, and other AGARICINEAE—the fleshy fruit-bodies contain certain long tubular hyphae filled with a dense and often highly refractive substance, which upon closer investigation may prove to be conducting elements. The fruit-bodies of *Lactarius* are furnished with [genuine] latex-tubes, in the shape of wide hyphae, with soft extensible walls; the finely granular latex contained in these tubes exudes from wounded surfaces in viscid drops, and coagulates on heating or upon treatment with alcohol. It is quite likely that the function of these latex-tubes is similar to that of the synonymous organs of Phanerogams, or, in other words, that they are concerned with the translocation of plastic substances. Morphologically considered, the laticiferous elements of *Lactarius deliciosus*—and probably of other species of *Lactarius* as well—correspond, according to A. Weiss, to articulated tubes (latex-vessels), since they are formed from rows of cells in which the transverse walls become obliterated. The fruit-bodies of *Lactarius* also contain specialised hyphae of a different kind. These latter are narrow and thin-walled, and are filled with clear liquid contents; they are surrounded by relatively large parenchymatous cells, which exhibit a rosette-like arrangement in transverse sections. It is uncertain whether these hyphae are also engaged in the transportation of plastic materials, or whether they possibly act as water-conducting channels.

## VI. ONTOGENY OF THE CONDUCTING SYSTEM.

Most **vascular bundles** are derived from primary procambial strands. In special cases, however, small bundles may be differentiated from fundamental meristem through the agency of secondary procambial strands; in the haulms of *Papyrus antiquorum*, for example, the small cross-connections in the diaphragms are formed in this way. The primary **conducting parenchyma**, comprising the parenchymatous bundle-sheaths and "nerve-parenchyma" of the leaf, and the medullary, cortical and interfascicular (medullary ray) parenchyma of the stem, but excluding the parenchymatous elements of the hadrome and leptome, arises out of the fundamental meristem.

The ontogenetic origin of the **endodermis**[182] is quite as variable as its phylogenetic development.    The author has demonstrated the procambial origin of the endodermal layers in the JUNCACEAE and CYPERACEAE, and a similar mode of development probably prevails among Grasses, except that in certain cases the endodermis represents a partially modified parenchymatous bundle-sheath.    Among Ferns the endodermis originates from fundamental meristem.    In this connection, Russow states that elements of the ground-meristem surrounding a procambial strand undergo one or more tangential divisions, thus giving rise to radial rows of cells. Either the outermost of the resulting layers, or one nearer the middle, is converted into the endodermis, while the inner layers give rise to the above-mentioned parenchymatous sheath (inner phloeoterma of Strasburger).    In the smaller bundles of Ferns, the genetic relation of the endodermis to the layers immediately interior to it is clearly recognisable even in the adult condition.    In roots, finally, the endodermis arises from the layer of fundamental meristem which immediately adjoins the axile strand.

Since the meristematic layer from which the endodermis originates, does not always become bodily transformed into endodermal cells, it cannot be simply regarded as the young endodermis.    The author, therefore, proposes to apply the term **coleogen** to this layer, and further to distinguish, according to the morphological character of the layer in question, between procambial and fundamental coleogen.    Just as the principal function of the phellogen is the production of cork, so the principal function of the coleogen is the formation of an endodermal sheath, as indeed its name implies.    But other tissues may also arise, in part at least, from coleogen.    As already indicated, this point is demonstrated by the development of the vascular bundles of Ferns, where the fundamental coleogen surrounding the procambial strand produces parenchymatous layers in addition to the actual endodermis. In the leaf of *Scirpus Holoschoenus* the procambial coleogen undergoes tangential division at points corresponding to the positions of the future hadrome- and leptome-groups, the innermost of the resulting segments developing into genuine bast-fibres ; in this case, therefore, the coleogen takes part in the construction of the mechanical system.

Passing on to consider the mode of origin of the individual conducting strands, we may first discuss what can be made out by means of transverse sections.    It will be found that the several elements of a strand do not arise simultaneously, but that their differentiation is, on the contrary, successive.    As a rule, the organisation of permanent conducting elements starts at the margin of the strand and gradually advances towards its centre.    By a slight modification of Russow's terminology, we arrive at the names of **protohadrome** and **protoleptome** for these first-formed vascular elements.

[Russow's own names were protoxylem and protophloem, while De Bary applied the term "Erstlinge" to both classes of elements.]

In the case of a collateral bundle, the protohadrome and protoleptome elements are situated at opposite poles in the cross-section of the strand; the differentiation of the two principal portions of the bundle progresses from these diametrically opposed points in the periphery towards the centre of the strand.   In the radial bundles of roots, the first-formed elements of the hadrome- and leptome-plates alternate with one another along the circumference of a circle; the further differentiation of the strand proceeds centripetally from these starting points.   In the concentric bundles of Ferns, finally, the protohadrome elements usually occupy the two extremities of the ribbon-shaped area corresponding to the future hadrome-mass.   Where the outline of the hadrome is circular or strongly curved, additional points of origin may be found; as a rule these are uniformly distributed over the cross-section of the strand.

With regard to the longitudinal differentiation of conducting strands, it may be remarked that acropetal development is characteristic of the bundles of roots, of cauline bundles in stems and also of the leaf-traces of certain plants (*Tradescantia albiflora*, species of *Potamogeton*, according to De Bary, and various other Monocotyledons according to Falkenberg).   In a very considerable number of Dicotyledons and Conifers, on the other hand, according to Nägeli, the leaf-traces develop basipetally.   Each vascular strand grows downwards in the stem from a node, and at the same time extends upwards into the leaf to which it belongs.

Turning, finally, to the development of the individual elements of vascular bundles, we may first of all consider the **vessels**, which arise by the fusion of longitudinal rows of meristematic cells.   Strasburger has shown, in the case of *Begonia dioica* and *Impatiens glandulosa*, that the transverse walls in those rows of cells swell up at an early stage of development, but that they do not become entirely obliterated until the thickening of the longitudinal walls is completed.   Indeed a marginal strip of the transverse wall always persists in the form of a narrow annular ridge.   The protoplasts of adjacent vessel-segments do not coalesce (Fig. 158).   As the thickening of the longitudinal wall progresses, each protoplast dwindles more and more, and finally disappears, nucleus and all.   Th. Lange [183] asserts that neighbouring protoplasts often fuse after the intervening septum is obliterated (e.g. in *Tilia, Malva, Hippuris, Fraxinus, Plantago, Cucurbita, Helianthus*); the same observer has shown that in many cases (*Cuscuta, Fraxinus, Secale, Hordeum, Triticum, Pinus laricio, Larix*, foliar pulvini of *Malva*, etc.)

the vessels (and tracheides) still contain living protoplasm at a very advanced stage of differentiation.

The development of **bordered pits** has been studied in detail by Sanio, Russow and Strasburger, especially in *Pinus sylvestris* and certain other Conifers. The formation of the border is preceded by the appearance of a wide "primordial pit," in the centre of which the torus is laid down as a circular thickened patch. At the same time the closing membrane of this pit bulges towards one side in a peculiar manner, its cross-section at this stage being compared by Russow to the shape of the Greek letter $\zeta$. Shortly afterwards the border is deposited on both sides of the primordial pit. As seen in radial section, the border-membrane first appears as a narrow, sharply defined ring, which rapidly grows in width. This annular membrane extends in an oblique direction towards the interior of the cell, thus gradually contracting the opening which constitutes the pit cavity, until the horizontal projection of its margin coincides with the edge of the torus. The development of the pit is then practically completed.

The ontogeny of **sieve-plates** [184] has been carefully examined by Russow, A. Fischer, Strasburger, Lecomte and A. W. Hill, among others. Each sieve-plate is derived from the closing membrane of a wide, shallow primordial pit, which occupies practically the whole of the transverse wall, if the latter is horizontal or only slightly oblique. The closing membrane of this pit becomes thickened in a reticulate manner. The unthickened portions or "sieve-fields" are pierced by protoplasmic connecting threads, as was first shown by Russow. Each sieve-field may be traversed by a single connecting thread (as in *Vitis vinifera*) or by several (as in *Wistaria sinensis* and *Cucurbita Pepo*). Both sides of the sieve-plate then become invested by callus, which stains readily with aniline blue. It is only in Angiosperms that the closing membranes of the sieve-fields disappear; according to Hill, the obliteration begins around the protoplasmic connecting threads, which are subsequently converted into "slime-strings." The narrow canals in which the slime-strings are contained gradually become dilated; if several are present to start with in the same sieve-field, they ultimately coalesce with one another. As a final result, the sieve-fields disappear altogether.

# CHAPTER VIII.

## THE STORAGE SYSTEM.

### *I. GENERAL CONSIDERATIONS.*

As a general rule, synthetic products—employing the term in its widest sense—are not made use of by the plant immediately after their formation. Even in the case of substances that are destined to be used up by the cells in which they are manufactured, consumption does not usually follow hard upon production. For the internal and external conditions controlling the activity of synthetic processes are by no means identical with the factors determining the utilisation of the product of synthesis, among which "growth" holds a prominent place; consequently a more or less extensive interval of time almost always intervenes between the production of plastic materials and their consumption. This fact may be readily verified even among simple filamentous Algae, such as *Spirogyra* or *Ulothrix*, which are photo synthetically active in the daytime, and undergo cell-division at night · evidently the plastic materials synthesised by day with the aid of sun light, are used up during the night in the formation of new cell-walls, or in other ways. Among the Higher Plants the condition of affairs is naturally more complicated. Here manufacture and consumption take place not only at different times, but also at different places. Now, it is the separation of these two processes in space that necessitates the translocation of synthetic products : similarly, their separation in time entails storage of these substances on a more or less extensive scale.

The greatest possible variety prevails with regard to the parts of the plant-body that are utilised for storage. The simplest plan consists in the deposition of plastic material at the place where it is formed: this condition is exemplified by the production of starch-grains in the interior of chloroplasts. In other instances the plastic substances are deposited, at points far removed from their places of origin, in cells of the most varied kind, including bast-fibres, certain elements of the conducting system—especially conducting parenchyma,—the subsidiary cells of stomata, etc. In the majority of such instances—for example, in the case of bast-fibres which contain starch-grains,—storage is most naturally regarded as a subsidiary function of the tissue or tissue-system involved. Sometimes the tissue or system may be considered as having undergone a temporary change of function; this is perhaps the correct interpretation of the presence of starch in the hadrome-parenchyma and the medullary rays. Strictly speaking, this aspect of storage does not fall within the scope of the present chapter. We shall confine our attention to the most typical form of storage, namely, that which involves the differentiation of a special **storage-tissue**; no tissue should therefore be termed a storage-tissue unless its *principal* function is the accumulation of a store of reserve materials.

It has already been stated, that the stored-up synthetic products of the plant consist in the main of **plastic materials**, reserved for consumption during succeeding phases of growth. These substances constitute **reserve-materials**, in the strict sense, such as may be met with more particularly in seeds, fruits, bulbs, tubers, rhizomes, and other organs of propagation or hibernation. In other cases, the accumulated synthetic products may be largely or exclusively utilised as **respiratory materials**. Further, storage is not always connected with strictly physiological requirements; ecological considerations may also lead to the deposition of reserve-materials in particular cells or tissues. Thus, various organic compounds are deposited in succulent fruits, in floral organs and so forth, in order to provide an attraction for animals, which in return perform some service on behalf of the plant in connection with seed-dispersal, cross-pollination or protection. Tissues and structural features which are concerned with ecological relations of this nature, may be included in a comprehensive definition of the storage system; they will be dealt with in a special section at the end of the present chapter.

Many propagative organs, such as bulbs and tubers, contain not only a supply of synthetic products but also a store of water, so that they are able to resume active growth even in a dry soil. Every gardener is familiar with the fact that many bulbs will sprout, even when they are hung up in a dry atmosphere. In such cases as these, the stored-up water is just as much a reserve-material as the accumulated supplies of starch or protein; for, in the building up of the plant-body, a sufficiency of water is no less indispensable than a proper supply of plastic material.

The acquisition, by one means or another, of a store of water may be of advantage to a plant in its fully developed state, as well as at the start of its vegetative life. Many xerophytes, in fact, depend for their very existence upon the possession of special water-reservoirs. "Water-tissues," and the other histological arrangements which subserve a similar purpose, are usually located in the immediate neighbourhood of the photosynthetic system, because this system is more liable than any other to suffer injury through failure of the water-supply. On the other hand, provision must also be made for the ready access of water from the conducting channels to partially or entirely depleted reservoirs.

After these preliminary remarks, we may proceed to characterise and define the **storage system** somewhat more formally. The storage system, then, may be said to comprise, in the first place, all those tissues which have as their principal function the storage of synthetic products, that will subsequently be utilised in connection with growth, or for other physiological or ecological purposes; it further includes the various structures which serve as reservoirs of water.

In view of the great diversity of reserve-materials, but little can be said concerning the general character of the storage system. Most storage-tissues, however, are, in accordance with their function, mainly composed of large-celled parenchyma; in addition they often display various features which tend to accelerate the entrance and removal of the reserve-materials.

## II. THE STORAGE OF WATER.

### A. WATER-TISSUES.[185]

In many plants of warm climates the uninterrupted activity of the photosynthetic organs is ensured by the presence of layers of water-storing cells, which constitute the so-called **water-tissue** of Pfitzer. This specialisation is particularly characteristic of xerophilous species that grow on rocky ground, or which lead an epiphytic existence, such as the BROMELIACEAE, many ORCHIDACEAE, and the members of genera like *Peperomia*, *Begonia*, and *Tradescantia*. In the tropics, the need for water-storing arrangements of this kind may even be felt by hygrophilous plants, especially if their leaves are large and delicate, in which case transpiration may at times attain to enormous proportions under the influence of the prevailing intense insolation. This condition is exemplified by *Canna*, *Maranta*, and other genera of SCITAMINEAE, by *Ficus elastica*, *Conocephalus ovatus*, *Euphorbia thymifolia*, and by many Palms. Many littoral plants, including the semi-aquatic Mangroves, are likewise provided with well-developed water-tissues. In these latter cases, the succulence of the leaves is regarded by Schimper as a safeguard against excessive loss of water by transpiration and consequent diminution of photosynthetic activity owing to accumulation of

*sodium chloride* in the leaves ; whether this is the correct explanation
or not, is still an open question.

Water-tissues always consist of living parenchymatous cells, which
may reach a very considerable size.  Apart from a thin peripheral
layer of protoplasm, in which are embedded the nucleus and a few
chloroplasts, the cell-cavities are entirely filled with a clear, watery—
or, in the case of deep-seated water-tissues, often somewhat slimy—
liquid.  The cell-walls are typically thin, and devoid of pits, but in
some cases develop a slight collenchymatous thickening along the
edges of the cells.  In the leaf of *Peperomia incana*, the layer of water-
tissue immediately above the photosynthetic tissue has the edges of its
cells thickened to a quite unusual extent.  These thickened strips of
cell-wall form an elegant, resistant framework, which completely roofs
in the thin layer of photosynthetic tissue.  The latter is thus preserved from deformation, even when a large amount of water escapes
from the leaf.

As regards their shape, the cells of water-tissues are in general
approximately isodiametric.  If, as sometimes happens, they are distinctly elongated (*e.g.* in *Hohenbergia strobilacea*), the orientation of their
longitudinal axes is such as to diminish the distance which the water
has to traverse on its way to the photosynthetic tissue.  In flat leaves
the cells of the water-tissue often assume a palisade-like form for
a similar reason (*Carapa moluccensis*, *Rhizophora mucronata*).

According to their situation, water-tissues may be described as
peripheral or internal.  **Peripheral water-tissues**, which may be regarded
as elaborations of the epidermal water-storing system (cf. p. 114), are
most frequently found in ordinary flattened leaves, where they are
situated close to one or other surface.  In dorsi-ventral leaves the
water-tissue is mainly or exclusively developed on the adaxial side.
**Internal water-tissues**, on the other hand, are characteristic of those
photosynthetic organs which are specialised, even as regards their
external form, for life under xerophytic conditions.  The CACTACEAE,
and the genera *Agave*, *Aloe* and *Mesembryanthemum*, provide excellent illustrations of this type of structure.  Both forms of water-
tissue are generally located in close proximity to the photosynthetic
system ; the surface of contact between the two sets of tissues
also invariably tends to be as large as possible.  In the leaf
of *Maranta arundinacea*, for instance, the water-tissue consists of
a layer of vertically elongated cells—situated immediately beneath
the adaxial epidermis—with obtusely conical lower ends, which project
into the underlying palisade-tissue.  The photosynthetic system and
the water-tissue are still more closely dovetailed together in the leaf
of *Hohenbergia strobilacea* ; the resulting enlargement of the surface of

contact between the two systems cannot fail to facilitate the emptying and filling of the water-reservoirs. A similar physiological significance doubtless attaches to the wedge-shaped masses of water-tissue observed by Lippisch in the leaves of *Ravenala madagascariensis* and of certain species of *Strelitzia* (*S. farinosa, S. Reginae, S. alba*); the masses in question consist of colourless palisade-shaped cells, and represent local intrusions of the adaxial water-tissue into the photosynthetic system.

The quantitative development of the water-tissue varies within wide limits. One end of this extensive and finely graduated scale is exem plified by certain epidermal layers, which reveal the accentuation of their water-storing function by an increase in height, or by a tendency to undergo tangential division; the opposite extreme is illustrated by those succulent photosynthetic organs in which the massive water-tissue enormously exceeds the photosynthetic tissue in thickness. It is an interesting feature of various leaves that the margin, which is the region most exposed to damage by excessive loss of water, is provided with a local water-tissue, composed of a varying number of layers, although water-storing arrangements are altogether absent, or, at any rate, far less conspicuously developed in the rest of the leaf. Such marginal water-tissues occur, according to Hintz, in various species of *Acacia* (*A. leprosa, A. salicifolia, A. longifolia,* etc.) and *Quercus* (*Q. pedunculata, Q. macrocarpa, Q. Ilex, Q. coccifera,* and *Q. Suber*), in *Ilex Aquifolium, Hakea eucalyptoides,* etc. Lippisch states that in *Musa* the margin of the young leaf is wholly composed of water-storing cells, but that this marginal tissue shrivels up later on, when the leaf-blade becomes broken up into numerous parallel strips. Schimper has made a remarkable observation regarding the quantitative development of the water-tissue in the epiphytic species of *Peperomia* and in certain GESNERACEAE. In these plants the water-storing cells elongate, and the whole water-tissue consequently increases greatly in thickness, as the leaves grow older. In *Codonanthe Devosii* (GESNERACEAE), for instance, Schimper found that, while the average thickness of a middle-aged leaf is 2·5 mm., old leaves, which are just beginning to turn yellow, may be as much as 5 mm. thick. This very appreciable difference depends entirely upon the greater thickness of the water-tissue in the older leaves. Schimper has further succeeded in proving experimentally that the older leaves actually serve as water-reservoirs, which are drawn upon by the younger leaves. The author has observed a similar relation in *Rhizophora mucronata*: on shoots of this Mangrove gathered by him on the shores of the coral island of Edam, near Batavia, the old, yellow leaves were exactly twice as thick as leaves which had just reached their full size. The following table of measurements shows

that the greater thickness of the old leaves is, in this case also, entirely
due to enlargement of the water-tissue.

| | Adult (green) leaf. | Old (yellowish-green) leaf. |
|---|---|---|
| Thickness of the photosynthetic tissue (including the lower epidermis - | ·426 mm. | ·426 mm. |
| Thickness of the water-tissue (including the upper epidermis) - - | ·355 ,, | 1·037 ,, |
| Total thickness of the leaf - | ·781 ,, | 1·463 ,, |

If, moreover, a severed shoot of *Rhizophora mucronata* is left to
transpire without any external supply of water, the old leaves contract
very noticeably and acquire a wrinkled surface in the course of a few
days ; the turgidity of the younger leaves, on the other hand—excepting those very young ones which are not yet fully developed—is
scarcely affected, and their surfaces consequently remain perfectly
smooth. The result of this experiment indicates very clearly that the
old leaves act principally as reservoirs of water.

Water-tissues are by no means restricted to leaf-blades and other
photosynthetic organs. In many epiphytic orchids, for example, storage
of water takes place in " pseudo-bulbs," which may consist of one or
several internodes. Water-storing tubers of varying dimensions are also
found among the *Rubiaceae, Vacciniaceae* and *Melastomaceae,* and in
species of *Gesnera.* One of the most familiar examples of a water-
storing tuber is the common potato ; this instance may also serve to
illustrate the fact that in tubers—as in bulbs—both water and plastic
materials are generally stored in one and the same tissue.

The largest known water-storing organs are represented by the
fleshy tuberous stems of *Hydnophytum* and *Myrmecodia,* two myrmeco-
philous and epiphytic genera of *Rubiaceae,* which inhabit the Malay
Archipelago, New Guinea and Southern Australia. In these plants
the tuber consists of the enlarged hypocotyl. In *Myrmecodia* it is
about 30 cm. long and 20 cm. thick, while in *Hydnophytum tortuosum,*
according to Beccari, it may attain a diameter of 60 cm. It consists
of succulent water-tissue traversed by a complicated system of cavities
and passages which are all continuous with one another. In the case
of *Myrmecodia* this system communicates with the exterior through a
large aperture situated among the roots on the side next the substratum ; smaller lateral openings are also present. The walls of the
cavities are clothed, like the outer surface of the tuber, with a layer of
cork, which is covered with numerous whitish tubercles, interpreted
by Trenh as organs of ventilation (lenticels). The cavities are occupied
by hordes of ants which strongly resent any intrusion upon the privacy
of their dwelling-place ; it is, however, difficult to imagine what enemies
of the plant could be kept at bay by these guardians, and there seems,

therefore, little or no justification for the widespread assumption that the plants in question are "myrmecophytes." Trenh has, in fact, shown by means of culture-experiments, that both the tubers, and the cavities and passages which they contain, develop quite normally in the entire absence of ants; he therefore quite properly regards these remarkable organs as enormous water-reservoirs, and interprets the system of passages as an arrangement for providing the massive and constantly growing structure with a sufficiency of oxygen [for purposes of respiration].

The physiological significance of water-tissues was recognised long ago by Pfitzer. Westermaier was, however, the first to approach the subject from the experimental point of view. The three succeeding paragraphs reproduce the principal results of Westermaier's researches, and at the same time summarise the leading physiological features of epidermal water-tissues in general.

1. If a leaf provided with a water-tissue is allowed to dry up gradually, the effects of the loss of water first become apparent in the water-storing cells themselves. These are found to have undergone a considerable amount of contraction, at a time when the photosynthetic tissue shows little or no signs of a shortage of water. This faculty of rendering up its store of water with ease and celerity is, of course, the most essential condition of efficiency in the case of any water-storing tissue.

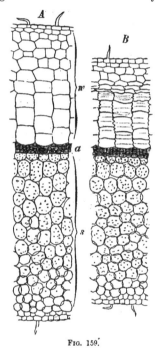

Fig. 159.

T.S. through leaves of *Peperomia tricho-carpa*. *A*. Leaf in the fresh condition. *B*. Severed leaf, which has been trans-piring for four days at 18°-20° C.; *w*, water-tissue; *a*, photosynthetic tissue; *s*. sPongy mesophyll.

2. As water is withdrawn from the storage-cells, they gradually contract and collapse, and their thin radial walls are thrown into folds or undulations (Fig. 159 B). This change of form is connected with the circumstance that the water lost is not replaced by air; the entrance of air would indeed be incompatible with the persistence of a living peripheral layer of protoplasm in the cells. The unthickened condition of the radial walls, which admits of this bellows-like contraetion and expansion of the entire water-tissue, is thus seen to be an anatomical feature of the first importance.

3. When the contracted water-tissue is once more plentifully

supplied with water, it quickly regains its former dimensions; the folded radial walls stretch out again under the influence of the turgor-pressure developed as a result of the reabsorption of water. This power of recovery after repeated depletion naturally also constitutes one of the important qualifications of an efficient water-tissue.

The above concise statements may be supplemented by a few explanatory remarks. The depletion of a water-tissue in time of drought involves two quite distinct processes. A comparatively small amount of the accumulated water escapes in the form of vapour, partly through the outer epidermal walls, and partly by way of the ventilating spaces which are present in all water tissues, although their development is there relatively feeble. A far larger proportion of the water removed is absorbed by the photosynthetic system, by virtue of the higher osmotic pressure which prevails in the green cells; the latter are thus enabled to cover the loss which they suffer by transpiration, at the expense of the water-storing cells, for a very considerable period. The protection which the water-tissue affords to the photosynthetic system, when the supply of water is insufficient, accordingly serves not only to preserve the latter from fatal injury through desiccation, but also to prevent any temporary interruption of photosynthetic activity through shortage of water. So long as plenty of water is available, the green parenchyma can obtain all the water and mineral salts that it requires from the vascular bundles; but if this normal source of supply fails for any reason, the photosynthetic cells can still fall back upon the store accumulated in the water-tissue.

Contraction of the epidermal water-tissue does not at first entail any folding of the radial cell-walls. To begin with, namely, the distended walls contract elastically, as the pressure in the protoplasts diminishes. If a suitable leaf, or a transverse section thereof, be immersed in a solution of common salt of sufficient strength, the turgor-pressure of the water-tissue cells will be completely neutralised, and the elastic tension in the cell-walls relieved; by means of this plasmolytic method, it is possible to determine, with a fair amount of accuracy, the extent to which the collapse of the water-tissue is referable to the mere elastic contraction of its radial walls. A number of experiments were carried out in this way by the author upon leaves of *Peperomia trichocarpa*; in every case, measurement revealed a radial contraction equivalent to nearly 6 per cent. of the total thickness of the water-tissue, when the turgor-pressure had fallen to zero. It is quite clear, therefore, that the mechanism of contraction is not the same for small variations of water-content, as it is for a serious loss of water.

According to Schwendener, a very unusual type of water-tissue is found in the MARANTACEAE, occupying the periphery of the heliotropi-

cally sensitive pulvini which form the distal termination of the petioles.
The tissue in question takes the form of a continuous layer of elongated
cells, with their long axes directed as a rule obliquely upwards [and
outwards]; the lateral walls of these cells show no signs of folding
when water is removed, but always remain fully extended.  Clearly,
the mechanism that permits of a diminution in volume, when water
escapes, cannot be the same here as in the case of ordinary water-
tissues.   In the present instance it is found, that the angle which the
longitudinal axes of the water-tissue cells forms with the longitudinal
axis of the whole pulvinus, diminishes when water is removed.  As a
result, the peripheral ends of these cells move a little nearer to the
centre of the petiole, and the radial diameter of the water-tissue is
correspondingly reduced.  At the same time the individual cells undergo
a certain amount of radial contraction.

Some Bryophytes possess water-tissues, which, however, often serve
for the storage of plastic substances as well.  Among Liverworts, the
MARCHANTIACEAE call for special mention in this connection; here the
colourless large-celled tissue situated immediately beneath the photosynthetic air-chambers is mainly utilised for water-storage.  In Mosses
the occurrence of water-tissues is restricted to the sporogonia.  In this
group of plants, the leafy shoots of the gametophyte generation have
adapted themselves to temporary failure of the water-supply in the
most straightforward manner; for they are able to undergo complete
desiccation, without suffering any permanent ill effects.  The sporogonial
water-tissue is sometimes sub-epidermal, in which case it may be located
either in the capsule-wall (*Funaria hygrometrica*), or in the apophysis
(*Webera nutans*); in other instances there is an internal water-tissue,
situated in the columella, which, however, often also contains starch,
especially when the capsule is young.

### B.  MUCILAGINOUS WATER-STORING CELLS.[186]

Vegetable mucilage subserves a great variety of physiological and
ecological purposes.  ‿ In the present section, however, we shall confine
our attention to such mucilaginous cells and tissues as can act as
water-reservoirs, owing to the fact that the more or less viscid mucilage
which they contain has a stronger attraction for water than ordinary
cell-sap.  Where the mucilage forms part of the cell-contents, as, for
example, in various succulents (*Aloe*, *Agave*, CACTACEAE, etc.) in bulbs
and in the tuberous roots of certain Orchids, it is difficult to draw a
sharp line of distinction between mucilage-tissues and typical water-
tissues, for the simple reason that the latter also frequently contain
mucilaginous cell-sap. ‿ Separate consideration must, however, be
given to those instances in which the mucilage is derived from

the cell-walls, certain of the secondary thickening layers being mucilaginous in character, and hence specially capable of retaining water. Nevertheless, mucilage-cells of this latter type are not treated separately here on account of the morphological relations of the mucilage, since these are of no particular importance from the anatomico-physiological point of view. The distinguishing characteristics of these tissues is, on the contrary, of a physiological nature, and consists in the fact that the absorption of water by the cells of typical water-tissues is directly dependent upon the presence of a living protoplast—or at any rate a plasmatic membrane—whereas this is not so in the case of typical mucilage-tissues (with mucilage derived from cell-walls).

Where a mass of mucilage consists of greatly thickened cell-walls, it is often very distinctly stratified ; the mucilaginous thickening layers exhibit their characteristic physical and chemical properties—as far as our observations extend—from the moment that they are laid down [i.e. they do not acquire these properties by secondary modification]. As a rule the primary layers of the wall do not become mucilaginous, unless the mucilage-cells are collected into groups or rows ; in this case the primary layers often break down and disappear, so that extensive cavities (TILIACEAE) or long ducts (*Fegatella*) are produced. The mucilaginous thickening layers may be laid down uniformly over the whole extent of the cell-wall, but in other instances are confined to a limited area; in the latter event, the cell-cavities often become reduced to mere slits, and are generally pushed very much to one side.

Attention has already been directed, on a previous occasion (p. 114), to the epidermal cells with thickened mucilaginous inner walls which occur in a variety of plants ; otherwise mucilage cells, in leaves, and in vegetative organs generally, are almost always idioblastic in character, whether they are solitary or collected into groups or rows. A few leaves (*e.g.* those of *Conocephalus ovatus* and *Rhizophora mucronata*) contain typical mucilage-cells *in addition* to a characteristic water-tissue devoid of mucilage. In *Rhizophora mucronata* elongated mucilage-cells, with the excentric type of thickening, are found at the boundary of the water-tissue and the palisade-layer, projecting for some distance into the latter.

The bark of a number of desert-plants (*Haloxylon, Eurotia, Calligonum, Halimodendron*) contains groups of mucilage-cells, which, according to their discoverer, Jönsson, arise from a phellogen, and hence receive the name of " mucilage-cork." [187] These cells sooner or later burst through the overlying layers of ordinary cork, and thus come to the surface ; thenceforth they act not only as water-storing elements, but also as organs of water-absorption.

A further instance of water-storage is furnished by the so-called mucilaginous cell-layers[188] which form a characteristic component of many pericarps and seed-coats. These mucilaginous layers are in the first instance designed to prevent the germinating seed from drying up, and are accordingly on the whole characteristic of xerophytes; they occur, for example, in a great many CRUCIFERAE and LABIATAE. As a rule, it is the outermost layer of the pericarp or testa that consists of mucilage-cells; this is the case, for instance, in *Linum*, *Salvia*, *Plantago*, etc., etc. This superficial location, with its attendant advantages as regards ready access of water and unlimited space for expansion, is, of course, highly appropriate in view of the functions which these cells perform. Occasionally, however, mucilaginous layers are also found in the interior of the testa, or on its inner surface.

The remainder of this section will be devoted to a more detailed consideration of the mucilage-tissues of pericarps and seed-coats in a few selected cases.

The mucilage-layer which forms the outer surface of the seed-coat of the Linseed (*Linum usitatissimum*), is often cited as a typical example of this form of mucilage-tissue. Each of the prismatic cells composing the mucilage-layer is separated from its neighbours by a thin, sharply-defined middle lamella, which terminates externally in the cutinised layers of the outer wall. The secondary thickening layers, both in the outer and the inner walls, possess the power of swelling to an enormous extent when wetted. If they are allowed to absorb water gradually, they at first exhibit a very evident stratification, which soon disappears again as the swelling proceeds. The thin inelastic middle lamellae cannot long resist the powerful tension to which they are subjected, owing to the pressure exerted by the swelling layers against the comparatively tough cutinised layers of the outer wall; sooner or later they give way, and the outer walls, which are intersected by numerous cracks, are thereupon lifted up by the prismatic masses of swelling mucilage. The outer walls thus form a sort of covering, which prevents the deliquescence of the mucilage from proceeding too rapidly.

The nutlets of *Salvia* are likewise provided with a superficial mucilage-layer, composed of prismatic cells, which are about twice as high as their width. In this instance the secondary thickening layers are deposited principally upon the lateral walls; hence, when swelling takes place, the results are somewhat different from those just described in the case of *Linum*. The middle lamellae separating adjacent cells do not break down, but form a delicate persistent framework, from the meshes of which the swelling secondary layers protrude in the form of long, tubular structures; these tubes of mucilage expand at their distal ends and finally become confluent with one another. The innermost (tertiary) thickening layer of the lateral walls takes no part in the swelling, but splits up, in each cell, into one or more spirally-twisted ribbons, which are drawn out by the expanding mucilage-tubes. The final result is a solid mass of mucilage containing numbers of these spiral ribbons with their coils drawn widely apart. The presence of such fibrous structures in mucilage-layers is by no means a peculiarity of the nutlets of *Salvia*; a similar arrangement prevails in the seed-coats of a great many plants, representing a variety of Natural Orders (*e.g.* in *Ocymum*, *Senecio*, *Collomia*, *Gilia*, *Iponopsis*, *Polemonium* [according to Unger]). The spiral fibres, therefore, presumably have some definite function to perform; in all probability they constitute a sort of skeleton or framework, which holds the mucilage together and thus prevents it from dissolving away too rapidly.

In addition to their principal function of preventing desiccation and of regulating the water-supply, such superficial mucilage-layers also assist in attaching the seed or fruit to the substratum.

### C. WATER-STORING TRACHEIDES.[189]

There is a special type of water-storing element, which is distinguished by the absence of a living protoplast, and by the presence of stiffening arrangements in the cell-wall, similar to those that are characteristic of vessels and tracheides. The reservoirs in question are filled with water injected into them from adjoining vascular bundles or from living parenchyma. When they are emptied, they do not collapse, like typical water-tissues, but become filled with rarefied air. These elements obviously resemble ordinary tracheides in many respects, though they differ markedly from the latter in their function —which is not conduction but storage—and in certain structural features correlated with this difference in function, such as their larger calibre and their frequent approximation to an isodiametric form. As a matter of fact, the cells in question have been termed **storage-tracheides** by Heinricher (Vesque's name of *reservoirs vasiformes* has the same significance).

In many plants, storage-tracheides only occur in the distal expansions of the bundle-ends; here the terminal portions of the water-conducting system are transformed into organs of water-storage (Fig. 160). The walls of the elements in question are more or less lignified, and are either strengthened with reticulate or spiral thickenings, or else are provided with numerous transversely elongated pits. Phylogenetically considered, these cells sometimes correspond to terminal tracheides of the vascular bundles, which have expanded and taken on a clavate ' or almost spherical form (*e.g.* in *Euphorbia biglandulosa*, *E. splendens*, *E. Myrsinites*, while in other cases, as Heinricher has shown (*Capparis spinosa, C. aegyptiaca*, etc., *Centaurea glomerata*), they represent modified parenchymatous elements pertaining to the parenchymatous bundle-sheaths, or even to the green mesophyll. At a more advanced stage of specialisation, the modification in question is not confined to the bundle-ends, but affects the whole extent of some of the minor vascular strands. In the leaves of *Astrolobium repandum*, for example, according to Heinricher, individual cells or cell-groups in the parenchymatous bundle-sheaths are transformed into storage-tracheides, the walls of which are furnished with transversely elongated

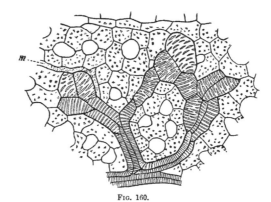

Fig. 160.

Bundle-ends with storage-tracheides in the leaf of *Euphorbia splendens*; *m*, portion of a latex-tube.

pits; the modified cells occur principally in the portion of the sheath adjoining the hadrome, or, in other words, in immediate proximity to the *conducting* tracheides. A similar state of things prevails in certain species of *Centaurea* (*C. scoparia, C. americana, C. senegalensis*). As regards the general distribution of storage-tracheides of this type, in leaves, it may be stated that they tend to occur more particularly in connection with the marginal veins.

Many plants that grow in dry, sunny situations are provided with storage-tracheides which have no immediate connection with the vascular system; the cells in question may be purely idioblastic in character, but in some cases form continuous layers, interpolated between the photosynthetic tissue and the epidermis. This latter condition is exemplified by species of *Pleurothallis* and *Physosiphon*. Examination of a transverse section through the leaf of *Physosiphon Landsbergii* (Fig. 161) reveals the presence of a typical water-tissue, situated beneath the adaxial epidermis, and comprising two layers of

cells with narrow pits on their walls.   Immediately below this water-
tissue lies a single layer of large palisade-shaped storage-tracheides,

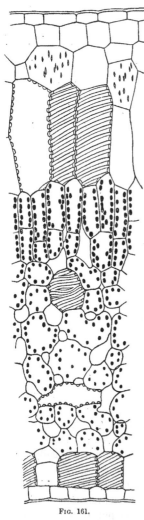

with spiral thickenings which alternate
on the two sides of each radial wall.
The continuity of this tracheidal layer
is frequently interrupted by water-tissue
cells, resembling the tracheides in shape,
and, like the latter, extending inwards
as far as the photosynthetic palisade-
layer.   The spongy mesophyll contains
numerous isodiametric or transversely
elongated water-storing tracheides.   Finally, there is a second continuous layer
of storage-tracheides situated next the
abaxial epidermis; like the adaxial layer,
it is interrupted, at a number of points,
by parenchymatous water-storing cells.

Most frequently, however, storage-
tracheides occur scattered singly throughout the mesophyll.   In the genus
*Reaumuria* (TAMARISCINEAE), for example,
both the palisade-tissue and the spongy
parenchyma of the needle-shaped leaves,
according to Vesque, contain numerous
" *reservoirs vasiformes* " which—as their
shape suggests—represent modified pali-
sade- or spongy cells.   Krüger has
described the storage-tracheides which
occur in the mesophyll or in the pseudo-
bulbs of certain epiphytic Orchids (*Liparis
filipes, Oncidium* spp.); the cells in question may be isodiametric or elongated,
and are usually provided with spiral
thickenings.   According to Kny and
Zimmermann, the characteristic tubular
" spiral cells " in the stem and leaf of

FIG. 161.

T.S. through a leaf of *Physosiphon
Landsbergii*, showing water-tissue and
storage-tracheides.

*Nepenthes*, which after active transpiration are filled with rarefied air or
water-vapour, likewise represent water-
reservoirs.   The " parenchymatous tracheides " discovered by Rothert in
the pith of *Cephalotaxus Koraiana* are probably of a similar nature;
they are furnished with irregular thickening fibres, and also with
bordered pits.

In vegetative organs, the intercellular spaces almost always serve for ventilation ; in a few cases, however, they are utilised for storage of water. This very exceptional condition is exemplified, according to Schimper, by the epiphytic Aroid *Philodendron cannaefolium.* The swollen spindle-shaped petioles of this plant contain large intercellular spaces which, during wet weather, are entirely filled—apart from minute air-bubbles—with a liquid of mucilaginous consistency. When the external water-supply fails, the liquid gradually disappears from these intercellular spaces, being transferred, as Schimper's experiments show, to the transpiring leaf-blades.

It is more usual for cell-layers containing numerous intercellular spaces to be employed for purposes of water-storage in pericarps and seed-coats. The hard central portion of the fruit of *Poterium spinosum,* for instance, is enveloped in a thick layer of tissue which in the dry state is full of air. According to Klebs, this layer consists of "stellate" parenchymatous tissue containing numerous large intercellular spaces ; when the fruit is wetted, both the cell-cavities and the intercellular spaces of the stellate tissue become filled with water. Heinricher states that, in *Adlumia cirrhosa,* the inner epidermis of the pericarp is converted into an open framework with wide intercellular meshes. The cell-walls are lignified and provided with numerous pits ; when this layer is wetted, all the interstices of the meshwork—intercellular spaces as well as cell-cavities—become filled with water.

## III. STORAGE OF PLASTIC SUBSTANCES.

### A. THE VARIOUS CLASSES OF RESERVE-MATERIALS.

1. *Non-nitrogenous reserve materials.*—The non-nitrogenous reserve materials that are deposited in storage-tissues, consist either of **carbohydrates** or of **fatty oils.** Reserve carbohydrates frequently occur in the solid form of starch or cellulose ; some of them, such as inuline or the various sugars, are contained in solution in the cell-sap.

**Starch**[191] generally takes the form of polyhedral or rounded granules, which vary in size from macroscopic to almost ultra-microscopic dimensions. The shape of the individual grains varies greatly in different plants, but is usually constant for any given species. The starch-grains of the Pea, the Bean, and other *Leguminosae* are ellipsoidal, with a central hilum ; those of Wheat and Rye are lenticular, while those of the Potato are ovoid in shape and excentric. These rounded shapes only occur where the starch-grains have room to expand freely in all directions, and thus do not interfere with one another's development.

Where the grains are closely crowded together, on the other hand, they tend to assume polyhedral shapes as a result of mutual interference. In the endosperm of Indian Corn, for example, the starch-grains of the " horny " endosperm are all polyhedral, whereas in the "mealy " portion rounded grains predominate.

In addition to the ordinary or simple starch-grains, various compound forms also exist ; these often resemble ordinary grains in shape, but on closer examination are found to be made up of two or more— sometimes of several hundreds—of partial grains. Even where the starch-grains are normally simple, they are almost invariably accompanied by a certain number of compound grains (e.g. in Potato, Wheat, etc.). Conversely, the compound forms are numerous, and the simple grains few and scattered, in the endosperm of Oat and Rice, and in many other cases. Compound grains may be further classified under the headings of completely compound and partially compound forms. In the former type the planes of separation between the component grains extend to the surface ; in the latter case, the partial grains are enveloped in a certain number of common layers.

As regards their chemical composition, starch-grains consist of several different carbohydrates—all represented by the empirical formula $(C_6H_{10}O_5)_n$—together with water and a small amount of mineral substance. According to Nägeli, every starch-grain is principally composed of *granulose*; this is the substance that is responsible for the familiar blue iodine-reaction of starch. The granulose is contained in the meshes of a delicate framework of " starch-cellulose "; this persists as a skeleton, when the granulose is removed by the action of solvents, such as saliva or dilute mineral acid. A. Meyer, on the contrary, believes that the majority of starch-grains consist of a single substance which he terms *amylose*; he distinguishes between two varieties of this body, namely, β-*amylose*, which deliquesces in contact with boiling water, and a-*amylose*, which remains solid under the same conditions. The β-compound corresponds to Nägeli's *granulose*, the a-variety to his *starch-cellulose*. Opinions differ as to the nature of the skeleton which is left, when a starch-grain breaks up under the influence of saliva or mineral acid. The starch-grains of certain forms of Rice and Millet (*Oryza sativa*, var. *glutinosa* and *Sorghum vulgare*, var. *glutinosum*) assume a reddish hue, in place of the normal blue coloration, on treatment with iodine. According to Meyer, these abnormal starch-grains consist principally of *amylodextrine*, a [hydrolytic] derivative of amylose ; Bütschli, on the other hand, assumes that a special amylaceous carbohydrate, *amyloerythrine*, is present in such cases.

With regard to the physical structure of starch-grains, it may first of all be stated that each grain consists of a number—usually a large number—of layers which are laid down either concentrically or excentrically around a morphological centre or hilum. In intact grains—even in those which are very excentric in structure—all the layers are complete. Interruption of one or more layers at any point is a sure indication of a preceding partial solution of the grain. The stratified appearance of starch-grains is due, as Nägeli first recognised, to differences in the density of successive layers, strata of low water content, and therefore of high density, alternating with layers of low density, which contain a larger proportion of water.

Different botanists hold very divergent views concerning the ultimate structure of starch-grains. The majority, including A. F. W. Schimper and A. Meyer, regard these bodies as sphaerocrystals (sphaerites) possessed of a radially fibrous structure, or in other words, composed of numerous exceedingly minute radially arranged needle-shaped crystals (trichites). In A. Meyer's opinion, the stratification of the grain is due to the varying thickness and number, in successive layers, of the more or less richly branched trichites. The alternation of layers of loose and dense texture is further regarded as a consequence of periodic fluctuations in the supply of material to the grain during its development. When the chromatophore (amyloplast) in which the grain is being laid down, is well supplied with sugar, it produces a large amount of starch-substance, and the corresponding stratum is therefore relatively dense; under the opposite conditions a loosely constructed layer is formed. Working with cuttings of *Pellionia Daveuana*, which were first starved [by darkening] and then once more exposed to bright light, A. Meyer endeavoured to keep count of the new layers laid down, during the second stage of the experiment, upon the starch grains which had been partially dissolved during the period of starvation. He came to the conclusion that a thick, dense layer was formed as the result of each diurnal period of illumination, while the nocturnal intervals of darkness were represented by thin and relatively watery strata. Whether this interpretation is the correct one or not, can scarcely be decided in the present state of our knowledge. It may, however, be stated that Hugo Fischer has recently contested Meyer's general position. He throws doubt upon the theory of the sphaerocrystalline structure of starch-grains, and suggests that some of the water in the layers of low density is contained in radial crevices, which are obliterated by contraction when the grain is dried. Bütschli, finally, believes that starch-grains have an alveolar or foam-like structure, similar to that which he ascribes to protoplasm.

It is probable that starch-grains never arise otherwise than in connection with chromatophores; the establishment of this important fact is mainly due to the researches of Schimper. In the case of storage-tissues, it is chiefly the amyloplasts that act as starch-formers. Every starch-grain develops in the *interior* of a chromatophore; whether it remains permanently enclosed in a layer of chromatophore substance, as A. Meyer supposes, is not so certain. In many cases, appearances suggest that a portion of the grain may burst through the chromato-phore-envelope. Schimper has shown that the shape of the grain and the character of its stratification vary, according as the envelope is uniformly distributed or not; in other words, they depend upon the position of the developing grain within the chromatophore. If the grain arises at the centre of a spherical amyloplast, the successive strata will be laid down concentrically, because the conditions of growth are the same at every point of the surface. If, on the contrary, the point of origin of the grain lies near the surface of the amyloplast, so that the chromatophore-envelope is distributed unevenly around the young grain, the deposition of starch will be favoured on one side, and the stratification will consequently be excentric in character.

The swelling of starch-grains which may be artificially induced by treatment with water at a temperature of 60° to 70° C., with caustic potash or soda, etc., need not be considered here. Some attention must, however, be given to the natural process of solution, which takes place when storage-tissues are emptied of their starchy contents, or, more generally, when starch is transported from one place to another. This so-called " solution " is really a process of hydrolysis, effected with the aid of diastatic enzymes. It results in the transformation of the starch by successive stages into *amylodextrine, dextrine, isomaltose,* and, finally, *maltose.* The changes of form which starch-grains undergo during solution, vary considerably in different plants. When the grains are large and excentric, as in the Potato or in *Lilium candidum* (in the bulb-scales), they dissolve gradually from the surface inwards. The rate of removal of the starch does not, however, stand in any direct relation to the orientation of the layers, but takes place, to begin with, at the same rate at every point of the surface. As a result, the successive layers soon begin to " crop out " at the thicker end of the grain; the more watery among those exposed then begin to dissolve more rapidly, so that annular grooves are formed. Except in the very earliest stages of the process, therefore, solution goes on more rapidly at the thicker end of the grain. A dissolving starch-grain finally acquires the shape of a rod, with constrictions at frequent intervals. The starch-grains of the Bean and other LEGUMINOSAE are traversed by radial fissures. Here solution proceeds uniformly all over the surface until the cracks are laid open; the latter then rapidly enlarge, owing to the entrance of diastase, and the more watery central portion of the grain thus quickly becomes exposed to the action of the solvent. The starch-grains in the endosperm of Maize, Barley, Rye and other GRAMINEAE, also dissolve from the surface inwards, but not simultaneously at every point. On the contrary, solution begins at a number of more or less widely separated spots. The resulting pits and channels sooner or later intersect one another and ultimately break up the grain into several fragments, which thenceforward dissolve independently. The corroded channels are not of uniform width throughout; their outline, as seen in optical (longitudinal) section, appears notched or jagged, owing to the fact that the channels remain narrower where they cross the denser layers than where they traverse those which are relatively watery, and hence more easily attacked by diastase.

It was stated above, that carbohydrate material may be stored in the form of greatly thickened cell-walls.[192] Where this is the case, the storage-tissue often acquires a horny or even a bone-like consistency; a familiar illustration is provided by the endosperm of *Phytelephas macrocarpa*, which constitutes the so-called "vegetable ivory." Various other Palms (*e.g.* the Date-palm), many LILIACEAE (*Ornithogalum, Fritillaria*) and IRIDACEAE and certain other Monocotyledons are provided with a similar type of endosperm. In certain LEGUMINOSAE carbohydrate substances are stored in the form of mucilaginous layers, in the walls of endosperm cells. In other members of the family the membranes of the cotyledonary cells are thickened for the same purpose; a similar arrangement is found in *Impatiens Balsamina, Paeonia officinalis*, and *Tropaeolum*. Finally, it has been shown by Schaar that the thick-walled cells of which the bud-scales of the Ash are composed, represent a store of carbohydrate material destined for the nutrition of the young shoots in spring.

The carbohydrates which are stored in the form of thickened cell-walls, may all be included in the comprehensive category of **reserve-cellulose**. Chemically, they belong, according to E. Schulze, to the class of *hemi-celluloses*, which are distinguished by the fact that they can be hydrolysed with relative ease (cf. p. 47). The most widely distributed of these hemi-celluloses are the *mannans* and *galactans*, the former being especially characteristic of Palm-endosperms, the latter of Leguminous seeds. A peculiar form of reserve-cellulose, which resembles starch in giving a blue reaction with iodine alone [*i.e.* without previous treatment with acid], is known as *amyloid*; it occurs, for example, in *Impatiens Balsamina*, in *Tropaeolum majus*, and in many PRIMULACEAE. The types of reserve-cellulose which have been mentioned, certainly do not include all the carbohydrates that belong to this category. It is most probable that such hemi-celluloses occur in thickened cell-walls in the form of mixtures, or combined to form compounds of a higher order of complexity.

The mobilisation of reserve-cellulose during germination is an enzymatic process, which presents many points of resemblance to the hydrolysis of starch-grains. In *Impatiens Balsamina*, according to Heinricher, the thickening layers disappear gradually, but are often removed much more rapidly from certain parts of the wall than from

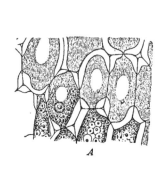
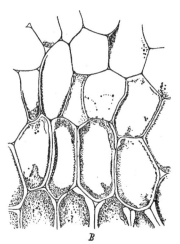

*A*

*B*

Fɪɢ. 162.

Cotyledonary storage-tissue of *Impatiens Balsamina*. *A*. Resting seed. *B*. Germinating seed. In *B* the secondary thickening layers of the walls, which consist of *amyloid*, have undergone partial solution. After Reiss (from Frank's *Text-book*).

others (Fig. 162 B). The middle lamellae remain intact, and finally form the thin partition-walls of the photosynthetic parenchyma into which the storage-tissue of the cotyledons becomes transformed, after it has been depleted of its reserve materials. In other cases (*Tetragonolobus, Lupinus, Goodia latifolia,* etc.), according to Nadelmann, the layers which are about to be dissolved, first of all acquire a radial striation. Then wedge-shaped fissures appear, which become enlarged by corrosion ; sooner or later the thickening layers dissolve, usually in a very irregular manner. Here also the middle lamella generally persists.

Among the carbohydrates which are stored in a state of solution in the cell-sap, the various sugars are by far the most important. *Cane-sugar*, for example, accumulates in the storage-tissue of the Beet-root at the end of the first vegetative season, while the fleshy bulb-scales of the Onion contain a mixture of *glucose* (grape-sugar) and other sugars. Almost all storage-organs, in fact, contain a certain amount of sugar.

The best known of the remaining soluble reserve carbohydrates is *inulin*,[193] which is so characteristic a chemical component in the COMPOSITAE; in the root-tubers of *Dahlia variabilis*, in *Helianthus annuus* and in *Inula Helenium*, this substance occurs in the form of a highly concentrated solution presenting the appearance of a mobile, pale yellow oil. When a freshly-prepared section of an inulin-containing tuber is treated with alcohol, the polysaccharide is thrown down as a finely granular precipitate. But if a whole tuber is kept in spirit for a long time, the slow entrance of the alcohol permits of the formation of large sphaerocrystals of inulin, which are usually aggregated into groups; when precipitated in this way, individual sphaerites often extend over a number of cells. Other carbohydrates which resemble inulin in their chemical relation to *fructose* (*triticin, graminin, scillin, irisin*), occur in various Monocotyledons.

The **fatty substances** which serve as reserve-materials, belong, like vegetable fats in general, to the class of glycerides; they are, in fact, glycerine-esters of various fatty acids (especially of *palmitic, stearic* and *oleic* acids). The majority of fats are liquid at ordinary temperatures (fatty oils); less frequently they are of the consistency of tallow or butter (*e.g.* cacao-butter and nutmeg-fat). They are readily soluble in ether, carbon disulphide and benzole; castor-oil is completely soluble in alcohol, even in the cold. In their natural condition the vegetable fats are never chemically pure substances; frequently they are accompanied by free fatty acids, and sometimes also by pigments (palm-oil), or by cholesterine (olive-oil and almond-oil).

When the amount of fat in a storage-cell is not very large, it is usually suspended in the protoplasm in the form of minute droplets or vesicles. In oily seeds, on the contrary, only a small quantity of the fat is present in this finely divided state; the greater proportion is contained in the interstices between the delicate meshwork of the cytoplasm on the one hand, and the so-called aleurone grains (see below, p. 417) on the other.

2. *Nitrogenous reserve-materials.*—Nitrogenous materials are almost always stored either in the form of **proteins** or of **amides** [and **amino acids**].[194] In the latter category are comprised *asparagin*—the most important of the vegetable amides—further, *glutamin, leucin, tyrosin*, etc. The reserve proteins include the various *globulins, vitellins* [*phytovitellins*] and *albumoses*, also certain *nucleoproteids*, and the members of the *gluten* group.

In succulent storage-organs—such as the majority of tubers, bulbs and rhizomes—the nitrogenous reserve-materials are for the most part dissolved in the cell-sap.[195] Thus, if a fairly thick section, taken from a mature potato tuber, is treated with alcohol, a bulky and somewhat granular precipitate is thrown down in the cell-sap of the starch-containing storage-cells; by means of appropriate reagents this precipitate may be shown to consist partly of amides (mainly in the form of *asparagin*) and partly of native proteins. In ripe potatoes, a large proportion (30-47 per cent.) of the total nitrogenous substance is present in the form of "amide-nitrogen." If a section from a bulb-scale of *Allium Cepa* is treated in the same way, only a slight flocculent or semi-granular precipitate appears in the majority of the cells; those parenchymatous storage elements, however, which immediately adjoin the leptome strands of the vascular bundles, become filled with a dense precipitate, which probably consists in the main of protein substance. The cells in question might in fact be regarded as members of a "protein-sheath" [comparable to the more widely distributed "starch-sheaths"]. The reserve amide-nitrogen of the Beet-root is principally made up of *glutamin*, which is accompanied by a small amount of *betain.* Generally speaking, a considerable proportion, and sometimes the greater part, of the reserve nitrogen of succulent storage-organs consists of soluble amides.

In dry storage-organs, on the contrary, such as seeds and many fruits, reserve proteins play so prominent a part, that non-albuminoid substances, as a rule, only make up between 2 per cent. and 10 per cent. of the total nitrogen. In these cases the protein-compounds occur in special forms, which constitute a very characteristic feature of quiescent storage-tissues. The most remarkable of these special forms of reserve protein are the so-called **crystalloids**.[196] These bodies are undoubtedly genuine crystals of protein, which possess all the essential features of ordinary crystals, but which in addition have the property of swelling in contact with water. According to Schimper, all protein-crystals belong either to the regular (*e.g.* the cubical crystalloids of the Potato) or to the hexagonal (*e.g.* the rhombohedral crystalloids in the endosperm of *Bertholletia excelsa*) system. While such crystalloids are principally found in dry storage-tissues, where they occur as inclusions of aleurone-grains, they are also quite frequently to be met with in succulent storage-organs (*e.g.* the Potato), or even in living tissues which are not specially concerned with storage. Zimmermann states that they are especially prevalent as inclusions of the nuclei in the OLEACEAE, SCROPHULARIACEAE and BIGNONIACEAE, and among Pterido-phyta; their frequent occurrence in chromatophores has been referred to on a previous occasion (p. 37). Crystalloids may also be freely suspended in the protoplasm, especially in epidermal cells and photo-synthetic elements. The wide distribution of these structures is scarcely surprising, in view of the fact that every living cell may accumulate a temporary store of plastic materials for its own use.

Various circumstances, in fact, point to the conclusion that protein crystalloids always represent reserve-materials, even when they are found in cells other than those belonging to the storage-system. From among the numerous observations bearing upon this question we may select those of Stock for special mention. Stock finds that crystalloids are dissolved in dying leaves, and that they also disappear from plants grown in nutrient solutions containing insufficient quantities of nitrogen.

**Aleurone-grains** (Fig. 163) constitute another characteristic form of reserve protein.[197] A typical aleurone-grain is a rounded granular body, the ground-substance of which, according to Tschirch and Kritzler, consists mainly of globulins soluble in water. In starchy seeds the aleurone-grains are usually small and lie crowded together in the interstices between the starch-grains and the protoplasmic meshwork. In fatty seeds, on the other hand, they are generally large, and often contain a variety of inclusions, such as crystals of *calcium-oxalate*, and the small rounded or botryoidal masses known as **globoids.** According to Pfeffer, a globoid consists of a *double phosphate of calcium and magnesium*, combined with certain organic substances. Very often one or more crystalloids also lie embedded in the amorphous matrix of the aleurone grains ; these become clearly visible when

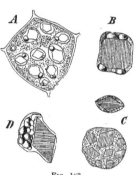

Fig. 163.

*A.* Cell with aleurone-grains from the endosperm of *Ricinus communis* (after treatment with alcoholic mercuric chloride solution). ×400. *B.* A single aleurone-grain from the endosperm of *Ricinus communis*, showing a crystalloid and four globoids. ×800. *C.* Aleurone-grain of *Elaeis guianensis*, mounted in oil. ×500. *D.* Aleurone-grain from the cotyledon of *Bertholletia excelsa*, fixed with alcoholic mercuric chloride and mounted in water. ×500. After Pfeffer.

the ground-substance is dissolved away by treatment with water (*e.g.* in *Ricinus communis* and other EUPHORBIACEAE).

Wakker and Werminski state that aleurone-grains are found ·in special vacuoles, which at first contain a highly concentrated solution of proteins, but dry up as the seed ripens ; during germination the aleurone-grains reabsorb water, and once more become converted into a corresponding number of vacuoles, which later coalesce to form a single sap-cavity.

How far other widely distributed plant-constituents, such as *glucosides* and *tannins*, may serve as plastic reserve-materials is a question which belongs to the field of pure physiology, and which consequently falls outside the scope of the present treatise.

### 1. *Tissues serving for the storage of plastic materials.*

In the tissues which serve for the reception of plastic substances, nitrogenous and non-nitrogenous compounds are deposited in the most varied combinations and proportions. The only further general statements that can be made with regard to this point are, first, that the non-nitrogenous reserve-materials usually exceed the nitrogenous substances in amount; and, secondly, that the various carbohydrates appear to be mutually interchangeable, while in certain cases they are altogether replaced by fats.

The equal value, from a nutritive point of view, of the various carbohydrates and fats is a point to which a little more consideration must be devoted; this equivalence may be deduced not only from the chemical changes attendant upon the germination of seeds, but also from a study of the comparative anatomy of storage-tissues. It frequently happens, for example, that different storage-organs of the same plant do not contain the same non-nitrogenous reserve-materials; in the same way totally different non-nitrogenous compounds may occur, in homologous storage-organs, in closely related species. Whereas, for instance, starch is stored in potatoes, inuline in the tubers of *Dahlia* and cane-sugar in Beet-root, the seeds of the two former plants contain oil, while those of the Beet are starchy. The grains of most Grasses contain starch; in a few instances, however, fatty oil is present instead (*Phragmites communis, Koeleria cristata*, etc.). In the case of *Impatiens Balsamina*, again, *amyloid* is stored in the cotyledons, in the form of enormously thickened cell-walls; in other species of *Impatiens* the cotyledonary tissue is quite thin-walled, and the place of reserve-cellulose is taken by oil. Similarly the bulb-scales of the Onion contain a large amount of glucose, whereas the Tulip and many other LILIACEAE accumulate starch in their bulbs.

From a purely physiological point of view, starch and oil are interchangeable; in given circumstances, however, one or the other may be preferable for ecological reasons. Fats contain a far higher proportion of carbon than starch or any other carbohydrate. Triolein, for instance [$C_3H_5(OC_{18}H_{33}O)_3$, the chief constituent of Olive oil], contains 77·4 per cent. by weight of carbon, as compared with the 44·4 per cent. of this element contained in the molecule of starch [$(C_6H_{10}O_5)_x$]. Since the specific gravity of starch is 1·56, while that of the fats lies between ·91 and ·96, a given volume of starch contains approximately the same quantity of carbon as an equal volume of fat, but is about 1·7 times as heavy as the latter. Fat thus represents a richer source of carbon than starch ; or, in other words, taking equal volumes and equal carbon contents, fat is much lighter than starch. The circumstance that the non-nitrogenous reserve-material of seeds consists far more frequently of fat than of starch, is probably to some extent connected with the fact that any reduction in ·the weight of a seed increases its chances of dispersal. This argument applies more particularly to seeds and fruits with parachute- or wing-arrangements, which are in fact almost invariably of the " oily " type. For the seeds of water-plants are mostly distributed by the aid of appropriate floating devices, which generally depend upon the presence of numerous large air-containing cells in the testa. The period of dispersal ends when these floats become water-logged ; such seeds will then most readily sink to the bottom (where, after a period of rest, their germination takes place), if their specific gravity is fairly high.

Bulky storage-organs, such as tubers, bulbs, rhizomes, and large seeds (spanish chestnut, horse-chestnut, acorn, bean), also generally accumulate their store of non-nitrogenous reserve-material in the form not of fat, but of starch. or some other carbohydrate. Here the underlying cause seems to be of a chemico-physiological nature. The quantity of oxygen absorbed during germination is much larger in the case of oily seeds than in that of seeds and fruits which contain starch. In the former case, oxygen is required, not merely for respiration, but also to a large extent for the oxidation of the fat, which has to be transformed into one or other of the more highly oxygenated carbohydrate substances before it can be translocated. As storage-tissues are, however, never provided with very abundant air-spaces, oxygen penetrates to their interior chiefly by diffusion from cell to cell ; the supply of oxygen to the deep-seated cells of a storage-tissue is thus most liable to become insufficient, where large quantities of this gas are used up at the time of germination. Hence, if a large proportion of the reserve-material of bulky storage-organs consisted of fat, it would be a difficult matter for the necessary oxygen to be absorbed with sufficient rapidity ; this would constitute a particularly serious disadvantage in the case of plants with rhizomes, tubers and bulbs, which are designed to produce their leafy shoots and flowers as soon as possible after the beginning of the vegetative season.

It has already been stated that carbohydrate reserve-material is most frequently stored either in the form of starch or in that of reserve-cellulose. The selection of one of these substances rather than the other, for this purpose, is again determined by ecological considerations. As a general rule, seeds which germinate quickly contain starch, which may either be present as such in the resting seed, or else may only appear during germination as a result of the oxidation of fat. Where the starch is stored in the endosperm, it is usual for the embryo to be attached to one side of the storage-tissue, or at most to be partially embedded in the latter. A familiar illustration of this relation of embryo to endosperm is furnished by the GRAMINEAE, and a similar state of things prevails in the CARYO-PHYLLACEAE, PORTULACEAE, POLYGONACEAE, CHENOPODIACEAE, etc. The peripheral, and hence exposed, situation of the embryo enables the latter to respond to comparatively slight external stimuli; like the starchy character of the reserve-material, the superficial location of the embryo seems to represent a device for hastening germination. In the case of seeds with a protracted period of germination, where the embryo withdraws the reserve-materials from the storage-tissue very gradually, the endosperm, when persistent, always surrounds the embryo completely, and moreover often possesses very thick cell-walls. In this instance, also, the advantages entailed by the character and arrangement of the reserve-materials are fairly evident. As compared with numerous starch grains, a thickened cell-membrane exposes a far smaller surface to the attack of hydrolysing enzymes. In seeds with thick-walled storage-tissues, therefore, the whole process of germination can be temporarily arrested and subsequently resumed with comparative ease; storage-organs of this type are also less likely to undergo decomposition, or to suffer from the attacks of insects and other animals, even when they remain buried for weeks and months. There can, in short, be little doubt that thickened cell-walls constitute a more appropriate form of carbohydrate reserve-material than starch-grains, where the process of germination is slow and liable to suffer frequent interruption. By taking into account the several methods of storage, and the various combinations of plastic materials belonging to different chemical classes, it is possible to distinguish a number of types of storage-tissue. These may be tabulated as follows ·

*A.* All the reserve-materials deposited within a single storage-tissue.

(i.) Reserve-materials restricted to the cell-cavities.

　1. Amides and proteins (dissolved in the cell-sap) *plus* sugar (dissolved in the cell-sap): Beet-root, Onion.

　2. Amides and proteins (dissolved in the cell-sap) *plus* starch: Potato. Many rhizomes.

　3. Aleurone-grains *plus* starch: Cotyledons of Pea, Bean and Lentil.

　4. Aleurone-grains *plus* fatty oil: Cotyledons of Soja Bean; endosperm of *Ricinus*.

(ii.) Reserve-materials contained partly in the cell-cavities and partly in the cell-walls.

    5. Aleurone-grains and fatty oil (in the cell-cavities) *plus* reserve-cellulose (forming the thickening layers of the cell-walls): Cotyledons of *Impatiens Balsamina*; endosperm of *Phytelephas macrocarpa*, *Phoenix dactylifera*, *Coffea arabica*, *Ceratonia Siliqua*.

*B.* Reserve-materials distributed between two different tissues.

    6. One portion of the non-nitrogenous reserve-material stored in the thickened mucilaginous walls of endosperm cells, the cavities of which are reduced to narrow crevices containing no appreciable quantity of plastic substance; the rest, comprising fatty oil—and sometimes also some starch —deposited in association with aleurone-grains in the cells of the cotyledons: *Trigonella Foenum graecum*, *Trifolium pratense*, *Medicago*.

In a number of seeds the outermost layer of the endosperm is specialised in a peculiar manner; it is composed of rather short, prismatic thin- or moderately thick-walled cells, which contain only aleurone-grains and oil, even when carbohydrates in the form of starch or reserve-cellulose predominate in the rest of the endosperm. Such an **aleurone-layer**, as it is termed, is particularly well developed in Grasses; here, however, in the author's opinion at anyrate, it does not, properly speaking, belong to the storage-system at all, but represents a glandular tissue which secretes diastase at the time of germination. Further investigation is required, in order to decide whether the aleurone-layer acts as an enzyme-secreting tissue in other families also, —notably among the LEGUMINOSAE—or whether it merely represents a specially differentiated portion of the storage-system.

When the various forms in which reserve food material is stored are set forth in tabular form, we see quite clearly that the several types of storage-system exhibit a gradually increasing perfection of adaptation, associated with an increasing division of labour. In the first of the types distinguished above (1), the elements of the storage-tissue do not differ morphologically from ordinary active parenchymatous cells. Storage-cells of the second type (2), may be said to combine the characteristics of active and of resting elements to a certain extent. In the third and fourth (3 and 4), all the plastic substances present have assumed forms which are associated with the quiescent condition of the cell. In all the cases so far considered the whole of the reserve-material is deposited in the cell cavities and within a single tissue. The two remaining types, on the contrary (5 and 6), are characterised by the segregation of the two main classes of reserve-materials, the protein compounds being separated—in great part at anyrate—from the non-nitrogenous substances. In the fifth type (5) nitrogenous and non-nitrogenous materials occur side by side in the same cell; in the sixth and most advanced type, however (6), they are deposited in two perfectly distinct tissues. As might be expected, the six principal types of storage-tissue that we have recognised, are connected with one another by a number of intermediate forms.

Some attention must now be given to the shape and arrangement of storage-cells, and to the characters of their walls.

In the case of starch-storing tissues, the cell-walls are frequently very thin, and quite devoid of pits (endosperm of Grasses, bulb-scales of *Oxalis esculenta*). Wherever, on the other hand, the walls are even moderately thick, numerous pits are present, which facilitate the filling and emptying of the storage-cells. These pits are very minute in the cotyledons of the Oak, but attain a considerable size in the Horse-chestnut, and also in the Bean and other Leguminosae. When the non-nitrogenous reserve-materials are stored in the form of oil, the walls of the storage-cells are generally thin and unpitted. Where the secondary thickening layers of a wall consist of reserve-cellulose or amyloid, they may be quite homogeneous, except for a highly refractive limiting pellicle, as in the endosperm of *Ornithogalum umbellatum* and *Phytelephas macrocarpa*; in other cases (*e.g.* endosperm of *Fritillaria imperialis*, Fig. 165 c) they exhibit very obvious stratification. Such thickened walls are, of course, always plentifully provided with pits which are deep and, as a rule, also fairly wide. Sometimes these pits are dilated at their lower ends in a funnel- or trumpet-shaped manner (Fig. 165 D); this arrangement serves to enlarge the area of the limiting membrane—which is often itself somewhat thickened—and thus to accelerate diffusion. In a few cases, according to Strasburger and Tangl, the same result is achieved in a different way, the limiting membranes being perforated by a number of protoplasmic connecting threads (*Ornithogalum umbellatum*, *Phoenix dactylifera*). Special interest attaches, in this connection, to the very narrow perforations of the endosperm-walls discovered by Tangl in *Strychnos Nux vomica*; these are not confined to the pit membranes, but also traverse the thicker portions of the wall. Whether such direct paths of communication serve for the rapid distribution of solvent enzymes, or whether they are utilised in the transportation of the products of solution, is still an open question.

In the less specialised types of storage-tissue, the component elements are isodiametric, and form a dense and irregular mass of cells, the arrangement of which seems to be devoid either of mechanical or of nutritive significance. This type of tissue—which may be compared to the simplest of the various forms of photosynthetic tissue —occurs, for example, in the cotyledon of the Bean, Pea, Oak and Horse-chestnut, and in tuberous storage-organs. Very often, however, the cells of storage-tissues are elongated and arranged in rows, which may be straight or curved, their exact disposition being determined with reference either to the mechanical requirements of the storage-tissue, or to the nutritive arrangements of the embryo [or other developing structure] which is destined to make use of the stored materials. In some cases a single system of curved cell-series serves both to increase the incompressibility of the tissue as a whole (mechanical function), and to facilitate the access of food-material to the embryo (nutritive function). In other instances these functions are distributed—in accordance with the principle of division of labour— between two distinct systems of curved cell-series, which are located in different parts of the storage-organ. A few examples will enable us to apprehend more clearly the precise significance of these statements.

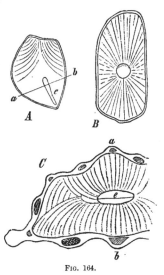

FIG. 164.

*A.* Tangential L.S. through a seed of *Iris* sp. ; *e,* embryo. ×5. *B.* T.S. through the same seed, taken along the line *a-b* (in *A*). ×14 *C.* T.S. through a mericarp of *Anethum Sova* ; *e,* embryo. In all three figures the course of the cell-rows of the endosperm-tissue is indicated by lines.

One of the adjoining figures (Fig. 164 c) depicts the greater portion of a transverse section across a mericarp of *Anethum Sova.* It will be noticed that, on the dorsal and ventral sides of the embryo—the cotyledons of which are seen in transverse section at *e*—the endosperm cells form curved series converging towards the centrally situated embryo. The lines of curvature correspond to the planes of compression in two dome-shaped structures placed back to back, an arrangement which must considerably increase the incompressibility of the fruit in the antero-posterior plane (*i.e.* in the direction *a-b*). As a matter of fact, it is precisely in this plane that the fruit is subjected to the greatest pressure when it swells through absorption of water. If a fairly thick transverse section of a mericarp is caused to swell, the antero-posterior diameter of the endosperm will be found to increase by

28 per cent., whereas the width (in the direction at right angles to *a-b*) will be augmented to the extent of 11 per cent. at most. The orientation of the two domes is thus closely correlated with the direction in which the surrounding soil exerts the greatest pressure when the seed swells under natural conditions. It may also be seen from the figure, that the lines of curvature of the radial cell-series indicate the paths along which the mobilised reserve-materials are conveyed to the embryo during germination. It is self-evident that this arrangement of the cells in converging rows must considerably accelerate the depletion of the storage-tissue. Evidently *Anethum* illustrates the case in which a single system of curved cell-series is concerned both with the mechanical and with the nutritive requirements of the organ in question.

The seed of *Iris* may be selected as an example of a different type of arrangement. In Fig. 164 A this seed is shown in tangential section. The pointed lower end of the seed contains the embryo (*e*); in the upper truncated portion the elongated and very thick-walled endosperm cells are arranged in curved series which trend obliquely outwards and downwards. The lines of curvature *intersect* the paths followed by the mobilised reserve-materials on their way to the embryo at various angles; they cannot, therefore, stand in any close relation to the nutritive requirements of the embryo, but must, on the contrary, have a purely mechanical significance. A transverse section across the narrow portion of the seed (cf. Fig. 164 B, *a-b*) reveals the fact that the endosperm cells in this region are disposed in straight rows radiating in every direction from the centrally situated embryo, and meeting the surface at all sorts of angles. Here there is clearly no mechanical adaptation. In order to produce increased resistance to radial pressure, the lines of curvature should form two dome-like structures placed back to back, each resting upon one of the flattened surfaces of the seed; as already stated, however, the radial cell-rows are perfectly straight. From the point of view of translocation, on the other hand, the arrangement of the endosperm-cells in the neighbourhood of the embryo is highly appropriate, since it brings every part of the storage-tissue into the closest possible relation with the embryo. The endosperm-cells of *Iris* are thus grouped in series belonging to two totally distinct systems; one of these, which can be made out in a tangential longitudinal section of the seed, is of purely mechanical significance, while the other, which is only visible in a transverse section, is developed in the interests of nutrition.

In many seeds, the cells of the endosperm are elongated at right angles to the surface, so as to resemble palisade-cells in form. Thus, a transverse section through the seed of **Polygonatum** *vulgare*

(Fig. 165 B), which is also a transverse section through the embryo, shows that the endosperm-cells are palisade-shaped near the periphery, but rapidly become isodiametric further in.    In this case the palisade-like form of the peripheral cells cannot possibly be connected with translocation, but must be a purely mechanical device.    For one thing, the radially elongated peripheral elements are those which have the smallest amount of translocation to per-form, whereas the inner part of the endosperm, which should be possessed of the highest conducting capacity, con-sists of isodiametric cells.    Further, the presence of numerous pits on the lateral walls of the palisade-shaped peripheral cells suggests that active translocation takes place, during germination, across the long axis of these cells.    The outer-most layer of all has no pits on its radial walls, or, at anyrate, very few. This feature likewise represents a mechanical adaptation, being clearly designed to prevent any weakening of the lateral walls in the layer which acts as the special mechanically protective tissue of the seed.

The four or five outermost layers on either of the flattened sides of the mericarp of *Siler trilobum* (Fig. 165 A) are palisade-like, whereas the inner por-tion of the endosperm is made up of cells which are either isodiametric or, more frequently, elongated in the tan-gential direction.    Since the fairly thick lateral walls of the peripheral cells are quite devoid of pits, there can be little

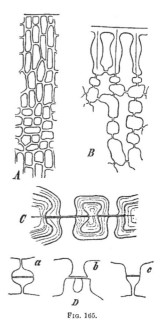

Fɪɢ. 165.

A. Part of a radial L.S. through the endosperm of *Siler trilobum.*  B. Peri-pheral portion of a T.S. through the seed of *Polygonatum vulgare.*  C. One of the thick, stratified walls of an endo-sperm-cell of *Fritillaria imperialis,* in section ; two pits are shown.  ×510. D. Various forms of pits from the endo-sperm of *Fritillaria imperialis* (semi-diagrammatic).  ×500.

doubt that, during the depletion of the endosperm, the mobilised reserve-materials of the outer layers travel, to start with, in the radial direction until they reach the inner region, where they are directed towards the developing embryo.    The path followed by the emigrating reserve-materials in this case is, therefore, closely similar to the course pursued by the synthetic products in the case of a leaf which has palisade-tissue on both faces.[198]

Hitherto discussion has been confined to tissues containing stores of plastic material which are destined to be utilised by the plant as a whole, or, at anyrate, by extensive portions of the plant-body.   A few words may be added with reference to **local storage-tissues,** which serve for the nutrition of individual organs or tissues.    According to Tschirch and Holfert, the so-called " nutritive layer " which is found in many seed-coats during the ripening of the seed, represents such a local storage-tissue.[199]    It usually consists of several tangential rows of cells containing abundance of starch, and is situated in the immediate proximity of the elements which later give rise to mechanical tissue or to mucilage-layers.   The plastic materials required, in either case, for the deposition of thickening layers upon the cell-walls, are derived from the nutritive layer.   The latter is, hence, entirely depleted when the testa is fully developed, and, as a rule, ultimately disappears altogether, its former position being indicated in the ripe seed by a zone of compressed and distorted cell-walls.   Excellent examples of these nutritive layers are to be found among the PAPILIONACEAE and CRUCIFERAE.

### 2.  *The storage of respiratory material.*

The plastic material deposited in storage-cells never becomes wholly incorporated in the growing tissues ; a certain proportion thereof is always destined to undergo katabolic change.   This portion is set apart to serve as **respiratory material,** to provide the fuel for the " physiological combustion " which results in liberation of chemical energy, and often also in evolution of heat.   Boussingault long ago noticed that seedlings of Indian Corn, when grown in the dark, lost 47·6 per cent. of their dry weight in the course of twenty days, mainly as a result of respiration.

So far as is known, " respiratory " reserve-materials are not, as a rule, deposited separately from those which are destined to be utilised in connection with growth ; in other words, there are no special storage-tissues set apart for the general mass of respiratory materials. In a few exceptional cases, however, where the **liberation of heat** is an essential, and not merely an incidental, feature of the respiratory process, storage-tissues may be developed for the reception of calorific material, in the form of starch or other carbohydrate substances.   It has long been known, that certain flowers and inflorescences evolve a very considerable amount of heat when they first open.   The most striking illustrations of this phenomenon are furnished by certain AROIDEAE.[200]   In the case of *Arum italicum,* for example, Gregor Kraus has recorded a temperature of 44·7°C. in the spadix, as compared with an external temperature of 17·7°C.; here there is a difference of 27°C. in favour of the plant.   The maximum temperature observed by Hubert during a series of observations upon *Colocasia odora* was as much as 49·5° C. Again, in a series of experiments performed by Kraus upon male cones of *Ceratozamia longifolia*, the highest temperature noted was 38·5° C., and the greatest excess over the temperature of the surrounding air, 11·7° C. The inflorescences of Palms and the flowers of NYMPHAEACEAE show a similar behaviour.

The metabolic changes correlated with such a pronounced rise of temperature, have been most thoroughly investigated in the case of *Arum italicum* by Kraus. The spadix of this plant is prolonged into a massive club-shaped appendage of a golden yellow colour, which protrudes from the cavity of the spathe for some distance. This appendage constitutes the calorific organ. It consists of a cylindrical core of water-tissue, traversed by vascular bundles and surrounded by a wide parenchymatous sheath, which is composed of a number of layers of thin-walled isodiametric cells containing large quantities of starch. In this peripheral parenchyma is deposited the store of respiratory material, consisting of numerous compound starch grains, together with a certain amount of sugar. Before the flowering period begins, carbohydrate substance makes up 77·8 per cent. of the total dry weight of the appendage, 66 per cent. being represented by starch, and 12 per cent. by sugars. In the course of the few hours during which the heating effect is noticeable, the dry weight of the appendage diminishes on an average by 74·1 per cent. This striking loss of weight is accompanied by the entire disappearance of the reserve carbohydrates; these substances are used up in respiration, except for a small fraction (3·7 per cent.), which is transformed into organic acids and other compounds that remain in the tissues of the appendage. The disappearance of the respiratory starch does not take place in any regular basipetal or centripetal sequence. On the contrary, the depletion of the storage-tissues follows a very irregular course in the present instance, patches, strips, and isolated spots retaining their starchy contents at a time when the intervening areas are already emptied of their reserve-material: ultimately, however, the entire storage-tissue becomes thoroughly depleted.

Both Kraus and Delpino regard the remarkable production of heat that takes place in the spadices of AROIDEAE, and in certain other inflorescences and flowers, as an ecological device which serves to attract pollinating insects. There can be no doubt that insects will be attracted to any plant which affords such an exceptionally warm habitation; the attraction will be greatest towards evening and in the early morning hours, when the difference of temperature between the flower and the outer air is most pronounced.

### 3. *Storage for ecological purposes.*

Many seeds and fruits, and also certain leaves and flowers, accumulate stores of plastic material for purely ecological purposes; in such cases, the reserve-materials—which most frequently consist of carbohydrates, but which may also include fats or protein-compounds—are employed to attract animals, the guests performing some useful office in return for the food with which they are provided. Storage-tissues of this kind may be termed **feeding-tissues**.

The most prevalent type of feeding-tissue is exemplified by the juicy pulp which forms so characteristic a feature of so many fruits and pseudo-carps; in all such cases the pulp is devoured by some animal—in most instances a bird—while the seeds are either scattered in the act of feeding, or subsequently ejected from the crop, or, finally, voided with the excrement. These pulpy tissues are generally composed of large thin-walled parenchymatous cells, with feebly developed protoplasts and large sap-cavities. The attractive material most often consists of one or more kinds of sugar (*glucose, fructose*), sometimes also of starch (*e.g.* in *Musa* and *Artocarpus*), or of oil (*Olea, Persea gratissima, Durio zibethinus*). The attractiveness of the fruit or seed is also frequently enhanced by bright coloration, due to the presence of chromoplasts, or of anthocyanin dissolved in the cell-sap, and by the development of a characteristic flavour or aroma which owes its existence to the formation of various metabolic by-products.

Oily storage-tissues are often developed in the special appendages of seeds which are known as **arils, caruncles** and **strophioles,** or in persistent floral organs subserving the same purpose, namely, the attraction of ants for purposes of seed-dispersal; the insects endeavour to carry off the seeds to their nests, but abandon a number on the way, and thus assist in the distribution of the plant.[201] Sernander unites all such oily appendages under the comprehensive category of **elaiosomes**. Among the commoner plants which have seeds provided with elaiosomes, may be mentioned *Luzula Forsteri, L. pilosa, Allium triquetrum, Helleborus foetidus, Chelidonium majus, Corydalis spp., Viola odorata* and *Melampyrum pratense*.

Similar in character to the oily appendages of seeds, are the small "**food-bodies**" borne by the leaves of certain myrmecophytes; these bodies afford a supply of food to the protective ants that regularly inhabit the plants in question.[202] In *Cecropia adenopus* and *C. peltata* (Trumpet-trees) the lower surface of each leaf-base is covered with a dense coat of hairs; interspersed among the hairs are numerous minute ovoid or pear-shaped structures, the **Müllerian bodies**, which, according to Schimper, consist essentially of a mass of parenchymatous cells containing a large amount of fat and protein. If these food-bodies are carried off by ants, or artificially removed, the leaf-base develops a fresh crop in the course of a few days. Morphologically considered, the Müllerian bodies undoubtedly correspond to trichomes; in Schimper's opinion, it is probable that they represent modified glandular hairs. Very similar structures (**Beltian bodies**), pear-shaped and of an orange yellow colour, occur at the apices of the pinnules in *Acacia sphaerocephala* (The Bull's Horn Thorn). These consist of a mass of thin-walled parenchyma (storage-tissue) filled with protein and oil, and are very readily detached from the leaf.

The food-supply which plants offer to insects in return for their valuable services as pollinating agents, generally takes the form of a sugary secretion termed **nectar**; among the less specialised flowers, a portion of the pollen may be sacrificed for the same purpose. Comparatively few flowers are provided with special feeding-hairs or -tissues.[203] Arrangements of this latter kind are particularly prevalent in the ORCHIDACEAE, and it is in connection with this family that they have received most attention. In every case these attractive devices occur in connection with the labellum, the perch upon which insect-visitors alight; and matters are always adjusted so that cross-pollination must result, when the appropriate insect proceeds to devour the feeding-tissue or -hair.

This feeding-tissue was first discovered by Ch. Darwin in *Catasetum*; his observations were subsequently confirmed by the author, while more recently Porsch has made a detailed study of quite a number of genera, such as *Stanhopea, Maxillaria, Cirrhaea, Oncidium* and *Odontoglossum*, with reference to these arrangements. The portions of the labellum containing the feeding-tissue, develop into warts, knobs or callosities of various shapes (*Oncidium, Stanhopea*), and may thus be recognised even with the naked eye. These excrescences consist of thin-walled cells which are filled with a variety of organic substances. Non-nitrogenous reserve-materials are represented by starch (*Catasetum*), amylodextrine (*Stanhopea*) or glucose, or sometimes by oil in globules or larger masses. In addition, a large amount of protein-substance is always present in the form of small granules, or more rarely in that of crystalloids (*Maxillaria lutescens, M. pumila*); in the case of *Maxillaria varia*, each cell contains a dense homogeneous mass of protein-substance in the form of a layer adhering to its inner wall, as well as numerous protein granules. Porsch mentions a very remarkable adaptive feature of feeding-tissues, which consists in the fact that the cuticle and the cutinised layers of the epidermal cells are cast off at any early stage of development; this arrangement naturally renders the attractive food-material more readily accessible.

Crüger long ago proved, by his observations on the pollination of *Catasetum* and other Orchids under natural conditions, that the feeding-tissues actually do perform the function which has been attributed to them.

In another group of tropical Orchids (spp. of *Maxillaria*, *Bifrenaria*, *Oncidium*, *Pleurothallis*, *Spiranthes*, etc.) the attractive apparatus consists of special uni- or multi-cellular hairs, which usually occur in dense masses on the labellum. According to Porsch, these hairs are crammed with fat and [amorphous] protein, while protein-crystals are also not infrequently present. In *Maxillaria rufescens* each feeding-hair consists of a single cell, which is very thin-walled except at the base, where the membrane is thick, strongly cutinised and brownish-yellow in colour; it is at the junction of the thick- and thin-walled regions that the hair is broken off by, the insects. In *Maxillaria ochroleuca* and *M. porphyrostele*, the slender basal cell of the [multicellular] hair is supported by short thick-walled accessory hairs, which keep the whole structure in the upright position. In certain cases the feeding-hairs are composed of short rounded cells, which ultimately separate from one another, and heap themselves up into aggregations resembling masses of pollen; this condition has been observed by Janse in *Maxillaria Lehmanni* and *M. venusta*, by Fritz Müller and Saunders in *Polystachya*, and by Penzig in one of the RUBIACEAE (*Rondeletia strigosa*).[204]

The alleged occurrence of floral feeding-hairs in members of a number of other Natural Orders (*e.g.* COMMELYNACEAE, *Aristolochia*, *Portulaca*, *Anagallis*, *Cyclamen*, *Verbascum*) requires confirmation.

## IV. THE STORAGE-SYSTEM IN THE THALLOPHYTA.[205]

Water-tissues comparable to those of land-plants are, naturally enough, unknown among the Algae. Even the littoral forms of Seaweed are sufficiently protected against desiccation by the water-retaining capacity of the mucilage with which they are coated. The gelatinous sheaths of *Gloeocapsa*, and the jelly-like substance in which the filaments of *Nostoc* are embedded, likewise represent water-storing arrangements which are essential to the existence of these terrestrial Algae.

Tissues for storage of plastic material were first observed in certain FLORIDEAE by Wille, while Hansteen and Hansen subsequently demonstrated their existence in the PHAEOPHYCEAE. In *Pelvetia*, *Fucus*, *Sargassum* and other FUCACEAE, plastic materials are deposited in the so-called inner cortex or primary cortex of the phycologists, that is, the zone which separates the peripheral photosynthetic system from the central strand of conducting elements.   This storage-tissue consists of numerous layers of relatively large isodiametric cells, with more or less thickened and pitted walls; the limiting membranes of the pits are traversed by fine perforations.

In the case of *Fucus serratus*, the contents of these cells consist, according to Hansteen, of granular masses of a special carbohydrate substance, which he terms *fucosan*.   In *Dictyota dichotoma*, the storage-tissue is represented by the central or medullary layer; this tissue consists of a single layer of cells in the flattened portions of the thallus, but is many-layered in the cylindrical stem-like region.   According to Hansen, each storage-cell contains a group of large oil-drops suspended in a central mass of cytoplasm, from which a number of protoplasmic strands radiate towards the cell-wall.   The depletion of the storage-tissue which takes place in connection with the production of adventitious shoots or reproductive organs, is preceded by emulsification of the large oil-drops.

Genuine water-tissues seem also to be unknown among the Fungi, unless the large vesicular **cystidia** that occur in the hymenium in certain Hymenomycetes (*e.g.* spp. of *Coprinus*), serve as reservoirs of water, besides performing the mechanical function—attributed to them by Brefeld and Von Wettstein [and also by Buller]—of keeping the gills apart while the spores are being liberated.   The **sclerotia** of Fungi are always provided with typical storage-tissues for plastic materials, in the shape of a medullary region, composed either of closely interwoven tubular hyphae or of pseudo-parenchyma (plectenchyma).   The non-nitrogenous reserve-materials consist of greatly thickened gelatinous cell-membranes or of masses of fat, and are accompanied by varying quantities of protein-compounds.   The depletion of these sclerotial storage-tissues, in connection with the development of fruit-bodies, closely resembles the corresponding processes in Higher Plants.   In the case of *Claviceps purpurea*, for example, the fat disappears completely, and the thickened cell-membranes are also largely used up. Finally, it may be mentioned that, according to Errera, *glycogen*, a polysaccharide closely related to the dextrines, which is widely distributed in the animal kingdom, also frequently occurs in large quantities both in the higher and in the lower Fungi, and probably always represents a reserve-material.

Protein-substance (? protoplasm) is stored for ecological reasons by a Southern Brazilian Agaric, *Rozites gongylophora*, in terminal swellings of special hyphae; according to Möller, these " Kohl-rabi-clumps " are devoured by the Leaf-cutter Ants which cultivate this Fungus in their nests.[206]

# CHAPTER IX.

## THE AËRATING OR VENTILATING SYSTEM.

### I. GENERAL CONSIDERATIONS.

ALL plants carry on a more or less active **gaseous interchange** with the surrounding atmosphere. During respiration, oxygen is used up and carbon dioxide excreted. The process of photosynthesis, which is characteristic of the green parts of the plant-body, involves the opposite type of interchange, carbon dioxide being absorbed and oxygen evolved. In certain cases other gases, such as hydrogen or sulphuretted hydrogen, are produced as the result of special katabolic changes. Large quantities of water-vapour escape in consequence of the transpiration which goes on in all the aërial organs. Many plants, finally, give off vapours consisting of " ethereal oils " and other volatile organic compounds.

Other things being equal, the activity of any of these forms of gaseous interchange depends upon the extent of surface through which gaseous diffusion can take place. In the case of microscopic unicellular organisms, the ratio of superficial area to volume is so large that special arrangements for increasing the free surface are entirely unnecessary. The larger a plant is, however, the greater does its need become for increase of superficial area. This demand can be met up to a point by more or less extensive ramification of the plant-body ; the additional free surface acquired in this way, may by itself enable a plant of moderate size to carry on and regulate all its various forms of gaseous interchange in a perfectly satisfactory manner. Among highly organised [land-] plants, however, the external surface alone can never be sufficient for the purpose, even when the plant-body is divided up to the utmost possible extent. In these circumstances, the plant resorts to the formation of an internal labyrinth of air-containing cavities and channels, which stand in the same relation to the adjoining tissues as does the outer atmosphere to the superficial cells. The ventilating system may thus be said to owe its origin to the influence of the fundamental **principle of maximum exposure of surface.**

The manner in which a living cell absorbs or liberates gaseous substances is always the same, whether the cell in question is situated in the interior of the plant body in contact with an air-space, or whether it lies at the surface, and is consequently directly exposed to the external atmosphere. In either case, the gaseous substances have to pass through membranes saturated with water of imbibition. In other words, gaseous interchange is essentially an osmotic process. In view of the fact, however, that we are regarding the ventilating system as an integral part of the plant-body, it will be necessary to consider gaseous interchange as a function of the whole plant rather than of the individual cell.

From this point of view we may—where the ventilating system is provided with special outlets or **pneumathodes**—distinguish between osmotic gas-exchange on the one hand, and direct intermingling of gases on the other.[207] The latter process generally depends wholly upon diffusion, as, for example, in the case of the gas-exchange attendant upon respiration and photosynthesis. It may, however, in part consist of a movement *en masse*, as, for instance, when variations of temperature or pressure occur in the outer atmosphere, or when the intercellular spaces expand or contract suddenly as the plant sways in the wind. These statements apply equally to movements of gases from one part of the ventilating system to another.

It was at one time generally assumed, in view of experimental results obtained by Boussingault, Mangin and others, that the gaseous interchange of leaves and other aërial organs, was not carried on solely through the intercellular spaces and pneumathodes, but that cutinised cell-walls were pervious to carbon dioxide and oxygen, and consequently that diffusion of these gases through the epidermal walls played an essential part in the normal processes of gaseous exchange. More recent experiments, performed by Blackman and by Stahl, have shown that, under natural conditions, this **cuticular diffusion** is practically negligible, so far as photosynthetic gas-exchange is concerned—and the same statement probably applies to respiration—in comparison with the **stomatic diffusion** which goes on through the pneumathodes. The normal condition in this respect is, in fact, closely comparable to the relation between cuticular and stomatic transpiration, which has already been discussed at some length (p. 105). The modern view thus attributes a greater physiological importance to the ventilating system, and to the pneumathodes associated therewith, than the early physiologists were disposed to concede, at any rate, in the case of aërial organs.

2 E

In the case of organs that are permanently submerged, or that grow in very wet soil, the possibility of gaseous interchange may be entirely dependent upon the continuity of their air-spaces with the ventilating system of the portions which are exposed to the outer atmosphere and provided with pneumathodes. Such organs may, on the other hand, carry on a direct exchange of gaseous material with the surrounding medium; but in this case the process is so slow, that the development of large internal air-reservoirs in the submerged or subterranean structures becomes an absolute necessity.

As a rule the ventilating system is entirely composed of intercellular air-spaces; in accordance with its function, it is a continuous system, and permeates all the other tissue-systems that have to carry out gaseous interchanges for purposes of respiration, or in connection with some special function. Another general feature of the ventilating system is the development of external outlets or pneumathodes, which exhibit great variety as regards the details of their structure. In the majority of cases, each individual tissue—the photosynthetic tissue, for example—produces the ventilating spaces which it needs by a partial separation of its cells. In certain instances, however, especially where large air-reservoirs are required, the formation of intercellular air-spaces becomes the sole, or at any rate, the principal function of a special tissue, which may consequently be termed **ventilating tissue** or **aërenchyma**.

## II. VENTILATING SPACES AND VENTILATING TISSUES.

### A. FORM OF THE AIR-SPACES AND CHARACTER OF THEIR WALLS.

If the ventilating spaces of the plant-body are to perform their functions satisfactorily, it is above all things essential that they should form a **continuous system**. The general arrangement of this system is such, that every cell of the tissue which it ventilates is actually in contact with one or more intercellular air-spaces. In other respects the extent and distribution of the ventilating spaces vary greatly, according to the requirements and special functions of the different tissue-systems.

Generally speaking, ventilating spaces may take the form of elongated passages, of wide cavities, or of narrow clefts and interstices. In ordinary parenchymatous tissues, composed of isodiametric cells, the intercellular spaces are developed as narrow passages, extending along the rounded-off edges of the cells, and collectively forming a dense meshwork which spreads out equally in all directions. In the case of palisade-tissue, every cell abuts directly against several passages, which run parallel to the long axis of the cell, communicating with one another at the surfaces of contact between successive layers of palisade-cells. Where a number of elongated cells are arranged in longitudinal series, it is usual to find air-passages of great length extending along the edges of the cells. This arrangement prevails, for example, in the leaves of *Elodea canadensis*, *Galanthus nivalis* and *Leucojum aestivum*, and in the inner region of the primary cortex of many roots. The wide air-passages which traverse all the vegetative organs in the case of many marsh- and water-plants, may be regarded as elongated passages of this type, which have become greatly distended.

In the spongy parenchyma of foliage-leaves, the ventilating system shows a marked tendency to assume the form of inter-communicating **cavities** or **chambers**, owing to the fact that the cells are usually produced into a number of arms or branches, which only abut against one another at their distal ends. Large polyhedral air-chambers, such as are found in the leaves of *Pistia* and *Pontederia*, and in the fronds of *Lemna*, may be compared to the wide air-passages included in the preceding class of ventilating spaces.

**Ventilating clefts** may arise between individual cells, just like air-passages or -cavities, or they may separate whole layers or sheets of cells from one another. The former condition is exemplified by the leaves of many MYRTACEAE, further by species of *Scirpus* and by *Cladium Mariscus*; the tubular photo-synthetic cells in the leaves of the last-mentioned plant are almost entirely separated from one another by such intercellular clefts; only a few small circular areas of the walls remain in contact with one another. Larger ventilating intercellular clefts are particularly prevalent in photosynthetic tissues; they occur, for example, in the leaves of *Pinus*, *Abies* and *Cryptomeria*, and also in many Monocotyledons.

The ventilating spaces in a particular tissue, whether developed as passages, chambers, or clefts, frequently present the appearance of isolated cavities. Nevertheless, the different spaces are, in the vast majority of cases, connected in such a manner that a free circulation of gases can take place throughout the tissue. The connecting channels are often exceedingly small, and hence may easily be overlooked.

The cell-walls which abut against intercellular spaces, generally remain unthickened. In close proximity to pneumathodes—and especially beneath stomata—they may be covered by a cuticle. Much more generally, however, intercellular spaces are lined by a delicate pellicle of pectic material, owing to the fact that the spaces arise by the splitting apart of the middle lamellae; the tubercles and rods, projecting into the spaces and traversing their entire width, which have been noted by Luerssen as characteristic of the parenchymatous tissues of many Fern-petioles, are likewise composed of pectic substances.[208]

## B. THE STRUCTURE OF VENTILATING SPACES AND VENTILATING TISSUES, CONSIDERED IN RELATION TO CERTAIN PHYSIOLOGICAL FUNCTIONS.

### 1. *The ventilating system in relation to respiration.*

The relations between the structure of the ventilating system and the function of respiration, are naturally most clearly revealed in the case of organs which are altogether colourless, or which at any rate contain very little chlorophyll, because gas-exchange is, in these circumstances, to a very large extent respiratory in character. Organs which respire actively, are invariably provided with a well-developed ventilating system; but it would not be correct to assume that there is always a close correspondence between the average intensity of respiration and the degree of development of the ventilating spaces.

In young organs which are growing rapidly, and hence also respiring actively, air-containing intercellular spaces generally make their appearance long before any appreciable tissue-differentiation takes place. In some instances—especially among roots—ventilating passages extend far into the primordial meristem of the apical region (Fig. 18). Occasionally the respiratory requirements of young organs are so great, as to necessitate the formation of special ventilating organs. An interesting illustration is furnished by the young leaves of *Nephrodium stipellatum*, Hook., a Fern which is not uncommon in moist wooded ravines in Western Java.[209] As long as they remain rolled up, the young leaves of this Fern are enveloped in a coating of mucilage, several mm. in thickness, which is secreted by special mucilage-hairs, and in all probability acts as a protective covering. This thick layer of mucilage would seriously impede respiration, were it not perforated, at frequent intervals, by certain conical or awl-shaped structures, which represent special breathing organs or **pneumatophores**. In the upper part of the leaf, these pneumatophores, which may reach a length of 5 mm., arise close beside the points of attachment of the pinnae, while further down, on the petiole proper, they are arranged in two longitudinal rows, and are, also, provided with shield-shaped appendages of unknown significance. The whitish external appearance of these pneumatophores at once suggests that they possess a spongy texture; as a matter of fact, they consist of a loose aërenchyma traversed by large ventilating spaces, which communicate with the outer atmosphere by means of numerous stomata. The walls of the aërenchyma-cells

are covered *externally* with innumerable thickenings in the shape
of tubercles or short rods, which project into the intercellular
spaces.

There can be little or no doubt, that these remarkable structures
actually represent special organs of ventilation, since their air-spaces
communicate freely with the ventilating system of the rhachis and
petiole.

It is probable, that a similar respiratory function should be attributed
to the so-called fore-runner-tip (*Vorläuferspitze*) of leaves, a structure
first observed by Crüger, and more recently investigated in detail by
Raciborski.[210] This is a highly characteristic feature of the very young
leaf in many lianes, and in certain other tropical plants ; it takes the
form of a long narrow terminal prolongation of the lamina, being
frequently marked off from the blade proper by a slight constriction.
The fact that it bears an unusually large number of stomata, suggests
that it carries on gaseous interchange, and particularly the respiratory
absorption of oxygen and evolution of carbon dioxide, on behalf of the
leaf-blade, while the latter is still in the embryonic condition. In
certain cases it seems to be more immediately concerned with the
secretion of (liquid) water. For the sake of convenience, the aërenchyma of the so-called "breathing-roots" which occur in many marsh-
and water-plants, will be discussed in a later section of this chapter.

### 2. *The ventilating system in relation to photosynthesis.*

The photosynthetic organs of land-plants are almost always covered
by a typical epidermis. Hence the characteristic gas-exchange of the
photosynthetic system, that is, the absorption of carbon dioxide and the
liberation of oxygen, is carried on almost entirely through the mediation
of the ventilating system. Even those palisade-cells which abut
directly against the epidermis, obtain their carbon dioxide from the
adjoining intercellular spaces, and not from the outer atmosphere *viâ*
the epidermal cells. It is for this reason that every photosynthetic
cell borders upon intercellular spaces at one or more points. In the
case of palisade-cells, it is usually strips of the lateral walls of varying
breadth which thus come into direct contact with air-spaces ; it will
be recollected, that the chloroplasts tend to congregate on these very
strips, while they are often altogether absent from the cross-walls.
The fact that chloroplasts favour those walls which abut against intercellular spaces, may be just as readily verified in subepidermal palisade-
cells as in more deep-seated layers. It furnishes histological evidence
in support of the conclusions drawn from experimental data. Stahl
and Blackman[211] both conclude, on independent experimental grounds,
that, under normal conditions, the photosynthetic tissues are entirely

dependent upon the ventilating system, so far as their gaseous interchange is concerned.   Stahl employed the following method in order
to investigate this point.   The stomatic lower surfaces of the experimental leaves (*Prunus Padus, Ribes petraeum, Lonicera tatarica, Philadelphus coronarius, Impatiens* spp.) are partially covered with cacao-wax,
so as to prevent the access of carbon dioxide to the underlying sections
of the ventilating system.   Leaves treated in this manner are exposed
to sunlight for several hours ; when subsequently tested by Sachs' iodine
method (cf. p. 292), the waxed portions appear yellow, and thus prove
to be devoid of starch ; the unwaxed areas, on the other hand, assume a
deep blue-black colour.   Again, if the entire lower surface of a leaf is
covered with cacao-wax, and incisions made on the upper side, so as to
provide artificial paths of communication between the ventilating
system and the outer air, a considerable quantity of starch is formed
in the immediate neighbourhood of each incision, after some hours'
exposure to sunlight. · Where the smaller veins are embedded in conducting parenchyma which extends as far as the epidermis on both
sides of the leaf, each of these zones of local starch-formation is limited
by the nearest veins ; here the veins evidently act as partitions, which
prevent the carbon dioxide from spreading for any considerable distance
through the mesophyll.

In many photosynthetic tissues, the intercellular spaces act as
barriers which prevent the synthetic products from diffusing in un-
desired directions (cf. p. 282); this subsidiary function has a large
share in determining the arrangement of the ventilating spaces within
the photosynthetic system.   The fact that photosynthetic organs so
often contain numerous transversely directed intercellular clefts, an
arrangement which renders it impossible for the products of synthesis
to travel longitudinally from the first, is really nothing more than an
expression of the principle of expeditious translocation ; this principle,
which, as we have already seen, plays such a prominent part in the
construction of photosynthetic tissues, thus extends its influence over
the ventilating system as well.

3. *The ventilating system in relation to transpiration.*

The transpiration-current does not represent the sole vehicle for
the transportation of the food-materials which green terrestrial plants
absorb from the soil, and hence scarcely possesses the fundamental
significance attributed to it by some physiologists; still, where the
climatic conditions are such that plants can transpire actively without
continually incurring the risk of excessive loss of water, transpiration
does provide a convenient means of *accelerating* the translocation of
nutrient salts.   We may therefore expect occasionally to meet with

features in the internal structure of foliage leaves and other transpiring organs, which tend to encourage transpiration. It is the ventilating system which most frequently exhibits hygrophilous adaptations of this kind. For the sake of the resulting increase of transpiratory activity, the intercellular spaces of a leaf may be enlarged to an extent which would seem quite inexplicable from the point of view of photosynthesis and respiration. In the ordinary dorsiventral type of leaf, it is the spongy parenchyma which performs the duties of a special transpiring tissue, besides performing other functions as a part of the photosynthetic system; by far the largest proportion of the water-vapour exhaled by a leaf is derived from the extremely well-ventilated spongy mesophyll. The degree of development of this tissue, therefore, at once shows whether any given plant is distinctly hygrophilous or xerophilous in character. In plants which are forced to reduce their transpiration in order to avoid the danger of excessive loss of water, the spongy parenchyma is comparatively small in amount and scantily provided with intercellular spaces. This diminution of the internal evaporating surface is often supplemented by a more or less far-reaching reduction of the external surface exposed by the transpiring organs. In the case of plants inhabiting damp, shady situations, or more generally, wherever the nature of the environment tends to lower the rate of transpiration, it is necessary that actively transpiring tissues should be plentifully developed. As Areschoug has pointed out, leaves in which the mesophyll is entirely made up of spongy parenchyma, should probably be regarded in the main as extreme cases of this hygrophilous adaptation.[212]

As a rule, adjustments of the quantitative development of the ventilating system, such as those described above, suffice to effect the requisite diminution or increase in the intensity of transpiration. A peculiar *arrangement* of the foliar intercellular spaces which is found in some Australian plants (*Hakea, Restio, Kingia*), and in *Olea europea*, also seems to represent a xerophytic adaptation. In the majority of leaves the palisade-cells abut directly against *longitudinal* air-passages, which open into the ventilating spaces of the spongy parenchyma; this arrangement makes for comparatively active gas-exchange, because the distances that have to be traversed by the diffusing gases are quite short. In the above-mentioned xerophytes, these longitudinal passages are replaced by tubular chambers, which encircle the cells in planes at right angles to their long axis. The short connecting passages which link up adjoining chambers, are all placed parallel to the surface of the leaf, so that there is no direct communication between the chambers in the radial direction. Tschirch[213] has pointed out, how greatly this unusual arrangement of the intercellular spaces adds to the distance which

water-vapour has to traverse in escaping from the interior of the leaf
to the outer atmosphere; instead of pursuing a straight or at most
a slightly curved course, the vapour has to make its way through
successive layers of palisade-tissue by a long and tortuous path.

### 4. *The ventilating arrangements of marsh- and water-plants.*

In **aquatic plants which are completely and permanently submerged,**
the entire gaseous interchange goes on without the aid of stomata, by
diffusion through the whole of the outer surface, just in the same way
as the respiratory gas-exchange of gill-breathing aquatic animals.

The disadvantage of this cuticular method of gas-exchange consists
in its extreme slowness. There are two prominent features in the
construction of submerged plants which tend to counteract this
drawback. Diffusion will obviously be accelerated if the superficial
area of the plant-body is increased. As a matter of fact, the external
surface is very often amplified in water-plants in a variety of ways.
Most frequently the leaves are subdivided into a number of narrow
segments (*Myriophyllum*). In *Ouvirandra fenestralis* the leaf-blade is
perforated so as to form a lattice-work, while in other cases again
(*e.g. Elodea canadensis*) the leaves remain undivided, but become
correspondingly thin and delicate. According to Goebel, the seedlings
of *Euryale* and *Victoria* are provided with localised surface-expansions,
which serve to accelerate gaseous interchange; here the resemblance
to the gills of animals is very pronounced.[214] In the case of
*Euryale*, the abortive primary root bears a four-lobed gill-organ,
which persists until the seedling has fully expanded and has come
into contact with water on every side; each of the lobes is branched
and thickly covered with hairs. Goebel believes that the organ is
concerned with the absorption of oxygen during the early stages of
germination; although this view has not been put to the test of
experiment, there is much to be said in its favour.

The inconvenience consequent upon the slowness of gaseous
interchange under aquatic conditions, may also be mitigated by the
development, in submerged organs, of large internal air-chambers. The
living tissues can thus maintain an active photosynthetic or respiratory
interchange with the "internal atmosphere" provided in this way. It
must be noted, that the entrance of carbon dioxide into the plant from
the surrounding water takes place more rapidly than the outward
diffusion of oxygen; the photosynthetic gas-exchange will therefore be
accelerated, if extensive internal air-spaces are provided for the
temporary accommodation of comparatively large quantities of oxygen.
If shoots of *Myriophyllum, Ceratophyllum* and similar plants are cut
across while they are engaged in photosynthesis, streams of bubbles,

consisting largely of oxygen, escape from the cut surfaces; this phenomenon has long been utilised by physiologists as a means of demonstrating the process of photosynthesis. In view of the slowness with which the $_o$xygen diffuses into the plant from the surrounding water, it is an advantage, from the respiratory point of view also, if submerged organs are provided with an internal store of oxygenated air; the possession of such a reservoir of oxygen further renders the plant in some degree independent of fluctuations in the oxygen content of the external medium. Those plants which possess the largest air-spaces will, of course, suffer the smallest amount of inconvenience from the difficulties by which the direct interchange of gases with the surrounding water is attended.

Air-spaces which are much larger than those of ordinary land-plants, are not restricted in their occurrence to *submerged* water-plants; on the contrary, they are also to be found in all the vegetative organs of floating aquatic species, and in the aërial organs of marsh-plants. Unger long ago calculated that air-containing intercellular spaces accounted for as much as 71·3 per cent. of the entire volume of the plant-body of the floating Aroid *Pistia texensis*, whereas the corresponding values in the case of two terrestrial plants, *Brassica Rapa* and [the more xerophilous] *Begonia manicata*, do not amount to more than 17·5 per cent. and 6 per cent. respectively. The aërial organs of such semi-aquatic plants are provided with stomata, and thus agree with land-plants as regards their mode of gaseous interchange. Nevertheless, there are obvious advantages to be gained, in their case also, from the development of large air-spaces. Such plants are liable to be temporarily submerged, or to be exposed to the action of surf; moreover, their stomata must often be in danger of becoming blocked by water. In the case of floating plants, the air-spaces also serve to increase the power of flotation, by lowering the specific gravity of the plant-body.

The characteristic large air-spaces of marsh- and water-plants often arise in connection with a special form of parenchymatous tissue, which is termed **aërenchyma,**[215] in allusion to its principal function. Even the single-layered plates of parenchyma which form the partitions between adjoining air-spaces and -passages in the stems of *Scirpus lacustris* and of species of *Potamogeton, Myriophyllum,* and *Papyrus,* in the fronds of *Lemna,* in the petioles of *Pontederia crassipes, Trapa natans,* various NYMPHAEACEAE, etc., may be included under the head of aërenchyma. There is a certain contrast between this lamellar form of ventilating tissue and the more prevalent stellate or spongy type; but the two extreme conditions are connected by numerous intermediate stages. Very typical aërenchyma is found in the "breathing roots" [pneumatophores] developed by mangroves and by certain other marsh-plants. Goebel, Jost, Schenck and Karsten are all agreed, that these breathing roots serve to supply the submerged parts of the plants—and especially those organs which are embedded in the very poorly oxygenated mud—with oxygen for purposes of respiration.[216] It is for this reason that breathing roots are negatively geotropic, and hence grow vertically upwards until they project for considerable distances into the air. Their ventilating system is

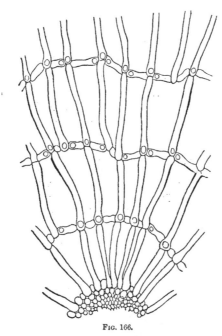

Fig. 166.

Aerenchyma in a breathing-root of *Jussiaea peruviana* (T.S.). After Schenck.

connected with the outer atmosphere by means of special pneumathodes, and they are accordingly able to supply the less favourably situated organs with air. Pneumatophores of this type occur in *Sonneratia, Avicennia, Laguncularia, Saccharum officinarum,* in some Palms and PANDANACEAE, etc.

The breathing roots of certain species of *Jussiaea* have been stüdied in detail by Schenck. These roots are clothed with a massive aërenchyma, made up of numerous concentric layers collectively representing the primary cortex. Most of the cells in each layer are prolonged radially into long arms or shanks; the shanks of adjacent cells belonging to the same layer are separated by large air-cavities, which collectively form a continuous cylindrical air-jacket (Fig. 166). The successive jackets corresponding to the concentric layers of aërenchyma, communicate with one another radially, by means of intercellular spaces which intervene between the tangentially expanded bodies of the aërenchyma cells. As seen in radial longitudinal section, therefore, each cell appears ⊢-shaped, the horizontal stroke representing the arm which traverses the air-jacket surrounding the particular layer to which the cell belongs, while the vertical stroke corresponds to the expanded body of the cell. Tangential connection between neighbouring cells of each layer is effected by those elements which do not become produced into radial arms. In these roots of *Jussiaea*, both the epidermis and the hypodermal layer (the cells of which do not undergo any radial extension) disappear at an early stage; the ventilating system thus soon opens directly into the surrounding water. As in other cases, the cells of the aërenchyma are not themselves full of air, but are, on the contrary, provided with a thin peripheral layer of protoplasm, which contains a nucleus and minute leucoplasts, and encloses a sap-cavity filled with a watery liquid.

In *Jussiaea*, and in certain other ONAGRACEAE and LYTHRACEAE, a very similar aërenchyma is also developed on the so-called "floating" roots, and on the submerged portions of the stem. In these latter cases, only a small portion of the aërenchyma consists of primary cortex, the rest being produced by a secondary meristem, which is homologous with the phellogen. Rosanoff long ago showed that the aërenchyma of the submerged stems of *Desmanthus natans* is a secondary tissue.

A different type of aërenchyma is exemplified by the felted stellate tissue which fills the air-passages in certain CYPERACEAE,—notably in species of *Scirpus*; the tissue in question also performs a mechanical function, by helping to prevent the collapse of the air-passages, and to preserve the cross-sectional outline of the entire organ (cf. p. 181).

### 5. *Diaphragms and internal hairs.*

Air-passages are often interrupted at intervals by **diaphragms,** composed of one or more layers of aërenchyma. Besides acting as stiffening plates (cf. p. 182), these diaphragms play an important part in connection with ventilation. They are always more or less lacunar, in structure, and hence do not in the slightest degree interfere with the movement of gases along the air-passages; in the haulms of *Papyrus*, they are actually responsible for effecting communication between adjacent ventilating chambers, since each diaphragm extends across several air-passages. Whereas, however, such diaphragms are quite pervious to air, they oppose, as Goebel has pointed out, a very considerable resistance to the passage of water. It is found, in fact, that where diaphragms are present, the air-spaces cannot be flooded with water, except by prolonged injection at a high pressure. When, therefore, a plant which is provided with diaphragms dies off at one end, or suffers some mechanical injury, flooding of the whole ventilating system is effectually prevented, since water will not readily penetrate beyond the first intact diaphragms. The chambered structure of the ventilating system in the inflated petioles of *Pontederia crassipes* and *Trapa natans*, in the fronds of *Lemna*, etc., should evidently be interpreted from the same point of view; but here, also, the walls of the water-tight compartments possess an accessory mechanical significance, just like the majority of diaphragms.

Aquatic plants which contain few or no diaphragms (*Rhizophora, Pilularia, Nymphaeaceae,* AROIDEAE, etc.), develop so-called **intercellular or internal hairs**[217] in their air-passages. The fact that these structures appear to take the place of diaphragms, in itself suggests that they perform a mechanical function; this view of their physiological significance is further supported by their arrangement, by their thick-walled character and by their shape, which is most frequently that of an **H** or **X**. The polyhedral air-passages in the petioles and peduncles of *Nymphaea* contain stellate hairs, which project from individual members of the cell-rows that occupy the narrow faces of the polyhedra; the thickened walls of these hairs are furnished with blunt external protuberances owing to the inclusion of numerous small crystals of calcium oxalate. In the vertical direction these hairs succeed one another at comparatively short intervals. The lamellar parenchyma in certain Aroids (*Monstera, Tornelia, Heteropsis, Pothos,* etc.) is traversed by intercelluar hairs which resemble bast fibres in appearance; they are branched in various ways, and are often hooked at their ends. The mechanical effect is probably similar to that of felted stellate tissue, such as is found in *Scirpus*. The lower (proximal) portions of the breathing roots of *Sonneratia acida*, which extend horizontally beneath the mud, are provided with aërenchyma containing characteristic slightly curved or **S**-shaped internal hairs. Westermaier[218] attributes a very curious function to these structures. He points out that the breathing roots of this mangrove are exposed to a varying pressure according to the state of the tides. When the water is high, the aërenchyma is compressed, and air rich in carbon dioxide is forced out. When the tide falls, the pressure upon the root diminishes; the aërenchyma expands once more, owing to the elasticity of the internal hairs—which thus act as springs—and oxygenated air is sucked into the intercellular spaces. Westermaier's account implies that these roots are provided with a genuine breathing mechanism, involving movements of inspiration and expiration. This suggestion certainly seems very plausible, but has not, so far, been put to the test of experiment.

In conclusion, it must be noted, that air-containing intercellular spaces do not necessarily always serve for purposes of ventilation. It frequently happens, that an axial or foliar organ, such as a Grass-haulm, a scape of *Leontodon* or a leaf of *Allium*, is traversed by a single wide air-passage; in this case the hollow type of construction is determined by mechanical considerations, and has no direct connection with ventilation. Mention has already been made of the fact that, in photosynthetic tissues, the intercellular clefts are arranged so as to ensure that the synthetic materials shall emigrate by the shortest path. Similarly, attention has been drawn to the fact that air-chambers and other ventilating spaces are responsible for the buoyancy of floating plants. Among those strand-plants, again, which depend upon ocean currents for the dispersal of their seeds and fruits, the testa or pericarp is often provided with a specialised floating tissue containing numerous inter-cellular spaces; the sole function of these air-spaces is reduction of the specific gravity of the seed or fruit (cf. Ch. XI.). In white petals, finally, and similar attractive organs, the ventilating function of the air-spaces is of minor importance, in comparison with their ecological value; for here the attractive white coloration is due to the total reflection of light by the air contained in the intercellular spaces.

## III. THE EXTERNAL OUTLETS OF THE VENTILATING SYSTEM.

Direct communication between the ventilating system and the outer atmosphere is rendered possible by the fact, that the dermal tissues are interrupted at certain points by open ventilating pores or **pneumathodes**,[219] through which gases can pass freely in both directions. There are three principal types of pneumathode, namely, the epidermal **stomata**, the **lenticels** which traverse the periderm, and the peculiar **ventilating pores** which occur in certain aërial roots.

### A. STOMATA.

#### 1. *Structure and mechanism of stomata.*[220]

The **stomatic pores** are the epidermal outlets of the ventilating system. They are never surrounded by ordinary epidermal elements; on the contrary, each is enclosed by a pair of specialised **guard-cells** (in Mosses the arrangement is slightly different), which constitute the stomatic apparatus in the strict sense of the term. [The whole structure, including the guard-cells and the orifice which they surround, is termed a **stoma.**] Not infrequently some of the epidermal elements immediately adjoining the guard-cells are also specialised as **subsidiary cells** of the stoma. Those cells, finally, which border upon the inter-cellular air-chamber that underlies the stoma, may also exhibit various peculiarities of form or structure. In such cases, the definition of the stomatic apparatus may be extended so as to include not only the guard-cells, but also the subsidiary epidermal cells, as well as any specially modified elements of the underlying tissue that may be associated with the stoma.

Every normal stoma has the power of expanding and contracting, of opening widely and of closing [more or less] completely; it is by virtue of this property that the stomata are able to regulate the processes of gaseous interchange, in accordance with the varying require-ments of the plant. The power of control is vested in the guard-cells, which, indeed, represent nothing more nor less than a mechanism for regulating the size of the stomatic orifice. The construction and mode of action of this mechanism is not the same in every case. There are, in fact, several distinct types of stomatic apparatus. All of them, however, resemble one another in one point: the opening and closing action can always be referred to the particular structure of the guard-cells. In almost every case the guard-cells take an active part in the process of opening and closing. It is therefore *not* correct to compare these cells to a pair of curved springs, which become passively com pressed as the stoma closes under the influence of an external force, but move apart elastically, and thus reopen the stoma, when the pressure is relieved.

Let us now examine the structure of a typical stoma, such as may be found in any ordinary Dicotyledon and Monocotyledon, with a thin-walled epidermis. The actual aperture of the stoma is enclosed by by two sausage-shaped **guard-cells,** placed side by side and separated from one another at either end by a thin partition. The average thickness of the inner or **ventral** face (*Bauch-seite*) of each guard-cell, *i.e.* of the side which is nearest to the pore, is considerably greater than that of the outer or **dorsal** face (*Rücken-seite*). As a rule, the ventral wall is provided with two heavily **cutinised** thickening **ridges** corre-sponding to its upper and lower edges; in a transverse section these ridges present the appearance of more or less sharply pointed horns or beaks (Fig. 167). Those ridges which project above and below, overarch two chambers, which may be respectively termed the **front cavity** (*Vorhof*), and the **back cavity** (*Hinterhof*); the two cavities communi-cate with one another through the **pore-passage** (*Zentralspalte, Spalten-*

*durchgang*). Both the cavities and the pore-passage are bounded by the ventral wall of the guard-cell. Collectively they constitute the **orifice** or **pore**, which is thus of unequal width at different points, often appearing roughly hourglass-shaped in vertical section. Tschirch terms the entrance to the front cavity the "eisodial aperture," and the exit from the back cavity the "opisthial" aperture.

In this widely distributed type of stoma, the dorsal walls of the guard-cells are entirely unthickened and usually bulge into the cavities of the adjoining epidermal (or subsidiary) cells.

A stomatic apparatus which changes its outline—as seen in surface view—when the stoma is opening or closing, obviously cannot be immoveably attached to the comparatively rigid outer epidermal wall; the latter is, in fact, specially thin in the immediate vicinity of the dorsal wall of each guard-cell, along a more or less well-defined strip, which thus acts as a dorsal hinge (the *Hautgelenk* of Schwendener) during the opening and closing of the stoma. In some cases the hinge only occupies an exceedingly narrow strip of the thickened outer wall (*Prunus Laurocerasus, Myrtus communis, Allium Cepa,*

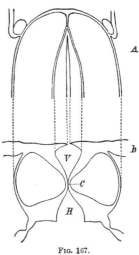

FIG. 167.

Stoma of *Narcissus biflorus. A.* In surface view. *B.* In T.S. *V.* Front cavity. *C.* Pore-passage. *H.* Back cavity.

Fig. 168 B), while in other instances it comprises a more extensive region of the cell-membrane (*Chlorophytum comosum* Fig. 168 A). Among the CYPERACEAE the hinges are represented by the entire thin outer walls of the subsidiary cells. In the first-mentioned case the movement of the guard-cells consists in a simple rotation around the hinge; in the instance last referred to, on the other hand, the expansion and contraction of the orifice involves outward curvature or straightening of the thin strip of membrane. In a number of plants, the epidermal cells, or, at any rate, the subsidiary cells of the stomata, are provided with more or less thickened inner walls. The author has shown that, where this is the case, these walls are likewise furnished with thin strips in the immediate neighbourhood of the guard-cells. The strips in question may be termed **inner hinges**, in contradistinction from the **outer hinges** which have been described above. In the leaves of *Chlorophytum comosum, Clivia nobilis, Uropetalum serotinum*, etc. the inner hinge takes the shape of a narrow, very well-defined strip, (Fig. 168). In *Linum usitatissimum*, and among the BROMELIACEAE, the thin and very convex inner walls of the subsidiary cells act as the inner hinges.

With regard to the contents of guard-cells, it may first of all be stated, that these elements are almost always characterised by the presence of chloroplasts, which often contain a large amount of included starch. The protoplast is well-developed in other respects

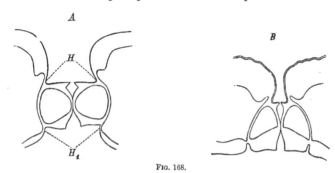

Fɪɢ. 168.

Stomata with inner and outer hinges ($H$ and $H_1$). *A. Chlorophytum comosum.*
*B. Allium Cepa.*

also, and the nucleus often lies near the centre of the central wall, especially in young guard-cells.

Attention must next be directed to the mechanism of a stoma of the ordinary type referred to above. In the first place, direct observation shows, that the pore of the stoma is opened or dilated when the guard-cells become more strongly curved, and that it is closed or contracted if the curvature of the guard-cells decreases. The width of the pore thus depends directly upon the degree of curvature of the guard-cells. The problem which we have to solve may therefore be restated as follows: how does the anatomical structure of the guard-cells lead, of mechanical necessity, to alterations in their degree of curvature [under certain conditions]? Von Mohl was the first to approach this problem along correct physiological lines. He pointed out, that when pieces of living epidermis are mounted in water, the stomata open or become dilated, whereas immersion in a solution of sugar produces the opposite effect. This simple experiment proved that changes of turgor provide most, if not all, of the motive power in the process of opening or closing. But it was reserved for Schwendener to show, first, that these turgor changes take place mainly in the guard-cells themselves; secondly, that the opening and the closing of the orifice are due to changes in the shape of the guard-cells; and thirdly, that turgor-variations produce these changes of shape owing to the peculiar distribution of thickened and unthickened areas in the walls of the guard-cells. Schwendener made a number of careful measurements, with a micrometer, of the dimensions of the different portions of the stomatic apparatus of *Amaryllis formosissima* and other plants, in the open and in the closed condition; in this way he was enabled to gain a complete insight into the changes of shape which the guard-cells undergo, and thus finally to explain the mechanism of the stoma. Closure of the stomata was brought about, in Schwendener's experiments, by plasmolysing the guard-cells with glycerine. The following table contains the measurements obtained in the case of *Amaryllis formosissima*:

|  | Stoma open. | Stoma closed. |
|---|---|---|
|  | $\mu$ | $\mu$ |
| Width of single guard-cell.    -    -    -    -    - | 37 | 33 |
| „    front cavity,    -    -    -    -    -    - | 16 | 7 |
| „    pore-passage,    -    -    -    -    -    - | 8 | 0 |
| Distance between the lines of attachment [220a] of the two adjoining epidermal cells,    -    -    -    - | 40 | 34 |
| Distance between each line of attachment and the dorsal wall of the corresponding guard-cell, | 19 | 16 |

Consideration of these figures at once shows, that the whole stoma is much wider in the open condition than when it is closed, whereas the length of the apparatus undergoes no appreciable change in the process of opening or closing. It follows that the volume of each guard-cell is greatest in the open condition of the stoma. The process of opening, therefore, depends upon endosmosis of water and consequent increase of the hydrostatic pressure in the guard-cells; this rise of pressure in turn causes a distension of the dorsal and ventral walls. The thin dorsal wall naturally stretches more readily than the partially thickened ventral wall; consequently the curvature of the guard-cells is increased and the orifice becomes wider. When turgor falls, on the other hand, the more active contraction of the dorsal wall diminishes the curvature of the guard-cells, and the size of the orifice is reduced. A guard-cell, constructed on the aforesaid lines, will, therefore, behave very much like a short length of rubber-tubing, the wall of which is considerably thickened along one side. Such a tube will curve when water or air is forced into it under pressure, the thick strip forming the concave side. The presence of *two* specially thickened ridges on the ventral wall of each guard-cell entails several advantages. In the first place it is obvious that the whole mechanism will act with greater precision when two longitudinal strips are "fixed," than it would do if the ventral wall were only provided with a single median ridge. Further, the interpolation of a thin strip between the two thickened ridges allows

2 F

the ventral wall to swing backwards and forwards around its median plane, so that it bulges towards the pore when the stoma closes. As Schwendener has pointed out, the two thickened ridges may be compared to the stiff boards of a portfolio, while the thin median strip corresponds to the flexible back. Finally, the stomatic pore can be closed more tightly, if the membranes which become opposed to one another are thin and pliant, instead of being thickened and hence comparatively rigid.

In curving under the influence of increased turgor, the guard-cells must, of course, overcome the resistance of the adjoining epidermal cells, which are themselves in a turgid condition. If this resistance is diminished—for example, at the edge of a severed strip of epidermis where the epidermal elements are laid open—the curvature of the guard-cells increases beyond the normal amount.

When a stoma is open, the internal tension of its guard-cells must, of course, exceed the osmotic pressure which prevails in the ordinary epidermal elements, and in the subsidiary cells, if such are present. According to Schwendener, the pressure in guard-cells amounts, under certain stated conditions, to about 5 to 10 atmospheres.

As already stated, the type of stoma which has just been described is the most widely distributed of all. Its action depends upon the fact, that the thickened strips are only developed on the ventral walls of the guard-cells, and are thus arranged asymmetrically with reference to the plane which passes through the middle of the guard-cell, at right angles to the surface of the leaf. In a considerable number of Dicotyledons, however, the walls of the guard-cells are thickened all round, in such a way that the cavities are reduced to narrow transverse slits. Here the massive semi-cylindrical thickened strips of the wall are distributed symmetrically on two sides of the median plane of the guard-cell. In these circumstances there is no appreciable difference between the dorsal and the ventral walls, with regard to the extension produced by the rise of turgor. Where stomata of this second type are functional at all, which is not always the case, the curvature of the guard-cells obviously cannot be brought about in the same manner as it is in the instance first described. It is possible that the guard-cells tend to elongate when their internal tension increases, but that they are forced to bend aside owing to the resistance of the epidermal cells to which their ends are attached, just in the same way as a [flexible] vertical column will be deflected to one side if too heavy a load is applied to the upper end. In order, however, that an *outward* curvature may result in every case, the guard-cells must be slightly curved in the same sense, even when they are not distended. Guard-cells of this symmetrical type occur in the

phyllodes of *Acacia leprosa, A. farinosa,* and *A. acinacea,* and in the
leaves of *Melaleuca uncinata, Hakea suaveolens,* etc.  It should be
stated that they are generally restricted to the older organs; in young
leaves and phyllodes the guard-cells are asymmetrical in structure and
display the ordinary or *Amaryllis*-type of mechanism.

The peculiar stomata of the GRAMINEAE and CYPERACEAE represent
a special modification of the ordinary symmetrical type.  Here each
guard-cell is shaped like a dumb-bell, the ends being dilated and
connected together by a narrow middle-piece, the upper and lower
walls of which are very thick-walled throughout, so that the cavity is,
in this portion of the guard-cell, reduced to an insignificant transversely

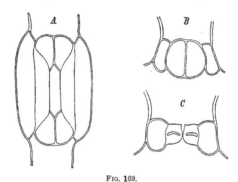

FIG. 169.

Stoma of *Poa annua*.   *A.* Surface view.   *B.* T.S. through the dilated ends of the
guard-cells.   *C.* T.S. through the middle of the guard-cells.

elongated slit (Fig. 169 c).   Towards either end, the thickened strips
rapidly become narrower and finally become reduced to the same
thickness as the thin walls enclosing the dilated terminal portions of
the guard-cell (Fig. 169 A).   The forces responsible for the movements
of a stoma of this type must obviously be altogether located in the
dilated ends of the guard-cells—except in so far as the well-developed
subsidiary cells, which are always present in such cases, may play a
part in the process.   Schwendener has, in fact, shown that a rise of
turgor in the guard-cells of a stoma is followed by an increase in the
volume of their dilated ends.   As a result, the total width of the two
ends of the stoma increases; points such as $m$ and $n$ in the adjoining
diagram (Fig. 170) draw apart from one another, *pari passu* with
more distally situated points, such as $a$ and $b$, because of the
comparative rigidity of the intervening stretch of thickened wall
($m$-$a$ and $n$-$b$).   Obviously a further consequence is an increase in the
angle $mon$, and hence in the width of the pore-passage $s$.   As seen
in surface view, the pore-passage is shaped like a hexagon, elongated in

one direction, but otherwise regular.   When the stoma opens or closes,
the two longest sides of the hexagon undergo the greatest displacement,
always, however, remaining parallel to their original direction, for
reasons which have been explained; as Schwendener remarks, the
mechanism is comparable to that of the adjustable slit of a
spectroscope.

In a few grasses (*Cynosurus echinatus*, *Aira capillata*, *Briza
maxima*) the pore-passage remains open, even after the guard-cells
have been killed and are therefore non-turgescent.   If such stomata

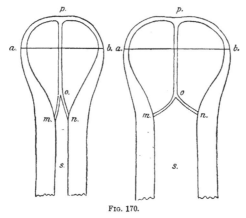

Fig. 170.

Diagram of one-half of a Grass-stoma in the open (right) and in the closed (left) condition.  Surface view.  (For explanation see text.)  After Schwendener.

are functional at all, closure must be effected by an increase in the
turgor pressure of the subsidiary cells.

All the forms of stomata that have been discussed so far, agree in
the fact that the width of the whole structure increases as the pore
opens.   The type which still requires to be described, on the other
hand, is characterised by the circumstance, that both the total width
of the stoma and the shape of the dorsal guard-cell walls remain
unaltered, either throughout the opening (or closing) movement or,
more frequently, during a considerable part of the movement.   The
changes in the shape of the guard-cells, which regulate the dimensions
of the pore in such cases, must be quite different from any that have
been described in connection with other types of stomata.

The simplest, and at the same time the most extreme, type of this
" constant-width " mechanism is exemplified by the stomata on the
capsule of *Mnium cuspidatum* (Fig. 171), which have been examined
from this point of view by the author himself.   The most striking
features revealed by a transverse section through such a stoma are, the

entire absence of hinge-areas, the thickness of the dorsal, and the thinness of the ventral walls of the guard-cells ; all these characters clearly indicate that the mechanism, in this case, is altogether different

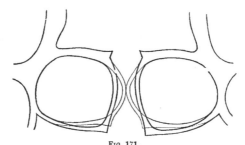

Fig. 171.

Diagram showing a stoma of *Mnium cuspidatum* in the open (heavy lines) and in the closed (faint lines) condition.   T.S.   For explanation see text.

from anything that we have previously encountered.   Attention may next be directed to the appended table of measurements ·

|  | | | | | | Stoma open. | Stoma nearly closed. |
|---|---|---|---|---|---|---|---|
|  | | | | | | $\mu$ | $\mu$ |
| Total length of the stoma, - | - | - | - | - | - | 51 | 51 |
| „    width    „    „   - | - | - | - | - | - | 42 | 42 |
| Width of a single guard-cell,  - | - | - | - | | | 17 | 20 |
| „    front-cavity,   - | - | - | - | - | - | 9 | 9 |
| „    pore-passage,   - | - | - | - | - | - | 8 | 2 |

As is quite evident from these figures, the total length and breadth of the stoma are always the same, whether the pore is open or closed.   Closure or contraction of the pore depends entirely upon an increase in the width of the individual guard-cells.   In the closed condition of the stoma, the cross-sectional outline of each guard-cell is approximately elliptical.   When turgor increases, this outline tends to become more nearly circular.   But since the thick dorsal and outer (upper) walls are practically immoveable, this change in shape can only be effected by a straightening out of the bulging thin ventral walls ; as these straighten themselves, they draw apart and thus enlarge the pore.   Here the alterations in the shape of the guard-cells, upon which the movements of the stoma depend, are not confined to the plane parallel to the leaf surface ; displacements also occur in planes which are perpendicular to that surface.

A mechanism very like that of *Mnium* had previously been described by Schwendener in the case of *Helleborus*.   The changes of form which take place in the latter instance, are indicated in the adjoining diagram (Fig 172).   Here, as in Fig. 171, the heavy lines correspond to the open, the lighter lines to the closed condition of the stoma). In *Helleborus*, according to Schwendener, "it is characteristic of the closing movement that the width of the front cavity remains constant, at any rate within the ordinary limits of internal tension. The back cavity, on the contrary, becomes much narrower, the contraction which results being roughly equal to twice the diameter of the pore-passage (when open); the curved dorsal wall of each guard-cell straightens itself and simultaneously revolves around its outer edge, as a door turns on its hinges. The character of these movements is perhaps most readily appreciated, if one expresses them in terms of the change of shape of the guard-cells. When the stoma is closed, the cross-sectional

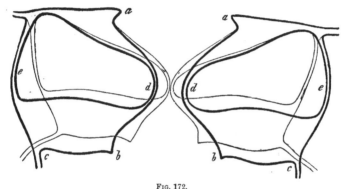

FIG. 172.

Diagram showing a stoma of *Helleborus* sp. in the open (heavy lines) and in the closed (faint lines) condition. After Schwendener. (From Sachs' *Lectures*.)

outline of the guard-cell cavity is roughly that of a scalene triangle. If the internal tension of the cell increases, this triangular figure becomes larger, and at the same time approaches an isosceles form, with the result that the ventral walls draw further apart and the pore-passage is enlarged."

Schwendener states that, in *Tradescantia discolor*, contraction of the pore-passage is also entirely brought about by an increase in the curvature of the ventral walls, which in its turn depends upon the fact that each guard-cell diminishes in height as its turgor falls. When the stoma opens very widely, however, the dorsal wall—and hence, also, the guard-cell as a whole—becomes more strongly curved. The stomata of *Tradescantia discolor*, and those of certain LILIACEAE and ORCHIDACEAE, are evidently constructed on a plan which is, in a sense, intermediate between the two extreme forms of mechanism that may respectively be termed the *Amaryllis* type and the *Mnium* type. The recent investigations of Copeland have brought to light still other types of stomatic mechanism.

The back cavity of every ordinary stoma opens into a more or less extensive air-space, which communicates at various points with the ventilating system. The regular occurrence of this **internal air-chamber** or ventilating chamber is readily understandable, in view of the fact that the stomata are the [principal and sometimes the only] outlets of the ventilating system. Not infrequently, special mechanical pro vision is made to prevent these air-chambers from collapsing, when the organ in which they occur is subjected to bending stresses. Where the stomata are placed in longitudinal series,—*e.g.* in the

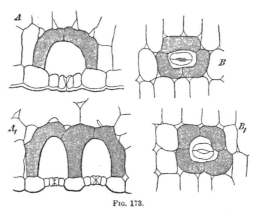

FIG. 173.

Mechanical protection of the internal air-chamber. *A*, *A₁*. Air-chamber overarched by green mesophyll-cells (T.S.). *A*. *Elymus*. *A₁*. *E. arenarius*. *B*, *B₁*. Air-chambers of *Iris germanica*, surrounded by a horizontal ring of green mesophyll-cells. (Surface view.)

majority of Grass-leaves, in Conifer-needles, etc.—all the air-chambers of a series generally coalesce to form a continuous passage; the cells that abut against these passages, are usually arranged in such a manner as to form a series of arches roofing in each passage. Most often each stoma is underlain by a pair of long, curved, photosynthetic elements, which converge so as to meet on the inner side of the air-passage (Fig. 173 A, A₁). In *Iris germanica*, where the stomata are irregularly scattered, each air-chamber is surrounded by a circular group of three or four tangentially-curved photosynthetic cells (Fig. 173 B, B₁).

The mechanical importance of the arrangements which have just been described, is purely local. Westermaier[221] ascribes a somewhat different significance to the rigid "gutters," composed of thick-walled cells, which enclose the hypostomatic air-passages (formed as in the pre-ceding instances by the longitudinal fusion of adjacent air-chambers) in the haulms of *Scirpus caespitosus*, *Eriophorum alpinum*, etc. (Fig. 174). Here the longitudinal seriation of the stomata weakens the tangential connection between neighbouring girders of the mechanical system. This difficulty is overcome by the interpolation of the aforesaid

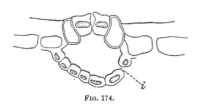

"gutters," which are firmly attached at both ends to the rigid epidermis by means of the thick radial walls of the epidermal cells. Free communication between the air-passages and the rest of the ventilating system is ensured by the fact that the "gutters" are traversed by a certain number of intercellular clefts (Fig. 174 $i$).

FIG. 174.

T.S. through a hypostomatic air-passage in the haulm of *Eriophorum alpinum.* The hypodermal cells enclosing the passage form a rigid gutter, which is interrupted by an intercellular cleft at $i$. After Westermaier

Some of the features in the structure of hypostomatic air-chambers are connected with the regulation of transpiration; in order to avoid needless repetition, these points are reserved for discussion at a later stage (cf. below III. A. 4).

### 2. *The importance of stomata as paths of gaseous interchange.*

Since the days of Dutrochet, it has been generally recognised that the stomata are [the principal] external openings of the system of intercellular spaces. This view has been subjected to a variety of experimental tests, in all of which the same criterion has been employed, namely, the escape from the cut ends of petiole or stem of air forced inwards through the stomata. Experiments of this kind devised by Dutrochet, Unger, Sachs and many others, are described in every text-book of Plant Physiology.

The interesting and valuable investigations of Brown and Escombe[222] have recently thrown fresh light upon the important question, as to how leaves and other aërial organs are able to carry on a sufficiently active gaseous interchange through such minute pores as the stomata. The experiments carried out by Brown and Escombe upon the diffusion of gases through minute holes in a thin septum, led to the discovery of a very remarkable physical law. It appears that, under the conditions described, the rate of diffusion depends upon the *linear* dimensions of the holes, and not upon their cross-sectional area. In other words, diffusion goes on more rapidly through very small pores than through larger apertures [with the same total cross-sectional area]. Hence, if a septum is perforated by a sufficient number of minute openings, it would seem as if diffusion should take place as rapidly as if the septum were absent; as a matter of fact, experiment shows that this is actually the case. Brown and Escombe's results thus form a remarkable tribute to the efficiency of the stomata as paths of diffusion.[223]

As the stomata are capable of opening and closing, they are not mere perforations in the epidermis, placing the inter-cellular spaces in communication with the outer air; on the contrary, they are the organs which *regulate* the activity of gaseous interchange in general, and of transpiration in particular. As a matter of fact, there is no need for the stomatic orifices to be adjustable, so far as respiration and photosynthesis are concerned; both these functions could be carried out quite satisfactorily, if the stomata were always open, so as to admit of rapid diffusion. For it is unlikely that plants would ever suffer injury, under natural conditions, through excessive photo-synthesis or respiration; hence gaseous interchange need never be restricted in the interests of either of these functions. Most land plants, on the other hand, are frequently threatened by the dangers attendant upon excessive transpiration; and it is for this reason that such plants must be furnished with the means of temporarily suppressing stomatic transpiration altogether, and thus of preventing, for the time being, all further escape of water-vapour, except for the small amount that is unavoidably lost by cuticular transpiration.

In the case of non-chlorophyllous organs [which are incapable of photosynthesis],—such as petals, stamens, the pneumatophores of *Nephrodium stipellatum* and the shoots of certain Phanerogamic saprophytes and parasites—and for which transpiration is a function of secondary importance, the stomata must be utilised mainly or exclusively for the purpose of respiration. The stomata of foliage-leaves, and of chlorophyll-containing organs in general, on the con-trary, play a very important part in connection with photosynthesis, since, as has already been explained (p. 437), the gaseous interchange incidental to that function is carried on almost entirely through the mediation of the ventilating system. Sachs long ago remarked that wilted leaves are unable to manufacture starch, because their stomata are shut; for a similar reason, no starch is formed, according to Stahl, in leaves the stomata of which have been blocked by artificial means.

It is in connection with the transpiratory process, that the control-ling action of stomata has been studied in the greatest detail. Garreau, Unger, Deherain, Boussingault and others have demonstrated, by a great variety of experiments, that, as a rule, the abaxial surface of a leaf evolves a far larger amount of water-vapour than the adaxial side. This difference is undoubtedly due to the unequal distribution of stomata on the two surfaces, the majority of leaves having all or most of their stomata on the abaxial surface. In these circumstances, the transpiration from the upper surface is purely cuticular, and hence comparatively insignificant, whereas in the case of the lower surface the far more active stomatic transpiration has to be taken into account as well.

Trustworthy and accurate quantitative data with reference to the regulation of transpiratory activity by stomata, can only be obtained by the method of weighing; but several investigators have achieved a considerable amount of success with the aid of methods which are more convenient, though less exact.[224]  Merget's plan is to place the experimental leaf in contact with paper previously coated with a mixture of the hypochlorites of iron and palladium.  This paper is yellowish-white when freshly prepared, but gradually becomes darker as it absorbs moisture; the stain produced [in a given time] by contact with a transpiring leaf will thus be darker or lighter accordingly as the rate of transpiration during that time was high or low.  Prints obtained in this way, thus give some indication of the regulatory power of the stomata.  Stahl's "cobalt-method" is quite similar in principle, and only differs in the nature of the substance employed as an "indicator." Filter-paper is impregnated with a dilute (1 to 5 per cent.) solution of cobalt chloride, a salt which is blue in the anhydrous state, but light pink when hydrated.  If a piece of blue cobalt-paper is placed upon the stomatic surface of a freshly-picked leaf—which should have been exposed to sunshine just before it was gathered—it frequently turns pink within a few seconds, whereas similar paper ·placed in contact with the astomatic surface of the same leaf may remain blue for hours (e.g. in the case of *Tradescantia zebrina*, *Salix caprea*, *Populus nigra*, *P. tremula*, *Pyrus communis*, etc.).  The fact that this striking difference in the behaviour of the two surfaces towards cobalt-paper can often be demonstrated in the case of young leaves still enclosed in the bud, shows that the cutinisation of the outer epidermal walls effectually restricts cuticular transpiration at a very early stage of development.  F. Darwin has utilised the hygroscopic curvature shown by thin strips of specially prepared horn ("Chinese sensitive leaf"), when placed in contact with the stomatic epidermis of a leaf, in order to estimate the relative activity of transpiration, and thus to determine whether the stomata are open or closed.

3. *The relations of external conditions to the movements of stomata.*[225]

It is obvious that the stomata would be incapable of regulating transpiration, and gaseous interchange in general, to the advantage of the plant, unless their movements were determined by external conditions.  It would seem, above all, to be essential, that the guard-cells should respond to any fluctuations in the intensity of illumination and in the relative humidity of the atmosphere.  As the width of the stomatic pore depends entirely upon the internal tension in the guard-cells, only those external factors which affect this internal tension will induce movements of the stomata. As a matter of fact, it was discovered long ago by Von Mohl, that the stomata of foliage-leaves open when they are illuminated, and that this movement takes place most rapidly in direct sunlight. Schwendener not only confirmed Von Mohl's observations, but also showed that the stomata close when a leaf is darkened.

The opening of stomata under the influence of light is readily comprehensible from an ecological point of view. Under normal conditions, it is to the interest of photosynthesis that the stomata should open as widely as possible, when the leaf is illuminated, so long as there is no danger of excessive transpiration ; a high rate of transpiration is in itself advantageous, within certain limits, because of the resulting increase in the supply of mineral salts to the leaves. It is not so clear what advantage is derived from the closure of stomata which takes place at night, or when a leaf is otherwise darkened. F. Darwin suggests, that this restriction of nocturnal transpiration serves to prevent the serious loss of heat which would result, if evaporation were as active at night as it is by day. Leitgeb, Stahl and Darwin all note that a number of plants, notably marsh- and water-plants, and also species with nyctitropic leaves, never close their stomata at night.

Opinions differ considerably as to the influence of temperature upon the movements of stomata. N. J. C. Müller maintains that a rise of temperature always causes the stomata to open, but Schwendener disagrees with this conclusion. Leitgeb states that the stomata of the pedicels and perianth segments of *Galtonia candicans* tend to open with rise of temperature, whereas those on the foliage-leaves of the same plant exhibit no such tendency. Kohl asserts that all stomata open more or less rapidly, not only when the surrounding air is warmed, but also when they are exposed to the action of (invisible) thermal radiations.

The majority of stomata are exceedingly sensitive to any fluctuations in the humidity of the atmosphere, and in the water-content of the soil ; it is this property, above all others, that enables these organs to exercise a beneficial control over the rate of transpiration. While Von Mohl and other even earlier observers were aware of the fact, that the stomata close as soon as a leaf begins to wilt, it was Leitgeb who first pointed out that the closing movement commences, before any other symptoms of withering make their appearance. If a leaf is transferred from the moist atmosphere of a greenhouse to an ordinary room, its stomata close even in bright daylight, a circumstance which proves that the relative humidity of the atmosphere has a more powerful effect than light upon the turgidity of the guard-cells. The ecological advantage of this high degree of sensitiveness of the stomata towards any decrease in the relative humidity of the atmosphere is sufficiently obvious. Doubtless the strong influence which dry air exerts upon the guard-cells, accounts for the fact that many stomata (*e.g.* those of *Berberis vulgaris, Syringa vulgaris, Ribes aureum, Saponaria officinalis,* etc., according to Leitgeb) remain closed in a high wind even after prolonged insolation.

There is a natural inclination to regard the fall in the turgor of the guard-cells which leads to the closure of the stomata in dry air, as a direct result of an increase in the transpiratory activity of these cells, especially as they occupy a relatively exposed situation, their outer, inner and ventral walls all bordering either upon intercellular spaces or upon the outer air. But active transpiration on the part of the guard-cells tends to *increase* the concentration—and hence also the osmotic strength—of their cell-sap ; these cells ought, therefore, to be able to withdraw water from the less actively transpiring ordinary cells of the epidermis, and thus to maintain their turgor-pressure at a more or less constant value. As a matter of fact, the relation between transpiratory activity and the condition of the stomata is not of a simple or direct nature at all ; the action of variations of humidity upon the turgidity of these guard-cells must rather be regarded—like the effect of light—as a process of stimulation.

We may conclude the present section with a brief reference to the chloroplasts and starch-grains which so frequently occur in the guard-cells of stomata. The turgor-variations that take place in these cells in response to external stimuli, are in their turn dependent upon alterations in the quantity and quality of the osmotically effective contents of the cells. A very plausible view is that which regards the chlorophyll-apparatus of the guard-cells as the source of these osmotically active materials ; according to this theory, the starch-grains represent a store of solid material which can easily be transformed into soluble and osmotically effective substances (most probably sugars), when such are required. This suggestion finds some support in Schellenberg's discovery of the fact, that plants which have been kept for some days in an atmosphere devoid of carbon-dioxide, and have thus been prevented from forming starch during that period, always keep their stomata tightly closed, even when they are exposed to bright light in a moist atmosphere.

4. *The structure of stomata in relation to climatic and edaphic conditions.*

In the preceding section, the anatomical structure of stomata has been considered solely with reference to the mechanism of opening and closing movements.  We have still to -discuss a number of structural features, which have no direct connection with the mechanism of the stoma, but which are nevertheless of great physiological importance. Most of these features have been developed in adaptation to particular climatic or edaphic factors of the environment.

Under what may be termed *average* conditions of water-supply, transpiration—and hence gaseous interchange in general—can be sufficiently regulated by the movements of the guard-cells alone.  In such

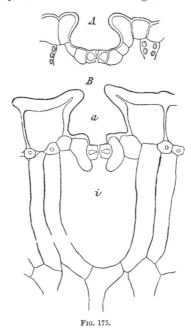

Fig. 175.

Sunken stomata.  *A. Amherstia nobilis* (upper side of leaf).  *B. Hakea suaveolens*;
ι, internal, *a*, external air-chamber.

cases the front- and back-cavities are never greatly developed; the entrance to the front-cavity is often very wide, and the stoma as a whole lies at the same level as the surrounding epidermal cells.

Among plants which grow in a dry climate, or in arid localities, the greater risk of desiccation renders it essential, that transpiration should be further restricted by special arrangements.  In such cases the stomata are generally sunk below the level of the surrounding epidermal cells, so that each appears to be situated at the bottom of a funnel, cup- or saucer-shaped depression, the so-called **external air-chamber** (Fig. 175).  Not infrequently, the entrance to this chamber is partially blocked by ridge- or wall-like outgrowths of the neighbouring

cells.   Each stoma thus opens into a cavity, the atmosphere of which
remains undisturbed, and hence highly charged with water-vapour,
however the surface of the leaf may be swept by air-currents.   Pfitzer
was the first who correctly interpreted the "sunken" stoma as an
arrangement for preventing excessive transpiration.   It was subsequently shown by Tschirch, that the various forms of stomata found in
members of the xerophilous Mediterranean and Australian floras mostly
conform to this general type of structure.   Many epiphytes have
stomata of a similar character, and an external air-chamber may even
be developed in a moist tropical climate, in the case of plants which
are liable to transpire very actively in bright sunlight, owing to the
fact that their leaves are both numerous and large (*e.g. Ficus elastica*).

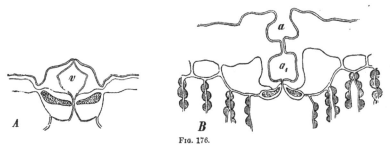

FIG. 176.

*A.* Stoma of *Cypripedium venustum* (upper side of leaf) (T.S.).   Wide front cavity (*v*)
overarched by the large cuticular ridges of the guard-cells.   ×400.   *B.* Stoma of
*Dasylirion filifolium* (T.S.).   External air-chamber divided into two compartments
(*a* and *a₁*) by projecting ridges.   ×670.

The hyperstomatic chambers containing stagnant air, which are of
such general occurrence among xerophytes, do not all originate in the
same way.   In *Cypripedium venustum, Clivia nobilis*, etc., these chambers
consist of the enlarged front-cavities of the stomata (Fig. 176 A), while
each of the very slightly sunken stomata on the photosynthetic stems
of *Euphorbia Tirucalli* is surrounded by a thick annular wall of wax
(Fig. 177).   According to Volkens, each of the stomata which occur
on the outer surface of the leaf of *Aristida pungens* is overarched by
four solid cuticular pegs ; these meet over the stoma in such a way as
to form a vaulted chamber, which only communicates with the outer
atmosphere through exceedingly narrow crevices.

As a rule, each stoma has its own external chamber ; certain plants,
however (*Nerium Oleander*, spp. of *Coscinium*), develop larger and
deeper chambers, with numerous stomata on their walls, which may be
termed collective or common external air-chambers.   In many steppe-
and desert-plants, in the genus *Casuarina*, etc., these common air-
chambers are replaced by longitudinal furrows bearing stomata on
their sides.

A great many of the special structural features of stomata are designed to impede the passage of gases—and thus to hinder the escape of water-vapour—through the internal and external air-chambers, or through the pore itself. All such arrangements act in the same general way, namely, by constricting or complicating the path followed by the diffusing gases. In a very large number of cases, the external air-chamber is partially occluded by wax, in the form of numerous granules or of a more compact porous plug. This particular arrangement was first observed in certain Conifers by Link and later by Wilhelm, while Wulff has recorded its occurrence in a number of Angiospermous families (GRAMINEAE, LILIACEAE, BETULACEAE, CASUARINEAE, SILENEAE, etc.). The external air-chambers of *Dasylirion filiforme* are

FIG. 177.

Stoma of *Euphorbia Tirucalli* (T.S.). External air-chamber surrounded by an annular Wall of wax.

constricted in a totally different manner; here a pair of stout flat-topped ridges project inwards from opposite sides of the chamber, which is thus divided into two superimposed compartments, approximately equal in size and communicating with one another by means of an exceedingly narrow slit (Fig. 176 B). According to Guttenberg, a very similar state of things is found in *Ruscus aculeatus*, where the front cavity of the stoma is divided into upper and lower halves by a pair of (sharp-edged) cuticular ridges, which only leave a very narrow fissure open (Fig. 178).

We are indebted to Bobisut for a description of the very peculiar stomata of *Nipa fruticans*. In this case the ventral wall of each guard-cell is furnished with several cuticular ridges of various sizes and shapes; the ridges of the two guard-cells fit into one another in the most exact manner, as a glance at the adjoining illustration (Fig. 179) will show.

Very frequently it is the internal air-chamber that is provided with arrangements which tend to restrict gaseous interchange. In *Kingia australis* and *Xanthorrhoea hastilis*, according to Tschirch, the chamber is lined by mechanical cells or cell-processes of peculiar shape. In the former case each air-chamber is shut off from the palisade-tissue by an irregularly twisted, knobby stereide. The lateral processes of this "protective cell" are all firmly attached to the neighbouring cells of the fibrous hypodermis. It is evident that the air-chamber is not

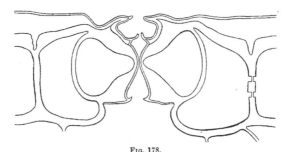

FIG. 178.

Stoma from a cladode of *Ruscus aculeatus* (T.S.). After H. v. Guttenberg.

hermetically sealed against the adjoining tissues. As Tschirch says, such a chamber may be compared to a cave with its mouth closed by a rock of very irregular outline; communication with the interior of the leaf is restricted, but not absolutely prevented; gases can pass freely

FIG. 179.

Stoma from the lower side of the leaf of *Nipa fruticans* (T.S.) After Bobisut.

between the protuberances of the protective cell. In *Xanthorrhoea* a similar partial occlusion of the internal air-chamber is effected with the aid of inward prolongations of the adjoining elements of the sclerenchymatous hypodermis (Fig. 180 A).

According to Pfitzer and Gilg, very remarkable arrangements for the restriction of transpiration are found, in connection with the internal air-chambers, in the RESTIACEAE, a xerophilous family of Monocotyledons peculiar to South Africa and the Australian Continent. In most cases (e.g. *Elegia nuda, Restio tectorum, R. paniculatus, Dovea mucronata*) each of the air-chambers, which project deeply into the photosynthetic tissue, is surrounded by a group of columnar "protective stereides," with cutinised walls. These stereides form an uninterrupted sheath around the upper or stomatic end of the chamber, but separate slightly from one another below; the narrow intercellular clefts produced in this way might almost be compared to a second, internal system of stomata (Fig. 180 B). In *Restio nitens* and *Lepidobolus Preissianus* the internal air-chamber is enclosed, except at its lower end, by a funnel-shaped sheath composed of inward prolongations of the epidermal elements immediately adjoining the subsidiary cells of the stoma.

Some plants contrive to block up the internal air-chambers of their stomata, either during prolonged periods of drought, or when the guard-

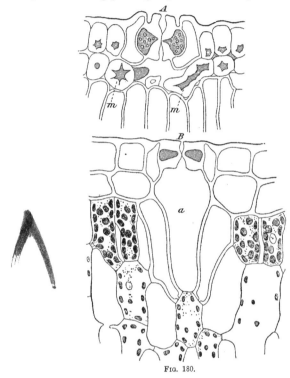

Fig. 180.

Partial occlusion or investment of the internal air-chamber by means of stereides.
*A. Xanthorrhoea hastilis*, after Tschirch ; *m*, stereides. *B. Elegia deusta.*

cells die or from some other cause lose the power of closing their pores effectually.   Most frequently this occlusion depends upon the fact that the immediately adjoining mesophyll cells grow out into the air chambers after the manner of tyloses.   This process has been described by Schwendener in the case of old leaves of *Camellia japonica* and *Prunus Laurocerasus* ; the lenticel-like developments observed by the author on the petioles of certain deciduous trees (*Aesculus, Acer, Tilia, Juglans*) are evidently of a similar nature ; here the air-chambers are filled with tightly packed cells—which finally die—derived from a special secondary meristem.   In *Pilea elegans* the air-chambers on the adaxial side of the leaf become invaded by parenchymatous cells, containing abundant protoplasm but few chloroplasts.   These cells show

2 G

a decided tendency to thicken their outer tangential walls. As a rule, one intrusive cell in each chamber develops its cellulose pad immediately below the stoma, which thus becomes completely occluded (Fig. 181). When plants of *Tradescantia viridis* are grown in the relatively dry atmosphere of an ordinary room, the epidermal elements

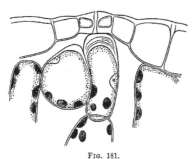

FIG. 181.

Occluded stoma from the upper side of the leaf of *Pilea elegans.* (T.S.)

adjoining the stomata put forth vesicular processes which become closely apposed to the guard-cells and thus obstruct the pore.[226] Another remarkable mode of stomatic occlusion has been observed by the author in *Dischidia bengalensis,* an epiphyte which is very common in Java. During dry weather the hypostomatic air-chambers of this plant become completely filled with a resinous substance, which is highly refractive and readily soluble in alcohol; this secretion appears to be derived from the subsidiary cells, which put forth vesicular processes towards the chamber; it generally extends outwards as far as the pore passage.

The thickness of the outer epidermal walls varies in different parts of the same plant; in the same way, the above-mentioned arrangements for the "protection" of the stomata may attain various degrees of development in different organs. As the accompanying drawing (Fig. 182) shows, even the two surfaces of the same leaf may differ in this respect. In *Populus pyramidalis,* most of the stomata of the adaxial leaf-surface are provided with a moderately deep, slightly funnel-shaped external air-chamber (Fig. 182 A); on the lower side of the leaf the stomata are scarcely depressed at all (Fig. 182 A). Similarly, in *Plantago major,* the front cavities of the adaxial stomata are on an average considerably wider than those of the abaxial stomata (cf. Figs. 182, B and B₁). In both the cases mentioned, the protective features of the stomata are more pronounced on the side of the leaf that is more exposed to the influence of factors—such as light and warmth—which tend to increase the activity of transpiration.

So far, attention has been mainly directed to those structural features of stomata, which serve to restrict transpiration, and which hence are characteristic of xerophytes. Some account must now be given of the peculiarities of the stomata of plants which thrive in a humid climate, or in moist localities. The leaves of certain floating water-plants have been examined by the author from this point of view.[227] Here the stomata often diverge widely from the ordinary "terrestrial" type; as a rule, closure of the stoma, instead of being brought about by apposition of the bulging ventral walls of the guard-cells, is entirely dependent upon the more or less close approximation of their external cuticular ridges, which are always well

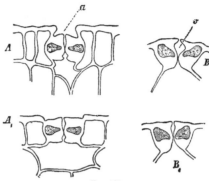

FIG. 182.

Unequal development of stomata on the two sides of a leaf. *A*. Stoma from the upper leaf-surface of *Populus pyramidalis*; guard-cells someWhat depressed, so that a shalloW external air-chamber results (*a*). *A₁* Stoma from the loWer leaf-surface; depression of guard-cells barely perceptible. *B*. Stoma from the upper leaf-surface of *Plantago major*; *B₁* stoma from the loWer leaf-surface. The former has the Wider front cavity (*v*). (All in T.S.)

developed (Fig. 183). The differentiation of front cavity, pore passage and back cavity is incomplete or altogether absent; the pore expands rapidly beneath the narrow fissure enclosed by the cuticular ridges, and thus opens directly by a wide aperture into the hypostomatic air-

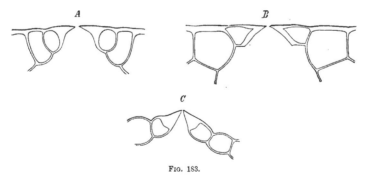

FIG. 183.

Stomata of the aquatic type. *A. Trianea bogotensis. B. Lemna minor.*
*C. Alsophila* sp. (All in T.S.)

chamber. Stomata of this type occur, for example, in *Lemna minor*, *Trianea bogotensis, Hydrocharis Morsus Ranae, Nymphaea alba, Victoria regia, Euryale ferox,* and *Trapa natans.* In certain other cases, on the other hand, in which the cuticular ridges are equally well developed

(*Potamogeton natans, Limnanthemum nymphoides* and *Alisma natans*),
the closure takes place in the ordinary manner by apposition of the
ventral walls.

With regard to the physiological significance of this "aquatic"
type of stoma, it is quite certain that the mere absence of any serious
danger of excessive transpiration does not suffice to explain the prevalence of such stomata among floating water-plants.  In the author's
opinion, the arrangement in question serves to diminish the risk of
capillary occlusion of the stomata by water; for liquid cannot lodge
between the sharp edges of the cuticular ridges, except in the form of
a thin film which readily bursts, and is quickly removed by evaporation.

Stomata of a very similar type have been observed by the author
in a large proportion of the plants inhabiting the constantly moist
wooded glens near Tjibodas, on the slopes of Gedeh in Western Java;
the species in question belong to the most diverse taxonomic groups
(*e g.* FILICALES, COMMELYNACEAE, ACANTHACEAE, GESNERACEAE).

### 5.  *The stomata of Gymnosperms, Pteridophytes and Bryophytes.*

Among the Archegoniatae, deviations from what has come to be
regarded as the normal type of stoma are even more frequent than

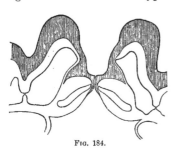

FIG. 184.

Stoma of *Juniperus macrocarpa.*  (T.S.)
After Klemm.

they are in the Angiosperms.  These
modifications are undoubtedly in part
adaptive in character, while in other
cases they merely represent cases of
variation of design.  A phylogenetic
classification of the various types of
stoma which occur in the Archegoniatae, has been attempted by Porsch.

From the researches of Kraus,
Tschirch, Mahlert, Klemm, Porsch and
others,[228] it appears that the stomata
of most Gymnosperms conform to a
type of construction which is principally characterised by the fact, that
the front cavity is reduced to a narrow fissure, of uniform bore or slightly
dilated in the middle (Fig. 184).  There are no internal cuticular
ridges, and the space enclosed by the ventral walls of the guard-
cells passes gradually over into the internal air-chamber, so that there
is in such cases nothing comparable to the pore passage and back cavity
of the normal stoma.  According to Porsch, feebly developed internal
ridges are present in a few cases (e.g. *Podocarpus nèriifolia*); here it is
permissible to speak of a back cavity which communicates with the
hypostomatic chamber through an unusually wide aperture.  As a rule,

the dorsal walls of the guard-cells are much thicker than the ventral walls, both as a rule being lignified and partially cutinised; a limited area of the dorsal wall always remains unmodified to permit of the inward and outward movements of water that accompany the turgor variations of the guard-cells. The cavities of the guard-cells are generally narrow, and have their long axes directed obliquely downwards and backwards (*i.e.* away from the pore). Freedom of movement of the stoma is always ensured by the development of hinge-areas. The stomata of Gymnosperms are almost invariably more or less deeply sunk. According to Copeland, the stomata of *Ginkgo* and *Larix* resemble those of Grasses in structure.

Among Pteridophytes the stomata display a considerable diversity of structure. Those of *Equisetum* approximate to the ordinary Gymnospermous type; the dorsal walls of the guard-cells are strengthened by transverse thickening ridges which project into the cavities of the subsidiary cells. Many hygrophilous Ferns have stomata of the "aquatic" type (Fig. 183 c). The guard-cells of *Salvinia natans* are thin-walled on all sides, and the pore is of the same width throughout. The guard-cells of the closely allied *Azolla caroliniana* are remarkable in more respects than one. They have an irregularly three- to five-sided outline; the pore is elongated at right angles to the plane of the septa between the two guard-cells, and these septa finally become partially or entirely obliterated. As in *Salvinia*, the guard-cell walls are entirely unthickened. In these two genera, the numerous hairs on the leaves afford sufficient protection against the risk of capillary occlusion of the stomata; that, in the author's opinion, is reason why these stomata are not constructed according to the usual aquatic type.

In the Mosses,[229] stomata are confined to the sporogonium [sporophyte], where they often occur in large numbers on the apophysis,—which is the principal organ of photosynthesis—and to a smaller extent on the capsule wall [where the latter is green]. The mechanism of the *Mnium* stoma has already been explained (cf. above, p. 452). A fully developed stoma of *Funaria hygrometrica* or of *Physcomitrium pyriforme* consists not of a pair of guard-cells, but of a single annular cell, which is provided with two symmetrically arranged nuclei and owes its origin to the complete coalescence of two primarily distinct guard-cells.

A single capsule of *Polytrichum* may bear three different types of stomata, namely: (1) "two-celled" stomata, with the normal paired guard-cells; (2) "one-celled" stomata, with fused guard-cells; and (3) four-celled stomata, in which each of the two primary guard-cells is divided in two by a median transverse wall.

The meaning of the fusion of the guard-cells which takes place in

*Azolla* and *Funaria* is quite unknown. It is conceivable that the obliteration of the septa serves to prevent any inequality of turgor-pressure on the two sides of the pore. The condition of *Polytrichum*, where the "one-celled" stomata are accompanied both by the four-celled and by the two-celled forms, seems inexplicable except as an instance of variety of design.

Among Liverworts, typical two-celled stomata do not occur, except on the sporogonium of *Anthoceros*. In the MARCHANTIALES the gameto-phyte is furnished with characteristic pneumathodes. But these differ

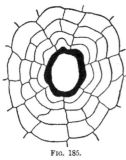

FIG. 185.

Air-pore of *Fegatella conica.*
(Surface view.)

so markedly from the stomata in other Bryophyta and Vascular Plants, that they are preferably distinguished under the name of air-pores.[230]

According to Leitgeb these air-pores may be either **simple** or **tubular**. A simple air-pore typically occupies the centre of an approximately hemispherical raised area of the epidermis, and is surrounded by several concentric series of epidermal cells. In *Fegatella conica* (Figs. 185 and 186) the walls of the innermost circle of cells are produced towards the circular pore in the form of a membranous border which thins out to a sharp edge, so that a vertical section of the pore recalls the structure of a stoma of the aquatic type. In this Liverwort, all the cells immediately underlying a pore are produced upwards into long, colourless, beak-like processes of unknown

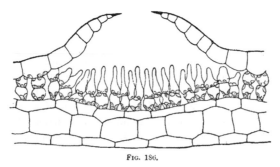

FIG. 186.

Air-pore of *Fegatella conica.* (Vertical Section.)

significance. Kamerling's suggestion that these cells represent an "evaporating apparatus" is purely speculative. Simple air-pores of various kinds are characteristic of the vegetative branches of the majority of MARCHANTIALES. The thalli of *Preissia* and *Marchantia*, on the other hand, and the gametophores in all cases, are provided with tubular air-pores; these are more or less barrel-shaped structures, consisting of several superimposed tiers of cells, enclosing a tubular passage (Fig. 187). The lowermost tier usually differs from the others, and was regarded even by Mirbel as a closing apparatus, witness his name of *anneau obturateur*. Experimental evidence as to its

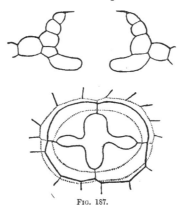

FIG. 187.

Air-pore of *Marchantia polymorpha*. Above, as seen in vertical section; below, as seen from below, in order to show the papillose cells of the lowermost tier.

function has been obtained by Kamerling, who succeeded in demonstrating that these lowermost cells can in many cases effect partial (*Fimbriaria Stahliana*), or even complete (*Plagiochasma Aitoniana*), closure of the pore. This basal tier usually consists of relatively large cells, and tends to contract more or less actively when turgor is reduced. The tubular pores of *Marchantia polymorpha* are immoveable. Here each of the pore-cells of the basal tier is provided with a large inwardly directed papilla-like process; as a result, the pore only communicates with the underlying air-chamber through a comparatively narrow cruciform fissure. It is not improbable that this arrangement is designed to retard gaseous interchange, and thus to restrict transpiration.

### 6. *Subsidiary cells.*

Very frequently, those cells of the epidermis which immediately adjoin the guard-cells of stomata, differ in structure from the typical epidermal elements; cells of this kind pertain both anatomically and physiologically to the stomatic apparatus in the wide sense [and are hence termed the subsidiary cells of the stomata]. The function of these subsidiary cells is not precisely the same in all cases. In the CYPERACEAE and BROMELIACEAE their thin tangential walls act as "hinge-areas." Where an external air-chamber is present, its sides and roof are often formed by the subsidiary cells. In some RESTIACEAE it is the subsidiary cells that send forth the thick-walled processes which enclose the internal air-chamber, while in *Tradescantia viridis* the same cells are responsible for the occlusion of the stomata which takes place under certain conditions.

Benecke has drawn attention to the fact that the presence of two or more subsidiary cells is particularly characteristic of succulent plants, and of xerophytes in general.[231] He suggests that these cells help to preserve the guard-cells from the injurious effects of the tensions that inevitably result from the shrinkage which leaves undergo through loss of water. Where the outer and inner tangential walls of the subsidiary cells are thin, they will undoubtedly undergo deformation when the epidermis contracts, and will thus protect the guard-cells from compression to a certain extent. Their action in this respect is essentially similar to their function as hinge-areas; in either case their effect is to preserve the stoma from being deformed owing to the rigidity of the epidermal walls.

### 7. *Distribution and location of stomata.*

The fact that the stomata serve to place the intercellular air-spaces of the plant in communication with the external atmosphere, sufficiently explains why these structures are restricted to aërial organs, and why they occur mainly where the subjacent tissue is at least moderately well provided with intercellular spaces, but are rare or altogether wanting on portions of the epidermis which are underlain by sclerenchyma or collenchyma.

The root-system is probably the only portion of the plant-body which never bears stomata. These structures are, however, usually present in the greatest numbers upon foliage leaves, in accordance with the fact that the leaves are generally the special photosynthetic organs, and also the most actively transpiring portions of the plant. On an average, the number of stomata borne on 1 sq. mm. of leaf-surface amounts to between 100-300. Much higher figures are recorded for special cases; thus the lower leaf-surface of *Olea europea* bears 675 stomata per sq. mm. (according to Weiss) and that of *Brassica Rapa* 716 (according to Unger).[232] The number of the stomata—like their detailed structure—is correlated with the transpiratory activity of the plant, and hence with the nature of its habitat; but the relation of structure to environment in this respect is far more complicated than might at first sight be supposed; for, as a matter of fact, the structure of the stomata varies greatly in different plants. The differences in the diameter of the pore passage are very considerable, the position of the guard-cells with reference to the general level of the epidermis is variable, and so forth.    Hence the amount of gaseous interchange that is effected by a single stoma in one plant, may require two or three stomata for its performance in another species. Consequently, we must not expect to find a *close* correspondence between the number of the stomata and the humidity of the surroundings.    All that can be stated in general terms is, that the stomata are on the whole least numerous, where the external conditions are most xerophytic.    This rule may be illustrated by means of the following examples, which have been selected from a more extensive list compiled by Tschirch:

| PLANT. | HABITAT. | NO. OF STOMATA PER SQ. MM. | |
|---|---|---|---|
| | | Upper side of leaf. | Lower side of leaf. |
| *Nymphaea alba* - - - | Water. | 460 | 0 |
| *N. thermalis* - - - | | 625 | 0 |
| *Quercus Robur* - - - | Humid forests. | 0 | 346 |
| *Q. pedunculata* - - - | | 0 | 288-438 |
| *Prunus domestica* · | Orchards. | 0 | 253 |
| *Pyrus Malus* - · | | 0 | 246 |
| *Triticum sativum* · | Fields. | 47 | 32 |
| *Avena sativa* - - - | | 40 | 27 |
| *Sedum acre* - - - | Stony places. | 21 | 14 |
| *Sempervivum tectorum* - - | Dry places. | 11 | 14 |

Such values are naturally most strictly comparable in the case of closely related species.    Thus, comparatively uniform results have been obtained by Spitzer with Grasses and by Ziegeler with species of *Carex*.

In leaves of the dorsiventral type, the stomata are principally located on the lower surface of the leaf, that is, in contact with the well-ventilated spongy parenchyma.    Not infrequently, however, a considerable number of stomata occur in addition on the upper side in connection with the palisade-tissue.    The circumstance that a current of air can traverse the whole thickness of the leaf, in such cases, is no doubt an advantage from the point of view of gaseous interchange.    In isobilateral leaves the stomata are as a rule equally numerous on both faces.    The fact that the stomata of floating leaves are all located on the upper side requires no explanation.

Stomata may be uniformly distributed among the ordinary epidermal cells, or they may be arranged in groups or in longitudinal series.    In the latter case, it is not unusual for all the members of a group or series to have a common internal and external air-chamber. .

As regards the position of stomata with reference to the surface of the leaf [or the organ] on which they are borne, it has already been stated that these structures usually stand at the same level as the ordinary epidermal cells, in the case of non-xerophilous plants. The significance of the "sunken" stomata which occur in xerophytes and in all plants that cannot afford to transpire actively, has likewise been discussed at some length (p. 461). At this stage, therefore, it is only necessary to consider the third possible arrangement, namely, that in which the stomata are raised above the general level of the epidermis. Great variety prevails with regard to the degree of elevation, which is sometimes scarcely visible, while in other instances (*e.g.* on the peduncle of *Cucurbita Pepo*, Fig. 188) it is so pronounced that each stoma comes to occupy the apex of a conical or cylindrical papilla, which is traversed by a passage corresponding to the internal air-chamber. The ecological importance of such raised stomata is not quite clear; from the fact that they not infrequently occur in plants which inhabit most shady spots (*e.g.* many ferns), it might be thought that the exposed situation represents a device for accelerating transpiration. Leaves which are readily wetted are also often furnished with raised stomata; here the elevation may possibly serve to provide protection against prolonged capillary occlusion of the pores by water. On the other hand, elevated stomata are also to be found on some very hairy leaves, in which case they are situated in common external air-chambers or furrows; in this latter instance no plausible explanation of the fact is forthcoming.

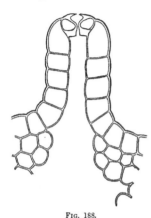

Fig. 188.

Raised stoma from the fruit-stalk of
*Cucurbita Pepo.*

### 8. *Loss or change of function on the part of stomata.*

In many plants, the stomata lose their power of adjustment more or less completely, or at any rate become incapable of closing tightly after a certain age. This physiological degeneration of the stomata takes place at a comparatively early age in floating and other aquatic plants, and also in a number of shade-loving hygrophytes, doubtless because members of these ecological classes never require much protection against excessive transpiration. It should, however, be explicitly stated, that neither Leitgeb nor the author himself have met with a single instance in which the stomata are devoid of the power of adjustment from the very first. In fully developed fronds of *Lemna minor*, however, the guard-cells are, according to the author's observations, practically immovable. In the case of *Limnocharis nymphoides*, treatment with glycerine causes the majority of the stomata to shut, but the closure is often incomplete. Similarly, most of the stomata of *Salvinia natans*— even in the case of young leaves—merely contract without closing properly, when the internal tension of the guard cells is diminished. In *Impatiens parviflora*, a plant with foliage which withers very readily, the stomata behave quite normally while the leaves are young. The same statement applies to the stomata on the upper side of adult leaves; if the abaxial epidermis of one of the older leaves is treated with glycerine, some of the stomata close completely, others merely contract to a certain extent, while others again remain widely open, having apparently lost all power of adjustment. In many other cases, also, different stomata on one and the same leaf may lose their power of movement at different times; this difference is most probably connected with the fact that all the stomata of the leaf are not of the same age.

Stomata which never functionate from their first formation onwards exemplify the process of phylogenetic loss of function. According to the author's observations, the capsules of certain Mosses illustrate various stages in the progressive reduction of stomata. The most complete atrophy of these organs occurs in the genus *Sphagnum*. A capsule of *Sphagnum acutifolium* bears more than 300 (rudimentary) stomata with well-differentiated guard-cells, but not a single one of these develops either a pore passage or an internal air-chamber. This far-reaching reduction of the stomata is correlated with the entire absence of photosynthetic tissue in the capsule of *Sphagnum*. According to Porsch, the stomatic apparatus is always more or less reduced among Phanerogamic saprophytes and parasites. It should be noted, however, that even where there is no photosynthetic gas-exchange, owing to the entire absence of chlorophyll, bulky and actively growing aerial organs can scarcely dispense with stomata altogether, if only in the interests of respiration.

The submerged foliar and axial organs of aquatic and amphibious plants are not infrequently provided with stomata, which may be regarded as vestigial structures, bearing testimony to the terrestrial ancestry of their possessors. According to Porsch, such stomata [which are, of course, always functionless] exhibit a variety of features that tend to prevent the entrance of water through them into the intercellular spaces.[233] Thus they may have fully differentiated guard-cells, but nevertheless always remain closed (*Callitriche verna, Hippuris vulgaris*); in other cases, where the pore is still developed, the external and internal cuticular ridges are closely appressed or interlocked (*Calla palustris, Menyanthes trifoliata*). Complete and permanent protection is ensured, when the guard-cells never separate entirely from one another, so that the stoma remains imperforate. In some of these cases the external cuticular ridges remain fused (*Potamogeton natans*); in others the guard-cells are completely joined together, with the sole exception of the internal cuticular ridges (*Polygonum amphibium, Schoenoplectus lacustris*). In either case, the stoma presents a very characteristic appearance in transverse section.

The pores of the stomata are the only perforations in the typical epidermis; hence it is hardly surprising that these structures are sometimes utilised as outlets for liquid secretions. Illustrations are provided by water-stomata, and by the so-called "sap-valves" of nectaries. The alleged secretion of wax by stomata, in the case of certain fruits (*e.g. Cydonia japonica, Rosa glandulosa, Prunus cerasus*, etc.), requires further investigation.

A very remarkable instance of change of function—discovered by the author and examined more recently by Bobisut—is furnished by the modified stomata which are found on the slippery region of the pitcher of *Nepenthes*. Goebel was the first to draw attention to the presence, on this slippery wax-covered tract of the pitcher, of small cells of crescentic outline with their convex sides facing upwards. The author himself next pointed out, that the shape and arrangement of these cells is such, that they assist insects to crawl down the inside of the pitcher, but offer no foothold to those that attempt to retrace their steps. It has now been shown [by Bobisut], that these peculiar cells are nothing more nor less than the raised guard-cells of transversely elongated stomata, which have [quite given up their original function] and have become transformed into structures accessory to the capture of insects. *Nepenthes Rafflesiana* illustrates various stages in this remarkable process of modification (Fig. 189), which depends essentially upon the fact that the upper member of each pair of guard-cells [*i.e.* that nearest the mouth of the pitcher] projects a good deal further than its fellow, so that the whole stoma acquires a very asymmetrical shape. As seen in a vertical section, the upper guard-cell or "clamp-cell" does not appear very greatly modified; its outer cuticular ridge is normal, or even slightly over-developed; the internal ridge, however, is entirely absent (Fig. 189 A, B, C). The lower guard-cell, which remains flush with the general level of the epidermis, undergoes a far greater alteration of shape. Here the outer cuticular ridge is rarely normal, being generally more or less obviously reduced or altogether suppressed; in the latter case, the "guard-cell" is barely distinguishable from an ordinary epidermal element (Fig. 189 c). The front-cavities of these modified stomata thus persist, though their shape is normally very much distorted ; occasionally the pore passage is also retained, but the back-cavity and internal air-chamber have always disappeared completely. This transformation of some of the guard-cells,

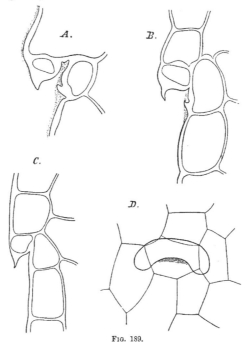

Fig. 189.

Modified stomata from the inner surface of the pitcher of *Nepenthes Rafflesiana*, in their natural positions. *A*. Both guard-cells still clearly recognisable. *B*. LoWer guard-cell greatly reduced, but still furnished With its external cuticular ridge. *C*. LoWer guard-cell entirely reduced. *D*. Front vieW of a completely modified stoma. After Bobisut.

in *Nepenthes*, into structures which assist in the capture of insects, is one of the most remarkable instances of secondary adaptation that is to be met with in the whole field of physiological anatomy.

## 9. *Lenticels.*[234]

The periderm, like the epidermis, is perforated at a number of points by pneumathodes, which in this case are termed **lenticels** (the name is due to De Candolle). The mode of development of these structures clearly indicates their physiological affinity with stomata; for, on the young shoots of our native trees and shrubs, every lenticel arises—as Unger discovered long ago—immediately beneath a stoma, the functions of which it assumes, when the epidermis is replaced by periderm.

A fully developed lenticel is a biconvex (lenticular) mass of special tissue, embedded in the periderm. Its inner boundary is marked by a

meristematic zone, which is continuous with the phellogen on every
side ; but the bulk of the lenticel is made up of so-called **complementary
tissue** containing a large number of intercellular spaces.   Stahl distinguishes between two leading types of lenticel, which differ in the
nature of their complementary tissue.   In the simpler of these types
(illustrated  by  *Sambucus*  [Fig.  190],  *Lonicera*,  *Euonymus*,  *Salix*

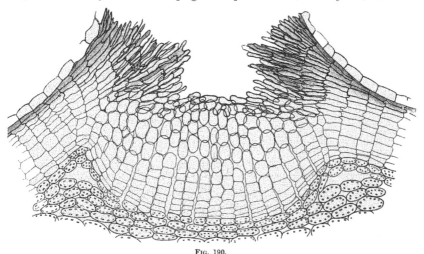

FIG. 190.

Lenticel of *Sambucus nigra* in transverse section (summer of the second year of growth).

*Ginkgo*, etc.), the complementary cells are somewhat closely united,
and form a tissue which is relatively compact and firm, in spite of the
numerous air-spaces that it contains.   The second type, which is much
more widely distributed, is characterised by the rounded form and
loose arrangement of the complementary cells.   This incoherent tissue,
which in addition is usually brittle or powdery, is prevented from
falling to pieces by the circumstance that the meristematic zone of
the lenticel occasionally produces so-called **closing layers**, composed of
one or more series of cells which are firmly united to one another.
These closing layers constitute relatively firm diaphragms, which give
cohesion to the whole complementary tissue ; they are traversed by
radial intercellular passages, and hence do not interfere with gaseous
interchange.   Sooner or later the closing layers burst, owing to the continued production of complementary tissue, and are replaced by freshly
formed layers of the same kind.   Examples of the second type of
lenticel are furnished by *Ulmus*, *Robinia*, *Sophora*, *Alnus*, *Betula*,
*Sorbus*, *Prunus*, *Aesculus*, etc., etc.   Both kinds agree in the fact that
the complementary cells are dead, more or less full of air, and provided with cell-walls which are thin and frequently brown in colour.

The perviousness of lenticels to gases might be inferred from their anatomical structure. Beneath every lenticel the intercellular spaces of the cortex are continued as passages right across the lenticellar meristem into the complementary tissue, where they ramify in every direction. Since the closing layers, if such are present, are also perforated, and since the epidermis becomes ruptured over every fully developed lenticel, an open communication is established between the cortical ventilating system and the outer air. Experimental proof of the pneumathode function of lenticels is also easily obtained in the following manner: One end of a severed twig, bearing lenticels, is occluded, by coating it with wax or in some other way, while the other is attached to the shorter arm of a J-tube; the whole apparatus is then placed in water so that the twig is completely immersed. If mercury is now poured into the longer arm of the J-tube, a rapid stream of air-bubbles will escape from the lenticels even under a small excess of pressure.

What has been stated regarding the significance of lenticels for gaseous interchange in general, naturally applies to the case of transpiration in particular. Lenticels are by no means as important as stomata in this connection; nevertheless, the author has thought it worth while to collect some numerical data concerning the value of lenticels as paths of transpiration. The method employed was that of direct weighing. Pieces of moderately old branches of various trees were rendered airtight at both ends by means of sealing-wax. The lenticels of some of these pieces were closed with asphalt varnish. On other pieces, which served as controls, an equal number of areas of periderm, of the same size as the lenticels, were similarly coated with varnish. The loss of weight by transpiration was observed daily for both sets of branches, and in this way a quantitative estimate of the influence of the lenticels upon transpiratory activity was obtained. A selection from the results of these experiments is reproduced in the accompanying table. The figures represent the loss of weight in a stated period of time, expressed as a percentage of the initial weight of the twig.

| Time. | Sambucus nigra. | | Triaenodendron caspicum. | | Morus alba. | |
|---|---|---|---|---|---|---|
| | Lenticels. | | Lenticels. | | Lenticels. | |
| | Open. | Closed. | Open. | Closed. | Open. | Closed. |
| After 5 days, - - | 10·6 | 7·66 | 5·35 | 3·58 | 9·76 | 9·26 |
| „ 10 - - | 19·65 | 15·9 | 11·1 | 7·69 | 19·84 | 17·47 |
| „ 15 „ - - | 28·02 | 23·71 | 16·41 | 12·18 | 27·75 | 24·62 |

It is quite clear that the rate of transpiration of the same branch varies considerably, according as the lenticels are open or closed. Klebahn obtained concordant results with branches belonging to other species of trees.

As regards the distribution of lenticels, the author has found that these structures are scattered uniformly over the surface in the case of erect branches, while they tend to be aggregated upon the lower sides of horizontal twigs; but the asymmetrical distribution in the latter instance becomes much less pronounced as the branches grow older. *Gleditschia triacanthos* illustrates the dorsiventral arrangement in its most extreme form; in one set of estimations a piece of yearling twig 20 cm. in length was found to bear, on an average, 210 lenticels on the lower side, as compared with 72 on the upper surface. In view of the close physiological relation between stomata and lenticels, there can be no doubt, that the preponderance of these pneumathodes on the lower surfaces of the organs that bear them, is determined by the same con siderations in both cases. According to Devaux, lenticels also tend to congregate in the vicinity of leaf-bases; the same author states that one or two lenticels invariably occur close to the insertion of adventitious roots.

The so-called **ventilating pits** (*Staubgrübchen*), which are characteristic of the stems and petioles of CYATHEACEAE and MARATTIACEAE, have certain features in common with lenticels. These organs have recently been studied in some detail by Hannig.[235] They consist of rounded or oval depressions filled with a loose complementary tissue, which is abundantly provided with intercellular spaces; they arise beneath stomata, and their development leads to the rupture of the overlying portion of the epidermis. In all these points they closely resemble ordinary lenticels. Although they are not provided with a continuous meristem, the immediately adjoining cells of the parenchymatous ground tissue give rise to the complementary tissue by irregular division. There can hardly be any doubt that these structures are really pneumathodes. In the MARATTIACEAE they only remain functional as long as the leaf is not completely unfolded, and close as soon as the lamina begins its photosynthetic activity.

## C. THE PNEUMATHODES OF AËRIAL ROOTS.

### 1. *The pneumathodes of breathing-roots.*

It has already been explained (p. 443) that the breathing-roots of *Jussiaea* ultimately throw off their epidermis, so that the aërenchyma comes into direct contact with the surrounding medium. As a rule, however, breathing roots are provided with special pneumathodes; in

*Avicennia officinalis* and *Laguncularia racemosa*, the periderm, which is several layers in thickness, is interrupted by structures agreeing in all essentials with ordinary lenticels. The breathing-roots of *Sonneratia acida* are surrounded, according to Goebel, by three-layered sheets of cork, which alternate with one- to two-layered zones of rounded cells; the latter resemble the complementary cells of lenticels, while the thin sheets of cork may be compared to the closing layers of those organs. As Karsten has remarked, the root of *Sonneratia* may be said to be invested by a continuous pneumathode; it is as if a single large lenticel had spread over the whole surface of the organ. According to Jost, the breathing-roots which occur in certain Palms, notably in the genus *Phoenix*, are provided with pneumathodes of a peculiar kind. When viewed with the naked eye, they appear as white powdery areas, which may be confined to the dilated root-tip, or may, on the other hand, be located on annular swellings further back. Microscopic examination shows, that at these points the subepidermal fibrous tissue is replaced by a mass of rounded sclerenchymatous elements, separated by narrow intercellular clefts, which in turn is overlain by a layer of thin-walled "spongy tissue," likewise richly provided with intercellular spaces. If the experiment described above in connection with lenticels is performed upon these roots, it is found that here also a very slight excess of pressure suffices to force air through the pneumathodes.

## 2. *The pneumathodes of aërial roots.*[236]

The pneumathodes which occur on the aërial roots of Orchids were described in detail by Leitgeb; but Schimper was the first to give a correct account of their function. To the naked eye they appear as fairly well-defined white spots, which are most conspicuous when the velamen is saturated with water. As a rule they are elongated parallel to the long axis of the root, and arranged in fairly regular longitudinal series; in cylindrical roots they are evenly distributed on all sides, whereas in flattened dorsiventral roots, such as those which constitute the photosynthetic organs of *Aëranthus fasciola* and *Taeniophyllum Zollingeri*, they are confined to the side which is next the substratum, and shaded.

Each of these pneumathodes comprises three distinct parts, which pertain respectively to the velamen, the exodermis, and the cortical parenchyma. The outermost portion consists of a wedge-shaped mass of velamen cells, distinguished by the fact that the spiral thickening fibres are unusually numerous, or that the cell-walls are uniformly thickened except for elongated pits. The outstanding physiological feature of these cells is the tenacity with which they

2 H

retain air in their cavities, when the rest of the velamen becomes
filled with water.   The inner or narrower edge of each wedge-shaped
mass is attached to a single, more rarely to two or three, thin-walled
air-containing exodermal cells (Fig. 191 L), which in their turn abut
against one or more rows of special rounded parenchymatous elements
(Fig. 191 L); these last-mentioned cells were termed " *cellules aquifères* "
by Janczewski on account of the colourless and watery nature of their

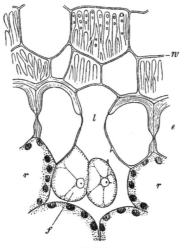

FIG. 191.

Pneumathode of a photosynthetic root of *Taeniophyllum Zollingeri* (T.S.) ;  *w*,
velamen ; *e*, exodermis ; *l*, exodermal cell containing air, its loWer right-hand Wall
perforated by a large hole ; *r*, green cells of the cortex ; *f*, complementary cells.

contents, but should more properly be compared to the complementary
cells of lenticels, in view of the fact that they are separated from one
another by wide intercellular spaces which extend as far as the aforesaid
thin-walled air-containing cells of the exodermis.   In *Taeniophyllum
Zollingeri*, the author has noticed that the inner wall of this cell breaks
down in one or two places where it abuts against an air-space ; in this
way large holes arise which permit of free communication between the
cortical air-spaces and the cavities of the thin-walled cells of the
exodermis.

Since the outer walls of the air-containing exodermal elements
always remain intact, the pneumathodes of these Orchidaceous roots
do not actually represent .open outlets of the ventilating system ; but
Schimper has shown that all the cell-membranes in such pneumathodes
are highly pervious to air, so that a sufficiently active gaseous interchange can go on by diffusion through the cell-walls.

## IV. ONTOGENY OF THE VENTILATING SYSTEM.

Though the ventilating system is always composed of air-containing intercellular spaces, it may develop in a great variety of ways, just like other tissue-systems. As a matter of fact, the majority of ventilating spaces arise by a process of separation which depends upon the splitting of a common primary cell-wall or middle lamella. The air-passages and -cavities of the NYMPHAEACEAE, and of *Papyrus antiquorum*, *Lemna*, *Trapa*, *Potamogeton*, *Ceratophyllum*, etc., have this schizogenous origin. In other cases ventilating spaces originate **lysigenously** (rhexigenously), *i.e.* by the collapse and disintegration of transitory tissues, which previously become pervaded by narrow schizogenous spaces; this case is illustrated by the air-passages of most CYPERACEAE, GRAMINEAE, and EQUISETALES. As regards the details of development, both lysigenous and schizogenous ventilating spaces display the greatest diversity.[237]

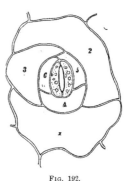

FIG. 192.

Stoma and subsidiary cells of *Sempervivum*, sp. The numerals indicate the course of segmentation.

Aërenchyma sometimes originates from fundamental meristem, but in other cases arises from a secondary meristematic layer, which is the phylogenetic equivalent of the phellogen

The cells which compose the stomatic apparatus, in the wide sense, pertain either to the protoderm or to the fundamental meristem. The guard-cells are, of course, always sister-cells, and invariably arise from a protodermal mother-cell. As a rule, however, certain preparatory divisions, which give rise to the subsidiary cells of the stomatic apparatus, precede the segregation of the actual guard-cells.[238] The details of these preliminary divisions vary considerably in different cases; it must suffice here to refer briefly to the development of stomata in two specific cases, namely in the CRASSULACEAE and in *Mercurialis* (Fig. 192). In both these instances the protodermal primordial mother-cell of the stoma divides after the fashion of an apical cell; in the CRASSULACEAE the segments are cut off in three planes, while *Mercurialis* adheres to the scheme of a two-sided apical initial.

Lenticels may arise either upon young stems, which still retain their epidermis, or upon older branches, after the formation of periderm has begun. Their development, in the former case, is described by Stahl as follows. The parenchymatous cells adjoining the internal air-chamber of a stoma increase in size and divide; the resulting daughter-cells give rise to a colourless complementary tissue which fills up the air-chamber. A curved cell-layer, convex towards the inside, undergoes tangential divisions and becomes the "lenticellar meristem," which continually cuts off additional complementary cells on its outer side. The pressure of the steadily expanding complementary tissue causes the epidermis to bulge outwards, and finally to burst; the complementary tissue then protrudes in places, hence the rough surface of the lenticel. On older branches, which are already covered with periderm, the lenticels arise from the phellogen, which, at certain points, produces complementary tissue with abundant air-spaces in place of the normal uninterrupted layers of cork, and thus becomes locally converted into lenticellar meristem. The layer of cork above a developing lenticel suffers the same fate as the epidermis in the other type of development, being distended and finally ruptured by the growing complementary tissue.

# CHAPTER X.

## SECRETORY AND EXCRETORY SYSTEMS.

### I. GENERAL CONSIDERATIONS.

ALL green terrestrial plants give off gaseous substances, in connection with the functions of transpiration, photosynthesis, and respiration. In the majority of cases various liquid compounds—or, it may be, solutions of solid bodies—are also liberated through the mediation of secretory organs composed of one or more specialised cells. The physiological importance of this process of secretion is very diversified, just as it is in the animal kingdom. Thus, many plants secrete liquid water, when transpiration is suppressed, an arrangement which prevents the hydrostatic pressure within the conducting system from becoming excessive, and hence protects the ventilating system against the danger of flooding. Such water - secreting organs, which vary greatly as regards the details of their construction, are termed **hydathodes**. Other plants, again, are furnished with special organs for the secretion of digestive enzymes; these structures, which may be compared to the digestive glands of animals, obviously play an important part in the general metabolism of the plant. Many secretions, such as *ethereal oils, resins, mucilage*, etc., serve for various ecological purposes, such as restriction of excessive transpiration, protection against animal foes, or, on the other hand,—as in the case of nectar—attraction of desirable insect-visitors. Finally, it is not improbable that some of the secretory organs of plants are entirely concerned with excretion of useless by-products of metabolism, like the kidneys or nephridia of animals.

Sometimes the products of the activity of secretory organs pass out directly from the organism. In other cases they are, to begin with, deposited within the confines of the plant-body, though outside the secretory cells; here exudation follows at a later stage, or may not take place at all. The escape of the secretion from the plant-body is therefore not an essential, though it is a common feature of vegetable secretory organs. The real distinctive character, which at once differentiates secretory organs from excretory reservoirs, consists in the fact that the secretion is removed from the organ or cell in which it is produced.

The above comprehensive definition takes no account either of the form of the secretory organ or of the mechanism of secretion. In the sequel the term **gland** will be applied to those localised secretory organs in which the production and liberation of the secretion is carried out by the living protoplasts of the secretory cells. The substance secreted by a gland may escape at once to the outside (as in the case of glandular hydathodes and nectaries), or it may pass into a **glandular cavity** of one kind or another. When the gland is superficial, the glandular cavity usually owes its origin to a local separation of the cuticle from the rest of the epidermal wall. In the case of internal glands, on the other hand, the cavity corresponds to an intercellular space, which may arise either by the separation or by the disintegration of the secretory cells. It is accordingly customary to discriminate between **schizogenous** and **lysigenous** glands, although the two types are connected by transitional forms.[239]

The principal difference between **excretory reservoirs** and **secretory organs** consists in the fact that the cells of which the former consist, or from which they develop, *contain* [and permanently retain] by-products of metabolism, *i.e.* substances which are valueless from a nutritive point of view. Excretory substances may secondarily acquire a considerable amount of ecological importance (for example, as agents of protection against animals); there are, however, good reasons for believing that many processes of excretion merely serve to remove useless waste-products from the photosynthetic and conducting system. It is, therefore, always easy to distinguish between a secretory organ and an excretory reservoir. There may, however, occasionally be some doubt as to whether a particular cell should be assigned to the excretory or to the storage-system ; one cannot, indeed, draw any sharp line of demarcation between useless metabolic by-products and substances of metabolic origin which are capable of further utilisation. Moreover, one and the same substance (such as *tannin*) may in some cases represent a plastic material, while in others it is purely excretory in nature.

It is, naturally, impossible to make many general statements concerning the structure of secretory and excretory organs. The simplest types of secretory organ are unicellular ; more often, however, as has already been stated, a larger or smaller number of secretory cells are closely associated together. Since the actual secreting elements are thin-walled and therefore exposed to injury, they are frequently provided with arrangements for mechanical protection. Secretory cells generally contain a well-developed protoplast and a large nucleus. The disposition

of secretory organs naturally varies according to the particular requirements in connection with which they are developed.

Excretory reservoirs may occur as isolated idioblasts, or they may
be associated to form rows, or even more or less massive layers of cells.
In a number of cases fusion of originally independent cells takes place.
Often the completion of the excretory process is followed by the suberisation of the walls of the reservoir, a modification which prevents
any subsequent effusion of noxious excretory substances; in this way
the excretion is as effectually withdrawn from the metabolic cycle as
if it had been actually ejected from the plant-body. The arrangement
of excretory reservoirs is determined by a variety of considerations.
The most influential factor, of course, is the location of the particular
tissue or organ that is ultimately responsible for the production
of the excretory substance. It must, further, be of advantage to a
plant, if the excretory reservoirs are situated where they are least
likely to interfere with the physiological activities of the several tissues,
and with their mutual relations; at the same time, it is important that
the excretory substances should be as far as possible eliminated from
the general metabolic cycle of the plant. Thus, excretory organs, especially when associated together in numbers, would be quite out of place, if
they were located at the surface of contact of two tissue-systems which
entertained an active intercourse with one another. Finally, the fact
that an excretory substance has an ecological value, may influence, or
even determine, the location of the organs in which it is produced.

## II. SECRETORY ORGANS.

### A. HYDATHODES.[240]

Many of our native plants are provided with organs which secrete
water in the liquid form. Such hydathodes are even more widely
distributed among plants inhabiting the humid tropics. They are
most frequently located on leaves. As a rule, their activity only
begins when the hydrostatic pressure in the water-conducting system
(the so-called "root-pressure" or "exudation-pressure") reaches a
certain intensity as a result of inhibited or reduced transpiration.
After a damp night, the leaves of hydathode-bearing plants are studded
with drops of water, each of which marks the position of a hydathode.
Formerly this secreted water was often confused with dew.

#### 1. *Structure and arrangement of hydathodes.*

The author has shown that the detailed structure of hydathodes
is subject to a considerable amount of variation. To begin with, we

may confine our attention to those **epidermal hydathodes** which do not
communicate directly with the water-conducting system ; these always
correspond either to modified epidermal cells or to multicellular
trichomes.

In *Gonocaryum pyriforme* (ICACINACEAE) numerous unicellular hydathodes are interspersed among the ordinary epidermal cells on both sides

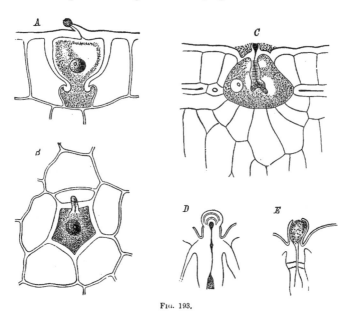

FIG. 193.

Unicellular hydathodes. *A* and *B. Gonocaryum pyriforme. A.* Hydathode in L.S.
*B.* Hydathode in surface view. *C-E. Anamirta cocculus. C.* Hydathode in L.S.
*D.* Papilla more highly magnified, showing the mucilaginous tip. *E.* Papilla with
tip open.

of the leaf.    Each water-secreting cell consists of three portions
(Fig. 193 A).    The thick outer wall is produced outwards into a
small oblique papilla, the extreme tip of which is of a mucilaginous
consistency.    This papilla is traversed by a narrow longitudinal canal,
which communicates with the cell-cavity at its inner end ; at its other
extremity the canal ends blindly in the distal cap of mucilage, or, if the
latter has been washed away by rain, opens directly to the exterior.
The papilla thus constitutes the outer opening of the hydathode.    The
central and largest portion of the hydathode is shaped like a four- to
six-sided funnel.    At an early stage of development, the internal
aperture of this funnel becomes surrounded by a stout projecting
flange of cellulose ; later on its lateral walls also undergo a considerable
amount of thickening; like the flange and the outer wall, they are strongly

cutinised. The third and innermost portion of the hydathode, finally, consists of a thin-walled chamber, which is sharply delimited from the thick-walled funnel. It is highly probable that this chamber regulates the volume of the whole hydathode in accordance with the large alterations of pressure which doubtless occur under varying conditions. The hydathodes of *Gonocaryum* contain a well-developed protoplast and a large nucleus, features which are typical of glandular cells in general. On an average, there are 58 of these water-secreting organs per sq. mm. of surface on the lower side of the leaf; the corresponding number for the upper side is 55.

No less remarkable is the structure of the unicellular hydathodes which occur on both sides of the leaves of *Anamirta Cocculus* (MENISPERMACEAE) (Fig. 193 c). Each lies in a shallow depression of the surface, and resembles an inverted funnel in shape. The inner and lateral walls, though unthickened, are strongly lignified. To the middle of the thickened outer wall is attached a peculiar "filtering apparatus," composed of a small external papilla, and a long peg-shaped internal process, which usually terminates in a number of knobs. The peg is traversed by a longitudinal canal, which is narrow for the greater part of its length, but expands to a varying extent in the papilla. At the tip of the papilla the cuticle is absent, or at any rate (Fig. 193 E) perforated by numerous pores. The cellulose layers in the papilla are modified in a mucilaginous manner. This part of the cell, therefore, really represents a funnel filled with mucilage, open above and continuous below with the narrow lumen of the peg. The outer layers in the wall of the peg are all strongly lignified; a delicate striation which often appears in this part of the structure, is due to the presence of numerous narrow, transversely elongated pits. In this case, also, the hydathode contains a massive protoplast and a large nucleus. According to Krafft, very similar water-secreting organs occur in the allied genus *Arcangelisia*.

From such unicellular hydathodes the transition is easy to the multicellular "water-glands" of the PLUMBAGINACEAE, the structure and action of which have been studied in detail by Volkens. Here the approximately hemispherical gland consists, as a rule, of four central and four peripheral cells. All these eight cells have abundant protoplasmic contents and unthickened walls. Even the outer walls are thin. Only those partitions which separate the gland from the surrounding tissues are somewhat thickened, besides being highly refractive and insoluble in sulphuric acid. Each gland arises by means of appropriate divisions from a single protodermal mother cell; it is usually enclosed by four epidermal subsidiary cells, which appear crescent-shaped in section (Fig. 194).

In a very large number of cases epidermal hydathodes take the shape of **multicellular trichomes,** which sometimes resemble ordinary tapering hairs, but are more often developed as clavate, capitate or scale-like structures.

The ternate leaves of *Machaerium oblongifolium* (PAPILIONACEAE), a Brazilian liane, are covered on both sides with long stiff hairs, which act as hydathodes. Each hair consists of a five- to six-celled basal portion or " foot," and of a two-celled "body" (Fig. 195 A). When the hair is fully developed, its long tapering terminal cell is devoid of living contents and provided with rather thick lignified walls; the partition which separates it from the short subterminal cell (the other cell of the " body ") is oblique, greatly thickened, strongly lignified, and furnished with a number of narrow slit-like pits. The lateral walls of the subterminal cell are also very thick, but are cutinised instead of being lignified; this cell contains a massive protoplast. The base or foot of the hair is composed of four or five very much flattened, disc-shaped cells placed one above the other; the whole column rests upon a cell which is somewhat less flattened. All the basal cells have massive protoplasts. Their transverse walls are thin, except for a thick marginal zone, which becomes more and more extensive as one passes upwards from the lowermost cell; the lateral walls, on the other hand, are greatly thickened and strongly cutinised, except in the case of the lowermost cell, which has both its lateral and its basal walls unthickened.

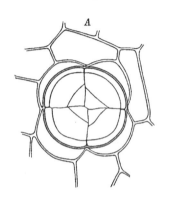

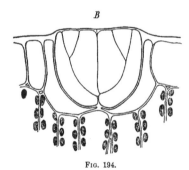

Fig. 194.

Hydathodes from the adaxial leaf-surface of *Plumbago lapathifolia.* *A.* Surface view. *B.* Vertical section. (Contents of the glandular cells omitted in both figures.)

The actual secretion of water is evidently carried on by the short subterminal cell and by the disc-shaped basal cells, which all contain abundant protoplasm. Since, however, these cells have cutinised lateral walls, the secreted water must escape from the living cells of the gland through the lignified oblique septum that separates the terminal from the subterminal cell; as a matter of fact, this wall is provided with numerous slit-like pits. The final exudation of the water must take place by filtration through the lateral walls of the dead terminal cell.

The leaves of *Phaseolus multiflorus* bear, especially on their lower side, curved clavate hairs, which act as hydathodes (Fig. 195 B). Each hair consists of a large, usually somewhat distended basal cell, super-imposed upon which are several (most often four) layers of smaller cells forming the club-shaped " body." As a rule the cells of the

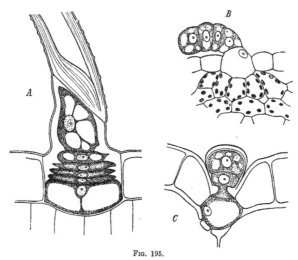

Fig. 195.

Trichome-hydathodes. *A. Machaerium oblongifolium. B. Phaseolus multiflorus. C. Piper nigrum.*

uppermost tier are divided in two by median walls. All the cells have thin lateral walls and abundant protoplasmic contents. Neither the cavities nor the membranes show any trace of oily, resinous or gummy secretion. In many PIPERACEAE (Fig. 195 C), BIGNONIACEAE and CONVOLVULACEAE, in species of *Artocarpus*, etc., the hydathodes take the shape of shortly stalked capitate or scutate hairs. While varying greatly in detail, hydathodes of this type always consist of three portions, namely, a head or scale, a stalk and a base or foot. The uni- or multicellular head is responsible for the actual secretion of the water. Here the external cell-walls are unthickened and covered by a very thin and pervious cuticle, which sometimes becomes distended and finally ruptured owing to the formation of a mucilaginous substance within the wall. In certain CONVOLVULACEAE (*Stictocardia tiliaefolia, Operculina Turpethum*), according to Svedelius, the cuticle is perforated by minute pores. The stalk-cell may be regarded as the mechanical component of the whole apparatus. Its lateral walls are often greatly thickened or at any rate strengthened by means of a stout flange, and are, in addition, nearly always strongly cutinised. As a consequence this cell acts as a rigid annular frame which prevents the "internal water-pore" (*i.e.* the aperture through which water passes from the general mesophyll into the actual secretory organ) from altering its dimensions. The unicellular or multicellular base, finally, which is often more or less distended, is responsible for all communication between the rest of the hydathode and the adjoining epidermal and sub-epidermal cells. It is, accordingly, always thin-walled; its form, moreover, is often such as to bring the greatest possible number of palisade-cells, or other sub-epidermal elements, into actual contact with the base of the hydathode.

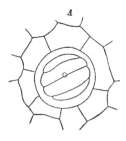

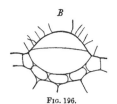

Fig. 196.

Scutate glands from a scale-leaf of *Lathraea squamaria*. *A.* Surface view. *B.* Vertical section; the small rod-shaped bodies adhering to the gland are bacteria.

The epidermal hydathodes which occur in the cavities of the scale-leaves of *Lathraea squamaria*, require separate consideration. The secretion of water by the scale-leaves of this plant was first observed by Charles Darwin, and has since been studied in detail by the author, by Goebel and by Groom. The interior of every scale-leaf of *Lathraea* is occupied by a continuous system of channels and clefts, which opens to the exterior through a single narrow outlet. The wall of this cavity corresponds morphologically to the abaxial surface of the scale-leaf; it bears glandular structures of two distinct types, which may respectively be termed **capitate** and **scutate glands**. Each capitate gland is composed of a one- or two-celled base, a stalk-cell, and two head-cells with abundant protoplasmic contents (Fig. 197).

The scutate glands (Fig. 196) have been studied in detail by Scherffel. Each of these consists, first, of a large lenticular cell, the bulk of which is sunk below the surface of the epidermis; and, secondly, of a tier of four cells superimposed upon the basal cell and collectively forming a lenticular or concavo-convex structure. The two middle cells of the group are not completely joined together, but enclose a groove which is roofed over by the cuticle. This cuticular film is perforated, at the exact apex of the whole gland, by a minute circular pore, not more than ·004 mm. in diameter, through which the aforesaid groove communicates directly with the exterior. The large lenticular basal cell rests upon a rosette-shaped group of eight cells,

separated from one another by more or less well-developed intercellular spaces.  As the study of development shows, this group of
cells represents the base or "foot" of the gland; the large middle cell
may be held to represent the stalk, while the apical tier constitutes
the head, which is the actual secretory apparatus.  The entire sub-
epidermal layer of the cavity-wall resembles spongy parenchyma in
structure; but its intercellular spaces, instead of being filled with air,
contain water which is pumped into them from the tracheides of the
bundle-ends that are present in great numbers in the scale (Fig. 197 I).
The glands are thus not directly connected with
the vascular system, but draw their supply of
water from the sub-epidermal system of intercellular spaces.

It is still uncertain, whether either the
capitate or the scutate glands are alone responsible for the secretion of water, or whether
both organs take part in the process.  Both
Goebel and Groom regard the scutate glands as
the actual hydathodes.[241]

We must now pass on to consider the
second type of hydathode, namely, that which is
characterised by the fact that the gland is in
direct communication with the water-conducting
system.  The simplest variety of this second
type is exemplified by the epidermal hydathodes

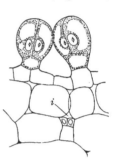

Fio. 197.

Capitate glands from a
scale-leaf of *Lathraea Squamaria* (vertical section); *i*, an
intercellular space in contact
with a group of tracheides
(bundle-end).

which occur on the leaves of many Ferns (*e.g.* spp. of *Polypodium*,
*Aspidium* and *Nephrolepis*), where they are either uniformly scattered
over the surface or else chiefly located in a continuous series along
the leaf margin.  In this case, each hydathode consists of a group of
thin-walled glandular epidermal cells sunk in a shallow depression,
immediately above an enlarged bundle-end, which is largely composed
of tracheides.  An even more widely distributed type of hydathode is
that which is provided with so-called water-pores.  In a large number
of cases, these hydathodes represent modified bundle-ends at the tip
or margin of the leaf; in the MORACEAE and URTICACEAE, many nodal
points of the general vascular reticulum become similarly modified.
The bundle-ends in question are usually enlarged to form a more or less
club-shaped structure, while the terminal tracheides generally diverge in
fan or pencil fashion, and insert themselves between the elements of
the **epithem**, a mass of thin-walled tissue interpolated between the
epidermis and the bundle of tracheides.  The innermost epithem-cells
are often elongated in the same direction as the adjoining tracheides;
all the elements of the epithem offer a marked contrast to the

surrounding mesophyll cells in several particulars ; besides being distinctly smaller, they contain abundant protoplasm and a large nucleus, but little or no chlorophyll. Typically they are more or less isodiametric, and either rounded or polyhedral ; not infrequently they are provided with spherical or irregular extensions (*Papaver, Geranium*), thus approaching the condition of spongy parenchyma (Figs. 198, 199 c).

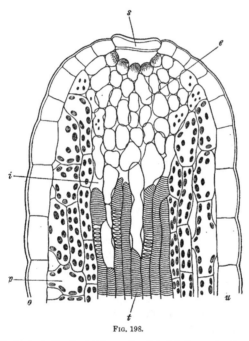

FIG. 198.

Hydathode of a leaf-tooth of *Primula sinensis* in R.L.S. ; *o*, adaxial ; *u*, abaxial epidermis ; *p*, palisade-cells ; *t*, terminal tracheides of a bundle-end, projecting freely into intercellular spaces (*ı*) of the epithem (*e*) ; *s*, guard-cell of the water-pore.

As was first pointed out by Volkens, every epithem-tissue is traversed by a more or less well-developed system of intercellular spaces. The author himself has demonstrated, in the case of *Fuchsia*, that the distal extremities of the terminal tracheides may abut directly against these intercellular spaces. As a rule, the spaces in question are filled with water, and open into one or more sub-epidermal chambers ; these chambers in their turn communicate with the exterior through open **water-stomata** that differ, more or less obviously, from the genuine ventilating stomata from which they are phylogenetically derived. The most important difference between the two structures consists in the fact that the guard-cells of water-stomata have lost their power of adjustment to a large extent. As a matter of fact, they usually become stationary at an early stage of development (*Fuchsia*), while not infrequently they are from the first incapable of movement. In accordance with this relative immobility, the cuticular ridges, which form so characteristic a feature of typical guard-cells—especially when seen in transverse section—are reduced (*Tropaeolum*, Fig. 199 D), or entirely absent. The guard-cells are sometimes short-lived (*Tropaeolum*, *Colocasia*, *Aconitum*), and occasionally disappear altogether (*Hippuris*, *Callitriche*). In certain cases (AROIDEAE, *Papaver*, *Tropaeolum*) the water-stomata are remarkable on account of their large size.

According to Tswett, the marginal epithem-hydathodes on the leaves of LOBELIACEAE are furnished with water-stomata of an unusual type. The wide pore of each stoma is divided in two by a cutinised septum, which evidently represents the middle lamella of the partition wall between the two guard-cells. The cuticle also remains imperforate, so that the pore is never open. The author has confirmed Tswett's observations in the case of *Lobelia syphilitica*. Tswett states that the permanent closure of these water-stomata does not prevent liquid water from exuding under suitable conditions. The cuticle must therefore be pervious to water at these points. What the advantages of this peculiar type of structure may be, is quite unknown.

The greatest diversity prevails with regard to the number of water-stomata that are associated with a single hydathode. In *Fuchsia*, *Aconitum*, *Delphinium*, *Primula*, etc., according to De Bary, the tip of each leaf-tooth is provided with a single large gaping water-pore. Groups comprising from three to six pores occur in *Ulmus campestris*, *Crataegus coccinea*, *Helleborus niger*, etc. In certain UMBELLIFERAE and COMPOSITAE, finally, and in *Potentilla*, *Alchemilla*, *Geum*, etc., numerous water-stomata are crowded together within a small area. The disc-shaped epithem-hydathodes which occur in spp. of *Ficus*, *Conocephalus* and other MORACEAE, are likewise provided with a large number of small water-stomata. In hydathodes of this type the epithem is usually shut off from the adjoining green mesophyll by a parenchymatous bundle-sheath, the cells of which frequently have their inner walls cutinised.

Some very simple hydathodes consist of water-pores without any associated epithem. Here the free ends of the terminal tracheides border immediately upon the intercellular spaces of the mesophyll, which in their turn communicate with the hypostomatic chambers of the water-pores. The leaf of *Vicia sepium* bears a solitary hydathode of this type, with from five to eight water-pores, at the tip of each pinna. In certain Grasses, again (*Secale ccreale*, *Triticum vulgare*, *Zea Mays*), the author has observed epithem-less hydathodes at the tips of the coty-ledonary sheaths and primary leaves; in these cases the water-stomata

are very typically developed and differ strikingly from the genuine or
ventilating stomata (Fig. 199 A, B).

The so-called **apical pores**, finally, which occur on the leaf-tips of
certain [mainly] Monocotyledonous water-plants, represent a peculiar
type of hydathode; the structures in question have been studied
more particularly by Sauvageau, Minden, and Weinrowsky. In these
instances the mid-rib of the leaf terminates in a little depression
(situate usually on the lower surface of the leaf), or in a cavity which
is roofed in by the cuticle (*Scheuchzeria palustris*). Numerous tufts of
tracheides project freely into this pit or cavity, nothing in the nature

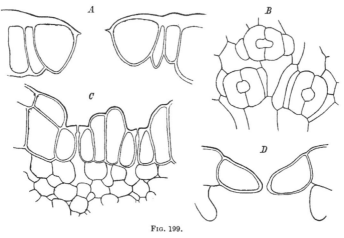

FIG. 199.

Water-pores. *A* and *B. Secale cereale* (tip of the cotyledonary sheath). *C. Cono-*
*cephalus ovatus. D. Tropaeolum majus. (B* in surface view; the rest in vertical
section.)

of an epithem being present. The depression owes its origin to the
degeneration of the guard-cells of water-pores (*Callitriche, Ranunculus*
*aquatilis*), or to the disintegration of groups of epidermal cells together
with adjoining sub-epidermal layers; as already stated, the cuticle alone
sometimes remains intact at these points. It is worthy of note, that
the distal ends of the tracheides often extend right up to the epidermis
even before the depression is formed. In *Aponogeton distachyum*, indeed,
Minden has observed that some of the epidermal cells undergo division,
become spirally thickened, and thus secondarily assume the characteristics
of tracheides. In this way the terminations of the water-conducting
system are pushed forward to the actual surface of the plant.[242]

## 2. *Function and importance of hydathodes.*

From a physiological point of view, it is convenient to distinguish between two different types of hydathode; in one case the secretion of water is a mere filtration under pressure, in the other the hydathode itself takes an active part in the process.

The category of **active hydathodes** comprises all epidermal water-secreting organs (including both the unicellular and the multicellular or trichome types among Angiosperms, as well as the water-glands of Ferns), and also the epithem-hydathodes of *Conocephalus* and *Ficus* (and perhaps also those of other MORACEAE and URTICACEAE). In all these cases the energy employed in the process of secretion is supplied by the glandular hydathode cells themselves. It is they that develop the force which causes water to exude. The exudation-pressure in the water-conducting system merely supplies the stimulus which sets the pumping action of the hydathodes going.

The author succeeded in proving experimentally that the hydathodes in question are really active water-glands—comparable to the sweat-glands of animals—by painting the surfaces of leaves provided with such organs with a weak ('1 per cent.) alcoholic solution of mercuric chloride, a procedure which kills the living cells of the hydathodes. If subsequently to this treatment the natural exudation-pressure in the water-conducting system rises, or if it is replaced by artificial pressure according to the methods of De Bary, Moll, etc. (the severed branch being attached to the shorter arm of a J-tube, and water forced into it under the pressure of 10 to 40 cm. of mercury), no liquid escapes from those leaves or leaf-regions on which the hydathodes have been killed. If only a portion of the leaf-surface is painted with alcoholic corrosive sublimate, an increased flow of water will be found to take place from the un-injured hydathodes. When, on the other hand, all the hydathodes on a leaf are poisoned, water-secretion stops altogether, and the inter-cellular spaces of the mesophyll become more or less completely blocked with water.

In the case of epithem-hydathodes which are provided with water-stomata, the process of secretion as a rule consists in a simple filtration, the requisite energy being provided by root-pressure, or, in more general terms, by the exudation-pressure which may be set up in stems and branches as well as in the root system. Here the hydathodes represent little more than points of least resistance to filtration. The anatomical structure of the epithem indicates the course followed by the water after it leaves the distal end of the terminal tracheides; some of the latter abut directly upon water-containing intercellular spaces, which in their turn communicate with the chambers underlying the water-

2 I

stomata. The path described evidently offers the smallest resistance to filtration, and is doubtless that by which the secretion actually escapes. The passive character of the epithem may be inferred from the fact that, if the tissue in question is poisoned or put out of action by anaesthesia or by exposure to extremes of temperature, artificial pressure will induce a flow of water quite as readily as it does when the hydathodes are in their normal condition. This conclusion was arrived at by the author on the basis of some experiments performed upon *Fuchsia globosa*; in view of the results previously obtained by Moll, who succeeded in forcing the red juice of *Phytolacca* and a 1 per cent. solution of tannic acid through the leaf-teeth in various plants, the same argument probably applies to the great majority of plants that are provided with epithem-hydathodes.

It now becomes necessary to explain the function of the epithem, which is usually a very well-defined tissue. Wilson and Gardiner noted that severed branches of *Fuchsia*, when kept in a moist atmosphere with the cut ends immersed in water, secreted a small amount of liquid, evidently as a result of active exudation on the part of the epithem tissue. Although this process of active secretion is not great, it no doubt suffices to keep the intercellular spaces *of the hydathodes* full of water, and thus to preserve the water-conducting system from contact with the outer air; this, in fact, most probably constitutes the special function of the epithem wherever the secretion of liquid water is merely a process of filtration under pressure. In *Conocephalus*—and probably in other MORACEAE also—the power of active secretion possessed by the epithem, which was originally feeble and of secondary importance, has become accentuated to such an extent as to become responsible for the entire water-secreting activity of the plant. It is not at all unlikely that further investigation may reveal the existence of various transitional stages between the types of hydathodes exemplified respectively by *Fuchsia* and by *Conocephalus*. In *Vicia sepium* and in the GRAMINEAE, where epithem is absent, the process of secretion must obviously consist entirely of filtration under pressure.

The quantity of water secreted by hydathodes is often very considerable. An adult leaf of *Colocasia antiquorum*, for example, was found by Duchartre to secrete, on an average, between 9 and 12 g. of water in the course of a night; on one occasion, however, it produced as much as $22 \cdot 6$ g. Molisch estimated the amount of water secreted in a single night by a young leaf of *Colocasia nymphaefolia* as varying from 48 to 97 c.c. In this plant the water exudes from a hydathode situated immediately behind the tip of the leaf. Here the liquid is actually *ejected* in drops, which follow one another in rapid succession. This curious fact was first recorded by Muntingh in 1672, and has since been verified by Musset and Molisch; the last-mentioned observer has seen as many as 163 drops ejected in a single minute. The significance of this process of "guttulation" is still obscure. According to an observation made by the author, an adult leaf of *Conocephalus ovatus*, weighing 13·02 g., secreted 2·76 g. of liquid in a night, a quantity equal to 26 per cent. of its fresh weight.

The liquid secreted by hydathodes is naturally not pure water. The solid matter present there is, however, as a rule quite small in amount; it comprises both organic and inorganic compounds. According to Unger, the liquid that escapes from the water-stomata of *Zea Mays* contains ·05 per cent. of solid matter, of which ·027 per cent. represents ash. The secretion of *Colocasia antiquorum* contains ·056 per cent. solids, but only ·008 per cent. of ash, that of *Brassica cretica* ·1 per cent. of solids and ·042 per cent. of ash. Van Romburgh finally states that the secretion of *Conocephalus ovatus*, on being dried at 100°, leaves ·045 per cent. of solid residue, which, when calcined, is reduced to ·02 per cent. of ash. Evidently the liquid secreted by leaves during the night contains no *appreciable* quantity of nitrogenous or non-nitrogenous by-products of metabolism.

In certain plants larger quantities of inorganic materials are liberated along with the water secreted by the hydathodes. Thus Wetter states that the secretion of the above-mentioned trichome-hydathodes of *Phaseolus* leaves nearly ·5 per cent. of solid residue, consisting principally of *potassium carbonate*. The same substance is present in the liquid secreted by the very similar hydathodes that occur on the leaves of various MALVACEAE. In such cases, when the water evaporates, the dissolved salts remain behind in the form of minute crystals. The salts in question are very **hygroscopic**, and hence deliquesce rapidly in moist air; as a result the hydathodes may seem to retain the power of secreting water even on severed leaves or leaf fragments—that is to say, independently of the general exudation-pressure of the root or shoot. A similar phenomenon had previously been noted by Volkens in the case of certain TAMARISCINEAE (*Reaumurea hirtella* and *Tamarix articulata*), and of some other desert plants. Here the saline efflorescence (consisting mainly of *chlorides of sodium, calcium* and *magnesium*, but also including *nitrates* and *phosphates*), forms a veritable incrustation, which actively absorbs moisture from the air at night. Volkens assumes, that the plant is able to withdraw water from the resulting saline solution. The absorption of water by living cells from a *concentrated* solution of salts, although not inconceivable, certainly seems somewhat improbable. Marloth, on the other hand, believes that the white saline crusts reduce the effects of insolation, and also that the cooling effect due to the evaporation during the day of water absorbed by night possesses a similar ecological significance, since it must tend to counteract the heating effect of the sun. Most probably, however, the secretion of saline matter in such instances mainly represents a device for obviating the accumulation of excessive quantities of mineral matter within the plant-body.[243]

More widely distributed even than these "salt-glands" are the "chalk-glands" which occur in a large number of families; here a varying amount of calcium carbonate is secreted together with the water and remains as a scaly deposit when the water evaporates. In certain species of *Saxifraga*, the "chalk-glands" simply represent epithem-hydathodes situated in depressions in which lime accumulates. In certain Ferns (*Lomaria attenuata*, species of *Polypodium* and *Nephrolepis*), and in many PLUMBAGINACEAE, the hydathodes similarly become encrusted with scales of secreted lime. A remarkable feature is the occasional provision of arrangements which prevent the scales of lime from falling off. In *Limoniastrum*, for example, the glands are situated, according to Volkens, at the bottom of a cavity, which is cruciform in section below and contracts to a simple tubular form above. Here it is quite impossible, in any circumstances, for the scale to fall out of its socket, just as a key cannot be removed from a lock in which it has been partially turned. In *Statice pruinosa* the epidermal cells adjoining the gland are produced into inwardly direct hook-shaped processes, which grip the scale from either side. Arrangements of this kind certainly suggest that these chalky scales — which are always present in considerable numbers — serve some useful purpose ; as the plants concerned are all desert-plants, or at any rate pronounced xerophytes, Volkens is probably right in supposing that these structures help to restrict transpiration.[244]

In many cases hydathodes may act secondarily as aërial water-*absorbing* organs, which are made use of when the normal water-supply is deficient. It is particularly the epidermal hydathodes (*e.g.* those of *Gonocaryum, Anamirta, Phaseolus* and *Machaerium,*(cf. above p. 488) that tend to assume this twofold responsibility. Minden has, however, also observed the entrance of water through epithem-hydathodes in various seedlings. Prolonged wilting of the leaves will, however, cause the intercellular spaces of the epithem to become full of air, whereupon any further entrance of water is, of course, rendered impossible.

In conclusion, we have to consider the question of the ecological significance of hydathodes. There seems no doubt that these organs play a very important part by regulating the water-content, and hence the turgor, of the foliage leaves and the whole plant-body. The activity of the hydathodes prevents that injection of the intercellular spaces with water which tends to result from any sudden increase of exudation-pressure ; such injection, though apparently not in itself injurious, would undoubtedly interfere with the nutritive metabolism of the leaves, by retarding the gaseous interchange attendant upon the photosynthetic process. The presence of hydathodes also ensures that, even when transpiration is entirely suppressed, a sufficient amount of water continues to flow through the plant, carrying with it mineral salts which are retained by the cells of the mesophyll. From this point of view hydathodes must be regarded as organs of great importance in relation to nutritive metabolism, especially in the case of plants inhabiting humid tropical regions. The same argument applies to the " apical pores " of waterplants, which, as Sauvageau, Minden and others have shown, likewise keep up a flow of water throughout the plant-body. This " hydathode-current " [or " secretion-current "] is, however, always feeble in comparison with the " transpiration-current " of land-plants, as indeed may at once be inferred from the prevalent reduction of the water-conducting system among aquatic plants.

In the genus *Lathraea*, the activity of the hydathodes in the leaf-cavities doubtless serves to draw large quantities of sugar-containing sap from the host-roots into the parasite, and particularly into the scale-leaves themselves. In this way the parasite can absorb a great amount of plastic material in a short space of time ; as a matter of fact, the scale-leaves do fill up rapidly with starch in springtime.

In certain cases hydathodes are adapted for very special purposes, and they may even become transformed into organs which perform functions other than the secretion of water. The first observations on such modified hydathodes were made by Treub upon the flowers of *Spathodea campanulata* (BIGNONIACEAE). The calyx of this plant forms a closed sac, which bears numerous trichome-hydathodes upon its inner surface ; these secrete a large quantity of liquid, within which the corolla and essential organs of the flower can develop without risk of desiccation. Koorders has since shown that such **water-calyces** are very widely distributed among tropical plants ; they occur, for example, in several other BIGNONIACEAE, and also among the SOLANACEAE (*Jochroma macrocalyx, Nicandra physaloides*), VERBENACEAE (*Clerodendron*), SCROPHULARIACEAE, etc. The " post-floral " water-calyces discovered by Svedelius in certain CONVOLVULACEAE probably serve to protect the developing fruit against premature desiccation.[245]

By a complete change of function hydathodes become converted into digestive glands or into nectaries.

### B. DIGESTIVE GLANDS.

#### 1. *The digestive glands of carnivorous plants.*[246]

Since Charles Darwin's classical investigation of the subject, it has been a matter of common knowledge that a number of species, representing very diverse families of Angiosperms, have acquired a highly characteristic method of nutrition, termed the "insectivorous" [or "carnivorous"] habit. The plants in question capture insects [and other small animals] by means of specially constructed foliar organs, hold them captive until they are dead, digest their carcases and absorb the soluble products of digestion. This is not the place to discuss the varied structure of the organs which serve as "traps." In the simplest cases they do not differ appreciably from ordinary leaves, while the trapping arrangement consists of the sticky digestive glands that occur in large numbers on the adaxial surface of the leaf (*Drosera*). Where specialisation and division of labour have progressed a little further, special capturing hairs may be developed (*Pinguicula*), or the leaf itself, or a portion thereof, may be converted bodily into a trap (*Dionaea, Nepenthes, Sarracenia*), in which case the digestive glands serve solely to produce a digestive solution (though in certain instances they perhaps also assist in absorbing the soluble constituents of the carcase).

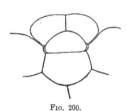

Fig. 200.

Digestive gland of *Pinguicula vulgaris* (vertical section).

Morphologically considered, all digestive glands correspond to trichomes of one kind or another. The leaves of *Pinguicula* bear glands of two distinct types on their upper sides. The capturing hairs are all glandular trichomes, with long stalks and discoid heads which secrete a sticky, mucilaginous substance. The other glandular structures are sessile, and consist of a discoid head or body, a short concavo-convex stalk cell and a basal cell, both the latter being sunk below the general level of the epidermis (Fig. 200). The body of these glands is usually made up of eight cells, which arise from a common mother-cell by means of a quadrant-division followed by the appearance of an anticlinal wall in each quadrant. All the cells contain abundant protoplasm, differentiated into a central mass enclosing a nucleus which contains numerous protein crystals, and a peripheral region distinguished by the presence of numerous vacuoles of various sizes. The lateral walls of the stalk-cell are cutinised. These sessile glands are the true digestive structures; their surfaces are perfectly dry until they are stimulated by contact with a dead insect [or other animal matter], whereupon they at once begin to secrete a digestive enzyme. The shortly-stalked shield-shaped glands of *Dionaea muscipula* behave in a similar fashion.

In the genus *Nepenthes*, the corresponding glands are sessile, spherical or cake-like structures, attached by a disc-shaped base to the floor of a depression with an overhanging upper margin (Fig. 201). The multicellular body of the gland consists of a large number of central cells surrounded by radially elongated, superficial secreting elements. A delicate bundle-end terminates in a tuft of short and rather wide tracheides immediately beneath each digestive gland, an arrangement which suggests that the glands are well supplied with water. As a matter of fact, these glands begin at an early stage of development to secrete a watery, mucilaginous liquid; this collects in the pitcher on which the glands are borne (they are situated on its inner surface near the bottom).

The most remarkable digestive glands of all are the tentacles of the genus *Drosera*, which not only serve as digestive organs but also represent capturing organs, endowed with special forms of irritability, and finally carry out the absorption of the soluble products of digestion. The tentaeles are multicellular, glandular villi which densely clothe the margins and adaxial surfaces of the orbicular or oblong-spathulate leaf-blades. Each consists of a stalk,—

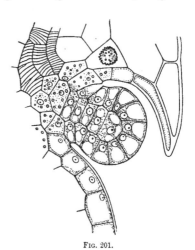

FIG. 201.

Digestive gland of *Nepenthes Phyllamphora* (vertical section).

which is longest in the case of the marginal tentacles—and of a swollen club-shaped head or body (Fig. 202). The whole structure is traversed by a bundle-end; in the stalk this consists of a single row of tracheides with closely wound spiral thickening fibres, while the head or body contains a row of short tracheides. Evidently the intense secretory activity of the tentacle necessitates an increase in the water-supply, which is ensured by the extension of a tracheidal strand into the tentacle. The large terminal tuft of tracheides is surrounded by three layers of parenchymatous cells, the innermost of which—the intermediate layer ("Mittelschicht") of Goebel—resembles a typical endodermis in having corrugated and suberised radial walls. The actual secretory cells form the outermost layer; they are radially elongated, especially at the apex of the gland, and their cuticle is very pervious. Both this superficial layer and the layer next to it contain a reddish-purple cell-sap. In the case of the marginal tentacles the preceding description only applies to the upper side, the lower surface of the gland being covered by an epidermis of the ordinary type.

In the allied genus *Drosophyllum* it can be shown by direct observation, that the remarkable permeability of the cuticle which covers the tentacles is due to the presence of innumerable microscopic pores. If the protoplasmic contents of the superficial cells are dissolved by treatment with Eau de Javelle, the cuticle appears minutely punctate in surface view, while in section it seems to be marked with delicate transverse striae.

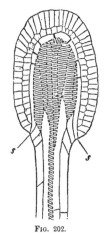

FIG. 202.

Digestive gland (tentacle) of *Drosera rotundifolia* (L.S ) ; *ss*, endodermis-like bundle-sheath. ×145. (From De Bary, *Comp. Anat.*)

The liquid secreted by the digestive glands [of carnivorous plants] contains a proteolytic enzyme, which, in its properties, closely resembles the pepsine of gastric juice, together with a certain amount of free organic acid (*propionic, formic, acetic* and *malic acids*) which assists the enzyme in its work of dissolving the protein constituents of the carcase in every case this process of solution results in the formation of peptones. In certain cases the secretion of digestive juice is an autonomous process, while in others it only takes place after chemical or mechanical stimulation.

From the phylogenetic point of view, the digestive glands of carnivorous plants may be regarded as modified trichome-hydathodes. This view, first put forward by Goebel, has been definitely established, as regards *Pinguicula*, by the author himself, who has also adduced strong evidence in its favour in the case of *Nepenthes*. In *Pinguicula* the lower surface of the leaf bears typical hydathodes, which agree closely in structure with the sessile digestive glands of the upper side. It seems probable, therefore, that the non-carnivorous ancestors of *Pinguicula* had both sides of their leaves furnished with water-secreting and -absorbing glandular hairs, with a structure approximating to that of the hydathodes on the lower leaf-surface of the living species of *Pinguicula*. For a reason which has been suggested by Goebel, the adaxial glands probably soon acquired the habit of secreting a somewhat mucilaginous fluid. A mucilaginous secretion adheres better to the leaf-surface and, perhaps, also evaporates more slowly, both of which properties obviously facilitate its reabsorption [at a later stage]. In this way the capture of insects became a possibility, though at first this process must have been purely incidental in relation to the ordinary activities of the leaf. Subsequently the adaxial hydathodes became transformed into digestive glands, and the plant was thus enabled to make better use of the advantages resulting from the carnivorous habit. As a final result there ensued the division of labour between the sessile and stalked glands to which attention has already been directed.

In all probability the digestive glands of other carnivorous plants have had a very similar evolutionary history.

### 2. *The digestive glands of embryos.*[247]

As Sachs has clearly explained, the mobilisation of reserve-materials which takes place during the germination of seeds, serves not only to render the various stored up plastic materials soluble, and thus suitable for translocation, but also to transform them into compounds which can be immediatelv utilised by the growing tissues. It is probable that this mobilisation is always brought about by the action of enzymes. Hansen is therefore, no doubt, justified in terming the whole process of mobilisation " diges-tion," by analogy with the corresponding processes which take place in animals. In the majority of cases the protoplasts of the storage-cells are themselves capable of secreting sufficient quantities of the appropriate enzymes. Sometimes, however, particularly when the endosperm of a rapidly germinating seed is being emptied, the storage-cells are assisted in

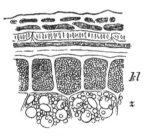

Fig. 203.

T.S. through the peripheral tissues of a resting Wheat-grain ; *s*, spermoderm (pericarp and testa) ; *kl*, aleurone-layer; *z*, cells of the starchy endosperm. ×300.

their work of enzyme-secretion by special digestive glands, which then become responsible for the solution of the bulk of the re-serve-materials. Researches upon this subject have hitherto been restricted to the GRAMINEAE. When germination first begins in these plants, the absorbing tissue of the scutellum secretes diastase, which attacks and dissolves the starch-grains in the adjacent layer of the endosperm. Soon afterwards, however, diastase also exudes from the so-called **aleurone-layer**, a tissue which was formerly assigned to the storage-system, but which, in the author's opinion, really represents a glandular digestive organ.

In the resting seed, the aleurone-cells, which constitute the outer-most layer of the endosperm, contain numerous aleurone-grains (Fig. 203, *kl.*), which, in spite of their minute size, possess the usual strue-ture (including the presence of a globoid). Each cell further contains a

somewhat oily protoplasmic matrix and a centrally situated nucleus.
In shape these cells are prismatic and often somewhat elongated radially
(*Secale, Triticum, Avena*). Typically the aleurone-layer is only one
cell thick, but in *Oryza sativa, Arrhenatherum elatum* and certain other
species it becomes two-layered, and in *Hordeum* even three- to four-
layered, owing to tangential division. The inner and lateral walls are
more or less thickened, and have been shown by Tangl to be traversed
by innumerable delicate protoplasmic strands.

When germination is in progress, the aleurone-layer presents a very
different appearance. While the starch-containing portion of the
endosperm becomes converted (in Rye, Wheat, Oat, etc.) into a soft

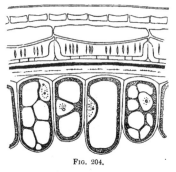

FIG. 204.

T.S. through the spermoderm and aleurone-
layer of a germinating Rye-grain.

pasty mass, the aleurone-layer remains
perfectly continuous, although it becomes completely separated from the
adjoining cells of the starchy endosperm. The reserve-materials stored
in the aleurone-cells are utilised in
the development of massive protoplasts, such as are characteristic of
glandular cells in general (Fig. 204).
The fact that these reserve-materials
of the aleurone-layer are utilised
locally in the manner described,
instead of being translocated to a
distance, clearly proves that the
layer in question cannot form part of the storage system, but must
be concerned with some special function demanding the presence
of massive protoplasts. Even in the later stages of its existence the
aleurone-layer never seems to give up any plastic materials to the
embryo. On the contrary, the protoplasts in this layer appear to
undergo fatty degeneration before they die, the cell-cavities becoming
filled with an increasing number of highly refractive globules, some of
which, at any rate, exhibit the reactions characteristic of fatty oils.
These globules are still to be found in the dead aleurone-cells, long
after the starchy endosperm has been entirely depleted, and the seedling
has attained to a state of nutritive independence.

That the above-described aleurone-cells, with their dense protoplasmic contents, do actually secrete diastase during germination, is
rendered highly probable by a fact which was known to Tangl among
others, namely, that corrosion and solution of starch-grains begin much
earlier in those endosperm cells that lie immediately beneath the
aleurone-layer, than they do in the central portion of the endosperm.
The aleurone-layer of Rye is imperfectly developed in the neighbourhood

of the ventral furrow, and it is precisely in this region that solution of starch is longest deferred.   The author has further shown that the aleurone-layer is capable of secreting diastase by the following experiment.   Fragments with a superficial area of several sq. mm. were removed, with the help of fine scissors, from the spermoderm of germinating grains of Rye, Wheat and Maize ; the adhering aleurone-layer was then separated from these fragments by careful washing with a weak (1-2 per cent.) solution of sugar, and covered with a paste of Rye-flour or starch.   After twenty-four hours most of the starch-grains were already deeply corroded or even broken up.   Starch-paste placed upon damp filter paper, to serve as a control, showed little or no corrosion in the same time ; the active corrosion that takes place in contact with the aleurone-layer cannot, therefore, be due to the diastatic activity of Bacteria.

The diastase secreted by the aleurone-layer is not derived at second-hand from the absorbing epithelium of the scutellum, as Tangl supposed, but is actually produced within the aleurone-cells themselves. This point can also be proved by experiment.   For this purpose, a shallow groove must be cut with a scalpel in a dry grain of Maize, parallel to, but just outside, the margin of the scutellum, so as to interrupt the continuity between scutellum and aleurone-layer.   When a grain which has been treated in this way germinates, it behaves in precisely the same manner as a normal grain, corrosion and solution of the starch-grains appearing first, as usual, in the cells immediately adjoining the aleurone-layer.

The diastatic activity of the aleurone-layer generally varies with the rate at which the growing embryo uses up the sugar produced by the hydrolysis of starch.   If a grain of Rye or Wheat which has been deprived of its embryo, is induced to germinate, no solution of starch takes place, as Sachs long ago remarked.   The aleurone-cells of Maize have some power of secreting diastase in the absence of the embryo, but even here this independent activity comes to end after a few days.[248]

How far the so-called " aleurone-layers," " albuminous layers " or " protoplasmic layers " which occur in various other families of Phanerogams, act as digestive glands, is a matter of uncertainty.

## C. NECTARIES.[249]

The sugary secretion [or " nectar "] that serves to attract pollinating insects to flowers, is produced by special glandular structures, the **nectaries** (nuptial or floral nectaries), which may be located on any of the various floral organs.   The nectar-secreting tissue is generally superficial, consisting, in fact, of modified epidermal cells, which are often papillose, or palisade-like in form.   Frequently the underlying thin-walled parenchyma is also more or less glandular in character. The superficial cell-layer is often furnished with stomata,—which, according to Behrens, often act as nectar-valves—and sometimes with arrangements for keeping the secretion in contact with the secreting surface. In many UMBELLIFERAE, for example (*Pastinaca sativa, Heracleum Sphondylium, Daucus Carota*), the outer walls of the nectar-secreting cells are provided with cuticular ridges, arranged in an intricate curvilinear pattern, which help to retain the nectar by adhesion. The "septal nectaries," which occur among the LILIIFLORAE and SCITAMINEAE, belong to the category of internal glands; according to Grassmann, they arise by the partial non-coalescence of the margins of adjacent carpels. When fully developed, they consist of branched or unbranched canals and crevices in the substance of the pericarp, lined by papillose or palisade-shaped secretory elements. Special outlets are provided for the escape of the nectar.

With regard to the mode of secretion of nectar, it is necessary to discriminate between two distinct processes namely, the exudation of some osmotically active substance, and the subsequent escape of water, the actual nectar being formed by the interaction of these two substances. The osmotically active material may be liberated by one of two methods: either certain layers of the superficial walls undergo an appropriate chemical modification ; or else the compounds that initiate the "osmotic suction" exude from the interior of the secreting cells. That the escape of water from nectaries depends upon "osmotic suction," has been proved experimentally by Wilson. The flow of liquid ceases immediately, if the osmotically active substances are removed by washing with water, and starts at once again if a minute fragment of moistened sugar is placed on the surface of the nectary. Nectar—that is, as already explained, the actual secretion diluted with water—always contains a large amount of sugar ; according to Bonnier, *gum, dextrine, mannite* and certain compounds of *nitrogen* and *phosphorus* may also be present in small quantities ; the same author states, that the water content of nectar varies between 60 and 85 per cent.

A great many plants are furnished with so-called **extra-nuptial** or **extra-floral nectaries,** which may occur within the floral region, on sepals, bracts, etc., or on purely vegetative parts, such as petioles and stipules. According to Belt, Delpino and others, these organs serve to attract protective ants, or, as in the case of *Nepenthes* and *Sarracenia,* to lure insect victims to their doom. Extra-nuptial nectaries exhibit considerable diversity as regards their histological features. Some of them closely resemble ordinary floral nectaries ; others are mainly or exclusively composed of thin-walled palisade-like secretory elements,

representing modified epidermal cells, or of groups of trichomes, which
are most frequently clavate or shield-shaped in form, though other
types also occur.    In *Vicia sepium*, for example, and in other species
of the same genus, every stipule bears, on its upper surface, a nectary
composed of densely-crowded club-shaped hairs (Fig. 205).    Each hair
is made up of a basal cell, a short stalk-cell, and three or four
glandular cells, which contain abundant protoplasm and numerous
tannin-vesicles.    Fehling's test gives a heavy precipitate of cuprous

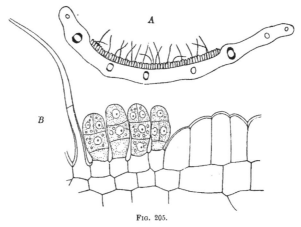

FIG. 205.

Extra-floral nectaries of *Vicia sepium*.  *A*. T.S. through a stipule, traversing the
nectary.  *B*. Margin of the nectary, more highly magnified.

oxide in the glandular cells, a circumstance which indicates the presence
of a considerable amount of sugar.    These clavate secretory trichomes
are accompanied by a smaller number of much longer pointed hairs
(three to six times as long as the club-shaped hairs), which are possibly
designed to retain the nectar by capillary action.    The extra-nuptial
nectaries of certain species of *Dioscorea* have been fully investigated by
Correns.    These are sunken, glandular structures which differ in shape
according to their location; they are ellipsoidal on the (abaxial) leaf
surfaces, but spindle-shaped on the petioles and internodes.    Each
nectary arises from a single protodermal mother-cell.    The superficial
layer of the fully-developed gland has suberised walls, and thus bears a
certain resemblance to an endodermis.

Even more remarkable nectaries occur, according to Zimmermann,
on the base of the petiole and also on the lamina, in the genus *Fagraea*.
Here the palisade-shaped (columnar) secretory cells enclose a cavity
which ramifies more or less extensively in the substance of the leaf, and
opens to the outside by a passage directed at right angles to the surface

(Fig. 206 B). The passage arises from a single tubular protodermal cell, which pushes its way actively into the tissue of the leaf for a time, but finally degenerates (Fig. 206 A). The cavity of this cell occupies the future position of the passage, but the latter also undergoes

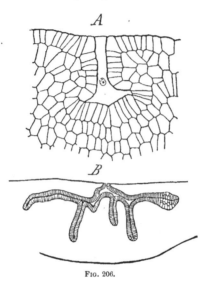

FIG. 206.

Extra-floral nectaries of *Fagraea*. *A*. Nectary of *F. fragrans* at an early stage of development, in vertical section ; the cavity of the enlarged tubular protodermal cell becomes the secretory passage of the nectary. *B*. Fully-developed nectary of *F. lanceolata*, in vertical section. (Both figs. after Zimmermann.)

a certain amount of independent enlargement at a later stage. This type of extra-nuptial nectary thus has some points in common with the septal nectaries which have already been described.

In accordance with their function, extra-nuptial nectaries continue to secrete nectar for a much longer time than nuptial nectaries, which are indeed generally quite short-lived structures. Moreover, washing with water does not affect their secretory activity to the same extent as it does that of floral nectaries. This point was demonstrated experimentally by Schimper in the case of *Cassia neglecta*, and the author has himself obtained confirmatory results with *Vicia sepium*. Thus, while the process of secretion is essentially the same in both types of nectary, the exudation of sugar continues for a longer period in the case of the extra-nuptial organs.

In all probability the majority of nectaries are phylogenetically derived from hydathodes, just like the digestive glands of carnivorous plants ; both the epithem- and the trichome-types of hydathodes have been modified in this direction. In *Vicia sepium*, for example, the clavate hairs of the extra-nuptial nectaries have exactly the same structure as the clavate trichome-hydathodes which occur on the upper side of the young pinnae.

### D. OIL-, RESIN-, MUCILAGE- AND GUM-SECRETING GLANDS.

#### 1. *External or dermal glands.*[250]

The glandular organs which secrete oil, resin, mucilage, or gum are frequently epidermal in a morphological sense, and are thus comparable to epidermal hydathodes and nectaries.

In the simplest cases secretion of such compounds is carried on by ordinary unspecialised epidermal cells. The so-called **glandular surfaces** produced in this way are more especially characteristic of such bud-scales as are, in addition, provided with glandular hairs (e.g. *Rumex, Rheum, Coffea, Alnus, Betula,* etc.). In certain species of *Silene* a sticky area is developed below each node. Within these areas specialised cells occur interspersed among the ordinary epidermal elements. In *Silene nemoralis,* according to Unger, these special cells are broader than the other epidermal cells, and are further distinguished by their papillose outer walls and abundant granular protoplasmic contents; it is highly probable that they represent the actual secreting elements. The modification of epidermal cells in relation to a secretory function is more far-reaching and conspicuous in the case of the so-called **glandular spots**, small, sharply-defined secretory areas which occur more parti-cularly on the leaf-teeth of various plants (spp. of *Prunus* and *Salix, Ricinus,* etc.). The secretory elements of which such a glandular spot is composed, are usually more or less columnar (prismatic) in form and are generally arranged in palisade-like layers; they are further distinguished from the neighbouring epidermal cells by their very thin walls and dense protoplasmic contents. Mention may finally be made, at this stage, of the so-called **intra-mural glands** of the genus *Psoralea.* Here the more or less spherical "body" of the epidermal gland is made up of a considerable number of tubular secretory elements, some of which are curved in a radial plane, while others are irregularly sinuate.

Glandular trichomes exhibit an extraordinary diversity of form. A typical **glandular hair** consists of two distinct portions—in addition to the base or foot, namely: first, a one- to several-celled stalk of varying length; and, secondly, the actual secretory organ or gland, which takes the form of a rounded head attached to the distal end of the stalk. The head may be unicellular, as in *Pelargonium zonale* (Fig. 207 A) and *Primula sinensis*; or it may be divided into quadrants, as in *Lamium, Plectranthus,* etc.; or, finally, it may be composed of a

large number of cells, as in *Sanguisorba carnea*. The stalk-cells sometimes contain chlorophyll, in which case they represent a localised
photosynthetic apparatus, belonging to the gland.    **Glandular villi** (or
glandular shag-hairs, " *Drüsenzotten* ")—like ordinary glandular hairs—

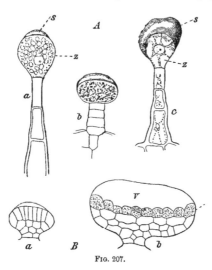

FIG. 207.

Dermal Glands.    *A*. Ordinary glandular hairs from a petiole of *Pelargonium zonale* ;
*z*, secretory cell ; *s*, secretion.    *B*. Young (a) and adult (*b*) glandular scales from a leaf
of *Ribes nigrum* (the secretion removed by treatment with alcohol); *z*, secretory cells ;
*v*, glandular cavity formed by distension of the cuticle.

are usually differentiated into a stalk, which here consists of an elongated mass of cells, and a secretory head, which is generally multicellular ; not infrequently, individual cells of the massive stalk are
produced into simple glandular hairs (e.g. *Begonia Rex*).    **Glandular
scales**, finally, likewise possess more or less well-developed stalks, but are

characterised, as their name indicates, by the fact
that the secreting elements are arranged in the
form of a flattened scale, or, in some cases, of an
almost basin-shaped cell-plate.    As examples of
glandular scale we may mention the well-known
lupulin-glands of the Hop, and the glandular scales
which occur in ·*Ribes nigrum*, and in species of

FIG. 208.

Glandular scale of *Py-
rethrum Balsamita.*

*Thymus, Mentha* and *Satureja*.    In *Pyrethrum Balsamita* (Fig. 208), each glandular hair consists of
a cell-plate, extended at right angles to the
surface of the parent organ and made up of four tiers of two cells
each.    The cells of the uppermost tier are colourless and represent the
actual secretory organs ; those of the two next tiers contain numerous

chloroplasts and constitute the photosynthetic apparatus of the gland.[250 a]

The process of secretion is not precisely the same in all dermal glands. The secretion of a mucilage gland, for instance, is from its first origin a constituent of the cell-membrane and not of the contents ; certain sub-cuticular layers of the wall become mucilaginous, with the result that the cuticle is distended in a vesicular manner and finally ruptured.

In the case of glands which secrete ethereal oil or resin, the secretion likewise generally originates within the cell-wall ; this view was first put forward as a suggestion by Hanstein, and was definitely enunciated by De Bary. After the cell-wall has become considerably thickened, the secretion appears in constantly increasing quantity between the cuticle and the cellulose layers. This deposition of secretion in the substance of the cell-wall sometimes begins, first of all, near the apex of the glandular head ; in other instances it commenees simultaneously at several points (Fig. 209). The vesicular distension of the cuticle, which takes place in this case also, may ultimately affect the walls of adjacent non-secretory cells, if the quantity of secretion produced is large ; according to Hanstein this distension is facilitated by a previous gelatinisation of the sub-cuticular layers of the wall. As a result a glandular cavity is formed, in which the secretion accumulates. The cuticle is incapable of indefinite distension, and finally breaks, whereupon the secretion escapes to the outside. Hanstein further states that this cuticle may be regenerated in certain cases ; when this occurs, the whole secretory process may be repeated several times.[251]

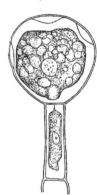

Fig. 209.

Glandular hair of *Pelargonium zonale*, plasmolysed and treated with alcohol.

More rarely the secretion, instead of being deposited in the outer walls of the glandular cells, appears within the partitions that separate adjacent secretory elements. This **intramural** type of gland is exemplified by the orbicular scales which cover the lower surface of the leaves of many species of *Rhododendron* and by the embedded multicellular foliar glands of *Psoralea* (cf. above, p. 511).

Complete uncertainty prevails with regard to the actual manner of formation of the secretion in the substance of the cell-wall, in all these cases. It is conceivable that the secretion owes its origin to a chemical metamorphosis of certain layers of the membrane, which are regenerated as fast as they are transformed. But it seems, on the whole, more probable that the raw material employed in the manufacture of the

2 κ

secretion is derived directly from the cell-contents. As a matter of fact, numerous highly refractive vesicles and globules of various sizes make their appearance in the protoplast both before and during the process of secretion; according to Tunmann, these bodies consist either of fat or of tannin. But in whatever fashion secretion may take place, the chemical changes involved in the process are undoubtedly controlled by the living protoplasts of the glandular cells. It is impossible, at the present time, to suggest how this influence is transmitted across the [apparently unaltered] innermost layer of the cell-wall.

In certain cases the secretion seems to originate in the cell-cavity, and to escape directly to the exterior without producing any rupture of the cuticle. According to Behrens, this case is exemplified by the glandular hairs of *Ononis spinosa*. Here the mobile ethereal oil which constitutes the secretion, first becomes visible in the protoplasm of the secretory cells, and later appears on the outer surface of the " head " in the shape of numerous small drops adhering to the cuticle; these drops continually increase in number and gradually coalesce to form a single large drop, which ultimately falls off. The glandular hairs of *Cicer arietinum* and those of certain ONAGRACEAE which secrete an acid liquid, probably belong to the same category; for Stahl states that fresh drops of the acid substance appear on the heads of these hairs even when the secretion has been repeatedly washed off with water. The "mealy" hairs found in the so-called " Gold " and " Silver Ferns " (*Gymnogramme* spp.) and in certain species of *Primula* (*P. Auricula, P. farinosa*, etc.), are probably of a similar nature. The mealy coating of the shortly-stalked capitate hairs consists of slender, often more or less curved, rods and needles of a resinous or fatty substance; in *Gymnogramme*, according to De Bary, these bodies are attached to the outside of the smooth cuticle.

The physiological and ecological significance of the dermal glands that have been described in the preceding paragraphs, varies according to the nature of the substance secreted. It is possible that in certain cases the secretion merely represents a useless by-product of metabolism, which must be got rid of by some means or other; apart from this possibility, however, most of the secretions in question probably serve to keep down transpiration, or to ward off the attacks of injurious animals. With regard to the first-mentioned function, Hanstein long ago pointed out that the mucilage hairs or " colleters " which occur on so many bud-scales, and on young foliar organs in general, afford protection against excessive transpiration and other injurious influences by virtue of the secretion (a mixture of gum and ethereal oil or resin) which they pour over the tender developing organs. The adult leaves of many xerophytes are similarly covered—sometimes on both sides, sometimes on the upper surface only—with a glistening coat of resin, which is generally secreted by dermal glands. Such "varnished leaves," as they are termed by Volkens, are very effectually protected against the danger of excessive transpiration by their resinous covering.[252]

Glands which secrete ethereal oils or resins, also frequently serve to provide a means of protection against the assaults of animals; but it is improbable that their importance in this respect is so great as Stahl is inclined to suppose. It is certain that many aromatic plants are readily devoured by a number of different animals. Kaltenbach states that the foliage of *Thymus serpyllum* is eaten by two beetles, by twenty-five different kinds of caterpillar, by the larva of a fly (*Trypeta serpylli*), by an Aphis (*Aphis serpylli*), and by a Mite (*Calycophthora serpylli*). No less than six of these animals are "specialists," which live exclusively on the Thyme.[253]

The well-known physicist, Tyndall, attributed a totally different ecological rôle to those external (and internal) glands which secrete volatile ethereal oils. He showed that a layer of air which is saturated with the vapour of an ethereal oil, is much less pervious to heat-radiations than an equally thick layer of ordinary air. The screening effect of such vapours is very considerable; hence plants which surround themselves with a layer of vapour produced by the vola-tilisation of ethereal oils are, to a certain extent, protected against rise of internal temperature—and consequent increase of transpiration —in bright sunshine, and, conversely, against excessive loss of heat by radiation under a clear sky at night. This theory, though attractive, is open to many objections.[254]

Floral organs are frequently provided with dermal glands secreting fragrant volatile esters, which help to attract insect visitors for the purpose of cross-pollination.

A coating of mucilage, which is generally secreted by glandular hairs, occurs on the vegetative organs of many water-plants. Stahl regards this slimy envelope as a means of protection against snails. The author himself, while accepting this interpretation, believes that the mucilaginous covering is also useful in preventing colonisation of the young organs by epiphytic Algae and Bacteria. Goebel assumes that the mucilage prevents the young organs from coming into direct contact with the surrounding water, thereby preventing the "extraction" (by exosmosis) of organic and inorganic materials; this hypo-thesis must be rejected in the entire absence of experimental evidence in its favour. Hunger attempts to explain the ecological significance of the mucilaginous coverings of young organs, in the case both of terrestrial and of aquatic plants, from a mechanical point of view; he regards the mucilage as a lubricant, which enables the organs concerned to glide smoothly, and without risk of damage, over opposing obstacles.    Such a lubricating arrangement would be especially useful in the case of unfolding buds, where it would greatly reduce friction [between the closely-packed organs].    As a matter of fact, mucilaginous secretions do occur very regularly in plants that possess more or less well-developed leaf-sheaths, which for a time completely envelop the younger portions of the bud (*e.g.* POLYGONACEAE, *Ficaria ranunculoides, Helleborus, Viola sylvestris, Valerianella* spp., etc.) ; the large surfaces of contact which occur in such cases, would experience great frictional resistance in the absence of special lubricating arrangements.    The frequent occurrence of numerous mucilage hairs on the young leaves of Ferns, while these are still tightly rolled up, is explained by Hunger from a similar standpoint.    There is certainly much to be said for this mechanical interpretation of the prevalence of mucilaginous envelopes on young organs.[255]

Finally, it is probable that mucilaginous secretions often help to prevent or retard desiccation.

### 2. *Internal glands (secretory sacs or reservoirs).*[256]

Axial and foliar organs generally, and foliage-leaves in particular, are often furnished with glands which are situated either immediately beneath the epidermis, or amid the more deeply-seated tissues.    Such **internal glands**, which are often visible to the naked eye as translucent spots or patches, occur, for example, in the leaves of species of *Hypericum, Lysimachia, Citrus, Amorpha,* etc. ; on the whole they display less variety of structure than the previously described dermal glands.    The differences which do occur, are very largely due to the various modes of formation of the walls of the gland, and also depend to some extent upon the presence or absence of special arrangements for the discharge of the secretion.

The intercellular cavity of an internal gland is generally more or less spherical, less frequently sac-like or tubular.    In the MYRTACEAE and HYPERICACEAE, and in the genera *Lysimachia* and *Amorpha*, it is formed by the partial separation of the secretory cells from one another ; in a fully developed gland of this type, the secretory cells form a continuous layer enclosing the **schizogenous cavity.**

In the RUTACEAE, on the contrary, the glandular cavity owes its origin to a precocious degeneration of the secretory elements, which takes place as soon as they have produced their secretion.    The walls of the secretory cells disappear altogether, while the secretion, which was originally present in the form of numerous small drops, gradually runs together into larger masses.    When such a **lysigenous cavity** is completed, no traces of the secretory cells remain apart from such fragments of their cell-walls as may have escaped destruction.

It should, however, be noted that in many cases the glandular cavity begins its development in a schizogenous manner, but later on undergoes lysigenous enlargement. The author was the first to draw attention to this composite mode of origin of glandular cavities, which was discovered by him in the leaves of *Ruta graveolens* (Fig. 210 C);

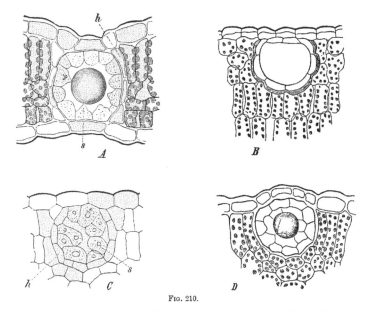

FIG. 210.

Internal glands of various foliage-leaves. *A. Hypericum perforatum.* The sheath of the gland (*h*) consists of thin-walled tangentially-flattened cells; the secretory cells (*s*) are more or less papillose. *B. Myrtus zeylanica.* The secretory cells are flattened and have thick outer walls. *C. Ruta graveolens* (young gland); *h*, sheathing layer. The glandular cavity is partially occluded by secretory cells which have become detached from the wall. *D. Dictamnus albus.*

it is, however, by no means a rare phenomenon, a fact which probably accounts for the frequent lack of agreement between different observers who have described the development of internal glands. Tschirch subsequently applied the term **schizolysigenous** to glands which have this composite origin. It must also not be forgotten, that the growth of the organ as a whole is liable to bring about a considerable secondary enlargement of glandular cavities, whether their early development is schizogenous or lysigenous in character.

The secretory tissue of a schizogenous gland generally consists of a single layer of glandular cells (Fig. 210 A, B), which are almost always readily distinguishable from the cells of the adjoining tissues by their form and contents, and often also by their smaller size.    In shape these secretory cells are either approximately isodiametric or else tangentially flattened ; in the former case, the " free " walls usually bulge more or less into the glandular cavity, and may actually be developed as protruding papillae (*Hypericum perforatum*).    In certain PAPILIONACEAE (*Lonchocarpus, Derris, Milletia*) the secretory cells are tubular and tangentially curved.    The contents of the cells usually consist of colourless granular cytoplasm, enclosing a nucleus which is often relatively large.    Opinions differ as to the mode of formation of the secreted substances and as to the manner in which they escape.    Many observers assume that the ethereal oil or resin which constitutes the secretion, arises first of all in the interior of the glandular cells, and subsequently passes through the cell-wall.    Tschirch, on the contrary, maintains that the secretion is produced within the cell-wall from a special mucilaginous stratum or " resinogenic layer " ; according to this view, the protoplasts of the glandular cells never contain any of the actual secretion, but merely supply the " resinogenic " material.

In the majority of plants with internal glands, the secretion is never actually expelled.    A gland may therefore also act as a secretory reservoir ; this is, in fact, its only function where the glandular cells become inactive through age and finally die, or where the cavity develops lysigenously and the secretory cells consequently disintegrate at an early stage.    In a few families, however, special arrangements are found which enable the contents of the glandular cavity to escape. A special discharging mechanism was first described by the author in connection with the sub-epidermal glands of certain RUTACEAE.    Here the gland is made up of a passive portion or " cover," and an active part or " body."    The cover generally consists of four cover-cells, which arise by the division of a protodermal mother-cell (Fig. 211 A).    The shape, structure and chemical composition of the membrane of the cover-cells is such that a glandular orifice arises in a predetermined region of the lateral walls (which may accordingly be termed the " pore-walls ").    The cover-cells are almost always considerably flattened vertically, in comparison with the adjoining epidermal elements ; in *Ruta graveolens* and *Pilocarpus pinnatifolius* they are sunk below the level of the general epidermis (Fig. 211 B).    This flattening of the cover-cells facilitates the formation of the pore.    It is also an advantage that the outer walls of the cover-cells—or, at any rate, their cutinised layers —are, as a rule, appreciably thinner than the corresponding portions of the neighbouring epidermal cells.    In *Agathosma pubescens* the cutinised layers of the cover-cells are, indeed, no thinner than usual ; but they are traversed by a deep furrow, which follows the line of the pore-wall and thus prepares the way for the formation of the pore.    Most im-

portant of all is the character of the lateral or pore-walls of the cover-
cells. These walls contain a middle lamella of varying thickness, which,
to judge by its staining properties, consists largely of pectic compounds,
though in *Ruta* and *Pilocarpus* some callose is also present. The pore
arises through the splitting of the walls along this relatively incoherent
layer. When this process of separation—which is more fully described

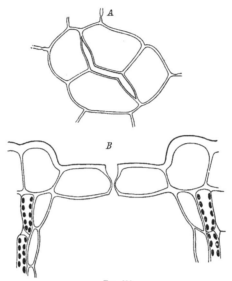

FIG. 211.

Discharging mechanism of the internal glands of *Ruta graveolens. A.* Cover and pore
in surface view. *B.* Cover and pore in vertical section.

below—is completed, the cross-sectional outline of the cover often
strikingly resembles that of a stoma.[257]

The one- to three-layered "body" of the gland, which represents
the active portion of the discharging apparatus, consists of flattened
cells arranged in a perfectly continuous layer. Its principal duty is to
exert pressure upon the contents of the gland by means of the turgor
developed in its constituent cells. The presence of a high internal tension
in these cells may be readily demonstrated, if a moderately thick
transverse section of a living leaf of *Ruta graveolens*—which must, of
course, traverse a gland—is examined in water. Under these conditions
the inner walls of the uninjured body-cells protrude to an extraordinary
extent into the glandular cavity; the cells in question then resemble
large vesicles, whereas in the intact gland they are quite flat. The
extension of the inner walls of the body-cells [in water] may amount
to between 25 per cent. and 80 per cent., and greatly exceeds their

limit of elasticity.   These cells must therefore contain large quantities
of osmotically active material.   The high pressure to which the contents of the gland are thus exposed, is nevertheless not in itself
sufficient to bring about the formation of the gland-pore and the
ejection of the contents.   The latter never become spontaneously dis-
charged.   On the contrary, the secretion only escapes when the leaf is
forcibly bent, so that the pressure upon the glands is suddenly increased ;
the tension set up in the cover-cells on the convex side of the bent leaf
assists in the formation of the gland-pore.   Vigorous shaking suffices
to bring about the discharge in the case of very turgid shoots of *Ruta
graveolens.*

The author's observations on this point have been extended by
Detto, who has shown that the shortly-stalked spherical or pearl-shaped
glands on the flowering shoots of *Dictamnus* are also furnished with a
special discharging mechanism, consisting essentially of a multicellular,
hair-like beak attached to the apex of the gland proper ; at the
slightest touch, the beak breaks off and the contents of the gland are
ejected.   Detto's figures suggest that the lowermost transverse walls in
the beak act as " pore-walls."

It is not only among the RUTACEAE that internal glands are
generally provided with a discharging apparatus ; a similar device also
occurs in all the MYRTACEAE that have been examined from this point
of view, though its construction is not quite the same as in the first-
mentioned family.   The author has given a brief account elsewhere of
the discharging arrangements of *Myrtus communis,* while Porsch has
made a detailed study of the corresponding structures in *Eucalyptus
globulus* and *E. pulverulenta.*   Here the " cover " consists typically of
a single pair of cells ; in *Eucalyptus* both the outer and the inner walls
of the cover-cells are very thin.   The septum between these two cells
is curved in an S-shaped manner, and is greatly thickened and
irregularly pitted.   The gland-pore again comes into being when the
leaf is forcibly bent ; in this case, however, it arises, not by fission of
the septum, but by rupture of the thin outer and inner walls of one
or both cover-cells.   Otherwise the discharging mechanism is the same
as in the RUTACEAE.   In the present instance, the functions of the
highly turgescent body-cells are vested in the persistent secretory cells,
which themselves exert a very considerable pressure upon the contents
of the gland.   According to Porsch, the internal glands which occur in
the petals of *Boronia megastigma* are provided with a discharging
mechanism of the *Ruta* type.

It has already been stated, that internal glands are frequently
enclosed in a special sheathing layer.   The functions of this sheath are
not always the same.   In the case of a lysigenous gland it serves prin-

cipally to shut off the contents of the gland from the surrounding tissues. Among schizogenous glands, on the other hand, the segregation of the contents is effected by the continuous (persistent) secretory layer itself; here the function of the sheath is probably in the first instance mechanical. The special part played by the sheath in relation to the discharging process in the RUTACEAE, has already been explained. The sheathing layers of glands may arise in a variety of ways. Thus, the outer walls of the secretory cells may become thickened where they abut against the surrounding tissues (*e.g.* in the smaller glands of *Myrtus zeylanica*, Fig. 210 B). The sheath may, on the other hand, consist of an irregular layer of collapsed cells, derived from the sur rounding tissues; these have thickened walls, and their contents do not differ noticeably from those of the neighbouring unmodified elements (e.g. *Eugenia australis* and *Eucalyptus cornuta*, according to Von Höhnel). The sheath may, lastly, take the form of a regular and continuous layer of flattened cells, with somewhat thickened walls and colourless contents (*Hypericum perforatum*). Occasionally, combinations of these modes of development occur (*e.g.* in *Myrtus zeylanica*, where the two first-mentioned methods of development are combined in the case of the larger foliar glands).

Chemically considered, the substances secreted by internal glands are generally in the nature of ethereal oils or resins. What has been stated above concerning the ecological value of dermal glands, applies equally to internal glands, where the secretion is of a similar nature in both cases. The discharging arrangements of the RUTACEAE and MYRTACEAE probably provide additional protection against the attacks of animal foes; presumably the animals take fright on suddenly encountering an intense odour which is not noticeable until the plant is roughly handled. An analogous method of protection is, of course, well known to be widely distributed in the animal kingdom.

### E. SECRETORY DUCTS AND PASSAGES (OIL-, RESIN-, GUM- AND MUCILAGE-PASSAGES).[258]

**Secretory ducts or passages** only differ from other internal glands in a single particular; they are always more or less elongated, and accordingly may, like latex-tubes, traverse whole organs, or even more extensive regions of the plant-body, from end to end. That there is no other difference between the two types of secretory organ, may be shown by the following consideration, among others: In the scale-leaves of *Thuja* and *Biota*, the resin-passages which are so characteristic a feature of the needles of *Pinus* and *Abies*, are replaced by glands of rounded outline. Structures intermediate between glands and secretory passages are also known. In the case of *Ginkgo biloba*, for example, it is difficult to decide whether the foliar secretory organs, which are about 1 mm. in length, should be regarded as elongated sacs or as short passages. All the features, accordingly, which have already been considered in connection with internal glands, may recur in the case of secretory passages. The glandular cavity is represented by a schizo-genous, lysigenous or schizolysigenous passage. Schizogenous passages are always lined by a sharply differentiated layer of secretory cells. Very frequently, again, the whole passage is enveloped by a special sheathing layer, which here assumes the characteristics of an endo-dermis.

The glandular cells of secretory passages are most frequently elongated in the same sense as the duct itself; more rarely they are transversely elongated (leaves of CYCADACEAE). The relative size of the secretory cells—as seen in transverse section—varies according to the character of the surrounding tissues. If the duct is one which traverses mesophyll (*e.g.* in *Pinus, Abies,* and other Conifers) or the cortical parenchyma of a stem, the secretory cells appear comparatively small. If, on the other hand, the passage is situated in the leptome strand of a vascular bundle, the glandular cells appear relatively wide. As a rule the walls of the secretory cells are thin and inclined to protrude into the cavity of the duct. The walls of the cells lining the mucilage-ducts of the MARATTIACEAE not only bulge inwards, but are furnished with conical papillae; in the foliar ducts of *Lycopodium* these papillae are represented by club-shaped outgrowths. In old passages the secretory cells may even grow out after the manner of tyloses and completely occlude the cavity. As in other glandular structures, the secretory cells usually form a single layer; in certain cases, however (*Hedera Helix, Philodendron*), this layer undergoes one or more tangential divisions at an early stage.

The substances produced in secretory passages are very diverse in character. In *Lycopodium,* among the MARATTIACEAE and CYCADACEAE, in species of *Canna* and *Opuntia* and in a few ARALIACEAE, a mucilaginous or gummy material is produced in these organs; among Conifers, in certain AROIDEAE and BUTOMACEAE, in the ALISMACEAE, UMBELLIFERAE, tubifloral COMPOSITAE and most ARALIACEAE, the secre-tion consists of ethereal oil or resin, or of an emulsion of gum-resin somewhat resembling latex in appearance. What has been said above with regard to the process of secretion in the case of internal glands, applies equally to secretory ducts. There is the same uncertainty as to whether the secretion arises within the cavity of the secretory cell, or in a "resinogenic" layer of the cell-wall.

It has already been stated, that secretory passages are often furnished with protective sheaths, which may be compared to those

endodermal layers that are principally mechanical in function. These
mechanical sheaths are very characteristically developed in the leaves
of *Pinus*, and in the roots of *Philodendron*. In *Pinus excelsa* the cells
of the sheath are flattened in the tangential plane, and their walls are
but slightly thickened. In *P. Cembra* there is no tangential flattening ;
the walls are thickened all round, and the cells are appreciably

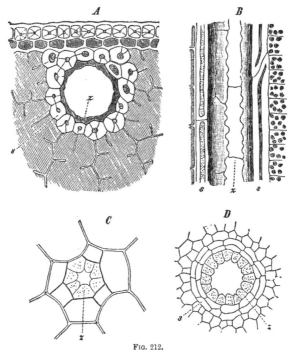

Fig. 212.

Resin- and Oil-passages.  *A*. Resin-passage from the leaf of *Pinus sylvestris*, in T.S.
*B*. Resin-passage from the leaf of *Pinus sylvestris*, in R.L.S. ; *s*, sheathing layer ;
*z*, secretory cells.  *C*. Oil-passage from the pith of the stem of *Heracleum sphondylium*
(T.S.) ; *z*, secretory cells.  *D*. Resin-passage from the primary leptome of the stem of
*Rhus Cotinus* (T.S.) ; *s*, sheathing layer ; *z*, secretory cells.

elongated (being six to eight times longer than their width), and here
and there show a tendency to become prosenchymatous. Finally, in *P.
Laricio, P. sylvestris* (Fig. 212 A) and *P. Strobus* the cells of the sheath
resemble typical stereides in all essentials, being very thick-walled
and strongly prosenchymatous. The sheaths may become two-layered
at certain places. A further point of agreement with endodermal layers
consists in the presence of isolated or longitudinally seriated " passage-
cells," which are conspicuous on account of the thinness of their walls.
The provision of passage-cells is quite comprehensible in view of the

fact that the raw material employed in the manufacture of the resin has to be conveyed to the secretory cells from the adjoining photosynthetic parenchyma.  As a matter of fact, the passage-cells may apparently undergo a considerable amount of thickening in old leaves, when the secretory cells have ceased to be active.  Similar thick-walled—often many-layered—mechanical sheaths are found surrounding the secretory ducts in the roots of *Philodendron*.  The oil-passages that traverse the primary leptome of *Rhus Cotinus* are likewise encased in a double sheath of cells, which are, however, thin-walled and flattened (Fig. 212 D).

Some account must next be given of the **course and arrangement of secretory passages.**  The simplest scheme is that in which an organ is traversed by longitudinal secretory passages which are closed at both ends ; this condition is exemplified by the acicular leaves of *Pinus sylvestris*, *P. montana* and *P. Cembra*, and probably, also, by those of many other ABIETINEAE.  A *Pinus*-needle contains a variable number of sub-epidermal resin-ducts.  There are always two principal lateral passages, and from two to twenty accessory ducts on the abaxial and adaxial sides of the needle.  The two lateral passages extend further towards the base of the leaf than the rest.  At a distance of 2·5 mm. above the leaf-base, the glandular tissue of each lateral duct is reduced to three or four cells (as seen in transverse section), which, however, still form a continuous cylinder ; half a millimetre from the base the secretory cells disappear altogether, and the resin-passage is replaced by a sharply-defined fibrous strand corresponding to the contracted thick-walled sheathing layer.  The median ducts terminate in a similar manner ; the intercellular cavity disappears first, and the secretory cells follow suit later, so that finally nothing remains but a subepidermal strand of fibres.  These accessory ducts, however, come to an end sooner than the lateral passages, their secretory cells disappearing at a distance of 5 to 7 mm. from the leaf-base.  The leaf of *Juniperus communis* contains a single median abaxial passage, which ends blindly close to its base ; the wide passages which occupy the three angles of each internode, likewise end blindly above and below.[259]

In the majority of cases, the secretory passages form a branched anastomosing system of tubes pervading the whole plant-body.  As a rule, this tubular system is most abundantly developed in the parenchymatous tissues, but not infrequently the vascular bundles also contain secretory ducts.  The mucilage-ducts of MARATTIACEAE and CYCADACEAE, the resin-passages of Conifers, and the oil-canals of the tubifloral COMPOSITAE and the UMBELLIFERAE, all occur principally in the cortical and medullary parenchyma of the stem.  In some species of *Araucaria*, in certain CLUSIACEAE, and in the ANACARDIACEAE (*Rhus*

*Cotinus, R. suaveolens, R. glauca,* etc.), the leptome-strands contain secretory passages; in *Pinus Laricio* and certain other Conifers, on the other hand, these structures are found in the primary hadrome-strands of the stem. In the majority of cases, however, the secretory ducts are not actually located *in* the vascular bundles, but are merely associated with them in a characteristic and constant manner. In Umbelliferous stems, for example, each vascular bundle (or, at any rate, each of the larger strands) is accompanied by a cortical oil-passage. In *Achillea millefolium, Cirsium arvense, Tanacetum vulgare* and other COMPOSITAE, a duct runs alongside of each leptome-strand; in *Solidago limonifolia* each leaf-trace bundle is provided with two oil-canals, one opposite the hadrome-, and the other opposite the leptome-strand; in *Helianthus annuus,* finally, every bundle is subtended by a crescentic group of oil-passages, both on its inner and on its outer side.

In conclusion, it will be necessary to consider the physiological and ecological significance of secretory passages. In view of the fact that these structures are so frequently associated with vascular bundles, and in particular with leptome-strands, it might at first sight seem probable that they serve for the reception of useless waste products, which travel towards the leptome-strands and other vascular tissues from the various tissues that are engaged in active metabolism; but the chemical character of the contents of secretory ducts hardly accords with this notion.

The reason for the frequent association of secretory passages with leptome-strands and other vascular tissues is therefore, in all probability, an ecological one, as has, in fact, been suggested by Stahl and Kniep. The substances contained in these passages are often of such a kind as to be capable of affording "chemical protection" against noxious animals; hence small assailants which have penetrated into the interior of an organ will be more or less effectually discouraged from attacking the conducting strands,—the continuity of which is so vital to the well-being of the plant—if the latter are protected by a series of secretory ducts (or excretory sacs).

Among woody plants, the resin- and gum-passages which occur in the cortical parenchyma (and sometimes also in the woody cylinder) of the stem, probably often serve to cover up wounds with an airtight layer of secretion; in this way the stem is rendered less liable to infection, and consequent decomposition, by fungoid parasites. This interpretation is strongly supported by the results that follow, when incisions are made in Coniferous stems for the purpose of collecting resin on a commercial scale. Here enormous surfaces become coated with the secretion, while the subsequent healing proceeds very satisfactorily. According to J. Moeller, balsam-ducts may actually be *formed* in the young wood of *Liquidambar orientalis* and *L. styraciflora* as an immediate consequence of mechanical injury.

### III. EXCRETORY RESERVOIRS.

It has already been explained (p. 486) that the distinctive characteristic of excretory, as opposed to secretory, organs consists in the fact that the metabolic by-products are not liberated, but on the contrary are permanently deposited within the cells. Excretory substances, in fact, never exude unless the plant is injured; they are accordingly frequently made use of for protection against animal foes.

#### 1. *Resin- and oil-sacs.*[260]

Sacs containing **resins** or **ethereal oils** are found in the parenchymatous tissues in the ZINGIBERACEAE, PIPERACEAE, LAURACEAE, MAGNOLIACEAE, ARISTOLOCHIACEAE, CANELLACEAE, and in many EUPHORBIACEAE; further, in the genera *Acorus, Aloe, Rheum, Lysimachia,* etc. They occur singly, or are aggregated in groups, or, more rarely, arranged in definite series. Most frequently they are relatively large, and isodiametric and rounded in shape. The contents consist mainly or exclusively of the excretion, which in many ZINGIBERACEAE and PIPERACEAE, and in *Acorus Calamus,* takes the form of a colourless or light yellow ethereal oil, but which is often more deeply coloured. The seriated sacs associated with the vascular bundles in certain species of *Aloe,* are filled with a dark-coloured—sometimes with a colourless—liquid, in which small drops of resin are often suspended. The "chrysophane-sacs" of the Rhubarb-root may contain a homogeneous orange-coloured sap, or a colourless liquid with bright red drops suspended in it. Occasionally (*e.g. Aloe* and *Acorus,* according to Johow) even fully developed resin- or oil-sacs retain a peripheral protoplasmic layer and a nucleus. The cell-walls of such cells are always thin and devoid of ornamentation. Zacharias states that they are very frequently suberised, or at any rate provided with a suberin-lamella.

Many excretions are certainly deposited in cell-cavities. Berthold, on the other hand, has observed, in a wide range of families, that excreted oil drops may be enclosed in a sac- or balloon-like internal protuberance of the cell-wall. The stalk of this vesicular process is cutinised; after treatment with sulphuric acid it persists in the shape of a little cup attached by its base to the wall, whereas the delicate membrane of the vesicle itself is dissolved. The author has not himself followed out the development of these organs in detail, but has confined his attention to the adult structure of the oil-cells of *Laurus nobilis* and *Asarum europaeum.* His observations entirely confirm

Berthold's statements. In thin transverse sections of the rhizome of *Asarum europaeum*, which have been treated with alcohol in order to remove the ethereal oil, the strongly cutinised cup-shaped stalk can be readily distinguished under a sufficiently high magnification; it is seen, on the one hand, to be continuous with the suberin-lamella of the cell-wall, while in the other direction it is prolonged into the extremely delicate vesicle which envelopes the oil-drop (Fig. 213 B). In alcohol

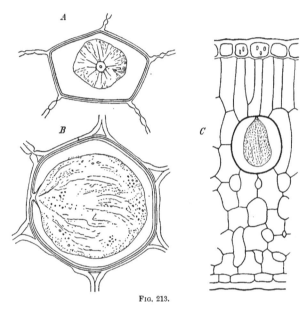

FIG. 213.

*A* and *B*. Oil-reservoirs of *Asarum europaeum*; *A*, from a scale-leaf (surface view); *B*, from the ground-tissue of the rhizome (T.S.). *C*. T.S. through a leaf of *Laurus nobilis*, showing an oil-reservoir. (Oil removed by treatment with alcohol in every case.)

material the vesicle is collapsed and wrinkled, and appears to be covered with minute granules. In order to obtain a surface view, it is best to examine the outer (adaxial) epidermis of the scale leaves (of the rhizome) which contains numerous oil-cells (Fig. 213 A). The cup-shaped stalk of the vesicle will then be seen, in every case, to be attached near the centre of the outer wall of an epidermal cell; the cavity of the stalk presents the appearance of a sharply defined circle, not unlike a minute pit, surrounded by a circular border, which corresponds to the wall of the cup. From the outer edge of this border there frequently arise numerous radiating striae; these correspond to the wrinkles which, as already stated, are produced in the wall of the vesicle, when the oil is removed with alcohol. The oil-cells in the leaves of *Laurus nobilis* have a similar structure, except that the cup-shaped stalk is exceedingly minute (Fig. 213 c).

The development of these "attached" oil-vesicles has been investigated by Rud. Müller, working in the author's laboratory, in the case of *Aristolochia brasiliensis*. Quite contrary to expectation, he found that the stalk of the vesicle does not, as Berthold believed, originate as a local expansion of the cell-wall. On the contrary, a large oil-vacuole arises in the cytoplasm—by the fusion of numerous smaller ones—and sooner or later puts forth a conical process towards the cell-wall. The plasmatic membrane of the vesicle then becomes transformed into the pellicle surrounding the oil-vesicle and its stalk, which is consequently continuous with the general cell-membrane. One is thus really dealing with an oil-vacuole, which originates within the cytoplasm, and only secondarily becomes connected with the cell-wall at a single point. The significance of this peculiar process of attachment is quite unknown.

## 2. *Tannin-sacs.*[261]

It has been remarked, on a previous occasion, that the compounds known as *tannins* or *tannic acids*, while possibly acting as plastic materials in certain circumstances, undoubtedly often represent by-products of metabolism. Compounds of this class are usually deposited in sacs arranged in long rows in close association with vascular strands. Such **tannin-sacs** are found in the parenchymatous tissues of the stems and petioles of many Ferns; they also occur in the ARACEAE and MUSACEAE in connection with the vascular bundles, in *Phaseolus multiflorus* and other LEGUMINOSAE in the leptome-strands of the primary axial and foliar bundles, and in *Phaseolus* also in the pith opposite each bundle. The most remarkable tannin-sacs, however, are those which are found in the cortical and medullary parenchyma of the stem of *Sambucus*. Dippel estimates the length of one of these sacs in the fully developed state at 18-20 mm. or more, and its average width at ·025-·164 mm. According to De Bary, however, it is probable that individual sacs may extend through an internode, that is, over a distance of 20 cm. or more. Each sac consists of a single enormously elongated spindle-shaped cell.

Closely related to tannin-sacs are the more or less tubular epidermal tannin-containing cells observed by Engler in *Saxifraga Cymbalaria* and allied species, and in *Sedum spurium*. In these plants the ordinary epidermal cells are isodiametric, with wavy radial walls, so that the elongated, scattered or seriated tannin-sacs are very conspicuous.

The ecological importance of tannin as a means of protection against snails and other animals has been pointed out by Stahl. G. Kraus and others, on the other hand, lay stress upon the antiseptic properties of these compounds.

### 3. *Enzyme-reservoirs*.[262]

In the CRUCIFERAE, CAPPARIDACEAE, TROPAEOLACEAE, RESEDACEAE and LIMNANTHACEAE, the ethereal oils which are responsible for the characteristic pungent odour and taste of all the vegetative parts (especially when these are cut or bruised) are never present in the free state in the uninjured plant; they are only produced as a result of mechanical injury by the action of an enzyme, *myrosine*, upon a glucoside-like compound, *potassium myronate*, which is hydrolysed with formation of *allyl isothiocyanate* (allyl mustard-oil) or a related body, *glucose* and *potassium sulphate*. In the intact tissues, according to Guignard, *myrosine* and *potassium myronate* are located in different cells, the former being contained in specially differentiated enzyme-reservoirs. These enzyme-cells, which are most often tubular in shape, though other forms also occur, were first discovered in the CRUCIFERAE and CAPPARIDACEAE by Heinricher, who termed them "protein-sacs" on account of their microchemical reactions. The fact that they contain *myrosine* was established later by Guignard. Heating with concentrated hydrochloric acid, containing a single drop of 10 per cent. watery solution of orcin per c.c., causes the cells in question to assume a violet coloration owing to the presence of the enzyme. The cell-sap of the enzyme-reservoirs is transparent in the living condition of the cell, but coagulates on heating to boiling-point; whether it contains ordinary protein-material in addition to dissolved enzyme or not, is still uncertain.

The distribution of *myrosine*-cells in the vegetative organs of the aforesaid plants is subject to considerable variation. Heinricher's detailed investigations of the CRUCIFERAE have shown that in some species (e.g. *Crambe cordifolia*) idioblastic *myrosine*-cells occur in almost every part. In other cases they are principally found in association with the vascular bundles, and in particular with the leptome-strands; in *Moricandia arvensis*, finally, they are all sub-epidermal both in stem and leaf. According to Guignard, these statements also apply to the other families, where, however, the organs in question are chiefly located in the cortex of root and stem.

Another enzyme that is sometimes contained in special reservoirs is *emulsin*, which hydrolyses the glucoside *amygdaline*—present, for instance, in bitter almonds—with the formation of *hydrocyanic acid*, *benzaldehyde* (oil of bitter almonds) and sugar. In *Prunus Laurocerasus*, according to Guignard, the *emulsin* is mainly contained in the

2 L

endodermis, but also occurs in individual cells of the pericycle, which may or may not be in direct contact with endodermal elements.

The fact that certain poisonous or pungent substances are only produced by enzyme action when the plant is injured, clearly indicates that the bodies in question serve for protection against animal foes. This argument was indeed put forward long ago by Nägeli. The usual peripheral location of enzyme-reservoirs, and their frequent association with vascular bundles, also point in the same direction.

### 4. Crystal-sacs.[263]

The vast majority of the genuine crystals that are found in vegetable tissues consist of *calcium oxalate*. Small quantities of this compound occur in the most diverse tissue-elements. But there are also special **crystal-sacs** or **-reservoirs**, characterised by the fact that *calcium-oxalate* crystals always form the bulk of their contents and often fill their cavities completely, or nearly so.

*Calcium oxalate* forms crystals which belong either to the monoclinic (clinorhombic or monosymmetrical) or to the tetragonal (quadratic) system. In either case the precise mode of crystallisation varies enormously. We may distinguish between the following principal types.

(a) *Solitary crystals.*—Solitary crystals of *calcium oxalate* are developed as quadratoctahedra, if belonging to the tetragonal system, but as hendyohedra when they are monoclinic. Numerous derivative forms, produced by twinning or otherwise, are also known. Such large perfect crystals, which are usually found singly in the cells, occur, for example, in the leaves of IRIDACEAE and PONTEDERIACEAE, in the phloem of the POMACEAE, *Robinia Pseudacacia, Ulmus campestris, Aesculus Hippocastanum* (Fig. 214 B), spp. of *Acer*, etc. Excellent specimens of twinned crystals are to be seen in the second layer of the testa of *Phaseolus vulgaris*.

(b) *Crystal dust (Krystall-sand).*—Crystal dust consists of exceedingly numerous minute crystals, the angles and edges of which can scarcely be distinguished, even under the highest magnification; this form of *calcium oxalate* occurs, for instance, in the foliar organs of many SOLANACEAE and in the primary phloem of the stem in the CUPRESSINEAE and in *Sambucus, Cinchona, Liriodendron*, etc.

(c) *Raphides.*—Raphides are long acicular crystals, usually aggregated in considerable numbers into a tightly-packed bundle or sheaf, in which all the crystals lie parallel to one another and are, as a rule, of approximately equal length. Typical raphides are particularly characteristic of the foliar organs of Monocotyledons; they occur regularly, for example, in the LILIACEAE, ORCHIDACEAE, COMMELYNACEAE, (Fig. 214 A), MUSACEAE, etc. Among Dicotyledons, also, they are found in a variety of genera, such as *Galium, Impatiens, Vitis, Phytolacca,* etc.

(d) *Spherical aggregates, sphaerraphides or drusy masses (Drusen).*— Sphaerraphides are rounded crystalline aggregates, the shape of which has been compared to the spiked mace-heads of medieval warfare ; such

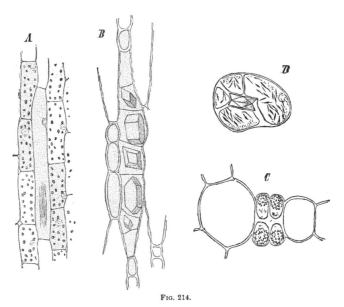

FIG. 214.

Crystal-sacs. *A.* Tubular raphide-sac from the stem-cortex of *Tradescantia zebrina.*
*B.* Crystal-cells from the secondary phloem of *Aesculus Hippocastanum* ⊂ in the
lamellar parenchyma of the petiole of *Trapa natans. B.* Cell from the pulp of·a
Rose-hip containing a twinned crystal suspended by cellulose trabeculae.

aggregates almost always occur singly in the crystal-cells. This form of *calcium oxalate* is of very frequent occurrence indeed ; a few of the many families in which it has been observed are the CHENOPODIACEAE, CARYOPHYLLACEAE, CACTACEAE, ARALIACEAE, MALVACEAE and TILIACEAE.

(e) *Genuine sphaerocrystals or sphaerites.*—Sphaerites of *calcium oxalate* are somewhat rare in the vegetable kingdom ; they are found in the pith of the twigs of *Terminalia Bellerica* and *T. paniculata* (according to Von Höhnel), in various CACTACEAE (according to Möbius), and in the testa of *Elisanthe noctiflora, Silene Cucubalus,* and certain other CARYOPHYLLACEAE (according to Hegelmaier).

In many plants, *calcium-oxalate* crystals occur in more than one of‐ the above-mentioned forms. According to Sanio, for instance, sphaerraphides occur mixed with solitary crystals in the secondary phloem of *Quercus pedunculata, Celtis australis, Morus alba, Fagus sylvatica*, etc.   The same remark applies to *Betula verrucosa* and *Alnus glutinosa* : but in these crystal dust is found in addition.   J. Möller concludes, as the result of repeated examination of various kinds of bark, that crystal dust, raphides and sphaerraphides are invariably contained in thin-walled cells, whereas well-developed solitary crystals occur principally in or near sclerotic cells.   Möller seeks to correlate this distribution with a difference in the velocity of osmosis in the various types of crystal-cell.   It is obvious, however, that the relative thickness of the wall is only one among a number of factors which influence the rapidity of osmosis, and consequently of crystallisation, in any particular case.   A factor of much more funda-mental importance is the varying rate of general metabolic activity ; as a matter of fact, Möller himself points out that sphaerraphides prepon-derate in the primary extra-cambial tissues of young, actively-growing internodes, whereas solitary crystals are more generally found at a later stage, after the formation of periderm and other secondary tissues. From the same point of view, it is quite comprehensible why sphaer-raphides always appear in connection with the autumnal depletion of deciduous leaves, a process which is certainly accompanied by very active metabolic changes.   Quite apart from such nutritive factors, the specific constitution of the protoplasts of the crystal-sacs undoubtedly often determines the mode of crystallisation ; the nature of this proto-plasmic control is, however, at present entirely unknown.

Crystal-sacs—like the crystals themselves—vary greatly in form. Not infrequently their shape stands in some relation to that of the crystals which they contain, though as a rule the connection is quite an indirect one.   Thus, sphaerraphides generally occur in isodiametric cells, while raphide-bundles are contained in tubular sacs.   The cor-relation is most marked in the case of the elongated crystal-sacs of many Monocotyledons (*e.g.* PONTEDERIACEAE), where the wall often fits closely around the dart- or needle-like crystalline enclosure.   According to Rothert, mechanical distension of the wall by the growing crystal is inconceivable.   Any correlation in respect of shape or mode of growth between cell-membrane and crystal must therefore be entirely due to the control exercised over both by the living protoplast of the crystal-sac.

All crystal-sacs, of course, contain living protoplasm to begin with. According to Johow and Fuchs, the raphide-sacs of many Monocotyledons and Dicotyledons permanently retain their living contents,—including the nucleus.   More often, however, fully-developed crystal-cells are dead structures containing very little beside the actual crystals.   Both solitary crystals and sphaerraphides are, in many instances, encased in a closely-fitting—occasionally suberised or lignified—cellulose membrane, which is connected with the cell-wall at one or more points (leaf of *Citrus*). Sometimes a crystal appears to be suspended in the cell-cavity by cellulose trabeculae (*e.g.* the so-called "Rossanoffian crystals," found in the pith of *Kerria japonica*, in *Ricinus communis*, in the petioles of the AROIDEAE, etc.). Each individual member of a raphide-bundle is likewise enclosed in a delicate sheath, which is at first of a protoplasmic nature, but which subsequently becomes converted into a very resistant substance of unknown chemical composition. The whole raphide-bundle, in its turn, is embedded in a mass of mucilaginous material, which is formed in the interior of a special vacuole; this substance has the property of swelling rapidly in contact with water. Zacharias and Rothert have shown that the walls of crystal-sacs are not uncommonly suberised, an inner suberin-lamella being laid down upon an outer unsuberised layer.

The arrangement of crystal-cells is determined by a variety of factors. Like other excretory organs, they tend to be associated with the parenchymatous conducting tissues, in the primary and secondary cortex, in the pith, in the "nerve parenchyma" and in the vicinity of leptome-strands and vascular bundles generally. On the other hand, they are also frequently located in the immediate neighbourhood of mechanical cells and strands; the advantage of this arrangement, no doubt, depends upon the fact that no interchange of material takes place between fibres and the adjoining conducting parenchyma, so that the presence even of large numbers of crystal-cells at the boundary between these two tissues is entirely unobjectionable. Specialised crystal-sacs rarely occur in the epidermis, although solitary crystals or spherical aggregates of *calcium oxalate* are not uncommon in epidermal cells. Möbius has noted the occurrence of multicellular scale-hairs on the pistil of *Cocos nucifera*, which contain raphide-bundles in their large marginal cells. Crystal-sacs may be isolated or arranged in rows; the latter condition is exemplified by the raphide-cells of most COMMELYNACEAE and AMARYLLIDACEAE and of many LILIACEAE. In the secondary phloem of many woody plants, the seriation of the crystal-sacs is due to the fact that cambial cells become divided by transverse walls into a number of vertically superimposed segments, each containing a single crystal or spherical aggregate. Such crystal-fibres, as Hartig has termed them, may comprise a small number of cells, or may, on the other hand, be made up of as many as 20 to 30 segments (Fig. 214 B).

As a general rule the *calcium oxalate* deposited in crystal-sacs represents an excretory product from the physiological point of view. *Oxalic acid* is formed in plants as a result of a variety of metabolic processes (particularly in connection with protein synthesis); but this substance is poisonous to the protoplasm, and is accordingly rendered innocuous by combination with *calcium* to form the very insoluble *oxalate* of that metal.    In certain cases, however, *calcium oxalate* deposits are redissolved and once more taken into the metabolic cycle; this process, which has been observed by De Vries, Tschirch, Schimper, and others, usually takes place when there is a deficiency of *calcium* in the plant.

There can be little or no doubt that *calcium oxalate*, especially when deposited in the form of acicular crystals or of raphides, frequently performs an ecological function by providing mechanical protection against noxious animals.    Stahl has been foremost in endeavouring to demonstrate this protective rôle of raphides with the aid of a variety of ingenious experiments.    Lewin, on the contrary, adduces experimental evidence to show that these structures are harmless to animals, though he admits that they may help to effect the injection of poisonous substances, just like the brittle points of stinging-hairs.

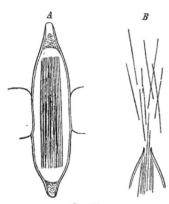

FIG. 215.

*A.* Raphide-sac from the leaf of *Pistia Stratiotes*, in the intact condition; the blunt tips of the cell are very thin-walled. *B.* Tip of an injured sac, more highly magnified; some of the acicular crystals have been forcibly ejaculated.

While it is probable that Stahl has exaggerated the protective value of raphides, there are a number of cases in which the shape of the raphide-sacs, the properties of the mucilaginous sheath investing the raphide-bundle, and the mode of thickening of the cell-wall, clearly indicate that the raphide-sac is specialised as a protective organ.    The following instance will serve to emphasise this point.    The single-layered plates of parenchyma that make up the ventilating tissue (aërenchyma) of the leaf of *Pistia Stratiotes*, are traversed at frequent intervals by spindle-shaped raphide-sacs which project into the air-chambers on either side (Fig. 215).    The walls of these sacs are fairly thick, except at the two ends, which are obtuse, and covered by an exceedingly thin membrane.    When a sac is mechanically injured (the mere access of water is not sufficient), the raphides are ejected with considerable force—generally one at a time—owing to the swelling of the mucilaginous envelope; the thin apical membrane is pierced by the needles as they pass out and soon disappears.    In such a case as this, the point of exit of the raphides is evidently predetermined by the structure of the cell-wall. The tapering of the sac at both ends prevents all the raphides from being ejected simultaneously; owing to the successive ejaculation of the raphides, an animal aggressor may be wounded in a number of different places by a single raphide-sac.

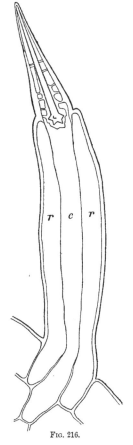

The protective function of acicular *calcium-oxalate* crystals is also very clearly illustrated by the so-called stinging-hairs of certain EUPHORBIACEAE (*Dalechampia*, *Tragia*, etc.), which were first described by Crüger, and have more recently been investigated by Rittershausen and by Knoll. The latter author describes the structure of a stinging-hair of *Dalechampia Roezliana* as follows: The axis of the hair is occupied by a "central cell" or stinging-cell proper (Fig. 216 E) which is sub-epidermal in origin; this cell is ·15 to ·17 mm. in height, and is surrounded for about three-quarters of its length by a sheath of three or four peripheral cells, which represent greatly elongated epidermal elements (Fig. 216 R). The free distal end of the stinging cell is shaped like a slender cone, and ends in a sharp point. It contains a sphaerraphide of *calcium oxalate*; one of the component crystals of this aggregate bears a long needle-shaped process which extends to the extreme tip of the stinging cell, where it lies close against the cell-wall. The whole sphaerraphide is enclosed in a cellular sheath which is connected to the cell-wall by a number of trabeculae, and is, therefore, held very securely in position.

The stinging-cells of *Dalechampia* agree with those of *Urtica* in containing a large amount of protein material dissolved in their cell-sap; it is probable, therefore, that the

FIG. 216.

Stinging hair from a bract of *Dalechampia Roezliana*; *c*, central or stinging-cell; *r*, peripheral cells.

chemical composition of the poison is similar in the two cases. As a matter of fact, the minute hairs of *Dalechampia* are incapable of wounding the human skin sufficiently to cause a stinging sensation. In *Tragia*, on the other hand, the hairs are much larger and produce quite an appreciable irritation. It is only the acicular crystal that penetrates the skin; the cell-wall slips backwards over the needle and is thrown into folds. The needle prepares the way for the entrance of the poison; it thus plays the part which Lewin attributes to raphides in general.

### 5. Cystoliths.[264]

The great majority of Urticaceae, Moraceae and Acanthaceae, many Combretaceae and Boraginaceae, and certain isolated genera, such as *Phlox*, *Klugia*, etc., develop on some of their cell-walls peculiar calcified thickenings which are termed **cystoliths**. In the Urticaceae and Moraceae (excepting *Pilea*, *Elatostema* and *Myriocarpa*), these bodies are restricted to the foliar epidermis; here the outer walls of certain epidermal cells—generally distinguished by their larger size —are provided with stalked spherical or ovoid processes covered

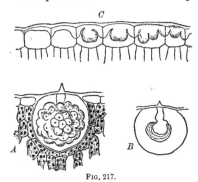

FIG. 217.

Cystoliths of *Ficus Carica*. *A*. Cystolith from the abaxial epidermis of the leaf. × 280. *B*. Decalcified cystolith from a fallen leaf (in autumn). *C*. Group of cells in the adaxial epidermis, with their outer walls thickened, impregnated with *calcium carbonate* and furnished with cystolith-like processes.

with projections, which may be blunt or more or less sharply pointed. The swollen part of the process may thus be compared to a mulberry suspended from the outer wall by its stalk. The cystoliths often fill the cell-cavities almost completely. These structures are particularly well developed in certain species of *Ficus* (*F. elastica*, *F. carica*, etc., cf. Fig. 217 a). Here the cystolith-cell arises directly from a protodermal cell, which remains undivided while the adjacent cells undergo several tangential divisions and become transformed into elements of the water-tissue. The cystoliths of the Acanthaceae are usually more or less spindle-shaped, with a laterally attached stalk, which is usually short and weak and often difficult to distinguish. They are not confined to the epidermis of the leaf, but occur also in the parenchymatous tissues of root, stem and leaf.

It has already been explained that cystoliths are strongly impregnated with *calcium carbonate*. If the mineral substance is dissolved by treatment with acid, an organic basis composed of watery cellulose-material is left behind; this skeleton always shows concentric stratification, to which, in the Moraceae, is superadded a system of radial fibrillar striae corresponding to layers that are relatively rich in cellulose. According to Zimmermann, the very similar striae in the spindle-shaped cystoliths of the Acanthaceae correspond to layers which contain *less* cellulose than the rest of the cystolith. Most probably the body of a cystolith always contains some *silica*; in the MORACEAE and URTICACEAE the stalk is strongly silicified.

There are a certain number of less specialised structures which are closely related to typical cystoliths. In the leaf of *Ficus carica*, for example, isolated groups of cells in the upper epidermis have greatly thickened outer walls, from which knob-shaped processes, impregnated with *calcium carbonate*, project into the cell-cavity (Fig. 217 c). The cystolith-like structures observed by Penzig in certain CUCURBITACEAE are of a similar nature.

It is probable that cystoliths are generally bodies of an excretory nature. For unknown reasons, plants provided with these structures seem to require a large supply of lime; special reservoirs are therefore needed for the reception of the quantities of *calcium carbonate* that become superfluous in the course of metabolism. In special circumstances, however, the lime deposited in cystoliths may be redissolved and utilised afresh. Thus the author has noted that some of the cystoliths of *Ficus carica* become completely decalcified when the leaves are emptied of their contents in autumn; the stalks of these cystoliths remain unaffected, but the bodies finally come to consist of nothing but a shrivelled brown cellulose skeleton (Fig. 217 B). Considerable numbers of such decalcified cystoliths may also be observed in old leaves of *Ficus elastica*, if the plant has been grown in a small pot for some time, a condition of affairs which is likely to result in a shortage of *calcium*. In such cases the cystoliths clearly undergo a change of function; excretory structures become transformed into repositories of reserve-materials, and the temporary surplus of lime is reintroduced into the metabolic cycle.

### 6. *Deposits of silica.*[265]

Silicification of cell-walls is a phenomenon of widespread occurrence in the vegetable kingdom. For the present it will only be necessary to consider those cases in which special cells—usually arranged in approximately spherical groups—have definite portions of their walls thickened and silicified. Generally it is the walls which face one another that become silicified in such a group of cells. The compact "silica-corpuscles" (*Kiesel-Körper*) that arise in this way, generally appear as characteristic translucent patches when the leaf or other organ in which they occur is examined with a hand-lens. Siliceous deposits of this nature are found in the leaves of many species of *Aristolochia*. They are generally composed of a small number of epidermal and hypodermal cells; less frequently they are deeply embedded in the mesophyll. In *Loranthus europaeus* spherical silica-corpuscles occur close to the tip of the leaf, and also along

its margins.  They either lie in the immediate neighbourhood of the
bundle-ends, or are actually intersected by them.  Very often the
distal ends of the terminal tracheides penetrate into the centre of such
a corpuscle (Fig. 218).

It is comparatively unusual for silica to be deposited in the *cavities*
of cells.  In various ORCHIDACEAE, SCITAMINEAE, and Palms, and also
in the HYMENOPHYLLACEAE, the fibrous strands are accompanied by
numerous small tongue-shaped cells, the stegmata of Mettenius.  In

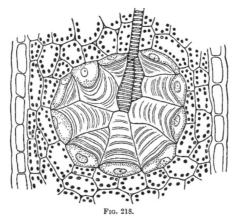

FIG. 218.

T.S. through a silica-corpuscle in the leaf of *Loranthus europaeus*.  The corpuscle is
penetrated by a bundle-end, composed of a few spiral tracheides.

these cells the walls next the fibres are more or less extensively
thickened; the opposite walls bulge outwards, and consist of a thin
median strip, which becomes gradually thicker towards either margin.
Each stegma contains a mass of silica, which, as a rule, fills the cavity
more or less completely, and seems to be devoid of any organic basis.

Möbius has described the small spherical concretions of silica which
occur in the leaves of *Callisia repens* (COMMELYNACEAE)..  The bodies in
question, which have a rough surface, are located in flattened cells cut
off on the outer side of ordinary epidermal elements by periclinal
walls.  Each cell contains several concretions, which are separated
from one another by outgrowths of the cell-wall; the cell-cavity is
thus transformed into a system of passages, in which the silica concretions are enclosed.

Space will not permit of more than a bare reference to the peculiar
silica-corpuscles that occur in the PODOSTEMACEAE, CHRYSOBALANEAE,
and in certain other families.

Silicified cells and silica-corpuscles have been included in the
category of excretory organs, mainly because nothing definite is known

concerning their physiological significance. As a matter of fact it is not improbable that they are actually repositories for the reception of any excess of silica that may be absorbed by the plant. Like so many other physiologically valueless structures, however, they may become secondarily adapted for ecological purposes. Warming believes that they play a mechanical part in the PODOSTEMACEAE, which all grow in swiftly flowing rivers and streams, and which, therefore, require to be specially protected against the shearing action of the violently agitated water.[266]

## IV. THE SECRETORY AND EXCRETORY SYSTEMS IN THE THALLOPHYTA.

The LAMINARIACEAE contain in all parts of their thalli mucilage-canals which have a very interesting structure. According to Guignard, they form a network of schizogenous clefts, lined by groups of cells, the glandular character of which is indicated by their abundant protoplasm and conspicuous nuclei. These secretory cells are not arranged as a continuous epithelium, but are disposed in separate patches, each of which may be regarded as a mucilage-gland. It is noteworthy that this extensive meshwork of mucilage-cavities occasionally sends forth passages or clefts towards the surface, where they end blindly immediately beneath the epidermis. Although these centrifugal passages do not actually open to the exterior, it is probable that they act as outlets for the mucilage.[267]

In many Fungi, resinous secretions are deposited in the interstices of interwoven masses of hyphae; but it is quite uncertain whether these substances are ever secreted by special glandular hyphae. Fungal hyphae are very frequently encrusted with crystals of *calcium oxalate*, but actual reservoirs for the reception of this substance are rarely found. In *Phallus caninus*, however, certain segments of the slender mycelial hyphae develop into large spherical vesicles, each of which is almost completely filled by a sphaerocrystal of *calcium oxalate*. The various hyphae with special contents recently observed by Istvanffy, Van Bambeke and others in the THELEPHORACEAE, and certain other groups, cannot be considered here, if only because it is not definitely known whether the substances that form the contents have any nutritive significance or not. Very possibly some of these specialised hyphae may actually represent excretory organs, while others contain stores of plastic materials, and others again serve as conducting tubes.

## V. DEVELOPMENT OF SECRETORY AND EXCRETORY ORGANS.

The organs which have been described in the preceding sections of this chapter may arise from any one of the three primary meristematic tissues.

The mode of origin of secretory organs from protoderm requires no special explanation, at any rate so far as dermal glands are concerned.

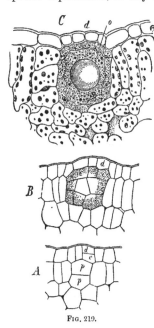

FIG. 219.

Development of a sub-epidermal gland from the upper side of the leaf of *Dictamnus Fraxinella. A.* Initiation of the gland; *p, c*, cells which give rise to the secretory tissue; d, cell which produces secondary epidermal elements. *B.* A later stage. *C.* Fully-developed gland; *o*, drop of ethereal oil; *d*, secondary epidermal cells.

But cases are also known in which glands are internal in a topographical sense, and yet are purely protodermal in origin. According to Von Höhnel, the sub-epidermal glands of *Amorpha, Myrtus*, and *Eugenia* belong to this category. In the case of *Amorpha*, a group of protodermal cells undergo radial extension ; then follow tangential divisions, resulting in the differentiation of secretory cells and secondary epidermal elements, separated from one another by one or two intermediate layers. The young gland bulges outwards, its protodermal origin being perfectly obvious at this stage. When fully developed, however, " the whole structure is pushed downwards into the mesophyll owing to the tension in the epidermis," and there is nothing to indicate its peculiar mode of development. From the ontogenetic point of view, such secretory organs would be classed among dermal glands (as Von Höhnel has indeed suggested), whereas they are undoubtedly internal so far as actual position is concerned.

According to Rauter, the sub-epidermal foliar glands of *Dictamnus Fraxinella* (Fig. 219) occupy an intermediate position between secretory organs which are protodermal, and those which originate in the fundamental meristem. The glands in question arise from paired primary mother-cells, of which one is protodermal, while the other belongs to the ground-meristem. The protodermal element, first of all, undergoes a tangential division. The outer of the resultant daughter-cells produces secondary epidermal elements, while the inner gives rise by further divisions to some of the secretory cells. The majority of the latter, however, are derived from the other primary mother-cell. Ontogenetically, these glands may be compared to the peripheral fibrous strands in the haulms of *Papyrus antiquorum*, which also arise partly from protoderm and partly from fundamental meristem.

Excretory organs rarely originate in the protoderm. The Eu-Crotoneae are furnished with epidermal oil-cells, which sometimes protrude in tubular fashion into the photosynthetic tissue (*e.g.* in *Croton cremophilus* and *Crotonopsis alutaris*, according to Froembling). Tannin-sacs and other elongated excretory elements occur in the epidermis of certain Crassulaceae, Saxifragaceae, Geraniaceae, etc. The cystolith-cells of the Urticaceae and Moraceae, and of some Acanthaceae, are also protodermal in character.

The origin of secretory and excretory organs from fundamental meristem hardly requires detailed consideration. As a matter of fact, the vast majority of internal glands, such as the various secretory passages, mucilage-, resin- and oil-cells, crystal- and tannin-sacs are derived from ground-meristem. The foliar crystal-cells of *Citrus*, though apparently belonging to the epidermis, are actually sub-epidermal in origin; as Guttenberg has shown, they attain their final position by active sliding growth, in the course of which they split apart the radial walls of the overlying epidermal cells. In this manner they may push their way as far as the cuticle, and are thus readily mistaken for protodermal elements. Rothert has described a similar state of things in the case of *Eichhornia speciosa*. According to Knott, another very remarkable instance of this invasion of the epidermis by sub-epidermal elements is furnished by the stinging-cells of *Dalechampia* and *Tragia* which have been described above. Rud. Müller states that the apparently epidermal oil-cells of *Aristolochia* are likewise sub-epidermal elements, which become superficial as a result of sliding growth.

It is the procambium, finally, which gives rise to the secretory cells of the resin- (or oil-) passages that occur in the primary leptome of the Araucarieae, Clusiaceae and Anacardiaceae, and in the primary hadrome (of the stem) of various Conifers (*Pinus*, *Larix*, etc.). According to Ambronn, the cortical oil-passages of Umbelliferae probably always originate together with collenchyma- or mestome-bundles from common procambial strands. In conclusion, it may be remarked that stegmata (cf. above, p. 538) are formed by the septation of peripheral components of procambial strands which for the rest become converted into bast fibres.

# CHAPTER XI.

## THE MOTOR SYSTEM.

### I. GENERAL CONSIDERATIONS.

MOVEMENTS analogous to those of animals, consisting in a change in the space-relations of the whole organism, or of some of its organs, are of widespread occurrence among plants; such movements are generally adaptive in character. An important distinction between the two classes of living beings consists in the fact that movement of the entire organism—that is, locomotion—is the rule among animals, but the exception among plants. Movements of individual organs, carried out in the interests of their own physiological activity, or for the benefit of other organs, on the other hand, are of general occurrence in the vegetable kingdom.

It is well known that a number of the lowest plants are freely motile throughout life, or, at anyrate, at certain stages of their life-history. Such cases show that, in spite of what has just been stated, we cannot invariably rely upon the prevailing type of movement in order to distinguish a plant from an animal. Several distinct modes of locomotion are known, and more than one form of special locomotor organ has been evolved for the purpose of carrying out these movements. Three principal types of active locomotion may be distinguished. These are exemplified respectively by the amoeboid creeping movements of Myxomycete plasmodia, the gliding movements of Diatoms, Desmids and OSCILLATORIEAE, and the swimming movements of BACTERIA, VOLVOCINEAE, and swarm-spores, which depend upon the action of vibrating or undulating flagella. Among Higher Plants, on the other hand, active locomotion is unknown; the only movements of translation that occur, are the gradual progression of growing parts of the plant-body, and the passive transportation which is almost entirely restricted to seeds and fruits. The last mentioned movements often depend upon the presence of special organs which provide points of application for the external agencies that actually effect the transportation; this type of adaptation is exemplified by the various contrivances which facilitate the dispersal of seeds or fruits by wind. Ejection of seeds or spores is often also brought about by the sudden release of tension previously set up by the action of turgor, or in consequence of the cohesive power of evaporating cell-sap. Many reproductive organs are furnished with hygroscopic arrangements, which assist in the distribution of the spores or seeds.

The various parts of the plant-body perform a great number of movements of different kinds, which are in no way concerned with locomotion or with the transportation of seeds or spores, but which serve to effect a suitable adjustment in space of the individual organs. Movements of this type are constantly performed by every highly organised plant. Some of them, such as the curling and uncurling of the leaves in certain xerophilous Grasses, are due to purely physical causes. As a rule, however, such adjustments are special manifestations of vital activity, and the energy required for their performance is derived from the metabolic activity of living protoplasts. External agencies act in general merely as **stimuli**, which may, however, not only induce an organ to move, but also determine the direction in which the movement takes place.

In a number of cases such active movements are carried out without the aid of special motor-tissues or -organs. The curvatures associated with heliotropic and geotropic response, for example, are generally performed by organs which are still engaged in longitudinal extension, and by tissues which will be concerned with totally different functions when they are fully developed. Frequently, however—more especially in connection with foliage leaves—special **motor-organs** are differentiated; these are usually capable of executing *repeated* movements. Such motor-organs, which may occur on stems, petioles, leaf-sheaths, or pinnae, are termed **pulvini**. In addition to the actual motor-elements, they naturally contain other tissues belonging to the dermal, skeletal, conducting and other systems, for the same reason that a photosynthetic organ is not exclusively composed of photosynthetic cells.

The motor system includes all tissues and anatomical features which are primarily concerned with the execution of passive or active movements. It will be found convenient to distinguish between **active** and **passive motor-tissues**. In the case of the latter, the movement is carried out with the help of external forces, whereas it is characteristic of the active type that the requisite energy is provided by the tissue itself. The flying-hairs and -tissues which assist in the dispersal of seeds and fruits by wind, thus belong to the class of passive motor-tissues, while the fibrous layer of an anther and the turgescent parenchyma of a pulvinus exemplify the active type. Active motor-tissues may further be subdivided into two groups. The action of the **inanimate** or **mechanical** variety is entirely dependent upon physical changes, which take place either in cell-walls (imbibition mechanisms), or in cell-cavities (cohesion mechanisms). The other group comprises the **living** or **physiological** motor-tissues, which functionate by virtue of the vital activity of their constituent cells.

On comparing vegetable and animal motor-tissues with one another, a striking contrast at once becomes apparent. The motor-tissues of plants exhibit great diversity of structure and mechanism. This diversity is mainly due to the fact that a variety of physiological processes and anatomical arrangements have been adapted and developed for the purpose of carrying out movements; moreover, this adaptation has not taken place at the same level of phylogenetic development in every case. In the animal kingdom, on the other hand, the structure and mechanism of the motor-tissues are comparatively stereotyped. There, movement is almost invariably performed with the aid of (smooth or striated) muscle-fibres which are derived, both phylogenetically and ontogenetically, from specialised protoplasmic fibrillae. This uniformity of the motor-tissues among animals is evidently correlated with the fact, that the specialisation in question appeared during the earliest stages of evolution, and has subsequently undergone a gradual and continuous development.

In the vegetable kingdom, special contractile protoplasmic organs corresponding to muscle-fibres are, so far as is known, entirely absent, principally, no doubt, because the prevalence of relatively rigid cell-membranes does not admit of the employment of delicate contractile protoplasmic fibrils for the purpose of performing movements. Plants have thus, as it were, been compelled from the very first to make use of cell-walls in the construction of their motor-organs.

## II. PASSIVE MOTOR-TISSUES.

### A. FLYING-HAIRS AND FLYING-TISSUES.

The flying arrangements of wind-distributed seeds and fruits consist either of air-containing trichomes, or of the thin sheets of tissue which are known as " wings." The principal function of these structures consists in increasing the friction between the air and the falling fruit or seed, so as to prolong the sojourn of the latter in the air ; as a result, the seed obtains a better chance of being carried to a distant point by the wind. Where, as often happens, their flying-organs expose a large surface to the wind, seeds may be carried over great distances.  The form and orientation of flying-organs and their relation to the mechanics of flight, cannot be discussed in detail here : the organography of the organs in question has been fully dealt with by Dingler.  In the present work attention must be concentrated upon the anatomy and histology of organs of flight.[268]

**Flying-hairs** sometimes envelop the whole seed uniformly, as in *Gossypium*.  In *Salix, Populus, Epilobium, Asclepias*, etc., they form a terminal tuft, which acts as a parachute ; the silky or woolly pappus of the fruits of COMPOSITAE illustrates this latter type of arrangement. The flying apparatus of a seed or fruit usually comprises a very large number of hairs ; the minute and exceedingly light seeds of *Aeschynanthus*, however, are only provided with three hairs apiece.

Typically, flying-hairs are unicellular structures of considerable size ; those of Cotton, for example, vary between 2 and 6 cm. in length.  The multicellular bristles of the COMPOSITAE are usually " feathered " in various ways.  The hairs which cover the rays of the pappus of *Tragopogon orientale*, etc., become interwoven into a network or web expanded in a single plane.  In *Centaurea calocephala* the two margins of the bristles are fringed with hairs, which are set so closely together as to form a continuous membrane, at any rate in the lower half of the bristle ; by this means the surface exposed to the wind is increased more than twofold (Fig. 220).

FIG. 220.

Margin of a pappus - bristle of *Centaurea calocephala*.

In accordance with their function, the structure of flying-hairs is always such as to combine lightness with rigidity ; the cell-walls are at most moderately thickened, but at the same time possess very considerable tensile strength.  The latter property is conspicuously developed in the case of Cotton, which has been tested with reference to this point in the following manner.  A single hair, 2 or 3 cm. in length, is cemented at both ends to narrow strips of paper.  One of the strips is fixed immovably, while to the other is fastened a tiny paper bag into which fine sand is cautiously poured ; in this way the absolute breaking load of a single hair can be determined to within $\frac{1}{100}$th of a gram. The effective cross-sectional area of the hair must then be estimated by careful measurement, when the load per sq. mm. can at once be calculated.  By this method the breaking strain of cotton " fibres " (*i.e.* the flying-hairs of the seeds) is found to lie between 18 and 22 kg.; the flying-hairs of *Gossypium* are therefore equal in strength to the toughest bast-fibres.

The general construction of the **flying-tissues** of winged seeds and

2 M

fruits is likewise arranged so as to reduce the specific gravity of the whole structure as much as possible, without impairing its inflexibility and shear-resistance.

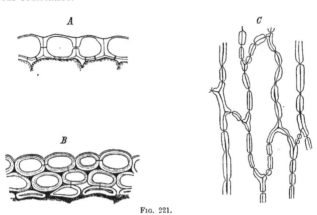

FIG. 221.

Flying-tissue of the winged seed of *Cedrus Libani*. *A*. T.S. through the thin part of the wing. *B*. T.S. through the thicker region. *C*. A portion of the thin region in surface view.

The simpler types of " wing " are entirely made up of one or more layers of flying-tissue. Among the ABIETINEAE the flying-tissue of the winged seeds is not, as might be supposed, an outgrowth of the testa, but represents a specialised portion of the ovuliferous scale, consisting mainly of epidermal cells, but to some extent also of sub-epidermal layers. In *Cedrus Libani*, for example, the thinner portion of the wing is composed of a single cell-layer, which, as already stated, corresponds to a portion of the epidermis of the ovuliferous scale; its component cells are elongated, and have fairly thick radial walls provided with numerous pits (Fig. 221 A). These radial walls form a rigid framework with its meshes closed by the thin tangential walls, which collectively constitute a continuous **flying-membrane**. Fragments of the disintegrated hypodermal layer adhere to the lower side of the wing in this region. The thicker portions of the wing comprise two or

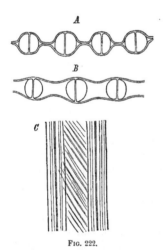

FIG. 222.

Flying-tissue of the winged seed of *Spathodea campanulata*. *A*. Part of a T.S. through the wing. *B*. Part of a T.S. through the wing after treatment with caustic potash. *C*. Small portion of one of the cells of the flying-tissue in surface view.

three sub-epidermal layers in addition to the epidermis (Fig. 221 B). Here the cells are oval in cross-section, and have their walls uniformly thickened and lignified; the cells of the innermost layer are usually much compressed.

The thin transparent wing of the seed of *Spathodea campanulata* is composed throughout of a single layer of flying-cells. These cells are prosenchymatous and arranged in radial rows; their lateral walls are so greatly thickened as to appear almost circular in cross-section.

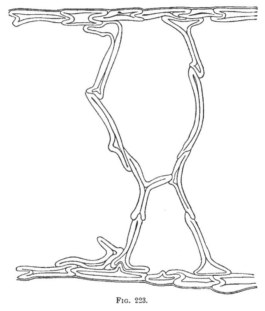

Fig. 223.

T.S. through the wing of the seed of *Zanonia macrocarpa* (traversing the thicker portion of the wing).

The actual flying-membrane consists again of the delicate tangential walls, which are stretched over the rigid framework of rods formed by the radial walls. In its natural state this flying-membrane is depressed between the rods to such an extent, that the outer and inner tangential walls meet (Fig. 222 A); the two halves of the membrane can be separated by treatment with caustic potash (Fig. 222 B). When examined in surface view, these tangential walls exhibit a delicate oblique striation, which is due to the presence of microscopic fibrillar thickenings. Both the radial and the tangential walls assume a greyish-violet coloration on treatment with chlor-zinc-iodine.

The large satiny wings of *Zanonia macrocarpa* are also exclusively composed of flying-tissue of a very remarkable type. The extremely delicate margin of the wing consists of very thin-walled elongated cells. A transverse section through the thicker central portion shows that the two epidermal layers which constitute the actual flying-membrane are linked together by loosely arranged, single-layered plates of tissue, which are also connected among themselves by cross-ties (Fig. 223). In a longitudinal section these buttressing plates are seen to consist of tabular cells, which are mostly very thick-walled; the terminal elements in each buttress expand so as to provide a large surface of attachment to the flying-membrane. The epidermal cells are likewise thick-walled and cohere very firmly, owing to the fact that their lateral walls are sinuous, and thus interlock; the cells of the buttresses are also frequently dove-tailed in a similar fashion.

The wings of fruits often have a more complicated structure than those of seeds, owing to the presence of vascular bundles (*Ulmus*), or stout mechanical strands (*Acer*), in addition to the actual flying-tissue. As Wahl has shown, such accessory tissues are always disposed in accordance with the mechanical requirements of the wing.[269]

Where the wing is the morphological equivalent of a leaf, it usually acts, to some extent, as a photosynthetic organ, at any rate while it is young; in such cases, the structure of the organ naturally embodies a compromise between the demands of two different functions.

### B. FLOATING-TISSUES.[270]

The seeds and fruits of water- and strand-plants are generally dispersed through the agency of water-currents. It is, therefore, most essential that the organs in question should be able to float for pro longed periods; this necessity has led to the development and specialisation of air-containing **floating-tissues**. Schimper discriminates between several types of floating-tissue, which differ in structure, and in their arrangement in the pericarp or testa. In some cases (*Cerbera Odollam, Laguncularia racemosa, Nipa fruticans, Aegiceras majus*) the tissue in question contains many large intercellular air-spaces, and thus resembles spongy parenchyma. More frequently, however, intercellular spaces are feebly developed or altogether absent, and the air is contained in cell-cavities. The cell-walls are thin, and often crowded with pits; they never consist of unmodified cellulose, but are often lignified, and sometimes impregnated with substances of uncertain composition. A very important property of such floating-tissues consists in the fact that they are pervious to air, but very impervious to water, so that they remain full of air for a long time, even in contact with water. Schimper found that fragments of the floating-tissue of *Barringtonia* *speciosa* and *Cynometra cauliflora* do not sink in a 3 per cent. solution of brine even after being immersed for twenty-two weeks.

Where the floating-tissue of the seed or fruit is superficial, as it is in many strand-plants (*Cocos nucifera, Barringtonia Catappa, Lumnitzera, Carapa,* etc.), it is usually traversed by numerous mechanical strands which prevent it from being easily abraded. In other cases (*Calophyllum inophyllum, Ximenia americana, Cycas circinalis, Excoecaria Agallocha*) the incoherent floating-tissue is protected by being enclosed within a hard endocarp or testa.

Kólpin Ravu has shown, that the seeds of many of our native marsh- and water-plants are provided with floating-tissues of a similar nature to those which have just been described.

## III. ACTIVE MOTOR-TISSUES.

### A. HYGROSCOPIC OR IMBIBITION MECHANISMS.[271]

The hygroscopic swelling and shrivelling of cell-walls involves changes in the volume of the membranes concerned, which, under suitable conditions, may produce a very appreciable amount of movement, in the shape of a curvature or torsion of some limited portion of the plant-body. It is an essential feature of every hygroscopic mechanism, that the two sides of the motor apparatus—which may consist of an extensive tissue, of a single cell or even of part of a cell— should be antagonistic in behaviour. The principle upon which the requisite antagonism usually depends may be explained as follows :

If the lateral walls of a cylindrical cell all have the same powers of imbibition, a loss or gain of water will always produce a simple contraction or extension of the cell. Curvature can only result, when two opposite longitudinal strips of the cell-wall have unequal powers of swelling, so that one of them undergoes a greater amount of elongation than the other when water is absorbed. The same statement applies —*mutatis mutandis*—in the case of a cylindrical mass of tissue. The *amount* of curvature obviously depends in the first instance upon the difference between the swelling capacities of the antagonistic sides, and is unaffected by the relative thickness of the cell-walls. The *energy* involved in the movement, on the other hand, varies directly as the thickness of the active membranes. As a matter of fact, hygroscopic curvatures of plant-organs always take place in opposition to more or less powerful resistances ; hygroscopic motor-cells are, therefore, usually more or less thick-walled.

The change in the volume of a cell-wall which accompanies a change of water-content, depends not only upon the quantity of water removed or introduced, but also upon the molecular or micellar structure of the membrane. The last-mentioned factor is responsible for the unequal distribution of the water of imbibition in different directions within the membrane. This fact, which is of fundamental importance for the theory of hygroscopic movement, was first established by Zimmermann. It is the differences in the power of imbibition in different tangential planes which are of the greatest importance. In the case of a spindle-shaped cell, the most active expansion and contraction of the walls may take place either in the transverse or in the longitudinal direction. In the former instance it is evident that water of imbibition is most readily introduced (or removed) in the transverse direction, whereas the longitudinal direction is favoured in the other case.

If we try to express this differential imbibition in terms of the molecular structure of the membranes, we are forced to assume that the molecular or micellar groups cohere with different intensities in different tangential planes. Where they form continuous longitudinal series water will be incorporated more readily between adjacent longitudinal rows,—that is, transversely—than between successive micellae of one and the same series—that is, in the longitudinal direction. If, on the contrary, the micellae are arranged in transverse series, water is most easily introduced between adjacent transverse rows, that is, longitudinally. For a little reflection will show, that the cohesion cannot be so great between adjacent micellar series as it is between successive micellae of the same fibrillar series, and that, consequently, the introduction of water between two series—that is, at right angles to their long axes—is opposed by a resistance smaller than that which has to be overcome before successive micellae in the same series can be pushed apart.

The orientation of micellar series in a membrane can be deduced from the arrangement of its slit-like pits or thickening fibres, as well as from any striations that may be visible, since, so far as is known, the two sets of features are always correlated in the same way. If, for example, an elongated cell is found to be furnished with longitudinal or very oblique series of slit-shaped pits, it may be safely concluded that the greatest swelling (and contraction) will take place in the transverse direction. Where, on the other hand, the pits are elongated transversely, or nearly so, the greatest swelling (and contraction) will take place in the longitudinal direction. In either case the maximum extension or contraction may vary, according to Eichholz, between 5 per cent. and 20 per cent. Where the histological features give no clue, the axes of maximum expansion and contraction may be determined by observing the optical behaviour of the cell-walls in polarised light.

This **principle of differential imbibition** finds a great variety of applications in the construction of hygroscopic mechanisms. In all such cases, the cell-walls on the two antagonistic sides of the motor apparatus differ in the manner described above in their molecular structure, and hence in their powers of swelling and contracting. The requisite antagonism may, however, also be produced in other ways, as will be seen later on in the case of anthers.

As regards their shape and detailed structure, hygroscopic motor-cells exhibit a good deal of variety. The prosenchymatous thick-walled forms are connected by various transitional types with genuine mechanical fibres. As a matter of fact, ordinary bast-fibres may assist in the production of movements, besides performing their principal mechanical functions. For cells of this composite type Eichholz suggests the name of "dynamo-static" elements. In the same way the term **dynamic cell** (Eichholz's actual phrase is "*specifisch-dyna-mische Zelle*") may be applied to any thick-walled prosenchymatous element which has its cell-wall molecules or micellae arranged in transverse rings, or very gently inclined spiral series, and which undergoes very well-marked longitudinal contraction on losing water. Parenchymatous hygroscopic cells are also far from uncommon; they may be isodiametric, or more or less rod-shaped. The thickness of their cell-walls varies greatly; very often local thickenings are developed, which play an important part in the mechanism of movement. The walls are also generally lignified, but this feature does not seem to be in any way directly correlated with the hygroscopic properties of the membranes.

The preceding generalisations, which are based upon elaborate researches by Zimmermann, Schwendener, Eichholz, and Steinbrinck (especially upon those of the last-mentioned author), may now be illustrated by a few concrete examples. The **peristome** of Mosses is an excellent instance of a hygroscopic apparatus formed by the persistence of certain definite portions of the walls of cells, which otherwise become disorganised. Where the peristome is double, hygroscopic properties are usually restricted to the teeth of the outer series. In damp weather, the hygroscopic teeth curl inwards, thereby closing the orifice of the capsule and preventing the escape of the spores, which, of course, cannot be effectively distributed during rain. When the air is dry, the teeth curl outwards, and spore-dispersal can proceed unchecked. Each peristome-tooth is made up of an inner and an outer cellulose lamella. Ontogenetically, these lamellae represent local thickenings of the partition-walls between two adjacent cell-layers; consequently they are separated from one another by a middle lamella. Now Steinbrinck has shown that the axes of maximum contraction of the two lamellae almost always cross one another at right angles; in this way the requisite antagonism between the two sides of the peristome-tooth is ensured.

Certain seeds and fruits are furnished with **hygroscopic flying-hairs**, which curl inwards from the base when wetted, so as to become closely appressed to the surface of the fruit (or to the style or pappus-ray, as the case may be), whereas in dry air they spread outwards. At the base of each hair there is a more or less extensive hygroscopic zone, in which the cell-wall consists of transversely seriated micellae on the outer or convex, and of longitudinally seriated particles on the inner or concave side. Not infrequently, the fact that the side of the hair which contracts most strongly on drying is made up of transverse micellar rows, can be inferred, without any examination of the optical properties of the membrane, from the presence of transversely elongated slit-shaped pits. In *Dryas Drummondi*, *Anemone Pulsatilla*, etc., a cushion-like enlargement of the more highly contractile side of the wall increases the intensity of curvature (Fig. 224).

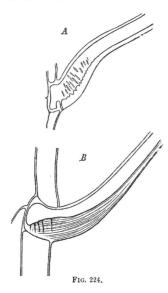

FIG. 224.

Flying-hairs. *A. Dryas Drummondi.* *B. Anemone pulsatilla.* (Only the basal portions of the hairs are shown.)

In other instances, the **antagonism between** the two sides of the hygroscopic organ is due to their being composed of **different tissues.** The involucral bracts of the ripe capitula of *Centaurea* and certain other COMPOSITAE curl inwards in damp, and outwards in dry air, thereby preventing the wind-dispersed fruits from escaping during rainy weather. The following description of the thick-walled motor-tissue of the bracts of *Centaurea* is based on Steinbrinck's account. Immediately beneath the outer surface—which, on drying, becomes the concave side—is situated a layer of prosenchymatous cells with transversely elongated pits; from what has been said above, it is evident that this layer will swell more actively in the longitudinal than in the transverse direction; conversely, on drying, the most pronounced contraction will be longitudinal. The second layer—reckoning from the outside—is parenchymatous in character; its walls are also transversely pitted, but have a smaller power of imbibition than those of the first layer. The following layer is likewise parenchymatous, but has somewhat oblique pits. Next the inner surface of the bract—which becomes the convex side on drying—there is a second prosenchymatous layer with very oblique or longitudinal pits, a feature which shows it to be the component antagonistic to the outermost layer. Evidently the axes of maximum contraction of the two prosenchymatous layers intersect one another at right angles; on drying, the outermost undergoes the greatest, the innermost the smallest, amount of contraction in the longitudinal direction, and the whole bract consequently curls outwards.

In the genus *Campanula* the valves of the capsule, which likewise curl and uncurl longitudinally, have a structure very similar to that of the bracts of *Centaurea*. Here there is an outermost zone of transversely pitted parenchymatous elements, a middle zone of somewhat more elongated cells with slightly oblique pits, and an innermost layer of parenchymatous cells with pits which are longitudinal or very oblique.

According to Steinbrinck, the branches of the familiar Rose of Jericho (*Anastatica hierochuntica*), which curl inwards when dry and outwards when wet, contain a mechanism of the same type. In the adaxial half of each branch the fibrous motor-cells are transversely pitted, while the corresponding elements in the abaxial half have very oblique pits; the antagonism is, therefore, of the same nature as in *Centaurea* and *Campanula*.

In the cases which have just been described, the cells of all the component layers are elongated in the same direction, namely, in the same sense as the whole organ. The requisite antagonism between opposite sides is therefore attained, not by any special orientation of the active cells, but mainly by differences in the micellar structure of the walls in opposite layers. In other words, the axes of maximum contraction of the cell-walls run in different directions on the two antagonistic sides. It is evident that the same effect might be produced in a different way, if cells all possessing the same micellar structure were arranged in rows which ran in different directions on the two sides of the organ. Let us, for example, consider the case of an organ in which the cells are all elongated transversely on one side, and longitudinally on the other. It is assumed that all the cell-walls consist of longitudinal or very oblique micellar rows—as shown by the orientation of the pits—so that they contract more strongly in the transverse than in the longitudinal direction. Such an organ, on drying, will tend to curve towards the side composed of transversely elongated cells. In order, however, that the elements in question may be able to overcome the resistance of the longitudinally extended elements, they must either form a thicker layer or be endowed with a greater power of imbibition.

This method of producing hygroscopic antagonism with the aid of crossed cell-layers ("woven structure," *gewebeartige Struktur*) was known to Kraus and Hildebrand. Steinbrinck was, however, the first to point out that the mechanism of many "explosive" dry fruits depends upon the same dynamical principle. Excellent illustrations are provided by Papilionaceous legumes, and by the capsules of LILIACEAE, RUTACEAE and ERICACEAE (tribe RHODOREAE). The dehiscence of most legumes appears to involve spiral torsion of each valve, but in reality the movement consists of a simple curvature around an axis which is inclined to the long axis of the valve. The inner side of the valve is occupied by a more or less extensive zone of thick-walled fibrous motor-cells (Fig. 225 $b$), which are elongated in the direction of the axis of curvature, that is, at an angle of $30°$-$40°$ with the long axis of the pod. The antagonistic tissue consists of the thick-walled outer epidermis (Fig. 225 $e_1$), which is sometimes supplemented by a few sub-epidermal layers. These antagonistic elements are elongated approximately at right angles to the long axes of the motor-cells.

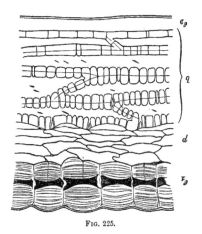

FIG. 225.

Oblique transverse section (taken parallel to the axis of curvature) through one of the valves of the pod of *Lathyrus latifolius*; $e_1$, outer epidermis; $p$, thin-walled parenchyma; $b$, thick-walled motor-tissue; $e_2$, inner epidermis. ($e_1$ and $b$ are the two antagonistic layers.)

The geniculate awns of many Grasses and the awn-like appendages of the mericarps of certain GERANIACEAE execute twisting movements of a hygroscopic character, which gradually force the fruits into the ground, owing to the presence of backwardly directed stiff bristles just behind their boring points. The mechanism of these torsions has been studied in detail by F. Darwin and by Zimmermann; both these authors conclude, that the twisting action depends upon the structure of the individual elements of the fibrous tissue, which forms the bulk of the hygroscopic organ in such cases. According to Zimmermann, it is the more peripheral bast-fibres that are mainly responsible for the movement. In these, as in many similar cells, the micellae are arranged in left-handed spiral series. The more internal stereides, on the contrary, do not twist actively, but serve to increase the torsion of the whole organ, owing to the marked contraction which they undergo on drying. The peripheral fibres may be isolated by treatment with potash, which reagent also causes the cell-walls to swell very considerably. In these circumstances, they exhibit a very well-marked right-handed torsion, which can be readily observed under the microscope; drying conversely produces a left-handed twist. The entire mechanism thus again depends essentially upon the fact, that the tangential imbibition is greatest at right angles to the long axes of the sinistral micellar series, and smallest in the direction parallel to those series.

The present section may conclude with a brief consideration of the **dehiscence-mechanisms of anthers**, a difficult problem which has formed the subject of numerous researches, but which still remains partially unsolved. As a general rule, each anther-lobe dehisces by means of a single longitudinal slit, which is common to the two pollen-sacs. In dry air, the two valves or flaps of the anther wall curl away from the slit, and thus allow the pollen to escape freely. If the anther is wetted, the valves at once uncurl and close over the slit. The hypodermal cells of the valves are provided with very characteristic fibrous thickening; they constitute the so-called **fibrous** or **mechanical layer**, which is altogether responsible for the movements of the valves, the epidermis taking no active part whatever in that process. According to Steinbrinck, the fibrous thickenings are generally arranged, in each cell, in such a way that "they traverse the whole length of the radial walls, and unite on the inner tangential walls to form a stellate or reticulate framework, a system of parallel bars, or sometimes even a massive plate, while the outer tangential wall always remains entirely unthickened." The fibrous layer therefore possesses two antagonistic sides, consisting respectively of the outer and the inner tangential walls, that differ appreciably as regards their rigidity, one being stiffened by means of special thickenings which are wanting in the other. During the tangential contraction of the fibrous cells which follows upon loss of . water, their delicate outer walls will tend to collapse more than their inner walls, since the latter are supported by thickening fibres (Fig. 226); the whole valve will consequently curl outwards in dry air. According to this view, which was first put forward by Leclerc du Sablon, the dehiscence of anthers depends upon a genuine hygroscopic mechanism, the active component of which is represented by the unthickened portions of the radial walls of the fibrous cells. Purkinje, Meyer, Chatin and Schinz, on the other hand, all attribute the movements of anther-valves to the hygroscopic properties of the fibrous thickenings themselves; this hypothesis has recently been revived by J. M. Schneider, who maintains that, in the case of *Tulipa*, at any rate, the inner side of each curved thickening fibre is more hygroscopic than the outer side, with the result that the curvature of all the fibres increases on drying.

Among recent investigators, Schwendener, Colling and Schneider all maintain that the dehiscence of anthers is a hygroscopic phenomenon. A strong argument in favour of this view is based by Schwendener upon the fact, that the valves do not begin to curl outwards, until every particle of liquid has disappeared from the cavities of the fibrous cells and the cell-walls have begun to lose water. The following experiment, performed by the author, also suggests that the hygroscopic properties of the cell-walls in the mechanical layer are at any rate partly responsible for the movement of the valves. Transverse sections of air-dried anthers of *Tulipa Gesneriana* or *Fritillaria imperialis* are mounted in a moist chamber and examined under the microscope; the valves, which were widely open in the dry condition, will be seen to curl rapidly inwards in the damp atmosphere, although they are not actually wetted; this movement is, however, not necessarily hygroscopic in character. If the sections are now exposed to dry air, the valves soon begin to curl outwards again. It should be stated that this opening movement never proceeds so far as to lead to a reversal of the curvature or even to a straightening of the valves. The sections appear quite opaque throughout this experiment, a fact which proves that the cells are filled with air, both during the inward curvature, and while the valves are uncurling. There can therefore be no doubt, that the partial opening of the valves, in this instance, is a purely hygroscopic movement.

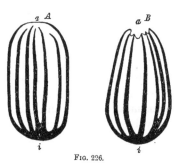

Fig. 226.

Isolated cells of the fibrous layer from the anther-wall of *Lilium candidum*. *A*, when moistened; *B*, in the dry state; *a*, outer; *i*, inner tangential wall. After Steinbrinck, semi-diagrammatic.

While Kamerling and Steinbrinck undoubtedly go too far when they relegate the whole mechanism of anther-dehiscence to the category of cohesion-movements (cf. the next section), it cannot be denied that co hesion plays some part in connection with the opening of anther-valves; it is the quantitative value of this factor which is still uncertain. Co-hesion may actually supply the entire motive power in certain instances. Colling states that this condition is realised in the case of *Tacca macrantha*, *Polygala grandis*, *Sagittaria natans*, and *Salvia officinalis*. The same investigator, on the other hand, found a hygroscopic mechanism to be present in more than one hundred of the species which he examined.

Hygroscopic mechanisms, as we have seen, depend entirely upon the swelling power of the cell-membrane.   Cohesion movements, on the contrary (which have been studied by Steinbrinck, Schrodt and Kamerling), are brought about by the water contained in the *cavities* of the motor-cells, the cell-walls at most undergoing passive deformation, which may result in a condition of strain.

The mode of action of a cohesion mechanism may be explained as follows : As the water contained in the lumen of a motor-cell diminishes in volume through evaporation, it draws the thin portions of the cell-membrane inwards owing to the adhesion between the liquid and the wet cell-wall.   The traction exerted in this way, depending as it does to a large extent upon the very considerable internal cohesion of water, is sufficient to produce deformation of any specially thickened cell-walls that may be present; these are accordingly brought into a condition of strain.   As evaporation continues, there comes a time when the cell-wall resists any further invagination or other deformation, or, in other words, overcomes the internal cohesion of the contained water; the previously continuous water-column thereupon collapses,— or, it may be, breaks away from the cell-wall—and a vacuum is suddenly formed in the cell-cavity.   The cell-wall is in all probability almost, if not entirely, impervious to air in the dry state, so that this vacuum may persist for some time.   As the membrane is readily permeated by water, renewed access of water quickly causes the vacuum to disappear again ; hence the entire process may be repeated again and again.

What may be termed the **primary movement**, in such cases, evidently depends upon contraction of the thin portions of the wall, the thicker regions undergoing a much smaller amount of deformation. This movement, of course, takes place quite gradually, its speed being determined by the rate at which water is evaporating.   When the internal cohesion of the water has been finally overcome, the cell-walls may, from causes which are still obscure, persist in their state of tension ; more usually, however, the elasticity of the strained thickened regions comes into play, and the collapse of the water-column is accompanied by an instantaneous release of tension, which results in a sudden jerky **secondary movement.**

The best-known illustration of a cohesion mechanism is provided by an ordinary Fern-sporangium.   In the POLYPODIACEAE, the single-layered sporangium-wall is furnished with an annulus, which extends over the back and apex of the sporangium, and as far as the middle of its anterior face ; this annulus consists of a row of cells with very thick inner and lateral, but with unthickened outer walls (Fig. 227 A).

When water is removed by evaporation from the cavities of the annulus cells, its cohesive power causes the thin outer walls to sink inwards, with the result that the thick, rigid, lateral walls are drawn closer together. The whole annulus consequently tends to straighten itself, and in so doing brings about the dehiscence of the sporangium. As water continues to escape, the annulus may gradually bend far backwards, so that the side which was formerly convex (*i.e.* the outer or

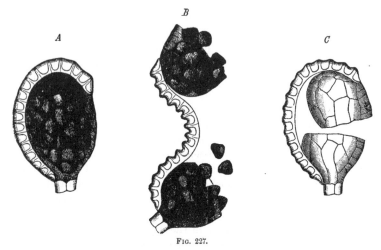

FIG. 227.

Dehiscence of a Fern-sporangium (*Pteris,* sp.). *A.* Ripe, but still intact, sporangium. *B.* Sporangium dehiscing. Annulus curved outwards and about to recoil; most of the spores still unshed. *C.* After the recoil of the annulus; all the spores ejected.

thin-walled side) becomes concave (Fig. 227 B). This primary movement takes place quite slowly. The drawing-in of the outer walls, and the deformation or lateral approximation of the radial septa, finally become so pronounced that the cohesion of the water is no longer able to withstand the elastic tension of the membranes; when this stage is reached, the water-columns in the cells suddenly collapse, the tensions in the radial and outer walls are instantly released, the whole annulus recoils with a violent jerk, and, striking against the surface of the leaf, throws the sporangium bodily upwards, or to one side. It is this sudden secondary movement, that actually brings about the dispersal of the spores. Repeated ejaculation may take place, if the water-columns do not give way simultaneously in all the cells of the annulus.

In the HYMENOPHYLLACEAE and CYATHEACEAE, where the oblique annulus completely encircles the sporangium, the mechanism is the same as in the POLYPODIACEAE. For an account of the variations in detail exhibited by other Pteridophyte-sporangia, the reader is referred to the writings of Leclerc du Sablon, Goebel, Steinbrinck, and Ursprung; the last-mentioned author states that hygroscopic action plays a part in certain cases.

According to Kamerling, the **elaters** of Liverworts also afford an excellent illustration of a cohesion mechanism. Typical elaters occur in most JUNGERMANNIALES and MARCHANTIALES. They are long, spindle-shaped structures, which, after the dehiscence of the sporangium, execute jerky movements, and so help to break up the spore-mass, and to disseminate the individual spores. Each elater has a thin wall provided with two parallel spiral thickening fibres. When water is withdrawn from the cavity of the elater, the thin strips between the spiral fibres are drawn inwards, owing to the internal cohesion of the

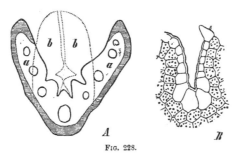

Fig. 228.

*A.* Diagram of a leaf of *Festuca glauca*, in T.S. ; *a-a*, leaf expanded ; *b-b*, leaf folded up. Sub-epidermal fibrous tissue of the lower side shaded. *B.* T.S. across one of the furrows on the upper side. Note the large epidermal cells at the bottom of the furrow. After Tschirch.

liquid; the radius of curvature of the fibres consequently diminishes, and the number of spiral turns is increased. In other words, the elaters undergo torsion, and the thickening fibres become wound up like the mainspring of a watch. When the tension in the membrane becomes greater than the internal cohesion of the water, the spiral fibres return with a sudden jerk to their original position.

Some of the movements of vegetative organs also depend upon cohesion. The leaves of many xerophilous Grasses (especially steppe-inhabiting species) become folded, or curl up, when they are insufficiently supplied with water, in order to avoid excessive transpiration (Fig. 228). Such **curling or rolling Grass-leaves** are furnished with a number of longitudinal grooves on the upper side. The epidermal cells which are situated at the bottom of these furrows, are distinguished by their large size (especially as regards their vertical diameter), and by the thin and flexible character of their walls. The abaxial half of the leaf contains sub-epidermal fibrous ribbons or strands (Fig. 228 A). Opinions differ with regard to the precise mechanism of the curling movement. Duval-Jouve believes that the curling is due to the active contraction of the large epidermal cells at the bases of the furrows, which takes place when their turgor is diminished owing to active transpiration. Tschirch, on the other hand, regards the abaxial fibrous strands as the active components; he assumes that the internal fibrous layers have a greater power of imbibition (and hence a greater tendency to contract on drying) than the more superficial layers. According to Steinbrinck, finally, the mechanism is a cohesive one, and depends principally upon the action of the large epidermal cells in the furrows and of the green mesophyll, and to a smaller extent, also, to that of the wide internal bast-fibres. In the author's own opinion, Steinbrinck's view is the most plausible.

There is still much uncertainty, in most cases, as to whether any particular mechanism depends upon imbibition or upon cohesion. The difficulty is due partly to the comparatively recent discovery of the cohesive type of mechanism, and partly to two accessory factors which complicate the main issue. In the first place, the action of a given mechanism may depend upon cohesion up to a certain stage of desiccation, but after that may be entirely a matter of differential imbibition; there is no doubt that the dehiscence of certain Fern-sporangia is a composite effect of this sort. Secondly, movements of homologous organs are often performed with the aid of totally different mechanisms in closely related species. Both these circumstances will have to be taken into account in future investigations of this subject.

### C. LIVING MOTOR-TISSUES.[273]

Among the various types of movement in which living tissues play an active part, we may, first of all, mention certain more or less violent movements performed by floral organs or fruits owing to **release of tissue-tensions.** Explosive movements of this kind are executed, in connection with the dispersal of pollen, by the staminal filaments in *Parietaria, Urtica* and other URTICACEAE, and in *Atriplex* and *Spinacia*, by the gynostemium of *Stylidium adnatum,* by the carina of *Indigofera* and by the stalk connecting the pollinia of *Catasetum* with their common adhesive disc. The different modes of explosive dehiscence of fruits exemplified by *Impatiens, Cardamine hirsuta, Cyclanthera explodens*, etc., likewise pertain to this first category of physiological movements. The external features of these movements, and the main principles of the underlying mechanism, are sufficiently well understood. The anatomico-physiological aspect of the subject, however, still awaits detailed investigation.[274]

The second group of physiological movements comprises all those various forms of curvature which are executed by means of asymmetrical growth on the part of the curving structure, but without the assistance of specialised motor-organs or -tissues. The curvature always takes place principally in that portion of the organ which is undergoing longitudinal extension at the time. Such **growth-curvatures** may be autonomous in character, in which case they are also known as "nutations"; "tropistic" growth curvatures, on the other hand (such as geotropic, heliotropic, hydrotropic curvatures, etc.), are induced [and directed] by external stimuli. Growth-movements of this type cannot be further considered here, because they are generally executed as it were incidentally by parenchymatous tissues which are possessed of principal functions of a totally different kind. The third and last class of physiological movements are characterised by the fact that, whether due to asymmetrical growth, to turgor-variation or to some other cause, they are invariably carried out with the aid of **special motor-tissues**, which may be located in special motor-organs. Here also, the movement almost always takes the form of a curvature, which may be either autonomous or induced.

We shall start by discussing a few cases in which the motor-tissues are *not* restricted to definite motor-organs. This simple condition is exemplified by "sensitive" stamens, styles and stigmatic lobes. The characteristic feature of the mechanism is the active longitudinal contraction which the motor-tissues undergo, as the result of mechanical stimulation.[275]

In *Centaurea* and other CYNAREAE the five staminal filaments of each floret are bent outwards in the resting condition; when touched, they undergo a contraction amounting to between 10 and 25 per cent. of their length. As a consequence, the filaments straighten themselves, and the whole anther-tube is drawn forcibly downwards. The pollen, which was previously contained in the anther-tube, being kept in place by a brush of hairs on the style, thus becomes exposed, when the filaments contract, in such a position that it is likely to adhere to the insect visitors which, under normal conditions, apply the mechanical stimulus to the filaments. The ecological significance of this particular mechanism is therefore perfectly clear. Pfeffer has shown that the contraction of the filaments is due to a sudden fall of turgor,—induced by mechanical stimulation—affecting the parenchyma which intervenes between the epidermis of the filament, and the central vascular bundle; this parenchyma, therefore, represents the special motor-tissue in this case. In the unstimulated filament, the longitudinal walls of the motor-cells are fully extended by turgor-pressure; the subsequent contraction will accordingly be proportional to the amount by which the turgor falls. As already stated, the net result is a very considerable decrease in the length of the filament. Since the total volume of the filament diminishes to a marked extent, contraction must be accompanied by an escape of cell-sap from the stimulated cells into the intercellular spaces which are present in large numbers. It is not known whether the fall of turgor depends upon a sudden diminution in the osmotic strength of the cell-sap, caused by the external stimulus, or whether the latter brings about an increase in the permeability of the ectoplast towards the osmotically effective constituents of the cell. Pfeffer favours the former alternative.

In *Centaurea jacea*, the structure of the motor-tissue in an unstimu lated staminal filament may be described as follows : The epidermal cells have thick, protruding outer walls. The space between the epidermis and the slender central vascular strand is occupied by a mass of fairly thick-walled parenchymatous tissue, which is traversed by numerous intercellular spaces of various sizes (Fig. 229 A). A longitudinal section shows that these motor-cells are somewhat elongated, and that it is only their lateral walls that are thickened and provided with large numbers of transversely elongated pits. From the orientation of the pits, it may be inferred that the micellar series also run transversely in these walls, or, in other words, that the tensile resistance of the membrane is smallest in the direction of the long axis of the filament, that is to say, in the direction in which the extension due to turgor-pressure is greatest. The dense protoplasm of the motor-cells adheres very firmly to the limiting-membrane of the pits, a fact which indicates the presence of protoplasmic connecting threads at these points. The regions of the walls that border upon intercellular spaces, likewise bear a few shallow pits with ill-defined margins ; whether these pits have anything to do with the escape of cell-sap into the intercellular spaces, or not, is uncertain. The protoplasm separates readily from the thin transverse walls of the motor-cells, which are smooth and unpitted. Both the longitudinal and the transverse walls exhibit the reactions of unmodified cellulose. As already stated, the protoplasts of the motor-cells are massive ; their nuclei are of moderate size, but contain unusually large nucleoli.[276]

Owing to the very considerable thickness of their longitudinal walls, the motor-cells require a greater expenditure of osmotic energy for their distension than is needed in the case of thin-walled cells. On the other hand, the contraction which follows upon the fall of turgor induced by mechanical stimulation, is correspondingly great, and a considerable amount of resistance can be overcome during the retraction of the anther-tube. As a matter of fact, the friction between the inner surface of the anther-tube and the mass of pollen enclosed therein—which is prevented from slipping downwards with the anther-tube by the circle of brush-hairs surrounding the tip of the style—is very appreciable.

In the case of *Centaurea Cyanus*, the longitudinal walls of the motor-cells are somewhat thicker than they are in *C. jacea*. According to the author's observations, the thickest walls of all are found in *C. montana*, where the motor-tissue closely resembles collenchyma when viewed in transverse section (Fig. 229 c).

The stout irritable staminal filaments of *Berberis vulgaris*, *Mahonia aquifolium* and *Opuntia vulgaris* are also provided with a more or less

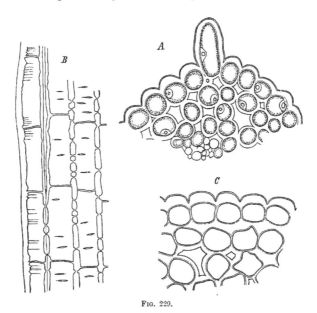

Fig. 229.

*A.* Part of a T.S. across a staminal filament of *Centaurea jacea* ; note the numerous intercellular spaces in the motor-tissue. *B.* Peripheral region of a L.S. through a similar filament. *C.* Small portion of a T.S. across a filament of *C. montana.* (All the sections prepared from spirit-material.)

thick-walled collenchyma-like motor-tissue, which is pervaded by narrow, (*Berberis*) or wide (*Opuntia*) intercellular spaces. Slender irritable stamens, on the other hand, such as those of *Portulaca grandiflora*, *Abutilon striatum* and *Helianthemum vulgare*, contain a thin-walled motor-tissue. In *Berberis*, where the stamens always bend in the same direction, *i.e.* towards the stigma, the motor-tissue is situated beneath the papillose sensory epithelium (cf. Chap. XII.) on the adaxial (concave) side of the filament. In other cases the tissue in question generally occupies the whole of the available space between the epidermis—or the sensory epithelium—and the central vascular strand.

Motor-tissues may be stimulated indirectly, through the mediation of sense-organs, as in *Berberis*, *Opuntia*, *Portulaca* and *Abutilon*, or directly, through the tension or compression induced within them by the stimulating agency. Longitudinal tension, in particular, often induces contraction when it is directly applied to a motor-tissue, even where stimulation is usually effected with the aid of special sense-organs.

It may be doubted whether all irritable stamens and styles (*Arctotis*), and stigmatic lobes (*Mimulus* and other SCROPHULARIACEAE, *Goldfussia*, some BIGNONIACEAE) conform to the *Centaurea* type of mechanism. In view of the fact that irritability of floral structure has certainly been evolved independently in a number of different families, it would seem *a priori* more probable that the motor mechanism should not be the same in all cases. This view is strongly supported by Pfeffer. It cannot therefore be regarded as proved, that contraction of motor-tissues is always brought about as it is in *Centaurea*, namely, by a decrease of osmotic pressure. Active contraction of protoplasts might equally well produce a fall of turgor. In view, finally, of the very thick-walled character of some active motor-tissues, it is not inconceivable that the walls might themselves be contractile and irritable, while the cells are alive, somewhat after the fashion of smooth muscle-fibres.

In the case of **tendrils**, such as occur in a number of Natural Orders (*e.g.* CUCURBITACEAE, PASSIFLORACEAE, PAPILIONACEAE, VITACEAE, etc.), the curvature which follows upon contact stimulation, depends upon acceleration of growth on the convex side. The exact proof of this important fact has only recently been furnished by Fitting, although Sachs and De Vries had long previously explained the main features in the behaviour of tendrils. According to Fitting, the concave side of a stimulated tendril undergoes no appreciable contraction. In accordance with their mode of curvature, the above-mentioned tendrils possess no specialised motor-tissue. In all the SAPINDACEAE that have been examined, on the contrary (*Urvillea ferruginea*, etc.), the hapto-tropic curvature of the tendrils—which are of the watch-spring type— depends, according to Ricca, upon a powerful contraction of the concave side. Here the motor mechanism resembles that which occurs in certain irritable stamens; it is not surprising, therefore, to find that the tendrils of *Urvillea ferruginea* contain a special motor-tissue, as the author had discovered previous to Ricca's researches. These tendrils are flattened at right angles to the plane in which they are coiled, and are markedly dorsi-ventral in structure (Fig. 230); irrita- bility is confined to the abaxial side. The most conspicuous feature in a transverse section is the very excentric position of the stereome ring, which is flattened like the tendril itself, and situated much closer to the adaxial or convex, than to the abaxial or concave side. At the time when the tendril first becomes irritable, the stereome elements are still in process of development. The small vascular bundles are situated just outside the stereome cylinder, on its abaxial side. The upper epidermis is separated from the mechanical tissue by a few layers of ordinary parenchyma; on the abaxial side, the space between the epidermis (or sensory epithelium) and the vascular strands is occupied by a bulky motor-tissue, consisting of elongated cells which contain

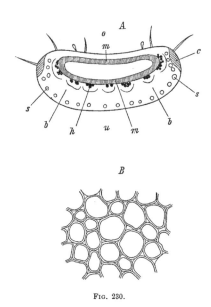

Fig. 230.

*A.* T.S. through a tendril of *Urvillea ferruginea*; *o,* upper (convex), *u,* lower (concave) side; *m,* stereome; *h,* hadrome-strands of the vascular bundles; *c,* collenchyma-strands; *b,* motor-tissue; *s,* tubular excretory sacs. *B.* Motor-tissue in T.S., more highly magnified.

no chlorophyll except in the immediate vicinity of the surface. The walls of the motor-cells are thickened in a peculiar manner (Fig. 230 B). The middle lamella is particularly well developed along the edges of the cells, and seems to possess very considerable powers of imbibition; the secondary thickening layers are highly refractive, and assume a blue coloration with chlor-zinc-iodine, thus contrasting sharply with the middle lamella, which remains unstained. As regards its general structure, therefore, the motor-tissue closely resembles that of many irritable stamens. That this tissue is really responsible for the curvature of the organ, may be inferred straight away from the fact, that it occupies the great bulk of the cross-section on the irritable and contractile side of the tendrils. There is, in any case, no other tissue

present that would be likely to possess the requisite contractile power.
The epidermis is not strong enough to produce the observed contraction; the vascular bundles are placed too near the centre of the
cross-section—to say nothing of other difficulties; the elongated sub-
epidermal excretory sacs, finally, which occur at intervals right across
the width of the motor-tissue, may also be safely left out of account.
It is probable that a motor-tissue of the same type is present in all
Sapindaceous tendrils; whether it occurs in other families as well, is a
matter for further enquiry.[277]

Attention may next be directed to the numerous cases in which
the motor-tissue is restricted to **special motor-organs** in the shape
of **pulvini**. The term pulvinus is applied to all pad- or cushion-
like swellings of stems and petioles which are directly concerned in
the execution of curvatures. Most often it is the basal ends of
internodes (COMMELYNACEAE, POLYGONACEAE, SILENEAE, etc ). or of the
primary petioles (LEGUMINOSAE), that are differentiated as pulvini; less
frequently the distal extremity of the petiole, close beneath the
insertion of the lamina (MARANTACEAE, many AROIDEAE), exhibits this
specialisation. Where the leaves are pinnate or bipinnate, pulvini
may be developed at the bases of the secondary petioles (sub-petioles)
and pinnules, as well as in connection with the main petiole (LEGU-
MINOSAE). Among Grasses the swollen leaf-bases act as pulvinoid
motor-organs.[278]

Most pulvini are cylindrical and capable of curving in all directions. A few, such as those on the pinnules of *Mimosa pudica* and other
LEGUMINOSAE, are more or less flattened, and accordingly only bend
in a single plane. As regards internal structure, cylindrical pulvini
may be either radially symmetrical (Grasses) or dorsi-ventral (e.g.
*Mimosa pudica* and *Biophytum sensitivum*, etc.); pulvini of the latter type
are, of course, also physiologically dorsi-ventral. Moreover, a pulvinus
may be anatomically radial and yet exhibit physiological dorsi-ventrality,
its two halves responding in different ways to the same external
stimulus, in spite of their apparent identity of structure. This type of
curvature, which depends entirely upon physiological differentiation, is
responsible for most " nyctitropic " movements.

In a strictly topographical sense, the motor-cells of a pulvinus
correspond to cortical tissue. They form a bulky zone, which is inter
polated between the epidermis and a central vascular strand, or a much
contracted central cylinder. The centralisation of the vascular and
mechanical tissues—the latter always consist of the relatively flexible
collenchyma—is connected with the fact, that an organ cannot bend
freely unless the mechanically resistant tissues are situated near the
centre. A peripheral disposition of the stereome, in particular, would

be highly unsuitable in the case of a motor-organ, which may have to undergo a large amount of curvature.

The motor-tissues of pulvini are usually composed of approximately isodiametric parenchymatous elements, and thus offer a marked contrast to the elongated or even prosenchymatous motor-cells of tendrils and irritable stamens. In the nodes of Grasses, the motor-cells are remarkably short, and sometimes approximate to a tabular form. In

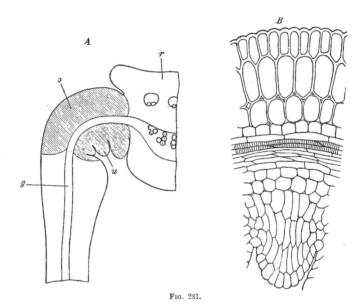

Fig. 231.

*A.* L.S. through the pulvinus of a pinna of *Biophytum sensitivum* (semi-diagrammatic); *o*, upper, *u*, lower side of the pulvinus ; r, rhachis (seen in T.S.) ; *g*, vascular bundle traversing the mid-rib of the pinna. *B.* Part of a L.S. through the pulvinus, more highly magnified (see text).

*Oxalis* and *Biophytum* the corresponding elements are also more or less flattened at right angles to the long axis of the pulvinus; in longitudinal sections their ends are usually seen to be more or less pointed (Fig. 231 B).

The walls of pulvinar motor-cells are generally thin, or, at most, slightly thickened. In *Mimosa pudica* and *Biophytum sensitivum*, where the pulvini are sensitive to shock, the upper (adaxial) and lower (abaxial) halves of each pulvinus represent two antagonistic zones, and the resulting movements depend upon their combined action. In the adaxial half the motor-cells have fairly thick walls, whereas they are quite thin-walled in the abaxial half (Fig. 231 B). The reason for the thickened condition of some of the cell-walls, in such cases, is quite unknown. It seems quite clear, however, that it forms an essential part of the whole mechanism ; otherwise, it is impossible to account for the occurrence of this feature in two genera such as *Mimosa* and *Biophytum*, which are widely separated in the natural system, but which have a striking physiological peculiarity in common with one another, inasmuch as the pulvini of both execute *rapid* movements when stimulated by shock. In the case of *Biophytum*, it is only the pulvini of the pinnae that respond in this way ; the motor-tissue of the petiolar pulvinus, which only executes nyctitropic movements, is thin-walled throughout. Hence thick-walled motor-cells are evidently confined to pulvini which carry out rapid movements in response to seismic stimulation.

Gardiner has found that the limiting membranes of the pits in the motor-cells of *Mimosa pudica* are traversed by delicate protoplasmic connecting threads ; it is probable that this feature is characteristic of pulvini in general. According to the author's own observations, all the longitudinal walls in the motor-tissues of Grass-nodes (*e.g.* in *Secale cereale, Festuca gigantea, Dactylis glomerata*), are provided with transversely elongated pits. From this circumstance it may be concluded, that the cohesion of these walls is smallest in the longitudinal direction, that is, in the direction in which growth-extension takes place [in response to geotropic stimulation]. The transverse walls, on the other hand, bear numerous *circular* pits.

In addition to a nucleated protoplast, the motor-cells of pulvini usually contain small chloroplasts, and frequently also irregular aggregations of tannin (*Mimosa pudica* and other LEGUMINOSAE).

Schwendener has drawn attention to an interesting point with regard to the development of intercellular air-spaces in pulvini ; as a rule, these spaces are feebly developed, or altogether absent, in the peripheral motor-tissue of pulvini that are sensitive to plastic stimulation. Such pulvini accordingly often present a peculiar translucent or glassy appearance. Schwendener explains this peculiarity on the ground that the motor-tissue must be translucent, if it is to be capable of responding to variations in the intensity of illumination, and that its transparency would be seriously impaired by the presence of numerous air-containing intercellular spaces. The ventilating spaces are larger and more numerous in the internal portions of such pulvini—*i.e.* in the neighbourhood of the central vascular strand—as well as in all parts of pulvini that move principally in response to geotropic stimulation (*Tradescantia*, GRAMINEAE).

Comparatively few pulvini (*e.g.* those of the lateral pinnae of *Desmodium* [*Hedysarum*] *gyrans*) perform autonomous movements ; paratonic responses are the rule in the case of these structures.

Perception of the mechanical (seismic), gravitational, photic or other effective stimulus, may take place within the motor-tissues itself, as in the case of nyctitropic movements. The perceptive faculty may, on the other hand, be located in special sense-organs; this more advanced condition is exemplified by geotropically sensitive organs in general, and also by *Mimosa pudica* and *Biophytum sensitivum*, though in these two plants the motor-tissues are also capable of direct stimulation. *Mimosa* and *Biophytum*, besides exemplifying the co-existence of direct and indirect stimulation of motor-tissues, also illustrate the fact that the same motor-tissue may respond to several different stimuli. In the case of heliotropically sensitive pulvini, the stimulus is often perceived by the leaf-blade, and transmitted to the motor-organ through a more or less extensive intervening region.

Curvature of pulvini may take place in several different ways. Heliotropic (phototropic) and geotropic curvatures are effected by means of the relative acceleration of growth on one side of the organ, whereas nyctotropic movements depend upon alterations of turgor; in the latter case the precise nature of the mechanism is not yet understood. Pfeffer believes that the two antagonistic halves of the pulvinus react in the same sense, but with unequal velocities or intensities. Schwendener, on the other hand, maintains that the two halves respond in opposite senses, a rise of turgor in the adaxial half being accompanied by a fall of turgor in the abaxial half, and vice versa. According to this latter view, when a leaf-blade which is horizontally extended in the diurnal position, sinks downwards in order to take up the nocturnal or "sleep" position (as happens, for instance, in the case of *Phaseolus, Amicia, Oxalis*, etc.), turgor increases in the adaxial half, and diminishes in the abaxial half of the pulvinus; the converse changes take place when the leaf returns to the diurnal position.

Changes of turgor in the motor-tissue must also be held responsible for the seismic and traumatic responses of the pulvini of *Mimosa pudica* and *Biophytum sensitivum*. The paratonic movements of the leaf of *Mimosa pudica* constitute one of the most interesting problems in the whole field of plant physiology, and have been studied by a whole series of investigators. The main external features of these movements are well known: the pinnules fold together in pairs (in the upward direction), the sub-petioles approach one another laterally, and the main petiole sinks suddenly downwards. All these separate movements are due to appropriate curvatures performed by pulvini. The pulvinus at the base of the main petiole is larger and more accessible than the others, and is therefore best adapted for experimental investigation; as a matter of fact the most important experiments have been performed upon these main pulvini. Brücke and Pfeffer agree that the paratonic movements of the main pulvinus are due to a decrease of turgor (amounting to at least 2 to 5 atmospheres) in the parenchyma of the sensitive abaxial half; the fall of turgor is accompanied by an escape of water from the motor-cells into intercellular spaces which previously contained air. The actual downward curvature of the pulvinus is partly due to a contraction of the walls of the motor-cells consequent upon the decrease of turgor, but is accentuated by expansion of the insensitive adaxial half of the pulvinus,—which was strongly compressed in the unstimulated condition of the organ—and also by the weight of the leaf. After stimulation, it is the collapsed abaxial half of the pulvinus that suffers compression.

In conclusion, attention may be directed to a morphological peculiarity, characteristic of many pulvini that execute large movements (OXALIDEAE, many LEGUMINOSAE). In such cases, the concave side of the pulvinus is often provided with transverse furrows (Fig. 231 A), which, as Schwendener acutely remarks, recall the folds of skin associated with the joints of the human finger. Similar folds sometimes occur on the convex side, where, however, they are generally less conspicuous. The presence of these furrows enables the pulvinus to undergo great curvature, without exposing the motor-tissue to the risk of undue compression or other serious deformation.

# CHAPTER XII.

## SENSORY SYSTEM.

### *I. GENERAL CONSIDERATIONS.*

ALL organisms, plants and animals alike, are possessed of the property of **irritability**, by virtue of which they react or **respond** to external or internal **stimuli.** The energy necessary for the carrying out of the response is not derived from the stimulus, but is furnished by the organism itself. The nature and final result of the reaction are, of course, dependent upon the structure of the particular organism concerned. Responses to stimuli generally tend to satisfy definite physiological or ecological requirements, and their importance to the organism depends upon this adaptive quality. Pfeffer terms the series of physiological processes which begins with the perception of a stimulus and ends with the final response, a "chain of stimulation" (*Reiz-Kette*). In the present chapter we shall deal with the first link in that chain, namely, the **perception of stimuli**, and with the anatomical features that are correlated with the perceptive function.

The perception of a variety of external stimuli by a plant presupposes the existence of different forms of irritability, and of correspondingly differentiated perceptive faculties. This sensitiveness towards external stimuli is vested in the living substance itself. It is obvious that there must be a structural basis underlying this sensitiveness; but this protoplasmic structure appears to lie beyond the limits of visibility.

Relatively unspecialised plants possess no arrangements for the perception of stimuli apart from this "irritable structure" of the protoplasm. In these circumstances, most if not all the living cells and tissues of an organ—or, it may be, of the whole plant-body—are sensitive. The structure of such unspecialised sensitive cells is primarily correlated with the performance of various other functions, and is only in a minor degree adapted for the perception of stimuli;

moreover, the perceptive faculty is often associated with certain stages in the development of the cells. In such cases, there are no histological features, apart from the ultra-microscopic irritable structure, to indicate the perceptive function of the cells.

This diffuse type of sensitiveness, in which the perceptive capacity may be distributed through an entire organ or over extensive portions thereof, was long thought to constitute an important distinction between plants and animals; while it does actually occur more frequently in plants than among animals, it can by no means be regarded as a universal characteristic of vegetable organisms.

At a higher level of specialisation, the faculty of perception is localised to the extent of being assigned to a definite tissue, instead of being diffused through an entire organ or distributed among a number of tissues; while the sensitive tissue is still primarily concerned with other functions, its anatomical structure does exhibit a certain amount of adaptation in connection with the subsidiary function of perception, some of the features of its component cells being designed to facilitate the perception of particular stimuli.

On account of its exposed position, the epidermis is particularly liable to experience a variety of external stimuli: as a matter of fact, epidermal cells do frequently exhibit histological features indicating a perceptive activity which is supplementary to their principal function of protection. Thus, in tendrils, the epidermis is often sensitive to contact, while in leaves it is frequently responsible for the perception of photic stimuli. When it is desired to discuss or emphasise the perceptive faculty of the epidermis, or the anatomical features correlated with perception, it is quite justifiable to apply the term **sensory epithelium** to the layer in question. This term is most appropriate, where perception of stimuli takes precedence over the protective function and, accordingly, finds permanent expression in the histological structure of the cells concerned. The sensory epithelium of the stamens of *Berberis* provides an excellent illustration of this grade of specialisation.

The highest stage of adaptation is characterised by the fact that the perceptive faculty is strictly confined to definite cells, cell-groups or cell-organs, which have perception as their sole, or at any rate as their principal function. Here the anatomical structure of the sensory organ is primarily arranged with reference to its perceptive function. In zoological nomenclature, organs concerned with the perception of external stimuli have always been known as **sense-organs**, even among the lower animals, and in other cases in which it is doubtful if the organs in question are responsible for sensation in the psychological sense. It is therefore not only permissible, but necessary in the interests of consistency, to apply the term sense-organ to the analogous structures in plants, especially as the latter often exhibit a close resemblance in plan to some of the perceptive organs of animals.[279]

So far as is known at present, plants have only developed sense-organs in relation to a few of the various types of external stimulation, namely, to the stimuli of contact and shock (mechanical stimuli, in the narrow sense), the gravitational stimulus, and, finally, photic stimulation. It is true that plants are also capable of perceiving chemical stimuli ; but special sense-organs comparable to the organs of taste and smell in animals have not as yet been discovered in the vegetable kingdom. Special arrangements for the perception of thermal stimuli likewise seem to be lacking.

In considering the general principles governing the construction of the sense-organs of plants, it is necessary to bear in mind that the real act of perception always takes place within the living substance. Noll was the first to point out, that the perceptive faculty cannot be distributed throughout the protoplasm of a sensitive cell, and, in particular, that it cannot be located in the streaming portion of the cytoplasm. Let us, for example, consider the case of mechanical stimulation, the perception of which depends upon deformation of the sensitive protoplasm. In this instance, it is quite obvious that the streaming granular cytoplasm (polioplasm) must be left out of account; for, since it is constantly undergoing ".spontaneous" [i.e. internally regulated] deformation, it cannot possibly be stimulated by deformations of external origin. By the same line of argument, we arrive at the conclusion that the rotating or circulating portion of the cytoplasm cannot possibly be entrusted with the duty of perceiving the direction of gravitation, or of the incident illumination. We are thus forced to assume, with Noll, that the perceptive faculties of a cell are mainly, if not entirely, vested in the stationary and relatively solid ectoplast. If the internal plasmatic membranes (vacuolar membranes or tonoplasts) should happen to be so firm, as to be able to resist displacement by the streaming protoplasm, they would also, of course, be able to receive the impress of external stimuli. The same argument applies to any other cytoplasmic structures that maintain a fixed position in the cell.[280]

The most characteristic structural feature of sense-organs in general, is the presence of arrangements which suitably direct and control the incidence of external stimuli upon the sensitive portions of the protoplasm ; it is these anatomical features that alone are accessible to direct observation. Hence, in attempting to discover the connection between the construction of any sense-organ and its function, we must always confine our attention to those structures and arrangements which facilitate or intensify the actual stimulation of the living substance. The protoplasmic stimulation itself is quite beyond the reach of direct observation.

It is not essential that every component cell of a sense-organ should be endowed with the perceptive faculty. Structures, for example, which serve to transmit the external stimulus to the seat of perception need not themselves be sensitive. The conical distal portion of the tactile bristle of *Dionaea muscipula*, which, from the mechanical point of view, represents the active part of the whole organ [but is itself insensitive to contact], illustrates this point very clearly. If we wish to apply a distinctive name to the actual perceptive elements, we may term them the **sensory cells.** In its narrow sense this term only refers to those elements of *sense-organs*, in which the actual protoplasmic perception takes place. It may, however, conveniently be extended, so as to include all cells which are shown by their anatomical structure to be concerned with the perception of stimuli.

It has already been explained that, when a stimulus acts upon a sense-organ, it starts a *series* of processes, which usually terminates with the execution of an adaptive paratonic movement. Sense-organs are very frequently situated in the immediate vicinity of the associated motor-tissues or organs. Thus, the sensory epithelium of the *Berberis*-stamen immediately overlies the motor-tissue of the filament; some of the tactile hairs and bristles of *Mimosa pudica* and *Biophytum sensitivum*, at any rate, are located upon the pulvini; finally, it is the rule for the statocysts or geotropic sensory cells of aërial organs to be situated in the internodes or pulvini that execute the geotropic curvature. In other instances, the sense-organs are further removed from the motor-tissues. One of the best illustrations of this separation in space of the perceptive and motor regions is furnished by ordinary roots, where the statolith-apparatus is largely or entirely confined to the extreme tip, whereas geotropic curvature takes place in the sub-apical growing zone.

A **transmission of stimuli** from the sense-organ or sensory cell to the motor-tissue must take place in all cases.

## II. TACTILE SENSE-ORGANS.[281]

A great many plants respond to shock, friction, contact and other stimuli of a "mechanical" nature, by executing movements which entail some ecological advantage. Instances are furnished by the foliage-leaves of *Mimosa pudica* (and other species of *Mimosa*) and *Biophytum sensitivum*, by certain insect-catching organs such as the leaves of *Dionaea muscipula* and *Aldrovandia vesiculosa* or the tentacles of *Drosera*, by the foliar or axial tendrils of various Angiosperms, and, finally, by those "irritable" stamens, stigmas, styles, gynostemia or perianth-segments which perform active paratonic movements in connection with pollination.

The perception of mechanical stimuli is often carried out through the mediation of special sense-organs, which are analogous to the tactile sense-organs of animals, and which indeed often resemble the latter in structure. According to their special mode of construction in different cases, these organs are termed **tactile pits, tactile papillae, tactile hairs** or **tactile bristles.**

It is important to note, that any given mechanical stimulus— whether presented as shock, friction or prolonged contact—cannot be perceived, and hence cannot induce a responsive movement, unless it produces a distinct deformation of the sensitive ectoplast. Moreover, in order to be effective, the deformation must be sudden, and in certain cases must take place several times in rapid succession. Nor is the actual mode of deformation a matter of indifference; for the author has shown that the sensitive ectoplast is principally affected by the *tangential* tensions (compressions or extensions) or deformations that result from shock or contact. In fact, the most characteristic general feature of the anatomical structure of tactile sense-organs is the presence of arrangements which facilitate or accentuate the requisite sudden deformation of the sensitive protoplasm, and which more particularly favour the production of tangential tensions within the ectoplast.

## A. TACTILE PITS.[282]

**Tactile pits** are sharply defined unthickened areas, more or less closely resembling ordinary pits, which occur in the outer walls of superficial sensory cells; they are occupied by special extensions of the sensitive protoplasm. The thinness of the cell-wall at these points reduces the amount of energy which has to be expended in deforming the insensitive cellulose membrane to a minimum, with the result that the greatest possible proportion of the total energy of the stimulus becomes available for the deformation of the sensitive protoplasm.

Tactile pits were first discovered by Pfeffer in the tendrils of the CUCURBITACEAE (Fig. 232). Most often they are confined to the morphologically lower side of the tendril (the concave side after curvature), which, as a rule, is alone sensitive to contact. In *Bryonia dioica* and *B. alba*, and probably in several other CUCURBITACEAE as well, the upper side (the convex side after curvature) is likewise provided with tactile pits. These structures are, however, never found on the insensitive basal region of the tendril.[283]

In *Cucurbita Pepo, Cucumis sativa, Lagenaria vulgaris, Cyclanthera explodens*, etc., the outer wall of each epidermal sensory cell is furnished with a single approximately central pit. The pit-cavity is nearly circular or elliptical in cross-section, and, as a rule, expands towards its outer end in a funnel- or basin-shaped manner. The diameter of these pits varies between 1·5 and 6 $\mu$. In *Cyclanthera explodens* there is no funnel-shaped expansion, and the pit-cavity consists of a shallow circular trough. The thin membrane which forms the roof of the cavity is flat, or bulges very slightly outwards (*Cyclanthera explodens*);

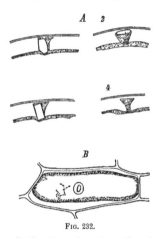

Fig. 232.

Tactile pits in the outer epidermal walls of Cucurbitaceous tendrils. *A. Cucurbita Melopepo*; pits in vertical section. *B. C. Pepo*; epidermal cell in surface view, showing a tactile pit in the outer wall. (All preparations from spirit-material.)

in *Cucurbita Pepo* it is only ·6 to ·8 $\mu$ in thickness. The cuticle is no thinner over the pits than elsewhere; it is underlain by a very thin cellulose-layer, which in *Cucurbita* is slightly cutinised.

The living contents of the sensory cells consist of a relatively massive peripheral layer of protoplasm and a large nucleus. As a rule, every tactile pit is completely filled with cytoplasm. The author has observed that in *Cucurbita Pepo* and *C. Melopepo* there are usually one or more minute crystals (? of *calcium oxalate*) embedded in the protoplasm of the pit (Fig. 232 A). It is probably not going too far to assume that these crystals form part of the perceptive mechanism. Any deformation produced by sudden pressure upon the end of the protoplasmic process in the pit, will necessarily be accentuated by the sharp angles and edges of the crystal, and the stimulation of the protoplasm will be correspondingly intensified; this argument is based upon the assumption that the sensory ectoplast actually comes into contact with the crystal, in which case one of the edges or angles of the latter will be driven into the protoplasm like a wedge, when stimulation takes place.

In *Bryonia alba*, each sensory cell on the lower surface of the tendril is provided with several tactile pits, which are of two kinds · the basin-shaped principal pits, which vary in number between one and three per cell, are accompanied by two or three accessory pits in the shape of minute transverse slits. All the pits in any one cell are arranged in an irregular row, with the principal pits in the middle.

Hitherto, tactile pits, like those of the CUCURBITACEAE, have not been discovered in any other family. In a solitary instance, however·

(viz. that of the Sapindaceous genus *Urvillea*), the author has described an arrangement which evidently must be interpreted in the same sense. Here, the more or less distinctly elongated epidermal (presumably sensory) cells on the lower side of the tendril contain abundant protoplasm, and are provided with remarkably thick outer walls. On either side of the longitudinally directed radial walls there is often present a narrow unthickened strip on the outer wall; the fissures or grooves produced in this way are filled by thin, flange-like projections of the protoplast. The whole structure, in fact, strongly resembles a very much elongated tactile pit. The transversely directed radial walls are usually flanked by several slit-like pits, while an irregular longitudinal series of similar pits is generally developed along the middle line of the outer wall.

Where the sensitive epidermis (sensory epithelium) of a tendril is devoid of tactile pits or analogous organs, it is still possible to point out various histological features that are most probably connected with the perception of mechanical stimuli. Thus, the small size and abundant protoplasmic contents of the epithelial cells, the thin and protuberant outer walls and the often very conspicuous longitudinal corrugations of the cuticle, are all features which directly or indirectly facilitate the process of perception.

Charles Darwin was the first to notice that tendrils can only be stimulated by contact with, or friction against *solid* objects, and not, for example, by the impact of drops of water; subsequently Pfeffer, with the aid of very elaborate experiments, arrived at the following more precise definition of the conditions under which tendrils perceive a mechanical stimulus. "In order that the stimulus may be perceived, it is necessary that separate circumscribed areas situated in the sensitive region of the tendril should be subjected to a shock or a tension of sufficient intensity, either simultaneously or in rapid succession. On the other hand, a tendril does not respond to a shock which affects every point within a considerable area with approximately equal intensity," a condition which is fulfilled, for example, in the case of contact with a liquid, or with a soft flexible rod of gelatine.[284]

The author, in his turn, has attempted to define, as exactly as possible, the difference in the character of the deformation which the outer walls of the sensory epidermal cells—and the ectoplast associated with them—suffer through contact with solid and liquid bodies respectively. A large, plastic, stimulating surface, which moulds itself to the shape of the tendril, produces a purely radial pressure; hence the impact of water, mercury, or recently solidified gelatine only brings about a *radial* compression of the ectoplast which is in contact with the outer wall of each sensory cell. But

2 o

experiment has shown that radial compression does not constitute a stimulus. When, on the contrary, very small stimulating surfaces, such as are provided by the innumerable small irregularities that are present on the surface of every solid body, come into contact with the sensory epithelium of a tendril, the protruding outer walls of the sensory cells become locally bent or depressed and the underlying ectoplast accordingly suffers the *tangential* compressions and extensions which constitute effective stimuli. It is obvious that the tangential tensions caused by contact with solid bodies, will be most likely to exceed the value corresponding to the lower limit of sensitiveness, if the cells of the sensory epithelium are small, and are furnished with thin and protuberant outer walls. Pfeffer's experiments have shown that these tangential tensions must be produced suddenly,—and often also several times in quick succession—if they are to be effective; hence, the presence of longitudinal cuticular ridges, which is a very conspicuous feature of many tendrils, will increase the chances of effective stimulation, by transforming the smooth gliding movement of the tendril over any object which it encounters in the course of its circumnutation, into a rapid succession of jerks. Each of the shocks produced in this way must be infinitesimally small; on the other hand, many tendrils are endowed with an extraordinary degree of sensitiveness.

With regard, finally, to the tactile pits of CUCURBITACEAE, it can readily be demonstrated, that the disc-like protoplasmic processes which extend into the pit-cavities, will be subjected to even greater tangential stresses than the rest of the ectoplast associated with the outer wall, when the tendril rubs against a solid object. Tendrils which are furnished with tactile pits, must accordingly be held to have attained a higher level of specialisation than those which are devoid of this complication.

The tentacles of *Drosera* respond both to mechanical and to chemical stimulation. Pfeffer has shown that the mechanical sensitiveness of these organs is closely similar to that of tendrils. Two conditions must be fulfilled, if stimulation is to be effective: In the first place, the outer walls of the glandular cells of the " head " must come into contact with a solid body; secondly, several " shocks " must occur in rapid succession. The response consists in a curvature of the stalk of the tentacle; the ecological aim of the whole performance is the capture, asphyxiation and digestion of insects.[285]

The sensory elements of the *Drosera* tentacle are represented by the superficial glandular cells of the head, which thus combine several quite distinct functions. They secrete a large amount of mucilage and smaller quantities of a digestive enzyme; they are responsible for the perception of mechanical stimuli; finally, they absorb the soluble products of digestion. The actual perception of mechanical (and perhaps also of chemical) stimuli seems to take place within minute papillose protoplasmic processes, which, in *Drosera rotundifolia* and *D. longifolia*, are restricted to the margins of the outer wall. These processes are contained in "marginal pits" (cf. above, p. 121), which arise in the following manner: The radial walls of the sensory cells are provided with small projecting ridges; these are thickest at their points of insertion on the outer wall, but rapidly become narrower as they pass downwards and inwards, and finally die away altogether (Fig. 233 c). These ridges enclose minute superficial pockets, which are prolonged inwards as short, and somewhat obliquely directed canals. The fact that these pockets contain protoplasmic processes, can be very convincingly demonstrated by mounting the head of a tentacle in dilute sulphuric acid for a short time, and then squeezing the preparation gently under the coverslip. By this treatment, the protoplasmic contents of the glandular cells are separated from the disorganised cell-walls and at the same time "fixed." The outer surface of the protoplast of each marginal glandular cell is then seen to be furnished with a number of minute papillae; in *Drosera rotundifolia* these processes are roughly isodiametric (Fig. 233 B), but in *D. longifolia* their height exceeds their width two- to three-fold (Fig. 233 A). In the palisade-shaped apical glandular cells the

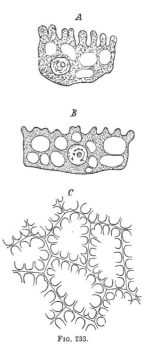

Fig. 233.

*A.* Protoplast isolated from one of the marginal glandular cells of a tentacle of *Drosera longifolia*. *B.* A similar preparation from a tentacle of *D. rotundifolia*. *C.* A few marginal glandular cells of *D. rotundifolia* seen in surface view.

processes are usually confined to the angles of the outer face, a single one being developed at each corner.

In other carnivorous plants, where the digestive glands and absorbing hairs are not sensitive to mechanical stimulation, the outer walls of the glandular cells are never furnished with tactile pockets. We are therefore justified in concluding that this structural peculiarity of *Drosera* is connected with mechanical sensitiveness, and not with the secretion of mucilage or enzymes, or with absorption of soluble products. This view is further supported by the strong resemblances of these marginal pockets to the tactile pits of Cucurbitaceous tendrils.

### B. TACTILE PAPILLAE.[286]

The term which forms the heading of this section may be applied to any mechanical sense-organ that projects above the general level of a sensory epithelium in the form of a more or less well-defined papilla. As a rule, it is only a limited area near the middle of the outer wall of the sensory cell that projects in this way; generally, this protruding part of the wall is specially thin. The tactile papillae on the tendrils of *Eccremocarpus scaber* are cut off from the underlying epidermal elements by cell-walls, so that each represents an independent cell.

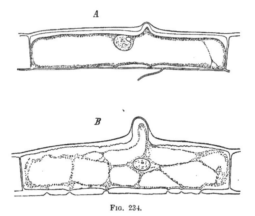

FIG. 234.

*A.* Epidermal cell from a staminal filament of *Portulaca grandiflora* in L.S., showing a tactile papilla. *B.* A similar preparation from a filament of *Opuntia vulgaris.* In both cases the protoplasts have shrunk away from the walls, the sections having been prepared from spirit-material.

Where the papilla consists of the whole outer wall, it often remains thin throughout; it may, on the other hand, be more or less thick walled, except for an unthickened marginal strip. In the latter case, only this marginal zone—or rather the ectoplast adhering thereto—suffers the deformation upon which stimulation actually depends. So far, tactile papillae have only been observed in connection with floral organs and tendrils.

The staminal filaments of *Portulaca grandiflora* are sensitive to contact at every point, except near the base. When touched or rubbed with a needle, they curve towards the stimulated side. Most of the narrow elongated epidermal cells of the filament bear, near the middle, or more rarely towards one end, a minute papilla, with a wall which is so thin as to consist of little beyond the very delicate cuticle (Fig. 234 A).

The thin-walled character of the papillae is rendered more conspicuous, by the fact that the outer epidermal wall is generally thickest around the base of each papilla. The cavity of the papilla is occupied by a small prolongation of the protoplast. Clearly these minute tactile organs hardly differ from the sensory pits of the CUCURBITACEAE, except in the fact that the thin membrane enclosing the protoplasmic process is developed as a papilla, an arrangement which must obviously facilitate stimulation considerably. The insensitive basal region of the filament is devoid of epidermal papillae.

In *Opuntia vulgaris* the staminal filaments are sensitive on all sides and throughout their length; here, however, the filament always curves inwards, *i.e.* towards the style. In this case, again, the outer wall of each of the elongated epidermal sensory cells is furnished with a small conical papilla, which is usually situated near its middle, and less frequently towards one end (Fig. 234 B). The papillae of the outer or longer stamens diminish in size towards the base of the filament, finally scarcely projecting at all, and thus approaching the condition of tactile pits. Around the bases of the papillae, the outer walls of the sensory cells are more or less strongly thickened, just as in the case of *Portulaca*. The walls of the papillae themselves are extremely delicate, never exceeding ·6 to ·8 $\mu$ in thickness; the cuticle is also very thin at these points. On coming into contact with a solid body, therefore, the papillae are readily pushed in, with the result that the underlying ectoplast suffers great deformation. Very similar tactile papillae occur on the irritable stamens of *Cereus spinosissimus*.

In that very remarkable genus of Orchids, *Catasetum*, the column of the male flower ends in a slender process, from which the single anther is suspended by means of a stout filament of considerable length. From the sides of the column project a pair of long, straight or slightly curved thick structures with tapering ends; each of these really consists of a ribbon-shaped prolongation of the margin of the column, which has become rolled up around its longitudinal axis into a tubular body. The name " antenna," applied to these appendages by Ch. Darwin, aptly expresses their function. For if the tip of one of the antennae is gently touched with a bristle or needle, the adhesive disc common to the two pollen-masses instantly separates from the rostellum, with which it was previously in organic connection; the curved stalk joining the disc to the pollen-masses straightens itself with a jerk, and the whole pollen-apparatus shoots forward with great violence. The actual propulsion of the pollinia is not a manifestation of irritability, but depends upon the instantaneous equalisation of tissue-tensions within the stalk, which behaves like a bent spring that is suddenly released. But this movement cannot be executed, until the adhesive disc of the pollen-apparatus is detached from the rostellum ; and this separation is a response elicited by the contact of a solid body with the sensitive antenna. Under natural conditions, stimulation of the antenna is effected by an insect-visitor, and the ejected pollen-apparatus becomes firmly attached to the back of the animal.[287]

In the species of *Catasetum* examined by the author (*C. Darwin-ianum* and *C. macrocarpum*), each epidermal cell in the apical region of the antenna bears a small tactile papilla, situated near the middle of the outer wall, or, more rarely, towards one end. The membrane of the papilla is thinner than the rest of the outer wall, although the difference is not so marked as in the case of *Portulaca* or *Opuntia*. According to Von Guttenberg, tactile papillae also occur in several other species of *Catasetum*.

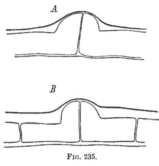

FIG. 235.

*A*. Tactile papilla from a staminal filament of *Centaurea Cyanus*. *B*. Tactile papilla from a filament of *Echenais carlinoides*.

Comparatively large, thin-walled tactile papillae are found on the staminal filaments of many CYNA-ROIDEAE, either as the only sensory organs (*Alfredia cernua*) or in association with tactile hairs (*Centaurea*, etc.; see below, p. 587). The most distinctive feature of these papillae consists in the fact that they are formed by *pairs* of sensory cells; the adjoining ends of two epidermal cells grow out together to form a common papilla, which is accordingly divided into two halves by a thin vertical partition (Fig. 235).

The tactile papillae of the stamens of *Berberis vulgaris* are constructed on a somewhat different plan to those which have been described so far. In this plant, sensitiveness is restricted to the adaxial side of the filament, and even there is absent at the base, and also just below the insertion of the anther. When any part of the sensitive region is touched, the filament curves suddenly towards the stigma. The sensitive adaxial surface is covered by a relatively large-celled and remarkably papillose sensory epithelium. The sensory cells, which, as usual, contain abundant protoplasm, are arranged in longitudinal rows ; their length and breadth are approximately equal. The cells usually alternate in adjacent rows, so that their cross-sectional outline is normally hexagonal. Each papilla, which is about as high as it is wide, arises by the protrusion of the entire outer wall of a sensory cell ; it has an obtuse apex, and is fairly thick-walled and provided with a stout, smooth cuticle. The papillae are almost always relatively thin for some distance above their base, a feature which recalls the characteristic hinge-areas so frequently developed in connection with the guard-cells of stomata (cf. p. 447). This thin region, which completely encircles the papilla, is quite narrow, and passes over insensibly into the thicker portion of the cell-wall. In the corners of the cell it frequently becomes modified so as to form a (marginal) pit occupied by an extension of the protoplast (Fig. 236 B).[288]

It can hardly be doubted, that the hinge-like, unthickened area in the outer wall of the sensory cells of *Berberis* serves to facilitate the

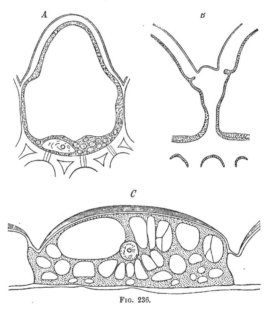

FIG. 236.

A. Sensory cell with tactile papilla from the ventral surface of a staminal filament of *Berberis vulgaris*, in vertical section. B. Another section passing through the adjacent corners of two cells, to show the hinge-areas which are present at these points. C. Sensory cell from a filament of *Abutilon striatum* in vertical section; outer wall pushed in.

perception of mechanical stimuli. For when the papilla is touched or pressed upon from above, the thin strip of membrane at its base will bend slightly outwards; if the papilla is pushed to one side, the hinge-action of the basal strip will again lead to local deformation of the ectoplast.

In *Abutilon striatum*, the free upper portions of the staminal filaments, which, further down, are united into a tube, are sensitive on all sides, especially towards their distal extremities. Curvature takes place towards the stimulated side. The somewhat elongated sensory cells are distinctly papillose, especially in the terminal portion of the filament (Fig. 236 c). The saucer-shaped outer walls of these cells are covered by an unusually thick cuticle, which is thrown into numerous longitudinal folds. The cuticle becomes much thinner towards the margins of each cell; since the cellulose layers of the outer wall are not very thick, this thinning down of the cuticle reduces the total thickness of the membrane very considerably. The thinnest portion of the outer wall of each cell is that which immediately adjoins the lateral walls; strictly speaking, this thin strip does not form part of the papilla. As in other cases, the sensory epidermal elements are distinguished by the density of their protoplasmic contents.[289]

After the preceding description of the structure of the sensory cells of *Berberis*, the actual process of stimulation in *Abutilon* scarcely requires explanation. The protruding portion of the outer wall, rendered thick-walled by the massive development of the cuticle, forms a comparatively rigid cover, which is attached to the lateral wall by means of a continuous hinge-membrane. This hinge is most perfectly differentiated at the acroscopic end of each cell. The pressure of a solid body upon this rigid cover produces a characteristic deformation of the hinge-area (Fig. 236 c). Owing to the more or less asymmetrical shape of the cover,—its centre being slightly displaced towards the basiscopic end of the cell—the mode of deformation differs at the two opposite extremities of each sensory element. At the acroscopic end, the hinge membrane becomes sharply folded inwards, whereas at the other extremity it bulges distinctly outwards. The underlying ectoplast accordingly suffers violent extension in the acroscopic portion of the hinge, and less pronounced compression in the basiscopic region.

In all the MALVACEAE with insensitive stamens that have been examined by the author, the epidermal cells of the filaments were found to be quite flat and smooth-walled; the outer walls (inclusive of the cuticle) were always moderately thick-walled throughout, the marginal region being in no way distinguishable from the rest of the wall.

Comparatively few tendrils are furnished with tactile papillae. On the lower and more sensitive side of the irritable petioles of *Adlumia cirrhosa*, *Corydalis claviculata* and *Clematis Viticella*, the ordinary narrow, elongated, flat-walled epidermal elements are accompanied by a considerable number of shorter and broader cells, with thin outer walls which protrude to a considerable extent. These projecting cells are particularly well adapted for the perception of mechanical stimuli. *Clematis Vitalba* is likewise provided with sensory cells which are isodiametric, papillose and specially thin-walled, and which thus differ markedly from the ordinary elongated epidermal elements. In the last mentioned plant, the sensory cells may be solitary or associated in groups to form prominent tactile cushions; each cushion is accompanied by a thick-walled and comparatively rigid unicellular hair, situated at its acroscopic end, which probably represents a "stimulator" (cf. below, p. 586).[290]

Very interesting and remarkable are the papillose tactile cells which occur on the repeatedly dichotomous foliar tendrils of a Chilian Bignoniaceous climber, *Eccremocarpus scaber*. Here the primary, secondary and tertiary branches of the tendril, and even its short terminal hooks, are much more sensitive to contact on the abaxial

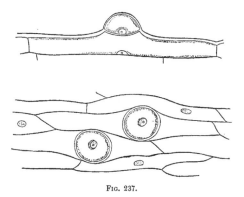

Fig. 237.

Tactile papillae of the tendrils of *Eccremocarpus scaber*. Above, in vertical section · below, in surface view.

than they are on the adaxial side. The numerous tactile papillae are likewise most abundant on that side; on the terminal hooks, indeed, they are always confined to the abaxial surface. These papillae represent entire cells, and take the form of hemispherical or lenticular outgrowths attached to the flattened central areas of the outer walls of ordinary elongated epidermal cells (Fig. 237). While the ten drils are still young and highly sensitive, the smooth outer walls of the papillae are distinguished by their extremely thin-walled character; their thickness only amounts to between 1·5 and 2 $\mu$, while the walls of the adjoining cells are about twice as thick. The protoplasts of these cells consist of a well developed peripheral layer of cytoplasm, and a nucleus which invariably lies in contact with the middle of the inner wall.

In all the preceding cases, the tactile papillae or cushions are specially adapted, on account of their exposed situation and their thin-walled character, to act as perceptive organs, even supposing that their protoplasts are not more sensitive than those of the ordinary epidermal cells.

### C. TACTILE HAIRS AND BRISTLES.

Trichome-structures, whether unicellular or multicellular, are particularly well adapted for the perception of mechanical stimuli. The superficial position of these structures renders them peculiarly accessible to the influence of such stimuli; they can also be readily utilised as levers, the function of which is to ensure that the stimulus is strictly localised, or, in other words, that all the available energy derived from the shock is concentrated at one point, so as to produce the greatest possible deformation of the sensitive protoplasm.

The simplest types of tactile hair or bristle merely serve to transmit the stimuli of shock or contact to the sensitive motor-tissue; these perform a purely mechanical function. Structures of this kind —which are not necessarily trichomes in every instance—are termed **stimulators** by the author, in contradistinction to sense-organs in the strict sense. The bristles which are found on the main pulvinus of the leaf of *Mimosa Spegazzini* may serve as a type of such ' stimulators. The bristles in question are composed of thick-walled, lignified mechanical elements; distally each terminates in a single cell, whereas the lower extremity is provided with a sharply conical and particularly thick-walled process, which is deeply embedded in the sensitive motor-tissue that composes the lower half of the pulvinus. Figuratively speaking, this thick-walled process constitutes a "thorn in the flesh" of the sensitive tissues, since even a slight displacement of the stiff bristle must result in a very pronounced deformation of the cells adjoining the embedded thick-walled process. The action of the process is probably accentuated by the fairly numerous large crystals of *calcium oxalate* that occur in the sensitive cells.[291]

In the case of *Mimosa pudica*, also, the majority of the bristles which occur on the under side of the main pulvinus serve only as stimulators.

Stimulators of a different type occur on the stamens of *Sparmannia africana*. These consist of peculiar reversed hooks, representing transverse folds of the epidermis, and are situated near the distal ends of the stamens and staminodes, on their inner side. The smooth inner surfaces of the filaments are quite insensitive to contact. When the outer side of a filament, on the other hand, is rubbed with a needle or bristle, the characteristic movement of the stamen at once follows, the filament bending outwards—*i.e.* towards the corolla—more particularly near its base. If the needle is drawn upwards—that is, contrary to the direction in which the barbs point, the movement of the stamen takes place much more promptly than it does if the filament is rubbed from above downwards. The barbs serve to convert a uniform friction into a series of jerks, and thus act as mechanical stimulators. It is only necessary to substitute an insect visitor for the needle or bristle used in the above experiment, in order to realise what happens under natural conditions.[292]

Some account must now be given of typical tactile hairs and bristles. We may first of all deal with the unicellular tactile hairs of *Mormodes*, a genus of orchids in which the pollen-apparatus is ejected much in the same way as in *Catasetum*. Here the twisted column is enclosed within, and compressed by the peculiarly formed labellum, which in *M. Buccinator* is more or less trumpet-shaped. When an insect alights upon the column, this pressure is increased, and the tactile hairs, which are present in large numbers upon the apical part of the column, are bent and consequently stimulated. The final result is the same as in *Catasetum*; the disc of the pollen-apparatus separates from the column, the tension in the stalk is relieved, and the whole pollen-apparatus is violently shot forward. In the distal region of the column, all the epidermal cells are prolonged outwards to form short tactile hairs. Each hair is from four to six times longer than its breadth, and has a blunt apex, and a more or less well-defined thin-walled basal zone (Fig. 238), which closely resembles the hinge of the tactile papilla of *Berberis*, and, like that structure, serves to produce a very intense local deformation of the ectoplast.[293]

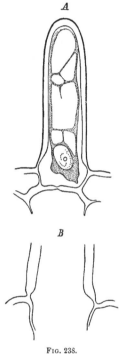

FIG. 238.

Tactile hair from the upper part of the column of *Mormodes Buccinator* (plasmolysed). *B.* Basal portion of another hair, showing the hinge-area very distinctly.

In many CYNAROIDEAE the irritable filaments of the stamens bear two-celled tactile hairs, which have been most thoroughly studied by the author in the genus *Centaurea*. The hairs in question, which are usually developed on all sides of the filaments, tend to be longest and most abundant towards the middle of the filament; near its base, on the other hand, they are altogether absent. In *Centaurea Cyanus* they form a continuous ring or collar around the middle of the filament, and on its outer side become united into a ciliated scale. Each tactile hair is composed of two elongated processes which

arise side by side from adjacent epidermal cells. In *C. jacea* and
*C. Cyanus* the hairs have thin walls, and hence are flexible throughout
their length. Where, on the other hand, the hairs are more or
less thick-walled (as in *C. montana*, Fig 239), they are usually
provided with a " hinge," in the shape of an unthickened basal zone. The
hinge-area never completely encircles the hair, and hence is only visible
in transverse sections through the filament; it can accordingly only
facilitate lateral displacement of the hairs. It can readily be shown
by experiment that these hairs actually represent sense-organs. For if
one rubs the hairs of a filament, which has been dissected out, with a

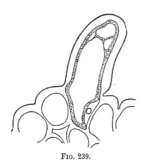

FIG. 239.

Tactile hair from a staminal filament of *Centaurea montana*; the
section being taken across the filament, only one of the two cells of
which the hair is composed can be
seen.

needle or bristle, without bending or even
touching the filament itself, the characteristic contraction at once takes place.
This experiment can be most satisfactorily
performed upon *Centaurea orientalis*. In
nature, when an insect visitor alights upon
a capitulum of *Centaurea*, and inserts its
proboscis into the corolla-tube of a floret
in order to reach the nectar concealed
therein, it cannot fail to rub against
several of the tactile hairs, and thus bring
about stimulation of the filaments.[294, 295]

The author has shown that the sensitive
foliage-leaves of *Biophytum sensitivum* and
*Mimosa pudica* are likewise provided with
tactile hairs or bristles. In these cases, the motor-tissues respond to
direct stimulation in the shape of shock or vibration ; they can also be
indirectly affected through the agency of the tactile hairs, which are
sensitive to a very light touch. From the ecological point of view, the
movement probably represents a method of protection against crawling
insects ; for such undesirable visitors will sooner or later brush against
one of the tactile hairs, and the sudden movement that ensues will
jerk them off the leaf, or perhaps cause them to fly away affrighted.[296]

The tactile hairs of *Biophytum* are located on the rhachis and on the
pinnae. On the rhachis they are arranged in transverse series; a
row, consisting of from four to seven hairs, extends between each
pair of secondary pulvini, while, in addition, three or four hairs occur
in the angles between the secondary pulvini and the rhachis.
Similar hairs are found on the upper sides of the pinnae, six to
twelve being situated on the mid-rib, and five to eighteen on the
lateral veins, and in the meshes of the vascular network. The lower
surface of the leaves, on the other hand, is entirely devoid of such
appendages. If the sensitiveness of the leaves represents a form

of protection against crawling insects, the above described distribution of the tactile hairs is obviously advantageous.

The histological structure of the tactile hairs of *Biophytum* is exceedingly interesting.   The hair proper (Fig. 240) consists of a single very thick-walled and lignified cell, ·9 to 1·2 mm. in length, and tapering to a sharp point.   The lower end of this cell is slightly curved and obliquely truncated.   While the convex side of the basal curved portion is freely exposed, the concave side rests against a projecting cell-cushion, which also envelops the two flanks of the hair cell.   This cushion projects far above the general level of the epidermis, and, in addition, fits into a shallow depression on the con-cave side of the hair; the wall of the hair is usually specially thickened at this point.   As far as the hairs on the rhachis are concerned, the cushion is entirely made up of several superimposed tiers of tabular epidermal cells. These cells, which are thinner walled and more abundantly pro vided with protoplasmic contents than the ordinary epidermal ele-ments, undoubtedly represent the sensory portion of the tactile organ.

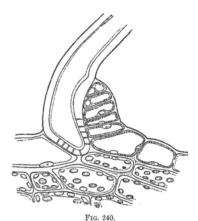

FIG. 240.

Median L.S. through one of the tactile hairs of the rhachis of *Biophytum sensitivum*; only the basal portion of the hair proper (which acts as a stimulator) is shown.

The hairs on the pinnae are some-what more complex in structure ; here a few palisade-cells, forming a conical prolongation of the photosynthetic tissue, are also included in the cushion; but there is no reason to suppose that these accessory cells are possessed of any perceptive capacity.

The stiff hair-cells act as levers; when they are bent downwards the sensitive basal cushion is forcibly compressed; and hence undergoes powerful stimulation.   The mechanism is comparable to that of the simple apparatus used in laboratories for compressing corks.   It is clear that stimulation will also take place if the hairs are bent back-wards, so that the cushion is subjected to tension.   Similarly con-structed hairs have been observed by the author in *Biophytum proliferum*.

In *Mimosa pudica* the sensitive lower half of the main pulvinus bears from ten to twenty-five upwardly directed tactile bristles. Each bristle is from 1 to 2·5 mm. in length and ends in a conical point ; it consists of a bundle of more or less thick-walled lignified cells. The structure of the base of the bristles varies considerably in different cases. In the simplest type of bristle the mechanical tissues extend right down to the base, the lowermost stereides being dove-tailed into the parenchymatous tissue of the pulvinus. Bristles of this type are mere stimulators and not independent sense-organs (cf. p. 586). In other instances, the bristle is seated on a small pediment composed of thin-walled elements resembling the cells of the sensitive parenchyma. If the pediment is regarded as an integral part of the bristle, the whole structure may be regarded as a tactile sense-organ. The third and most elaborate form of bristle, finally, is characterised by the presence of a well-defined basal cushion situated

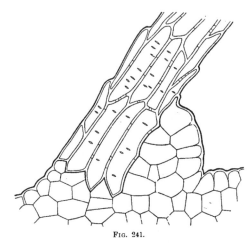

Fig. 241.

Basal part of a tactile bristle from the main pulvinus of *Mimosa pudica*.

in the axil of the obliquely inserted bundle of stereides (Fig. 241). The lower end of the cushion is encircled by a shallow but sharply defined transverse furrow representing a rudimentary hinge, which serves to accentuate the effects of slight upward or downward deflections of the bristle. The resemblance of these bristles to the tactile hairs of *Biophytum* is very striking, the plan of construction being identical in the two cases. There can be no doubt that the parenchyma of the basal cushion is sensory in function, since it is merely a local extension of the sensitive parenchyma of the pulvinus. The fact that these various forms of bristle act as stimulators or as sense-organs, is very easily demonstrated. If a bristle is slightly deflected by a gentle touch with a needle or bristle, the characteristic movement of the petiole instantly takes place.

Two carnivorous plants, *Aldrovandia vesiculosa* and *Dionaea muscipula,* possess very highly specialised tactile bristles. The sensory function of these structures was first recognised by Sydenham Edwards (1804) in the case of *Dionaea,* and by F. Cohn (1861) in that of *Aldrovandia.*

In *Aldrovandia vesiculosa,* which is a submerged aquatic plant, the two approximately semicircular halves of the leaf-blade close together instantaneously, when one of the tactile bristles situated on its upper surface is touched. About eighteen or twenty of these bristles occur on either side of the mid-rib, a smaller number being located on the concave surface of the lamina, especially towards its margin. Each bristle (Fig. 242) is made up of from five to seven superimposed tiers of cells; in the lower part of the bristle each tier is composed of four cells, but towards the apex the number of cells per tier is reduced to two. The lowermost tier represents the base or foot of the hair, and is made up of short cells. It is succeeded by from one to three tiers of elongated elements, with fairly thick outer, but quite thin inner walls. Next follows a tier of very short cells (they are only about twice as long as their width), with thin external walls. This hinge-like zone constitutes the sensitive region of the whole bristle. The remaining tiers again consist of elongated cells with thick outer walls, although this latter feature is less pronounced here than it is in the lower part of the hair. The two terminal cells generally diverge a little at their tips. Every cell in the bristle is furnished with a well

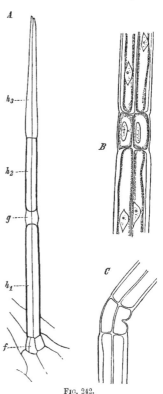

Fig. 242.

Structure of the tactile bristles of *Aldrovandia vesiculosa.* *A.* An entire bristle, highly magnified; *f,* base or foot ; $h_1, h_2, h_3,$ the tiers of elongated cells with thick outer walls ; *g,* the perceptive region or hinge. *B.* The hinge, and adjoining tiers, more highly magnified. *C.* Diagram showing the deformation which the hinge suffers, when the bristle is bent.

developed peripheral layer of protoplasm, and a fairly large nucleus, which is usually spindle-shaped with pointed ends. When one of these bristles is touched near its upper end, it does not simply bend or move to one side; for the regions above and below the hinge are stiff and inflexible, owing to the thickened condition of their outer walls. The bristle, therefore, bends sharply at the hinge, and the hinge-cells consequently become forcibly extended on the convex, and compressed on the concave side. If the deflection is considerable, the outer wall on the concave side of the hinge actually becomes pushed inwards, so that a transverse furrow is formed (Fig. 242 c). In any case, the protoplasts of the hinge-cells undergo very extensive deformation. The whole structure of the bristle is evidently designed so as to concentrate the mechanical effect of a shock or blow upon a limited portion of the organ, the obvious inference being that this portion is responsible for the actual perception of the stimulus.[297]

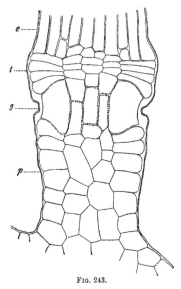

Fig. 243.

Basal portion of a tactile bristle of *Dionaea muscipula* in L.S.; *p*, pedestal; *g*, sensitive hinge; *t*, zone of tabular cells immediately above the hinge; *e*, prosenchymatous cells forming the terminal portion of the bristle.

The tactile bristles of *Dionaea muscipula* are much coarser than the corresponding structures of *Aldrovandia*, but are constructed on the same principle. Either half of the leaf-blade is provided with three bristles, which are placed at the corners of an approximately equilateral triangle. The two halves of the lamina will snap together, even if the upper epidermis is briskly rubbed, without any of the bristles being actually touched; but there is no doubt that the bristles are endowed with a much higher degree of sensitiveness than the rest of the surface, since a very gentle touch suffices to bring about the characteristic movement.

Each bristle is made up of four distinct portions, which may be described in basipetal order as follows:

1. The stiff, conical, pointed terminal region, which is about 1 mm. in length, acts as a lever, or, in other words, represents the stimulator, or mechanically effective portion of the entire apparatus. It is composed of living cells, which are prosenchymatous, and provided with moderately thick, but unlignified walls (Fig. 243, *e*).

2. Next follows a zone of tabular cells, comprising two or three tiers of central, and three or four of epidermal cells. The cells of this zone also invariably contain living protoplasts, in spite of the fact that their walls are more or less suberised.

3. The sensitive hinge of the bristle was first described by

Oudemans; its position may be detected from the outside by the presence of a furrow which completely encircles the bristle at this level (Figs. 243, g, 244). It consists of a central core of tissue surrounded by a ring of vertically elongated, tabular or wedge-shaped sensory cells, representing modified epidermal elements. These sensory cells contain a massive protoplast with a centrally situated nucleus (Fig. 244); their outer walls are thick, except for the portion immediately underlying the transverse furrow, and are covered by a well developed cuticle, which is furnished with numerous minute nodules and serrations upon its inner surface. The radial, longitudinal and inner walls of the sensory cells, which are also fairly thick, are traversed by delicate protoplasmic fibrils, so that the protoplasts of the sensory cells communicate not only with one another, but also with the living cells of the central core. This core usually consists of two superimposed tiers of cells; in a median vertical section each tier is seen to be three cells wide. The walls of these central cells

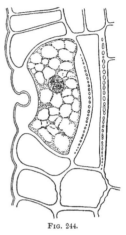

Fig. 244.

Part of a longitudinal section through the hinge of a tactile bristle of *Dionaea muscipula*, showing one of the sensory cells, with its protoplast. ×700.

are thick and highly refractive; their middle lamellae enclose numerous microscopic granules and rods of a cutin-like material (Fig. 244).

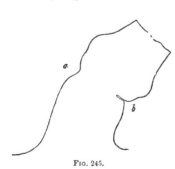

Fig. 245.

Diagram showing the effects of bending a tactile bristle of *Dionaea muscipula*. The furrow encircling the hinge is partially flattened out on the convex side (a), but is made deeper on the concave side (b).

In all probability, the principal function of the central core consists in the transmission of the excitation produced by deformation of the sensory elements to the pedestal of the bristle, and thence to the tissues of the lamina.

4. The lowermost portion of the bristle takes the form of a cylindrical pedestal composed entirely of parenchymatous cells. It is important to note that the pediment is widest above, and contracts somewhat towards its base. Consequently, when the bristle is violently deflected it bends to a certain extent at the base, so that the delicate hinge is preserved from excessive deformation.

As Goebel first remarked, it is the hinge-cells which suffer the

2 P

greatest amount of deformation, when a *Dionaea* bristle is bent; there can, therefore, be no doubt that the actual perception takes place within these cells. One can easily gain a clear idea regarding the distribution of deformation, if one bends a bristle under the microscope by pressing a needle against its stiff terminal portion. Under these conditions, one cannot fail to observe that curvature chiefly takes place in the region of the hinge, the transverse furrow being

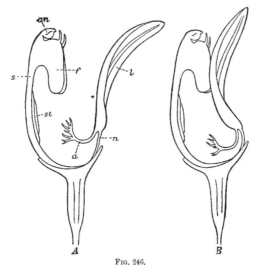

Fig. 246.

Semi-diagrammatic drawings of a flower of *Pterostylis* (type of *P. curta*). *A*, before ; *B*, after stimulation of the labellum ; *s*, column ; *an*, anther ; *f*, wing-like appendages of the column ; *st*, stigma ; *l*, limb of the labellum ; *n*, its claw (the motor-organ) ; *a*, appendage of the labellum (? sense-organ).

partially flattened out on the convex surface, and becoming narrower and deeper on the opposite side (Fig. 245). In a particular case observed by the author, the vertical extent of the hinge-area was found to be equal to nineteen divisions of the eye-piece micrometer, when the bristle was straight. On bending the bristle, the length of the convex side of the hinge increased to twenty-three divisions, an elongation of 21 per cent. The large tabular marginal cells may be seen to undergo great deformation, the obvious inference being that these elements represent the actual sensory organs.[208]

Various species of *Pterostylis*, a genus of Orchids inhabiting Australia, N. Zealand, and N. Caledonia, possess a sensitive labellum, furnished with a peculiar brush-like appendage ; as the latter probably represents a tactile organ, it may conveniently be discussed at this point. The labellum is made up of two parts—namely, a narrow

limb (Fig. 246 A, *l*), and a claw (*n*) of variable length.   Where the
limb joins the claw, it bears, on its upper side, an appendage (*a*), the
precise form of which varies in different species.   The characteristic
movement of the labellum takes place when a small insect alights
upon the limb and crawls upwards ; in these circumstances the claw,
which is the actual motor organ, curves suddenly inwards, throwing the
limb upwards against the column and imprisoning the insect in the
flower.   The animal can only escape by ascending the column, and
forcing its way between the two wing-like outgrowths (*f*), which crown
the distal extremity of that structure.   In crawling over the base of
the column, the insect transfers any pollinia that it may have brought
with it from other *Pterostylis*-flowers to the stigmatic surface (*st*), while
after emerging between the wings of the column it cannot fail to rub
against the rostellum, and thus to carry off the pollen-masses, if these
have not been previously removed.

So far as existing observations go, it seems highly probable that, in
*Pterostylis curta*, *P. striata*, *P. coccinea*, *P. truncata* and allied species, the
movement of the labellum takes place at the instant when the insect
touches the appendage of the limb.   In these species the appendage
consists of a ribbon-shaped basal portion which is bent upwards, and a
distal part composed of numerous lobes and lappets, which are densely
clothed with short thin-walled hairs.   The position and structure of
this remarkable appendage strongly suggests that it is the tactile sense-
organ of the labellum ; this hypothesis, however, still stands in need of
experimental verification.[299]

*Masdevallia muscosa*, an Orchid which grows in New Granada, likewise possesses an irritable labellum.   In this case F. W. Oliver has
shown, that the characteristic movement only takes place, when an
insect visitor touches a " tactile ridge " situated on the upper surface of
the limb.   The epidermis of this ridge consists of remarkably small
thin-walled cells, with papillose outer walls, and acts as a sensory
epithelium.   The ecological significance of the movement is the same as
in *Pterostylis*; insect visitors are imprisoned as a result of the sudden
curvature of the labellum, and bring about cross-pollination in making
their escape.

## III. GRAVITATIONAL SENSE ORGANS [STATOCYSTS].

### A. GENERAL CONSIDERATIONS.

The majority of plants possess the capacity of orientating their
several organs in an appropriate manner.   The most important
means of orientation is furnished by **geotropic irritability**, that is, the
power of perceiving the " direction of gravity," and of placing the various

members of the plant-body in some definite relation to this direction.
The trunks, or main stems, and the primary roots of Higher Plants tend
to assume a perpendicular position ; organs of this type are said to be
**positively geotropic** when they grow vertically downwards, and **negatively geotropic** when they grow vertically upwards, provided in either
case that they return to their normal position by means of appropriate
curvatures, if they are displaced.   Thus most primary roots are positively
geotropic, and most main axes negatively geotropic.   Lateral shoots and
roots, on the other hand, are generally **klinogeotropic,** that is to say, they
are in a geotropically stable position when they form an angle with the
vertical ; if this angle is one of 90°, or, in other words, if the horizontal
position is the stable one, the term **diageotropic** may be used.

The influence of gravity upon the vertical position of stems and roots
was first explained by Knight, in 1806, by means of his famous rotation
experiments.[300]   By attaching his experimental material—in most cases
seedlings—to a wheel revolving rapidly in a vertical plane, Knight exposed
the plants to two artificial conditions.   In the first place, the one-sided
action of gravity was eliminated [or rather equalised] by the rotation
around a horizontal axis ; secondly, the plant was subjected to the
action of centrifugal force, an agency which, like gravity, produces a
mass-acceleration.   As a result, the roots of the experimental plants
grew towards the periphery of the wheel, the stems, on the contrary,
towards the centre ; in other words, the direction of growth of both
organs was influenced by centrifugal force in the same way as it is
affected by gravity under natural conditions.   When a horizontally
revolving wheel was employed, so that gravity and centrifugal force
acted simultaneously but in different directions, a composite effect
was produced, roots growing outwards and obliquely downwards, and
stems inwards and obliquely upwards.   Evidently gravity and centrifugal force exert the same influence upon plant-organs ; these two
agencies are, in fact, interchangeable in this respect.   It follows that
the vertical orientation of stems and roots must be primarily determined by gravity.

Knight's experiments also give some indication of the manner in
which gravity affects organs that are sensitive to its influence.   This
particular external factor can only be felt by the sensitive protoplasm as a mass-acceleration, or, in other words, as an effect of *weight.*
[We now know that] the requisite weight-effect is produced by the
presence in the sensory cells of solid bodies which have a higher specific
gravity than the cell-sap or the semi-liquid protoplasm ; these heavy
bodies always come to rest upon the physically lower wall of the cell,
and exert a certain amount of pressure upon the corresponding region
of the ectoplast.   The author and Němec have independently attributed

this stimulus-producing function to the starch-grains, which have a relative density of about 1·5 ; but many other solid bodies, such as *calcium oxalate* crystals, particles of *silica*, etc., might equally well act as **statoliths**. This term may therefore be applied to all solid inclusions that serve, by virtue of their weight, to transmit gravitational stimuli to. the sensitive ectoplast.

According to the theory first suggested by Noll, and subsequently elaborated by the author and by Němec,[301] the mode of perception of gravitational stimuli in plants is analogous to the corresponding process in animals. The investigations of Ernst Mach, Breuer, Chun, Delage, Th. W. Engelmann, Verworn and others, long ago established the fact that the so-called otocysts which are found in so many animals—especially in the lower groups—have no connection with the sense of hearing, but represent balancing organs, or, more properly, organs for the perception of the "direction of gravity." For this reason Verworn suggested the term **statolith** in place of otolith, the name previously applied to the bodies which render gravitational stimuli effective by virtue of their weight.

Special gravitational sense-organs do not seem to be present in all plants; when they are absent, cells which perform totally different principal functions may in addition serve for the perception of gravitational stimuli, if they contain starch-grains or other bodies distinguished by relatively high (or, on the other hand, relatively low) specific gravity. At a higher level of adaptation, the geotropically sensitive members of the plant-body are furnished with special geotropic sense-organs, a striking instance of anatomico-physiological division of labour.

In the Higher Plants each gravitational sense-organ typically comprises several—usually a large number of—sensory cells or **statocysts**. Every statocyst consists of two essential portions, namely, first a statolith-apparatus composed of a variable number of readily movable or "falling" starch-grains, and, secondly, an ectoplast which is sensitive to the pressure of these statoliths. When an organ containing statocysts is in a condition of geotropic equilibrium, the pressure of the starch-grains against the physically lower portion of the ectoplast remains unperceived, or at any rate leads to no responsive movement. But as soon as the part under consideration is displaced from its stable position, the starch-grains fall against that portion of the ectoplast which is now on the physically lower side of the cell; a new and unfamiliar state of stimulation is thereby produced, with the result that a geotropic movement takes place, which brings the organ back into its former state of equilibrium. In this case the stimulus need not consist of a shock or jerk involving a sudden local deformation of the sensitive protoplasm, such as was seen to be essential in the case of tactile organs. The slow deformation due to continued or static pressure of starch-grains against the ectoplast constitutes an effective gravitational stimulus; for in geotropic experiments the characteristic curvature results even when precautions are taken to eliminate all vibration.

· Geotropic stimulation begins as soon as an organ is deflected from its stable position, so that a few starch-grains press upon the ectoplasts occupying the walls which are underneath in the new position; an actual rearrangement of the starch-grains is therefore not an essential condition of stimulation. As a matter of fact, the starch-grains do very soon migrate onto the physically lower walls, when a positively or negatively geotropic organ is placed horizontally, with the result that the intensity of stimulation gradually increases, attaining its maximum value when all the falling starch-grains have moved on to the lower region of the ectoplast. The time required for the complete rearrangement of the statoliths may be termed the **period of migration**; its average length varies from five to twenty minutes for different organs. The geotropic **presentation-time**, that is, the shortest period of time during which a geotropic stimulus has to be applied in order subsequently to induce an appreciable response, varies within similar limits. It is true, that if the first stages of geotropic curvature are determined with the aid of a microscope instead of by naked-eye observations, the presentation-time assumes a smaller value; this result, however, merely shows that the displacement of a portion of the statoliths suffices to produce effective stimulation.

In all specialised statocysts the starch-grains are highly movable; hence they are readily acted upon by gravity, and rapidly accumulate upon the lower walls of the sensory cells. As already explained, actual migration of the statoliths is not necessary. Even where the position of the starch-grains in the cell is practically unaffected by gravity, owing to the high viscosity of the cytoplasm, so that they are always uniformly distributed over the various cell-walls, those grains which lie against the physically lower walls must necessarily exert a definite pressure upon the corresponding portion of the ectoplast. Geotropic stimulation is therefore quite *possible*, even when the starch-grains are immovable. But mobility of the starch-grains, resulting as it does in a rapid accumulation of all the statoliths upon the lower wall, is for various reasons highly advantageous, and hence indicates a relatively specialised condition of the entire sense-organ.

Mobility of the statoliths first of all ensures that all the starch-grains in each statocyst are utilised for stimulation; in this way the total intensity of the stimulus is increased and the curvature correspondingly accelerated. In this sense, therefore, the evolution of movable statoliths may be regarded as an illustration of the principle of economy of material.

Mobility of the starch-grains entails a further advantage, inasmuch as it ensures that the ectoplasts associated with the walls which run parallel to the "direction of gravity," soon become free from statoliths, and hence are preserved from the disturbing influence of the accidental stimulation which would otherwise result from every slight vibration. Owing to their freedom of movement, the statoliths are also withdrawn from the orientating influence which the nucleus often exercises upon such structures. Otherwise, namely, leucoplasts and chloroplasts, with their included starch-grains, are frequently aggregated around the nucleus, and rendered incapable of responding to the influence of gravity by the high viscosity of the cytoplasm. In certain positions of the nucleus, starch-grains "anchored" in this way could not act as statoliths.

Before concluding this general discussion, it may once more be remarked that the statolith-theory agrees very well with all the experimental data which bear upon the subject of geotropism, or at any rate that none of these data are irreconcilable with the theory. Attention may be specially directed to the fact that geotropic stimulation attains its maximal intensity for a parallelotropic organ, when the latter is placed horizontally.[302] This result, which was assumed long ago by Sachs, confirmed by the researches of F. Darwin and Miss Bateson, and finally established beyond possibility of doubt by the recent work of Fitting, can be deduced a priori from the statolith theory; for it is only in the horizontal position of an organ that all the statoliths will be resting upon the physically lower portions of the ectoplast at the end of the period of migration. If the displacement from the stable position amounts either to more or to less than 90°,—if, in other words, the organ is placed obliquely—it is only a variable proportion of the statoliths that come to lie upon the lowermost of the longitudinal walls of each statocyst. Moreover only a limited area of the wall in question will become covered with statoliths in these circumstances; the rest of the falling starch will come to rest against the upper and lower transverse walls, and will be ineffective as far as stimulation is concerned. The maximum intensity of stimulation therefore coincides with the horizontal position.

After these introductory remarks, attention may be directed first to the anatomical characteristics of the statolith-apparatus in roots, stems and other geotropically sensitive organs, and subsequently to the experimental evidence upon which the statolith theory is founded.

### B. THE STATOLITH-APPARATUS IN ROOTS.

The positive geotropic curvature of ordinary roots always takes place in the sub-apical growing zone. Perception, on the other hand, is mainly located in the actual tip[303]; it is, however, not absolutely confined to this region, the growing zone being also sensitive,—in all cases which have been tested from this point of view—though to a much slighter extent. The very marked geotropic sensitiveness of the root-tip—which was first discovered by Ch. Darwin—is obviously a very useful property in view of the conditions to which roots are ordinarily exposed ; it enables the root to return as quickly as possible to its stable position, when it is diverted from its natural line of progression through the soil by the opposition of solid obstacles.

According to Němec's well-founded view, it is the central portion of the root-cap (the so-called "columella")—which always contains

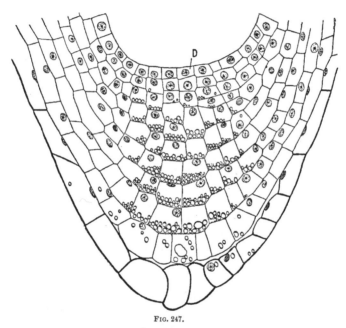

FIG. 247.

Median L.S. through the root-cap of an adventitious root of *Roripa amphibia* [*Nasturtium amphibium*]. All the starch-grains in the columella are resting on the (physically) lower cell-walls. After Němec.

numerous falling starch-grains (Fig. 247)—that represents the actual geotropic sense-organ of the root. This columella is shaped roughly like a truncated cone, and consists of several parallel or slightly divergent cell-layers arranged symmetrically on either side of the median plane. The individual cells are often slightly elongated ; apart from the youngest, which are still in the meristematic condition, they all contain numerous falling starch-grains. In many plants (*e.g.* in certain Conifers and CRUCIFERAE, in the aërial roots of *Monstera deliciosa*, etc.), the columella is sharply marked off from the rest of the root-cap ; in other cases the adjoining root-cap cells also contain some falling starch, and the limits of the columella are therefore not so well defined. The terminal—that is, the oldest—cells of the columella are either entirely devoid of starch, or else contain starch-grains which no longer respond to the action of gravity, but are irregularly distributed over the walls. It is highly probable that these cells are in a senile state, and have lost their faculty of perception.

The number of cells included in the statolith-apparatus of the root varies considerably in different species, and also according as the root under consideration is a primary or a secondary one. Němec estimates the number of statocysts in an aërial root of *Monstera deliciosa* at 1800-1900. As a rule the sensory cells are far fewer in number. Thus the main root of *Roripa amphibia* [*Nasturtium amphibium*] is furnished with about 216 statocysts, while the lateral roots of the same plant do not contain more than 96 of these cells.

Statolith starch usually consists of rounded grains, about 2-7 $\mu$ in diameter; the grains are generally of the compound type, but each is made up of a comparatively small number of partial granules. They arise in the ordinary way in the interior of leucoplasts, the stroma of which persists as a thin envelope around the mature grain. The starch-grains in the statocysts are sufficiently numerous to form from one to three complete layers over the whole extent of the lower wall of each cell. Thus every statocyst in the root-cap of *Roripa amphibia* contains about 26 starch-grains, a number sufficient to form two complete layers, one above the other.

The cytoplasm of statocysts is usually clear, hyaline and comparatively watery, so that the movements of the starch-grains are interfered with as little as possible. The nuclei, which are fairly large, may be unaffected by gravity, or they may respond like the statoliths, in which case they usually come to rest at the upper—less frequently at the lower—ends of the cells. The typical condition is illustrated by *Ceratopteris thalictroides, Hyacinthus orientalis, Allium Cepa, Canna indica, Salix viminalis, Phaseolus multiflorus, Cucurbita Pepo, Helianthus annuus*, etc., the exceptional case by *Equisetum arvense* and *Vicia Faba*. As long as the root is in the stable position, the cytoplasm is uniformly distributed over the walls of the statocysts; but, as Němec has shown, when the root is displaced, the distribution of cytoplasm in the sensory cells also changes, the peripheral protoplasmic layer becoming much thicker and denser, and staining more deeply with haematoxyline on the side which was lowermost, and hence covered with statoliths, in the stable position. This rearrangement of the cytoplasm is interesting, if only because it constitutes the first *visible* change produced in the sensory cells by stimulation; but it is improbable that the process forms a link in the chain of stimulation which culminates in geotropic curvature.

According to the author's own observations, the growing zone of the primary root is also sensitive—though less so than the root-tip—in *Vicia Faba, Lupinus albus* and *Phaseolus multiflorus*, and the same condition probably prevails in many other roots. In the aforesaid species, the greater sensitiveness of the root-tip is reflected in the more perfect development of the statolith-apparatus in the cap. In the less sensitive growing zone, the perceptive capacity is confined to the periblem, which contains numerous starch-grains; in *Vicia Faba* these are movable in the most actively growing region, but in other cases they are always irregularly distributed through the cells.

While the principal statolith-organ of the root is located in the root-cap in the great majority of cases, there are some exceptions to this rule. In *Selaginella Martensii*, for example, the root-cap is entirely devoid of starch, but falling starch-grains occur in the inner layers of the periblem, within a zone which begins at a distance of ·13 to ·16 mm. from the growing-point and extends over ·27 to ·34 mm. In *Trianea bogotensis*, again, the root-cap apparently contains neither starch-grains nor any other bodies of high specific gravity; but, here also, the cells of the inner periblem-layers are provided with falling starch-grains throughout the region of curvature. In accordance with this fact, Němec found, that if roots of *Trianea* are deprived of their tips and are then immediately exposed to geotropic stimulation, they exhibit appreciable curvature after three hours.

It was known to Sachs, that lateral roots of the second and higher orders are very feebly geotropic, or even quite indifferent to the influence of gravity. The author himself has found that the statolith-apparatus of these roots is always more or less obviously reduced; in some cases the root-cap is entirely devoid of falling starch-grains, while in others the statocysts are relatively few in number, and the individual starch grains unusually small. The fact that the ageotropic aërial roots of certain root-climbers (*Hedera helix, Marcgravia dubia, Hoya carnosa*) contain no starch-grains, or, at most, such as are immovable, also accords with the statolith theory. The grasping-roots of Aroids are likewise only slightly geotropic or indifferent; most of them accordingly possess a more or less extensively reduced statolith-apparatus, as Tischler and Gaulhofer have shown[304]; their statocysts are few in number, and their starch-grains small and sluggish or motionless. The close correlation which prevails, among roots, between the degree of geotropic sensitiveness on the one hand, and the development of the statolith-apparatus on the other, has been very fully investigated by Tischler.[305] This author states, that the root-caps of permanently ageotropic terrestrial roots (*e.g.* in *Arum maculatum, Salix* spp., *Epimedium alpinum, Carex arenaria*, etc.) contain either no starch-grains at all, or else only such as always remain irregularly distributed. This statement also applies to the geotropically insensitive roots of certain saprophytes and parasites (*Cuscuta, Orobanche, Pyrola*), and to various aquatic roots (*Eichhornia, Pistia, Pontederia*). In certain cases, where the roots are at first ageotropic but subsequently become positively geotropic (*Leontice Leontopetalum, L. Alberti, Festuca ovina, Poa*), the root-cap at first contains immovable starch-grains only, or none at all; the acquisition of geotropic sensitiveness coincides with the appearance of falling starch-grains. Negatively geotropic breathing-roots (*Jussieua, Phœnix canariensis*), again, are furnished with typical statoliths. The aërial roots of epiphytic Orchids, which in most cases are certainly ageotropic, contain no falling starch-grains. The positively geotropic, nutritive aërial roots of Aroids, on the other hand, are well provided with statoliths. The grasping roots of these plants have been dealt with above.

### C. THE STATOCYSTS OF STEMS AND LEAVES.

The author has shown that the statoliths of negatively geotropic organs (stems, inflorescence-axes peduncles, pulvini) are usually con-

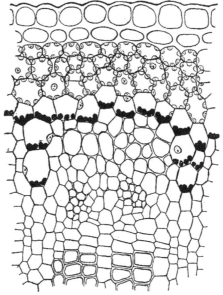

Fig. 248.

Part of a T.S. through an internode of *Linum perenne* (young enough to be capable of geotropic curvature) which has been lying on its side for some time. The endodermis contains large starch-grains (here stained with iodine), which are all resting on the (physically) lower cell-walls. The outer cells of the phloem-rays likewise contain falling starch-grains.

tained in the so-called "starch-sheath" [the endodermis].[306]  Formerly, this layer was regarded either as a carbohydrate-conducting tissue (Sachs), or as a local storage-organ (Heine), serving for the nutrition of the adjacent fibrous cylinder—or circle of mechanical strands—during its early development.

Morphologically considered, a typical single-layered starch-sheath represents the innermost layer of the primary cortex (the phloeoterma

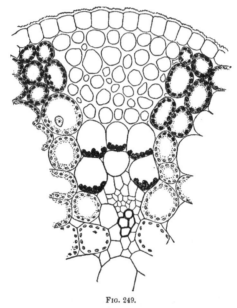

FIG. 249.

Part of a T.S. through a scape of *Arum ternatum*, showing a crescentic group of statocysts interpolated between a collenchyma strand and a vascular bundle.

of Strasburger).  Externally, therefore, it abuts against the cortical parenchyma, while internally it is contiguous to the vascular cylinder or circle of conducting strands, or to the fibrous tissue associated with the mestome (Fig. 248).  In many plants, each individual vascular bundle is enclosed in a separate starch-sheath.

Closely related to the above-described normal type of structure, is the condition in which the starch-sheath is interrupted by individual cells devoid of starch, or by a whole series of such elements.  In *Urtica dioica* the disruption of the starch-sheath has progressed still further, the leaf-traces being subtended by solitary statocysts, or rather by longitudinal rows of these cells.  Not infrequently the continuous "starch-sheath" is replaced by sickle-shaped groups of starch-containing cells, which are associated with the leptome- or hadrome-strands.  Such "starch-crescents" (*Stärke-sicheln*) occur, for example, in the nodes of Grasses, where they subtend the hadrome-strands; in the scape of *Arum ternatum*, on the other hand, they are located on the inner side of the leptome-strands (Fig. 249). Occasionally these groups of starch-containing cells are situated on the flanks of the vascular bundles; this occurs, for instance, in *Ranunculus acris*, where a transverse section shows two or three, or sometimes a single statocyst, on either side of each vascular bundle (Fig. 250). In *Chelidonium majus* all the vascular bundles are enveloped in complete starch-sheaths;

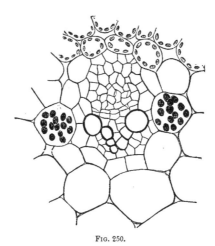

Fig. 250.

T.S. through one of the vascular bundles in the stem of *Ranunculus acris*; the stato-cysts are located on the flanks of the bundle.

but the starch-grains of these sheaths are only large and movable on the flanks of the bundles. In a few cases, finally (*Thalictrum flavum*), the functions of the starch-sheath are undertaken by the primary medullary rays, the cells of which are furnished with typical statoliths. It seems permissible, therefore, to assert, on the basis of the available data, that all geotropically sensitive axial organs are provided with a characteristic starch-sheath, or, where the latter is absent, with well defined groups of cells containing movable starch-grains.

The sensory cells of starch-sheaths, and the elements composing the groups of statocysts which frequently take the place of the sheaths, are always parenchymatous in character. They are not distinctly elongated, though they may be one and a half times or twice as long as their width. The advantage of this approach to the isodiametric condition consists in the fact, that it facilitates the rapid transference of a large number of starch-grains to the longitudinal walls, when the organ is

displaced from its stable position ; in order to achieve the same result
in a greatly elongated cell, the starch-grains would have to form a very
thick stratum in the stable position.    In *Fumaria officinalis* the sensory
cells are arranged in longitudinal rows composed of alternating long and
short elements ; the latter are one and a half to three times wider than
their length, and thus appear more or less disc-shaped.    In transverse
section the elements of the starch-sheath are usually four- to six-sided
and tangentially elongated ; more rarely they are nearly circular in outline (*Vinca minor*, Fig. 251 A).

    The starch-grains which act as statoliths, are sometimes simple and
spherical, but more often belong to the compound type, and consist
of two or more partial grains.    As a rule they are considerably
larger than the ordinary starch-grains of the medullary and cortical

*A*

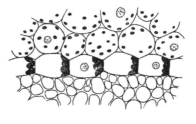

FIG. 251.

*A*, Small portion of a T.S. through a stem of *Vinca minor*, which has been lying
horizontally.   *B*. Small portion of a T.S. through epicotyl of *Phaseolus multiflorus* ; the
portion figured has been on one of the flanks, when the stem was lying horizontally.

parenchyma.    They are generally enclosed within pale-green chloroplasts, or more rarely within amyloplasts ; in either case the stroma of
the plastid persists as a thin pellicle around the statolith.    The number
of statoliths contained in each sensory cell varies much in the same
way as it does in the case of roots ; in the erect position of the stem
the starch-grains most often form a single layer on the lower transverse
wall of each statocyst, but two or three layers are found in certain cases.
According to Schwaighofer, the sensory cells in the hypocotyls of [307]
certain COMPOSITAE contain abnormally small numbers of statoliths.
In *Madia sativa*, for example, each statocyst usually contains five starch-
grains ; 5·6 per cent. of the statocysts examined actually contained
only a single starch-grain, which responded very readily to the influence
of gravity.    Cases of this kind militate strongly against the theory which
regards falling starch as a form of reserve-material.

The statoliths of stems are generally highly mobile, a deflection of the organ from the vertical of 10° or 11° usually producing a measurable displacement of the falling starch-grains (Fig. 252).

In respect of their living contents, the statocysts of stems resemble the corresponding cells of roots. The cytoplasm forms a uniformly thin peripheral layer, and is hyaline and apparently very liquid. The sap-cavity may be traversed by protoplasmic filaments. The nucleus is usually arranged without reference to any rule, but sometimes adheres to the upper transverse wall. The walls resemble those of ordinary parenchymatous elements.

A typical starch-sheath—containing normal statoliths—only occurs in those regions of the stem that are still engaged in longitudinal growth, and that hence are capable of executing geotropic [or other] curvatures. In ful' developed—and therefore no longer geotropically responsive—parts of the stem, the layer in question is usually empty, its starch having been dissolved and employed in the nutrition of the adjacent growing tissues ; this nutritive activity of the starch-sheath must, however, be regarded as a strictly

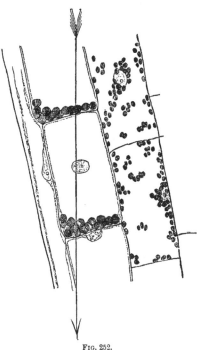

FIG. 252.

Small portion of a radial L.S. through a node of *Trades-cantia virginica*, which has been displaced from the vertical (indicated by the arrow) for some time. Note the position of the statoliths.

subsidiary function. When the cells of the sheath are about to lose their sensory capacity, the impending dissolution of the starch is often preceded by a scattering of these bodies.

According to the author's observations, the geotropically sensitive nodes [nodal pulvini] which occur in most RUBIACEAE, CARYOPHYLLACEAE, POLYGONACEAE, GERANIACEAE and COMMELYNACEAE invariably contain a typical starch-sheath. A similar layer has also been found by Němec in the foliar pulvini of certain LEGUMINOSAE (*Phaseolus, Lupinus*), the nyctitropic movements of which are, according to A. Fischer, affected by gravitational stimuli as well as by variations of illumination. Mention has already been made of the " starch-crescents " which are found in the highly reactive nodes (leaf-bases) of the GRA-MINEAE. Many Grasses (e.g. *Melica nutans*) are in addition provided with a well-developed one- or several-layered starch-sheath, which is situated immediately within the inner epidermis of the sheathing leaf-base.

F. Darwin has proved experimentally that, in certain Grass-seedlings, geotropic sensitiveness is confined to the tip of the so-called cotyledonary sheath; this localisation of the perceptive faculty is especially charac-teristic of the PANICEAE. Němec has found that, in all these cases, the parenchymatous ground-tissue of the apical region contains numerous starch-grains, which respond very readily to the influence of gravity.

The author's observations—which have subsequently been confirmed and extended by Schröder, Němec and Samuels[308]—have shown, that when floral organs exhibit geotropic sensitiveness, they are invariably provided with statocysts, which usually take the form of a starch-sheath surrounding the vascular strand or strands (*e.g.* in *Hemerocallis fulva, Funkia subcordata, Amaryllis robusta, Dictamnus Fraxinella, Azalea pontica, Epilobium angustifolium,* etc.).

From what has been stated above, it is quite clear that the distribu-tion, in space and time, of geotropic sensitiveness is correlated in a remarkable manner with the presence of falling starch-grains. This correlation might be regarded as accidental if it only occurred in a few instances. In point of fact, a very large number of cases have been described within recent years, which all support the statolith theory; indeed the anatomical evidence in favour of this theory may now be considered conclusive.

It is still unknown what structures, if any, act as statoliths among the Bryophyta and Thallophyta. Němec has, however, discovered falling starch-grains in the geotropically sensitive stems and setae of certain Mosses and Liverworts, and in the terminal cells of many Moss-rhizoids.[309] The positively geotropic rhizoids of certain species of *Chara* (*C. fragilis, C. foetida, C. aspera*) contain, near their tips, minute solid bodies of unknown chemical composition (*Glanz-Körperchen*), which have been shown by Giesenhagen to respond to the action of gravity.[310] Each group of these bodies, together with the asso-ciated cytoplasm, seems to represent the statolith-apparatus of the rhizoid. Possibly the terminal vacuoles, containing minute crystals, which occur in geotactically sensitive Desmids [*Closterium*, etc.,], likewise represent sensory organs. If plants of *Caulerpa prolifera* are darkened for some time, their " leaves " put forth slender branches which are very strongly negatively geotropic.[311] The peripheral cytoplasm of these branches contains rounded starch-grains at the exact point where geotropic curvature takes place. Although these starch-grains are not mobile, they may very well act as statoliths.

Since Fungi never form starch, other bodies of relatively high (or low) specific gravity must serve as statoliths in that group. This point still awaits investigation. In any case it is quite possible, that perception of gravitational stimuli does not always take place precisely in the same manner among Fungi and Algae, as it does in the Higher Plants.

### D. EXPERIMENTAL OBSERVATIONS REGARDING THE DEPENDENCE OF GEOTROPIC SENSITIVENESS UPON THE PRESENCE OF STATOLITHS.

The statolith theory has been subjected to the test of experiment in a variety of ways, so far without the discovery of a single fact which is irreconcilable with the theory; as a matter of fact, a whole number of data have already come to light which furnish more or less conclusive evidence in its favour.

1. Mention may first be made of certain amputation-experiments performed by Němec and by the author. All these experiments were arranged, so as to test whether a root or stem remains geotropically sensitive after its statocysts have been removed. Němec found, that a root is incapable of curving geotropically for a long time after its root-cap has been cut off, but that it returns to its normal condition when falling starch-grains reappear in the wound-callus. The temporary loss of the perceptive or reactive capacity is not, as might be supposed, merely due to shock ; for geotropic curvature is not interfered with to the same extent, if several transverse incisions are made in the root-tip which collectively produce a wounded surface equal to that created by the removal of the root-cap. Němec tried various modifications of this experiment, and always met with the same result; he therefore concludes that geotropic sensitiveness of roots is dependent upon the presence (ab initio, or as a result of regeneration) of falling starch-grains. Analogous amputations have been carried out by the author upon stems. In the COMMELYNACEAE, it is only the more or less distinctly pulvinoid basal region of each internode that is capable of perceiving gravitational stimuli and of executing geotropic curvatures. If the sub-epidermal collenchyma-strands and the greater part of the cortical parenchyma be removed from a node of *Tradescantia virginica*, while the starch-sheath and one or two of the immediately adjacent cortical layers are left intact, the geotropic reactivity of the organ is found to be almost as energetic as that of an uninjured node. If, however, the remainder of the cortex, including the starch-sheath, is likewise removed, the geotropic capacity is completely lost ; a feeble power of curvature is retained in the latter case, if, as sometimes happens, falling starch is present in the pith as well as in the endodermis.

2 Q

It must be admitted that such amputation-experiments are not very conclusive; the infliction of a serious wound cannot fail to affect the whole chain of stimulation, so that the interpretation of the results becomes a matter of doubt. Nevertheless, experiments of this type have a certain negative value, because their results have hitherto never come into conflict with the statolith theory.

2. A less drastic [and hence more reliable] type of experiment is that which does not involve the actual excision of the statocysts, but which aims at the removal of the statoliths by indirect means. For this purpose Němec embeds radicles of *Vicia Faba* in plaster of Paris, the effect of such treatment being to inhibit growth completely, and to bring about the solution of the starch contained in the root-cap. As a rule, the statoliths disappear entirely when the roots have remained in plaster of Paris for a week. If the roots are liberated at the end of this period, they at once resume growth, but are found to be incapable of curving geotropically until falling starch-grains are regenerated. This experiment may reasonably be criticised on the ground that the prolonged imprisonment of the root does not merely cause the statoliths to disappear, but also probably paralyses the perceptive faculty of the sensitive protoplasm for the time being. The following observation, however, is less open to objection. Onions were allowed to germinate, after having been stored for several years in a dry place. Vigorous roots were put forth, which, during the first few days, responded to hydrotropic, but not to geotropic stimulation: at this stage the root-cap was invariably found to be devoid of starch-grains. After several days, some of the roots exhibited geotropic curvature; in all the cases examined, these active roots contained falling starch in their root-caps. Other roots, which remained permanently ageotropic, never formed any statoliths. This experiment was repeated several years in succession, and always gave the same result.

Němec is, therefore, undoubtedly justified in his assertion that the correlation between the presence of falling starch on the one hand, and the power of perceiving gravitational stimuli on the other, can scarcely be considered accidental.

A closely related series of experiments was carried out by the author with stems in which the starch-sheath had been deprived of starch,—in common with all the other tissues—owing to prolonged exposure to low temperatures. Very suitable material presented itself in the shape of plants of *Linum perenne, L. usitatissimum, Capsella Bursa pastoris*, and *Ruta graveolens*, growing in the open during the mild and comparatively snowless winter of 1901-2. The experiments performed upon the two species of *Linum* may be dealt with first. Stems which were completely devoid of starch were laid horizontally for 2–2½ hours, at a temperature of 17°C.–20°C and were afterwards rotated upon the klinostat; not the slightest geotropic curvature could be detected. If the experimental shoots were then kept in the warm atmosphere of the laboratory for 24 hours, statoliths reappeared in the starch-sheath; when this had occurred, a sojourn of $2-2\frac{1}{2}$ hours in the horizontal position resulted in typical geotropic curvature after a few hours' subsequent rotation on the klinostat. It might, of course, be objected that the long exposure to a low temperature had resulted in a disturbance of the sensitive protoplasm, of the mechanism of curvature, or of some other link in the complicated chain of stimulation; this argument is partially disposed of by the fact, that shoots in which the statoliths had been regenerated under the influence of the warmth of the labora tory, responded *at once* to geotropic stimulation after a second period of exposure to temperatures between 2·5° C. and 5° C., lasting 17 hours. A few of the young inflorescence-axes of *Capsella Bursa pastoris* had either not become entirely deprived of starch through exposure to low temperatures, or else had regenerated their statoliths with unusual rapidity; these shoots displayed geotropic curvature in the course of a few hours, when subjected to higher temperatures, in contradistinction to control specimens entirely deprived of starch, which showed no trace of curvature under similar conditions. In the case of *Ruta graveolens*, the reappearance of statoliths was a slow process, occupying as much as five days; geotropic sensitiveness remained in abeyance for the same period. Shoots which are normally negatively geotropic, therefore, lose their capacity for geotropic response, if the starch-sheath becomes deprived of its starch-grains; this behaviour seems inexplicable, except on the assumption that the sensitive ectoplast of a statocyst cannot be effectively stimulated in the absence of the falling starch-grains which act as statoliths.

3. Further important evidence in favour of the statolith-theory is furnished by the vibration experiments which have been carried out by the author and by F. Darwin.[312] The argument is roughly as follows: If the effective stimulus actually consists in the *pressure* of the falling starch-grains upon the ectoplast into which they sink, the intensity of stimulation should be increased, when the deformation of the protoplasm is accelerated by the substitution of a rapid succession of vertical impacts in place of the single blow which each statolith inflicts under ordinary conditions. In the author's experiments, peduncles, inflorescence-axes or roots were placed horizontally and subjected to a series of shocks, or to rapid vibration, with the aid of special apparatus. The number of vibrations varied between five and fifteen per second, while their amplitude generally amounted only to fractions of a millimetre. The effect of this treatment was a more or

less pronounced shortening of presentation-time and reaction-time, a result which proves that stimulation was actually intensified under the conditions of experiment. F. Darwin made use of a large tuning-fork which was set in vibration by electrical means. By this method also a very pronounced acceleration of geotropic curvature was found to take place.

4. The following very elegant experiment, which strongly supports the statolith theory, has been devised by Buder.[313] The experimental material (consisting of radicles of *Lepidium sativum*) is placed horizontally for about 12–15 minutes, until all the falling starch-grains have taken up their position upon the physically lower walls of the statocysts. The experimental plants are now turned through 180°, and left in the new position for precisely the same length of time as that spent in the first position. The stimuli perceived in the two diametrically opposed positions neutralise one another; hence, at the end of the second period, the statoliths are all located on one side of the statocysts, but no tendency to curve geotropically has as yet been induced. During the third stage of the experiment, the radicles are subjected to intermittent stimulation, being placed alternately in the two diametrically opposed horizontal positions for periods of 8–10 seconds, or longer, the periods being in all cases made too short to allow of any rearrangement of the statoliths. The total duration of this stage amounts to about 20 or 30 minutes. During the fourth and last stage, the experimental plants are placed on the klinostat, whereupon geotropic curvature is found to result after 40–90 minutes; this curvature is always such that the side which faced downwards when the statoliths were also resting on the physically lower walls, during the period of intermittent stimulation, becomes convex. Concordant results have been obtained by Buder in analogous experiments performed upon inflorescence-axes of *Capsella Bursa pastoris*; in this case the intermittent stimulation was effected by means of rapid rotation (one revolution in 7–8 seconds) on the klinostat. These results seem entirely inexplicable except on the basis of the statolith theory.

5. It has been shown by F. Darwin and Miss Pertz,[314] and again, more recently, by Buder, that migration of the statoliths can be induced by the action of centrifugal forces of smaller intensity than the normal acceleration of gravity ($g$); in such cases the rearrangement naturally takes a considerable time. The amount of retardation agrees very closely with the prolongation of the presentation-time observed by Bach, when working with mass-accelerations of smaller value than $g$. This coincidence also seems to support the statolith theory, as has already been pointed out by Buder.[315]

## IV. OPTICAL SENSE-ORGANS.

The effect of light in inducing paratonic movements may depend either upon variations in the *intensity* of illumination, as in the case of nyctitropic (more properly nyctinastic) movements, or upon *unilateral incidence* of the light, as in all cases of heliotropic (phototropic) or helionastic (photonastic) movement.

Those foliar pulvini which respond to variations of light-intensity, do not seem to be furnished with special light-perceiving organs. In all probability, sensitiveness to the alternation of light and darkness is a property common to all the elements of the motor-tissue. The absence of intercellular spaces full of air from the peripheral regions of pulvini, must be regarded as an arrangement which facilitates the penetration of light into the interior of these motor-organs.

The question as to whether heliotropically reactive parts of plants are possessed of special optical sense-organs or not, is bound up with the more fundamental physiological question of the nature of the heliotropic stimulus; for it is not known with certainty whether heliotropic stimulation depends upon an *asymmetrical distribution* of light-intensity, or upon the *direction* of the incident illumination. Charles Darwin and Oltmanns, among others, maintain the former view, while the latter is supported by Sachs and by Noll. From the standpoint of physiological anatomy, the first of these alternative theories seems the more acceptable; for, given unilateral illumination, there will in all circumstances be appreciable differences in the intensity of illumination at different points in the interior of a light-perceiving organ, whereas the original direction of the incident rays must inevitably be more and more altered by refraction and reflection, as the light penetrates further into the organ under consideration.

### A. PERCEPTION OF LIGHT BY PARALLELOTROPIC ORGANS.

For the great majority of positively heliotropic axial organs, the region of heliotropic perception concides with the region of curvature (*i.e.* with the zone of longitudinal growth), just as in the case of geotropism. The question as to whether all the cells in this region have the same capacity of light-perception, or whether sensitiveness is more or less strictly confined to particular tissues, has not yet been properly investigated. But if it be assumed that sensitiveness is restricted to a single tissue, it is obvious that the epidermis possesses special qualifications for the purpose in question. For the rays of light that fall upon the epidermal protoplasts on the illuminated side of an organ, have undergone relatively less deflection from their original direction than those which reach the more deeply-seated cells;

similarly, it is opposite portions of the epidermal layer that must experience the greatest differences of light-intensity. It is nevertheless highly improbable that definite epidermal or cortical cells, or groups of cells, distinguished by special histological features, are set apart for the purpose of light-perception, in the case of ordinary stems; at any rate, there is at the present time no evidence in favour of the existence of any such specialisation.

Charles Darwin discovered that the apices of the positively heliotropic cotyledonary sheaths of certain Grass-seedlings (*Avena sativa, Phalaris canariensis, Panicum miliaceum, Setaria viridis*, etc.) are particularly sensitive to photic stimuli; more recently, Rothert has confirmed Darwin's observations by means of very elaborate and painstaking investigations.[316] The heliotropic stimulus travels from the apex towards the proximal end of the cotyledon, and in the case of the PANICEAE even into the hypocotyl—which is itself quite insensitive to direct illumination—and thus induces, or at any rate, accelerates, the heliotropic curvature of the basal region. In *Avena sativa*, the extreme apex of the cotyledon (comprising a zone from 1–1½ mm. in length) is specially sensitive; behind this region heliotropic sensitiveness rapidly decreases, being no greater 3 mm. from the tip, than it is at the actual base of the organ.

The tip of such a Grass-cotyledon might well be termed a heliotropic sense-organ; it has already been remarked (p. 608) that the same region of the cotyledon represents—or rather contains—a gravitational sense-organ as well. *A priori*, it seems somewhat unlikely that statocysts should also act as photic sense-organs. In the cotyledonary apex of *Phalaris canariensis, Panicum miliaceum, Eleusine indica*, etc., all the elements of the fundamental parenchyma are developed as statocysts; assuming that there is any division of labour at all, the only cells available for photic perception are the elements of the outer epidermis, which are entirely devoid of starch (the inner epidermis need hardly be taken into consideration). That the faculty of light-perception is confined to the epidermis, seems all the more probable in view of the fact that the underlying statocyst-tissue contains numerous air-spaces of considerable size, so that the light which penetrates beneath the epidermis must be repeatedly deflected from its original direction by reflection and refraction. Nevertheless, the epidermis displays no special histological features that could be regarded as adaptations in connection with light-perception. At the extreme tip of the cotyledonary sheath, the cells of the epidermis often, though by no means always, assume a palisade-like shape (*e.g.* in *Avena sativa, Phalaris canariensis*, etc.). All the epidermal elements of the specially sensitive region are provided with dense protoplasmic contents, a feature which is on the whole very characteristic of epidermal cells that are concerned with the perception of stimuli.

Kohl asserts, that in negatively heliotropic roots the faculty of light-perception is confined to the tip.[317]  There is no evidence, however, that sensitiveness is more strictly localised in the case of these organs, nor is a specialisation of this nature to be expected on theoretical grounds.

### B. THE OPTICAL SENSE-ORGANS OF FOLIAGE-LEAVES.[318]

A great many foliage-leaves are diaheliotropic; as Wiesner[319] has shown, their blades are in a condition of heliotropic equilibrium, when they are placed at right angles to the direction of the most intense diffuse illumination to which they are ordinarily exposed.  By assuming this " heliotropically fixed position " (*fixe Lichtlage*) the leaf is assured of the maximum amount of illumination.  Leaves which behave in this manner are termed " euphotometric" by Wiesner ; they are particularly prevalent among shade-plants.

Leaves assume their heliotropically fixed position by means of appropriate curvatures or torsions; the motor-organ may consist of the whole petiole or of a pulvinoid region thereof, or sometimes of a typically developed pulvinus.  Dutrochet long ago suggested, with reference to these cases of euphotometric adjustment, that the leaf-blade exerts a directive influence upon the actual motor-organ—represented by the petiole.  Vöchting[320] was, however, the first to prove experimentally, in in the case of *Malva verticillata*, that the light-perceiving lamina determines the behaviour of the petiole, forcing the latter, if need be, to execute movements in opposition to its own heliotropic tendencies, in order to bring the leaf-blade into a favourable relation to the available light. The author has since succeeded in demonstrating a similar directive influence of the lamina in several other plants.  Thus, the leaf-blade of *Begonia discolor* assumes the fixed position, even if the petiole is completely darkened by means of a sheath of tinfoil.  A similar result is obtained with *Monstera deliciosa*, if the distal region of the petiole, which is developed as a pulvinus, is covered with tinfoil; although itself shaded, the pulvinus executes the curvature or torsion necessary to restore the lamina to the position of heliotropic equilibrium, with the greatest possible precision.  In species of *Tropaeolum* (*T. majus*, *T. minus*, *T. Lobbianum*) the results of similar experiments lead one to conclude that it is only the coarse adjustment of the leaf-blade that is brought about directly by the positive heliotropism of the petiole, the fine adjustment being effected indirectly, through the influence of the lamina upon the petiole.  The primary leaves of *Phaseolus multiflorus*, on the other hand, take up their fixed position, as Krabbe[321] had already remarked, even when the blade is darkened, provided that the pulvinus is exposed to light. In any case, it may safely be asserted that the lamina is the principal, or even the only heliotropically sensitive part of the leaf in many, and perhaps in the majority of plants with diaheliotropic leaves, while the petiole chiefly represents a motor-organ. Here again, therefore, the perceptive and the reactive regions are separated in space.

The question next arises, as to whether the power of perceiving photic stimuli is common to all the tissues of leaf-blades, or whether it is confined to a particular tissue, or to special cells or groups of cells; *a priori* the latter alternative seems the more probable. The blades of euphotometric leaves respond to differences in the intensity of illumination produced by comparatively small deviations from the stable position. But in all the tissues internal to the epidermis, including even the palisade-cells, the inevitable reflexion, refraction and absorption collectively cause a great amount of dispersion and a serious diminution in the intensity of the incident illumination; these internal tissues thus seem very badly fitted for the task of perceiving the original direction of light which is generally not very intense in the first instance. It seems, therefore, highly probable that the power of perceiving photic stimuli is vested in the upper epidermis. As a matter of fact, the structure of the upper epidermis of euphotometric leaves is characterised by features which are correlated—in some cases without any doubt, in others with a high degree of probability—with the light-perceiving function of that layer. It is convenient in this connection to recognise several distinct types of light-perceiving epidermis (sensory epithelium), and to discuss each of them separately.

### 1. *Papillose epidermal cells.*

The most widely distributed, and at the same time the most perfect type of light-perceiving sensory epithelium, is represented by those forms of adaxial foliar epidermis, in which the outer walls are more or less papillose, while the inner walls are approximately flat and placed parallel to the leaf-surface. When a pencil of rays falls upon such a papillose epidermal cell, perpendicularly to the leaf-surface, the outer and inner walls, and the ectoplast attached to them, are illuminated in a particular manner, in accordance with well-known optical laws. The rays meet the middle of the protuberant portion of the outer wall at right angles, or nearly so, whereas, towards the margin of this wall, the angle of incidence becomes more and more acute; the underlying ecto-plast is therefore most brilliantly illuminated over a limited central area, while its marginal region remains dark in comparison. This differential illumination becomes even more pronounced on the inner walls. For every papillose epidermal cell acts as a plano-convex or

condensing lens. The rays falling upon the convex outer wall in the
direction parallel to the optical axis of the lens, are refracted so as to
converge upon the middle of the inner wall; hence, this central region
is brightly lighted, whereas a marginal zone of varying width is not
directly illuminated at all, but merely receives a small amount of
reflected light from the mesophyll. The accompanying diagram

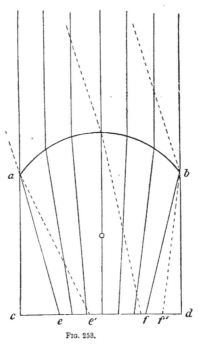

FIG. 253.

Diagram illustrating the lens-action of a papillose epidermal cell, *abdc*. In
vertical illumination there will be a bright central area, *ef*, on the inner tangential
wall, *cd*, surrounded by a relatively dark marginal zone. *ce*, and *fd*. If the light
falls obliquely from the left—as indicated by the broken lines—the bright area will
shift from the centre towards the right, to *e'f'*.

(Fig 253) represents the paths of the rays in an epidermal cell of this
type; for the sake of simplicity the outer wall is represented as part of
a spherical surface, and the mathematical construction is based upon
the refractive index of water (1·33).

The differential illumination of the inner walls of papillose
epidermal cells may also be directly observed by means of the following
very simple physical experiment, which has been termed the "lens-
experiment" by the author. A piece of the epidermis is carefully
removed with the aid of a razor, and at once mounted (with the
cut surface downwards) upon a *slightly* moistened cover-slip, care

being taken to keep the bulging outer walls dry. The cover-slip is
then inverted over a moist chamber placed on the stage of a microscope.
The preparation is illuminated by diffuse light, reflected from the plane
mirror of the microscope, so as to meet the epidermis approximately
at right angles. On focussing upon the inner epidermal walls, one at
once sees, in each cell, a bright central area surrounded by a dark
marginal zone (Fig. 254).

This well-marked differential illumination may also be demonstrated
in a somewhat different way, by placing the severed piece of epidermis,
with its outer surface uppermost, on a piece of sensitised paper moistened
with water, and by then exposing the preparation to vertical illumination. If the resulting photograph is examined, after fixation, with the
aid of a powerful lens, or as an opaque object under the microscope,
the contrast between the completely blackened central area and the
lighter marginal zone, in the image corresponding to each cell, may
often be seen with astonishing distinctness.[322]

When the incident light falls obliquely, instead of vertically, upon
the leaf-surface, the distribution of light intensities is no longer symmetrical with reference to the median radial plane of each cell. The
protruding outer walls are now more brightly illuminated on the side
facing the source of light, than they are at their apices or on the
opposite side. Consequently, the bright area on the inner wall shifts
from the centre towards the side furthest removed from the source
of illumination (Figs. 253 $e' f'$, and 255). This change can be readily
demonstrated in connection with the above-described lens-experiment,
by tilting the mirror to one side (after removal of the condenser, if
one is being used) so as to produce oblique illumination.

After what has been stated above, there can be no doubt that
a leaf-blade with a papillose epidermis is furnished with an excellent
optical apparatus, which enables it to orientate itself with reference
to the incident illumination. In the author's opinion, the action of
this apparatus is as follows: To begin with, it is necessary to assume
that the ectoplast lining the inner epidermal wall is sensitive to
photic stimuli. This sensitiveness involves a twofold power of discrimination; the ectoplast distinguishes not only between brightness
and darkness, but also between symmetrical and asymmetrical illumination of the cell (or, in other words, between concentric and excentric
distribution of light intensities on the inner wall). Heliotropic
equilibrium prevails, when the cells are vertically—and hence symmetrically — illuminated ; oblique — and, therefore, asymmetrical —
illumination, on the other hand, induces a heliotropic response of
the petiole, which continues until illumination is once more symmetrical,
or, in other words, until the lamina has regained its heliotropically fixed

position. The behaviour of a diaheliotropic leaf-blade has been compared by Ernst Mach to that of the human eye, which is in a state of heliotropic

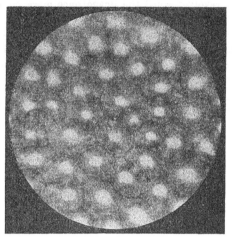

FIG. 254.

Lens-experiment with the adaxial foliar epidermis of *Anthurium leuconeurum*, in vertical light; symmetrical illumination of the inner epidermal walls. (Microphotograph.)

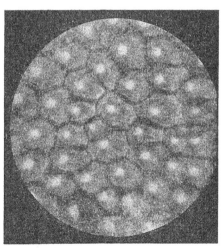

FIG. 255.

Another lens-experiment with *Anthurium leuconeurum*, in oblique light; asymmetrical illumination of the inner epidermal walls, the bright areas being all displaced towards the right. Same preparation as in Fig. 254. (Microphotograph.)

equilibrium when the image of the object under observation (*e.g.* a candle flame) falls upon the "yellow spot" of the retina. In this state the eye is comparable to a symmetrically illuminated epidermal cell. When the image shifts to the right- or left-hand side of the retina, the eye revolves, until the image once more coincides with the yellow spot.

The structure of the condensing portion of light-perceiving epidermal cells varies considerably in different cases. The simplest arrangement is that in which the whole outer wall protrudes, forming part of a more or less spherical surface. The degree of condensation naturally depends upon the radius of curvature. As a rule the curvature is not very great (Fig. 256 A); in this case, since the vertical height of

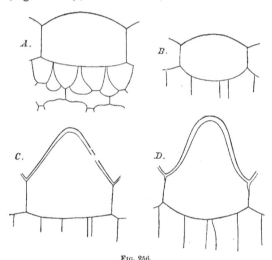

Fig. 256.

Foliar (adaxial) epidermal cells with bulging or papillose outer walls. *A. Oxalis acetosella. B. Trifolium incarnatum (cotyledon). C. Ruellia Daveauana. D. Anthurium Leuconeurum.*

the epidermal cells is small, the focus of the lens lies far below the inner walls, somewhere within the mesophyll. In the so-called "velvety" leaves, which are on the whole characteristic of plants inhabiting tropical rain-forests (MELASTOMACEAE, *Ficus barbata, Cissus discolor, Begonia Rex* and other species of *Begonia, Philodendron Lindeni,* spp. of *Anthurium,* etc.), the outer wall of each epidermal cell assumes the shape of a more or less steeply inclined conical papilla, with a rounded apex representing the actual condensing lens (Fig. 256 c, D). Here the focus generally falls within the epidermal cell itself, at a varying distance above its inner wall; in this instance, therefore, the bright central area seen in the lens-experiment represents the cross-section of a divergent pencil of rays. If the surface of·such an epidermis is covered with a thin film of water (a result which is easily produced in the lens-experiment with the aid of a fine brush), the rounded apices of the cells project above the water like so many islands, and continue to act as con densing lenses. Velvety leaves are thus able to perceive photic stimuli, even if their surfaces are constantly wet. The conical epidermal papillae of such leaves, therefore, most probably represent an adaptation to the frequent or continual wetting to which the leaves are exposed under natural conditions.

Where the bulging outer wall is thin, with its two surfaces parallel to one another, it is the cell-sap alone that constitutes the refractive medium. If, on the contrary, the outer wall is very thick, its high refractive index may become of importance, as Sperlich has shown ; the greatest amount of refraction will be obtained if this wall assumes a plano-convex, or even a biconvex shape (*e.g.* in *Faradaya, Paramignya, Albertisia*).

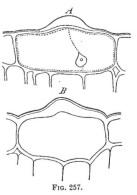

FIG. 257.

Local optical specialisation of outer epidermal walls. *A.* Adaxial foliar epidermal cell of *Colocasia antiquorum. B.* Adaxial foliar epidermal cell of *Aquilegia vulgaris.* See text.

In many cases, it is only a circumscribed area of the outer epidermal walls that protrudes, and this portion is often specially thickened so as to form a minute but powerful condensing lens. In *Colocasia antiquorum* (Fig. 257 A) and *Aquilegia vulgaris* (Figs. 257 B, 259), the central protuberance resembles a concavo-convex lens. The remarkably high refractive index of this lens, in the case of *Colocasia,* is probably due to the presence of pectic compounds. In *Vinca major* (Fig. 258, c) the very thick outer wall bears at its centre a minute, scarcely perceptible papilla; beneath this protuberance, the thick cutinised layers are invaded by a plug-like process of the cellulose layers, which are likewise very massive structures. The foliar epidermis of *Lonicera fragrantissima* is furnished with membrane-lenses of a somewhat flattened form ; here the cutinised layers are separated from the cellulose stratum of the outer wall by an intervening lamella composed of highly refractive material, which expands to form plano-convex or bi-convex lenses beneath the papillae. In this case the high refractive index of the lens is principally due to impregnation of the cell-membrane with wax. The outer walls of the adaxial epidermis of *Campanula persicifolia* contain strongly silicified bi-convex plugs (the "cell-wall plugs" of Heinricher), which act as condensing lenses. In *Petraea volubilis* also (Fig. 258 B), practically every cell of the foliar epidermis is

furnished with a silicified lens situated at the centre of its outer wall;
the point of special interest, in this case, is the fact that the lens
consists of an independent cell. While the inner wall of the lens-cell is
thin, its outer wall is bi-convex, and so thick that the cell-cavity
is almost entirely obliterated, appearing as a mere slit in transverse
section. The minute peg-shaped process attached to the middle of
the outer wall indicates that the lens represents a modified trichome.

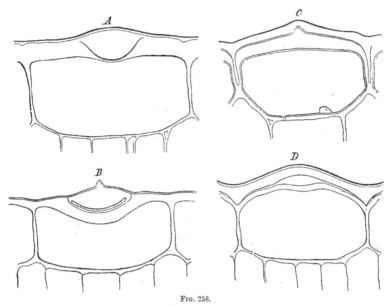

FIG. 258.

Local optical specialisation of outer epidermal walls. *A*. Adaxial foliar epidermal
cell of *Campanula persicifolia*. *B*. Adaxial foliar epidermal cell of *Petraea volubilis*.
*C*. Adaxial foliar epidermal cell of *Vinca major*. *D*. Adaxial foliar epidermal cell of
*Lonicera fragrantissima*. See text.

The efficiency of these minute membrane-lenses is illustrated by the
adjoining micro-photograph (Fig. 259), which shows the differential
illumination of the inner epidermal walls in the leaf of *Aquilegia*, as
seen in a lens-experiment.

Given a particular form of epidermal lens-cell and a suitable degree
of refraction, it may happen that the focal length of the lens coincides
very nearly, or exactly, with the vertical height of the cell; in this case,
more or less well-defined images of external objects will be projected
upon the sensitive ectoplast. The conditions referred to are realised,
for example, in the case of *Anthurium Warocqueanum*. One of the
accompanying illustrations (Fig. 260) is a reproduction of a micro-
photograph taken in the course of a lens-experiment with this plant;

inset in the bright circle (the image of the diaphragm-opening) on the inner wall of each epidermal cell, is a tiny image of a microscope-stand, which was interposed at a suitable height between the window that served as the source of illumination, and the plane mirror from which the light was reflected on to the epidermis. In such cases the *possibility* of a perception of external objects on the part of the plant cannot be denied. But there is no reason to suppose that any perception of this kind actually takes place. For leaves are concerned

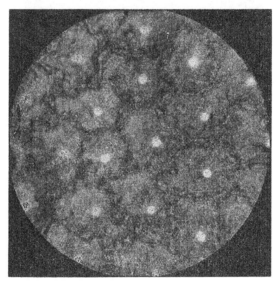

Fig. 259.

Lens-experiment with the adaxial foliar epidermis of *Aquilegia vulgaris*. (Micro-photograph, after Seefried.)

not with the perception of external objects, but with exact self-adjustment to their heliotropically fixed position; and this result can be attained, certainly with equal and perhaps with greater ease, if the focus of the epidermal lenses falls behind the sensitive ectoplast.

It is more particularly those leaves which are constantly exposed to feeble illumination, that stand in need of an upper epidermis with a well-developed power of light-perception; for it is leaves of this type that require to orientate themselves as accurately as possible in relation to the brightest available light. Hence, it might have been expected that, in one and the same species, shade-leaves would be more perfectly adapted in this respect than sun-leaves. As a matter of fact, Gaulhofer finds that this expectation is realised in a great many cases.[323] Thus, if epidermal papillae are present in both types of leaf (*Cercis siliquastrum, Prunus Padus, Fagus sylvatica,* etc.), their light-condensing action is more powerful in the case of the shade-leaves. In *Cydonia japonica,* the outer walls of the adaxial epidermal cells are parallel-sided in the sun-leaves, but distinctly plano- or bi-convex in the shade-leaves.

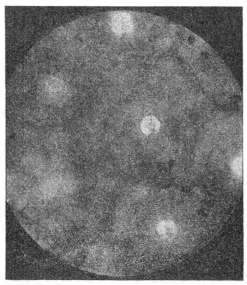

Fig. 260.

Lens-experiment with the adaxial foliar epidermis of *Anthurium Warocqueanum.* Note the image of a microscope in each bright area. See text.

### 2. Light-perception by epidermal cells with flat outer walls.

The papillose epidermis is not the only type of sensory epithelium that occurs in euphotometric leaves. There is another type,—which is neither so common nor so effective—characterised by the fact that the outer walls are flat or nearly so, or at any rate display no tendency to

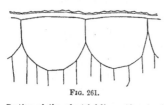

Fig. 261.

Portion of the adaxial foliar epidermis of *Franciscea macrantha,* in vertical section.

develop in a papillose manner. In this case the ectoplast adhering to the outer wall will always be uniformly illuminated, whether the light falls perpendicularly or obliquely upon the leaf; the intensity of this uniform illumina tion will, of course, be greatest in vertically incident light. As a rule, the inner walls of such epidermal cells are not flat, but bulge more or less into the underlying mesophyll. Each inner wall may appear as a continuously curved outline (Fig. 261) in

vertical section, or it may correspond to three sides of a polygonal figure
(Fig. 262); in the former event the lower half of each epidermal cell
is dome-shaped, while in the latter instance it resembles a truncated
pyramid. In either case, a central area of the inner wall will be most
brightly illuminated [just as happens in the papillose type of epidermal
cell] when light falls vertically upon the leaf, while a marginal
region (or the side-walls of the truncated
pyramid) will receive a smaller amount
of light under these conditions. In
oblique illumination the distribution of
light-intensities is altered, and the heliotropic equilibrium disturbed.

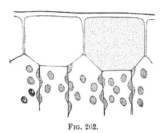

FIG. 262.

Portion of the adaxial foliar epidermis of
*Monstera deliciosa* in vertical section.

This second type of light-perceiving
foliar epidermis is exemplified by *Monstera
deliciosa* and many other Aroids, by
*Hedera Helix*, spp. of *Arabis*, *Franciscea macrantha*, etc.

Very frequently, the characteristic
features of the two opposite·types of light-perceiving epidermal cell
occur in combination, both the outer and the inner walls being more or
less protuberant. Cells of this type resemble bi-convex lenses in form.
In vertical illumination—that is, when the leaf is in the fixed position
—they are scarcely more efficient than the simpler plano-convex
form; in oblique light, on the other hand, the bright central area
shifts on to the inclined marginal region of the inner wall, where,
owing to the smaller angle of incidence, the local increase of light
is more pronounced than it would be if the inner wall were
quite flat.

In a number of diaheliotropic leaves, both the outer and the inner
walls of the upper epidermal cells are flat. In these cases there are,
according to Gaulhofer,[324] various special features which produce differential illumination—analogous to that which results from curvature of
the outer or inner walls—and which consequently enable the epidermis
to perceive changes in the direction of the incident light.

In all such special instances the plant makes use of the optical
property of total reflection. A very common arrangement (occurring, for instance, in *Banisteria splendens* [Fig. 263, A], *Aporrhiza
paniculata*, *Anomospermum japurense*, *Cocculus laurifolius*, etc.] is
that in which the outer wall of each epidermal cell is furnished
with so-called **marginal pits** (cf. above, p. 121); while varying considerably in shape, these pits invariably extend obliquely upwards and
outwards from the cell-cavity. Any light that falls from above upon
the wall surrounding such a pit, inevitably undergoes total reflection.

2 R

Hence, in vertical illumination, there will be a dark patch on the inner
wall beneath every pit. In oblique light these dark patches become
broader on the side of the inner wall facing the source of light, while
they simultaneously diminish in width or vanish altogether on the

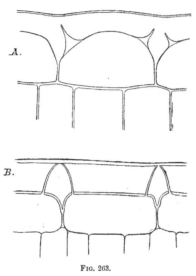

FIG. 263.

*A.* Vertical section through the adaxial foliar epidermis of *Banisteria splendens*,
showing marginal pits. *B.* Vertical section through the adaxial foliar epidermis of
*Hyperbaena laurifolia*, showing marginal fissures.

opposite side. The photic stimulus therefore once more consists in a
change from symmetrical to asymmetrical illumination of the sensory cells.

In a few plants, separate marginal pits are replaced by narrow
continuous **marginal fissures,** which are also directed obliquely outwards
(*e.g.* in *Abuta concolor, Hyperbaena laurifolia* [Fig. 263, B] and
*Anomospermum reticulatum*); in this way continuous reflecting surfaces
arise which produce differences of illumination in vertical and oblique
light identical with those caused by papillose protrusion of the outer
epidermal walls. Similar optical effects result if the radial walls are
twisted (*Hoya carnosa, Maranta setosa*).

### 3. *Localised organs of light-perception (ocelli, etc.).*

In the majority of diaheliotropic leaves the entire upper epidermis
is developed as a light-perceiving sensory epithelium. Not infrequently,
however, the margin, or some other circumscribed portion of the leaf-
surface, is specially adapted for the performance of this function
(e.g. *Tropaeolum majus, Campanula persicifolia, Begonia Rex*). A still

more far-reaching division of labour results in the development of optical sense-organs, composed of one or more highly specialised cells; where these cells differ anatomically from the ordinary epidermal elements, they may be distinguished by the name of **ocelli**.

In *Dioscorea quinqueloba* the ordinary epidermal cells have only slightly curved outer and inner walls; interspersed among them are large papillose cells, which may be solitary or collected into small groups. The papillae are conical with rounded tips; at the apex of each papilla the wall is specially thickened, sometimes to the extent of becoming biconvex. The lens-experiment convincingly proves, that the papillose cells condense light much more effectively than the ordinary epidermal elements.

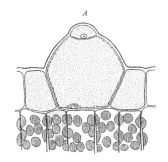

The best illustration of highly specialised ocelli is furnished by the leaf of *Fittonia Verschaffelti* (an Acanthaceous plant from Peru). Here the ordinary epidermal elements, which have flat outer walls, are accompanied by a considerable number (about 120-200 per sq.mm.) of much larger cells with strongly protruding outer, flat inner, and sloping lateral walls (Fig. 264). Each of these large cells is surmounted by a minute biconvex lens-cell, with a very convex outer and more gently curved inner wall. The contents of both cells are perfectly

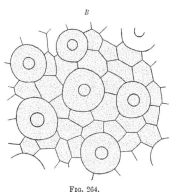

FIG. 264.

Ocelli from the adaxial surface of the leaf of *Fittonia Verschaffelti*. *A*. In vertical section. *B*. In surface view.

transparent, but the refractive index is slightly higher in the case of the small cells. In several respects this two-celled optical apparatus recalls the simple orientating eyes or ocelli of certain animals. The lens-experiment shows very clearly that the structure of the ocelli is such as to produce a very well-marked differential illumination of the inner walls in the larger of the pair of cells of which they are composed, the contrast between the bright central area and the dark marginal region being very marked indeed.

Ocelli of a very similar type occur also on the leaves of *Impatiens Mariannae*. Both here and in *Fittonia* they must be regarded as modified hairs.   In certain species of *Peperomia*, which have strikingly euphotometric leaves, the principal light-condensing organs are the large vesicular basal cells of special short trichomes.

4. *Experimental observations on the light-perceiving function of the upper epidermis of leaves.*[325]

The author himself has subjected his theory of the light-perceiving function of papillose (upper) epidermal cells to the test of experiment in various ways.   All the methods employed by him are the same in principle.   First, the sensory cells are artificially put out of action, so far as their light-perceiving function is concerned ; then, the petiole having been darkened, steps are taken to determine whether the leaf retains its power of perceiving the direction of the incident light and of taking up its heliotropically fixed position.

Assuming that the refractive index of the cell-sap is roughly equal to that of water, it should be possible to inhibit the light-perceiving activity by wetting the upper surface of the leaf ; for this purpose the leaf may be completely immersed in water, or its upper side may be moistened and covered with a sheet of mica, so as to produce a flat liquid surface.   It is clear that a total inhibition of the lens-action can only be effected by such means, where the refractive index of the cell-sap is not higher than that of water, where there are no highly refractive cutinised, silicified or wax-impregnated regions of the outer epidermal walls to act as condensing lenses, and where the inner walls do not bulge towards the mesophyll.   Even if all these conditions are fulfilled, reflection of the incident light often produces differential illumination of the inner walls, when the epidermis is wetted, although the contrast between bright and dark regions is not so striking as it is under normal conditions.   In these circumstances, a sufficiently sensitive leaf may adjust itself to changes of illumination, even when the normal lens-action of its sensory cells is inhibited.   Less sensitive leaves, however, lose their power of perception, and consequently become incapable of executing heliotropic movements, when the upper epidermis is wetted.   In order, therefore, that the results of experiments with wetted leaves may be correctly interpreted, it is, above all, necessary that the degree of sensitiveness of the leaves employed, towards differences of light-intensity, should be accurately known. The author has accordingly carried out experiments with the aid of incandescent gas-light, reflected from two mirrors of equal size placed on opposite sides of the plant or organ which it is desired to test ; by this method the following values were obtained for the minimum difference of light-intensity perceived by certain plants :

*Capsella Bursa pastoris* (inflorescence-axis),   -   -   $\frac{7.5}{}$

*Brassica Napus* (hypocotyl),   -   -   -   -   $\frac{1}{42}$

*Ipomoea purpurea*   „   -   -   -   -   $\frac{1}{38}$

*Avena sativa* (cotyledonary sheath),   -   -   -   $\frac{1}{38}$

*Lepidium sativum* (hypocotyl),   -   -   -   -   $\frac{1}{26}$

The corresponding value in the case of the human eye is originally about $\frac{1}{100}$, but rises in the course of life to $\frac{1}{30}$. It may thus be assumed, that the more sensitive types of plants are possessed of a capacity for perceiving differences of light-intensity which is not inferior to that of the human eye. It follows, that any local differences in the illumination of the inner epidermal walls that are detected by an observer, in such cases, are likewise within the range of perception of the sensitive protoplasm.

In view of the measurements that have just been cited, it was not to be expected that the experiments upon wetted leaves, carried out by the author and other physiologists, should give similar results with different plants. In the case of young plants of *Begonia semperflorens*, leaves with their upper surfaces wetted are quite incapable of perceiving the direction of the incident light; consequently, when such leaves are obliquely illuminated, they do not make the slightest attempt to return to their heliotropically fixed position. In quite a number of instances on the other hand, wetted leaves do return to their normal position, though not so rapidly as they do when dry; wherever this is the case, it will be found that the inner epidermal walls are curved, or else that reflection of light produces a differential illumination of the inner wall analogous to, though less pronounced than, that which prevails under normal conditions.

It is evident that such crude experiments with wetted leaves do not provide a satisfactory means of testing the importance of the light-condensing action of papillose epidermal cells, in relation to the perception of light by diaheliotropic leaves. For this reason the author has carried out experiments of a somewhat different kind upon leaves of *Tropaeolum majus*. A portion of the leaf surface is wetted and covered with a sheet of mica, the rest remaining dry; the two regions of the leaf are separated by a screen of black paper. The petiole having been darkened in a suitable manner, the dry and wet portions of the lamina are illuminated obliquely from opposite sides. The result is always the same; the leaf moves towards the light that falls upon the *dry* surface. This experiment succeeds even if the wetted area is from 2·2 to 4·8 times as large as the dry, both being equally illuminated, or if the two areas are equal, but the wetted one is caused to receive about twice as much light as the dry one. This improved " wet-leaf experiment," therefore, conclusively shows that the direction of the responsive movement is determined, under the prevailing conditions, by the light-condensing capacity of the cells of the upper epidermis. The "lens-action" of these cells is the deciding factor for one of two reasons: either it is actually responsible for that differential illumination of the inner epidermal walls which represents the first link in the chain of processes included under the heading of light-perception; or else it accentuates a contrast in illumination, already ensured by other structural peculiarities, to such an extent that the resulting stimulus is enormously intensified. In a certain sense, therefore, light-condensing epidermal papillae may be regarded as "optical stimulators." [326]

### C. EYE-SPOTS.[327]

A long familiar feature of many Flagellates (e.g. the EUGLENACEAE) of certain PERIDINEAE, and of the swarm-spores of most Green Algae, is the presence of a so-called **eye-spot** (or "stigma"); it is probable that this structure plays a prominent part in connection with the perception of light by the aforesaid organisms.

The eye-spot is a well defined cell-organ; it is discoid, lenticular, or rod-shaped, and is rendered very conspicuous by its red or brown coloration. Among Flagellates it consists—according to Klebs, Schilling and Francé—of a protoplasmic stroma, in which granules of colouring matter are embedded; in *Volvox* and in the swarm-spores of CHLOROPHYCEAE, on the other hand, it appears, from Overton's description, to be an entirely homogeneous structure. There is little reason to suppose that the colourless inclusions of the eye-spot (granules of starch or paramylum), observed by Francé, act as light-condensing organs. Sometimes the eye-spot is located in the polioplasm, not infrequently in close relation to a chromatophore (CHRYSOMONADINEAE, *Cryptoglena*, *Draparnaldia*). In other cases it is situated in the ectoplast, or rather represents a local modification of the external plasmatic membrane; this condition is exemplified by the EUGLENACEAE, in which group, moreover, the eye-spot abuts immediately against the principal vacuole. In *Cladophora laetevirens*, according to Strasburger, it represents an externally projecting local thickening of the ectoplast, underlain by a lenticular cavity in the cytoplasm, which is filled with some homogeneous substance.

There are various grounds for attributing a light-perceiving function to eye-spots. The location of these organs at the anterior ends of motile cells agrees with this view; it is, moreover, a significant fact that all Flagellates, PERIDINEAE and swarm-spores which possess eye-spots are also phototactic, though, on the other hand, some Flagellates and PERIDINEAE are devoid of eye-spots, and yet respond very actively to photic stimulation. It is also worth noting that in *Volvox*, according to Klein, brightly-coloured eye-spots only occur in the cells surrounding the anterior pole of the colony.

The only experiments so far carried out regarding the functions of eye-spots, are those of T. W. Engelmann. This observer states that if an individual of *Euglena viridis*, swimming in bright light, is partially shaded, cessation and subsequent reversal of the movement only takes place when the colourless anterior end—which contains the eye-spot— enters a dark region. " In *Euglena*, therefore, the light-perceiving faculty is strictly confined to the colourless anterior end of the cell." In this case, however, the response may begin before the eye-spot itself is darkened. The actual perceptive power must, therefore, be vested in the colourless cytoplasm, or in some portion thereof, perhaps, as Wager suggests, in the bi-convex thickened region of the flagellum, which lies just in front of the eye-spot. Similarly, in the case of swarm-spores, it is most probably the colourless cytoplasm in the immediate vicinity of the eye-spot which is responsible for the actual perception of light. The eye-spot itself is, in fact, in all probability nothing more than an auxiliary apparatus. The most obvious suggestion regarding the functions of the eye-spot is that it acts as a **light-screen**, which prevents the sensitive protoplasm from being illuminated equally on all sides, and which thus helps to indicate the direction of the incident light; if this view is correct, the eye-spot might be compared to the pigmented layers of animal eyes.

# CHAPTER XIII.

## STRUCTURES SERVING FOR THE TRANSMISSION OF STIMULI.

### I. GENERAL CONSIDERATIONS.[323]

BOTH in animals and in plants response does not follow immediately upon stimulation. The two processes are always separated by a number of intermediate changes. As already stated, the whole series may be collectively termed a chain of stimulation (*Reiz-Kette*); such a chain always begins with an act of perception and ends with the reaction or response induced, the most important of the connecting links being the **transmission of the stimulus** between the perceptive and the reactive regions of the cell or organ. Transmission of stimuli must occur, even where perception and response take place within the same cell; for the same part of the protoplasm, in a unicellular organism or in an individual cell of a higher plant, cannot simultaneously perform both these functions. When the sensory- and motor-organs are more widely separated in space, transmission of stimuli takes place over greater distances, and hence becomes a still more conspicuous feature of the chain of stimulation. The tactile sensory epithelium of the stamens of *Berberis* immediately adjoins the motor-tissue of the filament. Here the transmission of stimuli merely consists in a transference of the excitation aroused in the sensitive protoplasm to the adjacent motor-tissue by mechanical contact. In an ordinary root, the statolith-apparatus in the root-cap is separated by a considerable amount of meristematic-tissue from the sub-apical region that executes geotropic curvatures. Stimuli have to be transmitted over still greater distances in the leaves and stems of *Mimosa pudica*, where a single excitation may gradually extend to all the leaves on a shoot, or even over the whole plant. The three preceding examples sufficiently illustrate the diversity that exists in respect of the distances over which external stimuli may be propagated.

In considering whether special histological arrangements are required for the conduction of stimuli, it is necessary to take into account the **velocity of transmission.** The rate of propagation of stimuli is in general much lower among plants than it is in the case of animals. According to Czapek, Rothert and others, heliotropic and geotropic stimuli require five minutes in order to traverse a distance of one or two millimetres.[329]   Traumatic stimuli travel much more rapidly; thus, Kretschmar [330] states that the traumatic stimulus produced by cutting across a leaf or stem in certain water-plants (*Vallisneria, Elodea, Hydrocharis*), is propagated in the basipetal direction, under favourable conditions, at a rate of 1–2 cm. per min., judging by the starting of protoplasmic rotation, which constitutes the primary response. Fitting[331] has shown that an even more rapid transmission (1–2 cm. per *sec.* in *Passiflora coerulea*) takes place in the case of the traumatic stimulus which causes certain tendrils to curl up at the tip when they are cut off.   In the case of *Mimosa pudica*, finally, Dutrochet, Bert and the author all agree in estimating the maximum velocity of transmission at 1·5 cm. per sec. as regards the primary petiole, the corresponding value for the stem being usually somewhat smaller; Linsbauer, however, arrives at much higher figures for *Mimosa*—viz., 30–100 mm. per sec. *at least.*   Even the greatest of these velocities falls far behind the average rate of transmission of stimuli through the nervous system of animals, which amounts to at least 30,000 mm. per sec.   As a matter of fact, lower velocities are known to occur in the animal kingdom also; thus, Engelmann states that stimuli are transmitted through the heart-muscles, without the aid of nerve-fibres, at rates varying between 6·4 and 177 mm. per sec.   In any case, it may be confidently asserted that the actual velocity of transmission in any given case entirely accords with the nature of the requirements which the resulting response is intended to satisfy.

In approaching the question as to the existence of special arrangements for the transmission of stimuli in plants, it is further necessary to consider the various ways in which such transmission may be effected.

In the first place, stimuli may be transferred from cell to cell quite independently of any special transmitting arrangements; this statement applies more particularly to shock-stimuli, which can be propagated by purely mechanical means.   In *Mimosa pudica* the contraction of a single stimulated cell, or group of cells, in the irritable lower half of the pulvinus leads to a deformation of the neighbouring cells, whereby they are stimulated in their turn.   Pfeffer has already pointed out that a locally applied stimulus may spread in this way through the entire sensitive tissue of the pulvinus.

Where a shock- or wound-stimulus produces differences of hydro-

static pressure within a special system of tubular structures, transmission can also be effected in a purely mechanical manner through the
equalisation of these pressure-differences.   As will be shown later on
in detail, stimuli are actually transmitted in this way in *Mimosa
pudica* through the leaf-blades, petioles, and internodes from one
pulvinus to another.

Another way in which a stimulus may be propagated, is by the
diffusion of special substances—by osmosis from cell to cell—from
the region first stimulated.   If, however, the effects of the so-called
" formative " stimuli—which lie outside the scope of the present work—
are left out of account and attention is restricted to motor stimuli, it
will be found impossible to cite a single instance in which mechanical,
chemical, geotropic or heliotropic stimulation undoubtedly leads to the
formation of special substances that propagate the stimulus by their
diffusion.

It is probable that transmission of stimuli most often depends
upon the dissemination of some as yet incomprehensible excitation
or state of motion, or, to use more general terms, upon the propagation
of some physical or chemical change which takes place in the living
protoplasts (or in special local extensions thereof).[332]   The nervous
system of an animal represents a tissue-system specially adapted for
this mode of transmission.   Among plants, where physiological division
of labour has not progressed so far, there seem to be no stimulus-
transmitting organs analogous to the nerves of animals.   Here, on the
other hand, the power of transmitting stimuli is common to every
living tissue, provided that its component protoplasts are not completely isolated from one another by cell-walls but remain connected by
protoplasmic connecting threads.   There can be no doubt that the
protoplasmic connections are responsible for the transmission of stimuli
from cell to cell; the comparison so frequently instituted between these
structures and the nerves of animals is therefore perfectly legitimate.

## II. PROTOPLASMIC TRANSMISSION OF STIMULI.

That mode of stimulus-transmission  which is directly dependent
upon changes within the living protoplasm, may be termed **protoplasmic
transmission** in order to distinguish it both from the purely **mechanical**
and from the **osmotic methods.**   Where the stimulus is not merely transmitted  across  a  single  protoplast,  but  is  propagated—as  almost
invariably happens among higher plants—through a whole series of
cells, every path of transmission is made up of sections of two
distinct kinds, which alternate regularly with one another, the
protoplasts contained in the cell-cavities being responsible for **intra-**

cellular transmission, while the protoplasmic threads connecting adjoining protoplasts with one another represent the paths of intercellular transmission.

The differentiation of sensory and motor nerve-fibres which characterises the paths of stimulus-transmission in the animal organism, is correlated with the presence of a central nervous organ or reflex-centre. This organ is absent—or at any rate has so far remained undiscovered—in the vegetable kingdom. Hence, sensory and motor paths of transmission are not distinguishable in plants, although there must of necessity be a point somewhere in every path of transmission, at which the state of excitation produced by the stimulus becomes transformed into the impulse that initiates the responsive movement.

### A. INTERCELLULAR TRANSMISSION.

The protoplasmic connecting threads discovered by Tangl[333] (termed *Plasmodesmen* by Strasburger) generally take the form of extremely delicate protoplasmic filaments, which traverse the whole thickness of the cell-walls and thus place the protoplasts of adjoining cells in direct communication with one another (cf. above, p. 45). As a rule large numbers of these threads traverse the closing-membranes of pits (Fig. 265A); in the cortical parenchyma of *Viscum album*, for example, Kuhla often counted more than twenty connecting threads in each closing-membrane. Less frequently, the more or less thickened unpitted regions of the walls are perforated by solitary connecting threads. The two modes of distribution of the threads are, however, not sharply separable from one another; for the same cell-wall may be traversed both by groups of connecting threads and by isolated protoplasmic filaments.

It is highly probable that connecting threads consist of the same hyaloplasm that composes the outer plasmatic membrane or ectoplast; they may in fact be regarded as prolongations of the ectoplast. Evidence in favour of this view, which was first put forward by Noll, has been accumulated by Strasburger; this author points out that the diameter of the connecting threads, which are notoriously very slender, never exceeds the thickness of the ectoplast, so far at any rate as the latter can be measured; further, ectoplast and connecting threads are not only directly continuous with one another, but also consist of the same substance.

As regards the origin of the protoplasmic connecting threads, Strasburger has shown that they are not, as might be supposed, persistent spindle-fibres, but secondary formations, interpolated after the cell-wall has come into being, although they appear at a very early stage, before the deposition of the secondary thickening layers.

At the present time, the majority of investigators incline to the belief that the closing-membranes of all the pits developed between living cells of the plant-body are perforated by protoplasmic connections, while many hold the opinion that all the living cells of a plant are linked together by means of connecting threads, so as to form a single protoplasmic entity or "symplast"; but Kuhla's work on *Viscum album* represents the only attempt that has been made so far, to prove the truth of this assumption, in the case of a highly organised plant. Attention may therefore be drawn to the fact that, in all the cases hitherto examined, organs engaged in the perception of external stimuli are connected by means of protoplasmic threads with the associated reactive region or motor-tissue. Thus, Pfeffer[334] has demonstrated the presence of these structures in the parenchymatous tissues of tendrils, in which transmission of stimuli takes place, though not indeed over any great distance; connecting threads also occur in the walls separating the epidermal sensory cells of tendrils from the hypodermal parenchyma. The author has observed that the thickened inner walls of the sensory cells in the staminal filaments of *Berberis vulgaris*[335] are furnished with fairly numerous shallow —often very ill-defined—pits, the comparatively thick limiting membranes of which are traversed by protoplasmic threads (Fig. 236 A). Each sensory cell is thus able to transmit the excitation set up by the initial stimulus to the adjacent motor-tissue, the walls of which are also liberally provided with pits and protoplasmic connections. In the case of *Aldrovandia vesiculosa*,[336] the lower transverse walls of the sensory cells in the tactile hairs are devoid of pits; they are, however, relatively thin as regards the portion which lies next the central axis of the hair. By suitable treatment, the author has succeeded in demonstrating the presence of a small number of protoplasmic threads in the thin portions of these walls; as a rule, there are but two or three such threads in each wall, or even only a single one. Where the "hinge" consists of four cells, therefore communication with the neighbouring protoplasts is maintained with the aid of from four to twelve connecting threads. In the case of *Dionaea muscipula*, protoplasmic connecting threads have been found by

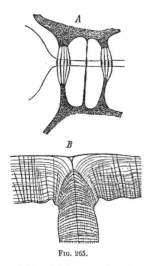

Fig. 265.

*A.* Protoplasmic connections between two cells of the mesophyll of *Viscum album* (after Kienitz-Gerloff). *B.* Curved protoplasmic connections in one of the radial walls of the aleurone layer of *Zea Mais*, after treatment with iodine and dilute sulphuric acid.

Macfarlane and by Gardiner, both in the tactile bristles and in the general mesophyll. Gardiner has also noted the occurrence of these structures in the lower, or sensitive half of the pulvinus of *Mimosa pudica*; they seem, however, to be quite as plentiful in the upper, or insensitive half of the pulvinus. According to Czapek and A. W. Hill,[337] root-tips, which are concerned in the transmission of geotropic, hydrotropic and perhaps also heliotropic stimuli, contain an unusually large number of connecting threads; moreover, the manner of their distribution in these organs distinctly supports the notion that the threads are responsible for the conduction of stimuli.

Direct experimental demonstration of the stimulus-transmitting function of protoplasmic connecting threads is naturally a matter of the greatest difficulty. Townsend,[338] working with Moss-protonema filaments and with hairs, has shown that the nucleus of one cell can induce an isolated non-nucleated mass of cytoplasm, contained in a neighbouring cell, to surround itself with a cell-membrane, if the two protoplasmic bodies are connected by protoplasmic filaments. While in this case it is not an external stimulus that is propagated, it is reasonable to suppose that the transmission of nuclear "impulses," and of internal stimuli in general, does not differ in principle from the propagation of external stimuli. Strasburger makes use of the fact that the majority of the protoplasmic connections are retracted or broken when a cell is plasmolysed, and fail to be regenerated when it is allowed to resume the turgescent condition. A root or stem is first completely plasmolysed by immersion in a suitable salt-solution; it is then thoroughly washed, allowed to become turgescent again, and finally tested with regard to its power of executing geotropic curvatures. Organs treated in the manner described are found to be no longer capable of geotropic response. It is certainly *possible* that this result is partly due to the rupture of the stimulus-transmitting protoplasmic connections. But it is just as likely that plasmolysis destroys the perceptive capacity or upsets the mechanism of curvature, or interferes with some other link in the chain of stimulation. Strasburger's experiments cannot therefore be regarded as conclusive, as he himself indeed admits.

It may be presumed that the protoplasmic connecting threads do not confine themselves to the propagation of external stimuli, or of the excitations produced by the latter, but that they are also concerned in the transmission of those internal stimuli which play so important a part in the mutual relations of different tissues and organs. It has, in fact, already been mentioned, that the influence of the nucleus upon the formation of cell-walls can be transmitted from one cell to the other by means of protoplasmic connections.

Whether one and the same connecting thread is capable of conducting different kinds of stimuli, or whether a division of labour prevails in this respect among the various protoplasmic connections which link up adjoining protoplasts, must, of course, remain an unsolved problem for the present. On the other hand, the question as to whether the protoplasmic connections serve for the conduction of different stimuli in different tissues, is open to experimental investigation.

The fact that protoplasmic connecting threads serve for the transmission of stimuli, does not exclude the possibility that these structures may in certain cases be partly or entirely engaged in translocation of plastic materials. As a matter of fact, protoplasmic connections were first discovered in the storage-tissues of certain endosperms, where they are often very well differentiated; in such cases it is quite possible that they are *principally* concerned with translocation.

### B. INTRACELLULAR TRANSMISSION.

Even if the protoplasmic connecting threads are regarded as the intercellular stimulus-transmitting paths, it is still necessary to account for the transmission within the individual protoplast. *A priori* several possibilities suggest themselves in this connection. In the first place, it is conceivable that the entire protoplast is endowed with the capacity for conducting stimuli; secondly, the faculty in question might be restricted to the ectoplast; finally, special fibrillar structures might be developed for the purpose. The last-mentioned possibility has been recently discussed at length by Němec.[339] Max Schulze, Apáthy, Bethe and others have shown that the nervous tissue of animals contains a system of delicate fibrillae, which are regarded by these investigators as the actual paths of transmission, on account of their unbroken continuity, and for other reasons. Starting from the observations and conclusions of the aforesaid animal physiologists, Němec succeeded in finding very remarkable fibrillar structures in the root-tips of a variety of plants (*Allium Cepa, Hyacinthus orientalis, Iris germanica, Cucurbita, Pisum, Aspidium*, etc.), more particularly in those rows of plerome-cells which later become converted into hadrome-vessels. The elements in question are traversed by longitudinal protoplasmic strands, which extend from one end of the cell to the other, and which correspond on the opposite sides of each transverse wall. In stained microtome-sections of suitably fixed material, and sometimes even in the living condition (*Allium Cepa*), these strands display a distinct fibrillar structure. The fibrillae form a reticulum, which, as already explained, extends from one transverse septum to the next. They do *not*, however, traverse the walls, nor are they anywhere continuous with protoplasmic connections; the fibrillar systems of different cells are thus not directly connected with one another. This circumstance alone throws grave doubts upon the validity of Nĕmec's assertion that these protoplasmic structures represent special paths of transmission analogous to the neuro-fibrillae of animal nerves. Moreover, the author has since demonstrated that protoplasmic streaming takes place within the fibrillar protoplasmic strands of the above-mentioned large plerome elements, and that the longitudinally fibrillar structure of these strands is due to the same morphological and physiological conditions that produce the longitudinally fibrillar structure previously described in connection with other forms of streaming protoplasm. It appears, in fact, that streaming protoplasm in general consists of elongated fibrous or lamellar plasmatic masses separated from one another by long and exceedingly narrow vacuoles.

If intracellular transmission of stimuli really depended upon the presence of special protoplasmic fibrillae, these structures ought to be present above all in those organs which are characterised by a high rate of transmission; they ought, therefore, to be readily observable in the cells that are interpolated between such sense-organs as the tactile pits or papillae of tendrils and irritable stamens, or the tactile hairs and bristles of *Centaurea*, *Aldrovandia* and *Dionaea* on the one hand, and the associated motor-tissues on the other. With regard to this point, however, the author's investigations have invariably led to negative results.

Intracellular transmission must, therefore, either be a function of the entire protoplast, or it must be principally located in the ectoplast. The latter seems the more probable alternative, especially in view of the fact that the protoplasmic connecting threads, which are held responsible for intercellular transmission, are merely local extensions of the ectoplast. It does not necessarily follow that the protoplasmic strands which traverse the sap-cavity are never utilised for transmission; indeed, where the nucleus is suspended near the centre of the cell, it is difficult to see how the impulses which it sends forth can travel towards the periphery of the cell by any other means. It is therefore quite possible that stimuli are actually propagated along the fibrillar protoplasmic strands observed by Nĕmec in root-tips; this possibility does not, however, justify us in assuming that these fibrillar strands are structures specially set apart for the work of transmission.

The question finally arises, as to how far the velocity of transmission depends upon the form of the cells in which transmission takes place. If stimuli travel in the ectoplast, or indeed in any other part of the peripheral cytoplasm, any elongation of the cells in the direction of transmission will naturally reduce the distance to be traversed, and hence increase the velocity of transmission; for every decrease in the number of transverse septa diminishes the loss of time caused by the relatively slow transmission through the protoplasmic connecting threads. The conditions are, of course, even more favourable for rapid transmission where the transverse septa are oblique, or in other words, where the transmitting elements are prosenchymatous. Stimuli might consequently be expected to travel more rapidly along vascular bundles than through parenchymatous tissues. As a matter of fact, Frank found that, in *Elodea canadensis*, the rearrangement of chloroplasts induced by a transmitted traumatic stimulus begins sooner in the vicinity of the midrib, than it does in other parts of the leaf. Kretschmar likewise states that transmission takes place most rapidly in the vascular bundles, in a number of water-plants. In a leaf of *Vallisneria* which was severed at a distance of 4 cm. from the apex, the traumatic stimulus that induces protoplasmic streaming was found by Kretschmar to travel to the apex in ten minutes along the midrib, and in twelve minutes along the two lateral veins; twenty-six minutes elapsed before all the mesophyll cells were engaged in streaming, and it was evident that the stimulus reached the mesophyll principally through the veins. Němec states that the traumatic stimulus which leads to a rearrangement of the protoplasm in the root-tip of *Allium Cepa* (cf. above, p. 23), is propagated most rapidly in the inner layers of the periblem and in the plerome. It is therefore quite to be expected that vascular tissues should be employed for the conduction of stimuli, where ecological considerations demand a high velocity of transmission. As a matter of fact, Batalin [340] states that stimuli applied to the tactile bristles of *Dionaea muscipula* travel towards the midrib mainly along the vascular bundles. In the case of *Biophytum sensitivum*,[341] the author's own observations render it probable that both mechanical and traumatic stimuli are likewise propagated along the vascular strands of the pinnae and rhachis. Transmission of stimuli in vascular bundles, in so far as it is effected by protoplasts and protoplasmic connections, must take place in the parenchymatous elements of the hadrome- and leptome-, and perhaps also in the living contents of the sieve-tubes, while in the case of leaves, the cells of the parenchymatous bundle-sheaths may also be utilised for this purpose.

The greater velocity of transmission in vascular bundles is perhaps not altogether due to the prevailing elongation of the component cells; it is possible that the protoplasmic paths of transmission are here endowed with a special molecular or micellar structure which renders them more efficient in this respect than those of the parenchymatous ground-tissue.

Where there is no special need for rapid propagation of stimuli, the transmitting capacity of the general parenchyma may be quite sufficient. In the cotyledonary sheath of *Avena sativa*, for example, heliotropic stimuli are propagated, according to Rothert,[342] with sufficient speed and without appreciable loss of intensity, even when both the vascular bundles are severed. The cases of *Mimulus* and *Martynia*, described by Oliver,[343] on the other hand, do not form real exceptions to the rule. Here the very rapid transference of contact-stimuli from one stigmatic lobe to the other takes place, not in the vascular bundles, but in the parenchyma. This so-called parenchyma, however, which consists of elongated cells, is really the motor-tissue; but every motor-tissue which executes rapid movements must be capable of transmitting stimuli with a corresponding velocity by means of ultra-microscopic, specialised, protoplasmic structures.

### III. *TRANSMISSION OF STIMULI IN MIMOSA PUDICA.*

At the present time we only know of a single instance, in the entire vegetable kingdom, in which a special tissue is set apart for the transmission of stimuli. This stimulus-transmitting system, which occurs in *Mimosa pudica,* was discovered and has been fully investigated from the anatomico-physiological standpoint by the author.[344] It should be pointed out at once, that the tissue in question is in no sense comparable to the nervous system of an animal, but merely serves as the pathway for a purely mechanical propagation of stimuli.

If a single pinnule of a sufficiently irritable plant of *Mimosa pudica* is stimulated by means of a gentle blow, the responsive movement takes place almost as soon in the opposite pinnule as in that which was directly stimulated, while the remaining pinnules of the same pinna fold up very shortly afterwards. A more intense local stimulus, such as that produced by wounding a pinnule, is transmitted to a greater distance. In this case, it is the rule for the main pulvinus to collapse first of all, whereupon the pinnules on each of the other leaflets fold up in acropetal succession. If the plant is in a highly irritable state, the pulvini of the sub-petioles may also become affected, in which case the stimulus is propagated through the stem to the neighbouring leaves. A very violent stimulus, such as results if a portion of a leaf is scalded or singed, may even be transmitted through the whole shoot-system of a moderate-sized plant.

Dutrochet long ago proved by experiment, that the transmission of stimuli through the stem of *Mimosa pudica* takes place neither in the cortical parenchyma nor in the pith, but only in the fibro-vascular system (according to Dutrochet the woody cylinder); the same observer further suggested that the transmission might depend upon movements of liquids contained in the conducting elements. As a matter of fact, if a sufficiently deep incision is made in the stem, a drop of liquid does exude from the wound, just before the neighbouring leaves perform

their characteristic movements.  The appearance of this drop of liquid
—believed by Sachs, Pfeffer and others to be a drop of water derived
from the woody cylinder—has been quite correctly regarded as a
phenomenon causally related to the transmission of stimuli.

In order to eliminate any influence which the living protoplasts
(including their intercellular connections) may exert over the process
of transmission, Pfeffer anaesthetised the middle portions of petioles
by means of chloroform or ether; he found that, under these conditions, traumatic stimuli never failed to traverse the narcotised

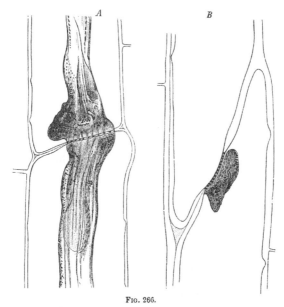

Fig. 266.

Stimulus-transmitting cells from the petiole of *Mimosa pudica* in L.S.  The contracted protoplasts adhere firmly to the limiting membranes of the large pits in the
transverse septa.   Sections  treated  with  dilute  sulphuric  acid  and  stained  with
picric-aniline blue.

region, while mechanical stimuli were also transmitted in a certain
number of cases.   It has since been shown by the author himself, that
traumatic stimuli continue to be propagated through parts of the petiole
which have been killed outright by scalding.   It follows, therefore, that
in *Mimosa pudica*, transmission does not depend upon the presence of a
continuous protoplasmic system, but consists in a disturbance of
hydrostatic equilibrium within the transmitting elements.   A disturbance of this kind can obviously travel through dead or anaesthetised
tissues; and there is no reason to suppose that the mechanism of transmission in the normal condition of the tissues is radically different from
that which prevails in scalded or narcotised organs.

The stimulus-transmitting elements of *Mimosa pudica* are not situated, as Dutrochet, Sachs and Pfeffer, among others, have supposed, in the woody cylinder, or in the hadrome portions of the primary bundles, but on the contrary occur in the leptome-strands, where they take the form of elongated tubular cells arranged in longitudinal series. In the main petioles and stems, these cells attain a length of ·6–1·2 mm., and an average width of ·018 mm. Their membranes are thin, soft and colourless, and give a deep violet reaction with chlor-zinc iodine. The longitudinal walls are well provided with pits. Each of the transverse septa, which may or may not be oblique, bears—not necessarily at its centre—a single wide circular pit, the closing-membrane of which is perforated by minute pores containing protoplasmic threads (Fig. 266). The cell-contents comprise a thin peripheral layer of cytoplasm, a large rounded or somewhat elongated nucleus, and cell-sap. The composition of this cell-sap is of interest; it contains, in addition to a gummy or mucilaginous substance, a large amount of a glucoside or glucoside-like compound in solution,—which is responsible for the intense reddish-violet reaction with ferric chloride and the reddish coloration with ferrous sulphate—and also suspended granules of a resinous character.

These stimulus-transmitting cells form a continuous system of tubes, which accompany the leptome-tissue throughout the stems, main petioles (Fig. 267), sub-petioles and pinnules; consequently they are to be found in the pulvini of petioles and pinnules, where they are separated from the motor-tissue in the sensitive half of the pulvini by abundantly pitted collenchymatous elements. In the adaxial marginal bundle of each sub-petiole, the transmitting elements form short-celled "nodal points" between the two members of each pair of pinnules; at these points stimuli can travel transversely between opposite pinnules, as well as along the sub-petiole. In the root-system, the transmitting cells are confined to the central cylinder in the main root, and are entirely absent from the lateral roots.

The effects of incision (see above, p. 641) show that stimuli are actually propagated in this system of highly turgescent tubes, and that the mode of transmission is a hydrodynamic one. If, namely, one or more of the tubular cells are laid open by an incision in the stem or petiole, their cell-sap instantly escapes in the form of a drop of transparent liquid; immediately afterwards the nearest pulvinus carries out its characteristic movement. The micro-chemical reactions of this drop of liquid—more especially the violet coloration with ferric chloride—prove that it is not derived from the woody cylinder, which indeed need not be wounded in order to produce the aforesaid consequences. The liquid that escapes is, in fact, not water at all, but *cell-sap* ejected from the stimulus-transmitting elements. The large size of the drop shows that this cell-sap can easily pass through the pores in the wide pit-membranes of the transverse septa. The rows of tubular cells therefore represent a continuous system, through which variations of hydrostatic pressure can be transmitted over considerable distances.

In the author's opinion, transmission of stimuli in this system of tubes takes place somewhat as follows: When the pulvinus of a pinnule moves upwards in response to a shock, pressure is exerted

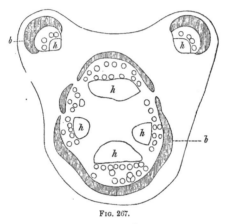

Fig. 267.

T.S. through the main petiole of *Mimosa pudica* ; *b*, stereome; *h*, hadrome.
The small circles indicate the position of the stimulus-transmitting elements.

upon the highly turgescent transmitting cells, partly owing to changes in the form and volume of the relaxed half of the pulvinus, and partly owing to the mechanical effects of the curvature; the local rise of pressure produced in this way will be propagated along the system of tubes, owing to the elastic tension of their walls, in the form of a pressure-wave, like the pulsations which travel through the arterial system of an animal. This wave of compression or positive tension acts like a shock-stimulus upon the nearest pulvinus, and so leads to an indirectly induced responsive movement. The initial rise of pressure which starts the wave is not large; hence the stimulus, while extending from one pair of pinnules to the rest, does not convey a perceptible shock to the comparatively insensitive pulvinus of the sub-petiole, and never penetrates as far as the main pulvinus.

Traumatic stimulation—such as may ,be caused by the severance of a pinnule—instantly destroys the turgor of the injured transmitting cells ; as a consequence a large local fall of turgor results, which is propagated through the tubular system as a wave of relaxation or negative tension. The initial change of pressure caused by such a mechanical injury is comparatively large; hence a much more violent disturbance is produced in the adjacent pulvini, than can possibly arise when the primary stimulus is due to shock. A traumatic stimulus can accord- ingly be transmitted over a relatively large distance; it will not only reach the main pulvinus, but may also travel through the stem to other leaves.

The rise of pressure produced by shock causes the cell-sap of the transmitting elements to travel in the same direction as the wave of compression. Traumatic stimulation, on the other hand, leads to a fall of pressure, and the cell-sap accordingly moves towards the region of reduced pressure, that is, in the direction opposite to that of the wave of relaxation.

It is impossible here to consider in detail, how the shock of the waves of compression and relaxation is transferred at various points to the sensitive parenchyma in the pulvini. It may, however, be noted that variations of pressure must result in alterations in the volume and form of the transmitting tissues, and that these latter changes in their turn affect the two or three layers of collenchyma which surround the transmitting tissue in the pulvini, and so indirectly bring about sudden deformation of the neighbouring sensitive parenchymatous elements. When the initial stimulus is feeble, the ultimate deformation will presumably be restricted to the immediate neighbourhood of pits, since the exceedingly thin and elastic limiting membranes will respond to the slightest variation of pressure by bulging towards the side of reduced pressure, the sudden protrusion constituting a fresh mechanical stimulus.

The velocity of transmission in *Mimosa* is naturally affected by a variety of circumstances, but depends, above all, upon the amount of friction between the liquid contents and the walls of the tubes, upon the resistance that has to be overcome during the filtration of the liquid through the transverse septa and upon the turgor-pressure in the transmitting system. According to Dutrochet and Bert, the rate of transmission through leaf and stem of the adult plant varies between 2 and 15 mm. per sec. Experiments performed by the author have shown that the stem usually transmits stimuli rather more slowly than the leaf, though the two organs do not always exhibit a marked difference in this respect. In a typical case, the velocity of trans- mission was found to be 8·5 mm. per sec. in the petiole, and 6·5 mm. per sec. in the stem. K. Linsbauer,[345] working with a kymograph, has estimated the average velocity of transmission of a traumatic stimulus, produced by making an incision in the primary petiole, at 31·2 mm. per sec. ; when the main petiole is severed altogether, the stimulus travels at the rate of no less than 100 mm. per second. Linsbauer's results stand in need of confirmation, especially as his methods involve the risk of a variety of errors.

If the author's theory of stimulus-transmission in *Mimosa pudica* is correct, a shock or wound-stimulus should only be propagated in organs which contain the special transmitting elements—that is to say, in the leaves and stems, and in the main root. According to the author's observations (which conflict with those of Borzi), section of lateral roots (even where these are of the first order) never provokes any movement of the leaves; this result is significant in view of the fact that the cells to which the transmitting function has been ascribed are entirely absent from the lateral roots.[346]

It may be presumed that the mechanism of transmission which occurs in *Mimosa pudica*, also prevails in the remaining sensitive species of the genus (*M. sensitiva, M. Spegazinii, M. casta, M. viva,* etc.), as well as in the other LEGUMINOSAE which exhibit the same peculiar form of irritability (*Smithia sensitiva, Aeschynomene sensitiva, A. indica, A. pumula, Desmanthus stolonifer,* etc.).

In conclusion, reference may be made to the fact that the stimulus-conducting cells of *Mimosa pudica*, though located in the leptome, do not stand in any direct phylogenetic relation to sieve-tubes; they are, on the contrary, homologous with certain excretory sacs which are of very general occurrence among the LEGUMINOSAE, where they frequently take the form of tannin-sacs associated with the leptome-strands (*e.g.* in *Phaseolus multiflorus, Robinia pseudacacia,* etc.).

# CHAPTER XIV

## SECONDARY GROWTH IN THICKNESS OF STEMS AND ROOTS.

### A. NORMAL SECONDARY THICKENING.

#### I. GENERAL CONSIDERATIONS.

IT is characteristic of all young stems and roots, that the region which is in process of conversion into permanent tissue is much thicker than the portion nearer the apical growing point, in which the primary meristematic layers originate from the primordial meristem. This increase of diameter, which accounts for the conical form of the apical region, is the result of the so-called **primary growth in thickness,** and does not depend upon any *special* growth-process; every cell in any cross-section takes some part therein, though all do not participate to an equal extent.

This primary growth in thickness, or "enlargement" (*Erstarkung*), as it may also be termed, determines the average diameter characteristic of the several organs of any given species, in their primary condition. When this thickness has been attained, the organ begins to grow in length. In the majority of Monocotyledons and Vascular Cryptogams, no appreciable growth in thickness takes place, once elongation has ceased and the primary permanent tissues. have become completely differentiated. In most Gymnosperms and Dicotyledons, on the other hand, in the Palms, PANDANACEAE and arborescent LILIACEAE, and in a few Vascular Cryptogams, the stems, and, as a rule also, the roots, are capable of further growth in thickness, even after longitudinal extension has come to an end and tissue-differentiation is completed.

According to Eichler and Barsickow,[347] the "secondary thickening" of certain Palm-trunks, which continue to increase in girth for a considerable time, really consists of a protracted "enlargement" of the stem. In such Palms, therefore, the diameter of the trunk increases steadily towards the base. The following measurements may be quoted from Martius: A trunk of the Coconut Palm (*Cocos nucifera*), 22·09 m. in length, had a diameter of ·46 m. at its upper end, and ·74 m. at its base; in the case of *Mauritia flexuosa* the corresponding measurements, for a trunk 25·5 m. in length, were ·52 m. and ·87 m. In these instances the increase of thickness is usually due to enlargement of the individual cells of the parenchymatous ground-tissue, which thus increases in bulk. The fibrous strands associated with the vascular bundles also augment their cross-sectional area, partly by dilatation of their cell-cavities, and partly by continued thickening of their cell-walls. In *Euterpe oleracea*, however, according to Kränzlin, the procambial elements which ultimately give rise to bast-fibres, gradually add to their number by division; the component cells of a given procambial strand are not formed simultaneously, but arise successively in centrifugal order. This multiplication of cells in the developing fibrous strands continues until the trunk has attained a diameter of about 10 cm. Strasburger has noted the formation of secondary vascular anastomoses, linking up primary water-conducting strands, in the peripheral part of the trunk of *Washingtonia filifera*. As a rule, however, the vascular tissues are not affected by the " secondary thickening " which a Palm-stem may undergo.

The above-described very simple type of secondary growth in thickness mainly serves to enhance the inflexibility of the stem; the tall stature of the trunk renders this increase of mechanical strength very necessary in the case of most Palms. Where secondary thickening depends altogether upon dilatation of the primary tissues, the leaf-output—and hence the development of photosynthetic tissue—is just as strictly limited, as it is in the entire absence of secondary growth in thickness. For when the paths along which water and minerals travel cannot be continually enlarged, by the addition of new conducting elements, the organs of transpiration and photosynthesis cannot be multiplied indefinitely. In Palms and other long-lived plants with secondary thickening of this primitive type, the leaves are constantly renewed, but their size and number remain constant after a certain stage.[348]

An unlimited development of the photosynthetic apparatus is, in fact, an impossibility, except where the conducting and mechanical tissues are also capable of indefinite expansion. This condition can only be fulfilled, when secondary growth in thickness consists in a continuous regeneration and amplification of the various tissue-systems of which the stem and roots are composed. In such cases the work of producing new cells is entrusted to a special meristem, which becomes secondarily interpolated among the permanent tissues, and which is termed the

**cambium.** Ordinarily the cambium takes the form of a hollow cylinder, which in transverse section, of course, appears as a complete ring.

## II. SECONDARY GROWTH IN THICKNESS IN GYMNOSPERMS AND DICOTYLEDONS.

### A. THE CAMBIUM.[349]

In the stems of Gymnosperms and Dicotyledons, the primary vascular bundles are intersected by the cambial cylinder at the junction of hadrome and leptome, the former being intra-, and the latter extra-cambial in position; the development of secondary tissue thus continually pushes the primary components of the bundles further apart in the radial direction.

The mode of origin of the cambium varies considerably in different axial organs. Frequently it stands in close relation to the primary

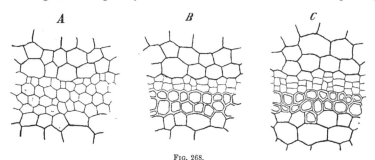

FIG. 268.

Successive stages in the differentiation of the primary procambial cylinder in the stem of *Salvia Horminum* (T.S.) See text.

differentiation that takes place in the apical region, while in other cases there is no connection between the two processes.

In a number of stems the cambium arises from a primary procambial cylinder. A couple of examples will render this statement clearer. A young internode of *Salvia Horminum* contains eight vascular bundles (leaf-trace bundles), a pair being situated in each angle of the four-sided stem. The two strands of each pair soon fuse to form a single compound bundle. At an early stage of development, four strips of procambial meristematic tissue arise between the compound bundles; these gradually extend parallel to the flat sides of the stem, and become connected at either end with the fascicular cambium. The complete primary procambial cylinder formed in this way, however, by no means represents the actual cambium. The interfascicular strips of procambial tissue comprise three or four layers of cells, which are arranged quite irregularly (Fig. 268 A), and not in the radial rows characteristic of a typical cambium. To begin with, these procambial strips give rise to small additional vascular strands. Later, the two or three innermost layers become transformed into mechanical elements; these, together with the stereides previously differentiated in the hadrome-strands of the vascular bundles, constitute a complete intracambial fibrous cylinder (Fig. 268 B). The outermost layer of the interfascicular procambium is ultimately converted into the interfascicular portion of the actual cambium, its cells becoming elongated radially and undergoing division principally by tangential walls (Fig. 268 c). The cambial cylinder is completed, by these interfascicular strips of serial cambium becoming continuous with the corresponding fascicular tissue.

A somewhat more complicated arrangement is found in the young internodes of *Pelargonium gibbosum*. Here, as in the preceding case, the young vascular strands become joined together by strips of primary procambial tissue, within which additional bundles are differentiated. In this instance, however, the procambial strips comprise as many as five to seven layers, and give rise to four distinct tissues, in place of the two produced by them in *Salvia*. The two outermost layers become transformed into fibres. The two or three next in order give rise to narrow conducting-parenchyma cells. The elements of the fifth or sixth layer, by elongating radially and undergoing tangential division, produce a serial cambium which, in conjunction with the fascicular cambium, constitutes the cambial cylinder. The innermost procambial layer, finally, becomes converted into narrow but elongated conducting-parenchyma cells. According to Hartig, Sanio and De Bary, the cambial cylinder originates in an essentially similar fashion in *Ephedra monostachya*, *Cheiranthus Cheiri*, *Rumex*, *Lunaria*, *Cobaea*, *Galium*, *Pyrethrum*, *Hieracium*, in the CARYOPHYLLACEAE, CRASSULACEAE, PLANTAGINACEAE, etc.

Where the cambial cylinder arises independently of the primary procambium, or where a primary procambial cylinder is never formed, the development of the actual cambium is comparatively straightforward. In this case, tangential divisions appear in the medullary rays, close to the margins of the strips of fascicular cambium; the rays thus become gradually bridged by strips of secondary meristem. Every cambial cylinder produced in this way is therefore composite in character, being made up of alternate strips of fascicular and interfascicular tissue, the former representing persistent portions of primary meristem, while the latter are secondarily derived from the fundamental parenchyma of the medullary rays. We have here another illustration of the fact—already mentioned on more than one occasion—that a tissue which is wholly concerned with a single physiological function, may yet be a heterogeneous structure when considered from a morphological point of view.

The last-described mode of cambial development shows some variation of detail in different cases. Typically (MENISPERMACEAE, *Casuarina, Aristolochia Sipho, Ricinus communis* [hypocotyl], *Begonia* [Fig. 269], *Cucurbita*, etc.), secondary meristematic activity is at once directed to the formation of the interfascicular cambium ; in a few cases, however (*e.g.* in *Clematis Vitalba*), the completion of the cambial cylinder is preceded by the differentiation of secondary cauline bundles, which appear in the medullary rays between the leaf-trace strands.

In the roots of Gymnosperms and Dicotyledons, the cambium originates in the central cylinder. Tangential division starts on the

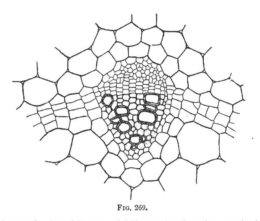

FIG. 269.

T.S. through the stem of *Begonia fuchsioides*, passing through a vascular bundle. The fascicular cambium is continuous with interfascicular cambial strips on either side. ×150.

inner side of the leptome-strands and thence extends laterally through the conducting parenchyma as far as the outer ends of the hadrome plates. The linking up of the separate strips of secondary meristem into a complete cambial cylinder is effected, by the cells of the pericycle (pericambium) outside each xylem plate becoming meristematic (Fig. 270). Here again the leptome is extra-cambial, and the hadrome intra-cambial. In other words, the orientation of the primary components of the vascular bundle with regard to the cambium is identical in the root and in the stem, and the subsequent course of secondary growth may also be the same in the two organs. On account of its mode of origin, the cambial cylinder of a root is, to begin with, curved inwards opposite each leptome-strand ; as the growth in thickness proceeds, however, the primary leptome is rapidly pushed outwards and the cambium acquires an approximately circular outline.

Attention may next be directed to the detailed structure of the cambium, and to the characteristics of its component cells. Our knowledge of the shape of cambial elements is mainly derived from Velten's detailed investigations. The form of a typical cambial cell is approximately that of a right prism, with its broader faces

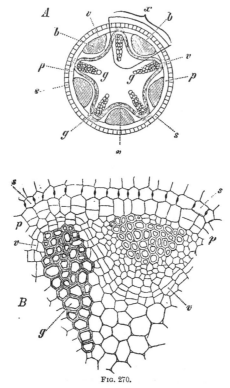

FIG. 270.

A. T.S. across the central cylinder of a primary root of *Vicia Faba*, after the commencement of secondary thickening (diagrammatic); *g*, hadrome-plates; *b*, fibrous strands, situated just outside the leptome-strands; *v*, cambium; *p*, pericycle; *s*, endodermis. *B*. Sector of *A* (*x*) more highly magnified; lettering as in *A*.

placed tangentially. The ends of the cells are gable-shaped, the terminal walls following an oblique tangential course. A tangential longitudinal section of the cambium, therefore, at once shows the prosenchymatous character of the cells, whereas in radial longitudinal sections the end walls are seen cut across, so that their sloping character is not apparent. While the tangential walls are thin and delicate, the radial walls are always—according to De Bary, Krüger and others—more or less thickened, and furnished with longitudinal rows of circular pits. The thickening of the radial walls is more pronounced in winter, *i.e.* during the resting season, than it is in summer. It is unlikely, however, that the decrease in the thickness of the radial walls which occurs when growth and division are resumed, is due to partial absorption of the thickened membranes. On the contrary, both Strasburger and Krüger are disposed to attribute this change to an extension of the walls, which constitutes the first indication of a resumption of radial elongation on the part of the cambial elements. The contents of cambial cells comprise finely granular cytoplasm and a nucleus, which is generally spindle-shaped.

It has already been mentioned on several occasions, that the cells of the cambial cylinder are arranged in regular radial rows. It is obvious, however, that all the elements in a given row are not exactly equivalent; for towards both ends of the series, the cambial cells become converted into elements of permanent tissue. Sanio accordingly assumes, that every row of the cambium (in *Pinus sylvestris*) contains a single, approximately median initial cell; each tangential division of this cell results in the formation of a new initial cell and of a "tissue mother-cell," cut off on the outer or inner side, according to circumstances. As a rule, each tissue mother-cell divides once more tangentially, the resulting daughter-cells becoming converted into elements of permanent tissue without further division (Fig. 272).

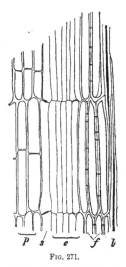

FIG. 271.

Portion of a radial L.S. through a twig of *Cytisus Laburnum*, showing the cambium and the secondary tissues immediately adjacent thereto; *p*, conducting parenchyma of the secondary phloem; *s*, young sieve-tube; *c*, cambium and the layers newly cut off from it; *f* and *l*, fibrous elements of the secondary xylem.

According to Sanio's view, the cambium proper comprises the single layer made up of the cells which are acting as initials at any given moment. But Raatz and Mischke have shown, that when the secondary growth in thickness is at all active, the primary segments cut off from the initial cells divide again more than once. Hence it is more convenient to consider the cambium as including all the cells in the radial rows, which are still undergoing tangential division. According to Raatz's conception, therefore, the cambial cylinder is a meristematic zone, in which the number of elements in each radial row is continually increased by successive tangential divisions. This extended definition of the cambium is, however, by no means incompatible with the existence of the initial cells postulated by Sanio. As a matter of fact, Sanio's conception has recently been revived by Schoute.

The cambium is gradually pushed outwards in the radial direction by the tissues which it produces on its inner side; its circumference must therefore continually increase as secondary thickening progresses. Up to a certain point, the individual cambial cells can keep pace with this increase in the circumference of the layer by means of tangential expansion; sooner or later, however, radial walls are interpolated, with the result that the number of radial cell-rows is increased. Since the ratio between the circumference of a circle and its diameter is constant ($= \pi$), the amount of tangential extension during any given interval of time can be readily deduced from the radial increase in thickness during the same time; in other words it is possible, in any given case, to calculate mathematically how many tangential divisions may take place in a radial row of the cambium, before the interpolation of a fresh radial wall becomes necessary. Nägeli was the first to perform calculations of this kind.

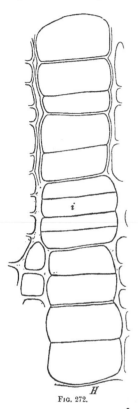

Fig. 272.

Small part of a T.S. through the cambial zone of *Pinus sylvestris*, showing a cambium-initial and the radial row of secondary elements derived therefrom; *i*, the cell assumed to be the actual cambium-initial. *H.* indicates the intra-cambial or xylem portion of the radial row. ×650. After Sanio (from De Bary, *Comp. Anat.*).

The secondary permanent tissues produced by the cambium in Gymnosperms and Dicotyledons may be classified as follows. The intra-cambial secondary tissue, consisting principally of mechanical and water-conducting elements, together with hadrome-parenchyma, may collectively be termed the **secondary xylem** or the **woody cylinder**; the extra-cambial increment, on the other hand, which is chiefly made up of leptome and conducting-parenchyma—often accompanied by mechanical strands serving for local protection—constitutes the **secondary phloem**. Both sets of secondary tissues are traversed by continuous radial strips of non-vascular tissue, known as **medullary rays**. That portion of a medullary ray which lies in the woody cylinder may be termed the "xylem-ray," the extra-cambial portion being distinguished as the "phloem-ray."

Additional or **secondary medullary rays** originate from one or more cambial cells, which undergo a definite number of transverse and oblique divisions, giving rise in this way to the initials of the several cell-rows of the ray. The author has followed out the development of the smaller medullary rays in *Cytisus Laburnum*.  Here certain elements of the cambium first undergo one or two preliminary divisions. Every primordial mother-cell of a medullary ray becomes divided by a radial longitudinal wall.  One of the two daughter-cells may thereupon be directly transformed into a definitive medullary ray mother-cell; it may, on the other hand, form a curved longitudinal wall, and thus cut off a small spindle-shaped cell, which then gives rise, by transverse division, to the initials of the several cell-rows of the medullary ray.

Medullary-ray initials behave essentially in the same way as ordinary cambial cells; by means of successive tangential divisions they add new elements, on their outer side, to the phloem-ray, and on their inner to the xylem-ray.  Krabbe has shown, for a number of woody plants, that the daughter-cells of the ray-initials never divide any more, but become directly converted into permanent tissue-elements, which accrue to the xylem-ray, or to the phloem-ray, according as they are cut off on the inner, or on the outer side.  In an eight-year old branch of *Cytisus Laburnum*, examined by the author, a particular ray-initial was found to have cut off two phloem-ray and eight xylem-ray elements in the course of each period of vegetation. If the duration of the latter be taken as approximately equal to 200 days, it is clear that successive divisions of the ray-initial must have taken place, on an average, at intervals of twenty days.

Translocation proceeds most actively in spring and in autumn; at these times the ray-initials are traversed by a constant stream of migrating carbohydrates, and thus for the time being act principally as conducting elements.  In some woody stems indeed (*Quercus pedunculata, Fagus sylvatica, Prunus Cerasus*, and especially *Cytisus Laburnum*), the change of function is so complete as to involve certain histological changes.  Almost all the ray-initials soon lose their meristematic character and acquire thickened and pitted cell-walls; in fact, they acquire all the characteristics of typical phloem-ray cells (Fig. 274, *m*).  In such cases as these, therefore, the xylem- and phloem-rays are directly continuous with one another from late autumn to early spring; when active growth in thickness is resumed, new ray-initials are formed by division of the innermost elements of the phloem-rays. Here, therefore, the cambial initials constitute a secondary meristem which is regenerated every year at a number of points.

## B. THE SECONDARY PHLOEM (EXTRA-CAMBIAL SECONDARY TISSUE).[350]

The extra-cambial products of the cambium are frequently included in the topographical definition of "bast"; naturally this usage cannot be accepted in the present work, since the term "bast" has already been restricted to a particular constituent of the mechanical system.

We must likewise reject the names "soft bast" and "bast-parenchyma," since the former is properly applicable only to the non-mechanical components of the secondary phloem, while the other should be employed to designate the conducting parenchyma—exclusive of the medullary rays—in the same region.

The various tissues that are comprised in the general category of secondary phloem are usually arranged in more or less distinct tangential layers, traversed radially by the medullary rays, which maintain the interchange of material between the woody cylinder and the extra-cambial region.

In secondary phloem, as elsewhere, the leptome is made up of sieve-tubes and of companion-cells—or of the rows of "albuminous cells," which take the place of companion-cells in Gymnosperms. It also includes some "cambiform" cells; but these are neither so numerous nor so typical as they are in the primary leptome, and are far less conspicuously developed than the ordinary leptome-parenchyma (conducting parenchyma). The sieve-tubes and companion-cells, on the one hand, and the conducting parenchyma on the other, are usually grouped in alternating tangential layers. The sieve-tube segments, being derived from cambial cells, have oblique end-walls, each sloping face bearing several superimposed sieve-plates; as has already been explained (p. 306), this is a feature which facilitates the translocation of protein compounds. The cells of the conducting parenchyma generally contain some chlorophyll, and at the end of summer become filled with large quantities of starch. This starch, however, disappears again completely in late autumn, before the period of "hibernation" begins; according to Russow, A. Fischer and others, it is mainly transformed into fat, to some extent also into glucose, and perhaps also into a third as yet unidentified compound, while a fourth and final portion in all probability passes into the woody cylinder, through the medullary rays. The first change in spring is the reappearance of starch; later this is transformed into soluble carbohydrates, which are conveyed to the opening buds along the secondary wood. The "ringing" experiments performed by Hartig, Hanstein and A. Fischer, show that the carbohydrates produced in the leaves travel downwards through the stem in the conducting parenchyma of the phloem, and thence pass into the woody cylinder along the medullary rays. Schellenberg states that, in some woody stems (*Vitis, Alnus, Aesculus, Betula*, etc.), the walls of the leptome-parenchyma cells become more or less extensively thickened during winter, but undergo partial solution in spring; the thickening layers involved consist of *hemicelluloses*.

The individual sieve-tubes, and their associated companion-cells—or albuminous cells—only remain functional for a comparatively short time; they lose their contents and become obliterated at the end of each vegetative period at the latest. The cells of the conducting parenchyma, on the other hand, may remain fully active for years.

The mechanical portion of the secondary phloem typically consists of bast-fibres and sclerenchymatous cells (sclereides). A considerable number of plants form no secondary fibrous elements (*e.g. Abies, Fagus, Platanus, Viburnum*); "stone-cells" (sclereides) are even more frequently absent (CUPRESSINEAE, *Taxus, Ulmus, Morus, Lonicera, Sambucus, Berberis, Tilia*, etc). It is, however, comparatively unusual for the secondary phloem to be entirely devoid of mechanical cells (examples are furnished by *Laurus, Nerium, Cornus, Ribes, Buxus*, etc.).

In typical cases the secondary bast-fibres are arranged according to one or other of the two following schemes. Either they are scattered singly or in small groups throughout the secondary phloem, or they form more extensive tangential layers or plates, alternating with similar masses of leptome-tissue. The second of these two modes of arrangement is exemplified by the CUPRESSINEAE, and by *Castanea vesca, Juglans nigra, Vitis vinifera, Clematis Vitalba, Quercus Robur*, etc.; where it prevails, the secondary phloem is often intersected by large masses of fibrous tissue. It would, however, be a mistake to suppose that this mechanical tissue contributes appreciably to the inflexibility of the stem as a whole. The strength of tree-trunks and large woody branches depends entirely upon the mechanical properties of the woody cylinder; the fibrous portion of the secondary phloem accordingly serves almost entirely to satisfy local mechanical requirements, or, in other words, to protect the leptome-tissue with which it is associated. It does not follow, however, that the presence of such fibrous layers is entirely without effect upon the total mechanical strength of the stem; their importance in this respect will naturally be greatest, when the woody cylinder is feebly developed to begin with. A twig of Lime, for example, may derive a large proportion of its mechanical strength from the massive fibrous strands in the secondary phloem during the first three years of its existence; but as continued cambial activity adds more and more to the growth of the woody cylinder, the mechanical significance of the bast becomes more and more strictly local.

In addition to mechanical elements, the secondary phloem very often contains crystal-sacs, and sometimes also secretory passages or latex-tubes. These structures, when present, exhibit no special pecnliarities, and hence demand no further consideration.

Finally, a few words must be devoted to the medullary rays of the secondary phloem (phloem-rays). Among Dicotyledons the constituent cells of these rays resemble the other conducting parenchyma elements of the secondary phloem as regards their contents and the character of

2 T

their walls. The two sets of cells are connected by means of numerous pits, through which an active interchange of materials is maintained; the companion-cells likewise enter into close relation with the medullary rays. Among the ABIETINEAE, and in some CUPRESSINEAE and TAXODINEAE, the phloem-rays contain in addition to starch-storing—or carbohydrate-conducting—elements, the so-called "albuminous cells" discovered by Strasburger; these cells, which generally occupy the upper and lower

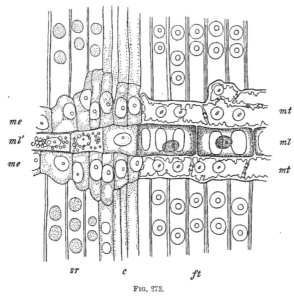

FIG. 273.

Secondary medullary ray of *Pinus Pumilio* (at the beginning of July). *me*, albuminous cells; *mt*, tracheidal cells; *ml-ml'*, row of conducting parenchyma cells; *sr*, sieve-tubes; *c*, cambium; *ft*, fibrous tracheides.

margins of the phloem-rays, differ from the starch-containing elements in their greater width and often also in the large size of their nuclei (Fig. 273, *m, e*). They adjoin the secondary sieve-tubes and communicate with them—in the ABIETINEAE, and probably also in the rest of the above-mentioned Conifers—by means of sieve-pores. They lose their contents at the same time as the sieve-tubes with which they are associated.

In the course of secondary growth in thickness, the circumference of the extra-cambial region naturally increases to a very considerable extent, since it is continually being pushed outwards. The (passive) tangential distension upon which this increase depends, is, of course, restricted to the living constituents and cannot, for example, take place in the fully developed periderm or bark. Most trees shed the outermost portions of the extra-cambial tissue—which usually consist of functionless, dead or moribund cells—every year. This process of desquamation is most conspicuous where the entire secondary phloem of the preceding year is cast off (*Vitis*); in such cases the large wastage of tissue is made good by a correspondingly generous annual addition of secondary phloem.

### C. THE SECONDARY XYLEM OR WOODY CYLINDER (INTRA-CAMBIAL SECONDARY TISSUE).[351]

1. *The component elements of the secondary xylem.*

The secondary xylem, woody cylinder, or intra-cambial secondary tissue comprises both mechanical and conducting elements. In their primary condition, Dicotyledonous stems are often strengthened by means of an extra-cambial fibrous cylinder (bast-cylinder); when secondary growth in thickness becomes active, the provision of mechanical strength largely devolves upon the intra-cambial wood fibres (libriform fibres). The reasons underlying this abnormal disposition of the mechanical elements are fairly evident. In the case of an organ which is undergoing secondary growth in thickness, the stereome cannot remain a permanent part of the organ, unless it is situated on the inner side of the cambial cylinder. But the cambium also produces the water-conducting elements, as well as a part of the carbo-hydrate-conducting tissue, on its inner side; hence the mechanical and conducting systems amalgamate to form the composite structure known as the woody cylinder. Since the two systems are no longer sharply separable topographically and anatomically, it is not surprising that their physiological individuality should also be lost to some extent. The leading anatomical features of the various tissues, in fact, no longer correspond exactly to their several functions; one tissue-system may undertake, as a subsidiary function, the work which properly belongs to another. Ultimately, indeed, it may become impossible to discriminate between principal and subsidiary functions; anatomically, this physio-logical compromise finds expression in the appearance of various cell-forms which are intermediate between typical skeletal and conducting elements.

Typically, the secondary wood is made up of the following diverse elements[352]: first, mechanical cells in the shape of **wood-fibres** (libri-form cells); secondly, relatively thin-walled **vessels** and **tracheides**, which represent the water-conducting tubes: and, thirdly, **xylem-parenchyma**, serving partly for the storage and partly for the conduction of non-nitrogenous plastic materials. Closely associated with these constituents of the wood proper are the cells of the **xylem-rays**.

Typical wood-fibres (libriform fibres or cells) are prosenchymatous cells, with thick walls which bear narrow oblique (sinistrorse) pits (Fig. 274, *lf*); they contain air and occasionally also shrivelled remains of the

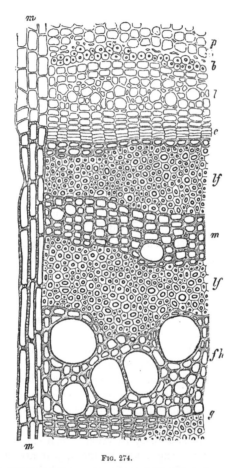

FIG. 274.

Part of a T.S. through the secondary xylem and phloem of a twelve-year-old branch of *Cytisus Laburnum* (at the end of October). *p*, conducting parenchyma ; *b*, plate of fibrous tissue ; *l*, leptome (the larger cells are the sieve-tubes); *c*, cambium ; *lf*, wood-fibres ; *m*, mestome (including intermediate cells, xylem-parenchyma, tracheides and narrow vessels); *g*, boundary between autumn and spring wood ; *m*, medullary ray.

former protoplasmic contents. Branched wood-fibres are of rare occurrence, but occur, for example, in *Tectona grandis* (Teak). Lignification is much more general—and also usually more intense—in the case of wood-fibres than in that of bast-fibres. Sanio has described a number of cases (*Cytisus Laburnum, Caragana arborescens, Gleditschia triacanthos,*

*Ulmus suberosa, Celtis australis, Morus alba*) in which the innermost layer
of the wall of the wood-fibres has a peculiar gelatinous or cartilaginous
consistency, and remains permanently unlignified. Leclerc du Sablon
and Schellenberg have shown that this layer consists of hemi-celluloses,
which are deposited late in the season and are converted into soluble
plastic material in the following spring.[353]

While wood-fibres are in general shorter than typical bast-elements,
they may nevertheless attain a length of ·3 to 1·3 mm. According to
Sanio, their average length is 1·26 mm. in *Prunus Laurocerasus,* 1·03
mm. in *Ulex europaeus,* ·8 mm. in *Quercus pedunculata,* ·53 mm. in *Salix
acutifolia* and ·46 mm. in *Tilia parvifolia.*

The *typical* vessels and tracheides of the secondary wood, that is to
say, those which are not utilised for mechanical purposes, are scarcely
more thick-walled than the corresponding primary elements. They are
frequently provided with delicate spiral thickening fibres (Fig. 295, D, E),
which are, however, generally too feeble to be mechanically effective.
Strasburger believes that these thickenings serve to "facilitate the
passage of water between the gas bubbles of the Jamin's chain and the
cell-wall, the liquid travelling along the spiral fibres." Diffusion goes
on in most cases through closely crowded bordered pits, or occasionally
through the unthickened areas in a reticulately thickened wall (*Crassulaceae, Opuntia, Cereus*). Where water-conducting elements abut against
conducting parenchyma (xylem-parenchyma or medullary-ray tissue),
they are furnished either with (one-sided) bordered or with large simple
pits.

Among the ABIETINEAE, and in other Conifers, as well as in certain
Dicotyledons (*Hippophaë rhamnoides, Salix fragilis,* etc.), the tracheides
are frequently traversed by radial trabeculae of cellulose. The development of these structures, which may extend through several annual
rings, is insufficiently known. It is likewise uncertain whether they
act as radial buttresses, or whether they merely represent abnormalities,
as Raatz maintains.

Xylem-parenchyma cells arise by the repeated transverse division
of cambial mother-cells; after differentiation is completed, it is often
possible still to recognise the genetically related elements, owing to
their being arranged in spindle-shaped groups. The individual cells
are always elongated and prismatic; in the vicinity of the larger vessels
they are often distinctly flattened. Their cell-walls are lignified, and
in general only moderately thickened. The pits are chiefly located on
the radial and transverse walls, and are always circular or elliptical in
outline. Xylem-parenchyma cells are always provided with living
contents, and in winter contain large quantities of reserve-materials in
the form of starch (most hard-wooded trees) or fatty oil (soft-wooded

trees, such as *Tilia, Betula, Pinus sylvestris,* etc.) ; occasionally a few chlorophyll corpuscles may also be present.

Mention has already been made (p. 334) of the ringing experiments carried out by Hartig, and more recently by A. Fischer,[354] which show that the carbohydrates formed in the leaves travel downwards in the stem exclusively in the extra-cambial conducting parenchyma and not in the xylem-parenchyma. One set of Fischer's experiments, in which branches of *Prunus Avium, Tilia* and *Betula* were ringed in two places, is particularly instructive ; no starch entered the (leafless) region between the two incisions, even after the lapse of several weeks. The upward transportation of carbohydrates in spring-time, on the other hand, takes place mainly in the wood-vessels, as we shall see later on. While, therefore, the xylem-parenchyma no doubt serves in the first instance as a repository of non-nitrogenous plastic materials, it would be going too far to regard it as a storage-tissue pure and simple, or to maintain that it has no carbohydrate-conducting capacity whatever. Like the medullary-ray tissue, the xylem-parenchyma certainly plays some active part in connection with translocation. The prevailing elongated form of its cells, and the relative abundance of pits on their transverse walls, further show that this tissue is not merely concerned with local interchange, but may also be responsible for conduction over greater distances.

If the cambial xylem-parenchyma mother-cells remain undivided,—a not infrequent occurrence—the resulting permanent elements, while agreeing closely with typical xylem-parenchyma cells as regards the character of their cell-walls and the nature of their contents, differ in their spindle-shaped form (Fig. 275 A). These elements, which frequently accompany or even entirely replace (*e.g.* in *Viscum, Caragana arborescens, Spiraea salicifolia*) the typical xylem-parenchyma, are the **intermediate cells** (*Ersatz fasern*) of Sanio. Each intermediate cell is homologous with a spindle-shaped group of ordinary xylem-parenchyma elements. The sole difference consists, as stated, in the fact that the cambial mother-cell remains undivided, in the one case, and undergoes transverse septation in the other. Since this distinction is manifestly of no value from the physiological point of view,—the difference between septate and unseptate bast-fibres being similarly regarded as unimportant—we shall follow the example of Troschel, and extend our conception of xylem-parenchyma so as to include the intermediate cells.

Having in the preceding paragraphs given some account of the typical elements of the secondary wood, or, in other words, of those which correspond to the normal components of the primary hadrome-strands, we must now turn our attention to the intermediate forms, which, as already stated, effect the transition from the mechanical to the con-

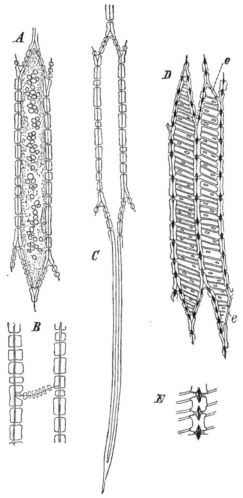

FIG. 275.

Various constituents of the secondary wood of *Cytisus Laburnum* (taken from a tangential L.S.). *A*. Intermediate cell, containing starch. *B*. Ends of two adjoining xylem-parenchyma cells. *C*. A transitional element which has the characters of a wood-fibre at its lower end, and those of an intermediate cell at its upper extremity. *D*. Two water-conducting elements; on the left a tracheide, on the right a narrow vessel; *e*, the perforations in the transverse septa of the vessels. *E*. Small portion of the wall separating two tracheides, showing bordered pits and spiral thickening fibres.

ducting and storing elements of the wood; such transitional forms in many cases actually make up the bulk of the woody cylinder, which thus acquires a very characteristic histological composition.

To begin with, we may consider the series of intermediate forms which link up typical wood-fibres, on the one hand, with vessels and tracheides on the other. The first indication which suggests that a wood-fibre has undertaken the transportation of water as a subsidiary function, consists in the appearance of small bordered pits on its walls ; the cavities of these pits are still narrow and oblique (*sinistrorse*), but the enlargement of the limiting membranes is sufficient to produce a small chamber (*Quercus, Fraxinus, Daphne* etc.). As the water-conducting activity of the wood-fibres becomes more pronounced, their bordered pits become larger and more typical, until finally the cells are almost equally adapted for performing mechanical and conducting functions. These strictly intermediate cells are prosenchymatous and thick-walled, and bear large circular bordered pits, with rounded, or with narrow oblique orifices. Cells of this type, which may be termed **fibrous tracheides** (*Fasertracheiden*), are typically developed in the Magnoliaceous genus *Drimys*, and even more so in the wood of Conifers. A still closer approximation to the structure of typical water-conducting elements is found in *Taxus*, and in other cases where the fibrous tracheides are furnished with spiral thickenings. From this condition it is only a small step to those thick-walled vessels and tracheides the mechanical function of which is purely subsidiary (*Fraxinus excelsior, Nerium Oleander, Convolvulus Cneorum*, etc.). Thin-walled vessels and tracheides represent the final link in the series.

For the sake of clearness, the above-mentioned transitional forms may be tabulated, with reference to their respective functions, as follows :

| Cell Form. | Principal Function. | Subsidiary Function. |
|---|---|---|
| 1. Wood-fibres with narrow simple pits. | Mechanical support. | |
| 2. Wood-fibres with small bordered pits (narrow pit-cavities). | Mechanical support. | Water-conduction. |
| 3. Fibrous tracheides. Prosenchymatous, thick-walled cells, with large bordered pits. | No differentiation of principal and subsidiary functions; the cells serve equally for mechanical support and for water-conduction. | |
| 4 Fibrous tracheides with spiral thickenings, otherwise as in 3. | | |
| 5. Thick-walled vessels and tracheides. | Water-conduction. | Mechanical support. |
| Thin-walled vessels and tracheides. | Water-conduction. | |

The second set of intermediate cell-forms effects the transition from typical wood-fibres to thin-walled xylem-parenchyma. This series is not so complete as the first, but is nevertheless quite clearly recognisable. The smallest amount of deviation from the normal type of wood-fibre, which is a dead structure, is illustrated by fibres that retain their living contents; storage or translocation of plastic substances can, of course, only take place in cells that are provided with living protoplasts. Such living wood-fibres are perfectly typical as regards their shape and the character of their walls; but the narrow, oblique (sinistrorse) pits show a tendency to congregate upon the radial walls, in many cases (*e.g.* in the genus *Acer*, according to Krah). The next step is exemplified by septate wood-fibres, such as occur in the Vine. Cells of a remarkable type, connecting mechanical elements with "intermediate cells," have been observed by the author in the wood of *Cytisus Laburnum*, where they occur in considerable numbers at the junctions of the broad plates of wood-fibres with tracts of xylem parenchyma; each of these special cells (Fig. 275 c) exactly resembles a typical thin-walled intermediate cell for half its length, while the other half agrees just as closely with a typical thick-walled prosenchymatous wood-fibre. This is evidently a case, in which two physiological functions find separate histological expression in two sharply differentiated portions of the same cell. A further link in the chain is supplied by those xylem-parenchyma cells which are distinctly thick-walled; in these, all the walls may be equally thick (*Convolvulus Cneorum*), or thickening may be confined to the radial walls (*e.g.* the autumn-wood of *Magnolia acuminata, M. tripetala, Liriodendron tulipifera, Gymnocladus canadensis, Amorpha fruticosa*, according to De Bary). In the case of these thick-walled parenchyma cells, as also in that of the intermediate cells with narrow pits which occur in a great many plants, the mechanical function is always of subordinate importance. These last-mentioned cell-types, therefore, effect the final transition to normal thin-walled xylem-parenchyma.

The second series of transitional cell-forms may be tabulated as follows ·

| Cell Form. | Principal Function. | Subsidiary Function. |
|---|---|---|
| 1. Ordinary dead wood-fibres. | Mechanical support. | |
| 2. Wood - fibres, with living contents— <br> (*a*) Non-septate. <br> (*b*) Septate. | Mechanical support. | Storage and conduction of carbohydrates. |
| 3. Composite elements, which are wood-fibres as regards one half, and intermediate cells as regards the other. | No differentiation of principal and subsidiary functions. | |

| Cell Form. | Principal Function. | Subsidiary Function. |
|---|---|---|
| 4. Thick - walled xylem - paren-chyma. | Storage and con-duction of car-bohydrates. | Mechanical sup-port. |
| 5. Intermediate cells with narrow pits. | Do. | Do. |
| 6. Thin - walled xylem - paren-chyma. | Do. | |

The problem as to the mode of evolution of the two anatomico-physiological transition series which have just been described, is no longer capable of exact solution. It is probable that, at the time when the Higher Plants acquired the property of secondary growth in thick-ness, primary differentiation had already led to the production of *distinct* elements serving, as the case might be, for mechanical support, for the conduction of water, or for the translocation and storage of plastic materials. It therefore seems permissible to assume, that these three principal components were represented in the secondary wood at the first inception of secondary growth in thickness, and that subsequently mechanical elements, where needful, undertook the work of conduction, and *vice versâ*, the change of function being attended by a corresponding modification of structure.[355]

The above general discussion of the longitudinally orientated constituents of the woody cylinder, may be supplemented by a few remarks concerning certain features which are restricted to particular families. In the PAPAYACEAE, the wood contains a well-developed network of latex-tubes, which extends throughout the xylem-parenchyma, and also comes directly into contact with the vascular tissue. As regards the occurrence of secretory or excretory organs in the woody cylinder, it may be noted that resin-passages are to be found in the xylem-parenchyma of Conifers, while crystal-sacs occupy a similar position in certain LEGUMINOSAE (e.g. *Haematoxylon*), in *Vitis,* and in other instances.

In the vast majority of cases, the tissue composing the **xylem-rays** is of the nature of conducting parenchyma. Its individual cells are as a rule radially elongated, in accordance with the fact that the medullary rays are specially destined to maintain an interchange of material in the radial direction. The cell-walls resemble those of the xylem-parenchyma, being more or less thickened and lignified, and furnished with very numerous pits on their transverse walls.

Among the ABIETINEAE, and especially in the genus *Pinus*, the parenchymatous cells of the medullary rays are accompanied by elements which agree with ordinary tracheides, both in structure and in function (Fig. 276, *mt*). The walls of these **tracheidal cells** are

strengthened by transverse bars, and bear bordered pits. These cells
occur more particularly along the margins of the medullary rays, and
serve for the conduction of water in the radial direction. In the
medullary rays of Dicotyledons, the central rows of cells are the most
typical; they are radially elongated, and serve principally for the
translocation and storage of non-nitrogenous plastic material. These
cells constitute the so-called procumbent cells of the medullary rays.
The elements occupying the upper and lower margins of the rays, on the

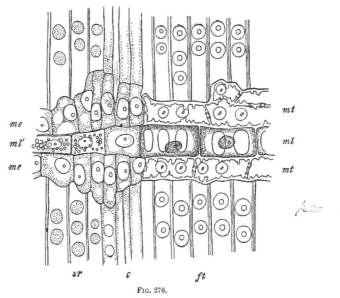

FIG. 276.

Secondary medullary ray of *Pinus Pumilio* (at the beginning of July). *me*, albuminous cells; *mt*, tracheidal cells; *ml-ml'*, row of conducting parenchyma cells; *sr*,
sieve-tubes; *c*, cambium; *ft*, fibrous tracheides.

other hand, are often comparatively short and high, and may be distinguished as the upright cells; they bring the medullary ray into
relation with the neighbouring vessels, which communicate with them
by means of one-sided bordered pits.

The so-called **medullary spots**, defined by De Bary as local
hypertrophies of the medullary rays, are of regular occurrence in the
wood of the genera *Alnus* and *Sorbus*; they are composed of irregularly
polyhedral cells, with pitted walls, and appear in the first instance to
constitute a form of storage-tissue. This view finds support in the fact
that the contents of the cells in question consist very largely of starch
and tannin. It is another question, to what extent such medullary
spots may act as secretory or excretory organs, or may represent

pathological developments.    Among Conifers resin-ducts often arise
secondarily within the medullary spots.

The present section may conclude with a few remarks with regard
to the organisation of the ventilating system in the woody cylinder.
Every portion of the conducting parenchyma of the wood is permeated
by a continuous reticulum of narrow air-containing intercellular spaces;
in the medullary rays, radial air-passages naturally predominate, whereas
the xylem-parenchyma is provided with narrow longitudinal passages.
In Conifers, air-spaces also occur between the fibrous tracheides;
according to Russow, these are particularly numerous, and also
relatively wide, in the CUPRESSINEAE and ARAUCARIEAE. The same
author[356] has described two features of purely physiological interest in
connection with the intercellular spaces of medullary rays.    Where, as
usually happens, the cells of the medullary rays are thick-walled,
they are provided with pits, even on those walls which border
upon intercellular spaces (*Larix, Quercus, Fagus Ulmus, Salix, Populus,
Cytisus Laburnum*, etc.).    It can hardly be doubted that those
pits which border directly upon air-spaces, serve for ventilation;
this is, however, the only instance so far definitely recorded, in
which pits appear to serve solely for *gaseous* interchange.    The second
point of physiological interest as regards the behaviour of the
intercellular spaces of the xylem-rays, consists in the fact that these
structures extend without a break right across the cambium and
thus communicate with the air-spaces of the phloem-rays, and through
them with the cortical air-spaces; since the latter open to the exterior
through the lenticels, the ventilating system of the woody cylinder is
really in free communication with the outer atmosphere.    This fact
has been demonstrated experimentally by Klebahn in the following
manner.    A twig of *Berberis*, about 10 cm. in length, is deprived of its
extra-cambial tissue for a distance of 2 cm. from the lower end; the
peeled surface and the upper cut end are then coated with some
air-tight material.    If now air is forced in at the lower end of the twig
under a pressure of 14 cm. of mercury, it escapes from the lenticels
in the unpeeled region.    Owing to the smallness of the pressure
employed, this effect cannot be attributed to diffusion through the
cell-membranes;    the experiment therefore proves that a direct
interchange of gases can take place between the outer air and the
ventilating system of the woody cylinder.    Klebahn has further shown
that, where no lenticels are present (*e.g.* in *Vitis spp., Lonicera
Periclymenum, Clematis Vitalba, Philadelphus coronarius*), the intercellular spaces of the medullary rays are continued radially right
through the periderm and to the surface, so that here also the wood is
placed in communication with the external atmosphere.

## 2. *The arrangement of tissues in the woody cylinder*

In considering the distribution and arrangement of the various elements of which the woody cylinder is composed, we are confronted with two questions; we have to decide, first, how far the arrangement is merely a geometrical or mechanical consequence of the process of secondary growth and its attendant changes; and, secondly, to what extent the arrangement depends upon, or is determined by, the functions of the elements concerned. The former of these two closely related questions may be dealt with first.

In the secondary wood, as in the secondary phloem, the newly differentiated elements are arranged in radial rows. If the conversion of these cells into permanent tissue is only attended by radial elongation, while their tangential dimensions undergo no change, the radial seriation naturally remains unaltered. This condition is exemplified by all ordinary xylem-parenchyma—at least when it forms large continuous masses—by the medullary rays and by tracheidal tissue. The radial seriation will likewise be maintained, if the ends of the elongating cells are all inclined in tangential planes, so that the cells can slide past one another without lateral deviation. It is for this reason that the fibrous tracheides of Conifers are always disposed in perfectly regular radial rows ; among Dicotyledons the wood-fibres very frequently exhibit a similar arrangement (*e.g.* in *Viburnum Opulus, Nerium Oleander, Laurus nobilis, Aesculus Hippocastanum, Paulownia imperialis,* etc.). In other cases the original radial seriation undergoes considerable modification, or even disappears altogether. This alteration may be brought about by the expansion of individual cells during their conversion into permanent elements ; every large vessel, for example, displaces, compresses or even obliterates the cells in its immediate neighbourhood. The disturbance may also be due to the circumstance that the elongating cells become dovetailed together in various ways ; this latter arrangement is exemplified by many wood-fibres (*e.g.* those of *Ulmus suberosa, Morus alba, Celtis australis, Cornus sanguinea, Cytisus Laburnum* and other LEGUMINOSAE, etc.). In either case a considerable amount of sliding growth must take place.

The morphological relations that have just been explained, are of less interest than the connection which exists, between the arrangement of the various constituents of the secondary wood, on the one hand, and their several physiological functions, on the other. In considering this second question, it must always be borne in mind that the woody cylinder is made up of elements belonging to two distinct tissue-systems. Obviously, therefore, any given arrangement of the component elements of the secondary xylem will not be satisfactory, unless it ensures that the two tissue-systems interfere with one another as little as possible.  Now, in the case both of the mechanical and the conducting systems, it is above all necessary that the component tissues should be *continuous*; it follows that the mutual interpenetration of stereome and mestome must not involve any interruption of either system.  The longitudinal strands of the stereome network must not be broken at any point; similarly, conducting elements must never become completely isolated.  In other words, every tracheide or vessel, and every cell of the xylem-parenchyma, or of a medullary ray, must have direct access to the stream of nutrient material.  Xylem-parenchyma or tracheidal tissue could not possibly remain active, if completely enveloped by mechanical tissue.

The continuity of the conducting system in the woody cylinder of Dicotyledons has been made the subject of detailed study by Troschel, and subsequently by F. W. Krah.[357]  Both these investigators come to the conclusion that the structure of the woody cylinder of deciduous Dicotyledonous trees provides in a very satisfactory manner for this continuity—which, as we have already explained, is demanded by fundamental anatomico-physiological principles.  As a matter of fact, the principle of continuity finds a three-fold application in the construction of secondary wood.  In the first place, the vessels and tracheides are not only longitudinally continuous, but also communicate with one another laterally; secondly, the medullary rays always abut directly against xylem-parenchyma on their flanks, as well as along their upper and lower edges; lastly, the vascular elements invariably have conducting parenchyma (xylem-parenchyma and medullary ray tissue) closely associated with them.

As regards the first-mentioned point, stress may be laid upon the existence of tangential communication between the water-conducting elements—their longitudinal continuity requiring no comment.  Anatomically, this tangential continuity is most marked, where the water-conducting portion of the wood consists largely or exclusively of fibrous tracheides; in this case the bordered pits are chiefly massed on the radial walls, so that diffusion in the tangential direction is facilitated as much as possible.  It is clearly to the advantage of a tree that this should be the case; for if the transpiration current is interrupted on one side of the trunk by some local injury, or if there is any serious irregularity in the development of the root-system, the ill effects of an interrupted water-supply upon the foliage are almost entirely obviated, owing to the ease with which water can move tangentially in the trunk.  In this respect the woody cylinder resembles the primary vascular system; for, in the latter case, the tangential vascular anastomoses act in the same way as the bordered pits on the radial walls of the secondary tracheides.

We must next turn our attention to the question of the close relation between xylem-parenchyma and medullary ray tissue. A few remarks must, however, be interpolated regarding the dimensions of medullary rays. The simplest form of medullary ray is only one cell in width vertically and horizontally, or, in other words, consists of a single row of cells. More frequently, however, the rays only *appear* to be uniseriate when seen in tangential or transverse section, but in reality consist. of several vertically superimposed cell-series (*e.g.* in the majority of Conifers). Generally speaking, both the height and the width of medullary rays vary within wide limits; rays which attain any considerable breadth, naturally comprise several or even numerous rows of cells, while a large number of cells may also be involved in the vertical direction. The number of rays is usually greatest, where the individual rays are narrow and of inconsiderable height, a relation which requires no further consideration from the physiological point of view.

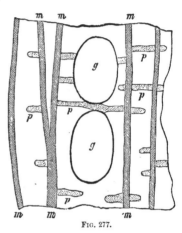

Fig. 277.

Diagram illustrating the continuity of the hadrome-tissue in the woody cylinder of *Casuarina.* *g*, vessels; *m*, medullary rays; *p*, strips of xylem parenchyma. All the intervening space is occupied by mechanical tissue (wood-fibres). After Wiesner.

The medullary rays—and more especially the procumbent cells of the rays — represent radial water channels which are linked together both tangentially and longitudinally by means of more or less extensive masses of xylem-parenchyma. Tangential commissures composed of this tissue connect every medullary ray with its neighbours (Fig. 277); the presence of closely crowded pits on the radial walls of the commissure-cells indicates that the flow of translocated material along these channels is not always longitudinal, but that it may, if necessary, follow a tangential course. In transverse sections of the woody cylinder, some of the shorter commissures may only appear to abut against medullary rays at one end (Fig. 277), or they may even seem to be entirely isolated; but, in view of the fact that the medullary rays occupy different radii at the various levels, it may be safely assumed, in such cases, that the apparently isolated commissures bridge the interval between adjacent rays at some level which is either above or below the plane of the section. In the same way, individual rows of xylem-parenchyma cells are never really isolated, although they often appear to be so; if their longitudinal course is carefully followed up, as has been done by Krah and Troschel, every series of xylem-parenchyma elements is found to intersect, or to become merged in, one of the medullary rays. In this way the xylem-parenchyma serves to effect a connection between medullary rays situated at different levels. It is, therefore, quite evident that the whole conducting parenchyma of the woody cylinder forms an inter-communicating system, even though its continuity may not be apparent in every transverse section.

The relations between the conducting parenchyma (xylem-parenchyma and medullary ray tissue) and the vessels has still to be discussed. In this respect, two different modes of arrangement of the xylem-parenchyma must be distinguished. This tissue may take the form of tangential commissures, in which case the vessels are opposed

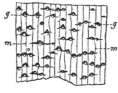

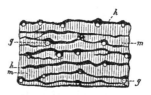

FIG. 278.

T.S. through the Wood of *Copaifera bracteata*, as seen under a simple lens. *m*, medullary rays; *g*, vessels, With externally apposed masses of xylem-parenchyma. After Wiesner.

FIG. 279.

T.S. through the Wood of *Pterocarpus santalinus*, as seen under a simple lens. *m*, medullary rays; *g*, vessels; *h*, xylem-parenchyma. After Wiesner.

to, or partially embedded in these plates. This "metatracheal" type of xylem-parenchyma, as Sanio calls it, occurs, for example, in *Acacia albicans*, *Celtis trinerva*, *Ficus elastica*, *Magnolia grandiflora*, *Castanea vesca*, *Juglans regia*, *Quercus pedunculata*, etc. (Fig. 279). In other cases the xylem-parenchyma envelopes the vascular tissue more or less completely, without forming continuous tangential plates; such "paratracheal" parenchyma is found in *Acacia Sophora*, *A. floribunda*, *Caesalpinia echinata*, *Capparis spp.*, etc. (Fig. 278). Since the two types of arrangement are connected by intermediate forms, and may, moreover, be combined with one another in a variety of ways, there is the greatest possible diversity as regards the appearance presented by different woods in transverse section. Krah has attempted to classify the known varieties by distinguishing a number of subtypes.

The connection between the medullary rays on the one hand, and the vessels and tracheides on the other, has been made the subject of special study by *P.* Schulz.[358] It has already been explained that the medullary rays of many ABIETINEAE are made up of two different kinds of cells, namely, ordinary medullary ray cells (conducting parenchyma) and water-conducting tracheidal cells. That the latter should communicate by means of bordered pits with the longitudinal fibrous tracheides, is quite in accordance with the general principle of continuity underlying the construction of the conducting system. The typical medullary ray cells are, however, likewise connected with the adjoining fibrous tracheides by pits, which are often of considerable size. These pits are particularly large in the genus *Pinus*, where each extends almost across the whole width of the medullary ray cell (Fig. 276, *ml*). The presence of such large pits greatly reduces the strength of the tracheidal walls; the tracheides accordingly develop remarkable transverse buttresses, which preserve them from compression at the hands of the turgescent medullary ray tissue. Each of these buttresses takes the form of a short rod bearing a disc-like expansion at either end; in longitudinal section, therefore, the buttress is shaped like an H, the upright arms, which correspond to the terminal discs, being attached to the lateral walls of the tracheide.

Among Dicotyledons also, the cells of the medullary rays communicate with the adjacent vessels by means of pits; this statement applies more particularly to the upright marginal cells, where these are specially differentiated. In the genera *Salix* and *Populus*, the pits are elliptical and slightly bordered. Among the Cupuliferae, and in a variety of other trees and shrubs, the pits are fairly large—in this respect agreeing with those of the Salicaceae—but show no trace of a border. They are larger still (·01–·012 mm. in diameter) in *Morus alba, Aristolochia Sipho, Vitis vinifera* and *Staphylea pinnata*; in surface view these exceptionally large pits no longer present the usual circular or elliptical outline, but are seen to be greatly elongated or irregular in cross-section. Among the Rosaceae—especially in certain Pomaceae—and Leguminosae, on the other hand, the corresponding pits are small, simple and irregularly distributed.

The question now arises as to the physiological significance of the above-described connection between conducting parenchyma and vascular tissue. The anatomical data certainly point to one of two conclusions: either the xylem-parenchyma and the medullary rays assist in the conduction of water, or else the vessels and tracheides take part in the translocation of carbohydrates. The two views are not mutually incompatible, and it is quite possible that they are both correct. The possibility that the conducting parenchyma participates in the transportation of water, has already been considered on a previous occasion (Chap. VII.). It may be added at this stage, that the production of exudation-pressure—which may be set up in the water-conducting tissues of the woody cylinder in axial organs as well as in

roots—is entirely dependent upon the close association of living parenchyma cells with the vascular elements.   When root-pressure is active
in spring, the sugar which arises from the starch stored in the xylem-
parenchyma and medullary rays, may diffuse into the water-conducting
elements and then travel rapidly in the transpiration-current towards
the growing organs.    It has long been well known that the sap of
certain trees contains considerable quantities of sugar in spring-time.
The sugar-content of the vernal sap is 3·57 per cent. in the case
of *Acer saccharinum*, according to Clark, and 1·15–3·44 per cent.
in that of *A. platanoides*, according to Schröder.[359]   This sugar can
only be derived from the xylem-parenchyma and the medullary ray
tissue, in both of which starch is stored during winter.   In summer-
time, water, with mineral salts in solution, is pumped by the
parenchyma of the absorbing roots into the conducting channels,
along which it travels (as the so-called transpiration-current) into
the photosynthetic organs; in precisely analogous fashion soluble
plastic materials are forcibly transferred in the spring from the xylem-
parenchyma and the medullary rays to the water-conducting elements,
and are thus conveyed to the developing leaves and floral organs much
more rapidly than could be done by osmosis alone.

While the transportation of plastic substances is evidently a subsidiary function of the water-conducting system, it is nevertheless of great
physiological importance, because it renders possible the very rapid unfolding of foliar and floral buds that takes place in spring.   The correctness
of this view—which had already been formulated in the first [German]
edition of the present work—was proved by the detailed investigations
of A. Fischer,[360] who showed that the vessels and tracheides of the
majority of our native deciduous and Coniferous trees always contain
considerable quantities of *glucose* in spring-time.   The *glucose* content
begins to rise during the bleeding period, and increases still further
when the reserves of starch are mobilised.   The concentration of sugar
in the vessels reaches a maximum about the beginning of May, and
falls off again gradually during the summer; but a certain amount of
*glucose* is present even in winter.   Fischer devised the following
experiment in order to demonstrate the ascent of *glucose* in the
transpiration-current.   Leafless portions were selected from four-year-
old twigs of *Acer dasycarpum* in the latter half of May, when the entire
reserve of starch had been transformed into sugar; these were cut off
and placed with their lower ends in water.   In such circumstances starch
appeared, immediately below the upper end, around the margin of the
pith and in the medullary rays, after 17 hours in a piece of twig
16 cm. in length, and in as little as 5 hours in a piece of half this
length.   Similar results are obtained if the pieces of twig are inverted,

the apical ends being placed in water. The explanation offered by Fischer is that the ascending water carries *glucose* in solution with it · but since this substance is not utilised at the upper end of the twig, it becomes converted into transitory starch.

Experiments performed by Hartig and Fischer indicate that carbo hydrates travel downwards, but not upwards, in the secondary phloem ; Fischer accordingly concludes, that all the synthetic products deposited both in the secondary phloem and in the woody cylinder are carried upwards by the transpiration current in spring-time, always excepting, of course, the portion which is used up locally in connection with respiration and growth. The possibility that the xylem-parenchyma participates in the longitudinal transportation of carbohydrates must, however, never be lost sight of; this matter, in fact, stands in urgent need of further experimental investigation.

### 3. *Annual rings.*

In the temperate zones, the growth in thickness of trees and shrubs is interrupted annually by the cold season. The addition which is made to the woody cylinder during a single vegetative season is often clearly distinguishable, owing to special anatomical arrangements, from the immediately preceding and succeeding zones ; these seasonal increments are usually known as the **annual rings**. The limits of each annual ring are formed by the **spring wood** and the **autumn wood** respectively. Similar annual rings may also develop in tropical woody plants, especially where vegetative activity is interrupted by a regularly recurring dry season.[361]

In erect trunks and branches, the annual rings are in general equally broad all round ; deviations from this rule are almost always referable to external influences. Mention has already been made in an earlier chapter (p. 195) of an experiment performed by Knight, in which the stem of a young Apple tree was fixed in such a way, that the wind could only bend it from north to south or *vice versa*; under these conditions, growth in thickness was found to be greatest in the plane of bending. Later investigators, such as Nördlinger, Metzger, R. Hartig and F. Schwarz, have also drawn attention to the connection between excentric secondary thickening and the direction of the prevailing winds. Ursprung has pointed out that excentric thickening of an erect stem, or asymmetrical development of individual annual rings, will be mechanically advantageous, if it accelerates the recovery of the stem from any accidental distortion.

Variations in the thickness of successive annual rings in one and the same transverse section of a stem are principally due to differences in the general conditions of nutrition in successive seasons. Thus, a summer in which the crown of a tree suffered serious injury through the ravages of insects, will be represented by a narrow annual ring in the stem, whereas seasons favourable to the development of foliage will be marked by the addition of wide annual rings.  In this way the fluctuations in the width of the successive annual rings in the trunk form a permanent record of the principal episodes in the history of the tree.

In oblique or horizontal branches and roots, the annual rings are usually all asymmetrically developed.  Thus, the upper half of each ring is thicker in *Acer*, *Alnus*, *Carpinus*, *Corylus*, *Cytisus Laburnum*, *Fagus*, *Tilia*, *Robinia*, and many other deciduous trees, while the lower half is favoured among Conifers and also in *Buxus sempervirens* and *Viscum album*.[362]  C. Schimper, who was the first to study excentric thickening in detail, applied the terms epinastic to the former, and hyponastic to the latter of these two modes of growth. Wiesner speaks of epitrophy and hypotrophy in the same connection.

The origin of such inequalities in the growth of the upper and lower sides of branches or roots is still imperfectly understood. Several physiologists associate this phenomenon with the influence of gravity ; others attribute it to the unequal incidence of temperature, light and moisture upon the two sides of the organ.   It is probable that the distribution of mechanical strains within an organ plays a leading part in producing asymmetrical thickening ; thus the upper side of a horizontal or oblique branch is mainly subjected to tension and the lower side to compression.   With regard to horizontally extended subterranean roots, Kny states that these organs only exhibit asymmetrical growth in thickness, when they are laid bare, and consequently exposed to the atmosphere ; the woody cylinder is then subject to the same influences as it is in the horizontal branches of the same plant, and accordingly assumes a somewhat similar anatomical structure.   Wiesner, finally, has shown that asymmetrical growth in thickness may also be brought about by internal influences. If, for example, a lateral branch of *Abies* is forced to develop in an upright position, the abaxial side of the woody cylinder still develops more strongly than the opposite face (exotrophy).

An epinastic or hyponastic tendency on the part of a twig or branch may result in very marked vertical expansion of the organ, a fact already noted by the author in his *Botanische Tropenreise*, in connection with a " candelabrum tree " belonging to the genus *Garuga* (BURSERACEAE). Here the basal curved region of each main branch is strongly flattened in the vertical plane.   In such a case as this the mechanical advantage of the asymmetry is fairly obvious.   As a result of the excentric development of the annual rings, the cross-section of the branch assumes the form of an ellipse, with its major axis vertical; in this way the bending moment is very considerably increased. The investigations of Hartig, Metzger, Sonntag and Ursprung have shown, that there is not only a quantitative but also a qualitative difference between the adaxial and the abaxial wood in branches with a distinct epinastic or hyponastic tendency. The adaxial wood (the "white" wood of *Picea excelsa* and other Conifers) is mainly inextensible in character, while the abaxial xylem (the "red" wood in Conifers) is incompressible. According to Metzger, Conifers have an inherent tendency to produce a relatively large proportion of incompressible wood, whereas deciduous trees are on the whole inclined to form inextensible woody tissue. It is, therefore, not surprising that Conifers concentrate their secondary xylem on the lower or "compression" side of the horizontal branches, whereas deciduous trees develop the bulk of the wood on the upper or "tension" side; in other words, Conifers tend to develop hyponastic, and deciduous trees epinastic lateral branches.[363]

The extreme epinasty exhibited by the flattened plank-roots which are characteristic of a number of tropical trees (spp. of *Ficus, Sterculia, Canarium commune, Parkia africana,* etc.) must be regarded as an adaptive phenomenon, since the lateral compression of the roots greatly increases their effectiveness as buttresses of the trunk. A transverse section of such a plank-root of *Parkia africana,* which the author brought back with him from Buitenzorg (Fig. 280), has a total vertical diameter of 104 cm.; but the organic centre is 92 cm. distant from the upper, and hence only 12 cm. from the lower surface of the root. Epinasty is not much in evidence in the zone represented by the first twelve annual rings (which amount *in toto* to 9 cm.

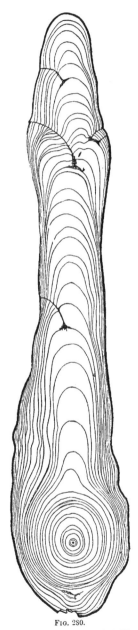

FIG. 280.

T.S. through one of the plank-like buttress-roots of *Parkia africana.* About one-sixth natural size. For description, see text.

on the upper, and 6·5 cm. on the lower side).   The thirteenth annual
ring initiates the subsequent prodigious development of the upper half;
this particular ring is 40 mm. thick above, 5 mm. on the flanks of the
root, and not more than 2·5 mm. below.   The greatest thicknesses
attained by individual rings—on the upper side—are 60 mm. (nineteenth
ring) and 57 mm. (fifteenth ring).   It will be noticed that a certain
amount of hyponastic thickening also takes place from the fifteenth ring
onwards, though this is always insignificant in comparison with the
permanent and extensive epinastic tendency, and is most probably
not a normal feature at all.   It is further impossible, in the case of
this particular cross-section, to determine with any certainty, whether
all the annual rings are complete or whether some of them die out
altogether on the lower side.   Ursprung has explained why it is an
advantage for plank-roots to be epinastic rather than hyponastic.   In
the first place this type of growth is unaffected by the pressure of the
soil ; and, secondly, the buttressing effect of the plank-roots is greater
if the latter are situated above ground.

The successive annual rings of a trunk or branch are more or less
clearly distinguishable in a transverse section, because of the anatomical
difference between spring and autumn wood.   This difference may be
due to an unequal distribution of the various elements within each
ring, or it may depend upon differences in the shape and structure of
homologous elements in the two regions of the ring ; very often the
two factors act in conjunction with one another.

As a rule, spring wood contains a larger proportion of vessels and
tracheides than autumn wood ; the vessels are also usually much wider
in the former type of wood, which hence has a looser and more porous
texture.   The elements of the spring wood are, moreover, as a rule,
relatively thin-walled, the contrast in this respect to the cells of the
autumn wood being accentuated by the fact that the latter are most
often tangentially flattened to a very considerable extent (Figs. 281,
282).   The aforesaid differences are most striking, where the composition of the wood is very homogeneous, as it is among Conifers.
Thus, Von Mohl found the average radial diameter of the fibrous
tracheides, in a well-grown 30-year-old trunk of *Pinus sylvestris*, to
be from three to six times as great in the spring wood as it is in the
autumn wood.   The flattening of the autumnal tracheides, measured
by the ratio between the radial and the tangential diameters, amounted
to $\frac{2}{5}$, while their walls were 1·6 times as thick as those of the spring
tracheides.   The tracheides thus primarily represent mechanical cells
in the autumn wood, and conducting elements in the spring wood.

There are a variety of histological features, which help to secure
radial communication between the conducting elements of successive

annual rings. In most Gymnosperms, the bordered pits (which are ordinarily confined to the radial walls of the tracheides) appear on the tangential walls as well, at the limits of each annual ring (Fig. 282) · in *Pinus*, where these tangential pits are absent, the tracheidal elements

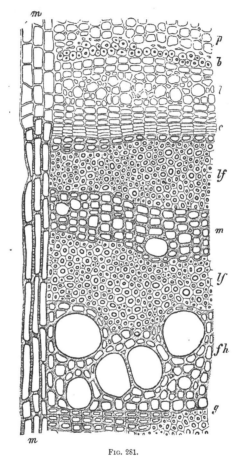

FIG. 281.

Part of a T.S. through the secondary xylem and phloem of a twelve-year-old branch of *Cytisus Laburnum* (at the end of October). *p*, conducting parenchyma ; *b*, plate of fibrous tissue ; *l*, leptome (the larger cells are the sieve-tubes) ; *c*, cambium ; *lf*, Wood-fibres ; *m*, mestome (including intermediate cells, xylem-parenchyma, tracheides and narrow vessels) ; *g*, boundary between autumn and spring Wood ; *m*, medullary ray.

of the xylem-rays are responsible for the radial conduction of water. Among Dicotyledons, according to Gnentzsch and Strasburger, the water-conducting elements of adjoining rays are often connected with one another by means of isolated narrow vessels, which develop in the autumn wood but abut directly against some of the vessels newly formed in the succeeding spring; in other instances, the vessels of successive rings are linked up by radial rows of tracheides, while the two methods of connection may also occur in combination with one another. In all cases communication is carried on through bordered pits, and *not* through open pores. Gnentzsch states that the xylem parenchyma of adjacent annual rings is likewise continuous.

Numerous attempts have been made to furnish a mechanical ex planation of the formation of annual rings.[364] Sachs and De Vries endeavoured to correlate the struc

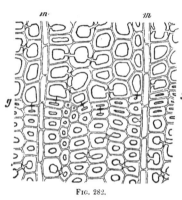

Fig. 282.

T.S. through the Wood of the Yew (*Taxus baccata*). *g*, boundary betWeen autumn and spring Wood; *m*, medullary rays.

tural characteristics of spring and autumn wood with differences in the tangential pressure exercised by the extra-cambial tissues; but this theory was refuted by Krabbe, who proved experimentally that this tangential pressure, and consequently also the compression exerted upon the wood, are approximately equal in spring and in autumn. Russow suggested that turgor-variations might be responsible for the structural differences between spring and autumn wood; but Wieler showed by means of plasmolytic experiments that the osmotic pressure in the young xylem elements is no greater in spring than in autumn. Hartig and Wieler have both attempted to correlate the formation of annual rings with conditions of nutrition, but in so doing have arrived at diametrically opposite conclusions; for, while Hartig maintains that autumn wood is better nourished than spring wood, Wieler regards poor nutrition and an insufficient supply of water as conditions which favour the formation of autumn wood. Lutz accepts Wieler's view to a certain extent, since he attributes the production of spring wood to a high water-content in the extra-cambial tissues and in the newly differentiated xylem. The author himself agrees with Krabbe and Jost, that it is at present impracticable to furnish any mechanical explanation of the origin of annual rings. In this respect the alternation of spring and autumn wood resembles other periodic growth phenomena, the mechanism of which has so far defied explanation.

It is, of course, quite another question, whether the difference between spring and autumn wood can be explained from the teleological point of view[365]; in view of the information that is available concerning

the function of the various elements of the woody cylinder, the answer
to this question would appear to be in the affirmative. The transpiring
surface exposed by the crown of a tree increases in size from year to
year. When, therefore, vegetative activity is resumed in spring-time,
one of the first requirements that has to be met is the demand for
additional water channels. This want is supplied by the inclusion of
numerous vessels in the wood produced during spring and early
summer. Hence, when the transpiratory activity of the crown reaches
its height in July and August, the additional water channels are there
ready for use. The water-conducting arrangements being thus perfected
so far as the current season is concerned, the tree can then concentrate
its attention upon increasing the mechanical strength of the trunk ;
strands of wood-fibres therefore appear, while the thick-walled character
and radial compression of the autumnal elements increases the amount
of mechanically effective tissue in the annual ring.

The foregoing anatomico-physiological explanation of the formation
of annual rings was first put forward by the author in the original
[German] edition of the present work. Subsequently similar opinions
were advanced by Hartig and Strasburger. On the basis of a large
number of measurements performed upon complete transverse sections
of the trunk of *Pinus sylvestris*, F. Schwarz comes to the significant
conclusion that " the distribution and quantitative development of
autumn wood do actually conform very closely to the prevailing
mechanical requirements." One-sided incidence of mechanical stresses
not only leads to an increase in the total growth in thickness on the
compressed side, but also raises the ratio of autumn to spring wood in
the individual annual rings.

Further evidence in support of the view, that the characteristic
differentiation of spring and autumn wood is determined by the general
ecological relations of the plant, is furnished by the occasional duplication of annual rings, a phenomenon known to Unger, and more recently
studied by Kny, Wilhelm and Jost.[366] This duplication, consisting in
the formation of two more or less sharply defined zones of secondary
wood in a single vegetative season, occurs, as a rule, when the young
foliage of the tree is destroyed by frost, the ravages of insects or some
other injurious agency, but is replaced during the same season, owing
to the unfolding of buds which would ordinarily have lain dormant
until the following year. As Jost has recently once more pointed out,
the phenomenon of duplication clearly demonstrates, that a correlation
exists between the production of leaves or the unfolding of buds on
the one hand, and the formation of annual rings on the other, and that
this correlation involves a possibility of self-regulation in respect of the
aforesaid processes.

Apart from the difference in the size of the component elements, which is primarily responsible for the contrast between spring and autumn wood in the individual annual rings, homologous elements increase gradually both in length and in width in successive rings. This statement applies more particularly to the mechanically effective elements, including fibrous tracheides as well as typical wood-fibres, the size of which goes on increasing steadily for a number of years, until a maximum is reached, but thenceforward remains comparatively constant. This general rule is applied to a very variable extent at different heights in the trunk, and in the various branches, twigs and roots of one and the same tree; such variations in detail cannot be considered in the present work. Most extensive and elaborate measurements with regard to this matter have been carried out by Sanio,[367] particularly in the case of *Pinus sylvestris*; the following data are quoted from that observer's results: In a 72-year-old transverse disc of a *Pinus* trunk, taken about 12 inches from the base, the average length of the fibrous tracheides was ·95 mm. in the first annual ring, 2·74 mm. in the seventeenth, 3·87 mm. in the thirty-seventh, 4 mm. in the thirty-ninth, 4·21 mm. in the forty-fifth and also in the seventy-second. The length of the tracheides had evidently attained a maximum value in the forty-fifth ring, and thereafter remained constant. In this case the mean maximum length is more than four times as great as the average length in the first season. The average width of the fibrous tracheides, which is ·017 mm. in the first ring and ·032 mm. in the seventy-second, likewise undergoes an appreciable amount of increase.

In Dicotyledonous wood, it is the fibres which display the greatest progressive increase in length; the vessel-segments and tracheides undergo less change in this respect, while the xylem-parenchyma cells do not deviate appreciably from their initial length. Sanio estimates the average length of the wood-fibres in a 130-year-old trunk of *Quercus pedunculata* at ·42 mm. in the first, and at 1·22 mm. in the three outermost annual rings. In the case of the tracheides, the initial average length was ·39 mm., the mean maximum ·72 mm. While, therefore, the wood-fibres had progressively increased nearly threefold in longitudinal diameter, the tracheides had not even doubled their length.

It is most probably this progressive elongation of the elements of the wood in the successive annual rings, that is responsible for the fact, that the strictly longitudinal orientation which prevails in the earlier rings, is gradually replaced by a tangentially oblique position of the cells.[368] The origin of this obliquity—which has been observed by A. Braun in 111 out of 167 woody Gymnosperms and Dicotyledons —is, however, not yet quite clear. According to Braun, the inclination of the fibres to the longitudinal axis of the stem, does not usually exceed 4° or 5°; in individual cases, however, it may be much greater, e.g. 10°–20° in *Aesculus Hippocastanum*, 30° in *Syringa vulgaris*, 40° in *Sorbus Aucuparia*, and 45° (the maximum value record) in *Punica Granatum*.

With regard, finally, to the advantage which the plant derives from the above-described increase in length of the xylem elements, and from the oblique "grain," which is probably a secondary consequence of that increase, attention may be once more directed to the fact that the increase in length is most pronounced in the case of the mechanically effective elements. That the resistance of the stereome increases with the length of its component cells, is a self-evident proposition. It is highly probable that the oblique course of the fibres also adds to the strength of the trunk; for, when the stem is bent, the tensile strength of the oblique fibres on the convex side is only tested by a portion of the total stress, the magnitude of this component depending upon the angle of inclination of the fibres; the other component acts at right angles to the course of the fibres, and thus really constitutes a shearing stress, so far as they are concerned. Thus, a woody cylinder with oblique fibres reproduces on a larger scale the structure of the individual mechanical cells, with walls composed of oblique molecular or micellar series.

Hitherto we have confined our attention to the width of the annual rings and the dimensions of their component cells in particular transverse sections, that is, at a given level in the trunk. Some consideration must now be given to the differences in the development of the annual rings at different heights in the trunk.[369] Von Mohl long ago concluded, on the basis of measurements made upon various Conifers, that the average width of the annual rings increases constantly from below upwards; the rings are also relatively wide right at the base of the trunk. F. Schwarz has carried out numerous measurements upon *Pinus sylvestris* with regard to this point; he records the following average widths of the annual rings at different levels in a Pinus trunk, 30 m. in height, for the period 1885-1894.

| Height from ground : in *m*. | ·3 | 1·4 | 5·5 | 19·9 | 24·1 | 26·2 | 28·4 |
|---|---|---|---|---|---|---|---|
| Average width of annual rings : in *mm*. | 1·14 | ·96 | ·83 | ·92 | 1·1 | 1·44 | 1·35 |

According to Schwarz, the narrowest annual rings in *Pinus sylvestris* are found, on an average, at a height of 1·3 to 3·4 m.; from this point onwards the width of the rings increases slowly, but steadily, towards the distal end of the trunk, where it finally diminishes once more.

The gradual increase in the average width of the annual rings from below upwards, is correlated with a tendency on the part of the tree to develop its trunk as a " girder of uniform strength " in relation to the bending action of the wind.[370]   This interpretation was first definitely put forward by Schwendener; subsequently Metzger proved, by means of numerous measurements of the radius at various levels, that the trunk of the spruce does actually conform very closely to the mathematical equation for girders of uniform strength of circular cross-section.   The elaborate researches carried out upon *Pinus sylvestris* by F. Schwarz led to similar results.   At the base of the trunk the annual rings considerably exceed the width necessary in the case of an imaginary girder of constant resistance, the greater thickness at the base depending upon the fact that this region of the trunk has to cope with the special demands which arise out of the necessity of anchoring the stem in the ground.   Metzger accordingly compares the base of the trunk to a rigid socket, in which the rest of the stem is embedded. The secondary diminution in the width of the annual rings, at the distal end of the trunk, is evidently correlated with the fact that the surface exposed to wind decreases (in the upper part of the crown) as one passes from older to more recent whorls ; the reduction of pressure permits of a corresponding weakening of the trunk.   Taking them all in all, the difference in the width of the annual rings, and consequently of the cross-sectional area of the trunk at various heights, is the result of very perfect adaptation to the mechanical demands that are made upon the trunk by the bending action of wind.

*4. Normal changes in the physical condition of the wood.*[371]

There are comparatively few deciduous trees (such as *Acer Pseudoplatanus, A. platanoïdes, Buxus sempervirens, Betula alba, Populus tremula*) the wood of which retains its original characteristics all over the cross-section of the trunk ; as a rule, the most recently added annual rings form a definite zone of **sap-wood** (*alburnum*), which is clearly demarcated from the older **heart-wood** (*duramen*).   The sap-wood presents the yellowish-white appearance that one ordinarily associates with lignified tissue, and is entirely composed of cells which are still functional.   Hartig, Wieler and Strasburger have, however, shown that, even in the case of the alburnum, every part is not engaged in the conduction of water to the same extent.   Under normal conditions, in fact, it is only the outermost annual rings that serve to convey the transpiration-current, the more internal portions of the sap-wood serving for the storage of water and synthetic products.   The dark-coloured heart-wood takes no part whatever in conduction or storage ; its functions are purely mechanical. All the cell-cavities in this central region are filled with various peculiar organic compounds, which likewise often impregnate the cell-walls. Among these substances special mention may be made of the pigments of dye-woods such as the *haematoxylin* of Log-wood (*Haematoxylon campechianum*), the brick-red *brasilin* of Brazil-wood (*Caesalpinia Sappan*), the dark-red *santalin* of Sandal-wood (*Pterocarpus santalinus*), the *morin* of *Maclura aurantiaca*, etc. ; resinous and gummy bodies are also frequently present, while in a variety of deciduous trees (*Ailanthus*, *Prunus*, *Xanthoxylon*, *Gleditschia*, *Sorbus*) Sanio found certain (at any rate originally) colourless substances, which are distinguished by their resistance to all ordinary solvents. According to Gaunersdorfer, *tannins* are often included among the substances with which heart-wood is impregnated.

In addition to these various organic substances, certain inorganic compounds are also not infrequently deposited in heart-wood. According to Crüger, for example, considerable quantities of amorphous *silica* accumulate in the cavities of the cells and vessels in *Hirtella silicea*, *Petraea volubilis* and *P. arborea*, and, in the vessels alone, in *Tectona grandis*. Molisch has demonstrated the occurrence of considerable deposits of *calcium carbonate* in the duramen—and especially in the old vessels—of a number of deciduous trees (*Ulmus campestris*, *Celtis orientalis*, *Sorbus torminalis*, *Fagus sylvatica*). The deposit of lime " is usually so heavy that, when the wood is burnt, the ash contains solid casts which reproduce not only the shapes of the vessel-cavities, but also every detail in the sculpturing of their walls." Molisch attributes the accumulation of calcium carbonate to the enfeebled conducting capacity of the heart-wood.

The changes above described naturally cannot fail to affect the physical properties of the wood. Heart-wood is, in fact, denser, harder and, as a rule, also tougher than the sap-wood of the same tree. These properties, which are, of course, responsible for the technical value of heart-wood, are also mechanically advantageous so far as the plant itself is concerned. From the ecological point of view, therefore, one cannot agree with De Bary that " the formation of hard and durable heart-wood is merely an episode of the early stages of degeneration." We must, on the contrary, regard the production of typical duramen, which is both mechanically superior to alburnum and better protected against decay and disintegration, owing to its being impregnated with the above-mentioned peculiar substances, as a perfectly normal and advantageous modification of the older wood. The complete loss of function, in respect of nutritive metabolism, is balanced by the greatly enhanced mechanical efficiency of the xylem.

In a few instances (*e.g.* in certain Willows and in the Canadian Poplar) the heart-wood does not differ appreciably from the sap-wood as regards either its strength or its durability. In such cases, the duramen is liable to decay, and its destruction is often hastened by the attacks of Fungi, so that trees of this type usually become hollow at a comparatively early stage.

The gummosis which often affects the wood of the AMYGDALEAE is a pathological phenomenon.

### III. SECONDARY GROWTH IN THICKNESS IN MONOCOTYLEDONS.[372]

It has already been pointed out, at the beginning of the present chapter, that, in the great majority of Monocotyledons, the entire growth in thickness takes place in the apical meristem. The stem does not begin to elongate until the growing point has undergone a definite amount of enlargement; in certain Palms elongation is succeeded by a second phase of enlargement (cf. above, p. 647). But genuine secondary thickening, with addition of entirely new tissues, only occurs among the arborescent LILIIFLORAE (*Aloe, Yucca, Dracaena, Cordyline, Aletris*), and in the tubers of the DIOSCOREAE; and even in these cases the process differs in certain essential features from the secondary growth in thickness of Gymnosperms and Dicotyledons.

In the trunks of the aforesaid LILIIFLORAE, the primary arrangement of tissues conforms to the ordinary Monocotyledonous scheme. The course of the bundles is of the same type as in Palm-stems. The cambial layer sometimes originates immediately behind the apex (*Yucca aloifolia, Aloe plicatilis, Beaucarnea tuberculata*), before the primary meristematic layers have become converted into permanent tissues; in other cases its appearance is deferred until all the permanent tissues have been differentiated at that level (*e.g.* in most species of *Dracaena*). In any case it always arises from one of the innermost layers of the cortical parenchyma—that is, on the outside of the vascular cylinder. The elements of this layer undergo active tangential division, and so give rise to a zone of meristematic cells (Fig. 283 A, *v*), which, however, differ in certain respects from genuine cambial elements; they are, for example, at most two to four times longer than their width, and are not prosenchymatous. Schoute states that at first this cambial layer is regenerated at the expense of the cortical layer immediately external to it; later on a definite initial layer is differentiated.

The cambial zone gives rise, on both sides, to a variety of secondary tissues; as in the case of Gymnosperms and Dicotyledons, these may be classified topographically under the heads of extra-cambial and intra-cambial tissues. The extra-cambial portion appears relatively late and is never voluminous; it consists of thin-walled parenchyma, containing a certain number of crystal-sacs. The woody cylinder, which is much more bulky and complex, represents a peculiar modification of the primary Monocotyledonous vascular arrangement, and thus differs very markedly

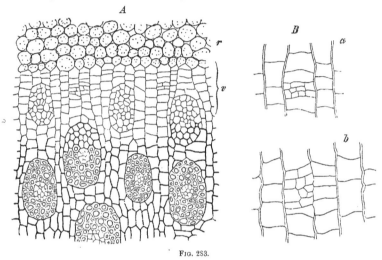

FIG. 283.

A. Part of a T.S. through the trunk of *Dracaena marginata*, illustrating the formation of secondary vascular bundles; *v.* cambium; *r*, cortical parenchyma.  *B a*, *B b*.
Early stages in the development of the bundles.

from the analogous structure in Dicotyledonous stems; it is com posed of numerous secondary vascular bundles embedded in a mass of secondary parenchyma (Fig. 283 A). The demarcation of the vascular bundles begins within the cambial zone, where repeated longitudinal divisions give rise to procambial strands of various sizes; as the cambial zone progresses outwards, these procambial strands are left behind and become transformed into permanent vascular tissue in centrifugal order. The smaller procambial strands may be seen to be derived from single cells in any particular cross-section (Fig. 283 B, *a*); the larger ones, on the other hand, originate from several mother-cells, belonging to at least two different radial series. In the formation of a procambial strand, the initial divisions are always tangential; these are first succeeded by radial divisions (Fig. 283 B, *a*, *b*), but, once sliding growth begins, walls may arise in any direction. The primary leaf-trace bundles present no peculiar features; ordinarily they are collateral, but in

*Cordyline*, according to Strasburger, become leptocentric (amphivasal) on entering the central cylinder. They are surrounded by typical bast-fibres, or by thick-walled parenchyma. The secondary bundles differ considerably from the primary strands, the functions of mechanical support and water-transport being both allocated to the same organic elements, just as in the woody cylinder of many Gymnosperms and Dicotyledons. The cells in question closely resemble the fibrous tracheides of Conifers, being thick-walled and prosenchymatous, and provided with narrow, oblique (sinistral) bordered pits. In each secondary bundle, the delicate central leptome is surrounded by a layer of these tracheides, which is interrupted at certain points, according to Strasburger, by parenchymatous passage-cells (*Cordyline*). The various secondary bundles are linked together by numerous anastomoses, so as to form a network with vertically elongated meshes.

The interfascicular secondary parenchyma, which is in a sense comparable to the medullary ray tissue of the Dicotyledonous woody cylinder, displays a distinct radial seriation, which is, of course, disturbed in the vicinity of the vascular bundles. The individual cells are furnished with moderately thick and pitted walls. From the physiological point of view, this tissue probably corresponds to conducting parenchyma.

In many species of *Dracaena* the older roots also undergo secondary growth in thickness. Here the cambial cylinder originates in the pericycle (pericambium); so far as is known the secondary tissues are precisely the same as in the stem.

The results which may be attained by means of the Monocotyledonous type of secondary thickening, cannot be more strikingly illustrated than by reference to the historical Dragons' Blood tree (*Dracaena Draco*), of Orotava (Teneriffe); this gigantic tree, first rendered famous by Von Humboldt, was destroyed by a storm in 1869. Over 70 feet in height, it had, in Von Humboldt's day, attained a girth, a short distance above the ground, of at least 45 feet. Its age was estimated at approximately 6000 years.

## B. ANOMALOUS FORMS OF SECONDARY GROWTH IN THICKNESS.

### I. GENERAL CONSIDERATIONS.

The mode of secondary growth in thickness which has been described in the preceding section, may be termed the **normal** type, so far as Dicotyledons and Gymnosperms are concerned. Its wide distribution may be attributed to the fact that it lends itself to the production of a massive woody cylinder more readily than any other method of secondary thickening.

In a considerable number of cases, however, secondary growth in thickness deviates more or less considerably from this normal type. Ontogenetically and morphologically these so-called **anomalous** forms of thickening are characterised by an unusual disposition of the cambium, and by abnormalities in the structure and arrangement of the secondary tissues.

From the physiological point of view all the anomalous forms of secondary growth fall under one of two categories. The first of these includes all those cases in which the deviation from the normal type is clearly connected with adaptation to definite external conditions. The bulk of this class is made up of the stems of climbing plants, which are widely different from ordinary vertical woody stems, in respect both of their mechanical requirements and of the conditions under which translocation has to take place. The first group of anomalous organs likewise includes many fleshy roots which serve principally for storage, as well as the submerged stems of certain woody marsh plants (*Aeschynomene aspera, A. indica, Herminiera Elaphroxylon, Pterocarpus Draco,* etc.), in which, according to Goebel, the cambium devotes its energies principally to the production of ventilating tissue in the shape of " floating wood." [373]

The second group of anomalous types comprises those instances— likewise a considerable number—in which the abnormal features are not adaptive but merely represent variations of design. Theoretically, namely, secondary growth in thickness might take place in a variety of different ways ; but the probability of any given type of structure actually appearing in nature largely depends upon its suitability. The determining influence of this consideration is greatest where bulky structures have to be produced. Where, however, a large woody cylinder is not required, modes of secondary thickening which are less efficient than the " normal" form of cambial activity, may suffice. Hence the mechanism of secondary growth exhibits a considerable amount of variation in the case of herbs and small shrubs. The resulting variations of design are all of approximately the same value from the physiological point of view ; in this respect they recall the variations in the cross-sectional outline presented by the stereome (cf. p. 170).

The extraordinary diversity of the anomalous modes of secondary thickening may now be illustrated by a series of examples, which will be classified in relation to the principles that have just been discussed. It need scarcely be said, that it is quite impossible to give anything like an exhaustive treatment of this subject in the space at our disposal.

## II. SECONDARY GROWTH IN THICKNESS OF LIANE-STEMS.[374]

The mechanical requirements of liane-stems are quite different from those of the trunks and branches of ordinary woody plants, which are almost always constructed with a view to inflexibility. A climbing stem requires above all things to be inextensible. When hanging freely, it has to bear its own weight; when wound around another living stem [or when attached thereto by tendrils], it is exposed to pulling and shearing strains, in consequence of the expansion of the support by growth, or its agitation by wind. In addition, long liane-stems have to be flexible, in order to escape the risk of their buckling or breaking, when they themselves or their supports are shaken by violent winds. If branches of the supporting tree are broken, or if the whole tree falls, the flexibility of the liane-stem is put to a still severer test. Moreover, every climber is itself constantly altering its relation to the support. For as the older grasping organs become useless, the stem gradually slips downwards, and may come to lie to some extent upon the ground; as a result, it is often thrown into a number of folds or loops, and also frequently becomes more or less extensively twisted. Twining stems are, in addition, often exposed to radial compression, owing to the growth in thickness of the supporting organ. These very special mechanical conditions are responsible for some of the most remarkable of the various anomalous types of secondary thickening; Fritz Müller was the first to examine such abnormalities from this physiological point of view.

All long inextensible liane-stems are constructed more or less after the pattern of a twisted rope or cable. This cable-like structure, however, which combines pliancy with a high degree of tensile strength, cannot come into being, if the woody cylinder is developed in the form of a solid mass. The wood must, on the contrary, consist of separate strands, which can slip past one another; this essential condition can only be fulfilled, when the woody cylinder is split up into more or less completely separated strands, by the interpolation of softer tissues. As a matter of fact, this splitting up of the wood is an anatomical feature common to all liane-stems, although it may be brought about in a great many different ways. Within recent years, Schenck has given a very careful and exhaustive account of climbing stems from this point of view.

In the less specialised forms of liane-stem, the cambium is normal to begin with, but soon produces xylem and phloem in different proportions at different points of its circumference. This relatively

simple condition prevails, for example, among Bignoniaceous tendril-
climbers, and also in various MALPIGHIACEAE (*Tetrapteris, Banisteria,
Stigmaphyllon*) and APOCYNACEAE, in the genus *Phytocrene*, etc.
Among the BIGNONIACEAE, the output of secondary wood is relatively
small from the first at four points, which are arranged in the form of a
cross, while the production of secondary phloem at these points is
correspondingly increased. As a result, the woody cylinder appears to
be provided with four longitudinal grooves. These grooves become
constantly deeper, as growth in thickness proceeds ; but since they are
filled with secondary phloem, the cross-sectional outline of the whole
vascular cylinder is not appreciably affected by this unequal development of the secondary xylem. The cambium itself becomes broken up,

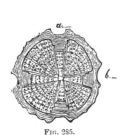

FIG. 284.

FIG. 285.

FIG. 286.

T.S. through a four-year-
old twig of *Bignonia capreo-
lata*. The Woody cylinder is
traversed by four furroWs or
grooves occupied by phloem.
×3. After De Bary.

T.S. through an undetermined
Bignoniaceous stem (? *Pleonotoma
sp.*). Nat. size. After Schleiden.

T.S. through a twig of *Melloa populi-
folia*. The lobed Woody cylinder is
shoWn in White, the phloem shaded.
×2. After De Bary.

as a result of its peculiar behaviour, into a succession of strips, the four
widest among which lie opposite the four projecting ridges of wood,
while the narrower portions are situated at the bases of the grooves.
In certain BIGNONIACEAE the furrows always remain four in number
(Fig. 284); in other species, on the other hand, additional grooves are
continually interpolated, as the result of a reduction of wood-production
at an increasing number of points in the circumference of the cambium.
The embedded plates of secondary phloem also often increase in width
towards their outer ends by a series of step-shaped increments
(Fig. 286). In this way the whole woody cylinder becomes split up
into a large number of separate lobes and strips.

In certain MALPIGHIACEAE, BIGNONIACEAE, etc., the splitting up of
the wood is carried a stage further, owing to secondary changes in
the xylem-parenchyma and pith. The primarily continuous, though
greatly lobed, woody cylinder becomes completely broken up, owing
to the active intercalary growth of interpolated masses of parenchyma. According to Schenck, this " dilatation-parenchyma " in

many cases arises from *unlignified* thin-walled xylem-parenchyma; in certain instances, however, it is lignified elements associated with vessels, or cells of the xylem rays, that undergo this secondary growth and division. Warburg, on the other hand, describes how, in a species of *Bauhinia*, the initial cells of the strips and wedges of

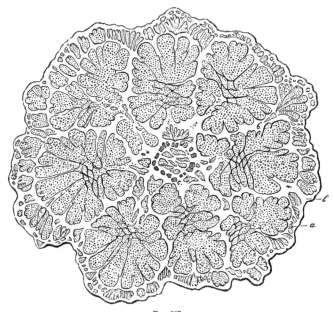

Fig. 287.

T.S. through a stem of *Bauhinia* sp. The irregular masses of xylem (*a*) dotted, the phloem and parenchymatous tissue (*b*) shown in white. Two-thirds nat. size. After Schleiden.

dilatation-parenchyma originate within the phloem, and only secondarily penetrate into the wood.

Once separate strands of wood have been formed, they may each grow in thickness by means of independent secondary meristems; additional strands may also be interpolated between the first-formed bundles. The whole stem thus comes to consist of a tangle of interwoven strands of wood, and accordingly often presents a highly complicated appearance in cross-section (Fig. 287). The resemblance of such a liane-stem to a thick rope is greatest, where the plates of tissue interpolated between the various strands of wood break asunder and are partly replaced by periderm. In such cases, mere external inspection shows the stem to be split into a number of longitudinal strands enveloped in cork, and interwoven or partially fused in a complicated manner.

In the instances that have been considered so far (except for *Bauhinia*), the secondary developments always start from a single cambial layer of normal character. In Sapindaceous tendril climbers, on the other hand (*Serjania*, **Paullinia**, *Thinouia*, etc.), several cambial layers are present from the first. These may be arranged in various ways in the cross-section of the stem. Sometimes the central portion of the organ is built up by a principal cambial cylinder, around which are grouped several minor peripheral cylinders; more rarely there is no central cambium, but only a circle of from five to seven peripheral cylinders, which are all of approximately equal strength.

Since this second type of secondary thickening was described by Gaudichaud, it has been carefully studied by a number of investigators,

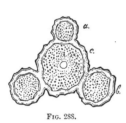  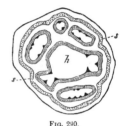

| Fig. 288. | Fig. 289. | Fig. 290. |
|---|---|---|
| T.S. through the stem of *Serjania sp.* (or *Paullinia sp.*). *c*, principal or central Woody cylinder; *a*, *b*, peripheral Woody cylinders. Nat. size. After Schleiden. | T.S. through the stem of another species of *Serjania* (or *Paullinia*). Nat. size. After Schleiden. | T.S. through a young internode of *Serjania caracasana*. *s*, within, the fibrous cylinder; *h*, principal or central vascular cylinder, surrounded by four peripheral cylinders. The primary bundles (leaf-traces) shown in black. |

among whom Nägeli and Radlkofer may be specially mentioned. Nägeli, who paid special attention to the ontogenetic development of the individual cambial layers, found that the circle of primary leaf-trace bundles is more or less deeply indented at certain points, so that individual groups of bundles tend to be abstricted from the circle even at this early stage. As a matter of fact, the first traces of this peculiar tendency are shown by the procambial strands; hence, when these become linked up to form a continuous cambial cylinder, the abstricted groups behave as independent circles of strands, and ultimately give rise to so many separate cambial cylinders (Fig. 290). Thenceforth each individual cambial layer behaves in normal fashion, cutting off secondary phloem on the outside and secondary xylem on the inside (Figs. 288, 289). The stem thus consists from the first of several separate woody cylinders; in other words, the cleavage of the wood is not a secondary process (as in the BIGNONIACEAE), but is predetermined by the behaviour of the apical meristem. It should, however, be stated that the secondary growth of some Sapindaceous climbers approaches the

Bignoniaceous type, the woody cylinder being continuous, but deeply lobed, or provided with prominent ridges.　It is further noteworthy, that in those SAPINDACEAE which are not climbers,—the bulk of the family—the structure of the stem is perfectly normal, so that the very aberrant construction of the climbing species really seems to be an adaptive feature.

A third type of anomalous thickening found in liane-stems, is characterised by the appearance of additional complete or partial cambial cylinders external to the primary cambium.　Each individual cambial layer produces secondary xylem and phloem in normal fashion for a time ; sooner or later its activity comes to an end, whereupon a new layer of secondary meristem arises further out in the fundamental parenchyma.　This process may be repeated several times.　Two different sub-types may be distinguished, with regard to the point of origin of these successive cambial layers.　The additional meristems may arise either in the primary cortex, as in various MENISPERMACEAE, in *Wistaria* and *Rhynchosia*, or in the secondary phloem, as in *Gnetum, Bauhinia, Machaerium, Phytocrene*, and certain BIGNONIACEAE (*Haplolophium, Glaziovia, Anisostichus*, etc.) and CONVOLVULACEAE (*Ipomaea, Argyreia, Convolvulus*).

A very remarkable type of anomalous thickening, finally, is exemplified by those lianes in which a secondary cambium arises on the inner side of the normal woody cylinder.　In *Tecoma radicans*, where this phenomenon was first observed by Sanio, the internal cambium gives rise both to xylem and to phloem, but in inverse order.　According to Scott and Brebner, a similar condition prevails in the stems of *Willughbeia firma* (APOCYNACEAE) and *Periploca graeca* (ASCLEPIADACEAE).

The most characteristic anatomical features of liane-stems are those which tend to produce a rope- or cable-like structure, and thus render the organs more pliable.　Tensile strength is most marked in the case of the young stem, where the first-formed or central secondary xylem (the "axile wood" of Strasburger) contains a remarkably large proportion of mechanical elements (wood-fibres and fibrous tracheides), whereas its vessels are unusually narrow and its xylem-parenchyma very feebly developed.　The later-formed portion of the secondary wood (the "periaxile" wood of Strasburger), on the contrary, contains large vessels and abundant xylem-parenchyma.　According to Schenck, this contrast between the central and the peripheral secondary xylem of liane-stems is almost invariably obvious and abrupt.　The younger stems are therefore distinctly inextensible in construction, after the manner of roots which have a stereome strand in the centre of the vascular cylinder.　As the stems grow older and thicker, however, their inextensible character becomes less pronounced, just as inflexibly constructed trunks and branches do not retain the peripheral disposition of their stereome, after secondary thickening has been in progress for some time.    In both cases, it is the numerical strength of the scattered mechanical elements that compensates for their failure to congregate at the centre or near the periphery, in accordance with approved mechanical principles.

The anomalous modes of thickening characteristic of climbing stems, besides leading to the rope-like construction based upon a splitting up of the woody cylinder, also involves another deviation from the normal arrangement of tissues; for the secondary leptome, instead of occupying a peripheral position, as it does in stems with normal secondary growth, shifts inwards—at any rate to a large extent —and becomes wedged in between the various strands or cylinders of secondary wood.   The formation of such **inter-xylary phloem** often depends upon the fact that the cambium, besides cutting off secondary phloem on its outer side, also gives rise on its inner side to isolated strands of sieve-tubes, accompanied by parenchymatous tissue; these intra-cambial masses of phloem (such as occur in *Dicella, Combretum, Mucuna, Entada*, etc.) thus become embedded in secondary wood.

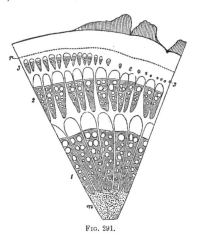

FIG. 291.

Sector of a T.S. through the stem of *Gnetum scandens*. 1, 2, 3 successive vascular cylinders; *m*, pith; *r*, mechanical cylinder situated at the inner boundary of the cortex. Xylem-strands shaded (except for the cavities of the large pitted vessels); phloem, cortex and medullary rays shown in white.  ×8. After De Bary.

According to Leisering, however, the relation of such internal phloem-strands to the cambium is indirect, since they become secondarily differentiated within masses of tissue resembling unlignified xylem-parenchyma.   In the genus *Strychnos*, leptome-strands of extra-cambial origin *subsequently* become enclosed in the woody cylinder, as a result of special growth-processes.

That inter-xylary leptome is particularly well protected against mechanical injury is an indisputable fact, and doubtless also an advantageous circumstance, since liane-stems are particularly liable to be subjected to violent flexure, to say nothing of the radial compression produced by the expansion of the living support.   It must, on the other hand, be noted that inter-xylary leptome-strands do occur in a number of woody plants which are not climbers, a circumstance which led Schenck to conclude that this peculiar feature is not an adaptation to the climbing habit, but is mainly a question of variety of design. The fact remains, that inter-xylary leptome is of particularly frequent occurrence in climbers belonging to the most diverse families.

The remaining histological peculiarities of liane-stems may be disposed of in a few words. Westermaier and Ambronn long ago drew attention to various features which represent anatomical responses to increased demands in the matter of translocation. The remarkable width of the vessels and sieve-tubes in the stems of climbing plants has been discussed in detail on a previous occasion (pp. 324, 334). Whether Westermaier and Ambronn are right in regarding the great development of xylem-parenchyma, and the often unusual vertical and tangential extent of the medullary rays, as further devices for increasing the rate of translocation is another question; in Schenck's opinion the last-mentioned anatomical feature serves to facilitate the bending and twisting of the stem.[375]

It has already been explained that, in their later stages of development, liane-stems often differ from ordinary upright woody stems, even as regards their external form. By far the most remarkable illustration of this point is provided by the greatly flattened ribbon-shaped stems of certain Leguminosae (*Rhynchosia*, *Dalbergia*, and *Machaerium* among the Papilionaceae; *Bauhinia* among the Caesalpinieae), which owe their characteristic shape to an unequal distribution of cambial activity, and in some cases also to the appearance of successive cambial layers. Such ribbon-shaped stems are sometimes thrown into folds in a most extraordinary manner; often the two edges take no part in the corrugation. The growth-processes responsible for this peculiar development have been studied and discussed by Crüger, De Bary, Von Höhnel, Warburg, and Schenck, but are still imperfectly understood. The ecological significance of the undulations is also not quite clear. Schenck supposes that the numerous abrupt curvatures assist the stem to gain a fresh hold, if it slips down through its tendrils letting go or the support giving way. With regard to those ribbon-shaped stems which are *not* corrugated, Schenck points out that a flattened organ will lodge more readily upon the supporting branches than a round one.

### III. SECONDARY GROWTH IN THICKNESS OF FLESHY ROOTS.[376]

The roots of a great many biennial and perennial Dicotyledons act as organs of storage [and hibernation] after the aërial structures have withered, and in accordance with this function assume a " fleshy " character. Such fleshy roots exhibit various anomalies in their secondary growth, all of which tend in the first place to ensure the development of a suitable parenchymatous storage-tissue; in addition, the anomalous cambial activity often leads to the multiplication and appropriate arrangement of the conducting strands required for the deposition and removal of the reserve-materials.

The morphological value of the parenchymatous storage-tissue varies greatly in different fleshy roots. In many cases there is a massive development of extra-cambial storage parenchyma, corresponding either to primary cortex (lateral roots of ASCLEPIADACEAE and PIPERACEAE) or to secondary phloem (*Taraxacum, Rubia,* UMBELLIFERAE). In either instance the secondary growth cannot strictly be termed anomalous, the deviation from the normal condition consisting solely in the relatively feeble development of the woody cylinder, as compared with the extra-cambial tissues. In another series of types the storage-tissue owes its origin to the massive development of the parenchymatous elements of the wood—that is, of the xylem-rays or xylem-parenchyma (*Urtica, Cucurbita, Symphytum officinale*); the latter condition is excellently illustrated by the napiform roots of *Brassica* and *Raphanus,* where the secondary xylem consists of a massive parenchymatous tissue, which is traversed by strands of vessels accompanied by a certain number of fibres. These fibro-vascular strands—the term is highly appropriate in the present case—are arranged in regular concentric circles, and form a network with vertically elongated meshes. In roots of this type, the anomaly consists essentially in the aberrant composition of the xylem; the characteristic peculiarity of the wood is often accentuated by the fact, that the distinction between xylem-parenchyma and medullary ray tissue is not strictly maintained (*Scorzonera hispanica, Raphanus, Brassica*).

J. E. Weiss has shown, that in a number of fleshy roots, **tertiary vascular bundles** may be differentiated within the storage parenchyma of the secondary xylem; in *Oenothera biennis* these tertiary strands consist solely of leptome, whereas in *Cochlearia Armoracia, Brassica Napus, B. Rapa, Raphanus sativus,* and *Gentiana lutea* they take the form of concentric bundles with central leptome. In *Bryonia dioica* the adventitious cambium surrounds the vascular strands of the woody cylinder, forming additional vessels in immediate contact with the pre-existing ones, and cutting off leptome tissue on its outer side.

The secondary growth in thickness exhibits even more striking anomalies in those roots which develop several cambial layers in succession. In the Beet-root (*Beta vulgaris*), a second cambium arises by tangential division in the parenchyma situated outside the primary cambial zone; it is only active for a limited period, and is in its turn succeeded by a third cambial layer, which originates at the outer limit of the phloem produced by its predecessor. This process may be repeated several times. The result is the formation of concentric zones composed alternately of xylem and phloem, the latter constituting the bulk of the storage-tissue. According to Trecul, the roots of *Myrrhis odorata* likewise develop several additional cambial layers, which, however, originate on the inner side of the primary cambium, and cut off xylem and phloem in inverse order.

The most complicated variety of this type of anomalous secondary thickening is found in certain CONVOLVULACEAE, and in the genus *Rumex*, where partial or complete additional cambial cylinders arise successively, sometimes within and sometimes outside the primary cambium, all forming xylem and phloem quite independently of one another. It is quite unnecessary to discuss these highly complicated cases in detail.[377]

Attention has already been drawn to the fact that such peculiarities of secondary growth do not *merely* serve to ensure the production of the requisite amount of storage-tissue ; another object of these " anomalous " processes is the thorough inter-penetration of storing and conducting tissues, an arrangement which renders the deposition and renewal of reserve-materials easier and more effective. The subsequent differentia tion of (tertiary) leptome-strands within the secondary xylem cannot be explained from any other point of view.

### IV.  NON-ADAPTIVE ANOMALIES OF SECONDARY GROWTH IN THICKNESS.

In conclusion, a few words must be devoted to those anomalous modes of secondary thickening which appear to be merely cases of variation of design. The anomalies in question occur more parti- cularly in certain herbs and small shrubs belonging to the CHENO- PODIACEAE, AMARANTACEAE, NYCTAGINACEAE, and TETRAGONIEAE, and in species of *Mesembryanthemum* and *Phytolacca*.[378] That we are really dealing with non-adaptive features, in such cases, follows at once from the circumstance that the abnormal features may vary considerably, not only within the limits of a single family, but even in different species of the same genus, quite independently of any differences in the ecological relations of the plants concerned.

The secondary growth in thickness of the CHENOPODIACEAE—and of the above-mentioned allied families—has been investigated by Morot, Hérail, and Leisering. In these plants one or more cambial layers arise in the pericycle—that is, outside the circle of primary leaf-trace bundles ; these extra-fascicular meristematic layers, which cut off secondary phloem externally and secondary xylem internally in perfectly normal fashion, are rarely complete; as a rule each only extends over a portion of the circumference, its edges intersecting the next older cylinder or curved strip of cambial tissue. According to Morot, each new arc of cambium arises opposite a phloem group, and extends laterally on either side until it meets an older cambial layer. Hérail, on the contrary, maintains that each new cambium begins to develop at one end as a lateral continuation of an antecedent cambial layer; thence it extends gradually across the leptome, and sooner or later rejoins an older cambial strip. Leisering believes that both possibilities may be realised. The individual strips of cambium are often small and very numerous, in which case they present the appearance of a dense network in a transverse section; since every part of this meristematic reticulum continually produces secondary xylem on its inside and secondary phloem on its outside, the final product of its activity is a homogeneous woody cylinder comprising vessels and xylem-parenchyma,—and usually also numerous wood-fibres—which encloses isolated leptome-strands. According to Leisering, this type of secondary thickening occurs not only in the above-mentioned families, but also among the HIPPOCRATEACEAE, PLUMBAGINACEAE, MELASTOMACEAE, LORANTHACEAE, COMBRETACEAE, etc.

Some account must, finally, be given of the secondary growth in thickness of Cycad-trunks.[379] Among Cycads the arrangement of the vascular tissues deviates considerably from the normal type, even in the primary condition of the stem. The broad leaf-bases contain numerous bundles; at the surface of insertion these are reduced by fusion to a single pair of leaf-trace strands, which at first follow a peripheral and nearly horizontal (tangential) course in the stem, but finally run obliquely downwards and inwards to join the circle of primary bundles situated between pith and cortex. Both these last-mentioned tissues are bulky, and consist of thin-walled parenchyma, which contains a large amount of starch, and in all probability mainly serves as a storage-tissue. The two members of each pair of leaf-traces are linked together by a horizontal commissure, the whole structure forming a transverse ring or girdle; further anastomoses appear, some of which connect neighbouring girdles with one another, while others place the girdles in direct communication with the primary bundles. In this way a richly branched cortical vascular reticulum is formed, a feature which is no doubt correlated with the great thickness of the cortical region. All these original vascular bundles—collectively termed by De Bary the "primary network of bundles"—are collateral in structure.

Secondary growth in thickness is initiated by the appearance, in the circle of primary bundles, of a cambium which produces secondary xylem and phloem in strictly normal fashion. The medullary rays, which are unusually broad and altogether massive, are traversed by the leaf-traces and by the radial anastomoses of the primary network. Since this network permanently retains its individuality, both radial and tangential anastomoses must constantly elongate as secondary thickening progresses; one result of the continued extension is, that the tracheides in these anastomoses ultimately undergo deformation and disruption. The genera *Zamia*, *Dioon*, and *Stangeria* never develop more than a single cambium. In *Cycas* and *Encephalartos*, on the other hand, a succession of more or less concentric cambial layers arise near the periphery of the secondary phloem (according to Constantin and Morot in the pericycle). Each of these successive "adventitious" cambial zones remain active for several seasons.

Regarding the products of cambial activity in the Cycads, it may be stated, that the secondary xylem consist of tracheides with scalariform-reticulate thickenings, or with transversely elongated bordered pits, together with a certain amount of xylem-parenchyma; the secondary phloem comprises bast-fibres as well as sieve-tubes.

In the genus *Cycas* itself, a secondary network of bundles—continuous with the primary network—appears in the cortical parenchyma. In *Encephalartos*, on the other hand, a dense weft of secondary vascular strands becomes differentiated in the pith; the peripheral strands of this medullary system become connected with the inner surface of the woody cylinder.

By way of a supplement to the preceding discussion of secondary growth in thickness in Phanerogams, a brief account may be given of the analogous processes which take place in certain Cryptogamic forms.[380]

In two genera of the OPHIOGLOSSACEAE, namely, *Botrychium* and *Helminthostachys*, the primary central cylinder becomes greatly dilated by the differentiation of a parenchymatous pith: in addition, a meristematic layer sooner or later appears between hadrome and leptome; the amount of secondary tissue produced by this cambium is, however, limited, and consists entirely of tracheides cut off on the inner side. The "cambium" which occurs in the tuberous stems of *Isoëtes* apparently only serves to increase the amount of storage-tissue; here the majority of the secondary elements are extracambial, only a small number being incorporated in the axile strand. Secondary tracheides have been observed in *Isoëtes lacustris* and in *I. Durieui*, but even here seem to be few and far between.

Among Algae secondary thickening is confined to the stipes of the LAMINARIACEAE. In some members of this family it is the superficial photosynthetic layers that cut off new tissues towards the inside by repeated tangential division of their cells; the layers in question thus undergo a well-marked change of function with advancing age. In *Thalassiophyllum*, according to Rosenthal, the "formative" layer of photosynthetic tissue (*Bildungsschicht*) gradually loses its meristematic activity and dies, whereupon a second cambial layer arises by tangential division in the peripheral portion of the "outer cortex." In several LAMINARIACEAE the secondary tissues are arranged in concentric strata, which recall the annual rings of a Dicotyledonous stem.

# ABBREVIATIONS.

*A.B.*   Annals of Botany (London).

*Abh. Amst.*   Verhandelingen d. koninklijke Academie van Wetenschappen (Amsterdam).

*Abh. bot. Ver. Brand.*   Abhandlungen d. botanischen Vereins d. Provinz Brandenburg.

*Abh. sächs. Akad.*   Abhandlungen d. königl.-sächsischen Akademie der Wissenschaften (math.-phys. Klasse).

*Abh. Senck. Ges.*   Abhandlungen d. Senckenbergischen naturforschenden Gesellschaft.

*Ann. agron.*   Annales agronomiques.

*Ann. Buit.*   Annales du jardin botanique de Buitenzorg (Leiden).

*Ann. Sci. Nat.*   Annales des sciences naturelles—Botanique (Paris).

*Arb. Würzb.*   Arbeiten des botanischen Institutes in Wurzburg.

*Arch. Anat. u. Phys.*   Archiv für Anatomie u. Physiologie (Leipzig).

*Arch. mikr. Anat.*   Archiv für mikroskopische Anatomie (Bonn).

*Arch. néerl.*   Archives néerlandaises (The Hague).

*B.G.*   Botanical Gazette (Chicago).

*B.Z.*   Botanische Zeitung (Berlin and Leipzig).

*Beih. bot. Centr.*   Beihefte zum botanischen Centralblatt (Cassel).

*Ber.*   Berichte d. deutschen botanischen Gesellschaft (Berlin).

*Ber. sächs. Ges.*   Berichte d. sächs. Gesellschaft d. Wissenschaften zu Leipzig.

*Bibl. Bot.*   Bibliotheca Botanica (Cassel and Stuttgart).

*Biol. Centr.*   Biologisches Centralblatt (Erlangen and Leipzig).

*Bot. Centr.*   Botanisches Centralblatt (Cassel).

*Bull. Acad. Crac.*   Bulletin internationale de l'Académie des sciences (Cracow).

*Bull. Mosc.*   Bulletin de la Société impériale des naturalistes de Moscou.

*Bull. Soc. Bot.*   Bulletin de la Société botanique de France (Paris).

*C.R.*   Comptes rendus hebdomadaires des séances de l'Académie des sciences (Paris).

*Cohn's Beiträge.*   Beiträge zur Biologie d. Pflanzen, herausgegeben von Dr. Ferd. Cohn (Breslau).

*Denkschr. Wien.*   Denkschrifte d. kaiserlichen Akademie d. Wissenschaften zu Wien.

*Engler's Jahrb.*   Botanische Jahrbücher für Systematik, Pflanzengeschichte u. Pflanzengeographie, herausgeg. v. A. Engler (Leipzig).

*Festschr. Halle.*   Festschrift d. naturforschenden Gesellschaft in Halle.

*Flora.*   Flora, oder Allgemeine Botanische Zeitung (Marburg).

*J.C.S.*   Journal of the Chemical Society of London.

*Jena. Zeitschr.*   Jenaische Zeitschrift für Naturwissenschaft.

*Journ. de Bot.*   Journal de botanique (Paris).

*Journ. Linn. Soc.*   Journal of the Linnean Society (London).

*Journ. Soc. imp. et centr.*   Journal de la Société impériale et centrale d'horticulture.

*K. Svenska Handl.*   Kongliga Svenska Vetenskaps-Akademienshandlingar (Stockholm).

*Kgl. Danske Vid. S.* Oversigt over det Kongelige Danske Videnskabernes Selskabs Forhandlinger (Copenhagen).

*Landw. Jahrb.* Landwirtschaftliche Jahrbücher.

*Landw. Versuchs-Stationen.* Die Landwirtschaftlichen Versuchs-Stationen (Chemnitz).

*Lund's Univ. Arsskrift.* Acta Universitatis Lundensis. Lund's Universitets Arsskrift.

*Mém. Bord.* Mémoires de la Société des sciences physicales et naturelles de Bordeaux.

*Mém. Soc. Cherb.* Mémoires de la Société des sciences naturelles de Cherbourg.

*Mem. Torr. Bot. Club.* Memoirs of the Torrey Botanical Club.

*Mitth. Graz.* Mittheilungen aus dem botanischen Institut zu Graz.

*Mitth. Steiermk.* Mittheilungen d. naturwissenschaftlichen Vereins für Steiermark.

*Monatsber. Berlin.* Monatsberichte d. königlichen preussischen Akademie d. Wissenschaften zu Berlin.

*Nachr. Gött.* Nachrichten von der k. Gesellschaft des Wissenschaften zu Göttingen.

*Naturw. Rund.* Naturwissenschaftliche Rundschau (Braunschweig).

*Nova acta.* Nova acta . . . Abhandlungen d. kaiserlichen Leopoldinisch-Carolinischen deutschen Akademie d. Naturforscher.

*Öst. bot. Zeitschr.* Österreichische botanische Zeitschrift.

*Pflüger's Archiv.* Archiv für d. gesammte Physiologie d. Menschen u. d. Tiere (Bonn).

*Phil. Trans.* Philosophical Transactions of the Royal Society of London.

*P.J.* Jahrbücher für wissenschaftliche Botanik, begründet von Dr. N. Pringsheim (Berlin).

*Proc. Camb. Phil. Soc.* Proceedings of the Cambridge Philosophical Society.

*Progressus.* Progressus rei botanicae (Jena).

*Q.J.M.S.* Quarterly Journal of Microscopical Science (London).

*Rev. gén. d. Bot.* Revue général de botanique (Paris).

*Sitz. Berlin.* Sitzungsberichte d. koniglich-preussischen Akademie d. Wissenschaften zu Berlin.

*Sitzb. Naturf.* Sitzungsberichte d. Gesellschaft naturforschenden Freunde zu Berlin.

*Sitzb. niederrh. Ges.* Sitzungsberichte der niederrheinischen Gesellschaft für Natur- und Heilkunde.

*Sitzb. Prag.* Sitzungsberichte der böhmischen Gesellschaft d. Wissenschaften in Prag.

*Sitzb. Wien.* Sitzungsberichte d. mathematisch-naturwissenschaftlichen Classe d. kais. Akademie d. Wissenschaften zu Wien (1te Abt.).

*Sitzb. Würzb.* Sitzungsberichte d. phys. med. Gesellschaft in Würzburg.

*Unters. Gött.* Untersuchungen d. botanischen Institutes d. Universität v. Göttingen.

*Unters. Tüb.* Untersuchungen aus d. bot. Institut zu Tübingen.

*Verh. Rheinl. u. Westf.* Verhandlungen d. naturforschenden Vereins d. Rheinlande u. Westfalens.

*Zeitschr. allg. Phys.* Zeitschrift für allgemeine Physiologie (Jena).

*Zeitschr. wiss. Zool.* Zeitschrift für wissenschaftliche Zoologie (Leipzig).

# NOTES.

1. That branch of natural science which treats of the adaptations of organisms is often termed " biology " ; some critics of the first [German] edition of the present work have remarked that the term " physiological plant-anatomy " should properly be replaced by " biological plant-anatomy." Although the question at issue is largely a matter of terminology, the author cannot assent to the change proposed, if only because different authors employ the term " biology " in very different senses. When used in its widest sense, biology includes the study of living organisms in general ; sometimes, however, biology is defined as the study of the adaptations of living organisms, while a certain number of investigators—including the author himself—limit the scope of biology, by excluding adaptations other than those which arise out of the relations of plants [or animals] to the various climatic and edaphic factors of the environment, or to animals and other plants. At the present day, biology in this limited sense is often termed " ecology." In the author's own opinion, biological or ecological anatomy merely constitutes a special section of physiological anatomy ; the study of function has always been regarded as the proper subject-matter of physiology.

Even in the first [German] edition of this book (p. 18) stress was laid upon the fact that physiological plant-anatomy—in the wide sense—has two objects in view. " On the one hand, it has to deal with the physiological processes that are responsible for the internal structure of the plant-body, while, on the other, it is constantly endeavouring to discover the relations between the structure and the function of the various organs." In the second edition (p. 2) the former of these objects was relegated to the science of ontogenetic or developmental anatomy (*entwickelungs-mechanische Anatomie*), the latter being thus regarded as the sole aim of physiological anatomy in the narrower sense. **Berthold** (Untersuchungen zur Physiologie der pflanzlichen Organisation, Vol. II. Pt. I. p. 8 [1904]), therefore, in defining the object of physiological anatomy as " the investigation of the mechanism of differentiation during development and in the adult condition," is only endorsing the author's own views. As a matter of fact, it is not quite evident how Berthold's various special investigations, which are in themselves of the greatest interest, can lead to an understanding of the " mechanism " of tissue-differentiation. This question must, however, remain open, until the available data have been worked up into a general theory.

It is gratifying to note that in recent times those branches of physiological plant-anatomy—in the wide sense— which concern themselves with the *causes* of differentiation, are showing signs of active development. **Vöchting's** " Untersuchungen zur experimentellen Anatomie und Pathologie des Pflanzenkörpers " form a notable contribution in this direction. The experimental aspect of Vöchting's work cannot, however, be regarded as its characteristic feature, since physiological anatomy, in the narrow sense, is also largely based on experimental evidence. The author would therefore prefer to term the subject-matter of Vöchting's work " ontogenetico-physiological " (*entwickelungsmechanische* or *entwickelungsphysiologische*) anatomy. This last-mentioned aspect of anatomy naturally cannot claim to be the only experimental branch of the science; in the same way the "phylogenetic anatomy" of **Strasburger** and his pupils cannot be regarded as the only legitimate " comparative anatomy." On the contrary, every anatomical investigation must follow comparative lines, if it is to lead to results of scientific value (cf. second [German] edition, p. 2).

2. The functional units are not always exactly identical with the formal units or cells. Thus, in the case of palisade-tissue, the functional unit (palisade-unit) *may*, and usually does, consist of a single cell; it may, on the other hand, represent a branch of a cell, every cell being made up of 2 to 4 palisade-shaped branches or "arms." The functional units of mechanical strands are the individual mechanical fibres. As a rule every fibre consists of a single morphological unit or cell. A septate bast-fibre, however, comprises more than one cell, while every collenchyma "fibre" is made up of the thickened edges of the walls of several adjacent cells. The functional units of the conducting system are the individual water-conducting tubes, but any given tube may consist of a single cell (tracheid) or of a row of fused cells (wood-vessel). These facts show that plants do not always maintain the individuality of the cell in constructing their functional units; in other words, the limits of functional unit are not always marked by cell-walls. Such exceptional cases do not, however, necessitate any departure from the long-established and, on the whole, fully justified view which regards the cell as the general functional unit of the plant-body.

3. **G. Haberlandt :** Sitzb. Wien, **111** (1902) [experiments with isolated cells].

4. The fact that, at the present time, the scientific conception of the structural units does not coincide with the traditional notion of the cell, occasionally causes some inconvenience, especially from a didactic point of view. Strictly speaking, for example, it is a contradiction in terms to call a naked swarm-spore a " cell." **Sachs** accordingly proposed (Flora, 1892) to restrict the term cell, so far as Botany is concerned, either to the cell-wall alone or to the cell-wall *plus* the cell-contents. The " morphological and physiological organic unit," consisting of a nucleus and the area of protoplasm controlled thereby, Sachs proposes to call an energid. So far as ordinary uninucleate cells are concerned, " energid " corresponds exactly to Hanstein's " protoplast " or Brücke's " cell-body " (*Zellenleib*). In the case of multinucleate cells and syncytes, on the other hand (latex-tubes and many bast-fibres), and among the non-cellular Siphoneae and Phycomycetes, Sachs regards each protoplast as made up of as many energids as there are nuclei. In such cases it is impossible to make out the hypothetical limits of the individual energids; hence the latter cannot be regarded as morphological units. It is also quite uncertain whether each nucleus, in a multinucleate protoplast, permanently exercises entire control over a definite tract of protoplasm (as Zimmermann has remarked, it is impossible to conceive of any such permanent relation in the case of a multinucleate cell which exhibits active protoplasmic circulation); the energid therefore appears an unsatisfactory unit from the physiological, as well as from the morphological, point of view. The difficulty is not restricted to the case of multinucleate cells; for, in most living tissues, the protoplasts of adjoining uninucleate elements are connected by means of protoplasmic filaments, and it cannot be assumed that the influence of the nuclei is never transmitted along these protoplasmic connections from one cell to another. Evidently, in defining the " organic unit " of the plant- or animal-body, it will not do to lay stress upon the still largely mysterious interrelations between any one constituent part of that unit and the remaining components, far less to make these interrelations the basis of our definition.

2 Y

5. The term " protoplasm " was introduced by **Von Mohl** (Über die Saftbewegung im Innern der Zellen, B.Z. 1846, p. 73). The fundamental importance of the protoplasm as the vehicle of the vital activity of the cell was recognised about the same time by **Nägeli** (Zeitschr. f. wiss. Bot. herausgeg. v. Schleiden und Nägeli, **3**, 1846). **Brücke** (Sitzb. Wien, **41**, 1861) was the first to lay stress upon the complex structure of the protoplast ; the term " organised," which Brucke used to emphasise this complex character of the living substance, was by Nägeli extended to purely physical matters, such as the micellar structure of substances which are capable of swelling, etc. (cf. **Nägeli** and **Schwendener** : Das Mikroskop, 2nd ed. p. 532). Nägeli accordingly includes starch-grains and cell-walls in the category of " organised " substances, whereas Brücke regards " organisation " of this sort as a property peculiar to living protoplasm.

Among later observers, Sachs and especially Pfeffer have upheld Brücke's theory of the highly complex structure of the protoplast (cf. **Pfeffer** : Abh. sächs. Akad. **16**, 1890 ; id. Studien zur Energetik der Pflanze, ibid. **18**, 1892, p. 156 *sqq.*). **Hanstein** (Das Protoplasma als Träger der pflanzlichen und thierischen Lebensverichtung, Heidelberg, 1880) takes up the same position.

The numerous attempts that have been made to demonstrate a definite structure in the living substance, may be grouped under two heads, according to the points of view of the investigators. Some authors rely chiefly upon direct observation, and lay much less stress upon theoretical considerations. This tendency is exemplified by the attempts of Max Schultze, Flemming and others to discover fibrillar structure in protoplasm ; similarly, Frommann and Heitzman on the zoological, and Schmitz and Strasburger on the botanical side, claim to have recognised a reticulate structure by direct observation. Bütschli, finally, concludes, after careful investigation, that protoplasm in general possesses an alveolar or " foamy " structure. In many cases, on the other hand, a definite structure has been assigned to the living substance on purely theoretical grounds, in order to render the phenomena of heredity, and the vital activity of protoplasm generally, more comprehensible. An instance is afforded by Nägeli's assumption that the solid " idioplasm "—*i.e.* that portion of the protoplasm which acts as the vehicle of the heritable properties—forms a continuous reticulum extending throughout each individual cell and through all the living regions of the plant-body. Some authors believe that the living substance is made up of equivalent ultimate particles, which are the actual vehicles of vital activity. The " plastidules " of Elsberg and Haeckel are still of the order of magnitude of chemical molecules ; but Altmann's " granula," De Vries's " pangens," Weismann's " biophores " and Wiesner's " plasomes " all represent very complex particles of the living substance. Altmann and Wiesner have endeavoured to demonstrate the existence of these ultimate particles by actual observation. As a matter of fact, the so-called " granula " of Altmann include a variety of different objects, such as protein-particles, oil-drops, pigment-granules, etc., which are preferably termed " microsomes," a term which has no theoretical connotation so far as plant-cytology is concerned. Most probably Wiesner's " plasomes," so far as they have been identified, are also by no means all equivalent particles.

All the attempts that have been made to resolve the living substance into a number of equivalent " ultimate particles," " living particles," " living elements," and so forth, are founded, consciously or unconsciously, on the assumption that every protoplast is made up of exactly equivalent particles, just as the various organs and tissues of animal and vegetable organisms are composed of homologous elements, or cells ; while it is possible that there is an analogy of this sort between multicellular organs and individual cells, there is no positive evidence in favour of such a comparison. The ultimate particles of which the living substance is made up, may equally well be of various kinds (as Pfeffer has already pointed out), just as the component parts of a clock or other complicated machine are by no means all alike. It should be remarked that those who propound or accept theories of protoplasmic structure are forced to associate all manifestations of vital activity—in the elucidation of which their theories are intended to assist—with these ultimate structural elements, although they are unable to make any definite statements concerning the internal structure of the elements themselves. But we learn nothing concerning the causes of vital activity by recognising ultimate protoplasmic particles, unless we understand the structure of these particles and its relation to the various functions of the living substance. The existing theories of protoplasmic structure therefore do not *explain* a single manifestation of vital activity ; all they do is to transfer the fundamental problem to another plane of argument. Such speculations nevertheless have their uses, inasmuch as they help to define the limitations of our knowledge, so far as the structure of living matter is concerned. Cf. **Flemming** : Arch. mikr. Anat. **18**, 1880. Id. ibid. **20**, 1881. Id. Zellsubstanz, Kern- und Zellteilung, Leipzig, 1883. **Schmitz** : Sitzb. niederrhein. Ges. 1880. **De Vries** : Intracellulare Pangenesis, Jena, 1889. **Altmann** : Die Elementarorganismen, Leipzig, 1890. **Weismann** : Das Keimplasma, Jena, 1892. **Wiesner** : Die Elementarstruktur, Wien, 1892. In recent times, it is especially Strasburger and his pupils who have published numerous papers on the subject of protoplasmic structure. The important question, as to the extent to which the granular, fibrillar, reticulate or other structures observed in fixed material represent artefacts, is dealt with in detail in **A. Fischer's** Fixierung, Färbung und Bau des Protoplasmas (Jena, 1899).

6. **Berthold** : Studien über Protoplasmamechanik, Leipzig, 1886. **Bütschli** : Untersuchungen über mikroskopische Schäume und das Protoplasma, Leipzig, 1892.

7. **Reinke** u. **Rodewald** : Unters. Gött. **2**, 1881 (Aethalium septicum). **Schwarz** : Cohn's Beitr. **5**, 1887. **Czapek** : Biochemie d. Pflanzen, **1**, pp. 20 *sqq.*

8. *Protoplasmic streaming.* **De Vries** : B.Z. 1885 (importance for translocation). **Ida Keller** : Über Plasmaströmung, Zürich, 1890. **Hauptfleisch** : P.J. **24**, 1892. **Kohl** : Wigand's Botanische Hefte 1. 1885, and **Noll** : Arb. Würzb. **3**, pp. 496 *sqq.* (arrangement of protoplasm in relation to curvature). **Tangl** : Sitzb. Wien, **90**, 1884 (continuity of protoplasm). **G. Haberlandt** : Über d. Beziehungen zwischen Funktion u. Lage d. Zellkernes b. d. Pflanzen, Jena, 1887, pp. 102 *sqq.* **Heidenhain** Sitzb. Würzb. 1897. **Nestler** : Sitzb. Wien, **107**, 1898 (streaming induced by traumatic stimulation). **Bierberg** : Flora, **99**, 1908.

9. *Ectoplast and vacuolar membrane.* **Pfeffer** : Osmotische Untersuchungen Leipzig, 1877. Id. Unters. Tüb. **2**, pp. 179 *sqq.* (absorption of aniline dyes). Abh. sächs. Akad. **16**, 1890, pp. 185 *sqq.* **De Vries** : P.J. **16**, 1885, pp. 465 *sqq.* **Went** : P.J. **19**, 1888, pp. 275 *sqq.* **Noll** : Naturw. Rund. 1888, 4 and 5 (action of light and gravity). Id. Arb. Würz. **3**, 1888, p. 532. **Němec** : Sitzb. Prag (experimental production of vacuoles). **Noll** : Biol. Centr. 1903.

10. *Structure and composition of nucleus.* **Strasburger** : Über Zellbildung und Zellteilung (3rd ed.), 1880. **Flemming** : Zellsubstanz, etc. **Zacharias** : B.Z. 1881 (chemical composition). Id. ibid. 1882, 1883, 1885, 1887.

For the voluminous modern literature cf. **Zimmermann** : Morphologie u. Physiologie d. pflanzlichen Zellkernes, Jena, 1896. **Häcker** : Praxis u. Theorie d. Zellenu. Befruchtungslehre, Jena, 1899.

11. **Strasburger** : Histol. Beitr. **5**, Jena, 1893. **Schwarz** : Cohn's Beitr. **4**, 1884. **Zacharias** : Flora, 1895 (Erg.-Bd.).

12. *Multinucleate cells*. **Schmitz** : Festschr. Halle, 1893 (Siphonocladiales). Id. Sitzb. niederrhein. Ges. 1879 (Aug.), 1880 (June). **Treub** : Arch. néerl. **15**. **Johow** : Unters. üb. d. Zellkerne in d. Sekretbehältern u. Parenchymzellen d. höh. Monokotylen, Inaug.-Dissert. Bonn, 1880. **Strasburger** : B.Z. 1880. **G. Haberlandt** : l.c. [8], Ch. 3, VII. (significance of the multinucleate condition).

13. *Functions of nucleus*. **Nägeli** : Mechanische Theorie d. Abstammungslehre, München u. Leipzig, 1884, p. 368. **Hertwig** : Das Problem d. Befruchtung u. der Isotropie d. Eies, Jena, 1885. **Strasburger** : Neue Unters. üb. d. Befruchtungsvorgang b. d. Phanerogamen als Grundlage f. eine Theorie d. Zeugung, Jena, 1884. **Weismann** : Die Kontinuität d. Keimplasmas als Grundlage einer Theorie d. Vererbung, Jena, 1885. **Kölliker** : Zeitschr. f. wiss. Zool. **42**, 1885. **Schmitz** : Festschr. Halle, 1879. **Klebs** : Biol. Centr. **7**, 1887. Id. Unters. Tüb. **2**, p. 489. **G. Haberlandt** : l.c. [8]. Id. Sitzb. Wien, **98**, 1889 (encysted protoplasts). **Palla** : Flora, 1890, p. 314, and Ber. **24**, 1906 (enucleate protoplasts). Palla, and later Acqua, found that enucleate portions of protoplasm from pollen-tubes, Marchantia-rhizoids and Urtica-hairs remained capable of forming new cell-membranes, so long as the organs from which they were derived showed active growth. It would be a mistake, however, to infer from this fact that cell-wall formation generally takes place independently of nuclear influence. On the contrary, as Palla points out in his first paper, the above-mentioned cases illustrate the after-effects of nuclear control ; presumably, this control depends upon the fact, that the nucleus of the growing cell secretes certain substances which induce the formation of the cell-wall, and that the substances in question are also found in the artificially isolated, enucleate pieces of protoplasm. The mere presence of a reserve-material, capable of conversion into cell-wall-substance, cannot enable enucleate portions of protoplasm to surround themselves with a new cell-membrane. **Van Wisselingh** : Beih. bot. Centr. **24**, I. 1908 (Spirogyra). **Gerassimow** : Bull. Mosc. 1890. **Acqua** : Malpighia, **5**, 1901. **Pfeffer** : Ber. sächs. Ges. 1896. **Townsend** : P.J. **30**. **Gerassimow** : Bull. Mosc. 1896. Id. ibid. 1900. Id. ibid. 1901. Id. Zeitschr. allg. Phys. **1**, 1902. **Magnus** : P.J. **35**, 1900 (mycorrhiza of Neottia). **Shibata** : P.J. **37**, 1902 (mycorrhiza). **Von Guttenberg** : Beiträge z. phys. Anat. d. Pilzgallen, Leipzig, 1905, pp. 22 *sqq*. **Küster** : Flora, **97**, 1907. In this paper, Küster cites a number of cases in support of the author's observations on the position of the nucleus in growing cells. At the same time he claims to have discovered a remarkable exception to the rule ; in a number of water-plants (Hydrocharis morsus-ranae, Trianea bogotensis, Potamogeton lucens, Stratiotes aloides, Vallisneria spiralis, Elodea canadensis, etc.), namely, the nuclei always remain at the proximal ends of the root-hairs. It should, however, be noted, that Küster has not proved that the root-hairs in question exhibit apical growth, like the hairs of terrestrial roots. It is quite possible that such aquatic root-hairs behave like the trichomes on the stems and petioles of many plants (Geranium sanguineum, etc.), which, as the author has shown, exhibit protracted basal growth,— after their apical extension has ceased—and accordingly always have their nuclei situated at the base. As regards zoological literature, attention may be specially directed to the papers of **Nussbaum, Gruber, Korschelt, R. Hertwig, Hofer, Verworn**, etc.

14. **Häcker** : l.c. [10], p. 114. The other views regarding the significance of nucleoli are exhaustively dealt with in this paper. **Von Derschau** : Ber. **22**, 1904.

15. **Zacharias** : B.Z. 1890. **Bütschli** : Über d. Bau. d. Bakterien u. verw. Organismen, Leipzig, 1890. **Palla** : P.J. **25**, 1893. **A. Fischer** : Unters. üb. d. Bau d. Cyanophyceen u. Bakterien, 1897. **Kohl** : Organisation u. Physiologie d. Cyanophyceenzelle, Jena, 1903. **Olive** : Beih. bot. Centr. **13**, 1905.

16. *Chromatophores.* **V. Mohl** : Unters. üb. d. anat. Verhältnisse d. Chlorophylls Tübingen, 1837. **Nägeli** : Zeitschr. wiss. Bot. 3, p. 110. **Schmitz** : Verh. Rheinl. u. Westf. 40, 1889 (Algae). **Schimper** : B.Z. 1880 (starch-grains). Id. ibid. 1883 (chloroplasts and chromoplasts). Id. P.J. 16, 1885 (chloroplasts). **Arthur Meyer** : Das Chlorophyllkorn, etc., Leipzig, 1883. **G. Haberlandt** : B.Z. 1877 (development of chloroplasts in cotyledons of Phaseolus). Id. Flora, 1888 (chloroplasts of Selaginella). **Mikosch** : Sitzb. Wien, 1878 and Öst. bot. Zeitschr. 1877 (development and division of chloroplasts). **Bredow** : P.J. 22, 1891. **Binz** : Flora, 1892 (Erg.-Bd.) (starch-grains). **Zimmermann** : Beitr. Morph. u. Phys. 1 (leucoplasts). Id. ibid. 2 (variegated leaves). **Senn** : Die Gestalts- u. Lagesveränderungen d. Pflanzenchromatophoren, Leipzig, 1908. Schimper's views as to the origin and genetic interrelations of the various types of chromatophores and the physiological importance of leucoplasts, have recently been combated by **Belzung, Eberdt** and **Koningsberger,** but the arguments so far brought forward by them are far from conclusive.

17. *Vacuolar membranes.* In addition to the papers of **De Vries, Pfeffer** and **Went,** already cited, see **De Vries** : P.J. 14, 1884. **Wakker** : P.J. 19, 1888. **Crato** : B.Z. 1893 (physodes). **Zimmermann** : Beitr. z. Morph. u. Phys. 1, p. 185 (elaioplasts). **Raciborski** : Anz. Akad. Krakau, 1893 (elaioplasts). For contractile vacuoles cf. **Pfeffer** : Pflanzenphysiologie, II. p. 398, and the literature there cited.

18. **Tangl** : P.J. 12. **Strasburger** : P.J. 36, 1901, and the literature there cited.

19. *Structure and growth of the cell-wall.* **Nägeli** : Bot. Mittheil. 2, 1866. **N. Pringsheim** : Unters. üb. d. Bau u. d. Bildung der Pflanzenzelle, 1854. **Crüger** : B.Z. 1855. **Dippel** : Abh. Senck. 10, 1876. **Schmitz** : Sitzb. niederrh. Ges. 1880. **Strasburger** : Üb. d. Bau u. d. Wachsthum d. Zellbäute, Jena, 1882. Id. Histol. Beitr. 2, Jena, 1889. Id. P.J. 31, 1898. **Leitgeb** : Üb. Bau u. Entwick. d. Sporenhäute, Graz, 1884. **Schenck** : Unters. üb. d. Bild. v. centrifugalen Wandverdickungen an Pflanzenhaaren u. Epidermen, Bonn, 1884. **Wille** : Üb. d. Entwickelungsgesch. d. Pollenkörner d. Ang. u. d. Wachst. d. Membranen durch Intussusception, 1886. **Wiesner** : Sitzb. Wien, 93, 1886. Id. Elementarstruktur, etc., 1892. **Krasser** : Sitzb. Wien, 94, 1887 (protein-content of cell-wall). **Noll** : Experimentalunters. üb. d. Wachst. d. Zellmembran, Würzburg, 1887. **Krabbe** : P.J. 18, 1887. **Klebs** : Biol. Centr. 6, 1886 (critique of Wiesner's 1886 paper). Id. B.Z. 1887. Id. Unters. Tüb. 2, 1886 (gelatinous membranes). **Zacharias** : Flora, 1891 (root-hairs). **Correns** : Flora, 1889 (intussusception). Id. P.J. 23, 1891. Id. ibid. 26, 1894 (critique of Wiesner's views). **Fitting** : B.Z. 1900 (megaspores of Isoetes). **Krieg** : Beih. bot. Centr. 31, 1907 (striation of coniferous tracheides).

20. *Chemical composition of cell-walls.* **Zimmermann** : Botanische Mikrotechnik, Tübingen, 1892, pp. 135 *sqq.,* and the literature there cited. **W. Hofmeister** : Landw. Jahrb. 1888. Id. ibid. 1889. **Gilson** : La Cellule, 9, 1893. **Reiss** : Landw. Jahrb. 1889 (reserve-cellulose). **E. Schulze** : Ber. 1880 (reserve-cellulose). **Wiesner** : Sitzb. Wien, 77, 1878 (phloroglucin-reaction). **Singer** : Sitzb. Wien, 85, 1882 (lignified walls). **Nickel** : Die Farbenreaktionen d. Kohlenstoffverbindungen, Berlin, 1890. **Von Höhnel** : Sitzb. Wien, 81, 1880 (cork). **Kügler** : Üb. d. Suberin, Strassburg, 1884. **Gilson** : La Cellule, 4, 1890. **Hofmeister** : Sitzb. sächs. Akad. 10, 1858 (mucilaginous walls). **B. Frank** : P.J. 5, 1865 (mucilage). **Nadelmann** : P.J. 21, 1890 (mucilaginous endosperms). **Walliczek** : P.J. 25, 1893 (mucilaginous walls in vegetative organs). **Mangin** : Bull. Soc. Bot. 109, 1889 (pectic compounds). Id. ibid. 111, 1890 (colour reactions of cell-wall constituents). **Kohl** : Anatomisch-physiologische Unters. d. Kalksalze u. Kieselsäure in d. Pflanze, Marburg, 1889.

Winterstein : Ber. 1895. pp. 65 *sqq*. (Fungus-cellulose). Schellenberg : P.J. 29, 1896 (lignified walls). Czapek : Hoppe-Seyler's Zeitschr. 27, 1899 (lignin-reaction). Id. Flora, 1899 (cell-walls of Mosses and Liverworts). Id. Biochemie, 1, pp. 506 *sqq*. Euler : Grundlagen u. Ergebnisse d. Pflanzen-chemie, Braunschweig, 1908, Vol. I.

21. Active growth of a cell-wall might take place in one of two different ways. In the first place, the growing membrane may be possessed of a definite molecular or micellar structure, which may be termed the "living structure" of the wall. Nägeli distinguished three different conditions of the cell-wall in respect of its micellar structure, namely, "life," "natural death" and "imbibition." The "living" condition is characterised by the fact that the membrane is capable of independent growth by intussusception ; in these circumstances, the underlying protoplasm merely keeps the membrane supplied with soluble plastic material. Under certain conditions, therefore, such a "living" cell-membrane should be able to go on growing, even though it is not in actual contact with living protoplasm, provided it is sufficiently supplied with soluble plastic compounds. Fitting (B.Z. 1900) alleges that this case is exemplified by the developing spore-wall in Isoetes and Selaginella.

The "living" state of the cell-wall, with which its power of active growth is correlated, might, on the other hand, depend upon the presence in the cell-wall of protoplasm continuous with the living contents of the cell. This view has found a strong adherent in Wiesner. There is undoubtedly much to be said in its favour. It dispenses with the necessity of assuming a mysterious "living" condition of the cell-wall, independent of the presence of protoplasm ; and thus maintains the fundamental physiological principle which regards the protoplasm as the sole living constituent of the cell. It is true that Wiesner does not explicitly state that the "membrane-plasma" is a part of the protoplast, but his remarks must undoubtedly be interpreted in this sense. As Correns has pointed out, the chief objection to Wiesner's theory is the lack of any proof that cell-membranes ever contain protoplasm ; in fact, Klebs, Fischer and Correns go so far as to deny the presence of protein-compounds—as alleged by Wiesner and Krasser—in cell-walls. Cf. Note 18 (Wiesner, Krasser, Klebs and Correns) ; also Strasburger : Hist. Beitr. 2, 1889, and Reinhardt : Festschr. f. Schwendener, Berlin, 1899 (plasmolytic method).

22. Sachs : Flora, 1893. Amelung : Flora, 1893.

23. Nägeli : Mechanisch-physiologische Theorie d. Abstammungslehre. 1884, pp. 357 *sqq*.

24. Sachs : Sitzb. Wurzb. 1878 (Nov.).

[24 a. The term "non-cellular," due to Sachs, is now chiefly of historic interest. In English botanical literature multinucleate "cells" are generally termed "coenocytes."]

25. Janse : P.J. 21, 1890, pp. 269 *sqq*.

26. The exceedingly voluminous literature of nuclear and cell-division cannot be considered in detail here. On the botanical side, special attention may be drawn to the work of Strasburger and his school (Strasburger : Hist. Beitr., P.J. 30, 1897, etc.), and to several papers by Němec. Cf. also Hacker : l.c. [10].

27. Warburg (B.Z. 1885, p. 29) maintains that "logically, one ought to distinguish several protective dermal systems, each corresponding to a different principal function of the dermal tissue ; the most important of these would be concerned with the reduction of transpiration, the others with the restriction of radiation, protection

against intense illumination, prevention of rapid change of temperature, etc."
Warburg evidently puts forward this counsel of perfection with the intention
of showing that the conception of a " dermal tissue " is inadmissible ; but he has
overlooked the fact that a dermal tissue does not perform *several* principal functions,
but, on the contrary, a single principal function which comprises several partial
functions.

28. **G. Haberlandt :** Eine botanische Tropenreise, Leipzig, 1893, p. 108. **Stahl :**
Ann. Buit. **11,** 1893 (rainfall and leaf-structure).

29. A discussion of the various ways in which the different tissues and organs of
the plant-body are brought into harmonious cooperation with one another would
be outside the province of this book. Very suggestive remarks on this subject will
be found in **Roux's** Kampf der Theile im Organismus, Leipzig, 1881 (second extended
version in Ges. Abh. **1,** 1895, pp. 135-422). **Weismann** also enters into the question
in his stimulating essay on Aussere Einflüsse als Entwickelungsreize, Jena,
1894.

30. Since the publication of **Schwendener's** Mechánische Prinzip, and **G. Haberlandt's** Entwickelungsgeschichte des mechanischen Gewebesystems, the theoretical and practical value of the anatomico-physiological classification of tissues
has repeatedly been the subject of vigorous discussion. A few words must be devoted
to this controversy at the present stage.

That the anatomico-physiological classification is sound in principle, is now
generally admitted. **De Bary's** Comparative Anatomy, though purely descriptive in character, lays stress upon the correlation between structure and function. Thus, on p. 2 of the introduction we find the following remarks:
" Investigation shows that the adaptation to, and participation in vegetative
duties, that is, the development into organs of definite function, and corresponding
structure, is far the most commonly and definitely carried out for members of
lower ranks, *i.e.* for cells, and groups of cells, or the products of their metamorphosis."
De Bary accordingly (l.c. pp. 24 *sqq.*) always treats the structure and arrangement
of tissues from the point of view of adaptation. He further distinguishes between
obvious adaptations, on the one hand, and unexplained anatomical features, on the
other ; the latter are also " derived from adaptations, which have happened in some
epoch or other of the phylogenetic development, but which cannot now be certainly
referred to their causes." In pointing out (l.c. p. 25) the desirability of arranging
the various chapters of his book, either according to different aspects of adaptation,
or else with reference to taxonomic groups, he recognises that the anatomico-
physiological classification is not only legitimate but also eminently practicable.
In fact, it was only the existence of a large number of unexplained anatomical
features, that led him to prefer a purely descriptive treatment. A considerable
time has elapsed since De Bary's book was written (it was begun in 1865), and
the unexplained characters are no longer so numerous as to place a serious obstacle
in the way of a rational anatomico-physiological classification of tissues.

It is interesting to consider what attitude one of the most noted of plant-physio-
logists, **Sachs,** has taken up with regard to the anatomico-physiological classification
of tissues. In the introduction to the eighth chapter of his last great work (Lectures
on Plant-Physiology) he defines tissues as " masses of cells, which in their growth
and other physiological relations, present a certain agreement, and are distinguished
from the other neighbouring masses of tissue." Similarly, a tissue-system is defined
as " a union of tissues " which " constitute a whole of definite physiological character."
In view of these statements it might have been expected that Sachs would have put
the anatomico-physiological system of classification into practice forthwith. As a

matter of fact, he did attempt to define the dermal and vascular systems from that point of view, though perhaps not with complete success. All the remaining tissues, on the other hand, are classed together as "ground-tissue" or "fundamental tissue," in contradistinction to the dermal tissue and the vascular bundles. But this ground-tissue includes green photosynthetic parenchyma, colourless water-tissue, storage-parenchyma, mechanical strands and cell-masses, endodermal layers and the multifarious tissues which make up pericaps and seed-coats. No one, therefore, will venture to maintain that "ground-tissue" constitutes a "whole of definite physiological character."

De Bary (l.c. p. 6) has pointed out that the conception of a "ground-tissue" is inadmissable even from a purely anatomical standpoint. "However much the sub-division of tissues into three main systems may be fitted to guide beginners, still, in my opinion, it does not answer its purpose, which is to serve as a basis for a uniform exposition of the various differentiation of plant-tissues. For the names Dermal and Fascicular [ =Vascular] Tissue indicate, in Vascular plants, systems of tissue, which are positively characterised by definite tissue-forms ; but the name Fundamental tissue implies the remainder, and this may just as much consist of different positively characterised tissue forms, and tissue-systems, which are equiva-lent to the Dermal and Fascicular systems."

In the author's opinion, successful application of the anatomico-physiological classification of tissues will provide a more satisfactory proof of the practical value of this system than any academic discussion of its merits ; it is hoped that the present work will be found to justify this belief. One point of detail may be specially referred to. Objection is frequently raised to the physiological classification of tissues, on the ground that most tissues perform more than one function, and must conse-quently be referred to one or another physiological tissue-system, according to the function which happens to be under consideration at the moment. But if the dis-tinction between principal and subsidiary functions—which is no mere theoretical conception but an actual fact—is kept in mind, there should be no difficulty or risk of error in assigning the various tissues to their proper systems. The principal function should be the sole guide in this connection. So far as the Higher Plants (from the Archegoniatae upwards) are concerned, there should rarely be any doubt as to which of the various functions of a tissue is the principal one ; where uncertainty exists, the systematic position of the tissue is indeterminate. It is surely also per-missible to remark, that the purely descriptive anatomists are also sometimes in doubt as to the correct systematic position of individual tissues, and that they have by no means always been in whole-hearted agreement with regard to the anatomical classification of tissues.

The need for a physiological classification was felt sooner by animal anatomists and histologists, than by botanists. The prominent histologist Leydig, for example, in his Lehrbuch der Histologie des Menschen und der Thiere (1857), sets up a physiological classification of animal tissues, justifying his action with the following words (l.c. p. 21): "In my opinion, the classification of tissues, more especially, should be carried out without reference, in the first instance, to morphological characteristics. . . . My own classification accordingly, is based upon the physiologi-cal relations of the structural elements." Cf. also **Hermann v. Meyer** : Biol. Centr. 1883, No. 12.

(The above note, which is reproduced almost word for word from the first [German] edition, is included chiefly on account of its historical interest.)

[31. The term "growing-point" has become so firmly established in English botanical literature, that I decided to retain it here instead of the more accurate "punctum vegetationis."]

32. **Sachs** : Flora, 77, 1893 (growth periods and formative stimuli). **Noll** : Biol. Centr. (embryonic substance).

33. **Krabbe** : Das gleitende Wachstum, Berlin, 1886. **Jost**: B.Z. 1901 (cambium). **Von Guttenberg** : Sitzb. Wien, 111, 1902 (crystal-cells of Citrus).

34. At the time of publication of the second [German] edition of this book, the author agreed with **Zimmermann**, in considering it improbable that neighbouring cells should become displaced relatively to one another by means of sliding growth. It seemed sufficient to assume, that when growing cells pushed their pointed ends or sharp edges between other cells, local surface-extension took place, which only affected the overlapping portions of the cell-membranes. More careful consideration, however, shows that this hypothesis is in reality far more open to criticism than **Krabbe's** theory of sliding growth. In order that the pointed end or sharp edge of a cell may insert itself between two adjoining cells, without any sliding of the cells over one another, the surface-extension must be strictly confined to the actual tip or edge. Surface growth must cease absolutely, close behind the tip,—or edge— so that the fresh layers of cell-wall-substance may be laid down in their final position upon the extended older parts of the membrane, otherwise sliding growth is bound to occur. It is, however, extremely improbable that surface-extension is always— or even commonly—so strictly localised.

The strongest argument against the theory of sliding growth consists in the universal occurrence of protoplasmic connections between adjacent cells. But there seems no reason to doubt that these connections may be formed secondarily, *i.e.* after sliding growth has come to an end. According to Strasburger, the connecting threads traversing the walls of latex-tubes are invariably secondary in origin. Cf. **Strasburger** : P.J. **36**, pp. 506 *sqq.* **Jost** : l.c. [33], pp. 8 and 10.

35. **Nägeli** : Die neueren Algensysteme, Zürich, 1847. Id. Beitr. **1**, 1858.

36. **Nägeli** and **Schwendener** : Das Mikroskop, 2nd ed. 1877, pp. 554-568.

37. According to the view embodied in the text, an apical cell represents a persistent initial cell. This conception of the apical cell is, in the author's opinion, the only reasonable one. Another view, which has received strong support in certain quarters, denies the existence of a single persistent rhythmically dividing apical initial ; at each division the apical cell is supposed to give rise to two daughter-cells, one of which represents a primary segment, while the other takes the place of the self-abolished apical initial. According to this theory, therefore, there is a constant succession of apical initials, each surviving only long enough to cut off a single primary segment. This second interpretation, however, amounts to little more than a play upon current phrases. It is customary, namely, to designate the two cells formed at each act of cell-division, daughter-cells with reference to their mother-cell, and sister-cells with reference to one another. This metaphorical terminology—for it is nothing more—is only appropriate when the two daughter- or sister-cells agree with one another morphologically and physiologically. But this condition is far from being fulfilled as regards the products of the division of an ordinary apical cell. At every division, one of the products, the segment, is a new formation, both from the morphological and from the physiological point of view ; the other retains all the features of the original apical cell unchanged. The only proper usage, in the present instance, therefore, is to describe the apical initial as a mother-cell, which cuts off the primary segments as daughter-cells ; all these segments are sister-cells with reference to one another, but none of them ought to be termed a sister-cell of the apical initial, which is their common mother-cell. The conception of a persistent apical cell is thus seen to be a perfectly legitimate one

The account of growth by means of apical cells given in the text is based upon the views of Nägeli. According to that author's definition, an apical initial is a cell from which all the tissues of an organ can be genetically derived. This definition obviously lays stress upon the individuality of apical cells.

Strongly divergent views regarding apical cells, and apical growth in general, were held by Sachs. Without entering into details, it may be noted that Sachs looks upon the apical cell as a mere hiatus in the system of intersecting surfaces formed by the cell-walls in the apical meristem ; this negative conception altogether obscures the cell-character of the apical initial and *a fortiori* ignores its special physiological activity in its rôle of primordial meristematic unit. Cf. **Sachs** : Arb. Wurzb. **2**, 1878 (arrangement of cells in meristematic tissues). Id. Lectures, pp. 431-459, where there is a very clear exposition of that author's views on the relations between growth and cell-division. (For a full and critical discussion of Sachs's conception of apical cells, cf. the second edition of the present work, pp. 87 *sqq.*)

In addition to Nägeli's papers, the following titles may be selected from the extensive literature dealing with growth with the aid of a single apical cell. **Cramer** : Pflanzenphys. Unters. **3**, 1855 (Equisetum). **Pringsheim** : P.J. **3**, 1863 (Salvinia). **Hanstein** : P.J. **4**, 1865 (Marsilia). **Geyler** : ibid. (Sphacelariaceae). **Reess** : P.J. **6**, 1867 (Equisetum). **Nägeli u. Leitgeb** : Beitr. wiss. Bot. **4**, 1867 (roots). **Leitgeb** : Sitzb. Wien, **57**, 1868 (Fontinalis). Id. Untersuch. üb. d. Lebermoose, **1-6**, 1874-81. **Russow** : Vergl. Untersuch. etc., St. Petersburg, 1872. **Pfeffer** : Hanst. Bot. Abh. **1**, 1871 (Selaginella). **Treub** : Mus. bot. Leide, **2**, 1877. **Strasburger** : Botanische Praktikum (4th ed.), 1902, pp. 311, 321, 340. **Rostowzew** : Flora, **73,** 1890. **Goebel** : Organographie, Jena, 1898-1901, p. 448.

38. Concerning growth by means of several apical cells, especially in Phanerogams, cf. **Schwendener** : Sitzb. Naturf. 1879. Id. Sitzb. Berlin, 1885 (apical growth and phyllotaxis). **G. Haberlandt** : Mittheil. Steiermk. 1881. **Dingler** : Üb. d. Scheitelwachstum d. Gymnospermenstammes. Id. Ber. 1886. **Korschelt** : P.J. **15**, 1884. **Groom** : Ber. 1885. **Koch** : P.J. **22**, 1891 (Gymnosperms).

Among the authors mentioned, Dingler and Korschelt confidently assert the presence of a single three-sided (pyramidal) initial in the stem-apex of Gymnosperms,—and of Phanerogams generally—while all the rest oppose this view with equal determination.

39. Schwendener's account of the apical growth of the roots of the Marattiaceae, has been confirmed, so far as essential facts are concerned, by Koch) P.J. **27,** 1895) ; but as regards the interpretation of these facts, the latter author follows in the footsteps of Sachs. Koch has shown that the four apical cells of Angiopteris evecta arise from a single initial by division in two planes at right angles to one another— a point of considerable theoretical interest.

40. *General remarks concerning growth with one and with several apical cells.* No doubt the presence of several apical cells is often a primitive arrangement, especially among Thallophyta ; for phylogenetic reasons, however, it is probably always to be regarded as a derivative condition among Phanerogams. For these plants are undoubtedly derived from Vascular Cryptogams, which almost invariably have a single apical cell in each growing-point. We are thus forced to conclude that the primordial meristem with several apical cells is secondarily derived from that with a single initial, so far as Phanerogams are concerned. In considering how the more complex condition has been evolved, we must pay special attention to any transitional cases that still exist. As a matter of fact, none of these cases (with the exception of one, which will be referred to again later on) illustrate the actual transition from the one·

celled to the many-celled condition; but they do exemplify certain intermediate states, and thus give some insight into the stages by which the transformation has in all probability taken place in the course of evolution.

Let us first consider the case of juxtaposed initials. There can be no doubt that the paired apical initials of Selaginella Wallichii, first described by Strasburger, have originated from a single four-sided (pyramidal) initial, owing to the appearance of a median longitudinal partition. It seems natural to assume that the four juxta-posed initials discovered by Schwendener in the roots of Marattiaceae have likewise arisen from a single apical cell, which has become divided into quadrants by two radial walls placed at right angles to one another. (This view was put forward tentatively in the first [German] edition of this book; it has since been proved to be correct by Koch in the case of Angiopteris evecta.)

The primordial meristem with superimposed initials can also be easily derived from that in which there is a single apical cell. This point has already been discussed by the author (l.c. [38]) in the following words: "If, for example, we imagine the three-sided (pyramidal) apical cell of a Fern-stem to become subdivided by two transverse walls into three superimposed compartments, and if we further suppose each of these compartments to go on growing and cutting off segments after the manner of the single original initial, we arrive at the type of growing-point charac-teristic of the lateral shoots of Ceratophyllum. If the original initial were subdivided into two superimposed portions, we should obtain the structure which prevails in the young dichotomous leaf-segments of this plant. In Ceratophyllum the appro-priate partition walls are there from the first. It is quite possible, however, that in other instances they are not laid down until the young leaves or shoots are becoming marked off from the main apex; it is in cases of this kind that should be particularly sought for."

After what has been said, there should be no difficulty in imagining how the growing-point with both juxtaposed and superimposed initials can be derived from that with a single apical cell.

In conclusion, we may enquire into the causes underlying the evolution of the complex from the simple type of growing-point; in regard to this question it is at present impossible to do more than put forward certain suggestions in very general terms. It will readily be admitted, for instance, that the individuality of the initial is much more pronounced where there is a single apical cell, than where there are several. This follows at once from the great regularity of the segmentation, and the constancy of the cell-shapes, in growing-points which possess a single initial. The transition from the simple to the complex type of structure thus involves a limita-tion of the independence of the individual cell in the interests of the whole organ. On the whole, this subordination of the cell becomes more pronounced as one ascends the scale of evolution; it is, therefore, not surprising to find the single apical cell of the Pteridophytes replaced by a group of initials among Phanerogams and especially in the Angiosperms.

41. **Hanstein** : Festschr. niederrhein. Ges. 1868. Id. Hanst. Bot. Abh. **1,** 1870.

42. *Apical growth of roots.* **Nägeli** und **Zeitgeb,** Beitr. **1,** 1867. **Hanstein** : Hanst. Bot. Abh. **1,** 1870. **Reinke** : Hanst. Bot. Abh. **1,** 3. **Janczewski** : Ann. Sci. Nat., sér. V., **20. Holle** : B.Z. 1876. **Treub** : Le meristeme primit. d. l. racine dans l. Monocotylédones, Leiden, 1876. **Eriksson** : P.J. **11,** 1878. **Flahault** : Ann. Sci. Nat., sér. VI., **6. Schwendener** : Sitzb. Berlin, 1882. **Van Tieghem** : Journ. de Bot. **1,** 1887. **Van Tieghem et Douliot,** Ann. Sci. Nat., sér. VII,, **8,** 1888. **Strasburger** : Botanische Praktikum (4th ed.), pp. 323 *sqq.* A number of interesting observations on the root-cap of aërial roots will be found in a paper by **Aladar Richter** : Bibl. Bot. **No. 54,** 1901.

43. It is an interesting fact, that axial organs may develop a protective and boring organ, resembling a root-cap, if they penetrate into the ground like roots. The so-called geocalycoid Jungermanniaceae—more appropriately termed " marsupiferous," as suggested by Goebel—illustrate this point. Here the apical region of each archegonium-bearing shoot becomes transformed into a pouch ; the shoot is positively geotropic and grows downwards into the soil, the archegonia, and later the sporogonia, being stowed away inside the pouch. The remarkable structure of the archegonial shoots of the Australian Acrobolbus unguiculatus has been described by Goebel. Here the pouch may be as much as an inch in length ; so long as the shoot is growing, the tip of the pouch is covered by a special boring and protective organ, which is surprisingly like an ordinary root-cap. The apical region of the pouch contains a subterminal meristematic zone, which is overlain by· a cap-like mass of resistant cells, that have ceased to divide. Cf. **Goebel** : Flora, **96**, 1906, pp. 155 *sqq.* **G. Haberlandt** : Flora, 1909.

44. The classification of primary meristems employed in the text, was first published in the author's Entwickelungsgesch. d. mech. Gewebesystems, where the reasons for its adoption will also be found. In that treatise the protoderm is called the "young epidermis," a term which no longer seems appropriate, since it is desirable, in an anatomico-physiological treatise, to reserve the name epidermis for the primary dermal layer in its fully developed condition. In the treatise mentioned, and in the first [German] edition of the present work, the prosenchymatous primary meristematic tissue is termed " cambium " in accordance with Nägeli's usage. At the present day, however, the term " procambium," first introduced by Sachs, has gained general acceptance, while most anatomists apply the name " cambium " only to the meristem which effects the secondary growth in thickness of the stems [and roots] of Dicotyledons and Gymnosperms ; in these circumstances, it seemed preferable to employ the term procambium in accordance with general usage. Finally, " fundamental meristem " ("ground-meristem ") is preferable to "fundamental parenchyma," so far as meristematic tissues are concerned, because the use of the former term prevents any uncertainty, as to whether embryonic or permanent tissues are under consideration.

45. **Hansen** : Abh. Senck. **12**, 1880. **Heinricher** : Sitzb. Wien, **78,** 1878. Id. ibid. 1881, and **Rostowzew** : Ber. 1894 [Gen.-Vers.-Hft.] (Ferns). **Goebel** : Biol. Centr. **22,** 1902.

46. **G. Haberlandt** : l.c. [44]. **Baranetzky,** Ann. Sci. Nat., sér. VII., 1897.

47. A very complete list of the early literature dealing with the epidermis will be found in **De Bary's** Comparative Anatomy. Note also : **Von Mohl** : Vermischte Schriften, pp. 260 *sqq.* (cuticle). **F. Cohn** : De cuticula, Warsaw, 1850. **Pfitzer** : P.J. **7,** pp. 532 *sqq.* Id. ibid. **8,** pp. 16 *sqq.* **Westermaier** : P.J. **14,** pp. 43 *sqq.* **Hiller** : P.J. **15,** (petals). **Solereder** : Systematic Anatomy of the Dicotyledons (Engl. ed. 1908).

48. **Pfeffer** : Physiology of Plants (Engl. ed.), I. pp. 116-9, 243-4, etc.

49. **Unger** : Sitzb. Wien, **44,** 1861, pp. 205 *sqq.* According to Unger, the ratio between the (total) transpiration of a leaf of Digitalis purpurea and the evaporation from a freely exposed water-surface varies between $\dfrac{1}{1 \cdot 4}$ and $\dfrac{1}{6 \cdot 9}$.

50. *Wax.* **De Bary** : B.Z. 1871, pp. 128 *sqq.* Wiesner : ibid. pp. 771 *sqq.* Id. ibid. 1876, pp. 225 *sqq.* **F. Haberlandt** : Wissensch.-prakt. Unters. **2**, 1877, p. 156. **Tschirch** : Linnaea, N.F. **9**, Nos. 3 and 4, 1881, pp. 147 *sqq.* **Tittmann** : P.J. **30.**

51. **Damm** : Beih. bot. Centr. **11**, 1902. **Emma Ott** : Ost. bot. Zeitschr. 1900.

52. **Stahl** : Jena. Zeitschr. **12**, 1888 (plants and snails). **Kohl** : l.c. [20].

53. *Calcium oxalate in epidermal walls.* **Solms-Laubach** : B.Z. 1871. **Heimerl** · Sitzb. Wien, **93**, 1886.

54. *Varnished leaves.* **G. Haberlandt** : l.c. [28], pp. 105 *sqq.* **Wiesner** : Festchr. zool.-bot. Ges. Wien, 1876.

55. **G. Haberlandt** : Ann. Buit. **11**, 1893, p. 114 (wettable velvety epidermis).

56. *Persistent epidermis.* **Damm's** (l.c. [51] ) results do not agree in all particulars with those of the author. At the end of the first year, according to Damm, the outer epidermal wall is thrown into folds over the radial septa, but only rarely develops any fissures. Careful re-examination, however, only confirms the author in his previous conclusion, to the effect that cracks are formed and *not* folds. The only possible explanation seems to be, that Damm and the author have examined different plants. Damm identifies his material as " Acer pennsylvanicum (striatum)." The author's own material was taken from specimens in the botanical gardens at Graz, labelled " Acer striatum, Lam., N. America." The Graz identification is undoubtedly correct. Possibly Damm's plant was really Acer striatum, hort. ( = A. palmatum, Thunb.). Cf. **Pax** : Aceraceae (in Engler's Pflanzenreich).

57. **Burns** : Flora, 1900.

58. **Von Höhnel** : Wissensch.-prakt. Unters. **1**, 1875, pp. 149 and 162.

59. *Mucilaginous inner walls.* **Radlkofer** : Monographie d. Gattung Serjania, 1875, p. 100.

60. *Chloroplasts in the epidermis.* **Stöhr** : Sitzb. Wien, **79**, 1879.

61. *Water-storing function of epidermis.* **Westermaier** : P.J. **14**, 1883, pp. 43 *sqq.*

62. *Vesicular (water-storing) epidermal cells.* **Volkens** : Flora, d. ägyptisch-arabischen Wüste, Berlin, 1887, pp. 52 *sqq.*

63. *Water-storing capacity of the ordinary epidermis.* The material used in these experiments consisted of leaves of Horse-chestnut, Hazel and Pear, for all of which the daily rate of transpiration had already been calculated (**G. Haberlandt** : Sitzb, Wien, **101,** 1892). The radial diameter of the two epidermal layers combined, amounts to ·025 mm. for Aesculus, ·02 mm. for Corylus and ·0325 mm. for Pyrus. The volume per sq. dcm. of both epidermal layers together, therefore, works out as follows :

| | |
|---|---|
| Aesculus - - - - - - | ·25 cc. |
| Corylus - - - - - - | ·2 ,, |
| Pyrus - - - - - - | ·325 ,, |

Assuming that the epidermis gives up one-half of its total water content to the transpiring mesophyll, when the external water-supply is cut off and transpiration goes on at the ordinary rate, the quantities of water available per sq. dcm. are as follows :

| | |
|---|---|
| Aesculus - - - - - - | ·125 g. |
| Corylus - - - - - - | ·1 ,, |
| Pyrus - | ·162 ,, |

Experiments carried out in the open air in the month of August gave the following values for the transpiratory activity of 1 sq. dcm. of leaf surface:

|  | Per 24 hrs | Per hr. |
|---|---|---|
| Aesculus - - - - - | 1·37 g. | ·057 g. |
| Corylus - - - - - | 3·33 ,, | ·138 ,, |
| Pyrus - - - - - | 5·97 ,, | ·248 ,, |

According to these figures the time for which the epidermis could cope with the loss of water by transpiration would be as follows:

| Aesculus - - - - | 2 hrs. 13 mins. |
|---|---|
| Corylus - - - - | 43 ,, |
| Pyrus - - - - - | 39 ,, |

The above estimate is based on the actual rate of transpiration exhibited by cut shoots standing in water, and hence with their stomata open. Since, however, the stomata close (in the plants experimented upon) when wilting begins, it is better to calculate on the basis of cuticular transpiration; the latter was calculated, for the aforesaid plants, by estimating the transpiration (during 24 hours, at 18° to 19° C., in air of relative humidity 70 per cent. to 75 per cent.) of leaves which were first allowed to wither slightly, and were then hung up with the cut end of the petiole waxed. Under these conditions the leaves withered very little more during the first 24 hours (except in the case of Corylus, where the marginal teeth shrivelled somewhat). The quantities of water lost by cuticular transpiration alone were as follows:

|  | Per 24 hrs. | Per hr. |
|---|---|---|
| Aesculus - - - - | ·18 g. | ·0075 g. |
| Corylus - - - - | ·407 ,, | ·017 ,, |
| Pyrus - - - - | ·562 ,, | ·023 ,, |

From these figures we get the following values for the ratio of cuticular to total transpiration:

| Aesculus - - - - - - | $\frac{1}{7·6}$ |
|---|---|
| Corylus - - - - - - | $\frac{1}{8·1}$ |
| Pyrus - - - - - | $\frac{1}{10·6}$ |

Assuming, as before, that the epidermis gives up one-half of its total water-content to the mesophyll, the time for which the combined epidermal layers could cope with the cuticular transpiration of the leaf is found to be:

| For Aesculus - | 16 hrs. 36 mins. |
|---|---|
| ,, Corylus - - - - | 5 ,, 52 ,, |
| ,, Pyrus - - - - | 7 ,, 2 ,, |

While these figures are far from exact, they give a rough idea of the water-storing capacity of the ordinary epidermis in certain cases.

64. *Anthocyanin.* **Von Mohl :** Vermischte Schriften, pp. 375 *sqq.* **G. Haberlandt :** Sitzb. Wien, **72,** 1876. **Kerner v. Marilaun :** Nat. Hist. of Plants. Vol. I. **Kny :** Atti. d. congr. bot. int. 1892. **Stahl :** Ann. Buit. **13,** 1896.

65. *Tannin.* **Warming :** Bot. Centr. **16,** 1883, p. 350. **Focke :** Kosmos, **10,** (protection against Fungi). **Stahl :** l.c. [52].

66. **Raciborski :** Flora, 1895 (Erg.-Bd.).

67. **G. Haberlandt** in Schenck's Handb. d. Bot. **11**, p. 579 (epidermal papillae).

68. **Stahl** : l.c. [64] (epidermal papillae).

69. **Ambronn** : P.J. **14**, (marginal pits).

70. **Damm** : l.c. [51].

71. *Hairs.* **Meyen** : Die Sekretionsorgane d. Pflanzen, Berlin, 1837. **Weiss** in Karsten's bot. Unters. Berlin, 1867. **Rauter** : Denkschr. Wien, **31**, 1871. **Uhlworm** : B.Z. 1873. **Solereder,** l.c. [47]. **Bobisut** : Sitzb. Wien, **113**, 1904. **Renner** : Flora, 1908.

72. *Screening action of hairs.* **L. Linsbauer** : Beih. bot. Centr. **10**, 1901. **Baumert** : Cohn's Beitr. **9**, 1907. As both these authors make use of methods which are open to many objections, it is not surprising that they arrive at very divergent results.

73. *Experiment with Stachys lanata.* In this experiment, no account is taken of the error introduced by the exposure of the uncutinised transverse septa of the hairs in the case of the shaven leaf. A simple calculation will show that this error is negligible. A leaf of Stachys lanata bears, on its upper side, 120 hairs per sq. mm., or for a leaf with an area of 12·92 sq. mm. 155,040 hairs on the whole of the upper side. In the case of a hair with a diameter of ·011 mm., the area of a single transverse septum works out at ·000095 sq. mm. As the hairs are unbranched, the total area of the exposed transverse walls in the aforesaid leaf is 155040 × ·000095 or 14·73 sq. mm. Even if we assume that the evaporation through these septa is equal to that of a freely exposed water-surface, the amount of water lost by the shaven leaf in this way in 24 hours cannot exceed ·01 g., and is thus small in comparison with the total loss (·915 g.) in the same period. The actual amount lost by evaporation through the transverse septa is probably much less.

74. **Goebel** : Pflanzenbiol. Schilderungen, **II.** 1891 (Espeletia).

75. *Stinging hairs.* **Schleiden** : Grundzüge d. wiss. Botanik (2nd ed.) **1**, p. 269. **Von Mohl** : B.Z. 1861, p. 219. **Duval-Jouve** : Bull. Soc. Bot. **14**, 1867. **G. Haberlandt** : Sitzb. Wien, **93**, 1886.

76. *Occlusion of hairs.* **Keller** : Nova acta, **55**, 1890.

77. *Structure of cork.* **Von Mohl** : Vermischte Schriften, pp. 212 *sqq.* **Sanio** : P.J. **2**, pp. 39 *sqq.* **G. Haberlandt** : Öst. bot. Zeitschr. 1875. **Von Höhnel** : Sitzb. Wien, **76,** 1877. **De Bary** : Comp. Anat. **J. Möller** : Anatomie d. Baumrinden. Berlin, 1882. **Kügler** : Über das Suberin, Inaug.-dissert. Strassburg, 1884. **Gilson** : La Cellule, **6**, 1890. **Van Wisselingh** : Arch. néerl. **26**, 1893. **Wiesner's** "Saft-periderm" (Öst. bot. Zeitschr. 1890) is practically nothing more than young cork which has not yet become dry; how far cork in this condition may act as a water-storing tissue, is a question which requires further investigation.

78. [The terms "cork-film" and "cork-crust" seemed to me the most convenient equivalents of "Korkhaut" and "Korkkruste."]

79. *Imperviousness of cork to water.* **Nägeli** : Bot. Mittheil. **1**, 1863, pp. 28 *sqq.* **Eder** : Sitzb. Wien, **72**, 1875.

80. **Wiesner u. Pacher** : Öst. bot. Zeitschr. 1875, No. 5 (transpiratory activity of woody stems at various ages).

81. **Mikosch :** Sitzb. Wien, **73,** 1876 (cork in bud-scales).

82. **Wiesner :** Sitzb. Wien, **79,** 1879 (imperviousness of cork to gases).

83. *Extensibility and elasticity of cork.* **Schwendener :** Abh. Akad. Berlin, 1882 (p. 39 of the separate).

84. *Phellogen.* **Sanio :** P.J. 2. **De Bary :** Comp. Anat., p. 109, **J. G. Weiss :** Denkschr. Regensbg. **6,** 1890. **Kuhla :** Bot. Centr. **71,** 1897. **Leisering :** Ber. **17,** 1899.

85. *Light-reflecting plates of Algae.* **Berthold :** P.J. 13, pp. 685 *sqq.* **Hansen :** Mitth. Neapel, **11,** 1893, Nos. 1 and 2.

86. **De Bary :** Comp. Morph. and Biol. of Fungi (sclerotia).

87. *Origin and regeneration of epidermis.* **Schwarz :** Sitzb. Wien, **77,** 1878 (Monstera). **Pfitzer :** P.J. **8,** pp. 40 *sqq.* (Peperomia). **Vöchting :** Unters. z. exper-Anat. und Path. d. Pflanzenkörpers, Tübingen, 1908, pp. 73 *sqq.* (Brassica).

88. Different botanists have attached—and still attach—different meanings to the term bast. It has already been explained in the text that, in its original sense, "bast" refers to a single tissue, namely, the extracambial fibrous tissue in the Dicotyledonous stem. Kurt Sprengel, for example, defines the term as follows (Von dem Bau u. der Natur d. Gewächse, Halle, 1812, p. 423) : "Bast is a whitish tissue, situated beneath the green rind, and distinguished by its fibrous structure and by its great elasticity, toughness and general resistant properties."

The further development of this primitive conception of bast took place along two divergent lines. A number of notable botanists, such as Link, Kieser and Meyen, continued to associate the term bast with one particular kind of tissue ; the advance of anatomical knowledge forced this school of observers to recognise that bast was not confined to the extracambial region of Dicotyledonous stems, but that a precisely similar tissue also occurred in the axial organs of Monocotyledons. Thus Meyen, in describing the figure of a longitudinal section through the leaf of Scirpus lacustris (Phytotomie, Berlin, 1830), writes as follows : "The wood-bundles (*Holz-bündel*) are of two kinds, larger ones situated in the middle of the leaf, and smaller ones placed nearer the epidermis ; bast-bundles (*Bast-bündel*) occur everywhere between the wood-bundles, some lying close beneath the epidermis." Strasburger is therefore mistaken when he alleges (Leitungsbahnen, preface, p. ix) that "the term bast-fibre has always been applied to cells of a particular morphological value and must accordingly be used only in this restricted sense." Similarly, the application of the term bast to mechanically specialised fibres by Schwendener and his school is in no sense an innovation ; all that Schwendener did, in this connection, was to set up a much more exact definition of the term than the earlier botanists were in a position to formulate, the increased precision arising largely as a consequence of the effect upon vegetable anatomy of a physiological mode of treatment.

The original definition of bast also became modified in a different direction altogether, the term being transferred, without much logical justification, from a particular tissue to a particular tissue-region. Bast, in this purely topographical sense, includes the whole of the extracambial secondary tissues of the Dicotyledonous stem (inclusive of the corresponding portions of the medullary rays), in which bast-fibres in the histological sense occur frequently, though not invariably. It then became necessary to discriminate between hard bast (equivalent to bast-fibres) and soft bast (comprising sieve-tubes and phloem-parenchyma). Later, the term bast was also transferred from the secondary phloem of Dicotyledons and Gymnosperms to the corresponding portions of the primary vascular tissue, which thus came to be known as bast both in Monocotyledons and in Dicotyledons.

There can be little doubt as to which of these modifications of the original significance of the term " bast " is the more precise and logical.

89. *Histology of bast-fibres.* **Schwendener :** Das mechanische Prinzip, 1874, pp. 3 *sqq.* **De Bary :** Comp. Anat. **Wiesner :** Mikroskopische Unters. pp. 24 *sqq.* Id. Rohstoffe d. Pflanzenreichs, Leipzig, 1912 (2nd ed.) Ch. XVIII. **Von Höhnel :** Über pflanzliche Faserstoffe, Vienna, 1884. Id. Mikroskopie d. technisch-verwendeten Faserstoffe, Vienna, 1887. **Krabbe :** P.J. **18.** The distortions exhibited by certain bast-fibres, described by **Von Höhnel** (P.J. **15,** p. 311), are attributed by **Schwendener** (Ber. **12,** pp. 239 *sqq.*) to the effects of preparation, a view which is shared by **Wiesner.**

90. **Nägeli :** Theorie d. Gährung, Munich, 1879 (micellar structure of bast-fibres).

91. *Collenchyma.* **Schwendener :** l.c. [89], pp. 3 *sqq.* **De Bary :** Comp. Anat. **Ambronn :** P.J. **12,** 1881. **Giltay :** Arch. néerl. **17,** 1883. **Van Wisselingh :** ibid. **Carl Müller :** Ber. **8,** 1890. **Jonas Cohn :** P.J. **24.**

Bokorny's interpretation of collenchyma as a water-conducting tissue, and C. Müller's views regarding the water-storing capacity of this tissue have already been disposed of by Strasburger (Leitungsbahnen, pp. 557 *sqq.*) and J. Cohn (l.c.). There was therefore no need to refer to those theories in the text.

92. Although sclereides and genuine fibres are connected by intermediate forms. these are not so numerous as to justify us in placing two such very different cell-types in a common category under the name of sclerenchyma. Where this is done (cf. Sachs, De Bary, etc.), the only distinguishing feature of a sclerenchyma-element is the thickness of its walls, a characteristic which is not only wanting in definiteness, but is also not directly connected with the mechanical function. Apart from their thick-walled character, however, there is little in common between bast-fibres and sclereides. Wiesner takes up the same attitude in his Elementen d. Anatomie u. Physiologie d. Pflanzen (Vienna, 1881), where he writes (p. 162): " Their unusual inextensibility and tensile strength, their flexibility, the uncoloured and altogether relatively unmodified condition of their cell-walls,—which usually exhibit the re-actions of pure cellulose—all these are features, in respect of which bast-fibres offer a very marked contrast to the hard, brittle, lignified, brown or yellow sclereides." The difference between the two types of cell could not be expressed in clearer language.

Concerning sclerenchyma (in the narrow sense) cf. **De Bary :** Comp. Anat. **Tschirch :** P.J. **16,** pp. 30 *sqq.* Id. Angewandte Pflanzenanatomie, Vienna and Leipzig, 1889, pp. 300 *sqq.* **Potonié :** Kosmos, **8,** pp. 33 *sqq.* **G. Haberlandt:** Sitzb. Wien, **75,** 1877 (testa of Phaseolus). Ibid. Mitth. Steiermk. 1887 (Begonia). **Solereder:** l.c. [47]. With regard to the filling in of the ruptured fibrous cylinder with sclerenchyma, see : **Tschirch :** l.c. [92]. **Schwabach :** Bot. Centr. **76,** 1898. **Devaux:** Mém. Bord., sér. V., **5,** 1899. **Schellenberg :** Schwendener-Festschrift, 1899. **Vöchting:** P.J. **34,** 1899. **Kny :** P.J. **37,** 1901. **Gaedecke :** Das Füllgewebe d. mechanischen Ringes (Inaug.-Dissert.), Berlin, 1907.

Strasburger maintains (Leitungsbahnen, p. 77) that the sclereides which occur in the bark of the Larch, Spruce and Fir, and in the flesh of the Pear, have no mechanical significance at all ; he believes that metabolic processes which go on in the starch-conducting tissues of the plants in question lead to the production of superfluous cellulose-material, which is then deposited in the form of secondary thickening layers on the walls of the sclereides. It seems, however, very unlikely that cellulose, which is ordinarily a plastic substance, should ever appear as a useless by-product of metabolism ; nor is there any good reason for supposing that translocation of carbohydrates involves the production and deposition of cellulose.

93. *Strength and elasticity of mechanical cells.* **Schwendener** : l.c. [91], pp.
9-16. **Weinzierl** : Sitzb. Wien, **76**, 1877. **F. Haberlandt** : Wollny's Forsch **1.**,
No. 5. **Firtsch** : Ber. **1**, 1883. **Sonntag** : Landw. Jahrb. **21**, 1892. **Schwendener** :
Ber. **12**, 1894, p. 239. **Wiesner** : Rohstoffe (2nd ed.), Ch. XVIII. **Sonntag** : Flora,
**99**, 1909.

[94. The translation, in the text, of *Tragmodul* as "modulus of elasticity,"
requires correction. In a foot-note to p. 151 of the original, Prof. Haberlandt defines
the *Tragmodul* as "the maximum load per unit cross-section, that can be supported
without the elastic limit being exceeded." The correct equivalent in English, therefore, is "load at the elastic limit," or (as it is more usually termed), simply, "elastic
limit."

In the same foot-note *Festigkeitsmodul* is defined as "the [minimum] load per
unit cross-section which causes rupture." The best equivalent for this term is
"ultimate strength."

Hence, on pp. 163 and 164 of the text, the terms "modulus of elasticity" and
"breaking strength" should everywhere be replaced by "elastic limit" and "ultimate strength" respectively. In the quotation from Schwendener, on p. 163,
"elastic limit" should also be substituted for "tensile strength."]

95. The mechanical principles underlying the arrangement of the stereome
were first explained by Schwendener in his classical treatise entitled Das mechanische
Prinzip. The objections raised by **Detlefsen** (Arb. Würzb. 3) were refuted by **Schwendener** in a paper Zur Lehre v. d. Festigkeit d. Gewächse (Sitzb. Berlin, **46**, 1884).
Schwendener dealt principally with the vegetative organs of Phanerogams. Interesting observations on the skeletal system of floral organs—particularly of sepals and
petals—have been made by **Herzog** (Uber. d. Systeme d. Festigung u. Ernährung i. d.
Blüte (Inaug.-Dissert.), Freiburg [Switzerland], 1902).

| ENGLISH. | | | GERMAN. | |
|---|---|---|---|---|
| Adjective. | Substantive. | Definition of Adjective. | Adjective. | Substantive. |
| Inflexible | Inflexibility | Offering resistance to bending | biegungsfest | Biegungs-festigkeit |
| Inextensible | Inextensibility | Offering resistance to stretching | zugfest | Zugfestig-keit |
| Incompressible | Incompressibility | Offering resistance to compression | druckfest | Druckfestig-keit |
| Shear-resisting | Shear-resistance | Offering resistance to shearing (*i.e.* mutual displacement of the component particles) | schubfest | Schubfestig-keit |

96. I have had considerable difficulty in finding suitable English equivalents for
the terms employed by German botanists in describing the various forms of mechanical
resistance. Terms such as "rigidity," "stiffness," etc., which are already to some
extent in use, are not sufficiently definite. My attempts to derive appropriate and
at the same time euphonious substitutes from Greek or Latin roots met with no
success. I therefore decided to make shift with words such as "inflexible," "inextensibility," etc., defining them in a particular sense for the purposes of the
physiological anatomist. When applied to a stem, therefore, "inflexible" must
be understood to mean, *not* "incapable of being bent," but "offering resistance

to bending "; similarly, "inextensibility" signifies the "property of offering resistance to stretching," not the property of being unstretchable. A list of all the terms of this kind used in the present work, their meanings and German equivalents, is given in the foregoing table.]

[96a. p. 165. The correct equivalent of *neutrale Faser* is "neutral lamina." p. 166. The designation of the axis common to the several I-girders in Fig. 55 as their "common neutral axis" (*gemeinschaftliche neutrale Achse*) is not strictly in accordance with British usage. I am indebted to Professor J. D. Cormack for the information upon which the present note and Note 94 are based.]

97. **Potonié** : Kosmos, **6**, 1882, pp. 181 *sqq.* (corrugated structure of stereome).

98. *Stereome in Palm leaves.* **Stahl** : Ann. Buit. **11**, 1893, p. 172. **Koop** : Beih. bot. Centr. **22**, I. 1907. Koop believes that the thinness of the lamina and the lowness of its girders are sufficient to account for the unusual flexibility of the leaves of Palms ; if this were the case, it would be difficult to understand why the fibrous strands are so often deeply embedded in the mesophyll in this group.

99. **Worgitzky** : Flora, **70**, 1887 (tendrils).

100. **Magocsy-Dietz** : Math. u. naturw. Ber. Ung. **17**, 1901 (septate pith).

101. *Protection of intercalary meristems.* **Westermaier** : Monatsber. Berlin, 1881, pp. 67 *sqq.* Id. Mitth. naturf. Freib. 1901.

102. **Westermaier** u. **Ambronn** : Flora, 1881 (inextensibility of climbing stems).

103. **Warming** : Engler's Jahrb. **4**, 1883 (buttress-roots of Rhizophora).

104. *Protection of leaf-margin.* The observations on this subject published by the author in the first [German] edition of the present work have been greatly extended by **Hintz** (Nova acta, **54**, 1889.) Cf. also **Lippitsch**: Öst. bot. Zeitschr. 1889 (Musaceae, etc.).

105. Sachs (Lectures, p. 50) attributes the function of protecting the leaf-margin against the risk of tearing more especially to the marginal commissures between the lateral veins ; in large or delicate leaves there are often several of these commissures one behind the other. "The leaf-margin is then comparable to a railway viaduct constructed of two or three storeys of arches ; a comparison which is by no means a merely superficial or formal one, but completely describes the fact itself, since this mechanical arrangement of the venation has a similar mechanical significance to that of the arch-piers of a bridge." As a matter of fact, the similarity upon which Sachs lays stress is purely superficial. For the arches of a viaduct have to resist the vertical pressure of the traffic passing over it, whereas, in the case of a leaf, the wind exerts its shearing action in a plane at right angles to that of the marginal commissures. If the leaf had to withstand pressure *against the margin,*—which is not the case—then its structure might indeed be justly compared to that of a viaduct.

106. **C. Müller** : P.J. **19**, pp. 497 *sqq.* (clamp-cells of Equisetum).

107. *Influence of mechanical stimuli on the development of stereome.* **Knight** · Phil. Trans. 1803, II. p. 280. Id. ibid. 1811, p. 217. **Treub** : Ann. Buit. **3**. **Pfeffer** Ber. sächs. Akad. 1891, pp. 638 *sqq.* **Wiedersheim** : P.J. **38**, 1902. **Vöchting** Nachr. Gött. 1902. Unters. z. experim. Anat. u. Path. d. Pflanzenkörpers, Tübingen, 1908. **Ball** : P.J. **39**, 1903. **Bücher** : P.J. **43**, 1906.

108. *Influence of moisture on the development of stereome.* **F. Haberlandt** Wollny's Forsch. **1**, pp. v *sqq.* **Kohl :** Die Transpiration d. Pflanzen u. ihre Einwirkung auf d. Ausbildung pflanzlicher Gewebe, Braunschweig, 1886, pp. 90 *sqq.*

109. **Wille :** K. Svenska Handl. **21**, 1885, No. 12 (stereome of Fucaceae).

110. *Ontogeny of stereome.* **G. Haberlandt :** Entwickelungsgesch. d. mech. Gewebesystems d. Pflanzen, Leipzig, 1879. **Ambronn :** l.c. [91].

111. **Karsten :** Bibl. bot. No. 22, 1891 (anchoring hairs of Avicennia).

112. *Anchoring hairs of epiphytes.* **F. Müller :** Ber. **13**, 1895 [two papers].

113. **Von Lengerken :** B.Z. 1885 (adhesive discs of Ampelopsis).

114. *Mucilage-hairs of Lythraceae.* **Klebs :** Unters. Tüb. **1**, 1885. **Correns :** Ber. **10**, 1892.

115. **G. Haberlandt :** Sinnes-organe im Pflanzenreich z. Perzeption mech. Reize Leipzig, 1906 (2nd ed.), pp. 117, 118 (eel-trap-hairs of Biophytum).

116. *Eel-trap-hairs of Aristolochia.* **Hildebrand :** P.J. **5. Correns :** P.J. **22,** 1891.

117. **Meyen :** Neues System d. Pflanzenphysiologie, **2**, 1839, pp. 9 *sqq.* **F. Schwarz :** Die Wurzelhaare, etc. Unters. Tüb. **1. Persecke :** Üb. d. Formveränderung d. Wurzel in Erde u. Wasser, Leipzig, 1877. **Mer :** C.R. **88**, 1879. **Westermaier** u. **Ambronn :** Abh. bot. Ver. Brand. **22** (Azolla). **Van Tieghem :** Ann. Sci. Nat.,sér. VII.,**6**, 1887 (twin-hairs). **Freidenfelt :** Bibl. bot. No. **61**, 1904. **Leavitt :** Proc. Boston Soc. 1904, p. 31. **Snow :** B.G. **40**, 1905.

**Kny** (Ber. **16**, 1898) found in the case of various roots (Zea Mays, Vicia Faba, Pisum sativum) that absorption (of nitrates and methyl violet) takes place, not only in the root-hair-bearing region, but also for some distance (a few mm.) in front of the latest-formed root-hairs. This fact does not, however, necessitate any modification of the statements already made in the text, regarding the location of the absorbing system in roots. No tissue enters suddenly upon its physiological duties at the moment when its differentiation is completed. On the contrary, its functional activity begins while it is still in process of development, although it usually becomes intensified as the tissue approaches the mature condition. Kny's results merely provide special illustrations of this general law.

118. **G. Haberlandt :** l.c. [8], pp. 54 *sqq.* (experiments on longitudinal growth of root-hairs). **Sokolowa :** Üb. d. Wachst. d. Wurzelhaare u. Rhizoiden, Moscow, 1897.

119. **Sachs :** B.Z. 1860, p. 117. Id. Experimental-Physiologie, 1875, pp. 170, 182, 188. **Czapek :** P.J. **29**, 1896. **Stoklasa** u. **Ernest :** P.J. **46**, 1908.

120. **Warming :** B.Z. 1883. **Briosi :** Sopra un organo di alcuni embryoni Vegetali, 1882. **Goebel :** Flora, 1901 (Malaxideae).

121. *Rhizoids of Bryophyta.* **W. P. Schimper :** Recherches anat. et morphol. sur les mousses, 1848, pp. 18 *sqq.* **Leitgeb :** Unters üb. d. Lebermoose, No. 2, p. 36 ; No. 3, p. 37 ; No. 6, (Marchantiaceae), p. 19. Id. Sitzb. Wien, **80**, 1879. **Zimmermann :** Arb. Würzb. **2**, p. 665. Cf. also **Sachs :** Vorlesungen (2nd ed.), pp. 36-39, and 528 *sqq.* **Goebel** in Schenck's Handbuch d. Botanik, **2**, Breslau, 1881 (Muscineae). **G. Haberlandt :** Öst. bot. Zeitschr. 1892 (geotropism). **Kamerling :** Flora, 1897 (Erg.-Bd.). **Correns :** Unters. üb. d. Vermehrung d. Laubmoose, etc. Jena, 1899.

**Vaupel :** Flora, **92,** 1903. **Paul :** Engler's Jahrb. **32,** 1903. In the last-cited paper, various facts and arguments are adduced in support of the view that the rhizoids of Mosses are always primarily organs of attachment, and have little to do with absorption. This view is no doubt correct, so far as epiphytic and aquatic species are concerned ; but all Mosses that grow rooted in the ground,—or rather all those that have a central water-conducting strand in their stems—must certainly absorb water through their rhizoids in the same way as a Higher Plant does through its root-system.

Fern-prothalli and Liverworts occasionally bear multicellular rhizoids ; these have been recorded, for example, in Danaea (**Brebner :** A.B. **10,** p. 120) and in Gottschea (**Goebel :** Flora, **96,** 1906, pp. 103 *sqq.*). In the latter case it is particularly the expanded and lobed ends of the rhizoids that are septate ; whether this point is of any physiological importance, is not known.

122. *Velamen.* **Unger :** Sitzb. Wien, **13,** 1854. Id. Anatomie u. Physiologie d. Pflanzen, 1855, pp. 306 *sqq.* **Duchartre :** Journ. soc. imp. et centr. 1856, p. 67. **Oudemans :** Abh. Akad. Amst. 1861. **Chatin :** Mém. soc. Cherb. 1856, p. 7. **Leitgeb :** Denkschr. Wien, **24,** 1864, pp. 179 *sqq.* Id. Sitzb. Wien, **49,** 1864 (two papers). **A. F. W. Schimper :** Die epiphytische Vegetation Amerika's, Jena, 1888, pp. 46 *sqq.* **Goebel :** Pflanzenbiologische Schilderungen, 1, Marburg, 1889, pp. 188 *sqq.* **Meinecke :** Flora, **78,** 1894, pp. 133 *sqq.* **Nabokich :** Bot. Centr. **80,** 1899.

With reference to the exodermis, cf. also **Von Höhnel :** Sitzb. Wien, **76,** 1877, and **Kroemer :** Bibl. Bot. No. **59,** 1903.

123. **Von Mohl :** Vermischte Schriften, pp. 294 *sqq.* **Oltmanns :** Üb. d. Wasserbewegung in d. Moospflanze (Inaug.-Dissert), 1884. **Russow :** Schriften d. Naturf.-Ges. zu Dorpat, **3,** 1887. **Lorch :** Flora, 1894, pp. 424 *sqq.*

124. For the earlier literature dealing with absorption of water by foliage-leaves, the reader is referred to the compilations of **Osterwald** (Wiss. Beilage z. Programm d. städt. Gymnasiums in Berlin, Easter, 1866) and **Burgerstein** (27ter Jahresber. d. Leopoldstädter Communalgymnasiums in Wien, 1891). Further note—**Lundström :** Pflanzenbiologische Studien, **1,** and **Wille :** Cohn's Beitr. **4,** 1887 (adaptations in relation to rain and dew). **Wiesner :** Sitzb. Wien, **86,** 1883. **A. F. W. Schimper :** l.c. [122], pp. 66 *sqq.* Id. Bot. Centr. **17,** 1884. **Volkens :** Die Flora d. ägyptisch-arabischen Wüste, Berlin, 1887, pp. 31 *sqq.* **Emily Gregory :** Comp. Anat. of the felt-like hair-covering of leaf-organs, Zürich, 1886. **Mez :** P.J. **40,** 1904, and **Steinbrinck :** Flora, **94,** 1905 (Bromeliaceae).

125. According to Mez, the thick outer walls of the head also have an important part to play in the absorption of water by the scale-hairs. When wetted, these walls swell and curve outwards. As a result, the cavities of the head-cells—which are almost entirely obliterated in the dry state—regain their proper dimensions ; being vacuous, they exert a powerful suction, and are thus able to draw in any water that lodges on the outside of the swelling hair. Steinbrinck raises various objections to this theory, and maintains that the action of the absorbing hairs of Bromeliaceae is entirely dependent upon the cohesive power of water. (For a discussion of cohesion-mechanisms see below, Ch. XI.).

126. **Treub :** Natur. Verh. Acad. Amst. **19,** 1879 (Orchidaceae). Id. Ann. Buit. 3, pp. 79 *sqq.* **Warming :** Engler's Jahrb. **4,** 1883, pp. 517 *sqq.* **G. Haberlandt :** Ann. Buit. **12,** 1893, pp. 91 *sqq.* (mangroves). **Koorders :** Engler's Jahrb. **21** (Tectona). **Balicka-Iwanowska :** Flora, 1899 (various Gamopetalae). **Goldflus :** Journ. de bot. **12** (1898) and **13** (1899) (Compositae). **Lloyd :** Mem. Torrey Bot. Club, **8,**

1899-1902 (Rubiaceae). - **Billings** : Flora, 1901 (Globularia, etc.). **Longo** : Annali di Botanica, 2.

127. **Sachs** : B.Z. 1862 (two papers). **G. Haberlandt** : Die Schutzeinrichtungen i. d. Entwickelung d. Keimpflanze, Vienna, 1877, pp. 39 *sqq.* and 87 *sqq.* Id. Ber. 1890. **Klebs** : Unters. Tüb. **1**, 1885, pp. 561 *sqq.* **Ebeling** : Flora, 1885. **Brown** and **Morris** : J.C.S. **57**, 1890, pp. 458 *sqq.* **Tschirch** : Ann. Buit. **9**, 1891, pp. 143 *sqq.* **Grüss** : P.J. **30**, 1897.

128. **Leitgeb** : Unters. üb. d. Lebermoose, No. 5, Pl. I. figs. 2 and 3 ; Pl. III. figs. 9, 10, and 11 ; Pl. IV. figs. 4c, 10b, and 11.

129. **Kamienski** : B.Z. 1881, pp. 457 *sqq.* (Monotropa). **Frank** : Ber. 1885. Id. ibid. 1887. Id. ibid. 1888. **Schlicht** : Landw. Jahrb. 1889. **Johow** : P.J. **16**, pp. 415 *sqq.* **Janse** : Ann. Buit. **14**, 1896. **Groom** : A.B. **9**, 1895 (Thismia). **Stahl** : P.J. **24**, 1900. In the last-cited paper, which contains a mass of interesting observations, Stahl endeavours to prove that mycorrhizic plants obtain part or the whole of the indispensable mineral substances in the form of organic compounds from the symbiotic Fungus. Hence it is particularly plants with feeble transpiration that enter into symbiotic association with Fungi for this purpose ; actively transpiring species can obtain a sufficient supply of mineral salts by their own exertions.

[129a. The term mycorrhiza, as used by Haberlandt, signifies a *root* in which the normal absorbing tissue is replaced by a symbiotic Fungus mycelium. In English botanical literature, it is used in a less strictly defined sense.]

130. Typical mycorrhiza, such as occurs among the Monotropaceae, Cupuliferae, Betulaceae and Coniferae, is distinguished by Frank as " ectotrophic " mycorrhiza. Somewhat different conditions prevail in the case of Alnus, Myrica, Burmannia, Apteria, Thismia, Voyria, Cotylanthera, Psilotum, the Ericaceae, Epacridaceae and Empetraceae, and the Orchidaceae (whether green or non-chlorophyllous). Here the Fungus inhabits the outermost layer—often also the rest of the cortex—of the host-root (or -rhizome), forming a dense tangle of mycelium in each infected cell, and sending forth comparatively few hyphae into the soil. This form of mycorrhiza is termed " endotrophic " by Frank, who believes that the living protoplasm of the infected cells ultimately digests the invading hyphae. Frank's interpretation has recently received considerable support through the investigations of Magnus and Shibata (see below). In the case of endotrophic mycorrhiza, *absorption* is probably at most a subsidiary function of the fungus mycelium ; this statement is supported by the fact that Orchids (even entirely colourless forms, such as Corallorhiza and Epipogon) are always provided with genuine root-hairs.

With reference to endotrophic mycorrhiza cf. **Frank** : Ber. **9**, 1891. **Nobbe** u. **Hiltner** : Landw. Versuchstat. **51**, 1899 (Podocarpus). **Magnus** : P.J. **35**, 1900 (Neottia). **Shibata** : P.J. **37**, 1902 (Psilotum and Podocarpus).

131. **G. Haberlandt** : P.J. **17**.

132. **Solms-Laubach** : P.J. **6**, 1868, pp. 509 *sqq.* Id. B.Z. 1876 (Pilostyles). **L. Koch** : Die Klee- und Flachsseide, Heidelberg, 1880. Id. Die Entwickelungsgesch. d. Orobanchen, mit bes. Berücksicht. ihr. Bezieh. z. d. Kulturpflanzen, Heidelberg, 1887. Id. P.J. **20** and **22** (Rhinanthaceae). **Leclerc du Sablon** : Ann. Sci. Nat., sér. VII. **6**, 1887. **Heinricher** : Cohn's Beitr. **7** (Lathraea). **Sperlich** : Beih. Bot. Centr. **11**, 1902. Id. Sitzb. Wien, **96**, 1907 (Balanophora). **Schaar** : Sitzb. Wien, **107**, 1898 (Rafflesia).

133. **Ch. Darwin** : Insectivorous Plants, London, 1875. **Batalin** : Acta Hort.

Bot. Petropol. 7 (Sarracenia and Darlingtonia). Goebel : Pflanzenbiologische Schilderungen, 2, Marburg, 1901-03.

134. Reinke : P.J. 10, 1876. Wille : Beitr. z. physiol. Anat. d. Laminariaceen, Christiania, 1897.

135. De Bary : Vergl. Morph. u. Biol. d. Pilze, Leipzig, 1884, pp. 18 sqq. Von Guttenberg : l.c. [13].

136. Treub : Ann. Buit. 13, 1895.   .

[136a. I have used the term photosynthesis throughout the translation as the equivalent of "carbon-dioxide assimilation." But cf. Pfeffer's Physiology, 1, p. 302, footnote.]

137. Schmitz : Verh. Rheinl. u. Westf. 40, 1881 (Algae).   Id. P.J. 15, 1884. N. Pringsheim : P.J. 12, 1881.    Arthur Meyer : Das Chlorophyllkorn, etc., Leipzig, 1883.   Tschirch : Ber. 1, 1883.   A. F. W. Schimper : P.J. 16, 1885.   G. Haberlandt : Flora, 1888, and Ber. 23, 1905 (Selaginella). Zimmermann : Beih. Bot. Centr. 4, 1894.   Kolkwitz : Festchr. f. Schwendener, Berlin, 1899 (Spirogyra). Senn : Die Gestalts- u. Lageseveränderungen d. Pflanzen-Chromatophoren, Leipzig, 1908.

138. The literature dealing with the pigments of chloroplasts and other chromatophores is very extensive. The following compilations and comprehensive treatises may be specially mentioned : Czapek : Ber. 20, 1902 (Gen.-Vers.-Heft). Id. Biochemie d. Pflanzen, 1, pp. 449 sqq. 1905. Id. Progressus, 1, 1906. Euler : Grundlagen u. Ergebnisse d. Pflanzenchemie, 1, pp. 191 sqq. Braunschweig, 1908.

139. Engelmann : B.Z. 1882, 1883 and 1884.   Id. Arch. Anat. u. Phys. 1902 (compl. chromatic adaptation). Timiriazeff : Bull. Congr. int. de bot. a St.-Pétersbourg, 1884. Id. Proc. Roy. Soc. 72, 1903. Gaidukow : Anh. z. Abh. Akad. Berlin, 1902, and Ber. 21, 1903 (compl. chromatic adaptation). Stahl : Zur Biologie d. Chlorophylls, Jena, 1909. Stahl's views are opposed by Wiesner (Der Lichtgenuss d. Pflanzen, Leipzig, 1907, pp. 241 sqq.).

140. Sachs : B.Z. 1862, p. 365.   Id. ibid. 1864, p. 289.   Godlewski : Flora, 1873, p. 378.   Boehm : Sitzb. Wien, 69, 1874.   Id. B.Z. 1883.   Dehnecke : Üb. nichtassimilierende Chlorophyllkörner (Inaug.-Diss.), Bonn, 1880. A. F. W. Schimper : B.Z. 1880, p. 881.   Id. P.J. 16, 1885.   Stock : Ein. Beitrag z. Kenntnis d. Proteinkrystalle (Inaug.-Diss. Tübingen), Breslau, 1892.

141. Boehm : Sitzb. Wien, 1856.   Frank : P.J. 8, 1872.   Borodin : Bull. Acad. St.-Pétersb. 1868.   Stahl : B.Z. 1880 (two papers).   A. F. W. Schimper : P.J. 16, G. Haberlandt : Ber. 4, 1886.   Senn : l.c. [137].   Lidforss : Lunds Univ. Arsskrift, N.F., 4, 1908.

142. Recently Senn (l.c. [137], pp. 265 sqq.) has attempted to prove experimentally, that the accumulation of chloroplasts on those cell-walls which abut against intercellular spaces, is not due to the chemotactic influence of carbon dioxide. Even if his views were correct, he has overlooked the fact that the *advantage* derived from this particular distribution of the chloroplasts consists in the improved carbondioxide supply, whatever the cause of the distribution may be. Senn inclines to the belief, that the preference on the part of the chloroplasts for the best ventilated portions of the cell-walls is connected with transpiration. "The chromatophores are attracted to the points of most intense evaporation, owing to the accumulation, in these portions of the walls, of the solid residuum—comprising mineral salts and

complex organic compounds—of the transpiration stream; this accumulation is sometimes so great that the walls become encrusted with solid matter." To the author, Senn's arguments do not appear convincing. It would be a very peculiar arrangement, if the chloroplasts had to obtain their supply of mineral salts from the cell-walls; what the importance of the "complex organic compounds" may be, it is impossible to guess, as Senn gives no further information as to their chemical nature. Even if we assume that the chloroplasts are attracted to the ventilated regions of the wall in the manner suggested by Senn, we are not justified in concluding that the physiological significance of this behaviour on their part has no connection with the intake of carbon dioxide. A carnivorous animal, though hunting by scent, does not pursue its prey for the sake of the odoriferous substances, but in order to obtain food-materials, which may themselves be quite odourless. As a matter of fact, Senn has himself demonstrated that the chlorophyll corpuscles in the leaf of Funaria display positive chemotaxis towards carbon dioxide (l.c. pp. 103-168).

143. **Knoll** : Sitzb. Wien, **117, 1908**

144. **Pick** : Beitr. z. Kenntn. d. assim. Gewebes armlaubiger Pflanzen, Bonn, 1881.

145. **Weber** : Arb. Würzb. **2**, p. 343.

146. **Brogniart** : Ann. Sci. Nat., sér. I., **21**, 1830. **G. Haberlandt** : P.J. **13**, 1881. Id. Ber. **4,** 1886. **Stahl** : B.Z. 1880. Id. Jen. Zeitschr. **16**, 1883 (sun- and shade-leaves). **Pick** : Bot. Centr. 1882. **Vesque** : Ann. agron. **9** and **10**, 1884. **Areschoug** : Engler's Jahrb. **2**, 1882. **Heinricher** : P.J. **15**, 1884 (isobilateral leaves). **Kohl** : Die Transpiration d. Pflanzen, etc., Braunschweig, 1886. **Volkens** : Die Flora d. ägyptisch-arabischen Wüste, Berlin, 1887. **Nilson** : Bot. Centr. **27,** 1886 (photosynthetic tissue in stems). **Eberdt** : Ber. **6,** 1888 (palisade-tissue). **Bonnier** : Bull. Soc. Bot. France, **35,** 1888, and C.R. 1890 (experiments on influence of alpine climate on plants). **Loebel** : P.J. **20**, 1889. **Leist** : Mitth. Naturf. Ges. Bern, 1889, and **Wagner** : Sitzb. Wien, **101,** 1892 (leaves of alpine plants). **Rikli** : P.J. **27,** 1895 (Cyperaceae). **Montemartini** : Atti. Ist. Bot. Pavia, **4,** 1895. **Jönsson** : Zur Kenntn. d. anat. Baues d. Blattes, Lund, 1896. **Roedler** : Zur Vergl. Anat. d. assim. Gewebesystems d. Pflanzen (Inaug.-Diss.), Berlin. **Warming** : Kgl. Danske Vid. S., sér. VI., **8,** 1897 (halophytes). **Rywosch** : Ber. **25,** 1907. Rywosch regards the tubular form of the typical palisade-cell as an adaptation to the prevailing direction in which *water* flows through the mesophyll. As a matter of fact, no appreciable flow of water takes place from the vascular bundles and spongy parenchyma towards the upper epidermis, for the simple reason that the latter is usually devoid of stomata and in addition well protected against loss of water by cuticular transpiration. Rywosch draws attention to the case of Asphodelus luteus, where the uppermost layer of palisade-cells is richer in chlorophyll and shorter (in the radial direction) than the more deeply seated layers; he has failed to realise that the more internal palisade-cells have to transmit the materials manufactured in the overlying layers, besides disposing of their own synthetic products. Such differences in the radial dimensions of different palisade-layers have no connection with transmission of water.

147. **Noll** : Arb. Würzb. **3**, 1888.

148. **Sachs** : Arb. Würzb. **3**, 1888. **A. F. W. Schimper** : B.Z. 1885. **Brown** and **Morris** : J.C.S. 1893. **A. Meyer** : Unters. üb. d. Stärkekörner, Jena, 1895. **Rywosch** : B.Z. 1908.

[148*a*. This term is explained in the next paragraph.]

149. Cf. the papers by **Stahl, G. Haberlandt, Heinricher, Volkens, Pick, Eberdt, Bonnier, Leist** and **Wagner** cited above [146].

150. When **Nordhausen** (Ber. 1903) objects that, in woody plants," the so-called ' sun ' and ' shade ' leaves may develop independently of the direct influence of light, either while still enclosed in the buds or immediately after the latter unfold," he is really confirming the conclusions arrived at in the text. Nordhausen further admits, that illumination has some direct influence upon the degree of development of palisade-tissue. The fact that the initiation of palisade-tissue takes place in the bud, and therefore in the absence of light, had been previously demonstrated by the author himself.

151. The not inconsiderable diminution in the absolute carbon-dioxide content of the atmosphere with increasing altitude, and the reduced vegetative period, may, as Wagner asserts, be partly responsible for the greater development of the photo-synthetic system in alpine plants ; but these factors are of minor importance in comparison with the intensity of illumination.

152. While the author refers the characteristic shape and orientation of palisade-cells to the physiological principles of maximum exposure of surface and expeditious translocation, Stahl attempts to correlate these two features with the intensity and direction of the prevailing illumination. By placing themselves on the radial walls of the palisade-cells, the chloroplasts assume the profile position towards vertically incident light, an arrangement which is decidedly advantageous, when the illumination is direct sunshine or intense diffused light (cf. p. 272). According to Stahl, the characteristic shape and orientation of palisade-cells are due to the disproportionate development of the radial walls, which in turn arises out of the advantage of enabling the greatest possible number of chloroplasts to assume the profile position in bright vertically incident light. The principal objections to Stahl's interpretation may be summarised as follows :

1. Under natural conditions, the sun's rays never strike the surface of any particular leaf at right angles for any considerable period. A leaf which has assumed a " heliotropically fixed position," [cf. p. 615] receives sunlight at different angles at various times of the day ; in our own latitudes a horizontally extended leaf experiences no vertical insolation at all. In Central Europe (Lat. N. 45° to 55°) the maximum altitude of the sun (on June 21st) never exceeds 68 5°. Diffuse illumination naturally meets the leaf at various angles, according to circumstances. Even those rays which are, to begin with, approximately vertical to the surface, undergo so much reflexion, refraction and absorption on their way through the leaf, that, even at a short distance below the epidermis it would be absurd to use the terms profile- and surface-positions in the sense which they bear, for example, in the case of a Moss-leaf or a Fern-prothallus.

2. Many plants form one or more layers of palisade-tissue even when growing in the deepest shade. This statement applies more particularly to evergreen plants. Stahl tries to evade this difficulty by arguing that long-lived evergreen leaves have to adapt themselves to several other special conditions (frost, increased mechanical requirements), and are therefore unable to adjust their organisation as perfectly as deciduous leaves to the prevailing conditions of illumination. But it is not obvious, why an increase in mechanical strength, and in the power of withstanding frost, should interfere with adaptations which are correlated with the conditions of illumination. Certainly, the differentiation of palisade-cells cannot increase either the mechanical strength or the frost-resisting capacity of a leaf. In any case, many deciduous plants also develop perfectly typical palisade-tissue in their shade-leaves

(*e.g.* Magnolia acuminata, Tropaeolum majus and Chelidonium majus, according to Eberdt).

3. Where palisade-cells are attached by their sides to the underlying cells, the inner tangential walls may lie nearly at right angles to the leaf-surface, but are nevertheless always free from chloroplasts, although any that might adhere to them would be approximately in the profile position; conversely, the lower curved regions of the lateral walls, which lie almost parallel to the surface, are crowded with chloroplasts, in spite of the fact that the latter are permanently exposed in the surface position. Similarly, if one end of a palisade-cell projects freely into an air-space, it usually bears chloroplasts over the whole of the exposed wall, although the corpuscles nearest the extremity have to assume the surface position.

4. Very frequently individual palisade-cells are more or less strongly curved, *e.g.* when their lower ends are inserted upon collecting-cells or attached to a bundle-sheath, or where they partly surround or over-arch a hypostomatic air-chamber; sometimes the curvature is so marked, that an L-shaped element results (Scilla bifolia, Fig. 107, *A*). As regards the distribution of the chlorophyll-corpuscles, such cells do not differ from ordinary, straight palisade-cells; all parts of the lateral walls are equally crowded with chloroplasts, although a varying proportion (depending upon the degree of curvature) of these must assume the surface position.

5. It has already been explained in the text (p. 291), that oblique orientation of palisade-cells has usually no connection with the direction of the incident light; the frequent occurrence of oblique palisade-cells is therefore unfavourable to Stahl's hypothesis. In the case of reclinate leaves [with oblique palisade-tissue] the palisade-cells slope upwards in the lower erect portion of the leaf, but downwards in the upper pendulous portion; their orientation with reference to the incident illumination is, therefore, opposite in the two regions of the leaf.

6. Stahl's interpretation does not apply to the girdle-type of photosynthetic system (and related types; cf. p. 284). Here, the photosynthetic cells which are situated between the vascular bundles and the leaf-surface, are elongated at right angles to the surface, and must therefore be regarded as typical palisade-cells; the cells on the flanks of the bundles, on the other hand, extend parallel to the surface, and every stage intermediate between these two extreme conditions is to be found in a single leaf. But there can be no doubt that the orientation of all the cells of a girdle depends upon the same cause; therefore the typical palisade-elements of the girdle cannot be orientated with reference either to the direction or to the intensity of the incident illumination. The same argument applies *mutatis mutandis* to those cases in which the photosynthetic cells are arranged in curved series converging towards the different bundle-sheaths (cf. Scabiosa ucrainica, fig. 121). Here, again, any hypothesis as to the arrangement of the photosynthetic cells must account for the orientation of every element in each series, from the outermost, which is a typical palisade-cell, to the innermost, which may be extended parallel to the leaf-surface.

7. Finally, it must not be forgotten that the palisade-cell is only a special variety of the elongated photosynthetic cell; other forms of this physiological unit may be orientated in a great variety of ways with reference to the surface of the organ in which they occur. A comprehensive theory of the shape and orientation of the photosynthetic cell must take all varieties of that cell into consideration; but if the direction (or intensity) of the incident illumination is regarded as the determining factor, this condition becomes impossible of fulfilment.

Another set of observers (Areschoug, Vesque, Kohl and Montemartini) have attempted to correlate the development of palisade-tissue with the intensity of transpiration, increased transpiration being supposed to favour the differentiation of palisade-cells. Areschoug, in fact, goes so far as to regard the formation of palisade-tissue as a device for restricting transpiration. As a matter of fact, the air-spaces between palisade-cells are by no means always narrow crevices; some forms of palisade-tissue are very loosely put together, and contain an exceedingly well-developed ventilating system. Further, Volkens has shown that desert-plants often possess a very loose palisade-tissue, although the structure of their dermal system and stomata shows them to be pronounced xerophytes. It is therefore quite clear that the shape and orientation of palisade-cells have no connection with the intensity of transpiration. In any case, narrow intercellular spaces can be developed quite as readily in an isodiametric photosynthetic parenchyma as in palisade-tissue; conversely, enlargement of the ventilating spaces—or, in other words, of the transpiring surface—can be achieved quite as readily within palisade-parenchyma as in any other form of tissue.

Warming regards the frequent convergence of groups of palisade-cells towards collecting-cells (cf. p. 285, and Figs 116 and 117), as an adaptation which serves to enlarge the intercellular spaces in proportion to their distance from the upper side of the leaf. This interpretation does not commend itself to the author's judgment. Even if such a progressive enlargement of the intercellular spaces were needed,—which remains to be proved—it could be equally well achieved by various other modes of arrangement of the palisade-cells.

153. **G. Haberlandt :** P.J. **17,** 1886. **Magdeburg :** Die Laubmooskapsel als Assimilationsorgan (Inaug.-Diss.), Berlin, 1886. **Roedler :** l.c. [146].

154. **Wille :** K. Svenska Handl. **21,** 1885. **A. Hansen:** Mitth. Neapel, **2,** 1893.

155. **G. Haberlandt :** Mitth. Steiermk. 1880.

156. **Pfeffer :** Abh. sächs. Ges. **18,** 1892. Id. Physiology, **1,** pp. 587 *sqq.,* **Czapek :** Sitzb. Wien, **106,** 1897.

157. **Von Mohl :** Verm. Schriften, p. 285. **Sanio :** B.Z. 1868, p. 113. **Caspary :** Monatsber. Berl. Akad. 1862 (July). **De Bary :** Comp. Anat. pp. 155 *sqq.* **Strasburger :** Üb. d. Bau. u. d. Verrichtungen d. Leitungsbahnen in d. Pflanzen, Jena, 1891, pp. 510 *sqq., et passim.* The terms "trachea" and "tracheide" were introduced at a time when anatomists generally believed, that the structures in question represented ventilating organs analogous to the tracheal tubes of insects. As it is now established, beyond any possibility of doubt, that the tracheae and tracheides of plants are water-conducting tubes, it becomes necessary to consider whether the modification of our views concerning the functions of these structures necessitates any change of terminology. Potonié has, in fact, suggested (Üb. d. Zusammensetzung d. Leitbündel b. d. Gefässkryptogamen, 1883), that the water-conducting-system as a whole might be termed the "hydrome," and its component elements "hydroids." The aptness of these terms cannot be denied; but the author has yet to be convinced of the need for a new terminology. Etymologically, the terms "trachea" and "tracheid" are derived from the Greek τραχυς, signifying rough, hard or uneven, and really imply that the walls of the tubular structures to which they are applied are furnished with devices for increasing their mechanical strength; as they may therefore be quite properly applied to water-conducting tubes, there is no need to depart from the old-established terminology.

158. **Rothert :** Bull. Acad. Crac. 1899. **Körnicke :** Sitzb. niederrhein. Ges. 1899.

159. *Bordered pits.* **Th. Hartig** : B.Z. 1863, p. 293. **Sanio** : P.J. **9**, 1873, pp. 50 *sqq.* **Russow** : Sitzb. Dorpat, 1881. Id. Bot. Centr. **12**, 1883, Nos. 1-5. **Strasburger** : Bau u. Wachstum d. Zellhäute, 1882, pp. 42 *sqq.* **De Bary** : Comp. Anat. pp. 158 *sqq.*

160. *Water-conducting tissues of Mosses.* **Unger** : Sitzb. Wien, **43**, 1861, pp. 497 *sqq.* **Lorentz** : P.J. **6**, pp. 388 *sqq.* **G. Haberlandt** : Ber. 1883. Id. P.J. **17**, pp. 372 *sqq.* **Oltmanns** : Cohn's Beitr. **4**, 1884. **Coesfeld** : B.Z. 1892. **Tansley** and **(Miss) Chick** : A.B. **15**, 1901. **Goebel** : Flora, **96**, 1906.

161. **Böhm** : B.Z. 1879, pp. 227 *sqq.* Id. ibid. 1881, Nos. 49 and 50. Id. Landw. Ver. **20**, 1877, pp. 375 *sqq.* Id. Forsch. Agrikulturphysik, **1**. **Von Höhnel**: Wissensch.-prakt. Unters. **2**, 1887, pp. 89 *sqq.* Id. P.J. **12**, 1879. **R. Hartig** : B.Z. 1883, pp. 250 *sqq.* (in part abridged from longer papers in Unters. Forstbot. Inst. München, **2**, 1882, and **3**, 1883). **Russow** : l.c. [159]. **Volkens** : Üb. Wasserauscheidung in liquider Form an d. Blättern hoherer Pflanzen (Inaug.-Diss.), Berlin, 1882. **Elfving** : B.Z. 1882, No. 42. Id. Acta Soc. Fenn. **14**, 1884. **Vesque** : Ann. Sci. Nat., sér. VI., **19**. **Scheit** : B.Z. 1884, p. 201. **Errera** : Bull. soc. bot. Belg. **25**. Westermaier : Ber. 1883. Id. Sitzb. Berlin, 1884. **Godlewski** : P.J. **15**, 1884. **Janse** : P.J. **18**, 1887. Id. ibid. **45**, 1908. **Schwendener** : Sitzb. Berlin, 1886, pp. 561 *sqq.* Id. ibid. 1892. Id. Vorles. üb. mech. Probleme d. Botanik (ed. Holtermann), Leipzig, 1909, pp. 63 *sqq.* **Pappenheim** : Ber. 1889. **Strasburger** : l.c. [157], pp. 537 *sqq.* Id. Hist. Beitr. No. 5, 1893. **Dixon** : Proc. Roy. Soc. Dublin, **10**, 1903. Id. Proc. Roy. Soc. (London), **79** B, 1906. **Dixon** and **Joly** : Phil. Trans. **186** B, 1895, and A.B. **10**, 1896. **Askenasy** : Verh. Heidelberg, 1895. **Pfeffer** : Physiology, **1**, pp. 220 *sqq.* **Copeland** : B.G. **35**, 1902. **Ursprung** : Beih. Bot. Centr. **18**, 1904. P.J. **42**, 1906, and **44**, 1907. **Steinbrinck** : Ber. **22**, 1904. Id. P.J. **42**, 1906.

In various publications (Arb. Würzb. **2**, pp. 291 *sqq.* ; Lectures, pp. 225 *sqq.*) Sachs upheld the theory of Unger, according to which water travels in the lignified *walls* of the water-conducting elements, and not in their cavities. This "imbibition theory" of the ascent of sap, which held the field for a considerable time, was first successfully attacked by Böhm ; it is now merely of historic interest.

162. Cf. **Schwendener** : Sitzb. Berlin, 1892, pp. 938 *sqq.* **Strasburger** : Hist. Beitr. No. 5, pp. 85 *sqq.*

163. **Jost** : B.Z. 1891 and 1893. **Kohl** : Die Transpiration d. Pflanzen, Braunschweig, 1886, pp. 90 *sqq.* **Schenck** : Ber. 1884. Id. Bibl. Bot. No. 1, 1886. **Constantin** : Ann. Sci. Nat., sér. VI., **19**. **Strasburger** : l.c. [157], pp. 929 *sqq.* **Westermaier** and **Ambronn** : Flora, 1881.

164. *Tyloses.* **Hermine von Reichenbach** : B.Z. 1845 (publ. anonymously). **Böhm** : Sitzb. Wien, **55**, 1867. **Unger** : Sitzb. Wien, **56**, 1867. **Reess** : B.Z. 1868. **G. Haberlandt** : Funktion u. Lage d. Zellkernes, Jena, 1887, pp. 71 *sqq.* **Molisch** : Sitzb. Wien, **97**, 1888. **Winkler** : Ann. Buit., sér. II., **5**, 1905. **Von Alten** : B.Z. 1909.

165. *Conducting parenchyma.* **Sachs** : Flora, 1863, p. 33. Id. P.J. **3**, 1863. Id. Experimentalphysiologie, 1865, pp. 574 *sqq.* Cf. also **Pfeffer** : P.J. **8**, 1872, p. 538. **De Vries** : Landw. Jahrb. **8**, 1879, p. 447. **A. F. W. Schimper** : B.Z. 1885. **Schubert** : Bot. Centr. **72**, 1897.

166. **Strasburger** (l.c. [157], p. 474 ; also Textbook of Botany [3rd Engl. ed.], p. 113) does not recognise the existence of a "cambiform tissue," as distinguished from ordinary leptome-parenchyma (phloem-parenchyma, *cribral-parenchym*). The author dissents from this attitude. It is true that the recognition of the separate

identity of the companion-cells (and the elements which take their place among Gymnosperms and Pteridophytes) has considerably restricted the connotation of the term " cambiform " ; nevertheless typical cambiform tissue can always be readily distinguished from typical phloem-parenchyma (conducting parenchyma). Strasburger attaches too much importance to the occurrence of transitional stages between conducting parenchyma and cambiform tissue. As a matter of fact, his whole conception of vascular structure is based upon the study of *secondary* tissues, where ordinary leptome-parenchyma (conducting parenchyma) bulks much more largely than cambiform tissue.

[Most British botanists follow Strasburger's example in this matter, recognising only three categories of leptome-elements, viz. sieve-tubes, companion-cells and leptome-parenchyma (phloem-parenchyma).]

167. Sieve-tubes were first described by Th. Hartig in 1857. Cf. **Nägeli** : Sitzb. München, 1861. **Hanstein** : Die Milchsaftgefässe, etc., Berlin, 1864. **De Bary** : Comp. Anat. pp. 172 *sqq.* **Wilhelm** : Beitr. z. Kenntn. d. Siebröhrenapparates dikot. Pflanzen, Leipzig, 1880. **Janczewski** : Mém. Soc. Cherb. **23**, 1891, p. 350. **Russow** : Sitzb. Dorpat, 1882, pp. 257 *sqq.* **Strasburger** : l.c. [157], p. 286. Id. Bau u. Wachstum d. Zellbäute, pp. 57 *sqq.* Id. P.J. **36**, 1901. Id. B.Z. 1901. **A. Fischer** : Unters. üb. d. Siebröhrensystem d. Cucurbitaceen, Berlin, 1884. Id. Ber. **3**, 1885. Id. Ber. sächs. Akad. 1886. **Zacharias** : 1884. **Lecomte** : Ann. Sci. Nat., sér. VII., **10**. **Poirault** : Ann. Sci. Nat., sér. VII., **18**, 1893. **A. W. Hill** : Phil. Trans. **194** B, 1901. Id. A.B. **22**, 1908.

168. **Frank** (Lehrbuch, 1), and, later, **Blass** (P.J. **22**) have suggested that the function of sieve-tubes is the *storage*, not the transportation, of protein materials. This theory was rejected in the second [German] edition of this work (p. 341).

169. As stated in the text, **Czapek** (Sitzb. Wien, **106**, 1897) has endeavoured to show that the leptome-strands—and especially the sieve-tubes—are responsible for the conduction of carbohydrates as well as of nitrogenous plastic compounds. Czapek asserts that conducting parenchyma—at any rate in petioles and stems— is incapable of transporting any appreciable quantity of carbohydrate material. His strongest argument is based upon certain extirpation experiments, in which the continuity of the tissues was interrupted in one-half of the cross-section of the petioles of various plants (by removing a half-disc at a convenient point). The result of this treatment was an entire cessation of the removal of carbohydrate materials from the affected halves of the experimental leaves. Now, the only tissues *completely* interrupted by the aforesaid operation were the vascular strands ; for the conducting parenchyma (parenchymatous ground-tissue) in the affected half of the petiole remains in lateral continuity with the corresponding tissue on the other side. Czapek infers that the paths along which carbohydrates travel must run straight up and down the petiole, and further concludes that the leptome-strands are the only conceivable conducting channels in the present case. The first of Czapek's conclusions is undoubtedly sound ; but the second is based upon the assumption that translocation in the conducting parenchyma takes place with equal readiness in the transverse and in the longitudinal direction. Czapek does, in fact, assume that this is the case. But the author himself holds the opposite opinion. Quite apart from the elongated form of the conducting parenchyma cells,—a feature which seems clearly to indicate the direction of prevalent translocation—it is highly probable that the plasmatic membranes lining the transverse septa are more pervious to plastic materials than those which clothe the longitudinal walls. In the author's opinion, indeed, Czapek's extirpation experiments provide an interesting demonstration of the fact, that, in conducting parenchyma, translocation takes place most

rapidly in the longitudinal direction, the transverse migration in that tissue being often so slight as to be practically negligible. *A priori*, this conclusion is quite as well founded as Czapek's assumption that translocation of carbohydrate takes place solely in the leptome-strands.

The above interpretation of Czapek's result is entirely in agreement with the rest of our information concerning the structure and function of conducting parenchyma. The author's anatomical observations upon the connection between the photosynthetic mesophyll and the conducting parenchyma of the bundle-sheaths, and Schimper's experimental results, both clearly indicate that transportation of carbohydrates takes place in the parenchymatous bundle-sheaths. Czapek does not, indeed, attempt to deny the truth of this statement (p. 139), but goes on to assert that "from the base of the leaf-blade onwards, the entire transportation of synthetic products is undertaken by the leptome-strands of the vascular bundles." But it would be a very singular circumstance, if the carbohydrates which have been conveyed out of the mesophyll through the parenchymatous bundle-sheaths were to depart from these channels after reaching the petiole, and to pass over into the leptome-strands. Tissues which are identical in structure always perform the same functions, no matter where they may be located in the plant-body. Hence, the sieve-tubes of the mesophyll must have exactly the same functions to perform as those in the petioles and stems. If, therefore, the former are solely or principally concerned with the transportation of protein materials, it is certain that the same statement also applies to the latter.

170. **Hanstein :** P.J. 2, 1860. In interpreting "ringing" experiments, it is necessary to bear in mind the polarity of woody twigs,—discovered by Vöchting which tends to encourage the production of roots at the basal ends of severed branches. But, as **Pfeffer** has pointed out (Physiology, **1**, p. 578), this influence of polarity does not invalidate such experiments. For it is found that roots are formed abundantly at the base of the long piece of each twig, and to a much smaller extent at the lower end of the shorter piece. This result clearly indicates the extracambial location of the protein-conducting tissues ; for, if ringing did not interfere with the conduction of proteins, one would expect the greatest number of roots to appear at the lower extremity of the twigs, as they do in the case of severed branches which are not ringed.

171. *Latex-tubes.* **Schultz-Schultzenstein :** Nova Acta, **18**, Suppl. II., 1841. **Von Mohl :** B.Z. 1843, p. 553. **Unger :** Ann. Wien. Mus. **2**, 1840. Id. Anatomie u. Physiologie d. Pflanzen, 1855, pp. 157 *sqq.* **Aron :** B.Z. 1846. **Schacht :** Die Pflanzenzelle, etc. 1852. Id. Monatsber. Berlin. Akad. 1856. **Hanstein :** l.c. [167]. **Sachs :** Experimentalphysiologie, 1865, p. 387. **Trécul :** Several papers in C.R. and Ann. Sci. Nat., 1862-8. **Dippel :** Entstehung d. Milchsaftgefässe, Rotterdam, 1865. **David :** Üb. d. Milchzellen d. Euphorb., Mor., Apoc. und Asclep. (Inaug.-Diss.), Breslau, 1872. **Schmalhausen :** Mém. Acad. St.-Pétersbg., sér. VII., **26**, No. 2. **De Bary :** Comp. Anat. pp. 183 *sqq.*, 432 *sqq.*, 487, 525. **Faivre :** Ann. Sci. Nat., sér. V., **6**, 1866, pp. 33 *sqq.* (Ficus elastica). Id. ibid., sér. V., **10**, 1869 (Morus alba). Id. C.R. **88**, 1879. **Schullerus :** Abh. Brandenburg, **24**, 1882 (Euphorbia Lathyris). **Scott :** Arb. Würzb. **2**, p. 648. **Emil Schmidt :** B.Z. 1882. **G. Haberlandt :** Sitzb. Wien, **87**, 1883. **Treub :** Ann. Buit. **3**. **Schwendener :** Sitzb. Berlin, 1885. **A. F. W. Schimper :** B.Z. 1885. **Chauveaud :** Ann. Sci. Nat., sér. VII., **14**, 1891. **Kny :** Sitzb. Naturf. Fr. Berlin, 1893, and **Zander :** Bibl. Bot. No. 37, 1897 (Latex-hairs of the Cichoriaceae). **Chimani :** Bot. Centr. **61**, 1895. **Molisch :** Studien üb. d. Milchsaft u. Schleimsaft d. Pflanzen, Jena, 1901. **Gaucher :** Ann. Sci. Nat., sér. VIII., **12**, 1900. **Kniep :** Flora, **94**, 1905. **Fitting :** Tropenpflanzer, **13**, 1909.

172. For the earlier literature on vascular bundles, and conducting strands generally, see **De Bary** : Comp. Anat. Note also **Von Mohl** : De structura palmarum, in **Von Martius** : Genera et species palmarum. Id. Vermischte Schriften, 1845, p. 129 (Palms). **Unger** : Üb. d. Bau u. d. Wachstum d. Dicotyledonenstammes, St. Petersbg. 1840. **Nägeli** : Beiträge Wiss. Bot. No. 1, Leipzig, 1859. **Nägeli** u. **Leitgeb** : ibid. No. 4, 1867. **Dippel** : Üb. d. Zusammensetzg. d. Gefässbündels d. Kryptogamen, Giessen, 1865. Id. Das Mikroskop, 2, 1869. **Russow** : Mém. Acad. St.-Pétersbg., sér. VII., **19**, 1872. Id. Betracht. üb. das Leitbündel- und Grundgewebe, Dorpat, 1878. **Schwendener** : Das mechanische Prinzip, etc. Leipzig, 1874. **Von Tieghem** : Ann. Sci. Nat., sér. V., **13**. Id. ibid., sér. V., **6**, 1866. Id. ibid., sér. VII., **3**, 1886 (polystely). Id. Bull. Soc. Bot. France, 1886 (Primula). Id. Traité de Botanique, 2nd ed. Paris, 1891, pp. 673 *sqq*. and 737 *sqq*. **Van Tieghem** and **Douliot** : Ann. Sci. Nat., sér. VII., **8**, 1888 (origin of endogenous members). **Kny** : Üb. einige Abweich. im Baue d. Leitbündels d. Monocot. Berlin, 1881. **G. Haberlandt** : Sitzb. Wien, **84**, 1881 (collateral foliar bundles of Ferns). Id. P.J. **17**, 1886 (Mosses). **Petersen** : Engler's Jahrb. **3**, 1882, and **Baranetzky** : Ann. Sci. Nat.. sér. VIII., **12**, 1900 (bicollateral bundles). **Potonié** : Jahrb. Bot. Gart. Berlin, **2**, 1883. **Leclerc du Sablon** : Ann. Sci. Nat., sér. VII., **9**, and **Zennetti** : B.Z. 1895 (Vascular Cryptogams). **Heinricher** : Ber. **1**, 1883, pp. 122 *sqq*. (Centaurea). **A. Fischer** : Ber. **1**. **Ross** : Ber. **1**, and **Reinhardt** : P.J. **16**, 1885 (anomalous root-structure in Monocotyledons). **Möbius** : Ber. **5**, 1887 (leptocentric bundles). **Strasburger** : l.c. [157]. **Perrot** : Journ. de Bot. **11**, 1897. **H. Fischer** : P.J. **35**, 1900 (pericycle). **Schoute** : Die Stelärtheorie, Jena, 1903. **Scherer** : Beih. Bot. Centr. **16**, 1904.

173. As is evident from the text, the terms " conducting strand " and " vascular bundle " are not precisely synonymous. The former is a physiological term, and hence has the wider application ; thus, from a physiological point of view, the simple central strand of a Moss-stem and the complex vascular bundle of an Angiosperm— composed of vessels, tracheides, sieve-tubes and hadrome- and leptome-parenchyma —are equivalent structures. As regards the significance of the term " vascular bundle," the author adheres to the definition given by **De Bary** (Comp. Anat. p. 316), who applies it to " strands which consist of tracheae [in the wide sense] and sieve-tubes, as their essential parts." The mechanical strands which so frequently accompany conducting strands, are not regarded as belonging to the vascular bundles by De Bary, although that author regards this separation as a convention, so far as descriptive anatomy is concerned (l.c. p. 400). In an anatomico-physiological treatise it is absolutely necessary to keep mechanical and conducting tissues strictly apart.

The author has found it necessary to substitute the terms "hadrome" (from ἁδρός, tough, coarse) and " leptome " (from λεπτός, thin, delicate) for the long-established Nägelian " xylem " and " phloem," because the older terminology is strictly *topographical* in its application, and hence not suited to the needs of the *physiological* anatomist. Other terms that have been employed [without gaining general acceptance] are *Gefässteil* and *Siebteil* [De Bary], and *Vasalteil* and *Cribralteil* [Strasburger].

[173*a*. The term " radial bundle," and the conception of vascular morphology which it implies, have fallen into disuse among British and American botanists since the general acceptance, on their part, in one form or another, of Van Tieghem's stelar theory. According to that theory, the vascular complex which De Bary called a radial bundle (cf. Comp. Anat. p. 319, pp. 348-366) is, properly speaking, not a bundle at all, but represents the entire stele or central cylinder of the root (or stem) in which it occurs. See also p. 359. ]

174. **A. F. W. Schimper** : Bot. Centr. **17**, 1884. Id. Die epiphytische Vegetation Amerika's, Jena, 1888. **Lierau** : Engler's Jahrb. **9**, 1888.

175. **Westermaier** : Sitzb. Berlin, 1884.

[175*a*. Cf. Note 173*a*.]

[175*b*. *Causes of medullation*. The problem of the exact phylogenetic origin of pith, though doubtless of great interest, in my opinion, hardly lies within the scope of the present work. The reader who may wish to pursue the subject, should consult the following recent papers (in which further references will be found): **Bower** : A.B. **25** 1911. **Lang** : A.B. **27**, 1913.]

176. **A. Fischer** : Ber. sächs. Akad. 1880. **A. Koch** : B.Z. 1884.

177. **Von Mohl** : B.Z. 1871. **Thomas** : P.J. **4**. **De Bary** : Comp. Anat. pp 378 *sqq.* **Zimmermann** : Flora, 1880. **Scheit** : Jena. Zeitschr. 1883. **Strasburger** : l.c. [157], pp. 102 *sqq.*

178. *The endodermis.* **Caspary** : P.J. **1**. Id. ibid. **4**. Id. ibid. **6**. **Schwendener** : Abh. Akad. Berlin, 1882 (abstract in Ber. **1**, 1883). Id. Sitzb. Berlin, 1890. **Russow** : l.l.c. [172]. **Van Wisselingh** : Versl. Med. Akad. Amst. 1884. **De Vries** : Maandbl. Naturwet. **13**, 1886. Id. B.Z. 1886, pp. 788 *sqq.* **Strasburger** : l.c. [157], pp. 105, 309, 344, 434, etc. **H. Fischer** : l.c. [172]. **Kroemer** : Bibl. Bot. No. 59, 1903. The paper last cited contains many interesting data with regard to the structure of endodermal layers, particularly with regard to the character of their cell-walls, although the author considers much of the new terminology introduced by Kroemer quite unnecessary. In the introduction, and again in his concluding remarks, Kroemer states that his investigation was based on a hypothesis framed by A. Meyer, to the effect that the cutinised or suberised strips in the walls of endodermal cells (and also in the walls of the exodermal elements and " Aufzellen " of roots) serve " in the first instance, to prevent as far as possible any diffusion, in the substance of the walls, of dissolved mineral salts or reserve-materials." This hypothesis does not appear to contain any novel idea. For all the numerous investigators who have insisted upon the relative imperviousness of suberised endodermal walls towards *water*, have undoubtedly regarded these walls as equally impervious to all substances dissolved in water. The author himself, indeed, has expressed his views on this point quite unequivocally in the second [German] edition of the present work in the following words : ". . . the framework of suberised strips in the radial walls of the endodermis prevents any diffusion of water [through the substance of the cell-wall] ; the dissolved materials conveyed by the vascular bundles are, of course, similarly prevented from escaping through the radial walls." **H. Müller** : B.Z. 1906.

179. The literature dealing with the course and arrangement of vascular bundles is very extensive. Cf. **De Bary** : Comp. Anat. Also the following—**Von Mohl** : De structura palmarum, Munich, 1831. Id. in **Von Martius** : Icon plant. crypt. Brasil. (tree-ferns). **Unger** : l.c. [172]. **Nägeli** : Zeitschr. Wiss. Bot. Nos. 3 and 4. Id. Beitr. Wiss. Bot. No. 1. **Cramer** in Nageli and Cramer: Pflanzenphysiol. Unters. No. 3. **Karsten** : Abh. Berlin. Akad. 1847 (Palms). **Hanstein** : P.J. **1**. Id. Sitzb. Berlin. 1857. **Von Ettingshausen** : Die Blattskelette d. Dicot., Vienna, 1861. **Hildebrand** : Anat. Unters. üb. d. Stämme d. Begoniaceen, Berlin, 1859. **Geyler** : P.J. **6** (Conifers). **F. Schmitz** : Das Fibrovasalsystem d. Piperaceen (Inaug.-Dissert.), Essen, 1871. **Vöchting** : Hanstein's Bot. Abh. **3** (Melastomaceae). Id. P.J. **9** (Rhipsalideae). **J. Weiss** : Flora, 1876. **Falkenberg** : Vergl. Unters. üb. d. Bau d. Vegetationsorgane d. Monokot., Stuttgart, 1876. **Westermaier** :

Flora, 1879 (Begoniaceae). Id. Monatsber. Berl. Akad. 1881. **Reinhardt** : l.c.
[172]. **Von Zalenski** : Ber. 1902. **Schuster** : Ber. 26, 1908 (Festschrift).

180. **Simon** : Ber. **26,** 1908 (Festschrift). **Freundlich** : P.J. 46, 1908.

181. **Klein** : Flora, 1877. **Ambronn** : B.Z. 1880. **Will** : B.Z. 1884. **Wille** :
Ber. 1885. Id. K. Svenska Handl. **21,** 1885. Cf. also Engler's Jahrb. 7. **Rosen
thal** : Flora, 1890. **Hanstein** : P.J. **24,** 1892. **De Bary** : Vergl. Morph. u. Biol.
d. Pilze, Leipzig, 1884, pp. 322 *sqq.* **A. Weiss** : Sitzb. Wien, **91,** 1885.

182. **G. Haberlandt** : Entwicklgsgesch. d. mech. Systems, Leipzig, 1879, pp.
21 *sqq.* 73. **Schwendener** : Die Schutzscheiden, pp. 63 *sqq.* **Strasburger** : Üb.
d. Bau u. d. Wachstum d. Zellhäute, Jena, 1882, p. 81.

183. **Lange** : Flora, 1891. **Nathansohn** : P.J. **32,** 1898.

184. Cf. Note 167.

185. *Water-tissues.* **Pfitzer** : P.J. **8. Westermaier** : P.J. **14,** pp 43 *sqq.* **Treub** ·
Ann. Buit. 3, 1883. Id. ibid. 7, 1888. **Volkens** : Die Flora d. ägyptisch-arabischen
Wüste, Berlin, 1887. **Heinricher** : Mitth. Steiermk. 1886. **Hintz** : Nova Acta, **54,**
1889. **G. Haberlandt** : P.J. **17,** 1886. **A. F. W. Schimper** : Die epiphytische Vegeta-
tion Amerika's, Jena, 1888. Id. Die Indomalayische Strandflora, 1891. **Goebel:**
Pflanzenbiologische Schilderungen, **1,** III. (Epiphytes), Marburg, 1889. **Lippitsch** :
Öst. Bot. Zeitschr. 1889. **G. Haberlandt** : Sitzb. Wien, **101,** 1892. **Schwendener** :
Sitzb. Berlin, **24,** 1896. **Holtermann** : Der Einfluss d. Klimas auf d. Bau d. Pflanzengewebe, Leipzig, 1907.

186. **Frank** : P.J. **5,** 1865.

187. **Jönsson** : Lunds Univ. Arsskr. **38,** 1902.

188. **Hofmeister** : Sitzb. sächs. Akad. **10,** 1858. **G. Haberlandt** : Die Schutzeinrichtungen in d. Entwickelung d. Keimpflanze, Vienna, 1887, pp. 11 *sqq.* **Klebs** :
Unters. Tüb. **1,** pp. 536 *sqq.*

189. **Vesque** : Ann. Sci. Nat., sér. VI., **13,** 1882 (two papers). **Krüger** : Flora,
1883, pp. 435 *sqq.* **Heinricher** : Bot. Centr. **22,** 1885. **Kny and Zimmermann** :
Ber. 3, 1885. **Rothert** : Ber. **17,** 1899.

190. **A. F. W. Schimper** : Die epiphytische Vegetation Amerika's, Jena, 1888,
pp. 41 *sqq.* **Heinricher** : Sitzb. Wien, **99,** 1890.

191. *Starch.* **Nägeli** in Nägeli and Cramer: Pflanzenphysiol. Unters. No. 2,
Zürich, 1858. Id. Bot. Mitth. 1863 (two papers). Id. B.Z. 1881. **Brown and Heron:**
Liebig's Ann. **199,** 1879. **A. F. W. Schimper** : B.Z. 1880. Id. ibid. 1881. **Stras-
burger** : Üb. d. Bau u. d. Wachst. d. Zellhäute, Jena, 1882. **Dafert** : Landw.
Jahrb. 1885. Id. ibid. 1886. **Arthur Meyer** : B.Z. 1886. Id. Ber. **4,** 1886. Id.
Unters. üb. d. Stärkekörner, Jena, 1895 (contains a very full list of literature).
**Mikosch** : Unters. üb. d. Bau. d. Stärkekörner, Vienna, 1887. **Krabbe** : P.J. **21,**
1890. **Binz** : Flora, 1882. **Dodel** : Flora, 1892. **Bütschli** : Verh. Heidelbg.
N.F. **5,** No. 1, 1893. Id. ibid. N.F. 7, 1903. **Salter** : P.J. **32,** 1898. **Winkler** : P.J.
**32,** 1898. **Hugo Fischer** : Cohn's Beitr. **8,** 1902. **Czapek** : Biochemie d. Pflanzen,
Jena, 1905, **1,** pp. 307 *sqq.*

192. *Reserve-cellulose.* **Sachs** : B.Z. 1862. **Godfrin** : Ann. Sci. Nat., sér. VII.,
**19,** 1884. **Heinricher** : Flora, 1888. **Reiss** : Landw. Jahrb. **18,** 1889. **E. Schulze** :
Ber. 1889. **Nadelmann** : **21,** 1890. **Schaar** : Sitzb. Wien, **99. Grüss** : Bot. Centr.

3 A

70, 1897. **Czapek** : l.c. [191], **1**, pp. 325 *sqq.* **Euler** : Grundlagen u. Ergebnisse d. Pflanzenchemie, Braunschweig, 1908, **1**, pp. 65 *sqq.*

193. *Inulin.* **Sachs** : B.Z. 1864, pp. 25 *sqq.* **Prantl** : Das Inulin, Münich, 1870. **Dragendorff** : Materialien zu einer Monographie d. Inulins, St. Petersbg. 1870. **Hugo Fischer** : l.c. [191].

194. Cf. **Pfeffer** : Physiology, **1**, pp. 457 *sqq.*, and the literature there cited.

195. The first and second [German] editions of this work contain the incorrect statement that the reserve-protein of succulent storage tissues is represented by the protoplasm in the storage-cells. As a matter of fact, the bulk of the nitrogenous reserve-material (amides, amino-acids and proteins) is contained in solution in the cell-sap of the storage elements. It is nevertheless quite likely, that when such storage-tissues are depleted, some constituents of the disorganised protoplasts are mobilised in order to serve as plastic materials ; but this statement applies equally to the dry storage-tissues of seeds and fruits.

196. *Crystalloids.* **F. Cohn** : Jahrb. Schles. Ges. 1858, pp. 72 *sqq.* **Nägeli** : Sitzb. Bair. Akad. **2**, 1862. **A. F. W. Schimper** : Üb. d. Proteinkristalle d. Pflanzen, Strassburg, 1879. **Heinricher** : Ber. 1891. **Zimmermann** : Beitr. z. Morph. u. Physiol. d. Pflanzenzelle, **1** and **2**. **Stock** : Cohn's Be.'+r. **6**, 1892. **Heinricher** : P.J. **35**, 1900. **Sperlich** : Beih. Bot. Centr. **21**, 1906.

197. *Aleurone-grains.* **Th. Hartig** : B.Z. 1855, p. 881. **Pfeffer** : P.J. **8**, pp. 419 *sqq.* **Wakker** : P.J. **19**. **Werminski** : Ber. 1888. **Tschirch** : Bot. Centr. **31**. **Tschirch** and **Kritzler** : Ber. Deutsch. Pharm. Ges. **10**, 1900. **Lüdtke** : P.J. **21**. **G. Haberlandt** : Ber. 1890.

198. **Hirsch** : Ber. 1890. Hirsch brings forward a number of fresh cases confirming the conclusions arrived at by the author, in the first edition of the present work, regarding the general principles of construction of endosperms and analogous storage-tissues, but makes no mention of the fact that his leading results had been anticipated six years previously.

199. **Tschirch** : Angewandte Anatomie, p. 459. **Holfert** : Flora, 1890.

200. **Kraus** : Abh. Naturf. Ges. Halle, **16**, 1882 and 1884. Id. Ann. Buit. **13**, 1896. **Knoch** : Bibl. Bot. No. 47, 1899.

201. **Sernander** : K. Svenska Handl. **41**, 1906, and the literature there cited.

202. **A. F. W. Schimper** : Die Wechselbeziehungen zw. Pflanzen u. Ameisen im trop. Amerika (Bot. Mitt. aus d. Tropen, No. 1), Jena, 1888.

203. **Ch. Darwin** : Fertilisation of Orchids, London, 1885. **Crüger** : Journ. Linn. Soc. **8**, 1865. **G. Haberlandt** : Sinnesorgane im Pflanzenreich zur Perzeption mech. Reize, Leipzig, 1901, pp. 64 and 65. **Porsch** : Öst. Bot. Zeitschr. 1905, 1906 and 1907. Id. Orchidaceae, in **Von Wettstein** : Ergebn. d. Bot. Exped. d. K. Akad. d. Wiss. nach Südbrasilien (Denkschr. Wien. Akad. **79**, 1908). Id. Text to Kny's Bot. Wandtafeln, **111** and **112** (Berlin, 1908).

204. **Janse** : Ber. **4**, 1886. **Fritz Müller**, in **H. Müller** : Befruchtung d. Blumen, 1873, p. 86. **Penzig** : Atti. Soc. Lig. **6**.

205. **Wille** : K. Svenska Handl. **11**, 1885. **Hanstein** : P.J. **24**, 1892. **Hansen** : Mitt. Neapel, **2**, 1893. **De Bary** : Vergl. Morph. u. Biol. d. Pilze, Leipzig, 1884,

pp. 31 *sqq.* **Errera** : Mém. Acad. Belg. **27**, 1885. **Clautriau** : Etude chimique du glycogene, 1895.

206. **Möller** : Die Pilzgärten einiger süd-amerikanischer Ameisen (Bot. Mitt. aus d. Tropen, No. 6), Jena, 1893.

207. **Pfeffer** : Physiology, **1**, pp. 176 *sqq.* **Wiesner** : Sitzb. Wien, **79**, 1879. **Wiesner and Molisch** : ibid. **98**, 1889. **Lietzmann** : Flora, 1887.

208. **Luerssen** : B.Z. 1873. Id. Sitzb. Naturf. Ges. Leipzig, 1875.

209. Cf. **Kühn** : Flora, 1889, p. 487. Kühn's plant is a species of Nephrodium, closely related to N. stipellatum, Hk.; his material was collected by Goebel near Tjibodas in Java. The author's specimens were collected in the gorge of the Tjiapus at the foot of Salak, near Buitenzorg. [Similar pneumatophores occur in various other species of Nephrodium. The translator is at present engaged upon an investigation of these organs in the case of several Jamaican members of the genus.]

210. **Raciborski** : Flora, **87**, 1900.

211. **Stahl** : B.Z. 1894. **F. F. Blackman** : Phil. Trans. **186 B**, 1895.

212. **Areschoug** : Engler's Jahrb. **2**, 1882.

213. **Tschirch** : Linnaea, N.F. **9**, pp. 154 *sqq.*

214. **Goebel** : Pflanzenbiologische Schilderungen, **2**, 1891, p. 268 (water plants). The tufted appendages of the leaves of many Podostemaceae are likewise compared to gills by Goebel ; but the structures in question contain numerous chloroplasts and abundance of starch, and must consequently be regarded as accessory photo-synthetic organs, as Warming points out and Goebel himself admits. Even in a botanical sense, a structure should not be termed a gill-organ, unless it is actually respiratory in function.

215. The term " aërenchyma " was introduced by Schenck, who restricted it to those ventilating tissues of secondary origin which are homologous with cork. The author agrees with Goebel (l.c. [214], p. 256), that it is preferable to define the scope of this term from the ecological or anatomico-physiological, rather than from the ontogenetic standpoint.

216. **Rosanoff** : B.Z. 1891 (Desmanthus). **Goebel** : Ber. 1886 (Sonneratia). **Jost** : B.Z. 1887. **Schenck** : Flora, 1889 (Avicennia and Laguncularia). Id. P.J. **20**, 1889. **Karsten** : Bibl. Bot. No. 22, 1891, pp. 46 *sqq.* **Wieler** : P.J. **32**, 1898. Wieler throws doubt upon the ventilating function of aërenchyma, on quite insufficient grounds.

217. Cf. **Gürtler** : Ub. interzell. Haarbildungen, insbes. üb. d. sog. inneren Haare d. Nympheaceen u. Menyanthoideen (Inaug.-Diss.), Berlin, 1905, and the literature there cited.

218. **Westermaier** : Zur Kenntniss d. Pneumatophoren, Freiburg i. d. Schw 1900.

219. The term " pneumathode," signifying anv external opening of the ventilating system, is due to Jost (l.c. [116]).

220. *Stomata.* **Von Mohl** : B.Z. 1856. **Unger** : Sitzb. Wien, **25**, 1857. **N. J. C. Müller** : P.J. **8**, 1872. **Schwendener** : Monatsber. Berl. Akad. 1881. **G. Haber-**

landt : P.J. 17, 1886.    Id. Flora, 1887.    **Leitgeb** : Mitt. Bot. Inst. Graz.    **Schaefer** : P.J. 19.    **Kohl** : Bot. Centr. 64, 1892.    **Schellenberg** : B.Z. 1896.    **F. Darwin** : Phil. Trans. 190 B, 1898.    Id. Proc. Roy. Soc. 63, 1898.    **Westermaier** : Festschr. für Schwendener, Berlin, 1899.    **Copeland** : A.B. 16, 1902.    **Porsch** : D. Spaltöffnungsapparat im Lichte d. Phylogenie, Jena, 1905.    This treatise contains many interesting observations and also a very complete list of the literature dealing with the structure, function and distribution of stomata.    **Von Guttenberg** : Engler's Jahrb. 38, 1907.

[220a. The "lines of attachment" (*Wandansätze*) are the *apparent* lines of insertion of the outer tangential walls of the adjoining epidermal cells on the dorsal walls of the guard-cells. *Actually*, they are—as was pointed out by Schwendener himself (Ges. Bot. Mitth. p. 44, foot-note)—the horizontal projections of the sharply curved edges of these tangential walls, where they bend inwards in order to insert themselves upon the dorsal guard-cell-walls.]

221. **Westermaier** : Monatsber. Berlin. Akad. 1881.

222. **Brown** and **Escombe** : Phil. Trans. 193B, 1900.

223. When considering stomata as minute perforations of a septum, it is necessary to bear in mind that the cross-section of the pore is elliptical.    Diffusion takes place with equal velocity through elliptical and circular apertures of the same cross-sectional area.    In order, therefore, to estimate the relative rates of diffusion through stomatic pores of different sizes, it is necessary to calculate the diameters of circular openings of the same cross-sectional area.    The velocities of diffusion will then be directly proportional to these diameters.

224. **Merget** : C.R. 87, p. 293.    **Stahl** : B.Z. 1894.    **F. Darwin** : Proc. Cambr. Phil. Soc. 9, 1897.    Id. B.G. 37, 1904.

225. [N.B.—This note applies to Section 4 of Ch. IX., not to Section 3 as inadvertently suggested in the text.]    **Pfitzer** : P.J. 7, 1870.    **Tschirch** : Linnaea, N.F., 9, 1881.    Id. Abh. Bot. Ver. Brandenburg, 23, 1881.    **Wilhelm** : Ber. 1883.    **Volkens** : Die Flora d. ägyptisch-arabischen Wüste, Berlin, 1887, pp. 49 *sqq.*    **Gilg** · Engler's Jahrb. 22, 1891.    **Wulff** : Öst. Bot. Zeitschr. 1898.

226. **G. Haberlandt** : l.c. [8].

227. **G. Haberlandt** : Flora, 1887.

228. **Kraus** : P.J. 4, 1866.    **Mahlert** : Bot. Centr. 24, 1885.    **Klemm** : P.J. 17, 1886.    **Porsch** : l.c. [220].

229. **G. Haberlandt** : P.J. 17, 1886, pp. 457 *sqq.*

230. **Leitgeb** : Sitzb. Wien, 1880.    **Voigt** : B.Z. 1879.    **Kamerling** : Flora, 1897 (Erg.-Bd.).

231. **Benecke** : B.Z. 1892.

232. **A. Weiss** : P.J. 4.

233. **Porsch** : Sitzb. Wien, 102, 1903.

234. **Stahl** : B.Z. 1873.    **G. Haberlandt** : Sitzb. Wien, 1875.    **Klebahn** : Ber. 1883.    Id. Jena. Zeitschr. 17 (N.F. 10), 1884.    **Devaux** : Ann. Sci. Nat., sér. VIII., 12, 1900.    Devaux erroneously regards the regulation of transpiratory activity as

the principal function of lenticels. Cf. Note 28 in the third [German] edition of this work (p. 427).

235. **Hannig :** B.Z. 1898, and the literature there cited.

236. **Leitgeb :** Denkschr. Wien, **24**, 1864, pp. 204 *sqq.* **A. F. W. Schimper :** Bot. Centr. **17**, 1884, p. 275. **Janczewski :** Ann. Sci. Nat. 2, 1885. **Jos. Müller ·** Sitzb. Wien, **109**, 1900.

237. **Frank :** Entstehung d. Intercellularräume, Leipzig, 1867. Id. Beitr. z. Pflanzenphysiologie, Leipzig, 1868.

238. **Strasburger :** P.J. **5**, pp. 297 *sqq.* **Pfitzer :** P.J. **7,** pp. 533 *sqq.* **Rauter :** Mitth. Steiermk. **2**, 1870. **G. Haberlandt :** Mitth. Steiermk. 1880.

239. Most of the early anatomists employed the term " gland " somewhat indiscriminately, often identifying structures as glandular organs on very superficial grounds. **Link** was the first to point out, that no structure ought to be termed a gland unless it is really secretory in function. A fairly precise definition of glands is to be found in **Meyen** (Üb. d. Secretionsorgane d. Pflanzen, 1837), although that author still includes guard-cells and stinging-hairs under that name. **Unger** (Anat. u. Phys. d. Pflanzen, 1855) reckons both uni- and multicellular excretory organs among glands.

Among more recent authors, **De Bary** proposes to restrict the term gland to dermal secretory organs, and to call internal glands "intercellular secretory reservoirs " (Comp. Anat. p. 92, pp. 133 *sqq.*). Evidently De Bary's standpoint is a purely morphological one ; and his suggestion, if carried out, would lead to the artificial separation of structures which are closely related both physiologically and anatomically. As a matter of fact, the term internal gland has been used in its original sense by several authors since De Bary's objection was published (*e.g.* by **Unger** [Anat. Unters. üb. einige Sekretionsorgane d. Pflanzen], and **Sachs** [Lectures].

240. **Unger :** Sitzb. Wien, **28**, 1858. **De Bary :** Comp. Anat. pp. 50 *sqq.* and 375 *sqq.* **Moll :** Versl. Akad. Amst., sér. II., **15**, 1880. **Volkens :** Jahr. Bot. Gart. Berlin, **2**, 1883. **Potonié :** Sitzb. Naturf. Berlin, 1892. **Gardiner :** Proc. Camb. Phil. Soc. **5**. **G. Haberlandt :** Sitzb. Wien, **103**, 1894. Id. ibid. **104**, 1895. Id. Ber. **12**, 1894. Id. P.J. **30**, 1897. Id. B.Z. 1898. **Nestler :** Sitzb. Wien, **105**, 1896. Id. ibid. **108**, 1899. Id. Ber. **17**, 1899. In the first of the above-cited papers, Nestler cast doubt upon the author's interpretation of the clavate hairs of Phaseolus as hydathodes, suggesting tentatively that the exudation of water from the leaves of this plant might take place through the stomata ; in his two later papers, however, Nestler records his entire agreement with the author's views. **Spanjer** (B.Z. 1898) denies the existence of epidermal hydathodes (trichome-hydathodes), and also attempts to disprove the active rôle of the protoplasts in various cases. (Cf. **G. Haberlandt's** remarks in B.Z. 1898, p. 177). **Von Minden :** Bibl. Bot. No. 46, 1899. **Lepeschkin :** Flora, **90**, 1902. Lepeschkin concludes, on quite insufficient grounds, that hydathodes are superfluous structures, " the existence of which at the present time is largely a matter of hereditary tendency." **Areschoug :** Bibl. Bot. No. 56, 1902. **Molisch :** Ber. 1903. **Krafft :** Syst.-anat. Unters. d. Blattstruktur b. d. Menispermaceen (Inaug.-Diss. Erlangen), Stuttgart, 1907.

241. **Scherffel :** Mitth. Graz. **G. Haberlandt :** P.J. **30**, 1897. **Goebel :** Flora, 1897. **Groom :** A.B. **11**, 1897.

241*a. Hydathodes of Lobeliaceae* (p. 495). **Tswett :** Rev. gén. **19**, 1907.

242. **Sauvageau** : Journ. d. Bot. 1890 (three papers). Id. ibid. 1894. Id. Sur les feuilles de quelques monocotyledones aquatiques, Paris, 1891. **Minden** : l.c. [240]. **Weinrowsky** : Fünfstück's Beitr. 1898.

243. **Volkens** : Flora d. ägyptisch-arabischen Wüste, 1887, pp. 27 *sqq.* **Marloth** : Ber. **5**, 1887 (Tamariscineae).

244. **Waldner** : Mitth. Steiermk. 1877 (Saxifragaceae). **Volkens** : Ber. **2**, 1884, and **Woronin** : B.Z. 1885 (Plumbaginaceae).

245. **Treub** : Ann. Buit. **8**, 1889 (Spathodea). **Lagerheim** : Ber. **9**, 1891 (Jochroma). **Kraus** : Flora, 1895 (Erg.-Bd.) (Parmentiera). **Koorders** : Ann. Buit. **14**, 1897. **Shibata** : Bot. Centr. **83**, 1900. **Svedelius** : Flora, **96**, 1906.

246. **Ch. Darwin** : Insectivorous Plants, London, 1875. **Pfeffer** : Landw. Jahrb. **6**, 1877. **De Bary** : Comp. Anat. pp. 100, 101, 374. **Goebel** : Pflanzenbiologische Schilderungen, **2**, V. 1891. **Arthur Meyer** and **Dewévre** : Bot. Centr. **40**, 1894. **G. Haberlandt** : Sitzb. Wien, **104**, 1895, pp. 92 *sqq.* **Vines** : A.B. **11** and **12**. **Lily Huie** : Q.J.M.S. **39**, 1897. **Rosenberg** : Physiologisch-cytologische Unters. üb. Drosera rotundifolia, Upsala, 1899. **Clautriau** : Mém. cour. Acad. Belg. **59**, 1900 (Nepenthes). **G. Haberlandt** : Sinnesorgane ım Pflanzenreich, Leipzig, 1901, pp. 94 *sqq.*

247. **Tangl** : Sitzb. Wien, **92**, 1885. **G. Haberlandt** : Ber. **8**, 1890. **Brown** and **Morris** : J.C.S. 1890. **Pfeffer** : Sitzb. sächs. Akad. 1893. **Hanstein** : Flora, 1894 (Erg.-Bd.). **Grüss** : Ber. **13**, 1895. Id. Landw. Jahrb. 1896. **Linz** : P.J. 1896.

248. The author's views regarding the diastase-secreting activity of the aleurone-layer of Grasses have met with opposition in various quarters. The principal objections have already been discussed in the third [German] edition of this work (Note 10, p. 476). It is unnecessary to repeat or supplement the arguments there put forward, as they have not yet been met by the author's opponents.

249. *Nectaries.* **Behrens** : Flora, 1879. **Bonnier** : Ann. Sci. Nat., sér. VI., **8**, 1879. **Wilson** : Unters. Tüb. **1**, 1881. **Stadler** : Beitr. z. Kenntn. d. Nektarien (Diss.), Zürich, 1886. **A. F. W. Schimper** : Die Wechselbeziehungen zw. Pflanzen u. Ameisen, Jena, 1888 (contains a detailed list of the literature dealing with extra-nuptial nectaries). **Correns** : Sitzb. Wien, **97**, 1888. **G. Haberlandt** : Sitzb. Wien, **104**, 1895, pp. 100 *sqq.* **Grassmann** : Flora, 1884. **Schniewind-Thies** : Beitr. z. Kenntn. d. Septalnektarien, Jena, 1897 (with many excellent figures). **Zimmermann** : Ann. Buit. sér. II., **3**, 1901. **Haupt** : Flora, 1902. **Schwendt** : Beih. Bot. Centr. **22**, 1907. **Elsler** : Sitzb. Wien, **116**, 1907.

250. **Meyen** : Üb. d. Sekretionsorgane d. Pflanzen, Berlin, 1837. **Hanstein** : B.Z. 1868. **Rauter** : Denkschr. Wien, **31**, 1821. **Reinke** : P.J. **10**, 1875. **De Bary** : Comp. Anat. pp. 88 *sqq.* **Behrens** : Ber. **4**, 1886. **Tschirch** : Angewandte Anatomie, Vienna and Leipzig, 1889, pp. 461 *sqq.* Id. Die Harze u. d. Harzbehälter, Berlin, 1900. **Tunmann** : Ub. d. Sekretdrüsen (Diss.), Leipzig, 1900. **Nestler** : Ber. **18**, 1900. Id. Sitzb. Wien, **109**, 1902. **Detto** : Flora, 1902.

251. In the second [German] edition of this work the author maintained (in agreement with N. J. C. Müller and Hanstein, but in opposition to De Bary) that secretions always arise in the lumina of glandular cells, and then pass through the cellulose layers of the membrane into the subcuticular glandular cavities. The work of Tschirch and Tunmann (l.l.c. [250]), however, led him to reconsider the question,

with the result that he now finds himself on the whole in agreement with the last-mentioned authors.

252. **Volkens** : Ber. **8,** 1890.

253. With reference to the case of Thymus Serpyllum, **Stahl** (Pflanzen u. Schnecken, p. 3) maintains that the aromatic secretion must at any rate repel many *omnivorous* animals. In his opinion, the success or failure of a single additional enemy may be a matter of the greatest moment to a plant—witness the case of Phylloxera and the Vine. Certainly a pest *may* become very dangerous to a plant, if it confines its attention to that particular species, though this very instance of Thymus shows that plants may continue to flourish in spite of the ravages of more than one specialised pest ; but it is highly improbable that the existence of any species of plant is ever seriously threatened by omnivorous animals.

**Detto** (l.c. [250]) considers that the preceding argument is weakened by the fact that only six of the numerous animal foes of Thymus Serpyllum enumerated in the text are specialised. The author, on the other hand, thinks that the existence of as many as half a dozen specialised pests is sufficient proof of the comparatively feeble protective powers of the oil-glands.

254. Tyndall passed air through cylinders of cardboard, soaked in various ethereal oils, and then determined its perviousness to thermal radiation in each case. By this method, air saturated with the vapour of ethereal oil of roses was found to absorb 36 times as much radiant heat as ordinary air. Other values obtained by Tyndall were as follows :

| | | | | | | | |
|---|---|---|---|---|---|---|---|
| Wormwood | oil | - | - | - | - | - | 41 |
| Cinnamon | ,, | - | - | - | - | - | 43 |
| Lemon | ,, | - | - | - | - | - | 65 |
| Rosemary | ,, | - | - | - | - | - | 74 |
| Camomile | ,, | - | - | - | - | - | 87 |
| Cassia | ,, | - | - | - | - | - | 109 |
| Anise | ,, | - | - | - | - | - | 352 |

Cf. **Focke** : Kosmos, **5,** p. 412. **Detto** (l.c. [250]) has advanced some weighty arguments against Tyndall's views.

255. **Goebel** : Pflanzenbiologische Schilderungen, **2,** pp. 232 *sqq.* **Schilling** : Flora, 1894. **Hunger** : Üb. d. Funktion d. oberflächl. Schleimbildungen im Pflanzenreich (Inaug.-Diss.), Leyden, 1899.

256. *Internal glands.* In addition to the above-mentioned treatises by **Meyen** and **Rauter** [250], cf. also **Frank** : Beitr. z. Pflanzenphysiologie, 1868. **Von Höhnel** Sitzb. Wien, **86,** 1881. **De Bary** : Comp. Anat. pp. 201 *sqq.* **Tschirch** : Angewandte Pflanzenanatomie, pp. 477 *sqq.* Id. P.J. **25,** 1893. **Sieck** : P.J. 27, 1895. **Lutz** Bot. Centr. **64,** 1895. **Solereder** : l.c. [47].

257. **G. Haberlandt** : Sitzb. Wien, **107,** 1898. **Detto** : Flora, 1903. **Porsch** Öst. Bot. Zeitschr. 1903. Id. Verh. Zool. Bot. Ges. Wien, **56,** 1906.

258. **Frank** : l.c. [256]. **Van Tieghem** : Ann. Sci. Nat., sér. V., **16,** 1872. **Trécul** : C.R. 1865-7. **F. Thomas** : P.J. **4. De Bary** : Comp. Anat. pp. 201 *sqq.* and 404 *sqq.* **Möbius** : P.J. **16,** 1885. **Tschirch** : Angewandte Pflanzenanatomie, pp. 477 *sqq.* Id. Die Harze u. d. Harzbehältes, Berlin, 1900. Id. Ber. 1901. **J. Moeller** Zeitschr. allg. Öst. Apotheker-Vereins, 1896. **Schwabach** : Ber. 1899-1900.

259. The statements of **De Bary** (Comp. Anat. p. 443) and **Tschirch** (Angew. Pflanzenanat. pp. 487 *sqq.*) concerning the course of the resin-passages in Pinus and Juniperus are incorrect.

260. **De Bary** : Comp. Anat. pp. 145 *sqq.* **Tschirch** : Angew. Pflanzenanat. pp. 472 *sqq.* Id. Festschr. f. Schwendener, 1899. **Molisch** : Grundriss einer Histo-chemie d. pflanzl. Genussmittel, Jena, 1891. **Zacharias** : B.Z. 1879. **Berthold** : Studie üb. Protoplasmamechanik, Leipzig, 1886, pp. 25 *sqq.* **Froembling** : Bot. Centr. **65**, 1896. **Biermann** : Arch. f. Pharmacie, **236**, 1898. **Solereder** : l.c. [47]. **R. Müller** : Ber. **23**, 1905. Further, as yet unpublished results obtained by Müller (working in the Graz Botanical Institute) are briefly summarised in the text.

261. **Dippel** : Verh. Rheinl. u. Westf. **Sachs** : Sitzb. Wien, 1859. **Engler** : B.Z. 1871. **De Bary** : Comp. Anat. p. 148, 153. **Berthold** : Unters. z. Physiol. d. pflanzl. Organisation, **1**, Leipzig, 1898 (gives many data concerning the distribution of tannin in the plant-body). **Solereder** : l.c. [47].

262. **Heinricher** : Mitth. Graz, 1888. **Guignard** : C.R. **111** (1880) and **117** (two papers). Id. Journ. de Bot. 1893. **Spatzier** : P.J. **25**, 1893. **Schweidler** : Ber. **23**, 1905.

263. *Crystal-sacs.* **Sanio** : Monatsber. Berl. Akad. 1857. **Hanstein** : ibid. 1859. **Holzner** : Flora, 1864 and 1867. **Hilgers** : P.J. **6**, 1867. **Schroff** : Beitr. z. näheren Kenntn. d. Meerzwiebel, Vienna, 1865. **Rosanoff** : B.Z. 1865 and 1867. **Pfitzer** : Flora, 1872. **J. Möller** : Anatomie d. Baumrinden, Berlin, 1882, pp. 420 *sqq.* 431-5. **Kny** : Ber. 1887. **Stahl** : Jena. Zeitschr. **22**, 1888. **Kohl** : Anat.-phys. Unters. d. Kalksalze u. Kieselsäure i. d. Pflanze Marburg, 1889 (with an exhaustive list of literature). Id. Bot. Centr. **67**, 1876. **Fuchs** : Öst. Bot. Zeitschr. 1898. **Rothert** : B.Z. 1900. Id. and Zalenski: Bot. Centr. **80**, 1899. **Von Guttenberg** : Sitzb. Wien, **111**, 1902. **Lewin** : Ber. **18**, 1900. **Möbius** : Ber. **23**, 1905. **Knoll** : Sitzb. Wien, **114**, 1905.

264. *Cystoliths.* **Schacht** : Abh. Senck. Ges. **1**, p. 133. **Weddell** : Ann. Sci. Nat., sér. IV., **2**, p. 267. **Richter** : Sitzb. Wien, **76**, 1, 1877, **Penzig** : Bot. Centr. **8**, 1881, p. 393. **Giesenhagen** : Flora, 1890. Id. Ber. **9**, 1891. **Zimmermann** : Ber. **9**, 1891. **Möbius** : Ber. üb. d. Senckenbergische naturf. Ges. in Frankfurt a. M. 1897. **Solereder** : l.c. [47]. **K. Fritsch** : Wiesner-Festschrift, Vienna, 1908.

265. *Silica-deposits.* **Mettenius** : Abh. sächs. Ges. **7**, 1864. **Rosanoff** : B.Z. 1871. **Cario** : B.Z. 1881. **Warming** : Études sur la famille des Podostemacées, Copenhagen, 1881-8. **Kohl** : l.c. [263, 1889]. **Marktanner-Turneretscher** : Sitzb. Wien, **91**, 1885. **Kölpin Ravn** : Bot. Tidsskr. **21**, 1897. Marktanner-Turneretscher regarded the spherical cell-masses discovered by him in the leaf of Loranthus europaeus as water-storing mucilaginous bodies ; their siliceous character was first recognised by Kölpin Ravn. **Solereder** : l.c. [47]. **Möbius** : Wiesner-Festschrift, Vienna, 1908.

266. Kohl's suggestion (l.c. [265]), that the stegmata of Palms, Orchids, etc., act as valves which allow water to pass from bast-fibres into the intercellular spaces in the adjacent parenchyma, but prevent it from travelling in the opposite direction, is entirely unsupported by evidence.

267. **Guignard** : Ann. Sci. Nat., sér. VII., **15**, 1892.

268. **Dingler** : Die Bewegung d. pflanzl. Flugorgane, Munich, 1889.

269. **Von Wahl** : Bibl. Bot. No. 40, 1897.

270. **A. F. W. Schimper** : Die Indomalayische Strandflora, Jena, 1891, pp. 164 *sqq*. **Kölpin Ravn** : Bot. Tidsskr. **19**, 1894.

271. The following is a selection from the extensive literature on *hygroscopic mechanisms*. **Kraus** : P.J. **5**, 1866. **Steinbrinck** : Unters. üb. d. anat. Ursachen d. Aufspringens d. Früchte (Inaug.-Diss.) Bonn, 1873. Id. Ber. 1888. Id. Verh. Nat. Ver. Preuss. Rheinl. **47**, 1891. Id. Flora, 1891. Id. Ber. 1895. Id. Bot. Jaarb. **7**, 1895. Id. Flora, 1897. Id. Festschr. f. Schwendener, Berlin, 1899. Id. und **Schinz** : Flora, **98**, 1908. **Zimmermann** : P.J. **12**, 1881. Id. Ber. 1883. **Tschirch** : P.J. **13**, 1882. **Rathay** : Sitzb. Wien, **83**, 1881. **Firtsch** : Ber. **1**, 1883. **Eichholz** : P.J. **17**, 1885. **Schinz** : Unters. üb. d. Mechanismus d. Aufspringens d. Sporangien u. Pollensäcke, Zürich, 1883. **Schrodt** : Flora, 1885. **Leclerc du Sablon** : Ann. Sci. Nat., sér. VII., **1**, 1885. Id. ibid. **2**, 1885. **Schwendener** : Sitzb. Berl. 1887. Id. ibid. 1899. Id. ibid. 1902. **Taliew** : Üb. d. hygròsk. Gewebe d. Compositen-Pappus, Kazan, 1894 (abstract by **Rothert** in Bot. Centr. **73**, 1895). **Ursprung** : P.J. **38**, 1903. **Hirsch** : Bewegungsmechanismus d. Compositen-Pappus (Diss.), Würzburg, 1901. **Colling** : Das Bewegungsgewebe der Angiospermen-Staubbeutel (Inaug.-Diss.), Berlin, 1905. **Schneider** : Der Öffnungsmechanismus d. Tulipa-anthere, Altstätten, 1908 (contains a very complete list of literature).

272. *Cohesion-mechanisms*. **Steinbrinck** : Ber. **15** (1897), **16** (1898), **17** (1899), **19** (1901), **20** (1902), **31**a (1908) (cf. **Tschirch** : l.c. [271]), and **32** (1909). **Schrodt**: Ber. **15** (1897). **Kamerling**: Bot. Centr. **72** (1897) and **73** (1898). Id. Flora, 1898. **Ursprung** : P.J. **38**, 1903.

273. **Gad** : B.Z. 1880.

274. **Ch. Darwin** : l.c. [203]. **G. Haberlandt** : Sinnesorgane im Pflanzenreich z. Perzeption mechanischer Reize, Leipzig, 1901, pp. 62 *sqq*. [2nd ed., 1906].

275. **Pfeffer** : Physiologische Untersuchungen, **1**, 1873. **G. Haberlandt** : l.c. [274].

276. **Pfeffer** (l.c. [275], p. 86) states that all the parenchymatous cells in the staminal filaments of Centaurea Jacea have thin walls (apart from the outer epidermal walls) ; the author's observations do not agree with this statement.

277. **De Vries** : Arb. Würzb. **1**, 1874. **Ch. Darwin** : Climbing plants, London, 1875. **G. Haberlandt** : l.c. [274]. **Fitting** : P.J. **38**, 1903. **Ricca** : Malpighia **17**, 1903.

278. **Brücke** : Arch. f. Anat. u. Physiol. 1848. **Pfeffer** : l.c. [275]. Id. Physiology, **3**, pp. 134 *sqq*. Id. Abh. sächs. Ges. **30**, 1907. **Ch. Darwin** : The Power of Movement of Plants, London, 1880. **Schwendener** : Sitzb. Berl. 1897 and 1898. **Jost** : P.J. **31**, 1898. **G. Haberlandt** : Ann. Buit. 1898 (Suppl. II.). **Möbius** : Festschr. f. Schwendener, 1899. **Pantanelli** : Atti. Soc. Nat. Modena, sér. IV., **2**, 1901. **Wiedersheim** : P.J. **40**.

279. **G. Haberlandt** : Biol. Centr. **15**, 1905. In this paper the author has demonstrated that terms such as "sense-organ," "sensitiveness," "sensation," "perception," etc., may be legitimately used in the comparative physiology of animals and plants, even when one is only concerned with the physical aspect of stimulation and response ; the possibility that psychical processes are also involved, can neither be proved nor disproved in the case of lower animals and plants, and is in any case of secondary interest to the physiologist. Cf. also **G. Haberlandt** : Almanach d.

K. Akad. d. Wiss. in Wien, **58**, 1908, and **Warwara Polowzow** : Methodologisches z. Problem d. Reizerscheinungen, Jena, 1909.

The psychology or psychobiology of plants recently developed on Lamarckian lines by Pauly, Francé, Wagner and others, involves a logical fallacy; for it is one thing to admit the possibility of a psychical aspect of animal and vegetable life, but quite another to suggest that the innumerable physiological and morphological adaptations of organisms, and all their self-regulating processes, are purposeful, in the sense in which this term is applied to the actions of an intelligent being.

280. **Noll** : Naturw. Rundschau, 1888. Id. Das Sinnesleben d. Pflanze (Lecture at the Jahresfest d. Senck. Ges. in Frankfurt a. M., 1896).

281. Cf. **G. Haberlandt** : l.c. [274], and the literature there cited.

282. **Pfeffer** : Unters. Tüb. **1**, 1888. **G. Haberlandt** : l.c. [274], pp. 126 *sqq.* **Borzi** : Contrib. alla Biologia Vegetale, **3**, 1903.

283. **Fitting** (P.J. **38**, 1903) has recently found, that the supposedly insensitive dorsal surface of Cucurbitaceous (and other dorsiventral) tendrils is sensitive to contact; stimulation of this side does not lead to curvature, but prevents curvature from being induced by stimulation of the ventral surface. Fitting believes that both surfaces possess the same kind of contact irritability. If this view is correct, it would seem rather strange that tactile pits are generally confined to the ventral surface ; indeed, this fact might be adduced as an argument against the perceptive function of the pits, were it not that Fitting himself (l.c. p. 560) points out that, in the case of Cucurbitaceous tendrils, stimulation of the dorsal side often only diminishes the curvature induced by stimulation of the ventral side, a result which suggests that this surface is the less sensitive of the two. Moreover, Pfeffer has found tactile pits on the dorsal side of the tendrils in Bryonia, and it is probable that the same state of things prevails in other Cucurbitaceae.

284. **Ch. Darwin** : l.c. [277]. **Pfeffer** : l.c. [282]. **G. Haberlandt** : l.c. [274], pp. 149 *sqq.*

285. **Ch. Darwin** : Insectivorous plants, London, 1875. **Pfeffer** : l.c. [282], p. 543. **G. Haberlandt** : l.c. [274], pp. 119 *sqq.*

286. **G. Haberlandt** : l.c. [274], pp. 17 *sqq.* Id. Physiol. Pflanzenanatomie (2nd ed.), 1886, p. 479.

287. **Ch. Darwin** : l.c. [274]. **G. Haberlandt** : l.c. [274], pp. 63 *sqq.* (2nd ed.). **Von Guttenberg** : Sitzb. Wien, **117**, 1908. The antennae of some species of Catasetum are constructed after the pattern of tactile bristles.

288. **Unger** : Anat. u. Phys. d. Pflanzen, 1855, p. 419. **Kabsch** : B.Z. 1861 **G. Haberlandt** : l.c. [274], pp. 24 *sqq.*

289. **G. Haberlandt** : l.c. [274], pp. 32 *sqq.*

290. **G. Haberlandt** : l.c. [274], pp. 154 *sqq.*

291. **G. Haberlandt** : l.c. [274], pp. 109 *sqq.*

292. **Morren** : Nouv. Mém. Acad. Sci. Brux. **14**, 1884. **G. Haberlandt** : l.c. [274], pp. 47 *sqq.*

293. **Ch. Darwin** : l.c. [274]. **G. Haberlandt** : l.c. [274], pp. 71 *sqq.*

294. **Kabsch** : l.c. [288], pp. 33 *sqq.* **H. Müller** : Alpenblumen, Leipzig, 1881. **G. Haberlandt** : l.c. [274], p. 34. **K. Linsbauer** : Sitzb. Wien, **114**, 1905. **Brunn**: Unters. üb. Stossreizbarkeit (Inaug.-Diss.), Leipzig, 1908. Both Linsbauer and Brunn regard the staminal hairs of the Cynareae as stimulators, and not as actual organs of perception. Cf. **G. Haberlandt** : l.c. [274], pp. 46-7.

295. [For staminal hairs of Centaurea Jacea cf. also Kny's Wandtafeln, Pl. 105 and text, pp. 480-482.]

296. **G. Haberlandt** : l.c. [274], pp. 101 *sqq.*

297. **Cohn** : Cohn's Beitr. **1**, 1861. **Goebel** : Pflanzenbiologische Schilderungen, **2**, 1891, p 72. **G. Haberlandt** : l.c. [274], pp. 129 *sqq.*

298. **Sydenham Edwards** : Curtis's Bot. Mag. **20**, 1804. **Oudemans** : Versl. en Med. Akad. Amst. **9**, 1859. **Ch. Darwin** : l.c. [285], pp. 259 *sqq.* **Munk** : Die elektrischen u. Beugungserscheinungen am Blatte d. Dionaea muscipula, Leipzig, 1876. **Batalin** : Flora, 1877. **Goebel** : l.c. [297], pp. 69, 201. **Macfarlane** : Contrib. Bot. Lab. Penn. **1**, 1892. **G. Haberlandt** : l.c. [274], pp. 133 *sqq.*

299. **Cheeseman** : Trans. N.Z. Inst. **5**, 1873. **Fitzgerald** : Australian Orchids, **1**, Sydney, 1882. **Haberlandt** : l.c. [274], pp. 85 *sqq.*

300. **Knight** : Phil. Trans. 1806.

301. **Noll** : Üb. heterogene Induktion, Leipzig, 1892, pp. 42 *sqq.* Id. l.c. [280]. Id. P.J. **34**, 1900. Noll was the first to suggest that plants might perceive gravitational stimuli in the same way as animals do, namely, with the aid of statocyst-like sense-organs ; but he believed that the organs in question were ultra-microscopic structures located in the ectoplast. The author and **Němec**, on the other hand, both arrived independently at the conclusion, that the statocysts of plants consist of entire cells, and that most gravitational sense-organs are made up of a number of statocysts. Cf. **Němec** : Ber. **18** (1900) and **20** (1902). Id. P.J. **36**, 1901. **G. Haberlandt** : Ber. **18** (1900) and **20** (1900). Id. P.J. **38** (1903) and **42** (1905).

302. **F. Darwin** and **Miss Bateson** : A.B. **2**, 1888-9. **Fitting** : P.J. **41** (1905).

303. The fact that a root with the apical ·5 to 1 mm. cut off does not curve geotropically, when laid on its side, though it continues to grow in length, led Ch. Darwin to conclude that geotropic sensitiveness is confined to the root-tip. This view, which is supported by other facts as well, has been the subject of a considerable amount of controversy. Wiesner found that decapitated roots, on being exposed to centrifugal force of 20 g. to 41 g. intensity, curved " geotropically " after some time in the region of most active growth ; he accordingly denies the localisation of geotropic sensitiveness. Czapek, on the other hand, supports Darwin's theory ; but the experiments with glass caps upon which his arguments are based, are not universally accepted as conclusive. Recently, Piccard has attacked the problem in a new way, by causing centrifugal forces to act in opposite directions upon the apex and the growing zone of a root made to revolve around a horizontal axis. This result was attained, by affixing the root in such a manner that it formed an angle with the axis of rotation, a point situated between apex and growing zone being centred on the axis. Piccard concludes, that not only the root-tip, but also the entire growing zone is capable of perceiving gravitational stimuli, and that the latter is actually the more sensitive of the two. As both Piccard's experimental method and his interpretation are open to criticism, the author has repeated his experiments

with a more satisfactory apparatus. He finds that, in Vicia Faba, Phaseolus multiflorus and Lupinus albus, both apex and growing zone are geotropically sensitive, the former being by far the more sensitive of the two, and the curvature of the growing zone being without a doubt largely induced by secondary stimuli transmitted from the apical region. Ch. Darwin's views were therefore in the main correct. Cf. **G. Haberlandt** : P.J. **45**, 1908, and the literature there cited.

304. **Gaulhofer** : Sitzb. Wien, 1907.

305. **Tischler** : Flora, **94**, 1905.

306. **G. Haberlandt** : P.J. **38**, 1903. **Schröder** : Beih. Bot. Centr. **16**, 1903.

307. From unpublished experiments carried out in the Graz Botanical Institute.

308. **Samuels** : Öst. Bot. Zeitschr. **55**, 1905.

309. **Němec** : Beih. Bot. Centr. **17**, 1904. Id. Flora, **96**, 1906.

310. **Giesenhagen** : Ber. **19**, 1901. **Schröder** : l.c. [306].

311. **G. Haberlandt** : Sitzb. Wien, **115**, 1906.

312. **G. Haberlandt** : P.J. **38** (1903) and **42** (1905). Id. Ber. **26a** (1908). **F. Darwin** : Proc. Roy. Soc. **71** B, 1903. **Bach** : P.J. **44**, 1907.

313. **Buder** : Ber. **26**, 1908 (Festschr.).

314. **F. Darwin** and **Miss Pertz** : Proc. Roy. Soc. **73** B, 1904.

315. **Jost** (B.Z. 1904, p. 279) exposed roots of Ervum Lens and cotyledonary sheaths of Panicum, for 2 or 3 hours, to centrifugal force of ·02 to ·05 g. intensity ; pronounced curvatures resulted, although the falling starch-grains remained evenly distributed. This result in no wise conflicts with the statolith-theory ; for when the starch-grains are evenly distributed, it is obviously only the grains adhering to the outer walls (i.e. the walls next the periphery of the revolving body to which the experimental plants are attached) that will exert pressure against the ectoplast under the influence of the centrifugal force. All the curvatures observed by Jost agree with this interpretation.

316. **Ch. Darwin** : l.c. [278], pp. 383 *sqq.* **Rothert** : Cohn's Beitr. **7**, 1896.

317. **F. Kohl** : Die Mechanik d. Reizkrümmungen, Marburg, 1894.

318. **G. Haberlandt** : Ber. **22**, 1904. Id. Die Lichtsinnesorgane d. Laubblätter, Leipzig, 1905. Id. Sitzb. Wien, **117**, 1908. **Von Guttenberg** : Ber. 1905. **Sperlich** · Sitzb. Wien, **116**, 1907. **Seefried** : ibid. **Wager** : Journ. Linn. Soc. 1909.

319. **Wiesner** : Biol. Centr. **19**, 1899.

320. **Vöchting** : B.Z. 1888.

321. **G. Krabbe** : P.J. **20**, 1889.

322. It is many years since the author first tentatively suggested that papillose epidermal cells might act as condensing lenses (cf. **G. Haberlandt** : Die physiol. Leistungen d. Pflanzengewebe, in Schenck's Handbuch, **2**, 1882) ; at that time, however, he believed that the sole advantage of this light-condensation consisted in the increased illumination of the photosynthetic cells.

323. **Gaulhofer :** Ber. 26*a*, 1908.

324. **Gaulhofer :** Sitzb. Wien, **117**, 1908.

325. **G. Haberlandt :** Die Lichtsinnesorgane d. Laubblätter, Leipzig, 1905, pp. 86 *sqq*. Id. Ber. 1906. Id. Biol. Centr. **27**, 1907. Id. P.J. **46**, 1909. **Kniep :** Biol. Centr. **27**, 1907. **Nordhausen :** Ber. 1907. **Gius :** Sitzb. Wien, **116** (1907). **Albrecht :** Ber. 1908. Id. Ub. d. Perzeption d. Lichtrichtung i. d. Laubblättern (Inaug.-Diss.), Berlin, 1908.

326. Mention should be made of a curious experiment carried out by **Kniep** (l.c. [325]). The upper surface of leaves of Tropaeolum minus, Begonia discolor and B. heracleifolia were wetted with paraffin oil, with a refractive index of 1·476. The resulting distribution of light on the inner epidermal walls was the converse of the normal arrangement, there being now a dark central area and a bright marginal zone ; the dark spot was, of course central in vertical illumination, but excentrically situated in oblique light. The illumination of the inner epidermal wall was therefore still differential, and, as a matter of fact, Kniep's leaves adjusted themselves more or less accurately to the prevailing light, even after being wetted with paraffin. Kniep's results by no means conflict with the author's theory of light-perception ; they merely show that the central portions of the ectoplast on the inner epidermal wall are not necessarily " attuned " to intense light, nor the marginal area to feeble light. Evidently, the power of perception depends not upon adaptation of the ectoplast to bright or dull light, but upon its power of distinguishing between symmetrical and asymmetrical illumination ; this power naturally presupposes a general capacity for recognising differences of light-intensity.

327. *Eye-spots.* **Zimmermann :** Beih. Bot. Centr. 1894, pp. 161 *sqq*. (Sammelreferat). **Engelmann :** Pflüger's Archiv, **29**, 1882. **Strasburger :** Über Reduktionsteilung, etc., Jena, 1900, p. 193. **Wager :** Journ. Linn. Soc. (Zool.) **27**, 1900. **Francé :** Zeitschr. f. d. Ausbau d. Entwickelungslehre, 1908.

328. The extensive literature dealing with the transmission of stimuli has been carefully compiled, and critically discussed by **Fitting :** Ergebn. d. Phys. **4** and **5** (also published as a separate work under the title: Die Reizleitungsvorgänge b. d. Pflanzen, Wiesbaden, 1907).

329. **Rothert :** Cohn's Beitr. **7**, 1896, p. 137. **Czapek :** P.J. **32**, 1898, pp. 217 *sqq*.

330. **Kretzschmar :** P.J. **39**, 1903.

331. **Fitting :** P.J. **39**, 1903.

332. **Pfeffer :** Physiology, **2**, 1901, pp. 90 *sqq*. and 199 *sqq*.

333. **Tangl :** P.J. **12**, 1879-81. Id. Sitzb. Wien, **92**. **Russow :** Sitzb. Dorpat, **6**. **Gardiner :** Arb. Würzb. **3**. **Kienitz-Gerloff :** B.Z. 1891. **Arthur Meyer :** B.Z. 1896. Id. Ber. 1897. **Kuhla :** B.Z. 1900. **A. W. Hill :** Proc. Roy. Soc. 67 B, 1901. **Kohl :** Ber. 1900. **Strasburger :** P.J. **36**, 1901.

334. **Pfeffer :** l.c. [282], p. 524.

335. **G. Haberlandt :** l.c. [274], p. 30.

336. **G. Haberlandt :** l.c. [274], p. 106.

337. **A. W. Hill :** Phil. Trans. **149** B, 1901, p. 103.

338. **Townsend** : P.J. **30**, 1897. Cf. also **Pfeffer** : Sitzb. sächs. Ges. 1896.

339. **Němec** : Die Reizleitung u. d. reizleitenden Strukturen b. d. Pflanzen, Jena, 1901. Id. Biol. Centr. **21**, 1902. Cf. also **Haberlandt** : Biol. Centr. **21**. Id. Ber. **19**, 1901. Id. l.c. [274], pp. 149 *sqq.*

340. **Batalin** : l.c. [298].

341. **G. Haberlandt** : Ann. Buit. 1898 (Suppl. II.).

342. **Rothert** : Cohn's Beitr. **7**, 1896.

343. **F. W. Oliver** : Ber. 1887.

344. **G. Haberlandt** : Das reizleitende System d. Sinnespflanze, Leipzig, 1890 (and the literature there cited). **MacDougal** : B.G. **22**, 1895. MacDougal adduces the following experimental evidence against the author's theory of stimulus-trans-mission in Mimosa.

1. If water is suddenly forced into the cut end of a stem, by means of a pump, under a pressure of 3 to 8 atm (?), no response follows. **Fitting** (see below) has pointed out, that the result of this experiment does not really militate against the author's theory, because the water forced in does not move in the secretory sacs, but in the wood-vessels. Fitting himself was unable to force water through the rows of secretory sacs under a pressure of 2 atm. As a matter of fact, when a stem of Mimosa is cut through, the ends of the intact secretory sacs next the cut surface quickly become occluded by plugs of resinous material. This point was explained by the author in his above-mentioned treatise (p. 19).

2. If the base of a cut stem is scraped, so as to expose the secretory sacs, and then immersed in a saturated solution of potassium nitrate,—after allowing the leaves to unfold—no response follows. In devising this experiment, MacDougal evidently started from the erroneous assumption that plasmolysis could produce a rapid fall of hydrostatic pressure in the system of secretory sacs. For stimulation only follows upon the sudden deformation of the sensitive cells in the pulvini produced by a *wave* of pressure ; the fall of pressure produced by plasmolysis is much too slow to act as a stimulus. Fitting repeated MacDougal's experiment in a modified form, and occasionally obtained positive results (*i.e.* a response) ; but microscopic examination showed that, in all such cases, the secretory sacs had died after being plasmolysed. These instances are therefore really special cases of stimulation by poisonous or corrosive substances. This point had also been discussed by the author in his above-cited treatise (pp. 61 *sqq.*).

**Fitting** : P.J. **39**, 1903. While Fitting arrives at no definite conclusion regarding the mode of stimulus-transmission in Mimosa pudica in this paper, his conclusions agree in many points with those of the author. In particular, he thinks it probable that the transmission depends upon the " movement of a liquid in living cells," though he does not clearly specify the cells that are concerned in the process ; in this connection it should be borne in mind, that the drop of liquid which suddenly exudes when a stem or leaf is cut, is certainly almost entirely derived from the secretory sacs.

It should be noted that, according to the author's observations (l.c. pp. 63 *sqq.*), the woody cylinder of Mimosa pudica is capable of transmitting stimuli to some extent ; for a traumatic stimulus is still transmitted by a stem in which the continuity of the secretory tissue has been interrupted. The author has endeavoured to give an explanation of this rather remarkable phenomenon in his above-mentioned treatise.

345. **L. Linsbauer :** Wiesner Festschrift, Vienna, 1908.

346. **Borzi** (Rivista di. Sci. Biol. **4,** 1899) states, as an objection to the author's theory, that the leaves of Mimosa respond when the roots are injured, although the latter contain no secretory sacs. According to the author's own observations, however, this statement only applies to the lateral roots, the characteristic secretory elements being undoubtedly present in the leptome of the main root ; a single experiment provides a complete answer to Borzi's objection. A vigorous young plant of Mimosa pudica was removed from its pot, and its root-system immersed in a vessel filled with tap-water, after the adhering soil had been carefully washed away as far as possible. On the following day (T = 22° C.), when the plant had completely recovered, several lateral (secondary) roots were carefully cut off, one at a time, with scissors (including the uppermost lateral root, the insertion of which was only 2·5 cm. from the lowermost leaf) ; no response resulted, even when a whole bunch of rootlets were severed at once. When, however, the main root was cut across, at a point about 3·5 cm. below the lowermost leaf, all the leaves, except the two uppermost, responded immediately. Response on the part of the leaves to traumatic stimulation of a root is therefore evidently correlated with the presence of secretory sacs in the latter, and failure to respond with the absence of these cells.

**Dutrochet** obtained an immediate response on pouring sulphuric acid over the root-system (the author's erroneous statements regarding this point are corrected by **Borzi**), the explanation, no doubt, being that not only the lateral roots, but also the main root suffered corrosion.

A further argument advanced by **Borzi** against the author's theory is based on the fact that transmission of stimuli takes place in the allied genus Neptunia, in which both Borzi and Fitting failed to find any secretory sacs. The author has so far had no opportunity of examining Neptunia oleracea himself ; but in view of his experiences with other Leguminosae, he remains sceptical regarding the absence of secretory sacs in the genus.

347. **Eichler :** Sitzb. Berlin, 1886. **Kraus :** Sitzb. phys.-mediz. Ges. Würzb. 1899. **Strasburger :** P.J. **43,** 1906. **Kränzlin :** Ber. **24,** 1906.

348. **Sachs :** Lectures, pp. 155-6.

349. **Nägeli :** Beitr. **1,** 1858, p. 4. **Sanio :** B.Z. 1863, pp. 357 *sqq.* Id. P.J. **9,** 1873. **Velten :** B.Z. 1875, p. 811. **G. Haberlandt :** Entw. d. mech. Gewebe-systems, 1879, pp. 39 *sqq.* Id. Ber. 1886. **Strasburger :** Bau u. Wachst. d. Zell-häute, 1882, pp. 39 *sqq.* Id. Leitungsbahnen, Jena, 1891. **Krabbe :** Das gleitende Wachstum, Berlin, 1886. **Mischke :** Bot. Centr. **44,** 1880. **Krüger :** B.Z. 1892. **Raatz :** P.J. **23,** 1892. **Wieler :** Ber. 1886. **Nordhausen :** Fünfstück's Beitr. **2,** 1898. **Schoute :** Verh. Akad. Amst., sér. II., 1902.

350. **Th. Hartig :** Forstliche Kulturpflanzen, Berlin, 1857. **Hanstein :** Unters. üb. d. Bau u. d. Entw. d. Baumrinde, Berlin, 1853. **J. Möller :** Anat. d. Baum-rinden, Berlin, 1882. **Schwendener :** Mechanisches Prinzip, pp. 143 *sqq.* **Strasburger :** Sitzb. Berlin, 1890. Id. Leitungsbahnen.

351. **Th. Hartig :** l.c. [350]. Id. B.Z. 1859. **Sanio :** B.Z. 1863. **Wiesner :** Die Rohstoffe d. Pflanzenreiches, 2nd ed. Leipzig, 1902, Ch. XVII. (Timber, by Karl Wilhelm). **J. Möller :** Denkschr. Wien., 1876. **De Bary :** Comp. Anat. pp. 475 *sqq.* **Strasburger :** Leitungsbahnen (contains a large number of fresh observa-tions on the structure of secondary wood).

352. The classification of the various components of the secondary wood employed in the present work is based on that of Sanio, which has not yet been surpassed for

convenience and precision. In addition to the medullary rays, Sanio recognises three classes of component elements, viz. : I. Parenchymatous system—(1) Xylem-parenchyma; (2) Intermediate cells. II. Fibrous system—(1) Non-septate wood-fibres (libriform cells); (2) Septate fibres. III. Tracheal system—(1) Tracheides; (2) Vessels.

353. **Leclerc du Sablon :** Rev. gén. d. Bot. 1904. **Schellenberg :** Ber. 23, 1905.

354. **Th. Hartig :** B.Z. 1858. **A. Fischer :** P.J. 22, 1891.

355. **Strasburger** (Leitungsbahnen, pp. 486 *sqq.*) only recognises two types of tissue in the woody cylinder, namely, tracheal tissue and parenchyma, and assumes that the mechanical elements are phylogenetically derived from members of these two other systems. It is, of course, quite possible that this view is correct, though the facts detailed in the text hardly seem to support it. Strasburger's statement (l.c. p. 468, Note) to the effect that the author erroneously derives fibrous tracheides from wood-fibres, on physiological grounds, although the two kinds of cell have really nothing to do with one another, is founded on a misconception. Strasburger forgets that the present work is written from the anatomico-physiological, and not from the phylogenetic point of view. The transitional series in the table on p. 664 might therefore just as well be read in the opposite direction. Moreover, while the author includes all the mechanical elements of the wood in the category of wood-fibres, whatever their phylogenetic origin may be, Strasburger restricts this term to those fibres which he believes to be derived from xylem-parenchyma; from Strasburger's point of view, therefore, it would naturally be absurd to attempt to derive fibrous tracheides from wood-fibres. As a matter of fact, the author stated quite plainly in the first [German] edition of the present work that his transitional series must not be interpreted in a phylogenetic sense.

356. **Russow :** Bot. Centr. 4, 1883. Cf. also **Klebahn :** Ber. 1, 1883.

357. **Troschel :** Unters. üb. d. Mestom im Holze d. dikot. Laubbäume (Inaug.-Diss.), Berlin, 1879. **Krah :** Ub. d. Verteil. d. par. Elemente im Xylem u. Phloem d. dikot. Laubbäume (Inaug.-Diss.), Berlin, 1883.

358. **P. Schulz :** Das Markstrahlengewebe u. seine Beziehungen z. d. leit. Elementen d. Holzes (Inaug.-Diss.), Berlin, 1882.

359. **Pfeffer :** Physiology, 1, p. 262.

360. **A. Fischer :** P.J. 22, 1891.

361. The formation of "annual" rings in tropical woody plants is evidently a somewhat complicated problem. Cf. **Reiche :** P.J. 30. **Ursprung :** B.Z. 1904. **Holtermann :** Des Einfluss d. Klimas auf d. Bau d. Pflanzengewebe, Leipzig, 1907.

362. **Kny :** Üb. d. Dickenwachstum d. Holzkörpers in seiner Abhäng. v. äuss. Einfl. Berlin, 1882 (and the literature there cited). **Wiesner :** Sitzb. Wien, 101, 1892. Id. Ber. 13, 1895. **R. Hartig :** Holzuntersuchungen, Berlin, 1901. **Ursprung :** Ber. 19, 1901. Id. Beih. Bot. Centr. 19, 1905. Id. Biol. Centr. 26, 1906. **Sonntag :** P.J. 39, 1904. **Metzger :** Naturw. Zeitschr. f. Forst- u. Landw. 1908.

363. According to R. Hartig and Sonntag, the white and red wood of Conifers differ also in histological characters. In the case of the Spruce, the red wood is more highly lignified, and the slit-like cavities of its bordered pits are more elongated and less inclined, than those of the white wood; the stratification of the walls is also

not the same in the two cases. It is highly probable that all these features are correlated with the physical characteristics of the two forms of wood.

364. **Sachs** : Textbook of Botany (2nd Engl. ed.), p. 813. **De Vries** : Flora, 1872, p. 241. Id. Arch. Néerl. **11**, 1876. **Krabbe** : Sitzb. Berlin, 1882, pp. 1093 *sqq.* **Russow** : Neue Dörptsche Zeitung, 1881. **Wieler** : P.J. **18**, 1887. **Gnentzsch** : Flora, 1888. **R. Hartig** : Unters. a. d. forstbot. Inst. z. München, **1**, 1880. Id. Das Holz. d. Nadelwaldbäume, 1885. **Jost** : B.Z. 1891. **Jahn** : Bot. Centr. **59**, 1894. **Lutz** : Ber. **13**, 1895. Cf. also Fünfstück's Beitr. **1**.

365. **G. Haberlandt** : Phys. Pflanzenanat. 1st ed. 1884, p. 371. Id. Ber. 1895. **Strasburger** : l.c. [355], pp. 945 *sqq.* **R. Hartig** : Allg. Forst- u. Jagdzeitung, 1889. Id. Forstl. naturw. Zeitschr. **3**, 1894. **F. Schwarz** : Physiologische Unters. üb. Dickenwachstum u. Holzqualität v. Pinus sylvestris, Berlin, 1899. **Ursprung** : Beitr. z. Anat. u. Jahresringbildung tropischer Holzarten (Inaug.-Diss.), Basle, 1900. **Metzger** : Naturw. Zeitschr. f. Forst- u. Landw. 1908.

Metzger, who is evidently unacquainted with the author's ecological explanation of the formation of annual rings, attempts to refer the structural differences between spring- and autumn-wood entirely to mechanical causes. "With a given expenditure of material, the maximum of inflexibility can only be attained, if the specialised mechanical cells are massed near the periphery of the annual rings; the weaker conducting elements thus naturally come to take up their position on the inner side of the rings, and are consequently produced at the beginning of each vegetative season." According to Metzger, therefore, the conducting tissue appears first, not because it is needed early in the season, but merely because it is necessary that the mechanical tissue should take up a peripheral position. This interpretation might pass muster, if the woody cylinder were composed of a few very wide annual rings. As, however, the converse is actually the case, the annual rings being numerous and comparatively narrow, the location of the mechanical tissue in the individual rings is a matter of very slight importance from the ecological point of view. This conclusion may also be directly arrived at from Metzger's own calculations. He considers the case of a stem 20 cm. in diameter, to which is to be added an annual ring of 3 mm. total diameter, made up of 2 mm. of spring-wood and 1 mm. of autumn-wood. Under certain conditions, the original strength of the stem may be expressed by the figure 4214·61. If a *homogeneous* ring 3 mm. in width is added, the value rises to 4596·93, an increase of 382·32 units. If the ring consists of 2 mm. spring-wood and 1 mm. autumn-wood, the new value is 4601·35, an increase of 386·74 units, or only 4·42 units more than in the first case. If the mechanical tissue were formed first and the conducting tissue later, the new value would be 4592·47, an increase of 377·86 units, or 8·88 units less than with the converse arrangement. But it obviously cannot be of much importance whether the strength of a stem is raised from 4214·61 to 4592·47, to 4596·93 or to 4601·35; the differences between the three new values only amount to ·1 to ·2 per cent., and are therefore of no ecological importance. Moreover, these differences diminish as the girth of the trunk increases, so that the location of the mechanical tissue in the annual rings becomes less and less important as the tree grows older; nevertheless, the most recently formed annual rings in an aged stem are just as sharply defined as the oldest ones. Metzger's attempt to arrive at a purely mechanical solution of the problem, cannot therefore be regarded as a success.

366. **Kny** : Verh. Bot. Ver. Brand. 1879. **Wilhelm** : Ber. 1883.

367. **Sanio** : P.J. **8**, pp. 401 *sqq.*

3 B

368. **A. Braun** : Üb. d. schiefen Verlauf d. Holzfaser u. d. dadurch bedingte Drehung d. Stämme, Berlin, 1854.

369. **Von Mohl** : B.Z. 1869. **R. Hartig** : Lehrb. d. Anat. u. Phys. d. Pflanzen, 1891. **Schwarz** : l.c. [365].

370. **Schwendener** : l.c. [350]. **Id.** Sitzb. Berlin, 1884. **Metzger** : Münchner forstl. Hefte, No. 3, 1893. **Id.** ibid. No. 5, 1894. **Schwarz** : l.c. [365].

371. **Nördlinger** : Die technischen Eigenschaften d. Hölzer, Stuttgart, 1860. **Wiesner** : l.c. [351]. **Molisch** : Sitzb. Wien, **80**, 1879. **Id.** ibid. **84**, 1881. **Gaunersdorfer** : ibid. **85**, 1882. **Temme** : Landw. Jahrb. **14**, 1885. **Praël** : P.J. **19**, 1888. **Wieler** : ibid.

372. **De Bary** : Comp. Anat. pp. 618 *sqq.* (and the literature there cited). **Kny** : Ber. **4**, 1886. **Röseler** : P.J. **20**, 1889. **Strasburger** : l.c. [355], pp. 393 *sqq.* . Kny's statement to the effect that the " tracheides " in the secondary bundles of the Dracaeneae and Aloineae are really vessels, is refuted by Röseler; Strasburger confirms Röseler's account. **Schoute** : l.c. [349].

373. *Floating-wood.* **Ernst** : B.Z. 1872, **De Bary** : Comp. Anat. p. 499. **Strasburger** : l.c. [355], pp. 178 *sqq.* **Goebel** : Pflanzenbiologische Schilderungen, **2**, 1891. According to Ernst, the large thin-walled prismatic cells in the dilated portion of the stem of Aeschynomene are " always full of water so long as the stem is submerged, and only contain air when the stem is above water." If this account is correct, Goebel is mistaken in regarding floating-wood as ecologically equivalent to aërenchyma.

374. **Gaudichaud** : Mém. prés. a l'Acad. d. Sci. **8**, 1841. **A. de Jussieu** : Arch. d. Mus. **3**, 1843. **Crüger** : B.Z. 1850-1. **Bureau** : Monographie des Bignoniacées, Paris, 1864. **F. Müller** : B.Z. 1866. **Netto** : C.R. **57**, 1863. **Id.** Ann. Sci. Nat., sér. IV., **20** and sér. V., **6**. **Nägeli** : Dickenwachstum d. Stengels b. d. Sapindaceen, Munich, 1864. **Radlkofer** : Monogr. d. Gattung Serjania, Munich, 1875. **Westermaier** u. **Ambronn** : Flora, 1881. **Von Höhnel** : P.J. 1882. **Warburg** : B.Z. 1883. **Schenck** : Beitr. z. Biol. u. Anat. d. Lianen, **2** (Anatomie), Jena, 1893 (contains an exhaustive list of literature). **Id.** P.J. **27**, 1895. **Gilg** : Ber. 1893. **Warburg** : ibid. **Schellenberg** : Festschr. f. Schwendener, 1899.

375. In the valuable treatise cited above, Schenck repeatedly attacks the theories of Westermaier and Ambronn, as well as the views put forward by the author, in the first edition of the present work, concerning the advantages of the anomalous structure of various liane-stems. The author, however, adheres in the main to his former statements, especially with regard to Westermaier's and Ambronn's explanation of the unusual width of the vessels and sieve-tubes of climbers (cf. the third [German] edition, p. 606, Note 36). Certain of Schenck's criticisms, which are accepted by the author, are mentioned in the text.

376. **Trécul** : C.R. **63**, 1866. **Schmitz** : Sitzb. Halle, 1874. **Id.** B.Z. 1875. **De Bary** : Comp. Anat. p. 516 *sqq.*, pp. 598 *sqq.*, 606 *sqq.* **Weiss** : Flora, 1880.

377. The splitting-up of rhizomes and roots observed by Koch in Crassulaceae, by Jost in Gentiana cruciata, Corydalis nobilis, C. ochroleuca, Aconitum Lycoctonum, Salvia pratensis and Sedum Aizoon, and by Arthur Meyer in the genus Aconitum, is quite a different phenomenon from the subdivision of the woody cylinder in lianestems. According to Jost, it depends upon the disorganisation of those tracts of tissue which are most directly connected with the (annual) aërial organs (leaves and

flowering shoots). The persistent portions of the root or rhizome generally become cut off from the disorganising tissues by layers of periderm. Cf. **Koch** : Unters. üb. d. Entw. d. Crassulaceen, Heidelberg, 1879. **Arthur Meyer** : Arch. f. Pharmacie, **219,** 1881. **Jost** : B.Z. 1890.

378. For the older literature concerning anomalous thickening in Chenopodiaceae, Amarantaceae, Nyctaginaceae, Phytolacca, etc., see **De Bary** : Comp. Anat. p. 590. Cf. also **Morot** : Ann. Sci. Nat., sér. VI., **20.** **Hérail** : ibid. sér. VII., **2.** **Leisering** · Bot. Centr. **80,** 1899.

379. **Brogniart** : Ann. Sci. Nat., sér. I., **16.** **Von Mohl** : Verm. Schriften, p. 195. **Miquel** : Linnaea, **18.** **Mettenius** : Abh. sächs. Ges. **7·** **De Bary** : Comp. Anat. pp. 608 *sqq.* **Strasburger** : l.c. [355], pp. 152 *sqq.*

380. **Von Mohl** : Verm. Schriften, p. 122. **W. Hofmeister** : Abh. sächs. Ges. **4.** **Russow** : Vergl. Unters. üb. d. Leitbündelkryptogamen. **Strasburger** : Das botanische Practicum, 2nd ed. Jena, 1887, pp. 178 *sqq.* **Rosenthal** : Flora, 1890.

# SUBJECT INDEX.

*Numbers in black type refer to illustrations of a particular plant or subject.*

# INDEX OF PLANT NAMES.

3 C

8.

199.

ERRATA.

P. 14, below headline    The Cells and Tissues of Plants, insert I. The Cell.
P. 45, lines 7 and 9, for plasmodesma read plasmodesms.
P. 458, line 35, delete 225.
P. 460, insert 225 at end of line 39.
P. 495, insert 241*a* at end of line 21.
See also Notes 94 and 96*a*

# Works on Botany.

A TEXT-BOOK OF BOTANY. By Dr. E. Strasburger, Dr. H. Schenck, Dr. L. Jost, Dr. George Karsten. Fourth English Edition. Revised with the Tenth German Edition by Dr. W. H. Lang. Illustrated. 8vo. 18s. net.

THE ORIGIN OF A LAND FLORA. A Theory based upon the Facts of Alternation. By Prof. F. O. Bower, F.R.S. Illustrated. 8vo. 18s. net.

PRACTICAL BOTANY FOR BEGINNERS. By Prof. F. O. Bower, F.R.S., and D. T. Gwynne Vaughan, M.A. Globe 8vo. 3s. 6d.

SYLVICULTURE IN THE TROPICS. By A. F. Broun. Illustrated. 8vo. 8s. 6d. net.

THE DISEASES OF TROPICAL PLANTS. By Prof. M. T. Cook Ph.D. Illustrated. 8vo. 8s. 6d. net.

THE COTTON PLANT IN EGYPT: STUDIES IN PHYSIOLOGY AND GENETICS. By W. Lawrence Balls, M.A. Illustrated. 8vo. 5s. net.

NOTES OF A BOTANIST ON THE AMAZON AND ANDES. By Richard Spruce, Ph.D. Edited and condensed by Dr. Alfred Russel Wallace, O.M. Illustrated. 2 vols. 8vo. 21s. net.

THE FUNGI WHICH CAUSE PLANT DISEASE. By Prof. F. L Stevens, Ph.D. Illustrated. 8vo. 17s. net.

DISEASES OF ECONOMIC PLANTS. By Prof. F. L. Stevens, Ph.D., and J. G. Hall, M.A. Illustrated. Crown 8vo. 8s. 6d. net.

BRITISH FOREST TREES AND THEIR SYLVICULTURAL CHARACTERISTICS AND TREATMENT. By J. Nisbet. Crown 8vo. 6s. net.

LONDON: MACMILLAN & CO., LTD.

# Works on Physiology. .

HUMAN PHYSIOLOGY. By Prof. LUIGI LUCIANI. Translated by FRANCES A. WELBY and edited by Dr. M. CAMIS. With a Preface by J. N. LANGLEY, F.R.S. Illustrated. In 4 vols. 8vo.

> VOL. I. CIRCULATION AND RESPIRATION. 18s. net.

> VOL. II. INTERNAL SECRETION—DIGESTION—EXCRETION THE SKIN. 18s. net.

A TEXT-BOOK OF PHYSIOLOGY. By Sir MICHAEL FOSTER K.C.B. Illustrated. 8vo.

> PART I. BOOK I. Blood : the Tissues of Movement, the Vascular Mechanism. Sixth edition. 10s. 6d.

> PART II. BOOK II. The Tissues of Chemical Action, with their Respective Mechanisms : Nutrition. Sixth edition. 10s. 6d.

> PART IV. REMAINDER OF BOOK III. The Senses. Sixth edition, revised. 10s. 6d.

> PART V. Appendix. By A. S. LEA. 7s. 6d.

A COURSE OF ELEMENTARY PRACTICAL PHYSIOLOGY AND HISTOLOGY. By Sir MICHAEL FOSTER, K.C.B., and J. N. LANGLEY, F.R.S. Seventh edition. Edited by J. N. LANGLEY and L. E. SHORE, M.D. Crown 8vo. 7s. 6d.

THE NERVOUS AND CHEMICAL REGULATORS OF META-BOLISM. By Prof. D. NOËL PATON, M.D., B.Sc. 8vo. 6s. net.

ELECTRO-PHYSIOLOGY. By Prof. W. BIEDERMANN. Translated by F. A. WELBY. 8vo. Vol. I. 17s. net. Vol. II. 17s. net.

HANDBOOK OF PHYSIOLOGY FOR STUDENTS AND PRAC-TITIONERS OF MEDICINE. By AUSTIN FLINT, M.D. Illustrated. 8vo. 21s. net.

A TEXT-BOOK OF THE PRINCIPLES OF ANIMAL HISTOLOGY. By ULRIC DAHLGREN and W. A. KEPNER. 8vo. 16s. net.

LONDON : MACMILLAN & CO., LTD.

Lightning Source UK Ltd.
Milton Keynes UK
UKOW06f1815061015

259981UK00005B/355/P